T0350633

ATOMISM
AND ITS CRITICS

THOEMMES

ATOMISM
AND ITS CRITICS

From Democritus to Newton

Andrew Pyle
UNIVERSITY OF BRISTOL

THOEMMES PRESS

Published in 1997 by

Thoemmes Press
11 Great George Street
Bristol BS1 5RR, England

US office: Distribution and Marketing
22883 Quicksilver Drive
Dulles, Virginia 20166, USA

ISBN 1 85506 502 9

© Andrew Pyle 1995

First published in 1995
First published in paperback in 1997

British Library Cataloguing-in-Publication Data

A catalogue record of this title is available
from the British Library

All rights reserved. No part of this publication may be
reproduced, stored in a retrieval system, or transmitted
in any way or by any means, electronic, mechanical,
photocopying, recording or otherwise, without the
written permission of the copyright holder.

Printed in Great Britain by Antony Rowe Ltd., Chippenham

For Peter, who taught me how the History of Philosophy should be done

CONTENTS

INTRODUCTION

This study is best characterised as a contribution to the history of Natural Philosophy during its first two millennia. Throughout this period – from Democritus to Newton – no sharp distinction existed between philosophy and natural science: figures like Aristotle, Descartes, and Leibniz may be found in histories of science as well as of philosophy. The Atomic Theory is essentially a philosophy of Nature: it is idle to ask, at least during the period in question, whether it belongs properly to metaphysics or to physics.

Qua philosophy of Nature, classical Atomism seems to involve four central claims, two of which might be described as *existential* (giving the foundations of its ontology), the other two as *explanatory*, i.e. laying down strict guidelines concerning legitimate patterns of physical explanation. These four claims can be summarised as follows:

1. A commitment to *indivisibles*, particles of matter either conceptually indivisible (i.e. such that one cannot conceive of their division) or physically unsplittable.

2. Belief in the existence of *vacuum* or 'Non-Being', purely empty space in which the atoms are free to move.

3. *Reductionism*: explanation of the coming-to-be, ceasing-to-be and qualitative alteration of sensible bodies in terms of the local motion of atoms which lack many (most) of the sensible properties of those bodies.

4. *Mechanism*. This is a thesis about the nature of physical agency: it claims in effect that no body is ever moved except by an external impulse from another, moving body.

These four assertions form the heart of classical or Democritean Atomism. (It is important to distinguish 3 and 4, since many Renaissance thinkers accepted the former but not the latter, insisting that the movements of minute bodies that constitute the generation and alteration of sensible bodies are guided by purposive, spiritual agencies of some kind. 3 is essentially a thesis about *qualities*; 4, about *agents*.) Even if no actual thinker subscribed wholeheartedly and unreservedly to all of 1-4, they still constitute, between them, an *ideal* Atomist position capable of serving as a sort of ideological landmark for future reference.

Our study concerns the reasons which led men to assert – or to deny – one or other of these four pillars of Classical Atomism, i.e. with the arguments employed on both sides. Although we shall range widely and with some freedom, these are the four key issues of which we must not lose sight: Indivisibles, Vacuum, Reduction and Mechanism. Under each of these headings, however, may be found questions belonging to several different fields of study: a chapter on indivisibles, for example, may contain issues chemical as well as mathematical or physical, while questions about the vacuum slide from physics into philosophy and even theology. An argument is pursued wherever it leads, with no respect of interdisciplinary borders.

The work divides naturally into three temporal periods, giving it an overall 3 × 4 structure, in which Chapters 1, 5 and 9 deal with indivisibles, 2, 6 and 10 with vacuum, and so on. The three time-periods are, roughly, Classical Antiquity (c. 500 B.C.–500 A.D.), the Middle Ages and Renaissance (c. 500–1600 A.D.), and the Seventeenth Century. These divisions, I feel, mark real changes in the direction of Western intellectual history, i.e. the break-up of the classical Greco-Roman learning, and the transition from Renaissance to modern, respectively. (The gulf that separates seventeenth from sixteenth century thought is at times quite striking.) Transitional figures such as Philoponus and Francis Bacon find themselves called upon as and where their ideas seem best to fit.

Why, one might ask, pursue the story to the year 1700, rather than, say, 1600 or 1800? This too, I would claim, is non-arbitrary: by the end of the seventeenth century there existed a conception of the physical world (e.g. in the writings of Boyle and Locke) not very far removed from that of Democritus: the debt of these 'corpuscularians' to classical Atomism is both transparent and profound. Around 1700, however, this world-view is already under threat: the ideas of, e.g. Newton and Leibniz are beginning, in their very different ways, to undermine the foundations of the whole edifice. To pursue the story to the year 1800 would involve, therefore, a host of new departures and themes. To call a halt around 1600, however, would be to leave too many loose ends waiting to be tied up, and would thus be profoundly unsatisfactory. I choose, then, to pursue the story as far as the year 1700 (give or take a few years), when the Atomistic world-view has been triumphantly re-established (and is just *beginning* to show signs of strain), rather than either to call a halt in the confusion of Renaissance thought or to pursue the tale on into pastures new. By around 1700, a *world-view* has been established which is clearly a revived version of classical Atomism: one of the themes of this study is, therefore, that of departure and return. Book One deals with the ancient Atomists and their critics; Book Two traces surviving Atomist ideas in the predominantly Platonic and Aristotelian (i.e. anti-Atomist) thought of the Middle Ages and Renaissance, and investigates developments during that period which facilitated, in one respect or other, the eventual re-emergence of one or other of the Atomists' claims; Book Three deals with the revival in seventeenth century thought of classical Atomism, often incorporating ideas from medieval and Renaissance thought. (It would be quite wrong to think of Book Two as merely a confused digression separating the weightier material of Books One and Three.)

The study argues for no single grand or universal thesis, and the reader is left, by and large, to draw his own or her own conclusions. This, I feel, is as it should be: any attempt to draw all the threads together into some great synthesis would produce either oversimplification and trivialisation, or else requires another work of comparable length: better, then, to tell the story and let the consequences take care of themselves.

In a work of this scope, there are bound to be faults: omissions, misinterpretations, fallacies, and plain errors of fact. I have eliminated many, and can only beg the reader's indulgence for those which, doubtless, remain.

INDIVISIBLES: 1

Part 1: The Constitution of the Continuum

The name of Zeno of Elea is commonly associated with the four notorious paradoxes of motion, which have bewildered and bedevilled philosophers and mathematicians through the ages, spawning a vast exegetical and critical literature. There is, however, another paradox, less well known perhaps, that is both logically prior to[1] and conceptually more profound than that fearsome quartet. Aptly characterised by Grünbaum as 'Zeno's metrical paradox of extension'[2] it concerns the very nature of extended magnitude as such, and is therefore of wide scope and applicability. (Although I shall speak throughout in terms of *spatial* extension, exactly the same principles will apply to other continuous magnitudes such as time.) Without following the details of Zeno's text – which can be found in Simplicius' commentary on Aristotle's *Physics*[3] – I shall attempt to spell out the basic line of thought involved.

We start with the concept of (mathematical) extension and divisibility. (We shall throughout this section be dealing entirely with conceptual and mathematical issues; the existence or otherwise of physical limits to division is of no concern to us here.) Extension, says Zeno, entails divisibility: any magnitude can, for example, be bisected. This thesis – which in due course becomes an axiom of Euclidean geometry – is put forward as something indubitable, either as a conceptual truth or as one accessible to mathematical intuition. If we conceive or intuit an extended magnitude, we can conceive of its division, which is therefore possible. Thus we can state, as a first premise:

Z1.   Every extended magnitude is divisible.

Now Z1 is already sufficient to guarantee infinite divisibility. Take a given spatial magnitude m and divide it n – 1 times to produce n equal parts. These n parts, x_1, x_2, x_3,

---

[1]   See Wesley Salmon's Introduction to his collection of papers on Zeno's Paradoxes, esp. p. 7.

[2]   Grünbaum, 2, p. 176 ff.

[3]   For accounts that adhere more closely to the text, see Barnes, Vol. 1, pp. 238-251, and the fine paper by Abraham.

... x_n, are each of magnitude m/n. Clearly, the sum of $x_1 + x_2 + x_3$ ... $+ x_n$ will equal m. This gives us Z2, the principle of the additivity of magnitudes:

Z2. $\Sigma\ x_1 + x_2 + x_3$ ... $+ x_n = m$.

i.e. the sum of the sizes of all the non-overlapping integral parts of a given magnitude is equal to the size of the whole. Z2 was deemed uncontroversial, even self-evident, by all the Ancients.

Now no matter what values one sets for m and n, each of these parts of magnitude m/n will, by Z1, be further divisible. For any finite value of n, therefore, it will be possible to divide m more times than n. No finite number of divisions of m will exhaust its 'potency' for further division. Hence m is *infinitely* divisible:

Z3.   Every extended magnitude is divisible *ad infinitum*.

Z3, as thus stated, masks a crucial ambiguity to which it will be necessary to return later. But let us now pursue the argument further.

Consider the following distinct types of division-process. In the first type of process one takes a unit magnitude, halves it, then halves *one* of the half-size parts produced, then bisects one of the two quarter-size parts produced by that division, and so on. This I call a P-division, since it yields proportional or P-parts of the original unit magnitude – in this case each P-part is half as large as its predecessor.

If, on the other hand, one were to divide *both* the half-size magnitudes produced by the first cut, than *all four* of the resultant quarter-size parts, and so on, one would produce, at any stage of the division-process, parts all of the same size. This I call an A-division, since it produces aliquot or A-parts.

After n phases of division, a P-process gives n + 1 resultant parts, whereas an A-process yields 2^n. But for all values of n greater than 1, $2^n > n + 1$. After the second stage of the respective processes, therefore, the A-process will always yield more parts than the P-process. If, then, a magnitude is divisible into an actual infinity of P-parts, it is *a fortiori* also divisible into an actual infinity of A-parts, since at no stage of the process is the number of A-parts less than the number of P-parts.

This, however, is where problems arise. How large are these A-parts, an infinity of which would reconstitute our original magnitude? One would like to say: 'as large as the last member of the set of P-parts 1/2, 1/4, 1/8, 1/16, etc. . . .'. But this set does not – and cannot – have a last member. It seems then that we cannot attribute any determinate size to the members of our set of ultimate A-parts. Perhaps 'vanishingly small' is the most apt description of them: it may have been considerations like these that motivated the postulation of the infinitesimal, that unhappy and unintelligible intermediate between something and nothing.

Now the metrical paradox, as Porphyry saw,[4] concerns A-division rather than P-division. It is not concerned with the old chestnut of the sum of an infinite series of P-parts; in this particular argument Zeno is not contending that:

$$\Sigma \ 1/2 + 1/4 + 1/8 + 1/16 + \ldots = \infty.$$

Aristotle's proof that this sum only approached one[5] (and can never exceed it) is therefore of no relevance to us.[6] If *both* the halves produced by each bisection are further divided at the next stage then at the end of each phase of the division-process there remains a set of equal or aliquot parts. Let us call such a set, following Barnes,[7] a Z^* set. Given Z3 there will be an infinite number of such Z^* sets for every non-zero magnitude m, thus:

		m			Z^*_0
	m/2		m/2		Z^*_1
m/4	m/4		m/4	m/4	Z^*_2

AND SO ON.

It is here, however, that difficulties arise. Although there are an infinity of Z^* sets, each of those sets has only a finite number of members.[8] For every integer n (the number of stages of bisections made), the value of 2^n is finite. One cannot by such means generate a Z^*_∞ set, consisting of an infinite number of A-parts of m. Yet it is this notion of a Z^*_∞ set that Zeno requires for his argument.

What Zeno plainly has in mind here is the idea of the *ultimate* constitution of the continuum. The ultimate constituents of m must, he feels, be A-parts: if one part is greater than another, it is further divisible, hence not an ultimate part of m.[9] Now if every possible bisection is made – i.e. if no part susceptible to further bisection remains – then, one might plausibly think, m has been divided into its ultimate constituents. Since no non-

---

4   See Abraham, pp. 42-43, and Barnes, Vol. 1, p. 247.

5   Aristotle, *Physics*, 3, 6, 206b 6-10.

6   Contrast the first two of the paradoxes of motion, the 'Stadium' and the 'Achilles', which do turn on the sum of an infinite series of P-parts.

7   Barnes, Vol. 1, p. 247.

8   *Ibid.*, p. 251.

9   See Salmon, Ed., p. 14.

zero extended magnitudes remain, m has been divided 'through and through' (henceforth, t & t). One might suppose that m had been divided at *every* point, but this clearly cannot be so, since no process of successive *bisections* could cut m at *irrational* points.

It is important that we do not think of the successive bisections of m as being temporally successive: so far as logic is concerned, they could all be simultaneous. The fact that temporally successive bisections would never come to an end,[10] would never yield our Z^*_∞ set, is therefore irrelevant to the argument.

Now after any finite number of phases of the A-process, extended parts will remain. But these, by Z1, will be further divisible. Now the actual dividedness of something divisible is possible. Something infinitely divisible, then, can be infinitely divided. Even if, for some reason, we cannot reach the Z^*_∞ set by successive or simultaneous divisions, facts about *divisibility* guarantee its existence. Infinite divisibility then involves the existence of a Z^*_∞ set of the ultimate component parts of the continuum. Even if infinite division never takes place, infinite divisibility into aliquot parts entails the existence of such an ultimate Z^* set, since the parts of a magnitude pre-exist and are not created by its division.[11] One can only divide a body if it already consists of distinct spatial parts. To divide a body into two, three, or four parts it must already be made up of (at least) that many distinct parts. Infinite divisibility into A-parts (Z^* sets) therefore guarantees the existence of a special Z^* set, the final or ultimate one Z^*_∞. The parts of Z^*_∞ are the ultimate constituents of the continuum, even if it is never divided into those parts. These conclusions are summed up in theses Z4 and Z5:

Z4.   Division presupposes the existence of distinct parts; it does not create those parts.
Hence
Z5: Infinite divisibility entails the pre-existence of an actual infinity of aliquot parts.

This is the notion which Zeno is leading up to, the conception of Z^*_∞ as the set of all the ultimate components of a magnitude m, such as would be produced by a t & t division of m. This vision is encapsulated in Z6:

---

[10]   Unless, like the 'infinity-machines' of Black and Thompson, each stage took less time than its predecessor, so that one division took one second; the next, half a second; the third, a quarter of a second, and so on.

[11]   See Furley, 2, p. 38.

Z6.   There are an infinity of equal or aliquot parts in every continuous magnitude (from Z3 and Z5).[12]

Now comes the terrible destructive dilemma. Are these ultimate parts – the members of Z^*_∞ – themselves extended or not? Do they have non-zero (> 0) size? If so, Z7 comes into play:

Z7.   For all $x > 0$, ∞ times $x = \infty$

So if the members of Z^*_∞ have > 0 extension, m must (by Z2) be infinite in magnitude. If, on the other hand, the members of Z^*_∞ are unextended, like mathematical points, nothing could be made up of them, and Z8 comes into play:

Z8. Σ $0 + 0 + 0 + 0 + 0 +$ . . . for *any* number of zeros, finite, or infinite, $= 0$

Or, alternatively, $n \times 0 = 0$ for all values of n.[13]

Z8 entails the utter impossibility of constructing or compounding a continuous magnitude out of unextended mathematical points. Even an infinity of them could not do the job. (The Greeks, of course, had no conception of the non-denumerable infinite of Cantor and Dedekind.[14]) A mere denumerable infinity of mathematical points will not constitute a continuum, and that was the only conception of infinity available prior to the nineteenth century. Imagine taking a line of unit length and marking one point at half-distance, another at three-quarter distance, and so on. There would always remain gaps or intervals between these points; they would not make up a continuum. The magnitudes of the gaps or intervals, moreover, would sum to unit length: one can therefore place a denumerable infinity of points (of the cardinality of the set of the natural numbers) on a line, and not even fill *part* of its extent. Hence Z9, universally accepted by Greek thinkers:

Z9.   No continuous magnitude can be composed of unextended mathematical points.

Since only a Cantorean, non-denumerable infinity of points will evade the force of Z8, Z9 remains orthodox – although by no means universally accepted[15] – until the nineteenth

---

12   Barnes, Vol. 1, p. 243.

13   If, says Aristotle (*Metaphysics*, B4, 1001 b 7–10), a point adds nothing to a line, it may without impropriety be called a 'nothing'. But of nothing, nothing can be made.

14   Grünbaum, 1, p. 296.

15   It seems that there always remained thinkers who were convinced, despite the apparently unshakeable Z8, that, *somehow*, a continuum must consist of points. William of Ockham was an example of such a thinker: for his – surprisingly modern – conception of the constitution of a continuum out of points see the papers by T. B. Birch and C. D. Burns.

century. (As Grünbaum shows,[16] the force of Z8 is only evaded by denying the applicability of the apparently self-evident Z2 to the sums of non-denumerably infinite sets.[17]) But if the continuum is not made up of unextended point-like elements, of what is it composed? Zeno's terrible dilemma is raised by Z6: as soon as one grants Z6 one is impaled on one or other of its horns. The members of Z^*_∞ either are, or are not, themselves extended magnitudes – *tertium non datur*. The strictness with which the Greeks took this dichotomy is a tribute to their logical rigour: to take the dichotomy seriously is implicitly to deny the intelligibility of the notion of the infinitesimal.[18]

To escape from Zeno's dilemma, then, one must find some way of rejecting Z6. But one cannot reject Z6 without also denying some of the apparently evident premises from which it was derived. Some denied Z1 and hence Z3, and thus opted for a theory of indivisible magnitudes;[19] advocates of such a view include Xenocrates the Academic and Diodorus Cronos as well as the Atomist school. Others, such as Aristotle, attempted to evade Z6 by so weakening the sense of Z3 that it does not entail Z5. The infinite, says Aristotle, is always something in process, never something actual and complete. Magnitudes, he argues, are divisible *ad infinitum* but are not made up of an actual infinity of parts.[20] Let us turn first to the former tradition, and the argument by which Democritus sought to establish his Atomism. The debt to Zeno is manifest at a glance.

The crucial text is that of Aristotle's *De Generatione et Corruptione*, Book 1, Chapter 2, where the Stagirite first expounds and then tries to refute an argument for mathematical indivisibles.[21] Aristotle speaks of this as 'the argument which is believed to establish the necessity of atomic magnitudes'.[22] The chapter discusses two varieties of

16   Grünbaum, 1, p. 301.

17   Z2 does apply, of course, to the sums of denumerably infinite sets: ordinary geometry and analysis depend on the summation of infinite series.

18   See Abraham, p. 43; Evans, p. 549.

19   See Aristotle, *Physics*, 1, 3, 187a 3, for the postulation of atomic magnitudes as an answer to Eleatic puzzles about divisibility. See also Barnes, Vol. 1, p. 245.

20   See Edel, pp. 51–57, for a clear account of the Aristotelian conception of the potential infinite.

21   I am heavily indebted to Dr. Christopher Williams, whose edition of *De Gen et Corr* is forthcoming in the Clarendon Series, for much of the following material.

22   *De Gen et Corr*, 1, 2, 317a 1–2.

Atomism, that of the *Timaeus* and that of Leucippus and Democritus. The former theory is briefly dismissed, then Democritus is praised as follows:

> Democritus would appear to have been convinced by arguments appropriate to the subject, i.e. drawn from the science of nature. Our meaning will become clear as we proceed.[23]

Since Aristotle now proceeds to state the crucial Atomist argument, and since it is clearly 'physical' rather than 'dialectical' in nature (it resembles in form and content much of the argumentation of Aristotle's own *Physics*), it is surely not too daring to attribute this argument to Democritus as a possible response to Zeno.[24]

The context of the argument is as follows. Aristotle is discussing the nature of coming-to-be and passing-away and has come across the theory of Leucippus and Democritus, which explains these phenomena in terms of the association and dissociation of abiding atoms. He then immediately raises the following question:

> The fundamental question, in dealing with all these difficulties, is this: 'Do things come-to-be and "alter" and grow, and undergo the contrary changes, because the primary "reals" are indivisible?' For the answer we give to this question makes the greatest difference.[25]

How does the truth or falsity of Atomism – the existence or non-existence of indivisibles – affect the 'generation = aggregation' thesis? A refutation of Atomism will not refute the aggregation theory, since a compound body could come-to-be by the aggregation of *divisible* particles. (Empedocles and Anaxagoras both accept that coming-to-be involves the coming-together of pre-existing stuffs, but neither accepts indivisibles.) If anything, Atomism entails the 'generation = aggregation' thesis rather than *vice versa*. Perhaps the Atomic theory is introduced here because it is thought to provide grounds for accepting such a theory of generation; a refutation of the theory of indivisibles is therefore not out of place here.

---

23 *Ibid.*, 1, 2, 316a 13–14.

24 This has been commonly believed by scholars since Philoponus. See Furley, 2, p. 83, and Abraham, pp. 45–46, for examples of two modern scholars who attribute this argument to Democritus.

25 *De Gen et Corr*, 1, 2, 315b 25–29.

The argument for Atomism runs from 316a 15 to 316b 16, and takes the form a *reductio ad absurdum*. The proposition to be reduced to absurdity is Q:

Q   Some body is everywhere divisible (divisible at every point).

Now Q is ambiguous between Q1 and Q2, the difference being best brought out by a little formalisation. For simplicity, let us take as our universe of discourse a line L and the set of points $x_1, x_2, x_3 \ldots$ situated on L. Using N and P for the modal operators, we can now formulate Q1 and Q2 thus:

Q1.   P (X) Dx

Q2.   (X) P Dx

where Dx means 'there is a division (of L) at x'. The Atomists' *reductio*, it will become clear, works against Q1 and thus establishes its contradictory, Q3:

Q3.   N (∃ x)∼Dx.

Q3, however, asserts only that there must be some points on L at which it is undivided. But 'it is necessary (given the existence of matter at all) that there are undivided particles' does not entail 'there are indivisible particles'. The Atomists need to claim that there are points on L at which it cannot be divided; i.e. they require Q4, the contradictory of Q2:

Q4.   (∃ x) N∼Dx.

In fact, even Q4 is too weak to give atomic magnitudes and indivisibles. Suppose L to be indivisible only at its mid-point. Then L would still be infinitely divisible, and could be cut at points closer and closer to that mid-point, producing ever-smaller 'slices' of L containing that point. The Atomists therefore need a thesis to the effect that L is only divisible at a finite number of points (more exactly, that there are only a finite number of points at which it is possible to divide it) rather than Q4.

This reconstruction, however, will not prove of great value to us. If the Atomists simply proceeded by *reductio* of Q1 to Q3, confused Q3 with Q4, and assumed that Q4 was sufficient to establish the existence of indivisibles, not much can be said in their defence. It transpires, however, that deeper and more problematic issues are at stake. The Atomists' argument, as set out by Aristotle, begins thus:

. . . to suppose that a body (i.e. a magnitude) is divisible through and through, and that this division is possible, involves an absurdity. What will there be in the body which escapes the division?[26]

[26]   *Ibid.*, 316a 15–16.

The target of the *reductio* is clearly the idea of a magnitude as t & t divisible or divisible at *every point*. (There is no suggestion here that the divisions occur by bisections, so no problems arise about cutting L at its irrational points.) If we suppose L to have been divided at every point, what remains? What has L been divided into?

> What, then, will remain? A magnitude? No: that is impossible, since then there will be something not divided, whereas *ex hypothesi* was body was divisible *through and through*.[27]

If L has been divided into smaller magnitudes l_1, l_2, l_3, etc., these in turn must (by Z1) contain points at which they are undivided. But if l_1 were a part of L, any point on l_1 was a point on L. Hence L was not divided at every point, contrary to the original hypothesis:

> But if it be admitted that neither a body nor a magnitude will remain, and yet division is to take place, the constituents of the body will *either* be points (i.e. without magnitude) *or* absolutely nothing.[28]

The latter alternative is abruptly dismissed:

> If its constituents are nothings, then it might come-to-be out of nothings, and exist as a composite of nothings: and thus presumably the whole body will be nothing but an appearance'.[29]

Ex nihilo nihil fit. If things come-to-be out of their constituent parts, then on this theory a body could come-to-be out of nothing. Similarly, it would pass away into nothing by t & t division. (This is reminiscent of Epicurus' claim that indivisible minima are necessary 'if all is not to be dissolved into nothing'; the Epicurean arguments for indivisibles make much better sense if their Zenonian origins are borne in mind.[30])

Points, however, fare no better than nothings as candidates for the ultimate constituents of the continuum. A mere mathematical point adds nothing to the extension of a line; however many points one adds, one does not generate a

---

[27]   *Ibid.*, 316a 23–26.

[28]   *Ibid.*, 316a 26–28.

[29]   *Ibid.*, 316a 28–29.

[30]   See Epicurus's 'Letter to Herodotus', 56–57 in Bailey, 1, p. 33, and also Abraham, pp. 45–46.

magnitude. Even a (denumerable) infinity of points will not suffice (Z8). Hence no continuum is compounded of mathematical points (Z9).

The argument thus works by setting up a destructive trilemma. What, Democritus asks, remains after a t & t division? Is it:

a. Magnitudes?
b. Nothing?
c. Points?

None of these answers escapes absurdity. One or two half-hearted attempts are made to evade the trilemma (what remains, it is tentatively suggested, might be like 'sawdust' or 'a separable form or quality') but these do not bear scrutiny.

The supposition that L might be divided t & t is therefore refuted. Q1 is reduced to absurdity. What, according to Democritus, will follow from this? His answer seems to be a flat assertion of Q4:

> Since, therefore, it is impossible for magnitudes to consist of contacts or points, there must be indivisible bodies and magnitudes.[31]

As we have seen, however, a *reductio* of Q1 yields only Q3 and not Q4. Does Democritus' argument then turn on a mere conflation of Q3 and Q4, an elementary modal fallacy? Or is there more to be said?

Like Zeno, Democritus seems to have had in mind the question of the ultimate constitution of the continuum. The Aristotelian notion of an infinite *potential* for division, never actualised, was alien to him. The Aristotelian effectively *denies* that a continuum has a final or ultimate constitution at all; one can, on the peripatetic view, create as many Z^* sets as one wishes, but one will never arrive at the Z^*_∞ set. For each Z^* set Z^*_n there is a successor, Z^*_{n+1}, but there is and can be no ultimate or final Z^* set, Z^*_∞, that marks the end of the division-process. Not only could one not reach such a Z^*_∞ set by any conceivable process of division, the Aristotelian will argue, but the very idea of it is incoherent. There is and can be no set of ultimate component parts of the continuum. Infinite divisibility, far from guaranteeing the possibility of infinite dividedness, ensures that it can never occur.

But, Democritus can reply, his point does not turn on the concept of *division*. If at every point on a line it can be divided, this entails the actual existence of a corresponding

31   *De Gen et Corr*, 1, 2, 316b 15–16.

(i.e. infinite) number of parts. The Democritean argument rests on Z4 and Z5 and concerns, first and foremost, the *constitution* of the continuum rather than its division. Aristotle's contention that infinite divisibility does not entail the possibility of actually infinite dividedness will not therefore impress an Atomist. An infinitely divisible body, he insists, can be conceived as made up of infinitely many A-parts. But then Zeno's Z6 comes into play, and his terrible dilemma becomes inescapable. By eschewing Aristotle's escape-route (the potential infinite) and committing himself to Z4 and Z5, the Atomist can only evade Z6 by denying Z3. But to deny Z3 entails denial also of the apparently intuitive Z1 and assertion of the existence of *indivisible magnitudes*.

This, I contend, is what lies behind the apparent equivocation in the Atomist argument. Acceptance of Z4 and Z5 makes Aristotle's strategy unworkable. Z5 entails the existence of a Z^*_∞ set for any infinitely divisible magnitude, and Democritus' trilemma applies to these members of Z^*_∞. If they are magnitudes, the original magnitude has not been divided t & t. If they are nothings, *ex nihilo nihil fit*. If they are unextended points, Z8 and Z9 come into play. Hence, says the Atomist, no magnitude is infinitely divisible. But if Z3 is false so is Z1: there must be indivisible magnitudes.

The hypothesis of mathematical indivisibles, however, is not – as Aristotle saw – without its undesirable and counter-intuitive implications. In the first place, the Democritean must deny Z1, which is, at least *prima facie*, one of the fundamental and self-evident truths of mathematical intuition. As Aristotle says, if we do postulate indivisibles,

> we are confronted with equally impossible consequences, which we have examined in other works.[32]

Such a theory is, he insists, 'in conflict with mathematics',[33] notably with the principles of geometry:

> Admit . . . the existence of a minimum magnitude, and you will find that the minimum which you have introduced, small as it is, causes the greatest truths of mathematics to totter.'[34]

[32]   *Ibid.*, 316b 17–18.

[33]   *De Caelo*, 3, 4, 303a 20–22.

[34]   *Ibid.*, 1, 3, 271b 11–13.

The accusation is developed further in Book 6 of the *Physics*, where Aristotle opposes his own theory of the continuum to the rival theory of indivisibles, and in the small treatise *On Indivisible Lines*, which may not be the work of Aristotle himself but is certainly peripatetic in content and origin. 'It is not hard,' says Aristotle, 'to destroy the theory of indivisible lines.'[35] The author of the treatise – perhaps a pupil? – sets out to substantiate this claim. The existence of such minima, he has no difficulty in showing, is incompatible with many of the theorems of Greek geometry. For the sake of brevity, let us cite only two of the anti-geometric consequences of the postulate of minimal magnitudes:

1.   All magnitudes will be commensurable with one another, since all will consist of integral numbers of *minima*.[36]

2.   Not all magnitudes will be susceptible to bisection: lines consisting of an odd number of *minima* will not be divisible into two precisely equal parts.[37]

I cite only two of a multitude of examples used by the author – the reader can construct many more *ad lib*.

This opposition between indivisibles and geometry accounts for much that seems strange in the writings of the Atomist tradition. Epicurus, for example, seems to have been frankly contemptuous of geometry and geometricians,[38] while Zeno of Sidon, an Epicurean contemporary of Posidonius, wrote attacking the principles of geometry.[39] The Epicurean needs his indivisible magnitudes to answer Zeno, and is therefore forced to deny the truth (or physical applicability) of the theorems of the geometers. Of course, those theorems remain *approximately* true (given the limitations of our perceptual acuity, we may never discover any detectable deviation from them), and they may remain of great practical *utility* for various purposes; but nevertheless, considered strictly and literally, they are *false*. Euclidean geometry is a mere thought-creation inapplicable to the concrete physical world.

Fundamentally the same attitude can be found in many later writers in the Atomist tradition. The Mutakallimun

[35]   *Physics*, 3, 6, 206a 17.

[36]   *On Indivisible Lines*, 970a 1–3.

[37]   *Ibid.*, 970a 26–28.

[38]   See Furley, 2, pp. 155–156. For the Epicurean attitude to mathematics, see also Mau and Bailey, 2, p. 234.

[39]   See Long, p. 221.

school of the kalām, for example (a school of Islamic philosophical theology), were led by their theory of indivisibles to deny altogether the existence of incommensurables. The diagonal of a real square (as opposed to a geometrical one) is, they insist, commensurable with its side.[40] In the fourteenth century Nicholas of Autrecourt argues vigorously in favour of Z4 (against the Aristotelian conception of the potential infinite) and hence against infinite divisibility.[41] An infinitely divisible magnitude, he insists, must consist of an actual infinity of parts, and hence can only be of infinite extent.[42] This would be nonsense if Nicholas had P-parts in mind, but is a perfectly intelligible position if he is thinking of the A-parts produced by an A-division process.

Even as late as the seventeenth century a fictionalist philosophy of mathematics can be discerned in the Epicurean Walter Charleton, disciple of Gassendi. The naturalist, says Charleton, is not concerned with the abstractions of mathematics. The geometer may assume infinite divisibility, and hence implicitly accept an infinity of parts in each continuum (see Z5):

> not that he doth, or can really understand it so, but that many convenient conclusions, and no considerable incongruities, follow from the conclusion thereof.[43]

Suppose one were to ask bluntly whether the theorems of geometry were true or false. If true, such a one might say, they must be applicable in physics; if false, necessary and certain conclusions could not be drawn from them. To this dilemma Charleton replies:

> Our expedient is, that, though we should concede those suppositions to be false, yet may they afford true and necessary conclusions: every novice in logick well knowing how to extract undeniable conclusions out of the most false propositions.[44]

Charleton therefore applies to geometry the sort of fictionalism that had long been influential in astronomy (cf. of course

---

[40] For the rejection of incommensurables by the Mutakallimun, see Maimonides, p. 122.

[41] See Weinberg, 1, p. 158; O'Donnell, p. 110.

[42] Weinberg, 1, pp. 158–159.

[43] Charleton, p. 95.

[44] *Ibid.*, p. 96.

Osiander's notorious preface to *De Revolutionibus*). The principles of the mathematicians may be useful in natural philosophy, but they are, he insists, *not true*.[45]

The postulation of indivisible magnitudes is then indeed, as Aristotle saw, 'in conflict with mathematics', and Atomists through the ages had to deny the strict and literal truth (or applicability to the physical world) of the principles of geometry as a result of their acceptance of such *minima*. But let us return to the text of *De Generatione* to see how Aristotle attempts to evade the force of the Democritean argument.

His first attempt involves the use of his own favourite distinction between potentiality and actuality, or 'potency' and 'act'.[46] A body may of course be divisible (potentially divided) at a point where it is actually undivided. By contrast to other problem-situations, however, this distinction does not seem to help here. For to say that a body is potentially divided at every point is to say that no contradiction would result from its being divided at every point. But this is precisely what Democritus denies:

> Hence (it is urged) the process of dividing a body part by part is not a 'breaking up' which could continue *ad infinitum*; nor can a body be simultaneously divided at every point, for that is not possible; but there is a limit, beyond which the 'breaking up' cannot proceed. The necessary consequence – especially if coming-to-be and passing-away are to take place by 'association' and 'dissociation' respectively – is that a body must contain atomic magnitudes.[47]

Simple application of the potency/act distinction therefore achieves nothing, and only leads us back to a brisk reassertion of the thesis of atomic magnitudes. Aristotle's next attempt goes deeper:

> . . . since point is not 'immediately next' to point, magnitudes are 'divisible through and through' in one sense, and yet not in another. When, however, it is admitted that a magnitude is 'divisible through and through' it is thought there is a point not only anywhere, but also everywhere, in it: hence it is supposed to follow, from the admission, that the magnitude must be divided

45   *Ibid.*, p. 97.
46   *De Gen et Corr*, 1, 2, 316b 20–27.
47   *Ibid.*, 316b 29–35.

away into nothing. For – it is supposed – there is a point everywhere within it, so that it consists either of contacts or of points. But it is only *in one sense* that the magnitude is 'divisible through and through', viz. in so far as there is one point *anywhere* within it and all its points are everywhere within it if you take them singly one by one. But there are not more points than one *anywhere* within it for the points are not 'consecutive': hence it is not simultaneously 'divisible through and through'. For if it were, then, if it be divisible at its centre, it will be divisible also at a point 'immediately-next' to its centre. But it is not so divisible: for position is not 'immediately next' to position, nor point to point – in other words, division is not 'immediately next' to division, nor composition to composition.[48]

Two lines of thought are detectable in this passage. One is the familiar any / every distinction, and the difference between Q1 and Q2. Our hypothetical line L may be divisible at any point, but not at all points 'simultaneously'. Thus Q1 may be false but Q2 is true, and Q2 is still sufficient to refute the thesis of atomic magnitudes. Aristotle can therefore accept Democritus' argument as a proof of Q3 by *reductio* of Q1, while continuing to reject the confusion of Q3 with Q4.

What lies behind Aristotle's rejection of Q1? According to the Stagirite, Q1 is, strictly, a senseless proposition. To show this he borrows a thesis from the *Physics*[49] to the effect that point cannot succeed point on a line. If two points p_1 and p_2 are non-identical, p_1p_2 will give a line – which, by Z1, will be capable of being bisected, e.g. at p_3. But then p_1p_3 constitutes a line susceptible to further division, and so on. Hence, p_1 *has no successor point*. (Points, Aristotle shows, cannot touch or be in contact without *coinciding*.) But for the nightmare possibility of the division of a line into nothing but points to be realisable, p_1 must have a successor. Otherwise, no matter how many times one cuts L one will not resolve it into points.

Aristotle could have gone on to show that division of a line into nothing but points would require a more-than-infinite number of cuts. If, for example, we start with the series of P-parts 1/2, 1/4, 1/8, 1/16, etc., we can divide all these fractions in turn and still have magnitudes left, thus:

1/4 + 1/8 + 1/16 + 1/32 + . . .
1/8 + 1/16 + 1/32 + 1/64 + . . .

---

48   *Ibid.*, 317a 3–13.
49   *Physics*, 6, 1, 231b 6–9.

$$1/16 + 1/32 + 1/64 + 1/128 + \ldots$$

We can thus take all the parts left by an infinite division (into P-parts) and divide them again and again. Each successive division leaves us with ever-smaller magnitudes, not with points. To divide a line into points, therefore, a more-than-infinite number of cuts would be needed, which is absurd. (Given Aristotle's reservations about the actual existence of even a denumerable infinite, he would hardly be sympathetic to the ideas of Cantor and Dedekind.) Q1, therefore, makes no sense for an Aristotelian. It is unintelligible to suppose that L could be simultaneously divided t & t: this is quite simply *inconceivable*.

Aristotle concludes the argument of Book 1, Chapter 2, of *De Generatione* with the following statement, expressing his own position:

> Hence there are both 'association' and 'dissociation', though neither (a) into, and out of, atomic magnitudes (for that involves many impossibilities), nor (b) so that division takes place through and through – for this would have resulted only if point had been 'immediately next' to point: but 'dissociation' takes place into small (i.e. relatively small) parts, and 'association' takes place out of relatively small parts.[50]

For Aristotle, any continuous magnitude may be said to be divisible *ad infinitum*, but all this means is that it is divisible as many times as one might wish to divide it. The infinite, he contends, 'has a potential existence' only.[51] This, he insists, is all the geometers need for their work – that a line will always have a 'potency' for further division.[52] Each continuous magnitude is composed of smaller ones, and those of still smaller ones, and so on without end. There is and can be no Z^{*}_{∞} set of *ultimate* parts, either points or magnitudes, that make up a given magnitude. A continuum is made up of any number of proximate constituents, but no answer can be given, in principle, to the question of its ultimate constitution. (The Stoics made this denial of ultimate parts still more explicit.[53])

---

50   *De Gen et Corr.*, 1, 2, 317b 13–17.

51   For the Aristotelian theory of the potential infinite, see *Physics*, Book 3, Chapter 6.

52   See Boyer, p. 40.

53   See Hahm, 1, p. 210, and Plutarch, Vol. 13, p. 817, for Chrysippus' denial of ultimate parts.

This, then, is the problem-situation. Zeno accepts all the theses Z1–Z6, but by means of a destructive dilemma and the subsidiary theses Z7, Z8 and Z9 succeeds in reducing Z6 to absurdity. But Z6 follows from Z5 and Z3. Now Z5 follows from Z4 (divisibility presupposes distinct parts and does not create them), and Z3 follows from Z1 (all magnitudes are divisible). But Z4 seems evident, and Z1 is part of our everyday conception of extended magnitude. Hence, Zeno can conclude, Z6 cannot be avoided, in which case our ordinary notion of a finite magnitude is incoherent – implicitly self-contradictory.

Democritus, on my reading, accepts Zeno's *reductio* of Z6. Since he accepts Z7, Z8 and Z9 he too must find Z6 untenable. But he also accepts Z4, and hence Z5 – if a body is infinitely divisible, it must contain an actual infinity of A-parts.[54] Now if one accepts Z5 but wishes to deny Z6 and hence escape the horns of Zeno's dilemma, one must deny Z3 and hence, ultimately, the apparently intuitive Z1. Denial of Z1 and Z3 allows Democritus to accept Z2, Z4, Z5, Z7, Z8 and Z9 while still rejecting the perilous Z6.

By denying Z1 and Z3, Democritus managed to avoid asserting Z6, but instead was left with the curious and anti-geometrical theory of atomic magnitudes. Aristotle, convinced both of Z1 and of the truth of geometry, was not prepared to take such a route. Instead he attempted so to weaken Z3 as to be able to deny Z5, the actual existence of an infinity of A-parts in every continuum. But to deny Z5 Aristotle had tacitly to deny the apparently evident Z4 (divisibility presupposes distinct parts). A point on a line, says Aristotle, is only *actualised* when a cut is made there.[55] Aristotle, then, accepts of the Zenonian Z1–Z9 the following subset: Z1, Z2, Z3 (in an attenuated form consistent with the peripatetic doctrine of the potential infinite), Z7, Z8 and Z9. Between them, Z7–Z9 reduce Z6 to absurdity. To evade Z6, Aristotle finds that he has to deny Z5 and hence ultimately Z4.

The eventual consequence of all this is as follows. The Atomists have a viable theory of the ultimate constitution of the continuum, but a theory which flatly contradicts the very foundations of contemporary Greek geometry. The Aristotelians, by contrast, can develop a conception of the continuum compatible with geometry, but only at the cost of

---

54   See Furley, 2, p. 86.

55   See Evans, p. 548; Edel, p. 49.

having no view about its *ultimate* make-up. In the final analysis, they would have to dismiss the question as unintelligible. To make this position defensible however, they have to make the counter-intuitive move of denying Z4.

If we turn to Adolf Grünbaum[56] for a clear exegesis of the modern (post Cantorean) view we find a quite different strategy in operation. Whereas both Atomists and Aristotelians accepted Z7, Z8 and Z9, and hence saw Z6 as a thesis to be avoided at all costs, Grünbaum is prepared to accept Z6. He then boldly grasps the second horn of the Zenonian dilemma, denying that is leads to disaster. To accept the constitution of a continuum out of points he must of course reject Z9 and hence Z8. His rejection of Z8 depends upon the distinction between denumerable and non-denumerable infinities, and on his denial that the apparently innocuous Z2 (the principle of additivity) can meaningfully be applied to the 'sums' of non-denumerably infinite sets. Thus Grünbaum will accept Z1, Z3, Z4, Z5, Z6, and Z7, but by restricting the range of applicability of Z2 will be able to deny Z8 and hence Z9.

This gives us three quite distinct responses to Zeno. Zeno himself, in antinomanian mood, asserted all of Z1–Z9, saw that contradictory consequences ensued, and simply concluded that our everyday conception of finite magnitude is self-contradictory. Democritus denied Z1, and hence Z3 and Z6, giving him his theory of indivisible magnitudes. Aristotle denied Z4, and hence Z5 and Z6, giving rise to the theory of the potential infinite. By contrast, the modern theory, as outlined by Grünbaum, *accepts* Z6 (as entailed by Z3 and Z5) but by restricting the scope of Z2 is enabled to deny Z8 and Z9.

One attractive compromise position would be to separate altogether questions about division from questions about constitution. One could then be an Aristotelian about the former – no process of division that we could conceive would divide a line into points – and a Cantorean about the latter – nevertheless, the line *is* made up of points. But to attain this insight, and to produce a viable theory of the continuum that is compatible with infinite divisibility, a great part of the resources of modern mathematics is required. Small wonder that the ancients never developed such a theory but quarrelled instead over the relative merits of the rival Atomist and Aristotelian positions (neither of which was really

56 Grünbaum, 1, p. 288 ff.

satisfactory, since both involved denial of evident truths) and that the problem of the constitution of the continuum was long regarded as a 'labyrinth' of complexities and paradoxes, far surpassing the wit of mere human intellects to unravel.

Part 2: Democritus, Aristotle, and Epicurus

A. The Democritean Beginnings

It is uncontroversial that all the ancient Atomists, from Leucippus to Lucretius, believed in corpuscles that were at least *physically* indivisible. The atom was conceived as perfectly solid,[57] as an absolute plenum of pure Eleatic 'Being' or homogeneous matter*. (I introduce the term matter* here to stand for the uniform, featureless, homogeneous stuff of which the atoms consist; all macroscopic bodies, for the Atomist, are compounded of matter* or 'Being' and vacuum or 'Non-Being'.[58]) In virtue of its perfect solidity (lack of internal void) the atom cannot be divided by any physical process. For a body to be susceptible to physical division (cutting, shearing, splitting, etc.) it must contain internal vacuum: lacking such inner void, the atom is immune to such processes.

An atom, according to all the Greek Atomists, is eternal, immutable, homogeneous and physically indivisible. Is it then an Eleatic 'One', as Barnes suggests?[59] If so it must be a *unitary* substance, a genuine unit in a sense sufficiently strong to meet the rigorous demands of the Eleatic dialectic of Parmenides and Zeno. Now a true unit, according to Democritus, cannot consist of substantial parts. In Aristotle's words:

> If a substance is a unit, it cannot consist of inherent substances in this way, as Democritus rightly says; for he says that it is impossible for one thing to come from two or two from one (for he makes the atomic magnitudes his substances).[60]

---

57 Cf. Barnes, Vol. 2, p. 43, who stresses the influence of Melissus on the Atomists' conception of their atoms as absolutely solid, full of matter*. For more on physical indivisibility see Appendix 1.

58 According to Aristotle (*Metaphysics*, A4 985b 4) the 'elements' of the atomists are 'the full' (= matter*) and 'the empty' (= void).

59 '. . . the first atoms,' says Barnes (Vol. 2, p. 44), 'came from Elea.' For the Eleatic roots of the Atomic theory see Aristotle, *De Gen et Corr*, A8 325a 2 ff.; Kirk & Raven, pp. 401–407, and Burnet, pp. 330–349.

60 Aristotle, *Metaphysics*, 13 1039a 7–11. See also Pancheri, p. 139.

To be a true unit, something must not consist of distinct parts. Anything made up of parts is a plurality compounded out of those parts, not a genuine unit.[61] An atom which is merely physically indivisible, however, can always be divided – in thought – into its constituent parts, and can therefore be thought of as compounded out of those parts. As such, it is really a plurality, even if in some respects it behaves as a unit; it is a stable cluster or complex rather than a true unit. By Zeno's logic, if parts can even be conceived in it, it is many and not one. According to Zeno, extension entails possession of distinct parts, which in turn entails plurality.[62] Mathematical points would count as true units, since one cannot even conceive of their division, but of them nothing can be compounded (see Z9 of Part 1). As for a physically indivisible corpuscle, it will count, by Zenonian standards, as a compound entity made up of spatially distinct parts, and hence as many rather than one.

I conclude therefore – with Furley[63] and Pancheri[64] – that physically indivisible atoms do not suffice to answer Zeno. So long as atoms are theoretically/conceivably divisible into distinct parts, Zeno's arguments will get a grip (cf. especially the 'metrical paradox' discussed in the first part of this chapter). As Furley says:

> a physically unsplittable atom which is still theoretically divisible will not meet the Eleatic arguments at all.[65]

If we take seriously the claim that Atomism emerged as an attempt to reconcile Eleatic metaphysics with sense-experience, we *must* attribute to Leucippus and Democritus a stronger indivisibility thesis. We must, that is, claim that they postulated theoretical indivisibles, bodies that cannot be conceived to be divided; i.e. *minimal magnitudes*. If there are, *pace* Zeno, magnitudes without distinct parts, they will truly be units. In the words of Furley again:

> It seems to be impossible to understand the development of atomism out of Eleatic metaphysics without considering theoretical indivisibles as well as physical ones.[66]

---

61   Barnes, Vol. 1, p. 48.

62   See Vlastos, 6, p. 120, for Zeno's 'paradox of plurality'.

63   Furley, 2, p. 86.

64   Pancheri, p. 140.

65   Furley, 2, p. 86.

66   *Ibid.*, p. 3.

When we come to discuss the Epicureans, we shall find substantial evidence to support our attribution to them of a doctrine of geometrical minima. In the case of Leucippus and Democritus, however, the evidence is less compelling. The most powerful argument in favour of such a view is that only by positing mathematical indivisibles could Democritus have answered Zeno. (This is the argument on which Furley lays most emphasis in his admirable book.) It is possible, however, to piece together a more direct line of argument from the following fragments:

1. 'Some people,' says Aristotle, 'gave way to the Eleatic argument from dichotomy by positing "indivisible magnitudes".'[67] This may refer to Xenocrates or Diodorus Cronos – known advocates of mathematical minima – but more probably concerns the Atomists. (The same people, says Aristotle, answer another Eleatic argument by admitting the existence of 'non-being', which makes it almost certain that the Stagirite has Leucippus and Democritus in mind.) But to answer the Zenonian 'dichotomy' one must posit mathematical minima and not merely physical ones.[68]

2.   Atomism, Aristotle alleges, is 'in conflict with mathematics',[69] a remark explicitly directed against Leucippus and Democritus. Barnes[70] attempts to show that Aristotle might have said this of a theory of merely physical indivisibles, but his argument is far from convincing: a strong case can be made, as Furley[71] shows, for the claim that Aristotle treats Atomism as committed to mathematical indivisibles in all his counterarguments.

3.   Chrysippus' reading of Democritus' famous dilemma of the cone involves, according to Hahm (and others), the attribution to Democritus of a theory of geometrical minima.[72] I cannot go into this issue in detail, but the claim does seem a plausible one.

4.   According to a scholiast on Euclid, 'the Democriteans' posited the existence of a smallest magnitude[73] (*elachiston megathos*).

---

67 *Physics*, 1, 3, 187a 1–4.

68 Furley, 2, pp. 81–83.

69 *De Caelo*, 3, 4, 303a 20–22.

70 Barnes, Vol. 2, p. 56.

71 Furley, 2, pp. 89–90.

72 See Hahm, 1, pp. 219–220, and, for a more cautious discussion, Barnes, Vol. 2, p. 55. For the text, see Plutarch, Vol. 13, p. 821.

73 Barnes, Vol. 2, p. 53. D.K., 68A 48a.

5.   To estimate the areas of surfaces, Archimedes would cut them into thin strips, each approximately rectangular, and then sum the areas of these 'rectangles', each of unit width and n units length. Archimedes himself used this merely as a heuristic method for discovery of theorems to be proved by more rigorous methods. The technique was derived, however, from Democritus, who may well have taken its implicit geometrical atomism quite seriously: he, unlike Archimedes, may have regarded areas as quite literally made up of a mosaic of geometrical minima.[74]

6.   We must also cite the already-quoted remark from the *Metaphysics* to the effect that Democritus 'makes the atomic magnitudes his substances'.[75] One could hardly imagine Aristotle speaking of a merely physical indivisible as an 'atomic magnitude' since, quite clearly, it would not be such. *Qua* magnitude, that is, it would not be atomic (indivisible).

7.   If, as we have suggested in Part 1 of this chapter, the argument from *De Gen et Corr* really is Democritus', this would clinch the case for seeing him as a mathematical atomist; that argument clearly concerns the constitution of the continuum as such and is independent of questions about the physical divisibility of matter. It applies – as a mater of pure mathematics – to all extended magnitudes (space, time and motion) as such.

None of these seven points is conclusive. Individually, indeed, they are fairly weak – each involves a certain amount of conjecture and/or tendentious interpretation. Collectively, however – and in combination with Furley's argument that only a theory of mathematical minima could have been seen as an answer to Zeno – they yield a powerful case for seeing Democritus as a mathematical atomist.

Immediately, however, a very serious problem emerges. Are we to think of Democritus' *atoms* as geometrical minima, as the minimal magnitudes? If so, surely, they must all be of the same size – one minimum cannot be greater than another. But we know that Democritus' atoms have a great (infinite) variety of sizes and shapes. Atoms with hooks and eyes, moreover – or indeed with any determinate shapes – must have distinguishable parts. Democritus' physics requires attribution to atoms of a variety of shapes and sizes, but this makes nonsense of the claim that they are mathematical minima.

[74] See Boyer, p. 21, and pp. 50–51; Evans, pp. 449–450.

[75] *Metaphysics*, Z13, 1039a 11.

This leaves us with three options. We can (a) retract the claim that Democritus believed in mathematical minima. This seems a wrong-headed and retrograde step: it ignores the powerful arguments that have been marshalled, by Furley and others, in favour of attributing such a theory to Democritus. If this seems unpromising, we could (b) simply allege that Democritus was confused, that he failed to notice the discrepancy between his conception of atoms as minimal conceivables and his physical theory to the effect that they varied in size and shape. There seems to be some slight evidence favouring (b) – the Democritean atom is referred to by Simplicius as 'partless',[76] and is often conceived as a true Eleatic unity – but it is hard (*pace* Furley[77]) to credit that Democritus could have missed so blatant a confusion. The other and more speculative proposal (c) is that of Luria, who would father the Epicurean 'minimal parts' theory on Democritus. According to that theory, the atom is only physically indivisible (so atoms can vary in size and shape without absurdity), but is composed of serried ranks of *minimal parts*, the smallest conceivable units of extension. While this is the most coherent and attractive of the three possibilities, direct evidence for it is thin. The only direct positive evidence comes from the peripatetic Alexander, who remarks[78] that Democritus' atoms have 'partless parts', 'conceptually present' in them. This may be no more than a mistaken identification of the views of Democritus and Epicurus – no other ancient author attributes such a theory to Democritus[79] – but one hopes it is not. Since option (a) would make nonsense of the Eleatic roots of the Atomic theory, only the unattractive (b) remains if (c) is rejected. Like Konstan,[80] I tend therefore to favour (c), but the grounds for this preference are weak and largely negative.

If the argument from *De Gen et Corr* is Democritus', he is committed – if only implicitly – to a theory of minimal units of space, time, and motion as well as of matter. The argument against infinite divisibility is topic-neutral and applies to all extended magnitudes as such. He may not, however, have

---

76 Simplicius, *Physics*, 925 17, quoted from Furley, 2, pp. 111.

77 Furley, 2, pp. 94-97, p. 111.

78 Alexander, *Metaphysics*, 36, 25-27, quoted from Furley, 2, p. 98. See also Barnes, Vol. 2, pp. 50-51.

79 Furley, 2, p. 98. Aristotle and Theophrastus certainly find no trace of a theory of 'minimal parts' in Democritus.

80 Konstan, 2, p. 402.

seen these implications: there is no evidence that he posited time-minima, for example. We need not attribute to a thinker all the implications of views he holds explicitly.[81] The theory of minima of space, time, and motion was worked out by Epicurus in response to the Aristotelian critique of the indivisibilist position.

We now come to a further point which must at least be mentioned. If we think of Democritean Atomism as a reasoned response to Zeno, we must assume that the third and fourth paradoxes of motion (the 'Arrow' and the 'Moving Rows') were not – as several scholars have suggested[82] – directed against indivisibles of time and space. (There is no indication that the Atomists felt under any obligation to answer those paradoxes.) In particular we should show that they were not directed against an alleged Pythagorean theory of 'unit-point-atomism'.[83]

There is, fortunately, a growing scholarly consensus on these issues, a consensus which may be spelt out by the following four theses:

1.   This is insufficient evident to attribute to the Pythagorean school any doctrine of spatial or temporal indivisibles, and in particular no warrant to see them as proponents of a doctrine of 'unit-point atomism'.[84]

2.   The first pair of paradoxes, the 'Stadium' and the 'Achilles', clearly do turn on the concept of infinite divisibility: it is therefore credible that indivisibles were first posited *after* Zeno to escape his unpalatable conclusions.[85]

3.   The 'Arrow' paradox does not depend on any assumption about indivisible time-minima. All it requires is the commonsense notion of a moment or instant of time. It then argues as follows: if at every moment throughout its flight the arrow is at rest, it is at rest throughout the duration of its flight (c.f.: if at every moment last night I was asleep, I was asleep all night[86]).

---

81   See Furley, 2, p. 101.

82   Owen, Lee, and Heath all seem to accept this interpretation of the 4 paradoxes – the first pair (the 'Stadium' and the 'Achilles') directed against infinite divisibility, the latter pair (the 'Arrow' and the 'Moving Rows') against indivisibles.

83   For an account of the supposed Pythagorean theory of 'unit-point atomism' see Kirk and Raven, pp. 245-249.

84   Booth, 2, pp. 90-103; Furley, 2, pp. 44-55; Barnes, Vol. 1, p. 234.

85   Booth, 1, p. 4.

86   Furley, 2, pp. 71-72; Owen in Salmon, Ed., p. 158; Barnes, Vol. 1, p. 285; Vlastos, 5, p. 11.

4. The 'Moving Rows' paradox can be construed as a sophisticated argument against a theory of indivisibles,[87] but there is no evidence that any ancient commentator saw it as such. If, says Owen, the 'Moving Rows' is a powerful and telling argument against indivisibles, 'the Greeks seem to have missed the point by a wide margin'.[88] Perhaps then the paradox really is, as Aristotle thought,[89] merely an elementary sophism about relative motion, into which later scholars have read a depth and significance it did not originally possess.[90] As Barnes says, it is tempting to see the paradox as an ingenious attack on a developed theory of indivisibles, but this sophisticated reading 'has no historical support'.[91] We have no right to assume that all of Zeno's arguments were equally profound and well articulated: the great '*antilogikos*' of antiquity was not above the occasional fallacy or sophism.[92]

It is time now to draw together a few conclusions. The most important of these is that Democritus posited indivisible magnitudes – geometrical minima – as an answer to the Zenonian (Eleatic) demand for true units or 'ones'. Anything which can be divided – even in thought – is subject to the dichotomy argument: only something indivisible in a very strong sense will count as an Eleatic 'One'. These true units are either the atoms – in which case Democritus is severely muddled, since atoms vary in size and shape – or the 'minimal parts' of which atoms are composed. The latter view – that of Luria and Konstan – is preferable in that it attributes to Democritus a coherent and credible position, but has insufficient evidential support to rate as more than a plausible conjecture.

B. The Aristotelian Critique

Book 6 of Aristotle's *Physics* contains an extended critique of the contention that continuous magnitudes are composed of indivisibles. Most of the arguments turn on the concept of indivisibility in general, and are therefore applicable indifferently to points or minima. The first chapter moves swiftly

---

87   Kirk and Raven, pp. 295–296; Lee, p. 83 ff.

88   Owen in Salmon, Ed., p. 149.

89   *Physics*, 6, 9, 239b 33–240a 18.

90   See Barnes, Vol. 1, pp. 290–291; Furley, 2, p. 74, and Salmon's introduction to his collection of papers on Zeno, p. 12.

91   Barnes, Vol. 1, p. 291.

92   For this vision of Zeno, see Barnes, Vol. 1, pp. 231–234.

from definition to argument. It opens[93] with definitions of the terms 'continuous', 'in contact' and 'in succession' – things, says Aristotle, are continuous 'if their extremities are one', in contact 'if their extremities are together' and in succession 'if there is nothing of their own kind intermediate between them'. Given these definitions, says the Stagirite, 'nothing that is continuous can be composed of indivisibles: e.g. a line cannot be composed of points'.[94]

According to the above definition, two things are continuous if their extremities are one. Suppose this to be the case for two points: since a point has no other part distinct from its extremity, it follows that we should then have not two points but one.

Perhaps, then, the points which constitute a line are not continuous but in contact with one another. Contact, Aristotle replies, is either (a) part to part or (b) whole to whole. Indivisibles, having no distinguishable parts, cannot touch in sense (a). As for (b),

> if they are in contact with one another as whole to whole, they will not be continuous: for that which is continuous has distinct parts and these parts into which it is divisible are different in this way, i.e. spatially separate.[95]

Nor, by the above definition, can points (or instants) be in succession, since between any two non-identical points there is always a line, and this will contain further points. If, for example, p_1 and p_3 are distinct points, there is a *line* p_1p_3, divisible at p_2. But then $p_1 \neq p_2$, so p_1p_2 gives us a line which is in turn further divisible, and so on . . . No point can have a successor point 'immediately-next' to it. (Exactly the same argument applies to unextended instants of time.)

The upshot of this argument is the general conclusion that no continuum is composed of indivisibles. For indivisibles to constitute a continuum they would have to be continuous or contiguous or at least successive. Points, however, can satisfy none of these relations. What of *indivisible* magnitudes? These, clearly, would fall foul of the argument concerning contact. *Either* they touch 'part to part', in which case they have distinct parts distinguishable by the mind (in which case they are not mathematical minima) *or* they cannot make up anything greater than one minimum by aggregation.

93 *Physics*, 6, 1, 231a 20–23.

94 *Ibid.*, 231a 24–25.

95 *Ibid.*, 231b 3–6.

The same reasoning, Aristotle continues, 'applies equally to magnitude, to time, and to motion: either all of these are composed of indivisibles, or none'.[96] He set about showing this as follows:

> If a magnitude is composed of indivisibles, the motion over that magnitude must be composed of corresponding indivisible motions: e.g. if the magnitude ABC is composed of the indivisibles A, B, C, each corresponding part of the motion DEF of X over ABC is indivisible.[97]

If 'a motion' here means simply the traversal of a magnitude, the argument is plainly valid – and would, if pursued, also entail the existence of minimal discrete time-quanta. But, Aristotle alleges, this consequence is absurd.[98]

It is surely analytic that one cannot *both* be in the process of doing something (for the first time) and have finished doing it at the same time. It can never therefore be true to say of X both (1) that it *has passed* over A and (2) that it *is passing* through A, at the same time. Nor of course could (1) ever be true before (2). If, then, (2) is ever to be true, it must be so before (1) is. But:

> if X actually passed through A *after* being in the process of passing through, the motion must be divisible: for at the time when X was passing through, it neither was at rest nor had completed its passage but was in an intermediate stage.[99]

A stubborn defender of indivisible motions cannot admit that (2) is true either before, after, or at the same time as (1); i.e. he must deny that (2), for an indivisible motion, is ever true. This is the position toward which Aristotle is forcing him:

> if it is not in motion at all over the partless section A but has completed its motion over it, then the motion will consist not of motions but of starts, and will take place by a thing's having completed a motion without being in motion.[100]

Motion over indivisible magnitudes must be discontinuous – there can be no time at which X has *half* traversed A. At one

96   *Ibid.*, 231b 18–19.

97   *Ibid.*, 231b 20–24.

98   See Furley, 2, p. 117, for an account of this train of thought.

99   *Physics*, 6, 1, 232a 2–4.

100   *Ibid.*, 232a 7–9.

moment t_0 it has not begun to traverse A; at the next moment t_1 is has finished crossing A: there is no moment at which it is (in the process of) crossing the indivisible magnitude A. This vision of motion as essentially *cinematic* – a rapid succession of discrete 'stills' of sufficient frequency to deceive the eye – became, as we shall see, the Epicurean position, in spite of Aristotle's objections. One consequence of this theory, Aristotle alleges, is that the moving object, like Zeno's arrow, can truly be said to be at rest throughout the period of its flight.[101]

It is easy to show that indivisible spatial magnitudes entail indivisible time-minima. If the time-period T in which X traverses the geometrical minimum A is divisible, then in time T/2 X would have crossed half of A. But A is indivisible: the concept of half a minimum makes no sense. Therefore T must also be a minimum – a smallest discrete quantum of time. It is useless to suggest that at time T/2 X has either not started to cross A or has completely crossed it, for this simply makes the time taken by X to cross A T/2, and then one can ask where X is at time T/4 or 3T/4. One can by such manoeuvres vary the mesh of the grid, as it were (i.e. the actual values of the minima), but one cannot circumvent the demand for indivisible time-quanta of some magnitude.

We now move on to the second chapter of Book 6 of the *Physics*, in which Aristotle presents what is clearly regarded as the conclusive proof of the (potentially) infinite divisibility of space and time. Imagine, says Aristotle, two moving bodies α and β, of which α is the faster. In virtue of its greater speed, α will traverse a given distance AM in less time than β. Let α traverse AM in a time TZ, thus:

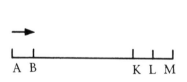

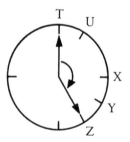

101 *Ibid.*, 232a 13–15.

If α traverses AM in TZ. then in the same time β will cover a distance AL < AM. But α will cover AL in a time TY < TZ, in which time β will cover a distance AK < AL, a distance α will traverse in a time TX < TY, *and so on*:

> The quicker will divide the time and the slower will divide the length. If, then, this alternation always holds good, and at every turn involves a division, it is evident that all time must be continuous.[102]

And of course the distance must be continuous – i.e. infinitely divisible – if the time is.

It is easy to show that the process of division described above cannot come to an end. Let α traverse a distance AB in a time TU. AB, let us suppose, is a minimum magnitude. But in time TU β must cover some distance (or it would be at rest). It cannot, however, cover a distance as great as, or greater than, AB, since it is *ex hypothesi*, the slower of the two bodies. It must therefore cover a distance AA¹ < AB in TU, in which case AB is not a minimum magnitude. An exactly parallel argument holds against anyone who tries to postulate a minimal time unit TU: find how far β travels in TU and enquire how long it would take α to travel that distance – it must be less time than the proposed minimum TU. All Aristotle needs for his argument is the existence of *different, uniform* velocities in Nature. To avoid Aristotle's conclusion, as we shall see, Epicurus found it necessary to deny that apparently obvious premise.[103]

Aristotle also argues that an indivisible cannot move – or alter in any other way – except 'accidentally', i.e. as a limit or surface of some moving body:

> For suppose that it is changing from AB to BC – either from one magnitude to another, or from one form to another, or from some state to its contradictory, and let D be the primary time in which it undergoes the change. Then in the time in which it is changing it must be either in AB or in BC or partly in one and partly in the other.[104]

All these options are inadmissible: in one case the indivisible simply hasn't changed at all, in another it has changed without *changing*, and on the third and final option it would not be indivisible: if one part is F and another non-F, it has distinct parts and is therefore divisible, at least in thought.

---

102   *Physics*, 6, 2, 233a 7-10.

103   See Furley, 2, p. 121.

104   *Physics.*, 6, 10, 240b 20-25.

This argument, Aristotle admits, could be met, but only if time were composed of indivisible minima and motion of 'starts', a theory which, he feels, he has already sufficiently refuted. His argument clearly tells only against mathematical indivisibles – there could be no objection to a physically indivisible atom being partly in one place and partly in another, for example.[105] To answer Aristotle, the Atomist must either opt for a conception of motion as discontinuous or posit atomic parts; Epicurus, as we shall see, does both.

It is clear that Book 6 of the *Physics* is as much concerned to refute the constitution of continua out of indivisible but extended minima as it is out of unextended mathematical points. In common with all the ancients, Aristotle accepted Z8 and Z9 (see Part 1) and so felt himself in a position decisively to refute any theory of the composition of a continuum out of points. Even an actual (denumerable) infinity of them will not suffice – as building blocks, they are plainly inadequate. Much of the argumentation of Book 6 is, however, directed against indivisible but extended minima of space, time, and motion: a finite number of these could constitute an extended magnitude. The arguments of Book 6 may, I venture to suggest, be directed against Democritus as well as such known advocates of indivisible magnitudes as Xenocrates. In any event, as we shall see, the argumentation of this book had a great influence on the subsequent direction of Atomist thought.

C. The Epicurean Response

In order to make sense of Epicurus, it is necessary to presume that he was familiar with the basic Aristotelian texts and was endeavouring, at a number of places, to argue against peripatetic doctrines.[106] If we bear in mind both the fundamental Atomist argument for indivisible minima and Aristotle's counterarguments from Book 6 of the *Physics*, much that seemed obscure in Epicurus will become clear.

A continuum, says Aristotle, cannot be composed of indivisibles. They cannot touch 'whole to whole' without coinciding, nor 'part to part' without having distinct parts. This is the most fundamental of all the peripatetic objections to indivisibles. In answer, Epicurus[107] proposes the following analogy.

---

105   Furley, 2, p. 113.
106   See A. E. Taylor, p. 40; A. A. Long, p. 10; Furley, 2, p. 121 ff.
107   Epicurus, 'Letter to Herodotus', 58–59, from Bailey, 1, p. 35.

The existence of a least perceivable magnitude is uncontroversial. Let this 'minimum perceivable' have the value of q^*. Every perceptible magnitude, Epicurus continues, must consist of an integral number of such minima – a magnitude $3q^*/2$ would be *perceived as* q^*. It could only be perceived as greater than q^* if, *per impossibile*, $q^*/2$ were itself perceptible. All perceived magnitudes consist therefore of integral multiples of q^*.

Now these minima do combine to form a continuum, the perceptual continuum. But what is actual is possible. Therefore (*pace* Aristotle) minima can associate to form an extended magnitude. (A strikingly similar argument can be found in Hume's *Treatise*.) If we look at the very edge of a perceptible body, we see its outermost minimum or 'point'; a 'point', for the Epicurean, is simply another term for a minimum – the term denotes a real physical entity, not a mathematical abstraction. Adjacent to this extreme 'point' on the outermost surface of the perceived body there must, says Epicurus, be another, similar point, its *successor*.[108]

> And when we look at these points in succession starting from the first, not within the limits of the same point nor in contact part with part, but yet by means of their own proper characteristics measuring the size of bodies, more in a greater body and fewer in a smaller.[109]

This immediately informs us that a *finite* number of these so-called 'points' make up an extended perceptual magnitude. (The terms 'more' and 'less' were considered inapplicable to infinite sets.) They are, therefore, minimal *magnitudes* rather than true mathematical points. They touch, says Epicurus, *neither* 'part to part' *nor* 'whole to whole', thereby evading Aristotle's dilemma, but rather serve as *measures* of the body's magnitude. This will hardly satisfy an Aristotelian – *how*, he will demand, do these minima touch if neither 'part to part' nor 'whole to whole'? Is there a third possibility? All Epicurus can do is what Hume does; i.e. insist that what is actual (in the case of perceptual minima) is possible, even if it is difficult to explain how it is possible.

The minimal parts of atoms, Epicurus continues, are analogous to these perceptual minima:

> Now we must suppose that the least part in the atom too bears the same relation to the whole; for though in

108   See Pancheri, p. 141.

109   'Letter to Herodotus', 58, from Bailey, 1, p. 35.

smallness it is obvious that it exceeds that which is seen by
sensation, yet it has the same relations. . . .we must
consider these least indivisible points as boundary-marks,
providing in themselves as primary units the measure of size
for the atoms, both for the smaller and the greater, in our
contemplation of these unseen bodies by means of
thought.[110]

Just as every perceptible magnitude is made up of an integral
number of perceptual minima, so every conceivable or
thinkable magnitude is composed of an integral number of
'minimal conceivables'. (The Epicurean notion of thinking or
conceiving an analogous to mental picturing facilitates this
analogy.[111]) Atoms have different sizes and shapes, and are
therefore to be thought of as constituted out of different
numbers and arrangements of tightly-packed 'minimal parts'.
By implying the existence of minimal conceivable magnitudes,
the perception-conception analogy provides the ground for an
entire theory of mathematical minima. (Fortunately, that
theory can be assessed on its own merits – one suspects that
the perception-conception analogy would not bear much
scrutiny.) There will be spatial minima of size q, the smallest
possible size, that of a single minimal part of an atom. But
then, by Aristotle's arguments, there must also be minima of
time and motion. Given a time-minimum t and a space-
minimum q, it will follow that no body can move at a velocity
other than $v = q/t$ (or zero). A velocity less than q/t involves
splitting the space-minimum; a speed in excess of q/t either
divides the time-minima or involves the possibility of a body
'leaping' from one place to another without passing through
the intermediate places, a possibility not seriously considered
before the rise of quantum physics.

This cinematic theory of motion follows by Aristotle's
arguments from the postulation of a minimal spatial
magnitude q. That such a theory can be attributed to
Epicurus is clear from Themistius, who writes as follows:

But our clever friend Epicurus is not ashamed to use a
remedy more severe than the disease – and this in spite of
Aristotle's demonstration of the viciousness of the argument. The moving object, he says, moves over the whole

110   *Ibid.*

111   See Furley, 2, p. 9. Barnes (Vol. 2, p. 540) dismisses the idea of a
minimal conceivable magnitude as a 'wretched muddle', but finds the
theory of a granular space made up of geometrical minima perfectly
conceivable.

distance, but over each of the indivisible units of which the whole is composed it does not move but *has moved*.[112]

Space therefore is composed of indivisible minima, over which motion is discontinuous. Epicurus is prepared to countenance 'X has crossed A' being true – where A is an indivisible – without 'X is crossing A' ever being true.

If we turn to the pages of Sextus Empiricus we find further evidence warranting our attribution of such a theory to Epicurus. Sextus quite clearly thinks that Epicurus – as well as Diodorus Cronos – is committed to such a theory, and raises a number of perceptive objections to it. Imagine, says Sextus,[113] two bodies separated by an odd number of space-minima, and let both bodies be moving with a velocity of q/t towards one common space-minimum. At some moment these converging bodies will be separated only by that single space-minimum. What happens next? They cannot both move into that space without either interpenetrating or dividing the indivisible. But if one or other of the bodies swerves aside or stops the implications are equally absurd. Why one body rather than the other? What becomes of the principle of sufficient reason here? And why should *either* body change course at all without a proper physical cause, such as a push? Is one body supposed somehow to 'know' of the presence of the other? Whatever the Epicurean says will raise immense – perhaps insuperable – difficulties.[114]

It is clear that Epicurus did accept the cinematic theory of motion.[115] Given the Democritean argument for indivisibles and the logical scaffolding provided by Aristotle, he could hardly have done otherwise. But now if all atoms really move at a velocity of q/t Epicurus can pre-empt one of Aristotle's most powerful arguments for infinite divisibility. If Nature really presents no different uniform velocities, the argument of *Physics*, 6, 2, will fail – its crucial premise is that α travels faster than β.[116]

How can the denial of differences in velocity be squared with sense-experience? The answer is surprisingly easy. An atom, for Epicurus, always moves at q/t: in every time-minimum t it moves a space-minimum q. If it always moves in

---

112 Themistius, *Physics*, 184.9, quoted from Furley, 2, pp. 113–114.

113 Sextus Empiricus, *Against the Physicists*, 142, p. 283.

114 Sextus (S 149–154) has further, and equally damaging, objections to the 'cinematic' theory of motion.

115 See Long, pp. 36–37; Konstan, 2, p. 398.

116 See Furley, 2, pp. 121–122.

the same direction, its speed is 'as quick as thought' – the passage of the atoms through resistanceless void space is like this.[117] Normally, however, atoms are involved in frequent collisions with one another, causing changes in the direction (though not the speed) of their motion. If, then, the minimal perceptible spatial and temporal magnitudes q^* and t^* are orders of magnitude above the absolute minima q and t, the problem is readily solved. Different net velocities at the perceptual level are perfectly compatible with constant atomic *speeds* (note the use of the scalar term) at the microscopic level.[118]

The Epicurean atom divides – mathematically of course rather than physically – into discrete minimal parts. These minima are necessarily incapable of separate existence:[119] they are essentially parts of atoms in the same way that a surface is essentially part of a body – the notion of a surface that is not the surface of some body is incoherent. Lucretius describes them as 'without parts' and 'endowed with the least nature'. A greater (finite) number of them make up a large atom; a lesser number constitute a smaller atom. Just as unity was thought of by the ancients not as a number but as the measure of number, so a minimum magnitude was conceived not as being itself extended but as the measure of extended magnitude. (Nevertheless, since a finite number of them constitute a measurable magnitude, elementary arithmetic will suffice to attribute to each one a minimum value of extension.)

By what arguments do Epicurus and Lucretius seek to establish the existence of these minima? If, says Epicurus, a body is infinitely divisible, it must be composed of an actual infinity[120] of parts:

> For if one says that there are infinite parts in a body or parts of any degree of smallness, it is not possible to conceive how this should be, and indeed how could the body any longer be limited in size? (For it is obvious that these infinite particles must be of some size or other; and however small they may be, the size of the body too would be infinite).[121]

---

[117] 'Letter to Herodotus', 61, 46b, from Bailey, 1, p. 37.

[118] See Furley, 2, p. 118.

[119] Lucretius, Book 1, 599–600 (p. 207 in the Bailey edition). See also Konstan, 2, p. 403 ff.

[120] For Epicurus's rejection of the potential infinite, see Furley, 2, p. 38.

[121] 'Letter to Herodotus', 56–57; Bailey, 1, p. 33.

If we remember the process of A-division – dichotomous division into aliquot parts – discussed in the first part of this chapter, this argument becomes intelligible. We need not simply assume that Epicurus misunderstood Aristotle's discussion of the sum of an infinite convergent series at *Physics*, 3, 6, 206b 6–10. If the actual infinity of parts are A-parts, there is 'some size or other' that they all share – (∃ s) (X) Sx rather than (X) (∃ s) Sx. If a body is infinitely divisible into P-parts, it must also be infinitely divisible into A-parts, and the question then immediately arises – as it did for Zeno – about the size of those A-parts.

Lucretius first introduces the concept of minimal parts in the ninth of his eleven arguments for physically indivisible atoms. The reasoning runs as follows:

> Since there are extreme points, one after another on that body which our senses can no longer descry,[122] each point, we may be sure, exists without parts and is endowed with the least nature, nor was it ever sundered apart by itself nor can it be so hereafter, since it is itself but a part of another, and that the first single part: then other like parts and again others in order in close array fill up the nature of the first body, and, since they cannot exist by themselves, it must needs be that they stay fast there whence they cannot by any means be torn away. The first-beginnings then are of solid singleness; for they are a close, dense mass of least parts.[123]

The 'first beginnings' are of course the atoms, the physical indivisibility of which is guaranteed by their composition out of a densely-packed mass of minima. One must not, of course, read Lucretius' '*cacumen*' as referring to a mathematical point in the sense of the geometers: mathematical points have no successors and cannot combine to yield magnitudes; minima have successors and, in aggregation, produce bodies. Atoms, then, are made up of 'a close, dense mass' of these 'least parts' or minima. Moreover, Lucretius continues,

> if there were not a least thing, all the tiniest bodies will be composed of infinite parts, since indeed the half of a half will always have a half, nor will anything set a limit. What difference then will there be between the sum of things and the least of things? There will be no difference; for however

---

122   Is there a hint here of the analogy with the minimal perceivable magnitude? Furley (2, p. 350) certainly thinks so.

123   Lucretius, Book 1, 599–601, p. 207.

completely the whole sum be infinite, yet things that are
tiniest will be composed of infinite parts just the same. And
since true reasoning cries out against this, and denies that
the mind can believe it, you must be vanquished and
confess that there are those things which consist of no parts
at all and are of the least nature. And since these exist, the
first-beginnings too you must needs own exist, solid and
everlasting'.[124]

Lucretius here produces a mathematical argument against
infinite divisibility and in favour of indivisibles. But, he
concludes, minima are essential parts of atoms – they cannot
be separated from atoms, nor can an atom be physically
divided into its constituent minima. If, therefore, minima
exist, there must also exist solid and external atoms, of which
those minima are, necessarily, integral parts.

Why should he think that infinite divisibility precludes size-
differences? The answer will turn on the process of A-division
and especially on the notion of the *ultimate* constitution of a
magnitude. If two bodies of different sizes are both divisible
into an actual infinity of A-parts they must, Lucretius can
argue, be made up of the *same number* of *equal* parts. Since
nothing can be greater than infinity, no infinite number can
be greater than another – this was a commonly accepted
premise of mathematicians and philosophers for many
centuries. But the ultimate A-parts of the larger body cannot
be greater than those of the smaller – all *ultimate* constituents
of all magnitudes are equal. Infinite divisibility, therefore,
would seem to entail the *equality* of all magnitudes: if all
bodies consist of an infinity of equal ultimate A-parts, none
can be greater than another. Only given finite numbers of
ultimate A-parts will it be meaningful to say of one magnitude
that it is larger than (i.e. consists of more minima than)
another.

Minima are also crucial to Lucretius' proof that (*pace*
Democritus) there cannot be an infinite variety of atomic
shapes. For such infinite variety to exist, says Lucretius, some
atoms would have to be infinitely large:

For, within the same tiny frame of any one single seed, the
shapes of the body cannot be very diverse. For suppose the
first-bodies to be made of three least parts, or if you will,

---

[124] Lucretius, Book 1, 615–627, p. 207. For the possible targets of this
       argument (the Stoics?) see Furley, 1, pp. 14–15. For a strikingly similar
       argument see Plutarch, Vol. 13, pp. 813–815.

make them larger by a few more; in truth when you have tried all those parts of one body in every way, shifting top and bottom, changing right and left, to see what outline of form in that whole body each arrangement gives, beyond that, if by chance you wish to make the shapes different, you must needs add other parts; thence it will follow that in like manner the arrangement will ask for other parts, if by chance you still wish to make the shapes different: and so greater bulk in the body follows on newness of forms. Wherefore it is not possible that you can believe that there are seeds with unbounded difference of forms, lest you constrain certain of them to be of huge vastness, which I have taught above cannot be approved.[125]

This argument quite clearly presupposes that a finite number of minima can only be arranged in a finite number of ways. This is evident if the minima are mathematical minima, units than which nothing smaller can be conceived. Such units – which we shall represent for simplicity as cubes – can only combine full-face to full-face. Two minima, then, could only combine in one way; three could join in two ways; four in nine ways, thus:

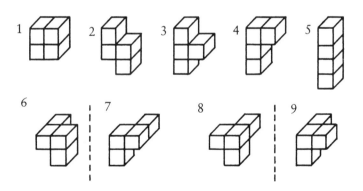

6 and 7, 8 and 9 are mirror-image forms of one another. For any finite number of minima, there will be a corresponding finite number of distinct atomic shapes. Lucretius' argument goes through in a perfectly straightforward manner

125 Lucretius, Book 2, 483-499, pp. 261-263.

– an infinite variety of shapes would require gargantuan atoms, composed of an infinity of minima.

The argument only works, however, if one takes the minimal parts to be *minimal conceivables*, indivisible even in thought. Unless this is assumed, one can generate an infinite number of atomic shapes with any number of minima – at a pinch, with only two – as follows:

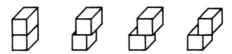

AND SO ON.

So long, then, as the 'minimum' is divisible in thought, Lucretius' argument will not get off the ground.

We have now built up a very strong case for construing Epicurean minimal parts as *minimal conceivable units of extension*. The case might be itemised as follows:

1. The minima are explicitly called 'partless' and 'least parts' by Epicurus and Lucretius.

2. They must be indivisible in a stronger sense than the merely physical if their postulation is not to be a sheer redundancy: the atom is already physically indivisible.

3. The analogy with the perceptual minimum and the whole cinematic theory of motion support the conception of minimal parts as mathematical minima.

4. Epicurus and Lucretius both argue quite explicitly against infinite (mathematical) divisibility for what are basically the (Zenonian) reasons outlined in the first part of this chapter.

5. The Lucretian argument just discussed turns on the existence of minimal conceivables.

I conclude, with Furley, that:

> The *minimae partes* of atoms in Epicurean theory were theoretically indivisible portions of matter: that is to say, they were such that no parts could be distinguished within them by the mind.[126]

---

126 Furley, 2, p. 4.

The only scholar who has attempted to account for all the relevant textual material while still resisting this conclusion is Gregory Vlastos.[127] He alleges that the minima serve as units of measure for extended magnitudes, but are not themselves mathematically indivisible. Instead, he argues, the Epicureans put forward the following pair of theses, T1 and T2, as *contingent*, empirical hypotheses:

T1. 'Atoms are so constituted that variations in atomic lengths occur only in integral multiples of the smallest atomic length.'

T2. 'Atoms are so constituted that variations in their shapes occur only in permutations of a modular unit of invariant size and shape.'[128]

T1 accounts for the rôle of the minima as measures of extension; T2 accommodates Lucretius' argument against infinite variability of atomic shapes. There remains, however, something profoundly unsatisfactory about Vlastos' interpretation, its ingenuity notwithstanding. He has to give an attenuated sense to the use of the terms 'least part' and 'partless' of the minima. He fails to see and to account for the Zenonian roots of the Epicurean arguments against infinite divisibility, arguments which seem to demand the postulation of mathematical minima. He cannot deal adequately with the Epicureans' cinematic conception of motion (although of course he could postulate further empirical theses T3, T4, etc., to fill this deficiency). Above all, T1 and T2, on Vlastos' interpretation of them (i.e. as contingent propositions) have an arbitrary and unfounded look: how could an Epicurean ever have discovered them or sought to justify them? On Furley's view, which we have advocated, T1 and T2 follow as *necessary* corollaries of the theory of conceptual minima. Furley's view captures all that is best in Vlastos', but not *vice versa*.

Vlastos' only real *argument* against attributing mathematical indivisibles to the Epicureans is the argument from geometry. The discovery of incommensurables, he claims, conclusively refuted – and was *seen* to have conclusively refuted – any such theory. Epicurus, coming after Euclid and Eudoxus, must have accepted the infinite divisibility of the continuum.[129] We have, however, already effectively refuted this argument by pointing out the existence of an

---

127   Vlastos, 3.

128   *Ibid.*, p. 138.

129   *Ibid.*, pp. 125–126.

anti-geometric Atomic tradition from antiquity to the seventeenth century. If Ash 'ari, Nicholas of Autrecourt and Walter Charleton can all deny the truth (or physical applicability) of the principles of geometry, why not Epicurus?[130]

There remain, however, serious problems about the minima. Do they, for example, have *shape*(s)? If finite size entails shape, one assumes that they must. Earlier I depicted them as cubes: this has the advantage that all their sides are equal. But a cube has diagonals: lines which, intuitively, would be longer than one minimum but less than two, which is impossible.[131] Are the minima then spheres? All the diameters of a sphere are equal. But spheres still have chords . . . True spheres, moreover, touch only at mathematical points and do not pack tightly. (An atom made up out of spherical minimal parts would not be perfectly solid but would contain internal vacuum.) More generally, anything with a determinate shape must have distinguishable parts (e.g. centre and extremity) and could not therefore be a minimal conceivable magnitude.

In the final analysis, then, one must deny that the minima have shape and insist that one should not try to picture or imagine them, nor apply even rudimentary and intuitive geometrical ideas to them.[132] While there is a significant positive 'Analogy of Nature' between an atom and a macroscopic, compound, perceivable body, below the level of the atom this analogy breaks down completely. Advocates of mathematical minima are committed to a profound and radical *disanalogy* between the microstructure and the macrostructure of space. Ordinary geometry and mental images are quite inapplicable to the minimal conceivables. The physics of indivisibles is not without surprising implications – and profound difficulties.[133]

---

[130] For the possibility that the Epicureans had their own rival geometry, based on indivisibles, see Mau.

[131] If the lost work of Epicurus on 'The Angle in the Atom' had survived, we might know more about his views on these difficult matters. See Diogenes Laertius, Vol. 2, p. 557, for a list of Epicurus' writings.

[132] For a similar theory, and the resulting objections to 'the tyranny of the imagination' in attempting to visualise the least things or 'perfect parvitudes', see Henry More, 1, Vol. 2, pp. 3–4. The least things, More insists, *have no shape at all*; reason must overcome the demands of 'fancy' and insist on this conclusion.

[133] For some of these problems, see Konstan, 2, pp. 394–407.

VACUUM: 1
The Ancients

A. The Eleatic Challenge

What is, says Parmenides in the 'Way of Truth', is, and cannot not be. Likewise, what is not is not, and cannot be.[1] But void or empty space (*to kenon*) is not (anything); it is a privation of being. Hence it does not and cannot exist – a thesis with momentous consequences.

In the first place, 'what is' (Being) must exist in the form of a single plenum: being could only be separated from being by non-being. Since there can be no empty or void space dividing one being from another, all being is one.[2]

This single, unique Being must moreover be perfectly homogeneous, since it is absolutely full of being. It could only be heterogeneous, the Eleatics contend, if there were more being in one part of it and less in another, which could only be the case if being could, as it were, be in places 'diluted' by non-being. But this, Parmenides insists, is impossible:

> neither is there anything that is not, so that there might be of what is here more and there less . . ..[3]

Melissus too quite clearly presupposes that differences in density and rarity could only be due to the admixture of different proportions of void (non-being); since void cannot exist, he argues, being must be homogeneous. Melissus' 'Being' is, says Barnes,

> a solid, physical body; it is 'full', a material occupant of space; and it is impenetrable, refusing to countenance any co-occupants.[4]

The non-existence of void therefore precludes (a) the existence of a plurality of beings and (b) the internal

---

[1] For more detailed discussion of the 'Way of Truth', see Barnes, Vol. 1, p. 155 ff; Kirk & Raven, p. 269 ff, and Cornford, 2, p. 30 ff.

[2] This argument is most clearly found in Melissus. See Sweeney, p. 128.

[3] Quoted from Barnes, Vol. 1, p. 179.

[4] Barnes, Vol. 1, p. 226. I follow Barnes in rejecting Simplicius' testimony to the effect that Melissus' 'One' is incorporeal.

differentiation of the 'One' into distinct parts. It will only
have the latter implication, however, if it is assumed that
being as such is perfectly *homogeneous*. Why, we might ask,
should not the 'One' consist of qualitatively different parts?
Must all qualitative differences depend on differences of
density and hence ultimately on the admixture of non-being
with being? Such does seem to be the Eleatic position.

One reading of Parmenides, that of Taran, provides at least
a hint of an Eleatic argument for this homogeneity thesis.
According to Taran's Parmenides:

> Being is being and nothing but being. Only being is real.
> There can be no difference in what exists, so nothing can be
> distinguished inside Being . . . Being is all alike because it
> can only be without restriction.[5]

Parmenides, on this view, treats 'Being' as the name of a
substance rather than as an abstract noun.[6] Since all that is is
being, it must have the essential nature of being – all
properties not properties of being as such can safely be
dismissed as unreal. Being, then, lacks altogether those
qualities such as hot, cold, red, blue, etc., which might be
thought capable of differentiating one part of it from another.
Qua being and nothing but being, it can only be
homogeneous.

Alternatively, the Eleatics could have argued from monism
to homogeneity. What is, says Parmenides, is *homoion*, all
alike. This, according to Barnes, may indicate only
'existential' homogeneity (all parts of being exist equally,
without gaps or intervals of non-being) and not qualitative
homogeneity.[7] It is only in Melissus that we find a clear
argument from monism to qualitative homogeneity:

> And being one, it is in all respects homogeneous
> (*homoion*); for if it were heterogeneous (*anhomoion*),
> being several things it would no longer be one but many.[8]

The idea seems to be that qualitatively different parts could be
distinguished by the mind, and thus seen to be 'many',

---

[5]   Quoted from Sweeney, p. 102.

[6]   Cf. Aristotle, *Physics*, 1, 3, 186a 21–26. Parmenides' monism rests, says
     the Stagirite, on the unargued and false premiss 'that things are said to be
     in only one way . . .'. On this reading, Parmenides slides from ' "Being" is
     univocal' *via* ' "Being" names one thing' to 'There is one and only one
     being'.

[7]   Barnes, Vol. 1, pp. 210–211.

[8]   Quoted from Barnes, Vol. 1, p. 208.

whereas qualitatively identical parts are not a threat to monism. (Or perhaps the Eleatic 'One' is 'one' in such a strong sense as to be altogether without parts?)

In any case, Melissus' 'Being' is qualitatively homogeneous. So too, I suspect, is that of Parmenides: this seems clearly to be his intention,[9] even if no explicit statement to that effect can be found in the 'Way of Truth'. Since Eleatic being is perfectly homogeneous, the mere non-existence of non-being (void) will preclude both plurality (the existence of many distinct and separate things) and internal differentiation within the 'One Being' that remains.

Another consequence of the non-existence of the void is the impossibility of local motion, an implication clearly seen by Melissus:

> Nor is it [Being] empty (*keneon*) in any respect; for what is empty is nothing, and it will not be nothing. Nor does it move; for it has no way to retreat, but it is full. For if it were empty, it would retreat into what was empty; but not being empty, it has not any where it may retreat.[10]

Melissus thus quite clearly accepts the following principle V:

V. No body can move unless there is empty space (void) for it to move into.

If empty space is indeed, as V asserts, a necessary condition of local motion, then the Eleatic denial of void gives them what seems a conclusive refutation of the existence of locomotion. According to Aristotle:

> Melissus, indeed, infers . . . that the All is immovable; for if it were moved there must, he says, be void, but void is not among the things that exist.[11]

Whether or not the above argument (and principle V) can be found in Parmenides is a matter of controversy. Some commentators[12] deny that Parmenides is the author of the above argument and contend instead that his (undoubted) denial of local motion rested solely on his general objection to all forms of change and coming-to-be on ontological grounds. Since both local motion and qualitative alteration (*alloiosis*)

---

9 *Pace* Barnes, Vol. 1, p. 212.

10 Quoted from Barnes, Vol. 1, p. 214.

11 *Physics*, 4, 6, 213b 12-13.

12 Kirk and Stokes, p. 1; Barnes, Vol. 1, p. 222.

involve the coming-to-be of something new, both fall foul of Parmenides' attack on *genesis*.

Bicknell, however, dissents from this view and re-affirms the more traditional interpretation. The argument of Parmenides' fragment 8, lines 29–33, can, he claims, be paraphrased as follows:

> What-is could only change its position if there were vacancies into which it could move. If, however, there were such vacancies then what-is would be incomplete. But what-is could not be incomplete for if it were then it would fall short of completeness by what is not (literally, 'what is not would be lacking from the whole') and what is not is inconceivable and unutterable.[13]

This may not explicitly mention *to kenon*, as Melissus does, but is nevertheless, Bicknell insists, essentially the same argument – *from* V and the impossibility of the existence of void space *to* the non-existence of local motion.[14]

For our purposes it is not necessary to settle the above exegetical controversy. We can envisage Leucippus as replying to an *Eleatic* argument based on V without worrying overmuch about the precise source of that argument. Suffice it to say that there existed an Eleatic argument (certainly in Melissus, and possibly in Parmenides as well) from V and the denial of void space to the non-existence of local motion.

B. The Atomist Response: Leucippus and Democritus

If one identifies Parmenides' 'Being' with homogeneous matter* and his 'Non-Being' with void, one arrives at the ontology of Leucippus and Democritus, who, according to Aristotle,

> say that the full and the empty are the elements, calling the one being and the other non-being – the full and the solid being being, the empty non-being (whence they say being no more is than non-being, because the solid no more is than the empty); and they make these the material causes of things.[15]

The Eleatics' denial of non-being had led them to dismiss all differentiation, plurality and motion as merely illusory, no

---

13   Bicknell, p. 3.

14   *Ibid.*, p. 3.

15   *Metaphysics*, 1, 4, 985b 5–10. See also Diogenes Laertius, Vol. 2, p. 441.

part of true reality, a conclusion which, as Aristotle saw,[16] would undermine physics altogether. Although the Eleatic conclusions appear to follow dialectically, he says, 'yet to believe them seems next door to madness when one considers the facts'.[17] Our senses reveal a world of many, qualitatively different, moving objects: it is the business of physics to account for these appearances in an intelligible manner. (If sense-experiences are to be deemed illusory, one requires at least an explanation of how the illusion arises.)

By identifying 'Being' with their featureless, homogeneous matter*, the neo-Eleatics,[18] Leucippus and Democritus, left themselves no effective option: to 'save' the phenomena (plurality, differentiation, local motion) they were forced to posit the existence of non-being or void. The Atomists, says Aristotle, concede too much to Parmenides' arguments:

> Some people gave in to both arguments: to the argument that if 'is' means only one thing, all things must be one, when they said that there is that which is not, and to the argument from dichotomy when they posited indivisible magnitudes.[19]

Here it is to 'save' plurality that the existence of the void is posited: without such empty space, all would be a single homogeneous Parmenidean plenum. (More aphoristically, 'Democritus minus void equals Parmenides'.) Elsewhere, of course, vacuum is admitted to 'save' local motion:

> Leukippos . . . thought he had a theory which harmonised with sense-perception and would not abolish either coming-to-be and passing-away or motion and the multiplicity of things. He made these concessions to the facts of perception: on the other hand he conceded to the Monists that there could be no motion without a void. The result is a theory which he states as follows: 'The void is a "not-being", and no part of "what is" is a "not-being"; for what "is" in the strict sense of the term is an absolute *plenum*. This plenum, however, is not "one": on the contrary, it is a

---

[16] See *Physics*, 1, Chapters 2 and 3.

[17] *De Gen et Corr*, 1, 8, 325a 18-20.

[18] For Leucippus' debt to Parmenides, see Kirk & Raven, pp. 400-401. According to Simplicius (*Physics*, 28, 4) Leucippus has 'associated with Parmenides in philosophy'; we need not believe this to be a literal historical truth to accept the enormous Eleatic influence on early Atomism.

[19] *Physics*, 1, 3, 187a 1-4. See also Stokes, p. 219.

"many" infinite in number and invisible owing to the minuteness of their bulk. The "many" move in the void (for there is a void): and by coming together they produce "coming-to-be"; while by separating they produce "passing-away" '.[20]

This is reminiscent of Melissus' remark to the effect that 'if there were many things, they would have to be of the same description as I say the one is'.[21] Each Leucippan atom is a perfectly full plenum of matter*, a homogeneous and indivisible 'One' – eternal, immutable, absolutely solid being.[22] All Leucippus has done is to *multiply* these Eleatic 'ones' and give them *room* in which to move: their mere local motion will then suffice to save the appearances.

Space, for Leucippus and Democritus, is *to kenon*, 'the empty', i.e. *unoccupied* place or void.[23] It is complementary to matter* and mutually exclusive with it – matter does not *occupy* but rather *excludes* space.[24] Space was not conceived as an abstract three-dimensional framework 'containing' all things, but as the much more concrete notion of an *ontological* void or emptiness.[25] It was, says Grant, this identification of space with nothing (*inane, nihil*) that guaranteed paradoxes and difficulties for subsequent ages.[26] How can *nothingness* exist and have magnitude?

What Leucippus and Democritus need, clearly, is a viable distinction between 'reality' and 'existence'. Only the atoms are real beings – only they could be said to exist in the fullest sense. Space, however, although it may be called 'non-being', nevertheless exists ('. . . being no more is than non-being, because the solid no more is than the empty'[27]). Tangibility ceases to be the criterion of existence, although it remains the touchstone of reality. Although the void exists, it is not something positive but rather a privation of reality, a *lack* of (Eleatic) being, an *emptiness*.[28]

20   *De Gen et Corr*, 1, 8, 325a 24–33.

21   Quoted from Furley, 2, p. 57.

22   See Bailey, 2, p. 71.

23   But see *Physics*, 4, 5, 213a 15–19, where Aristotle attributes to the early vacuists a more sophisticated theory of space.

24   Jammer, 1, p. 9.

25   Bailey, 2, p. 18.

26   Grant, 9, p. 3, p. 284.

27   Aristotle, *Metaphysics*, 1, 4, 985b 8–9.

28   Bailey, 2, p. 75.

Before moving on to Aristotle's critique of 'vacuum', let us briefly mention two further aspects of that theory. One doctrine associated with Atomism from its beginnings was that of the infinite extent of the universe. This thesis follows from the notion of matter* and vacuum as mutually delimiting principles, combined with the conception of void as a mere privation of matter*. If every portion of void is delimited by matter* and *vice versa*, there can be no limit or extremity to the physical universe, no absolute boundary of 'the all' (*to pan*). Since our material cosmos is finite and bounded, it must make sense to ask what lies beyond that boundary – the answer can only be 'void'. Space must be infinite in extent (a limit or bound is inconceivable), so even if our cosmos exhausted all the matter* there is there would have to be an infinite extracosmic void outside it. But, if there were only a finite number of atoms in infinite space, they would become so sparsely scattered as to be incapable of aggregating to form a world at all. It is therefore more probable that the universe consists of an infinity of atoms moving, combining to form worlds (*cosmoi*) and separating to dissolve worlds, all in a void space of infinite extent. Such is the basic cosmological vision of the Atomists.[29]

The second point involves an abrupt change of scale from the macro to the micro-realm. If all compound bodies are made up of different proportions of matter* and void, it should be possible to explain their physical properties in terms of this constitution. Bodies with more internal vacuum will tend on average to be softer, more liable to cutting and crushing, than those containing less. Condensation can be explained in terms of the compression of the atoms of a body into its interstitial vacua, rarefaction in terms of the separation of the atoms leaving more internal void.

Given the homogeneity of matter* and the immutability of the atoms, Atomism will also entail the existence of a physical limit to the compression of any given body. More precisely, it will entail the existence of three different types of limit, namely:

L1.   *The Theoretical Limit*: A body cannot in principle be condensed into a smaller volume than the total bulk of its matter*, i.e. the sum of the volumes of its constituent atoms.

---

[29]   I find it far more plausible to try to derive the Atomists' cosmology from their ontology (matter* and void) than to suggest, with Cornford (3, p. 12) that the Atomists arrived at their vision of infinite space by attributing physical existence to the abstract space of the geometers.

(Matter*, being already absolutely *full*, is not itself liable to condensation.)

L2.   *The Ideal Practical Limit*: Unless the constituent atoms of a body will pack perfectly so as entirely to exclude all internal vacuum, L1 will not be practically attainable. A set of perfect equal cubes could be packed so as to attain L1, but Democritean atoms vary in size and shape; in all probability a given set of them, chosen at random, will not be capable of such tight packing. Given the invariant sizes and shapes of all the atoms, there will be a minimum volume into which they could conceivably all be packed, with the smallest amount of space left empty. This 'ideal practical limit' (L2) could, however, only be attained if the atoms happened to fall into the ideal arrangement prior to or during compression.

L3.   *Actual Practical Limits*: If one were to take a set of Democritean atoms at random and compress them as far as they would allow, one would obtain a volume V_1. If the same atoms were then jumbled and recompressed, one would, in all probability, arrive at a different volume V_2. Sooner or later – perhaps much later – one should by this random process arrive at L2. (One will not, however, know that one has done so.)

The existence of condensation and rarefaction supplies the Atomist with a further argument for the existence of the void. Given perfectly homogeneous matter*, there must be interstitial vacua if condensation is to take place: since the immutable constituent atoms of a body cannot themselves undergo condensation, there must be empty space in a compressible compound body for them to withdraw into under pressure.

Another advantage of the postulation of interstitial vacua is that it enables one to explain the *permeability* of compound bodies to heat, cold, sound, etc.: if these are conceived as carried by some material bearers, permeability will (given V) need to be accounted for in terms of pores. This became a stock argument in favour of that theory.

Furthermore, the Atomist can claim, we are now in a position to give a viable account of the differences in specific gravity (S.G.) between observable bodies. Since matter* is perfectly homogeneous, we can expect either all of it to gravitate, or none. The perceived differences in S.G. between observable substances can readily be explained on the plausible supposition that the S.G. of a given substance depends on its ratio of matter* to void – the higher the proportion of matter*, the greater the S.G. While all matter*

as such has gravity and tends to fall,[30] bodies containing a high degree of inner vacuum will have a weaker downward tendency and may therefore tend to be forced upwards.

Enough has been said by now to give a sufficiently clear picture of the rôle of the vacuum in the system of Leucippus and Democritus. It is evident that that rôle is an absolutely fundamental one: without the void, Democritus' universe would collapse into a Parmenidean 'One', an eternal, immutable, motionless *block* of matter* or Being. A refutation of the existence of the void would therefore be a refutation of Democritean physics – whatever one might say of the views of the sphinx of Elea, they do not constitute a credible physical theory.

C. Aristotle Against the Void, 1: Refutation of the Pro-Vacuist Arguments

What grounds, Aristotle asks in Chapter 6 of Book 4 of the *Physics*, have been given for belief in the existence of the void, of 'place bereft of body'?[31] There have, he says, been arguments given based on the following phenomena:

    a.   Locomotion.
    b.   Condensation and Rarefaction.
    c.   Two bodies (apparently) occupying the same place.
    d.   Plurality.[32]

a.   The argument from locomotion to the existence of void or empty space is the familiar one of Leucippus, based on principle V. If Aristotle is to accept locomotion and still reject the void, he must show that V – common ground to Eleatics and Atomists alike, is false. The void, he insists,

> is not in the least needed as a condition of movement (*kinesis*) in general, for a reason which, incidentally, escaped Melissus; viz. that the full can suffer *qualitative* change.[33]

This may seem irrelevant to our discussion, but in fact turns out to be crucial to it: among the varieties of *kinesis* which

---

[30] I shall not discuss here whether the gravity of the atoms is innate and intrinsic to them, or a product of the 'cosmic whirl'. The issue is difficult and controversial, and does not need to be resolved in this place – my point will hold on either view.

[31] For Aristotle's characterisations of void space, see *Physics* 4, 1, 208b 26; 4, 7, 213b 31.

[32] See *Physics*, 4, 6, and Solmsen, 4, p. 136.

[33] *Physics*, 4, 7, 214a 26-28.

Aristotelian matter can undergo, one must include (intrinsic) condensation and rarefaction. How is this possible?

Aristotle's sophisticated treatment of the concept of 'Being' in the *Metaphysics* allows him to escape Eleatic pitfalls and permit qualitative differences (and changes) within Being. 'Being' becomes for Aristotle a highly abstract and formal notion, applicable – with different but analogous senses – in each of the categories, rather than a sort of super-substance endowed *per se* with the unchanging properties of unity and self-identity. There can, for Aristotle as for common sense, be a plurality of substances qualitatively distinct one from another without engendering the contradiction 'Being is F and non-F'. There are (many) beings, but no Being-as-such. Hence Aristotle can deny the homogeneity of matter demanded by the Eleatics (and granted by the Atomists) without self-contradiction.

Among the forms of change accepted by Aristotle are the intrinsic condensation and rarefaction of matter. Since Aristotle can deny the homogeneity thesis, he can reject homogeneous matter*. His matter can undergo condensation without interstitial vacua: it can, as it were, occupy space more or less *intensely*. (The void, says Aristotle, might be said to have 'potential' existence, in so far as matter can be rarefied without limit.[34]) Thus, he says,

> things can increase in size not only by the entrance of something but also by qualitative change; e.g. if water were to be transformed into air.[35]

Dense differs from rare, says Aristotle elsewhere, 'in containing more matter per unit area'.[36] Once one denies the existence of homogeneous matter*, the existence of such matter susceptible to intrinsic condensation and rarefaction becomes intelligible. But, if matter can be condensed and rarefied in this way, local motion will be possible without the existence of a void.

Even without such intrinsic condensation and rarefaction, says Aristotle, V would not hold, since:

> bodies may simultaneously make room for one another, though there is no interval separate and apart from the

---

[34]   *Metaphysics*, 9, 6, 1048b 9–17.

[35]   *Physics*, 4, 7, 214a 34–b2.

[36]   *De Caelo*, 3, 1, 299b 9.

bodies that are in movement. And this is plain even in the rotation of continuous things, as in that of liquids.[37]

This theory of cyclical mutual replacement of bodies was adopted by Plato in the *Timaeus*,[38] and of course was in time to become the Cartesian solution to the problem of reconciling the existence of local motion with plenism. (Since Descartes, unlike Aristotle, *did* accept homogeneous matter*, the cyclical replacement theory was the only solution open to him; for Aristotle, it is of less significance.)

Aristotle therefore has a two-pronged answer to the argument, based on V, from the phenomenon of local motion to the existence of void space. In the first place, he argues, a body can move in a medium which, although full, is susceptible to real (i.e. intrinsic) condensation and rarefaction. Furthermore, he can add, the Platonic theory of cyclical replacement is sufficient to show the falsity of V.

b.   The phenomena of condensation and rarefaction had been cited by the Atomists as evidence of the existence of interstitial vacua. It should now have become transparently clear that this argument rests on the assumption of the existence of particles of matter*, themselves immune to condensation and rarefaction. Deny this assumption, as Aristotle does, and the Atomists' argument will seem a blatant *petitio principi*, This alleged proof of the existence of interstitial vacua will only work, as Aristotle sees, if one first grants the (unproved) Eleatic premiss of the homogeneity of matter.

c.   The Atomists' third argument for interstitial vacua starts from the (allegedly) observed fact that a vessel full of ashes can absorb as much water as the same vessel when empty. This, according to the Atomists, would involve the interpenetration of bodies (two bodies coming to occupy the same place) *unless* the ashes and water both contained interstitial vacua. On that assumption, the water-particles can fill the pores or vacuities in the ashes without the question of interpenetration arising.

Once again, however, it is clear that the argument presupposes the homogeneity thesis. On an Aristotelian view, the two substances can interact to form a qualitatively different (and *denser*) product without any actual interpenetration of bodies taking place. Aristotelian matter must fill

---

[37] *Physics*, 4, 7, 214a 29-32.

[38] *Timaeus*, 79-80, p. 106 ff. The theory dates back to Empedocles (see Barnes, Vol. 2, pp. 95-97).

some space,[39] but is not to be quantified in terms of the amount of space it occupies at any given moment; we must, says Joachim, think of *hyle* as 'capable of filling space with all possible degrees of intensity'.[40]

d. The argument from plurality to the existence of the void also presupposes the homogeneity thesis. Given a homogeneous stuff like Eleatic 'Being', *only* non-being or void could differentiate distinct entities. For Aristotle, however, matter is differentiated by *forms* and *qualities*: one thing can be distinct from another in virtue of these without any need for empty space to separate them.

D. Aristotle Against the Void, 2: Conceptual Objections

What, asks Aristotle in Book 4 of the *Physics*, is *place*? The observed fact of the mutual replacement of bodies testifies to its existence, but leaves us in the dark about its nature. Can the place of a body be identified with either its matter or its form? Surely not: 'The form and the matter are not separate from the thing, whereas the place can be separated'.[41]

Just as a body can leave a vessel or container, so can it quit its place. This suggests to Aristotle that place is like a vessel or container, something which is of course no part of the body contained.[42] With some refinements, this will emerge as his final position: place, he says, is a 'non-portable vessel':[43] more precisely, '*the innermost motionless boundary of what contains is place*'.[44] The place of a body is thus analogous to an infinitely thin 'skin' around it and containing it, a two-dimensional (2D) surface marking out the bounds of the body yet – unlike its own form or shape – separable from it and capable of holding a quite different body.

Of the theories of place which Aristotle rejects,[45] one is of special importance. This is the view that the place of a body can be identified with the interval or gap between its

---

[39] Sokolowski, p. 277.

[40] Quoted from Hesse, 1, p. 65.

[41] *Physics*, 4, 2, 209b 22-24.

[42] *Ibid.*, 209b 28-29.

[43] *Ibid.*, 4, 4, 212a 14.

[44] *Ibid.*, 20-21.

[45] At *Categories*, 6, 5a 5-14, Aristotle seems to *accept* the 3D 'interval' theory of place. This passage has always been a thorn in the side of peripatetic commentators; many have followed Averroes in the suggestion that Aristotle was only outlining a common or 'vulgar' view, the falsity of which he exposed in the *Physics* (see Grant, 9, p. 9).

extremities. Proponents of such a theory, says Aristotle, assume that, when one body replaces another in a given vessel, 'the extension is something over and above the body displaced',[46] i.e. that there exists a (3D) extension which is not the magnitude of any body, 'an extension which were such as to exist independently and be permanent'.[47] On this view, place is a 3D incorporeal reality, the dimensions of which interpenetrate with those of (successive) bodies.

Let us call such a 3D incorporeal reality vacuum$_1$. It is important to realise that vacuum$_1$ need never actually be empty or void (= vacuum$_2$): it could be contingently true that vacuum$_1$ is always full, that every part of it is always occupied by some body. Vacuum$_1$ is of such a nature that, existing as it does independently of body, it *could* exist as void or vacuum$_2$. Vacuum$_2$ therefore implies vacuum$_1$ but not *vice versa* – any believers in void space, such as the Atomists, are implicitly committed to a conception of place as vacuum$_1$.

Aristotle objects to the 'interval' or vacuum$_1$ theory of place that 'there is no such extension'[48] separate from those of bodies. For Aristotle there can be no *incorporeal dimensions*: 'three-dimensional', he insists, entails 'corporeal', making the notion of vacuum$_1$ 'really vacuous'.[49]

How does Aristotle argue for this fundamental thesis, '3D ⟶ corporeal'? At the back of his mind there resides, of course, the doctrine of categories as elaborated in the *Metaphysics*. 3D extension belongs of course to the category of quantity. But a quantity must inhere in a substance: it must be a quantity *of something*.[50] A quantity of nothing (or, worse still, of nothingness) is quite unintelligible.[51] But the only 3D extended substances are material ones. Therefore 3D extension can only inhere in a corporeal substance. This thesis has momentous implications, and was accepted by good peripatetics for many centuries.[52] It lies, as we shall see, at the back of the famous 'cube' argument, which runs as follows.

---

46 *Physics*, 4, 4, 211b 15-16.

47 *Ibid.*, 19-20.

48 *Ibid.*, 16. See also Henry, p. 561, for Aristotle's rejection of the theory of place as vacuum$_1$.

49 *Physics*, 4, 8, 216a 27-28.

50 For the primacy of substance, see *Metaphysics*, Z1.

51 See Grant, 9, p. 17.

52 See *Ibid.*, p. 6.

Suppose, says Aristotle, one places a solid cube into a fluid medium such as air or water; clearly, it *displaces* its own bulk or volume of that medium. But in a void, he continues,

> this is impossible; for it is not body; the void must have penetrated the cube to a distance equal to that which this portion of void previously occupied in the void, just as if the water or air had not been displaced by the wooden cube but had penetrated right through it.[53]

The wooden cube, however, still retains its own characteristic bulk or volume in virtue of which,

> it will occupy an equal amount of void, and fill the same place, as the part of place or of the void equal to itself. How then will the body of the cube differ from the void or place that is equal to it? And if there can be two such things, why cannot there be any number coinciding?[54]

For a body to occupy an equal volume of vacuum$_1$ would require the *interpenetration* of dimensions; but this, says Aristotle, is absurd: if dimensions can interpenetrate, all bodies could collapse together.

This objection tacitly presupposes the '3D →corporeal' premiss, or at least that the impenetrability of a body depends on its dimensionality.[55] Advocates of vacuum$_1$ will have to distinguish incorporeal from corporeal dimensions, and insist that the impenetrability of bodies is due to their solidity rather than to their dimensionality as such. Aristotle, perhaps for the metaphysical reasons mentioned earlier, seemed to find the notion of incorporeal extension hard to grasp: the cube argument tacitly treats void as a subtle material medium, and then argues that as such it would have to interpenetrate with bodies.[56]

Even if such interpenetration were possible, Aristotle continues, the existence of vacuum$_1$ would be *superfluous*; why, he asks, 'need we assume a place for bodies over and above the volume of each . . .?'[57] To posit an extra set of (incorporeal) dimensions is a mere redundancy. Each body has, one might say, its internal space, but this is the space *it* fills, that is occupied by the dimensions of that particular

---

53   *Physics*, 4, 8, 216a 33–b2. See also Grant, 7, p. 551.

54   *Physics*, 4, 8, 216b 8–11. See also Wolfson, 1, p. 147.

55   See Wolfson, 1, pp. 59–60, p. 187. This is of course the premiss that advocates of vacuum$_1$, such as Philoponus and Crescas, must deny.

56   See Grant, 7, pp. 551–552.

57   *Physics*, 4, 8, 216b 13–17.

body. Why should we be tempted to posit a further (and unintelligible) set of incorporeal dimensions?[58]

But now, if the conception of place as vacuum$_1$ is rejected, the non-existence of vacuum$_2$ or void follows as an immediate consequence. As Aristotle says, 'if separate place does not exist, neither will void . . .'[59] Once one has grasped the Aristotelian definition of place as 2D surface of containing *body*, it becomes plain that 'on this showing void does not exist, either unseparated or separated'.[60] A body in a vacuum$_2$ would have no (Aristotelian) place at all – it would be nowhere: 'If then a body has another body outside it it is in place, and if not, not'.[61] In a void, moreover, there would be no places, hence *a fortiori* no termini '*a quo*' and '*ad quem*', hence – for Aristotle – no possibility of motion.[62]

Aristotle thus presents a number of arguments of a conceptual nature against the existence of vacuum$_2$ or empty (void) space. Most of those arguments turn, however, on the '3D ⟶ corporeal' premiss. He argues, for example, that the advocates of vacuum$_2$ must presuppose the 'interval' (vacuum$_1$) conception of place, and attacks that theory as incoherent. Furthermore, he alleges, dimensionality belongs to the category of quantity; quantity must inhere in substance; but only material substances are extended in three dimensions . . . Both these arguments – that against the vacuum$_1$ theory of place and that from quantity to corporeality, presuppose the unintelligibility of incorporeal dimensions. At the back of all Aristotle's thought on this topic we find the single principle, '3D ⟶ corporeal'; if that principle is true, the very notion of the void is incoherent.

One curious consequence of Aristotle's definition of place as 2D surface of containing body should at least be mentioned here. It is this – that the cosmos as a whole, or, more precisely, the outermost sphere of the heavens, cannot be said to have a place:

> The heaven . . . is not anywhere as a whole, not in any place, if at least, as we must suppose, no body contains it.[63]

58 See Grant, 7, p. 552, and Grant, 9, p. 6.

59 *Physics*, 4, 8, 214b 28.

60 *Ibid.*, 4, 7, 214a 18-19.

61 *Ibid.*, 4, 5, 212a 32-33.

62 Wolfson, 1, p. 44; Furley, 3, p. 88. For termini *a quo* and *ad quem*, see Wolfson, 1, p. 141.

63 *Physics*, 4, 5, 212b 8-10.

But whatever exists, as Aristotle says elsewhere,[64] exists *somewhere*; only non-entitles are nowhere. This remained an anomaly – and a headache – for Aristotelian commentators over the ages: I shall not weary the reader with the tortuous dialectic and fine distinctions employed to try to evade the appearance of flat self-contradiction here.[65]

E. Aristotle Against the Void, 3: Dynamical Objections

Not only, says Aristotle, is void not a necessary condition of the existence of local motion, but in fact, 'not a single thing can be moved if there is a void'.[66] He has a number of arguments for this claim, based either on the tenuity (lack of resistance) of the void or on its homogeneity (lack of differentiation). Let us begin with the latter type:

> . . . as with those who for a like reason say the earth is at rest, so, too, in the void things must be at rest; for there is no place to which things can move more or less than to another; since the void in so far as it is void admits no difference.[67]

The reference to 'those who for a like reason say the earth is at rest' leads us to Anaximander,[68] according to whom the earth rests because there is no sufficient reason for it to move in one direction rather than any other. Aristotle is therefore claiming that a body situated in a void would be at rest, and for the same reason: in a uniform and homogeneous void there could be no sufficient reason for it to move in any particular direction rather than another.

While this argument may suffice to show that no body would *begin* to move *in vacuo*, unless for example it were pushed, it surely does not show the impossibility of motion in a void. The Atomists could reply by reiterating their thesis of the eternity of atomic movement: while every particular atomic motion has as its 'sufficient reason' (in a very attenuated sense) a proper mechanical cause or impulse, the whole series of atomic motions and collisions has no first member, and no sufficient reason. The eternal existence of

---

64   *Ibid.*, 4, 1, 208a 30.

65   See Wolfson, 1, p. 43 ff.

66   *Physics*, 4, 8, 214b 30–31.

67   *Ibid.*, 31–34.

68   Anaximander's explanation of the stability of the earth is discussed by Aristotle in *De Caelo*, 2, 3, 295b 10 f.

atoms moving in void space is for Democritus simply a brute fact.

The Democritean can therefore explain why, for example, one atom A moves towards another, B, rather than away from it – A moves in the direction it does because of a prior collision with a further atom C. And, although the void lacks absolute differentiation, it still permits relative differentiation: it makes a real and discernible difference whether A moves towards B or away from it.

The next argument rests on Aristotle's all-important distinction between 'natural' and 'forced' (or 'violent') motions. A motion is natural, according to Aristotle, if it results from an 'internal principle'[69] in the moving body: in such a case no external force need be exerted on the body – it will move, as it were, spontaneously in accord with its 'nature'. By contrast, a 'forced' motion, being contrary to nature, requires an external force; lifting a weight would serve as a paradigm of forced motion, while free fall would exemplify a natural one.

For Aristotle, natural motions are directed towards *natural places*. Fire, for example, 'naturally' rises until it reaches the 'sphere of fire', where it stops, while earth naturally falls until it reaches its natural place. Such a doctrine seemed to Aristotle merely a generalisation and systematisation of common sense experience, but was in fact pregnant with momentous consequences. From his doctrine of natural motions and natural places Aristotle derives the following:

a. A finite and unique cosmos.
b. A fixed and central earth.
c. No void.
d. Irreducible qualitative differences between elements.
e. 'Simple bodies' to correspond to the simple natural motions.[70]

Much of peripatetic cosmology, physics, and even chemistry is therefore consequent on the 'natural motions' theory. It will be discussed in more detail elsewhere: at this point we are concerned only with the third of the above implications, the argument from the phenomena of natural motions to the non-existence of the void.

A motion is constrained, says Aristotle, if it is contrary to a natural motion – all constrained motions therefore logically

---

69   For 'nature' as an internal source of motion, see *Metaphysics*, 5, 4, 1014b 19-21, 1015a 13-15.

70   See Cherniss, p. 196.

presuppose prior natural motions.[71] But in an infinite and homogeneous void, there can be no natural motions, for:

> how can there be natural movement if there is no difference through the void or the infinite? For in so far as it is infinite, there will be no up or down or middle, and in so far as it is a void, up differs no whit from down; for as there is no difference in what is nothing, there is none in that void (for the void seems to be a non-existence and a privation of being), but natural locomotion seems to be differentiated, so that the things that exist by nature must be differentiated. Either, then, nothing has a natural locomotion, or else there is no void.[72]

This argument seems to be cast in the following form:

P1.   All motion is either natural or constrained (215a 1).
P2.   Constrained motion presupposes prior natural motion (215a 4).
P3.   Natural motion is impossible in a undifferentiated void or vacuum$_2$ (215a 5–13).
P4.   Constrained motion is also impossible *in vacuo* (from P2 and P3).
P5.   There can be no motion at all in a void (from P1, P3 and P4).

How might Democritus have replied to this apparently valid and conclusive argument? Suppose one grants Aristotle his characterisation of natural motion as due to an internal principle; such motion, one might also accept, is impossible *in vacuo* (P3). The problem for the Aristotelian is that only one of P1 and P2 can be treated as analytic. If P2 is analytic (definition of constrained motion), P1 becomes synthetic and questionable. This is so because 'natural' and 'constrained' would cease to be jointly exhaustive of all the possibilities – conceptual space would be left for a motion which was *neither* in accord with nor in opposition to the 'nature' of the moving body: a body without a 'nature' (= internal source of motion) could have a motion of this third kind.[73]

If on the other hand P1 is taken as analytic, 'constrained' becomes a blanket-term covering all motions not due to the inner principles of 'nature'. But then P2 becomes dubious: it need not be the case that a body moving non-naturally is

---

71   *Physics*, 4, 8, 215a 1–5.

72   *Ibid.*, 5–13.

73   Cf. of course the later concept of a 'neutral' motion, discussed at length by Galileo.

moving in a direction contrary to its natural motion. Once again one could imagine a body without any natural motion, moving non-naturally (but *not* 'contrary to nature') in the void. The motion of the Democritean atoms may be of this kind.[74]

But Nature does seem to present us with 'natural motions': this is the crux of Aristotle's argument. The above manoeuvre would therefore not have impressed him. Fire, Air, Water and Earth all have their natural places to which they move spontaneously and from which they can only be wrested by force; how, Aristotle can demand, could Democritus explain these undeniable facts on the basis of the random and non-natural motions of atoms in an infinite homogeneous void? (To evade the force of this argument, as we shall see, Epicurus attributes to *his* atoms a natural motion – downwards in an *anisotropic* void space.)

Everything that is in motion, says Aristotle, must be moved by something.[75] In the case of the constrained motion of a projectile, a sustained external push is required:[76] without such an impulse, it would revert to rest or to its natural motion. Thus:

> things that are thrown move though that which gave them their impulse is not touching them, either by reason of mutual replacement, as some maintain, or because the air that has been pushed pushes them with a movement quicker than the natural movement of the projectile wherewith it moves to its proper place. But in a void none of these things can take place.[77]

Here are presented two distinct versions of the notorious 'motor-medium' theory of peripatetic physics.[78] On the one hand, the projectile may be given a further forward impulse by the mutual replacement of the parts of the medium – air displaced by the forward motion rushes round behind the projectile and urges it on. Alternatively, the projector (sling,

---

[74] For the suggestion that the Democritean atom has no natural motion, see Barnes, Vol. 2, p. 65.

[75] *Physics*, 7, 1, 241b 24.

[76] In the case of a projectile its original *mover* is the *projector* (hand, sling, bow, etc.), but the *instrument* by which the projectile is kept in motion is the *medium* through which it moves. See Weisheipl for a fuller discussion.

[77] *Physics*, 4, 8, 215a 13–18.

[78] See Koyré, 7, p. 7; Shapere, pp. 45–46.

hand, or bow) could somehow impart to the air a capacity or power to impel the projectile; this specially empowered medium then becomes the motor cause of the continued motion of the projectile after it has left the projector.

It hardly matters for our purposes which theory of this type Aristotle favoured.[79] Both serve to preclude non-natural motion *in vacuo*: if such motion requires a sustained push or impulse which can only be provided by a physical medium of some kind, then in the absence of any such medium there can be no constrained motion.

Here Aristotle trades once again on the ambiguity between 'non-natural' (i.e. *not due to* an internal principle) and 'violent' (i.e. *opposed to* an internal principle). It is precisely this equation of 'non-natural' with 'violent' which the Atomists must deny – clearly what they require is a conception of motion which is non-natural and yet not 'violent' or 'constrained' in Aristotle's sense. Equally clearly, they must deny that such motions require a sustaining push: this would lead them in the direction of a different (and superior) theory of dynamics from that of the Stagirite. Unfortunately, there is little or no evidence to suggest that such a theory of dynamics ever emerged from ancient Atomism.[80]

Just imagine for a moment what the Atomists *should* have said in response to the following Aristotelian argument:

> Further, no one could say why a thing once set in motion should stop anywhere, for why should it stop here rather than here? So that a thing will either be at rest or must be moved *ad infinitum*, unless something more powerful gets in its way.[81]

If, contrary to the previous argument, there could be some non-natural motion which did not depend on a sustaining impulse from the medium, a body with such a motion would go on moving for ever, thus:

$$A \longrightarrow \qquad \cdot \quad \cdot \quad \cdot \quad \cdot \cdot \cdot$$
$$P_1 \quad P_2 \quad P_3$$

Why, asks Aristotle, should arrow A stop at P_1 rather than P_2 or P_3? What (sufficient) reason could be given why it should come to rest at any particular point? Given the

79   He does seem to have a preference for the second, 'empowered-medium' type of theory. See J. F. O'Brien, pp. 350–351; Dijksterhuis, p. 28.

80   See Balme, 2.

81   *Physics*, 4, 8, 215a 19–22.

perfectly uniform, homogeneous, and *resistanceless* nature of void space, no such reason could be given.

We have here the following three options, O1–O3:

O1.   A does not move at all.

O2.   A stops at P_1 or P_2 or P_3 or . . .

O3.   A moves *ad infinitum*. (We assume that this will be with uniform velocity – there is after all nothing to cause A to decelerate or change course.[82])

Now Aristotle uses the principle of sufficient reason to refute O2. But for him O3 is equally absurd: in a finite Aristotelian cosmos no infinite rectilinear motion is possible. Besides, the motion of A must be non-natural (there can be no natural motions *in vacuo*), but only natural motions (like those of the heavenly spheres) can be everlasting. O1 remains therefore the only intelligible possibility.

Aristotle alleges in effect that the modern principle of inertia is implicit in ancient Atomism, and uses it as part of a *reductio* of that theory. It seems clear, as Balme[83] and Meyerson[84] insist, that Democritus could never have *explicitly* accepted such a principle: there is no indication that he was in possession of a theory of dynamics far in advance of those of his contemporaries. Aristotle's use of the above argument as a *reductio* only makes sense if he presumed that Democritus would have been unwilling to accept O3.

Nevertheless, given (a) immutable atoms, (b) a conception of locomotion as mere change of place (and *not* an inner process of development, the actualisation of a 'potency') and (c) resistanceless void, the law of inertia would seem – with a few extra, but not implausible, assumptions – to follow. The Atomist too must reject O2 above, and for much the same reasons as Aristotle. Since he is firmly committed to the denial of O1, it seems he is left with O3. Add the extra conditions of rectilinear progression and uniform speed, and we have the modern law of inertia. And why *should* the moving atom change speed or direction without a proper cause, such as an impulse given by another body, something of course the void can never supply?

The law of inertia may then have been implicit in Democritean Atomism, in the weak sense that a plausible derivation of it can be made from the basic principles of that theory. It seems, however, never to have been derived or

---

82   See Grant, 9, p. 266.

83   Balme, 2, pp. 24-25.

84   Meyerson, p. 113 ff.

explicitly stated by Democritus, and was considered so implausible by Aristotle as to be made part of a *reductio* of the very notion of motion *in vacuo*.

The next argument runs from 215a 24 to 215b 20 and divides into two parts. Consider two freely falling heavy bodies, B_1 and B_2, where B_1 falls faster than B_2. Two different explanations, says Aristotle, might be given of this fact:

E1. B_1 is falling through a less dense medium than B_2, a medium which offers less resistance to motion.

E2. B_1 is heavier than B_2.

Consider first E1. The velocity of a freely falling body is, according to Aristotle, inversely proportional to the density or resistance of the medium in which it falls, according to the following formula:

$v_1/v_2 = R_2/R_1.$

But, he continues,

> there is no ratio in which the void is exceeded by body . . . Similarly, the void can bear no ratio to the full, and therefore neither can movement through the one to movement through the other, but if a thing moves through the thickest medium such and such a distance in such and such a time, it moves through the void with a speed beyond any ratio.[85]

Given the truth of the $V \propto 1/R$ formula, a body falling freely *in vacuo* ($R = O$) should indeed have a velocity 'beyond any ratio', i.e. an *infinite* velocity. But this is absurd: all motion is in time; an infinitely fast motion, since it would take no time at all, is a self-contradictory notion.

Suppose then that the freely falling body B_1 does take some small amount of time to fall a given distance through the void. Let B_1 cover a distance D of void in a time T_1. Then it will cover the same distance in air in a time $T_2 > T_1$. (For convenience, let $T_2 = 2T_1$, although the argument would work for any finite ratio.) Then in T_1, by the $v \propto 1/R$ formula, B_1 would cover D/2 in air. But in a corporeal medium half as dense as air, it would cover $2 \times D/2$, i.e. D, in T_1. Hence, Aristotle concludes:

> . . . if there is a time in which it will move through any part of the void, this impossible result will follow: it will be found to traverse a certain distance, whether this be full or void, in an equal time; for there will be some body [i.e.

[85] *Physics*, 4, 8, 215b 12–23.

medium] which is in the same ratio to the other body [medium] as the time is to the time.[86]

This argument therefore presupposes that all motion takes time, then assumes that all finite time periods stand in finite ratios to one another, and works the *reductio* with the aid of the ubiquitous v α 1 / R formula. It is that formula which *must* be flatly rejected if motion *in vacuo* is to be 'saved': given that formula, and the resistanceless (R = O) nature of void space, Aristotle's conclusions are inescapable.

We have not yet, however, completed the argument. B_1 might fall faster than B_2 in virtue simply of being heavier. But, says Aristotle:

> We see that bodies which have a greater impulse either of weight or of lightness . . . move faster over an equal space, and in the ratio which their magnitudes bear to each other. Therefore they will also move through the void with this ratio of speed. But that is impossible; for why should one move faster? (In moving through *plena* it must be so; for the greater divides them faster by its force. For a moving thing cleaves the medium either by its shape, or by the impulse which the body that is carried along or is projected possesses.) Therefore all will possess equal velocity. But this is impossible.[87]

Heavier bodies must fall faster than lighter ones. But *in vacuo* they would not do so – they could not cleave the medium more rapidly. Hence *in vacuo* all bodies would fall equally fast; but this, Aristotle feels certain, is impossible.

This whole battery of dynamical arguments rests firmly on the general peripatetic dynamical formula, 'if F > R, v α F/R'.[88] For a projectile, Aristotle argues, F = O *in vacuo*, so v = O. But allow F any finite value (e.g. perhaps for natural motions?[89]) and v becomes 'beyond any ratio', since R = O.

86   *Ibid*, 216a, 3-8.

87   *Ibid*., 13-21.

88   For Aristotle's 'laws of motion' see Drabkin. Aristotle himself may not have proposed his formulae as more than rough and ready rules of thumb (he was certainly no mathematical physicist in the modern sense) but his disciples certainly took them as more, and developed a system of quantitative dynamics on the basis of Aristotle's hints. See Schramm, pp. 101-102; Crombie, 3, p. 152.

89   Koyré (1, p. 412) suggests that for Aristotle all forced motions in a vacuum₂ would have zero velocity, and all natural motions infinite velocity.

Where R = O, differences in weight (F) make no difference to the speed of fall. And so on – the whole family of arguments stand – or rather, eventually, *fall* – with the v α F/R formula. In a subsequent chapter, we shall follow their refutation more closely.[90]

Before moving on, let us add one further Aristotelian argument against the void. It derives from *De Anima*, and turns on Aristotle's theory of light and vision. Light, says the Stagirite, is the actualisation of the potential transparency of the medium. In the absence of such a transparent medium, therefore, light – and hence vision too – would be impossible:

> Democritus misrepresents the facts when he expresses the opinion that if the interspace were empty one could distinctly see an ant on the vault of the sky; that is an impossibility. Seeing is due to an affection or change of what has the perceptive faculty, and cannot be affected by the seen colour itself; it remains that it must be affected by what comes between. Hence it is indispensable that there be something in between – if there were nothing, so far from seeing with greater distinctness, we should see nothing at all.[91]

One need not accept Aristotle's curious theory[92] of light and vision to make use of this argument: *any* theory in which the medium plays an essential rôle will provide grounds for a similar point. The Stoics were to make much of a similar argument,[93] as in time to come were advocates of 'species' and wave theories. If *any* such mediumistic theory of light is true, there is no vacuum – at least not within those parts of the universe perceptible to us.[94]

F. Epicureanism

Epicurus' brief remarks on the void in the 'Letter to Herodotus'[95] are, to say the least, disappointing. In effect he is

[90] The dynamics of Philoponus, although temporally they belong in this chapter, will be discussed in its successor, since ideologically they belong with later views.

[91] *De Anima*, 2, 7, 419a 15-21.

[92] For more on Aristotle on light, see Ronchi, pp. 14-15.

[93] Sambur, 3, p. 41.

[94] The Epicureans of course had a rival theory of 'eidola' – the ancestor of corpuscular theories – which did not essentially depend on a corporeal medium.

[95] 'Letter to Herodotus', 39-40; Bailey, 1, p. 23.

content to reiterate the argument of Democritus. The senses testify to the existence and motion of bodies; but (principle V) without void or vacuum$_2$ bodies could have neither positions nor motions; therefore there is a void. Since the whole 'proof' rests so firmly on V, it is surprising that Epicurus does not attempt some defence of this principle against those (peripatetics and Stoics) who would deny it.[96] Perhaps elsewhere – in one of the (lost) lengthier works – he did give a fuller defence of V such as we can find in Lucretius. The poet begins his argument for the void with an explicit assertion of V, thus:

'There is then a void space, mere space untouchable and empty. For if there were not, by no means could things move; for that which is the office of body, to offend and hinder, would at every moment beset all things; nothing, therefore, could advance, since nothing could make a start of yielding place.'[97]

Before moving on to deal with objections to this argument, Lucretius adds the following, largely familiar, points:
1.   The seeping of water through rock, the assimilation of nutrients by organisms, and the passage of sounds through apparently solid walls all testify to the existence of pores or microvacua in even the most solid-seeming bodies.[98]
2.   If we postulate interstitial vacua we can explain differences in S.G. between different substances: we can explain, for example, why a ball of wool weighs so much less than an equal volume of lead. High S.G. corresponds to high matter*/void ratio; bodies of low S.G. consist largely of void space.[99]
   Neither of these arguments is of course conclusive: the first implicitly relies on V (if movement is possible in a plenum, bodies could be permeable without being porous, or could be porous and yet have their pores filled with a subtle fluid matter), while the second rests firmly on the homogeneity thesis. If matter is not homogeneous, no difficulty arises in accounting for differences in S.G.: some substances simply have a more intense '*gravitas*' than others. (This need not correlate with their 'quantity of matter'.) Both these arguments serve rather as *illustrations* of the *explanatory power* of

96   Bailey, 2, pp. 279-280.
97   Lucretius, Book 1, 334-339, p. 193.
98   *Ibid.*, 346-357, pp. 193-195.
99   *Ibid.*, 358-365, p. 195.

the Atomist theory: by positing microvacua, says Lucretius, we become able to give plausible and convincing explanations of phenomena of the above kinds.

Lucretius then proceeds to rebut the two outstanding attempts to deny V and thereby evade the force of his pro-vacuist argumentation, namely (a) the theory of cyclical mutual replacement from the *Timaeus* and (b) the theory of condensation and rarefaction borrowed by Aristotle from the Milesians. The former is discussed first; its proponents:

> say that the waters give place to the scaly creatures as they press forward and open up a liquid path, because the fishes leave places behind, to which the waters may flow together so they yield: and that even so other things too can move among themselves and change place, albeit the whole is solid. In very truth this is all believed on false reasoning. For whither, I ask, will the scaly creatures be able to move forward, unless the waters have left an empty space? Again, whither will the waters be able to give place, when the fishes cannot go forward? Either then we must deny motion of every body, or we must say that void is mixed with things, from which each thing can receive the first start of movement.[100]

This is not a mere reiteration of the argument – it introduces a new consideration, explicitly stated only at the end of the quoted passage. Very well, Lucretius concedes, we can conceive how there could be motion in a plenum, but how could there be *stops* and *starts*? How could there be 'a first start of movement'? And the motions we observe in Nature are not (with the significant exception of those of the heavenly bodies) continuous cyclical motions. A fish will, on occasions, *start* to move, will initiate a series – perhaps even a cycle – of motions. But in a plenum this origination of motion could not occur unless *all* the bodies in the cycle were to move simultaneously. This, however, leads to dynamical difficulties; it is the fish that moves the water and not *vice versa*. If we attribute to Lucretius the plausible-sounding premiss that causal priority involves temporal priority, we can get some mileage out of the above argument.

What of the theory of condensation and rarefaction? This issue comes up during discussion of Lucretius' final pro-vacuist argument, which runs as follows:

100   *Ibid.*, 372–383, p. 195.

Lastly, if two broad bodies leap asunder quickly from a meeting, surely it must needs be that air seizes upon all the void, which comes to be between the bodies. Still, however rapid the rush with which it streams together as its currents hasten round, yet in one instant the whole empty space cannot be filled: for it must needs be that it fills each place as it comes, and then at last all the room is taken up.[101]

However fast the air flows in to occupy the empty space, it can only move with finite velocity; in which case it will successively occupy one part of that place after another. But in that case there will be a time-period – however short – in which part of that space is perfectly void.[102] If, Lucretius adds,

> anyone thinks that when bodies have leapt apart, then this comes to be because the air condenses, he goes astray; for in that case that becomes empty which was not so before, and again, that is filled which was empty before, nor can air condense in such a way, nor, if indeed it could, could it, I maintain, without void draw into itself and gather into one all its parts.[103]

I find the exact purport of this argument elusive. Its gist, however, seems to be as follows: suppose one were to say (following up a hint from Aristotle's De anima[104]) that the two surfaces did not actually come into contact but by coming together caused a thin film of air to *condense* between them; when they spring apart, of course, this air then rarefies. Lucretius seems to have a theory of this kind in mind; both his counterarguments, however, turn on the homogeneity thesis. When the two plates come together and cause the air between them to condense, some other space must, he argues, become empty; when they separate, it will again become filled. In any case, he alleges, condensation presupposes internal void: it cannot therefore be invoked by an anti-vacuist. Neither of these arguments would impress a Stoic or an Aristotelian, prepared as they were to countenance the intrinsic 'thickening' and 'thinning' of matter.

After Aristotle's careful analysis of the concept of 'place' in Book 4 of the *Physics*, we can expect greater theoretical

---

[101]   *Ibid.*, 384–390, p. 195.

[102]   See Grant, 9, p. 87.

[103]   Lucretius, Book 1, 391–397, pp. 195–197.

[104]   See Grant, 9, p. 92.

sophistication in subsequent conceptions of space and place. For Leucippus and Democritus space was a concrete ontological concept, *to kenon*, 'the empty', contrasted with 'the full': the notion of space as a uniform abstract 3D mathematical framework was alien to them.[105] In Epicurus and Lucretius, however, the situation is different. Although the fundamental conception of space remains that of *empty* space or vacuum$_2$ (*inane*, *nihil*[106]) there is a growing awareness that the existence of vacuum$_2$ presupposes a conception of place as vacuum$_1$, as the *interval* between the extremities of a body. Body *excludes* vacuum$_2$ (void) but *occupies* vacuum$_1$ or 'intangible nature'. Thus, says Sextus Empiricus:

> according to Epicurus, 'of the intangible nature one part is named "void", another "place", another "room" ', the names being varied according to the different applications, since the same nature is termed 'void' when destitute of any body and is called 'place' when occupied by a body and becomes 'room' when bodies pass through it. But the general designation 'intangible nature' is given to it by Epicurus owing to its lack of resistant touch.[107]

There is therefore such a thing as mere space or 'intangible nature' (i.e. vacuum$_1$) which when occupied by a body is called its *'place'* (a place is, analytically, the place of some body) and when empty is called 'void'. It is in its own nature, however, quite indifferent to occupation, perfectly impassive. The Epicureans clearly saw that in order to defend void or vacuum$_2$ against Aristotle's conceptual objections they needed the notion of space as vacuum$_1$.[108] This conception of vacuum$_1$ was then generalised to give the notion of an all-embracing 3D 'intangible nature', parts of which are occupied by bodies (and are therefore called 'places'), and parts of which are empty (and may therefore be called 'void').[109] Vacuum$_1$ is in an important sense prior to body: without vacuum$_1$ in its rôle as place, atoms could not be anywhere – could not even have *positions*, far less motions.

This remains the orthodox Epicurean theory of space in the seventeenth century, as is clear from the writings of Walter

---

[105]   But see *Physics*, 4, 5, 213a 15-19, where Aristotle attributes to the early vacuists a theory of place as vacuum$_1$.

[106]   See Grant, 9, p. 3.

[107]   Sextus Empiricus, Book 1, 1-2, p. 211.

[108]   Simplicius attributed such a theory to Epicurus (see Bailey, 2, p. 295).

[109]   See Bailey, 2, pp. 294-296.

Charleton, disciple of Gassendi. In Chapter 6 of Book 1 of his 'Physiologia' he turns his attention to 'place'. 'Inanity' and 'Locality', he states,

> bear one and the same notion *essentially*, and cannot be rightly apprehended under different conceptions, but *respectively*; or more expressly, that the same *space*, when possessed of a body, is a *place*, but when left destitute of any corporeal tenant whatever, then it is a *vacuum*.[110]

Space therefore is essentially vacuum₁ or 'intangible nature'; it is *accidental* to any given portion of vacuum₁ whether it is filled (and therefore to be called 'place') or empty (and therefore 'void'). This *Epicurean* definition of place is, Charleton insists, superior to that of Aristotle.[111] It is also, clearly, an ancestor of the absolute space of Newton.[112]

The whole universe, Epicurus insists, must be infinite in extent.[113] Since only void can delimit matter*, and only matter*, void, the concept of a boundary beyond which there is neither of these mutually exclusive and jointly exhaustive realities is unintelligible. Since our material cosmos is finite, there must be infinite empty space surrounding it, in which no doubt further worlds are present. This is orthodox Atomist doctrine, with little or no departure from Democritus. Lucretius makes essentially the same point: the whole universe, he insists,

> is bounded in no direction of its ways; for then it would be bound to have an extreme edge. Now it is seen that nothing can have an extreme edge, unless there be something beyond to bound it, so that there is seen to be a spot farther than which the nature of our sense cannot follow it. As it is, since we must admit that there is nothing outside the whole sum, it has not an extreme point; it lacks therefore bound and limit.[114]

The very idea of a limit or extremity to 'the all' is absurd. But, adds Lucretius, even if one were to suppose:

---

[110]   Charleton, p. 62.

[111]   *Ibid.*, p. 63.

[112]   See the introduction to Čapek, Ed., p. xx. For the influence of Gassendi (possibly *via* Charleton) on Newton's conception of space and time, see McGuire, 5.

[113]   'Letter to Herodotus', 41–42; Bailey, 1, p. 23.

[114]   Lucretius, Book 1, 958–964, p. 225.

that all space were created finite, if one were to run on to the end, to its farthest coasts, and throw a flying dart, would you have it that that dart, hurled with might and main, goes on whither it sped and flies afar, or do you think that something checks and bars its way? For one or the other you must needs admit and choose. Yet both shut off your escape and constrain you to grant that the universe spreads out free from limit.[115]

If the spear flies on, there must be space ('room') in which it can move. If, on the other hand, it is arrested, there must be something solid (matter*) impeding it. But matter* occupies vacuum$_1$ ('place'). If this impeding boundary of matter* is infinitely thick, the infinity of the universe is conceded. If it is only of finite thickness, go to its very edge and release the spear again. Since *ex hypothesi* no impeding solid remains, the projectile must fly on into the infinite void space outside the cosmos.

But an infinite void, the Aristotelians will object, can sustain no natural places and hence *a fortiori* no natural motions. *Qua* infinite, it can possess neither centre nor extremity; *qua* void, it must be perfectly homogeneous.

In answer to these objections Epicurus developed a very curious and idiosyncratic theory. Instead of trying to make do *without* natural motions, he attributed a natural motion – viz. downwards – to his atoms. In virtue of their *weight*,[116] says Epicurus, all atoms naturally *fall* through the infinite void. But in *De Caelo* Aristotle had argued at length that use of such terms as 'up' and 'down' presupposes a finite and structured space composed of a nested set of spheres, the 'natural places' of the 'simple bodies'. In an infinite void there can be no such natural places. What, then, do 'up' and 'down' signify for Epicurus? He answers as follows:

> . . . in the infinite we must not speak of 'up' or 'down' as though with reference to an absolute highest or lowest – and indeed we must say that, though it is possible to proceed to infinity in the direction above our heads . . . the highest point will never appear to us – nor yet can that which passes beneath the point thought of to infinity be at the same time both up and down in reference to the same thing: for it is impossible to think this. So that it is possible

---

115 Lucretius, Book 1, 968–676, p. 225.

116 Weight was for Epicurus (along with shape and size) one of the three 'primary qualities' of the atoms.

to consider as one single motion that which is thought of as the upwards motion to infinity and as another the downward motion, even though that which passes from us into the regions above our heads arrives countless times at the feet of beings above and that which passes downwards from us at the head of things below; for none the less the two motions are thought of as opposed, the one to the other, to infinity.[117]

According to Aristotle,[118] 'up' and 'down' should be defined respectively as away from and towards the centre of the universe. This of course fits with his description of natural motions as motions tending towards natural places. The Epicurean universe, however, can have neither centre nor periphery, hence no natural places; so 'up' and 'down' cannot be defined as 'towards the top' and 'towards the bottom' of the universe, since it can have no such extremities. If we are to accommodate natural motions into the Epicurean universe, they must be divorced from their supposed link with natural places. In Epicurus' universe, 'up' and 'down' can be defined absolutely (if only ostensively) for *directions*, but only relatively (i.e. as 'above' and 'below') for places. The direction of my feet from my head is down; hence my feet are below my head.

According to David Konstan,[119] the passage quoted above constitutes an extended reply to the Aristotelian cosmology of *De Caelo*. Imagine a body B in an Aristotelian world moving from P_1 to P_2 through the centre C of the cosmos, thus:

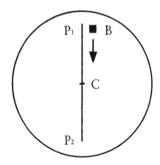

---

[117] 'Letter to Herodotus', 60; Bailey, 1, pp. 35–37.

[118] *De Caelo*, 1, 8, and 3, 2.

[119] Konstan, 1.

According to Aristotle, B is first falling, then rising, in passing from P_1 to P_2. This Epicurus considers absurd. The only sense we can make of 'down', he feels, is 'the direction of my feet from my head' (or something similar) and all lines parallel to that and taken in the same sense. In the diagram, P_1C and CP_2 are parts of the same line taken in the same sense; if then P_1C is 'down', so must CP_2 be.

In the Aristotelian world, says Epicurus, to travel from P_1 to P_2 one would first have to move 'down' and then 'up'. This for Epicurus is equivalent to saying that P_2 is both above and below P_1. As Konstan explains, the Epicurean theory is the product of the following confusion:

> 'up' and 'down' as descriptions of a direction of motion with respect to a point or points of natural attraction are identified with 'up' and 'down' in the sense of 'above' and 'below' with respect to any given observer.[120]

Epicurus thus accepts only one concept of 'up' and 'down', treated fundamentally as ostensively defined directions, 'opposed to infinity' (i.e. through all space) in terms of which 'above' and 'below' will be defined. His rival cosmology will look like this:[121]

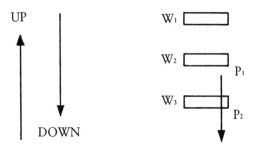

$W_{1 \ldots n}$ are worlds in descending order, and the movement of a body from P_1 to P_2 (through W_3) is uniformly downward. The upward and downward arrows at left can be regarded, as Epicurus says, as single motions, 'opposed to infinity'.

120   Konstan, 1, p. 271.

121   That Epicurean cosmology is really of this level of primitiveness is clear from Lucretius' indignant rejection (at Book 1, 1052-1113) of the theory of gravity as directed towards a central point, and of the *ludicrous* notion of the antipodes!

Epicurean space is therefore homogeneous (one part is like any other) but *anisotropic*:[122] it has a built-in directionality which affects the behaviour of bodies. It resembles an electric or magnetic field of infinite extent and uniform strength, but *without poles*. A body has exactly the same downward tendency anywhere in the universe (the void is truly homogeneous) but it is a tendency with a definite direction (space is *not* isotropic).

In this void space, Epicurus continues, all bodies move with equal speed. We have already seen (in Chapter 1) that for Epicurus all atoms move one space-minimum q in one time-minimum t: atomic velocity is always q/t. A body moving in purely empty space free from any resistance would move with atomic velocity i.e. 'as quick as thought'; its motion 'accomplishes every conceivable distance in an inconceivably short time'.[123] In the absence of either external resistance or internal agitation even perceptible bodies would move with this fantastic velocity. In fact, however, all sensible bodies are slowed by the two aforementioned causes: we have no direct familiarity with bodies moving at atomic speed, at a velocity which, according to Lucretius:

> . . . must needs . . . surpass in speed of motion . . . the light of the sun, and rush through many times the distance of space in the same time in which the flashing light of the sun crowds the sky.[124]

It is important for Epicurus that the atoms always move at 'atomic speed' q/t – it enables him to rebut one of the key Aristotelian arguments for the infinite divisibility of space and time. The above theory of atomic motion can, however, be seen as a reasoned reply to more than one peripatetic argument. Aristotle contended that there could be no natural motion in a void; Epicurus attributed weight to his atoms as a primary quality and anisotropy to his space, giving the atom a natural downward motion. In a vacuum$_2$, says Aristotle, bodies would move with a speed 'beyond any ratio'; the atoms move, replies Epicurus, with a velocity that 'accomplishes every conceivable distance in an inconceivably short time'. *In vacuo*, Aristotle objects, all

---

122   See Jammer, 1, p. 11.

123   'Letter to Herodotus', 61–46b; Bailey, 1, p. 37.

124   Lucretius, Book 1, 160–164, p. 245.

bodies would fall with equal velocity; this, reply Epicurus and Lucretius, is precisely the case![125]

Epicurus thus simply incorporates a number of undigested Aristotelian premisses and inference-patterns into his system of the world. Konstan[126] lists the following four consequences drawn by Aristotle in *De Caelo* from the assumption of an infinite universe without natural places:

1. One direction of motion for all things.
2. All bodies would have infinite velocity.
3. No body would accelerate or decelerate.
4. No body would naturally be at rest.

Epicurus, according to Konstan, simply accepted all of 1–4. Even infinite velocity?! What Epicurus says is that his atoms move 'inconceivably' fast. Bailey[127] misses the point altogether when he sees Epicurus' intuitions as being confirmed by modern science, e.g. by the discovery that the hydrogen molecule moves at c 1,800 m/sec, i.e. very quickly. 1,800 m/sec stands in a definite ratio to 1 m/sec; it is in no way 'inconceivable', nor will it answer Epicurus' definition. In fact, Epicurus seems to assume that Aristotle's 'speed beyond any ratio' is something incredibly, amazingly, perhaps staggeringly, fast – that is, he confuses a logico-mathematical point with a psychological one. What he should have done was to argue as follows: atoms move in the void; however quickly they move, their velocity must be finite;[128] therefore there is something wrong with Aristotle's dynamics and in particular with his v α 1/R premiss. One can only regret that Epicurus tried to take on too much unsifted peripatetic doctrine here: had he argued *against* the v α 1/R formula instead of lapsing into psychologism and fudging the issue, Epicureanism might have developed a superior theory of dynamics to any other than classical antiquity could boast.

G. Developments in Late Antiquity, 1: Philoponus and the Concept of Place

The conception of place as vacuum$_1$ rejected by Aristotle was adopted in late antiquity by a number of thinkers of Platonic

---

[125]   Heavier bodies, Lucretius insists (Book 2, p. 225 f.), would *not* fall faster than lighter ones in a resistanceless void. His argument for this is the same as Aristotle's.

[126]   Konstan, 1, p. 275.

[127]   Bailey, 2, p. 315.

[128]   The Aristotelian doctrine of the 'potential infinite', always a position of dubious coherence and intelligibility, is particularly unhappy in its application to the velocity of such a motion. For attempts to apply such a notion to motion, see Furley, 3, p. 91, and Grant, 9, p. 48.

leanings,[129] among them the important figure of Philoponus. The quantity of space or vacuum$_1$ occupied by a body is, he says, the proper measure of its bulk. A pitcher thus has a fixed capacity (of pure space or vacuum$_1$) independent of its particular contents (air, water, or whatever) at any given time. It is in virtue of these *independent* dimensions that the vessel can hold some particular quantity or bulk of material contents. These independent incorporeal dimensions are, according to Philoponus, part of an absolute, all-embracing 3D space, a 'receptacle' in which all things are contained.[130] Space, he insists,

> is not the boundary of the containing body, as one can well conclude from the fact that it has a certain extension in three dimensions, different from the bodies placed into it, incorporeal according to its proper nature and nothing but the empty interval of a body – in fact, space and the void are the same by their nature.[131]

Of its nature, then, space (vacuum$_1$) is empty and void (vacuum$_2$); being filled or occupied by bodies is quite inessential to it.[132] It possesses its property of 3D extension quite independently of its occupation by bodies: it is indeed in terms of the volume of space it fills that we characterise the bulk of a given body. There are therefore, *pace* Aristotle, *incorporeal dimensions*; indeed, if Philoponus is right, incorporeal dimensions are a necessary precondition of corporeal ones! Place is a 3D portion of the whole of vacuum$_1$, the essentially incorporeal space coterminous with the physical universe. (Interestingly, Philoponus still believes in a universe finite in extent.[133]) Since Aristotle's '3D —→ corporeal' premiss is *false*, vacuum$_1$ need not be conceived as analogous to a fluid medium, in which case the cube problem fails to disturb. The dimensions of bodies can – and always do – interpenetrate with those of vacuum$_1$: the impenetrability of bodies is due rather to their solidity than to their dimensionality.[134]

---

[129] See Samburphy, 4, p. 3.

[130] Grant, 9, pp. 19–20.

[131] Quoted from Samburphy, 4, p. 6.

[132] Grant, 7, pp. 558–559. For the influence of this theory on later writers such as Pico, Telesio, Patrizi and Bruno, see Grant, 9, p. 19.

[133] See Duhem in Čapek, Ed., p. 40. Beyond the finite material cosmos, says Philoponus, is space 'conceived by reason only', devoid of physical reality (cf. the 'imaginary' extracosmic space of the scholastics).

[134] Grant, 9, pp. 19–20.

But if space is conceived as vacuum$_1$ – an independent 3D reality quite distinct from matter – it could possibly exist unoccupied, i.e. as void or vacuum$_2$. Given the conception of space as vacuum$_1$, that is, all the conceptual objections to vacuum$_2$ collapse, leaving the issue of its existence an empirical one subject to scientific enquiry. Since vacuum$_1$ is in essence independent of body it could in principle exist in a void state.[135] In fact, however, says Philoponus, all vacuum$_1$ is always full or occupied by body:

> I do not maintain that this extension either is or can be empty of every body. This is never the case, for though the void in its proper sense is different from the bodies placed into it, as I said before, space is never devoid of bodies, just as we say that matter differs from form but is never devoid of form.[136]

For Philoponus to argue that vacuum$_1$ is never in fact empty of matter is perfectly intelligible, but for him to allege – as he seems to here – that vacuum$_1$ *cannot* exist as vacuum$_2$ (no more than matter can exist without form) is perverse. If vacuum$_1$ is of its own nature independent of matter – if occupation by matter is *accidental* to it – it should be capable of separate existence as vacuum$_2$. What physical or metaphysical reasons may have led Philoponus to try to back down from this apparent consequence of his position I do not know.

H. Developments in Late Antiquity, 2: Empirical Pneumatics

The discovery that air is a material substance may be attributed to the Presocratic '*phusikoi*' (natural philosophers) as a group. Air may be compressed or rarefied; it exerts resistance (try pushing an upturned beaker down into a vessel full of water); in rapid agitation, it produces wind,[137] which is tangible and capable of exerting a considerable mechanical impulse on bodies, and so on – the phenomena could be multiplied *ad lib*. One of the most significant of the early pneumatic experiments was that of the *clepsydra* or water-clock, an experiment often attributed to Anaxagoras. Water will not flow out of the *clepsydra* unless an air-hole is opened in the top. (Many of us have had similar experiences with beer

---

135   See Duhem in Čapek, Ed., p. 39.

136   Quoted from Sambursky, 4, p. 7. See also Grant, 7, p. 558, and 9, p. 15.

137   For a convincing demonstration of the corporeality of wind (and hence of air), see Lucretius, Book 1, 271–297, pp. 189–191.

barrels.) Thus, although water 'naturally' falls (has an innate downward tendency) and is impeded by no material obstacle, it still does not flow out of the vessel. From this experiment – and the phenomena of *suction* – there emerged the notorious theory of the '*fuga vacui*' (or, more anthropomorphically, the '*horror vacui*'). 'Nature', it was said, 'abhors' a vacuum$_2$ – a maxim liable to a variety of interpretations. On the one hand, one could say, with the orthodox Aristotelians, that 'Nature' acts so as to *forestall* vacuum-formation; alternatively, one might say, with the 'pneumatic' school, that there are (small) vacua but that they exert a force of suction which draws bodies into them.

One possible source of belief in interstitial microvacua was Plato's *Timaeus*. Although Plato flatly denied the existence of the void,[138] the cubes, icosahedra, octahedra and tetrahedra (= particles of earth, water, air and fire respectively) from which he constructs the material world will not pack tightly together so as to exclude all empty space. Whether he likes it or not, he seems committed to interstitial vacua: a contradiction in his thought he seems not to have noticed.[139]

One of Aristotle's successors as head of the Lyceum, Strato of Lampsacus, may be regarded as the great pioneer of empirical pneumatics. Strato accepted the vacuum$_1$ theory of space[140] and was therefore able to treat the existence or otherwise of void space as an empirical matter subject to experimental enquiry. On balance, Strato was inclined to accept the existence of interstitial microvacua (and used them to give an Atomist-style account of S.G.) but felt that the results of pneumatic experiments (e.g. the *clepsydra*) told against the existence in Nature of a 'collected' or 'coacervate' vacuum$_2$.[141]

In this position Strato was followed by Philo of Byzantium and Hero of Alexandria. All opt for a particulate theory of matter with interstitial vacua, and explain condensation, rarefaction, S.G., etc., along conventional Atomist lines.[242] Let us briefly examine the opening pages of Hero's *Pneumatics*. Some, he begins, hold that there is absolutely no void; others say that:

138 *Timaeus*, 79, p. 106.

139 See Vlastos, 7, p. 90; Duhem in Čapek, Ed., pp. 23-24.

140 Sambursky, 4, p. 3.

141 Farrington, p. 182.

142 See Marie Boas, 1, for a vision of Hero as one of the spiritual fathers of the 'corpuscular philosophy' of the Renaissance.

while no continuous vacuum is exhibited in nature, it is to be found distributed in minute portions through air, water, fire, and all other substances: and this latter opinion, which we will presently demonstrate to be true from sensible phenomena, we adopt.[143]

There next follows a proof of the corporeality of air from the resistance encountered in attempting to force an upturned beaker into a vessel of water; it is manifest, says Hero,

> that the air, being matter, and having itself filled all the space in the vessel, does not allow the water to enter. Now, if we bore the bottom of the vessel, the water will enter through the mouth, but the air will escape through the hole.[144]

This escaping air, Hero adds, may be felt as a tangible draught – further proof, if any be needed, of its corporeality. Air-particles, he continues,

> are in contact with each other, yet they do not fit closely in every part, but void spaces are left between them.[145]

Hero likens air-particles to grains of sand: although they touch, they leave small interparticulate spaces. In compression, they are pressed into these spaces; in rarefaction, they 'spring' back again. The fact that more air can be forced into a vessel already apparently full of it is for Hero the clinching argument for the existence of such interstitial vacua: if, he argues, the space of the vessel had really been quite filled by the air already present in it, no more could be forced in. (Hero's arguments clearly presuppose the incompressibility of the air-particles themselves; they turn, that is, on the existence of some homogeneous matter like that of the Atomists.)

Although a 'collected' vacuum$_2$ does not exist in Nature, Hero continues, one can be made artificially; no insuperable obstacle stands in the way of this. One can, for example, suck the air out of a vessel until it will hang

> suspended from the lips, the vacuum drawing the flesh towards it that the exhausted space may be filled. It is manifest from this that there was a continuous vacuum in the vessel.[146]

---

[143]   Hero, p. 1.

[144]   *Ibid.*, p. 2.

[145]   *Ibid.*, pp. 2–3.

[146]   *Ibid.*, p. 3.

The vacuum$_2$, says Hero, exerts a force of *suction*: there is no trace here of a theory of (external) air-pressure (not surprisingly, since the phenomenology of the experience, as Linus was to urge against Boyle millennia later, is of suction from within the vessel, not of pressure from without[147]).

The 'pneumatic' school of Strato, Philo and Hero therefore accept the existence of a 'fuga vacui', as a force of suction exerted by the vacuum$_2$ or void. Since this force is only of *finite* strength, however, it can in principle be *overcome* by 'art' and a 'collected' vacuum$_2$ made.

Very similar theories were accepted by the medical school of Erasistratus. These physicians accepted a corpuscular theory of matter,[148] emphasised the importance of 'pores' and 'channels' in physiology, and accepted the existence of interstitial vacua and a finite *fuga vacui*, a force of suction exerted by the vacuum$_2$. This force explains, for example, the operation of cupping-glasses, the action of purgatives, etc.,[149] as well of course as the standard pneumatic phenomena of siphons, straws, water-clocks, etc.

I. Developments in Late Antiquity, 3: The Extracosmic Void of the Stoics

The Aristotelian cosmos is finite in extent: beyond it, says Aristotle, is 'neither place nor void nor time'.[150] This, however, entails that the outermost sphere of the heavens has no (Aristotelian) place and is therefore *nowhere*! What happens, moreover, if someone at the very edge of the finite material cosmos were to extend his hand? Would it not pass into a space beyond; if not, why not?[151] This argument (closely related of course to that of Lucretius' spear) was first proposed by Archytas of Tarentum,[152] a contemporary of Plato, but was to become known as the argument of the Stoic's hand.

To cope with these difficulties the Stoics posited an *infinite extracosmic void space*. Thus Cleomedes:

147 See Boyle, *Works*, vol. 1, p. 127.

148 Farrington, p. 208.

149 See Heidel, 1, p. 147. For a contemporary critique of the view of Erasistratus, see Galen, p. 97 ff.

150 *De Caelo*, 1, 9, 279a 12-13.

151 For this Stoic argument, see Simplicius' commentary on *De Caelo*, discussed in Grant, 9, p. 322.

152 See Grant, 9, p. 106.

> Outside the cosmos there is the void, extending from all sides into the infinite. That which is occupied by a body is called place and that which is not occupied is void. Every body must necessarily be in something. That in which it is must be different from the body occupying and filling it, and thus is incorporeal and so to speak insipid. That reality which can receive the body and be occupied by it we call the void . . .[153]

The Stoics therefore accept the vacuum$_1$ theory of space; when full, it is called 'place', when empty, 'void'. We can, Cleomedes continues, imagine the world moved from one place to another. In the periodic conflagrations (*ecpyroses*) in which the whole cosmos is destroyed in fire, it actually expands to fill a place 'a thousand times greater' than that which it currently occupies. The extracosmic void is then no merely 'imaginary' or fictitious thought-creation, but a physical reality.

Even during the *ecpyrosis* phase of the cosmic cycle there is, the Stoics insist, no danger of the world's disintegrating into the external void: the '*tonos*' (tension) of the all-pervading *pneuma* (spirit) serves to maintain its unity and cohesion, the 'universal sympathy' of all its parts. This universal sympathy also precludes the existence of interstitial microvacua; although conceptually possible, they would disrupt the divine order and harmony of the cosmos. For the Stoics, then, there is physically real extracosmic void, but no *intracosmic* vacuum$_2$ is allowed to mar the perfect unity and organic cohesiveness of the divine cosmos.

J. Developments in Late Antiquity, 4: The Divinisation of Space

If one accepts the Aristotelian conception of place as analogous to a container one will immediately encounter problems about the relation of God to space. Where, the unbeliever asks, is your God? Is He in any place? If so, He is contained in or circumscribed by something else, some reality greater than Himself (in at least the spatial sense). But this is flagrantly heretical, blatantly at odds with the Judaeo-Christian conception of the deity as the being than which nothing can be greater (Anselm's definition, of course). 'The heaven, and the heaven of heavens, cannot contain thee,' says

---

[153] Quoted from Sambursky, 3, p. 128. See also Plutarch, Vol. 13, p. 581, for the (similar) views of Chrysippus.

Solomon.[154] Rather than being Himself *in* a place, God is more like the place that contains all things: if God fills the space outside the cosmos it is literally true, as St. Paul says, that 'in Him we live and move and have our being'.[155] In rabbinic lore, one of the names of God is māqôm[156] ('place'): He is (like) the place *in* which all things exist.

I mention these obscure and enigmatic remarks not so much because they represent any systematic doctrine concerning the relation of God to space as for their *later* relevance, when they were taken as expressions of the wisdom of the ancients, the '*prisci theologi*'. The nearest the ancients came to the explicit divinisation of space was perhaps in the Hermetic corpus, a body of predominantly religious writings of a Platonist tenor which acquired during the Renaissance an almost scriptural authority. The finite material cosmos, according to the Hermetics, is *in* (contained by) an *ideal* incorporeal space.[157] This is most emphatically *not* void: in the first place, the very idea of void or nothingness existing is incoherent; in the second place, it would violate the Platonic 'Principle of Plenitude' for God to leave gaps or vacancies in His Creation – it would mean that He had neglected certain possibilities of creation. But God, says Plato in the *Timaeus*, is not 'envious': he grudges existence to no type of being.[158] He therefore *fills* all of space with His creatures, leaving no void, an argument accepted by many Platonists until Leibniz's day. What then is the extracosmic space if it is not void? It is:

> Mind, entire and self-encompassing, free from the erratic movement of things corporeal, it is imperturbable, intangible, standing firm – fixed in itself, containing all things, and maintaining in being all things that are: and in it is the light whereby soul is illuminated.[159]

Its functions, clearly, are theological rather than physical – the above remark is less than clear about the *physical* properties of this quasi-divine space. At one point, the author comes close to identifying God with space, thus:

154 Quoted from Locke, 1, 11, 13, 27, Vol. 1, p. 144. Their presence in such a sober philosopher as Locke testifies to the seriousness with which these ideas were treated.

155 As 154.

156 See Copenhaver for the later developments of this tradition.

157 W. Scott, Ed., p. 135.

158 *Timaeus*, 29–30, p. 42. See also Lovejoy, pp. 50–52.

159 W. Scott, Ed., p. 141. There is more than a trace of Neoplatonic 'light-metaphysics' in the final sentence.

> For thou art the place in which all things are contained; there is no other place beside Thee; all things are in Thee.[160]

Elsewhere, however, it is insisted that this usage is metaphorical: while all things are 'in' God, they 'are not situated in God as in a place'.[161] A place is, on the Aristotelian definition, the limit of a surrounding *body*, but what surrounds and contains the whole material cosmos is *incorporeal*: it is, says the author, only in a quite different sense that things can be said to be 'situated in' something incorporeal.

In the *Asclepius*, the tone is a little more agnostic. The cosmos, of course, is a material plenum. What lies beyond it? Certainly not void or nothingness – the very idea of such an entity is absurd.[162] If, says the author, *anything* 'extra-mundane' exists it is certainly not void but rather

> filled with things apprehensible by thought alone, that is, with things of like nature with its own divine being.[163]

Extracosmic space is full of the divine spirit. It is idle to read too much physics into this spiritualistic Neoplatonic-Hermetic rhapsody; I quote these passages only because of their significance for later parts of our story.

---

160   *Ibid.*, p. 165.
161   *Ibid.*, p. 219.
162   *Ibid.*, pp. 317–321.
163   *Ibid.*, pp. 317–319.

MATTER, FORMS, AND QUALITIES: 1

Part 1: The Presocratics on Change

A striking feature of Presocratic philosophy is its inability to come to terms with the phenomena of change in all its varieties, whether the coming-to-be and ceasing-to-be of substances (*genesis* and *phthora*) or their local motion or qualitative alteration (*alloiosis*). This inability gave rise on the one hand to (one aspect of) the Atomic Theory of Leucippus and Democritus and, on the other, to the theory of Real Qualities. In Part One of this chapter I wish in particular to discuss this second theory and its origins.

We begin with Parmenides and his monumental but oracular 'Way of Truth'.[1] The one premiss from which the entire Parmenidean system is to be deduced consists of one word, viz:

1.   '*Esti*' ('it is').

The reader is left to his own devices to discover the logical subject of this proposition: what, it is natural to ask, is here being said to be? Perhaps the most plausible suggestion is that of Cornford[2] to the effect that '*esti*' should be read as 'that which is, is'. This reading, as we shall see, squares with Parmenides' development of the argument. What is, he continues, cannot not be:

2.   '<That which is> is, and it is impossible for it not to be'.[3]

This thesis (2) leaves two crucial ambiguities unresolved. Not only does the distinction between the predicative and existential senses of 'to be' arise (the ambiguity is as natural in Greek as in English, and is endemic in early Greek philosophy), but also the question of the scope of the modal operator. Since these two questions are independent, we immediately find ourselves faced with four variants of thesis (2), from the trivial (2,1) to the very strong (2,4):

2,1.   Necessarily: What exists, exists.

2,2.   Necessarily: What is F, is F.

---

[1]   See Barnes, Vol. 1, p. 155; Kirk and Raven, p. 264 ff.

[2]   Cornford, 2, p. 30.

[3]   *Ibid.*, pp. 30–31.

2,3.   What exists, exists – necessarily (has necessary existence).

2,4.   What is F, is necessarily-F.

While (2,1) and (2,2) are trivial and innocuous, (2,3) and (2,4) are very powerful – and almost certainly false – metaphysical claims. Cornford[4] and Barnes[5] both suggest that Parmenides conflates (2,1) with (2,3), or (2,2) with 2,4), and is therefore guilty of a modal fallacy in his defence of (2,3) and 2,4). By the same principle, one could take:

3.   'What is not, cannot be'

and generate four possible readings, namely:

3,1.   Necessarily: What does not exist, does not exist.

3,2.   Necessarily: What is not F, is not F.

3,3.   What does not exist, is necessarily-non-existent.

3,4.   What is not F, is necessarily non-F.[6]

Only, therefore, by first conflating the existential and predicative senses of '*esti*' and then committing a modal fallacy could Parmenides derive (2,4) and (3,4) from the trivial (2,1) and (3,1). Given (2,4) and (3,4), however, it is easy to account for Parmenides' denial of all forms of change. While (2,3) and (3,3) rule out the possibility of all substantial coming-to-be and ceasing-to-be, (2,4) and (3,4) equally prohibit all qualitative alteration and local motion. This can easily be shown.

By (2,3), what exists, at any time, exists necessarily. But what exists necessarily exists for all time: it cannot have come-to-be, nor again can it pass away when its time is up. Similarly, by (3,3), what does not exist at any particular time cannot exist at any time. The contents of the universe are all *eternal*: what is real cannot undergo generation and corruption.

Neither, if we accept (2,4) and (3,4), can they (or it) undergo qualitative change or even local motion. If 'F' stands for possession of a quality or occupation of a place, then if whatever is F is necessarily-F (2,4), and whatever is not F is necessarily-non-F (3,4), then if anything is ever F it will always be F, and if something is ever non-F it will never be F. Qualitative change and local motion are therefore just as illusory as coming-to-be and passing-away.

This is a highly conjectural reading of Parmenides' thesis (2), one that goes far beyond the textual evidence. While it

---

4   *Ibid.*, p. 31.

5   Barnes, Vol. 1, p. 162.

6   *Ibid.*, pp. 166–167.

may have provided one strand in Parmenides' argument
against (all forms of) change, it does not seem to have been his
main line of attack. To this we must now turn our attention.

The argument is explicitly directed against the notion of
genesis or coming-to-be, but can easily be extended to cover
ceasing-to-be as well. If an object O comes-to-be at a time t_1,
there was a time t_0 at which O did not exist. All coming-to-be,
therefore, is from 'what is not'.[7] But from what-is-not (the
non-existent) nothing can come to be:

> Nor will the force of conviction allow anything to arise out
> of what is not besides itself (viz. what is not).[8]

The argument can be found in a more articulate form in
Melissus; nothing, he insists, can come-to-be:

> For it if came into being, it must have been nothing before it
> came into being; if it were nothing, nothing would ever
> come into being out of nothing.[9]

What the Eleatics object to, as Barnes[10] sees, is any statement
of the form 'O did not exist but now exists'. But if no
proposition of this form can be true, *alloiosis* and locomotion
are prohibited along with *genesis* and *phthora*. When
Socrates becomes musical, 'musical-Socrates' comes-to-be;
when bricks and mortar are moved and arranged in a certain
manner, a house comes-to-be. All local motions and
qualitative changes, Parmenides contends, involve the
impossible processes of coming-to-be and ceasing-to-be. But
nothing can come-to-be from what is not:[11]

> For it is not to be spoken or thought that it is not.[12]

Since 'what is not' cannot be intelligibly thought of or referred
to, it is inconceivable that things should come to be out of it.
Furthermore,

> What need would have driven it to grow later or sooner,
> beginning from nothing?[13]

This is most naturally read, as by Stough, as implying that
coming-to-be out of what is not involves a violation of the

---

7   See Stough, pp. 100–103.
8   Quoted from Stough, p. 91.
9   *Ibid.*, p. 103.
10   Barnes, Vol. 1, p. 185.
11   See Cornford, 2, p. 36 ff.
12   Quoted from Stough, p. 92.
13   *Ibid.*, p. 93.

principle of sufficient reason, since there could be no
explanation of why O came-to-be at one time t_1 rather than at
another time t_2. (Barnes, it must be noted, has an alternative
reading.[14])

> Nor will the force of conviction allow anything to arise out
> of what is not besides itself.[15]

Even if – *per impossible* – 'what is not' had some measure of
existence, nothing could be generated out of it.

The thesis 'Nothing can come-to-be out of that which is
not' is therefore for Parmenides an undeniable axiom, the
contradictory of which is not even thinkable, since 'what is
not' can be neither conceived nor referred to. But all coming-
to-be is *from* that which is not, and all ceasing-to-be is *into*
'non-being'. Hence, says the sage of Elea:

> coming into being is extinguished and perishing not to be
> heard of.[16]

What is, is

> . . . beginningless, endless; since coming into being and
> destruction have wandered far away and true trust has
> driven them off.[17]

Ordinary people, says Parmenides, think that things can
move or alter without loss of identity. Every such change,
however, involves the coming-to-be or ceasing-to-be of a state
or position of a body.[18] The argument therefore runs as
follows:

4.   *Genesis* and *phthora* are unintelligible, and therefore
impossible.
5.   *Alloiosis* and locomotion involve *genesis* and *phthora*.
6.   *Alloiosis* and locomotion are impossible.

Or, in the words of Parmenides:

> . . . all things are mere names which mortals laid down
> believing them to be true, coming into being and perishing,
> . . . change of place and variation of bright colour.[19]

14   Barnes, Vol. 1, p. 188.
15   Quoted from Stough, p. 93.
16   Quoted from Barnes, Vol. 1, p. 178.
17   *Ibid.*, p. 179.
18   See Barnes, Vol. 1, p. 169, and Kirk and Stokes, p. 2.
19   Quoted from Kirk and Stokes, p. 2.

Nothing real, for an Eleatic, undergoes any sort of change: complete immutability is a necessary condition of reality. Melissus, for example, is arguing along classical Eleatic lines in the following syllogism:

7.   The objects of sense undergo changes.
8.   What is real is immutable.
9.   The objects of sense are not real.[20]

All genuine entities are, for the Eleatic, sempiternal: a thesis with a long and fascinating history in Western thought.[21]

The Atomists, as we know, responded to the challenge of Elea by postulating eternal and immutable atoms, the mere local motion of which gives rise to what men call *genesis*, *phthora*, and *alloiosis*.[22] The absolute realities (the atoms) do not come-to-be or cease-to-be, but only move in the void; those (compound) things that come-to-be, cease-to-be and undergo qualitative alteration are of an inferior level of reality.[23] Compound bodies, says Leucippus, can come-to-be and cease-to-be by the aggregation and dissociation of their constituent simples, and can undergo alteration by their rearrangement, but simples of course can undergo none of these processes: they are eternal and immutable. Complexes – such as atomic arrangements – can therefore, *pace* Parmenides, come-to-be and cease-to-be: the Atomic theory can give an intelligible account of this.[24] If one atomic arrangement gives bread, and another, flesh, we can *explain* intelligibly how the bread we eat is transformed into the flesh of our bodies.

But I digress. My main concern here is with a quite different response to the Eleatic challenge. The fundamental principle bequeathed by Parmenides to the philosophy of the fifth century B.C. was that only something abiding, something conserved, can be truly real.[25] To reconcile Eleatic metaphysics with common sense, there is an alternative strategy to

20   See Gomperz, pp. 167–168.

21   See Barnes, Vol. 1, p. 197.

22   For Leucippus' answer to Parmenides, see Aristotle, *De Gen et Corr*, 1, 8, 325a 33–35, and 1, 1, 315b 6–9.

23   See Barnes, Vol. 2, pp. 142–143, and Bogaard for the inferior ontological status of compound bodies in Democritean Atomism.

24   This presupposes of course the intelligibility of local motion itself, and hence assumes that the Atomists felt themselves able to answer Zeno's paradoxes.

25   See Kirk and Raven, p. 319, for general features of post-Parmenidean philosophy.

that of the Atomists: it is to this that we must now turn our attention.

The great post-Parmenidean natural philosophers – not only Leucippus and Democritus but Empedocles and Anaxagoras too – accept Parmenides' thesis (4) that there is no coming-to-be or ceasing-to-be of absolute realities. Mere mortals, according to Empedocles, are

> . . . fools; for their thoughts are not deep, since they think that what before did not exist comes into being, or that something dies and is completely destroyed.[26]

In truth, however,

> It is impossible for anything to come into being from what is not; and it is unattainable and unaccomplished for what exists to perish.[27]

Anaxagoras too lends his voice to the Eleatic choir. 'No thing,' he tells us bluntly, 'comes into being or is destroyed.'[28] Everything fundamentally real is eternal: no new substances come into being; no old ones cease-to-be.

The great problem for such natural philosophers is of course to reconcile this Eleatic thesis (4) with the facts of experience. To do so they have to deny Parmenides' (5), which asserts that alteration and locomotion involve coming-to-be and ceasing-to-be.

When an object O becomes F, says Parmenides, O's Fness comes-to-be *ex nihilo*: it had no prior existence. This is Parmenides' reason for accepting (5) and hence, given (4), (6) follows. Now an Atomist will retort that O's possession of F is *reducible* to the arrangement of its constituent atoms, and hence that O's becoming F can be explained in terms of the local motion (rearrangement) of those atoms – the only intelligible form of change – without the *genesis* of any real substance.[29] But if we reject this reducibility claim – never widely accepted, for many values of F, in antiquity – can we still find an answer to Parmenides? It seems that we can. According to the Atomist, what comes-to-be is always something complex, a compound of abiding simples. A similar analysis, it is clear, could be applied at the level of qualities. O's Fness does *not* come-to-be; it is transferred to O

---

[26]   Quoted from Barnes, Vol. 2, p. 134.

[27]   *Ibid.*, p. 134.

[28]   *Ibid.*, p. 135.

[29]   See Meyerson, pp. 93–94.

from a distinct object O* which was previously F. Schematically,

O*F plus O ─────────────→ O* plus OF.

Here O has become F without any trace of absolute *genesis* creeping into the picture. We have instead, as the diagram plainly shows, a rearrangement of abiding realities.

Let us pursue this line of thought to its most extreme conclusions. An immediate corollary seems to be what Barnes calls the 'Principle of Synonymy', namely that if a brings it about that b is F, a must be F. Or, as Barnes puts it,

> since I cannot give you what I do not myself possess, causes must themselves be endowed with the properties they impart.[30]

Barnes discovers this principle in both Plato and Aristotle, and does not doubt that it is Presocratic in origin, but fails to see it as part of a possible response to the Eleatics. If we think of qualities as thing-like, the principle will seem self-evident: one cannot impart (give) a property (thing) one does not oneself possess.

Complementary to the 'Principle of Synonymy' we might postulate a 'Principle of Transmission' to the effect that something only ceases to be F by imparting its Fness to another object. The two principles together will constitute a 'Principle of Qualitative Conservation' (PQC): any quality which obeys both principles can never come-to-be nor cease-to-be, but is conserved and transferred from one object to another.

We might also mention what might be called a 'Principle of Emanation', to the effect that an object with a quality F will tend to radiate it, to impart Fness to neighbouring things. Strictly, all that will follow from our account so far is that F objects can, while non-F ones cannot, radiate F-ness: to claim that they do or must impart that quality to other things is to go one step further. It is here, I suspect, that we find the roots of the theory of *species*, which rests firmly on the idea that bodies naturally impart their forms and qualities to a surrounding medium.

It is immediately apparent, however, to the most casual observer, that not all qualities obey PQC. When Socrates becomes musical, his previous unmusicality is not transferred to another bearer – it simply ceases-to-be. Perhaps, one might reply, that is because unmusicality is a mere privation, not a

---

[30]   Barnes, Vol. 1, p. 119.

positive reality. But are all positive realities conserved? What, for example, of colours – when a green apple becomes red, what becomes of its greenness? Does it cease-to-be? When a punctured football ceases to be spherical, do we even think of asking what becomes of its sphericality?[31]

This last question reminds us of an important point. For a vast range of values of F, common sense is Democritean. We do not ask what becomes of the sizes, shapes, and other structural properties of bodies: we *all* believe these to be *reducible* to the arrangements of smaller parts. Where the properties of a body depend simply on the arrangement of its parts – as is clearly the case for many artefacts[32] – there is no temptation to 'substantialise' or reify those qualities.

No philosopher, therefore, is going to propose that PQC holds for all values of F. A vast class of properties will turn out to be reducible à la Democritus and hence not subject to PQC. There remain, however, many qualities that uneducated common sense does not unhesitatingly consign to the Democritean. What, for example, of colours? Do they obey the principles of synonymy and transmission?

There are, it must be granted, many positive instances. Paints and dyes impart their colours to other bodies, stained glass tints the light passing through it, brightly coloured awnings may seem to possess an 'aura' of their own hue,[33] etc. These are the sorts of phenomena which might be thought to warrant the reification of colours, the claim that colours are conserved. It is easy, however, to think of counterexamples: an apple does not require the presence of red objects in order to ripen, nor does it (so far as we know) impart its greenness to anything else. How then can such a simple phenomenon as a change of colour be accounted for by a philosopher who accepts Parmenides' (4) but rejects a Democritean reduction of such qualities?

It is time now to turn the clock back a little. From the Greek medical tradition – and notably from the Hippocratic corpus – arose the element-theory of Empedocles, Plato and Aristotle.[34] The four elements are of course Fire, Air, Water and Earth (hereafter FAWE): their associated qualities, Hot, Cold, Dry and Moist (HCDM). To call HCDM *qualities* of

---

[31] See Meyerson, p. 334.

[32] Hence the importance of machine-analogies to the corpuscular philosophy of the seventeenth century.

[33] For this example, see Lucretius, Book 4, 75–82, p. 367.

[34] See Lloyd.

FAWE is, however, a tendentious and anachronistic reading of the early Hippocratics. More traditionally, HCDM were conceived as the primary elements of the physical world, the 'blood and guts', as it were, of bodies. Far from being mere qualities, they are 'powers', active or passive,[35] constitutive of the natures of things: the human body, for example is conceived as quite literally made up of HCDM.[36] As late as Aristotle the view is still current: the Stagirite consistently sees HCDM as prior to the 'so-called' elements FAWE. In 'The Parts of Animals' this view is spelt out quite unequivocally:

> The moist, the dry, the hot, and the cold are the matter [substratum] of composite bodies.[37]

In some Hippocratic texts, then, HCDM are conceived as the basic stuffs of which all things are composed: they were thought of, says Meyerson, as 'real substances', the 'elements of the elements'.[38] In other – mainly later – Hippocratic treatises HCDM are treated as qualities of FAWE. As Lloyd explains:

> In the late fifth and early fourth centuries the hot and the cold, the dry and the wet, whether alone or in conjunction with other pairs of opposites, were used in at least two types of physical theories, (1) where they figure as elements themselves, i.e. as the primary substances of which other things are composed (as in On the Nature of Man) and (2) where they are associated with the elements, when these were conceived in the form of other substances (as in On Fleshes).[39]

Now, according to the very influential medical theory proposed by Alcmeon (and adopted in Plato's Timaeus[40]), a correct balance or harmony (isonomia) of the powers HCDM constitutes health, while the dominance (monarchia) of any one gives imbalance and hence sickness. If a person is, e.g., feverish (too hot) application of the contrary power (cold) will restore balance and hence health.[41] The patient's illness

---

35   For the terminology of active and passive powers, see Aristotle, Meteorology, 4, 1, and Kahn, p. 129.

36   Meyerson, p. 326.

37   De Partibus Animalium, 2, 1, 646a 14–17.

38   Meyerson, p. 326.

39   Lloyd, p. 93.

40   Timaeus, 82, pp. 109–110.

41   Lloyd, p. 94.

ceases-to-be, and he comes-to-be healthy. But the cold (stuff) added is not itself *healthy*: the principle of synonymy is violated. The drug confers health without possessing it.

Here, it is plain, we have the germ of an alternative reductive account of qualities to that of Democritus. If qualities such as colours, smells, and tastes are supervenient on HCDM (or FAWE) ratios, they can come-to-be and cease-to-be without obeying any conservation law. HCDM, on such a theory, are conserved, and by their various mixtures give rise to all manner of other qualities.

There are, then, let us suppose, four fundamental stuffs in Nature, viz. H-stuff, C-stuff, D-stuff and M-stuff.[42] The ratio or proportion of these in a compound body determines its other properties, such as colour, smell, etc. On such a theory, *only* HCDM need obey PQC: all other qualities may violate it. If, for example, $H_3C_3D_2M_3$ gives a green apple, and $H_4C_3D_2M_2$ a red one, the ripening of an apple can be explained in terms of gain of H and loss of M without supposing that the greenness need be conserved or the redness pre-existent. Already we have the beginnings of a powerful explanatory theory, involving the explanation of the proper-ties of a body in terms of its *material constitution*.

HCDM, moreover, seem well-chosen to abide by PQC: by and large, hot bodies do impart their heat to others, while cold bodies cool (and are themselves warmed); likewise for moist and dry. Of course, as Boyle was later to show,[43] HCDM are not always conserved (the mixture of two cold substances may generate a lot of heat), but it is significant that his examples are somewhat sophisticated, drawn from the chemical laboratory rather than from everyday experience. As far as common sense goes, fire warms, ice cools, water wets and towels dry – what could be simpler? The idea of a heat-stuff, a *caloric* fluid (*sic*) that, as it were, flows from one body to another, is a natural response to the challenge of the Eleatics. As Meyerson shows,[44] *both* the kinetic theory of heat *and* the caloric theory can be seen as responses to the demands made by Eleatic metaphysics: both postulate some abiding identity beneath the superficial flux of qualitative change manifest in our everyday experience of the world.

---

[42]   For a keen diagnosis of the Presocratics' tendency to think of qualities as *ingredients* of things, see Heidel, 2.

[43]   Cf. especially his treatise 'On the Imperfection of the Chemist's Doctrine of Qualities' in *Works*, Vol. 4, p. 273 ff, esp. pp. 278–279.

[44]   Meyerson, p. 93, p. 334.

In the philosophy of Empedocles (the father of the 4-element theory[45]) HCDM are conceived as qualities or powers of the 'roots', FAWE, but a similar argument-form remains valid. Since FAWE, the four 'roots', never themselves undergo *genesis* or *phthora* (any given portion of W is always W), the qualities possessed by the elements themselves are conserved. F always contributes H to compounds; W, M, and so on. The FAWE ratio characteristic of a given compound explains its other qualities. Bone, for example, is said to be defined by the ratio $E_2A_1W_1F_4$.[46] (This theory, as Aristotle sees,[47] has the consequence that no true *alloiosis* is possible: no body can alter in quality without accession or loss of one or other of the four 'roots'.)

This theory will yield an important dichotomy of qualities. Those qualities of compound bodies that are shared by one of the elements must be attributed to the presence in the compound of that particular element. (The alchemical principle 'One quality – one material bearer' survived a vigorous critique at the hands of Boyle,[48] and re-emerged in the phlogiston theory of the eighteenth century.) By contrast, qualities of compounds (e.g. health) not possessed by any element must be attributed to the particular ratio of elements which constitutes that compound. Only the former set of qualities need be conserved: if, for example, colours are dependent on specific FAWE ratios, they need not obey PQC.

For an example of a thinker who accepts Parmenides' denial of coming-to-be and ceasing-to-be, but is generally antipathetic towards reductions of either the Democritean or the Empedoclean variety, we have only to turn to Anaxagoras. Nothing, he insists, can come-to-be from what is not: hair cannot arise from non-hair, flesh from non-flesh, etc.[49] How, then, can any changes take place? Only, says Anaxagoras, by the *separation* of *pre-existent* stuffs.[50]

This immediately entails that any quality or material present in our cosmos must have been present in the 'primal soup' from which all things emerged. In this original mixture, therefore, we find:

[45]   Pederson and Pihl, p. 142; Paneth, Part 1, p. 10.

[46]   Aristotle, *De Anima*, 1, 5, 410a 1-5.

[47]   Aristotle, *De Gen et Corr*, 1, 1, 314b 13-27; Charlton, Ed., pp. 64–65.

[48]   Boyle, Vol. 4, p. 273 ff.

[49]   Cf. Aristotle, *Physics*, 1, 4, 187a 27-28.

[50]   Separation, of course, implies pre-existence of what is separated – see Aristotle, *Physics*, 1, 4, 187a 32-b7.

Fire, Air, Dry, Hot, Cold, Bright, Dark, Hair, Flesh, Thin, Thick, Cloud, Water, Stone, Gold, Blood, Lead, Black, White, Sweet, . . .[51]

This mixture, then, consists of *stuffs* and *qualities* rather than things. Now, according to Vlastos,[52] a given stuff (gold, hair, flesh) is *constituted* by its characteristic properties. Flesh, says Vlastos, is red, soft, warm, moist, etc.: given all these qualities, what need is there for any *extra* quality of 'fleshiness'? The primitive constituents of the 'primal soup' are therefore, he argues, the opposite pairs of qualities, hot and cold, dry and moist, light and heavy, bright and dark, sweet and sour, etc. Since the 'primal soup' contains all these qualities, it contains the qualities of all material stuffs. Hence, by a simple process of separation, any given material can be extracted from it.

This, according to Vlastos, is all that Anaxagoras means when he says that every (type of) substance contains 'seeds' of every other. It is not that bread actually contains discrete *particles*[53] of hair, flesh, and bone, but rather than it contains all the necessary *ingredients* (i.e. qualities) from which those tissues can be formed. Since, according to Anaxagoras, sense-evidence teaches that any given substance can be made from any other, it follows that every substance contains at least a little of every quality, but manifests to sense only its predominant ones.[54] Nothing, says Aristotle, is for Anaxagoras

> purely and entirely black or sweet, bone or flesh, but the nature of a thing is held to be that of which it contains the most.[55]

In the beginning, then, there was a uniform, homogeneous mixture of all the contrary qualities. Then, under the formative influence of 'Mind', distinct stuffs began to emerge, to separate out of the 'primal soup'. Each particular stuff will be characterised by a specific blending of qualities, some more intense, others less so, none totally absent, and will appear to

---

[51]   See Barnes, Vol. 2, p. 16.

[42]   Vlastos, 2, p. 483. See also Barnes, Vol. 2, p. 16 ff.

[53]   For the denial that Anaxagoras' 'seeds' are discrete particles, see also Barnes, Vol. 2, pp. 21-23.

[54]   For the argument that each stuff must contain a portion of every other, see Aristotle, *Physics*, 1, 4, 187a 32.

[55]   *Physics*, 1, 4, 187b 4-6.

our senses to possess only those qualities that are its predominant constituents. Each of these materials can be itself *homoiomerous* (= homogeneous) and yet still contain 'seeds' of every other. 'X contains seeds of Y', on Vlastos' reading (endorsed by Barnes) will mean no more than 'Y can be produced from X' and will be true so long as X contains – even in the slightest degree – the constituent qualities of Y.

This, I contend, is the best current reading of Anaxagoras. On this theory, as Aristotle saw,[56] the fundamental constituents of Nature are opposed pairs of quality-stuffs – better conceived, perhaps, as forms of energy or power [57] (*dunamis*) – The Hot and The Cold, The Dry and The Moist, The White and The Black, The Sweet and The Sour, etc. On Anaxagoras' view, says Aristotle, qualities are conceived as *ingredients* of things,[58] which involves the absurd supposition that 'modifications and accidents could be separated from substances'[59] and exist independently of them.

None of these basic qualities will ever, on Anaxagoras' view, come-to-be or cease-to-be, although they may become manifest and disappear. Hence the notorious syllogism:

Water is black
Snow is (frozen) water.
Therefore, snow is black.[60]

The blackness of the water cannot cease-to-be when it is frozen; instead, it ceases to be manifest. To save his Eleatic presuppositions, Anaxagoras has to deny that qualities – even perceptible qualities such as colour – are always perceived as such. The presence of much 'whiteness' in the snow masks the presence of the blackness.

As for material stuffs, in one sense they do, and in another sense they do not, come-to-be and cease-to-be. We have – albeit tentatively – attributed to Anaxagoras a sort of reduction of stuffs to qualities: there is nothing more, on this view, to a given material than a particular combination of qualities. Flesh, blood and bone, therefore, can come-to-be and cease-to-be, but only by the commingling and separation of their constituent quality-stuffs. Hair does not come-to-be

---

56 *Ibid.*, 1, 4, contains Aristotle's critique of Anaxagoras.

57 Vlastos, 2, p. 471; Cherniss, p. 52, p. 61.

58 For objections to the idea of qualities as ingredients, see *Metaphysics*, 1, 9, 991a 8f.

59 *Metaphysics*, 1, 8, 989b 2-3.

60 See Kirk and Raven, p. 380.

from non-hair, but from the 'hair-seeds' present in bread, meat, etc. – i.e. from its own constituent principles present, albeit latent and hidden, in the food we eat. In this weak sense fire is in wood, butter in milk, and so on. This account I think best resolves the apparent contradiction between the theory of *homoiomers* and the thesis that every stuff contains seeds of every other.

We can now distinguish two possible meanings of the term 'real qualities', and two possible theses that one might wish to assert. The stronger theses I label RQ1, the weaker RQ2:

RQ1.   (Some) qualities are quasi-substantial, capable of independent existence, and are the basic elements out of which things are constituted.

RQ2.   (Some) qualities fall under PQC; i.e., obey the principles of synonymy and transmission.

RQ1 will – given the Eleatic denial of coming-to-be and ceasing-to-be – entail (and explain) RQ2, but not *vice versa*. One could reject quality-stuffs as metaphysical nonsense, yet accept, as a physical thesis, the conservation of certain basic qualities. It has been our suggestion that Empedocles adopted this latter stance, accepting RQ2 while denying RQ1.

If our reading of Anaxagoras is right, he is firmly committed (as Aristotle alleged) to the stronger thesis RQ1. The world consists, on such a view, of a great number of basic and irreducible quality-stuffs which commingle and separate to form and dissolve perceptible things and stuffs. From RQ1, with the aid of an Eleatic premiss or two, one derives RQ2 as a corollary: all these basic quality-stuffs are conserved. To evade obvious empirical counterexamples recourse must then be had to the all-important distinction between the possession of a property and its manifestation: without this distinction, the theory simply would not stand up.

Empedocles, then, accepts RQ2 only, while Anaxagoras adopts the stronger thesis RQ1. Where does Aristotle stand? The previously quoted remark from the 'Parts of Animals' would see to favour RQ1 and a vision of HCDM as the true elements of matter. Against this isolated remark, however, stands the clear and explicit doctrine of the *Metaphysics*:

> . . . heat and cold and the like are modifications of these [FAWE], not substances, and the body which is thus modified alone persists as something real and as a substance.[61]

[61]   *Metaphysics*, 3, 5, 1002a 1–3.

Affections and qualities, Aristotle insists, require a substratum in which to inhere, and are hence inseparable from substances.[62] Thus Aristotle objects to Anaxagoras that:

> If then colours and states had entered into the mixture, and if separation took place, there would be a 'white' or a 'healthy' which was nothing but white or healthy, i.e. was not the predicate of any subject.[63]

The tendency to reify qualities – to assume that they could have independent existence – makes no metaphysical sense. (Seventeenth century critics of the doctrine or real qualities – men such as Digby and Hobbes – often represent themselves as restorers of the pure Aristotelian doctrine that had been corrupted by scholastic accretions.) According to Lacey,

> In Aristotle's mind substances and qualities were clearly different things, and the question whether qualities were 'real' could be answered tolerably easily. They were not, just because and in so far as they were not substances, though of course it did not follow that they were not 'really' possessed by their possessors.[64]

Only in a material substrate, Aristotle insists, can HCDM be *active* and *passive*.[65] H does not act on C to convert it to H – that would violate Eleatic constraints on intelligibility – rather, a hot thing heats a cool thing (and is itself cooled in the process). Without a common material substratum, action and passion would be impossible.

Aristotle is committed, therefore, to a flat rejection of RQ1 and the idea of quality-stuffs. Nevertheless, the quotation from the 'Parts of Animals' cannot be dismissed as a mere aberration – Aristotle consistently seems to think of the 'powers' HCDM as quasi-substantial and prior to the 'so-called' elements FAWE. It seems inadequate to say that an undigested residue of older medical theory persisted in his thought: some superior explanation must be sought to try to resolve this appearance of paradox.

I suggest then that Aristotle accepted RQ2 for HCDM – i.e. believed them to fall under a conservation principle – and tended to slide from RQ2 to RQ1, despite the opposing

---

62   See also *Physics*, 1, 7, esp. 189a 25–b10.

63   *Ibid*, 1, 4, 188a 7–9.

64   Lacey, p. 461.

65   *De Gen et Corr*, 1, 6, 322b 16–17. See also Sellars in McMullin, Ed., 1, pp. 262–263.

pressure of his metaphysics. If HCDM are *conserved* it is after all *natural* to think of them as quasi-substantial. (RQ1 would provide a satisfying *explanation* of why RQ2 was true.)

Aristotle's commitment to the doctrine of real qualities seems therefore to have been a minimal one. He explicitly denies RQ1, as he must, and can accept RQ2 unreservedly only for HCDM. Although inseparable from a material substratum they are, as Aristotle says, the active and passive 'powers' at work within Nature.[66] These powers are of course 'contraries' in the proper Aristotelian sense (i.e. opposite ends of a spectrum of intermediates) and are therefore, once embodied, ideally suited to the rôles of agent and patient, which for Aristotle must be generically identical but specifically contrary. Thus, he says:

> We can now understand why fire heats and the cold thing cools, and in general why the active thing assimilates to itself the patient. For agent and patient are contrary to one another, and coming-to-be is a process into the contrary: hence the patient must change into the agent, since it is only thus that coming-to-be will be a process into the contrary.[67]

Fire must heat and ice must cool: it is a *necessary* proposition of Aristotelian physics that the patient is assimilated to the agent. The thesis will, however, only hold for elementary qualities: we have already seen that a non-elementary quality need obey no conservation law. Now all other qualities, says Aristotle, are explicable in terms of the 'primary' qualities HCDM.[68] It has been suggested that in *De Generatione*, 2, 2, Aristotle is arguing only for the reducibility of all other *tangible* qualities to HCDM, but this is belied by what he says elsewhere, e.g. in the *Physics*:

> Further, some pairs of opposites are prior to others, and some, like sweet and bitter, pale and dark, arise from others.[69]

Colours, tastes, etc., are therefore reducible to some prior qualities – presumably, to the 'primary' qualities HCDM. If a given HCDM ratio constitutes something white and sweet, there is no *a priori* reason to suppose that whiteness and

66   For the conception of HCDM as active and passive powers, see esp. *Meteorologica*, 4, 1.

67   *De Gen et Corr*, 1, 7, 324a 10–14.

68   *Ibid.*, 2, 2, esp. 329b 9–11, 330a 25–26.

69   *Physics*, 1, 6, 189a 17–19.

sweetness fall under PQC: objects can become white or cease to be sweet without any real *genesis* or *phthora*.

Aristotle's commitment to real qualities is therefore a *minimal* one. His metaphysics commit him to a firm rejection of RQ1, while he can accept RQ2 only for the primary qualities HCDM. (Unfortunately, as the occasional lapse shows, he was not able with perfect consistency to keep RQ1 and RQ2 separate.) Other qualities, such as colours, smells, etc., will turn out to be reducible to HCDM ratios, giving Aristotle a reductive theory of qualitative explanation in terms of material constitution. (Contrast the scholastic tendency to posit a new 'form' or 'real quality' for every new phenomenon.)

There are then, as we have seen, two basic sources for the conception of real qualities. We have Parmenides' denial of coming-to-be and ceasing-to-be, which put an end to cosmologies based on the biological concepts of birth and death, and gave rise to a new tradition of cosmological thought based on the mixing and separating of abiding constituents;[70] and we have the early Hippocratic tradition of the essay 'On the Nature of Man' and its conception – by no means unusual at the time – of qualities as the basic ingredients of stuffs. It is clear that to a generation of natural philosophers struggling to reconcile the strictures of Eleatic metaphysics with the commonplaces of everyday sense-experience the latter tradition will seem to offer an escape route, a natural philosophy compatible with Eleatic constraints. Now the conception of qualities as abiding stuffs clearly involves PQC. Equally clearly, many (most) observed qualities do not obey a conservation principle. Size, shape and structure can be dealt with along Democritean lines. What of colours, scents, tastes, etc.? Here we can adopt either a reductive explanation in terms of a set of allegedly primitive qualities (HCDM) or the idea of Anaxagoras that qualities can exist unperceived.

Gradually, however, Greek philosophy developed a sharper distinction between the categories of substance and quality, and the notion of quality-stuffs lost currency. In Empedocles and Aristotle the simplest bodies (the simplest things capable of *independent* existence) are FAWE. HCDM are powers or qualities of FAWE; other qualities are explicable in terms of FAWE (or HCDM) ratios. All that remains of the conception of real qualities in Aristotle is the

[70]   Kahn, p. 155.

idea of the conservation of the four primitive qualities HCDM. These, I would contend, are the only 'real qualities' – and that in the weaker sense RQ2 – in the natural philosophy of Aristotle.

Part 2: Reductionism and its Critics

A. Leucippus and Democritus: The Programme of Mechanistic Reductionism

As we saw in Part 1 of this chapter, a measure of mechanistic reductionism is implicit in common sense. We all believe that some of the qualities of things depend essentially on the spatial arrangement of their constituent parts. A house consists of bricks and mortar put together in a certain way; a written syllable, of letters enscribed in a particular order, and so on. It was only in Leucippus and Democritus, however, that this idea was generalised and developed into a systematic programme for the explanation of qualities. How did this programme arise?

In the first place, we must remember that the Atomists' matter* can be identified with the 'being' of the Eleatics, and the Leucippan atom (with some reservations[71]) with the Parmenidean 'One'. The atoms are eternal and immutable:[72] they provide an *abiding Eleatic substratum*[73] 'beneath' the ever-shifting phenomena. By positing such atoms, Leucippus felt able to reconcile Eleatic metaphysics (the demand for abiding identity) with the Heraclitean 'flux' of sense-experience.[74] The demand for an unchanging reality 'behind' the appearances is honoured without having to dismiss the appearance of change as merely illusory. Reason (intelligibility) demands a reality which retains its self-identity; the senses reveal an ever-changing phenomenal world: the Atomic Theory *explains* the latter on the basis of the former.[75]

[71]   E.g. with respect to its theoretical indivisibility.

[72]   Kirk and Raven, p. 405.

[73]   For the Atomists' debt to Elea, see Meyerson, p. 95; Barnes, Vol. 2, p. 43.

[74]   See of course Aristotle, *De Gen et Corr*, 1, 8, 325a 24–26.

[75]   If, as Meyerson does, one sees the Atomic Theory as essentially an answer to Eleatic identity-problems, one will see the core or essence of that theory as substantially unaltered over many centuries: the Atomism of the Greeks, says Meyerson (pp. 85–86), was 'absolutely complete' and

The genuine reals are, then, as the Eleatics demanded,[76] eternal, immutable and, most importantly, *impassive*.[77] Since the atoms move (and collide) in the void, this requires a particular conception of local motion. Displacement, says Meyerson,[78] is the change that is not a change: it is the only ultimately intelligible variety of *kinesis*. The moving body remains perfectly self-identical – no process of alteration occurs *in* it, its *intrinsic* nature remains constant – and merely 'changes' in respect of its *purely external* spatial relations to other things. Local motion is therefore the only form of change that raises no awkward Eleatic identity-questions, thus motivating the programme of reducing all other forms of change (*genesis*, *phthora*, *alloiosis*) to varieties of locomotion.[79]

The atoms are of course all composed of one and the same homogeneous stuff, matter* or 'being'.[80] As Aristotle says, atoms:

> are distinguished, we are told, from one another by their figures; but their nature is one, like many pieces of gold separated from one another.[81]

Their substance, then, is perfectly uniform:[82] two atoms of the same size and shape are perfectly qualitatively identical. (Not the slightest difference would be made if they were to be instantaneously swapped around.[83])

What is this common nature of the atoms? They are, say the doxographers, '*apoioi*', usually translated as 'quality-less'.[84] (Plutarch cites 'colourless' in illustration.) Three sound reasons can be given for this claim:

1. If atoms had such qualities as heat, cold, colour, etc., they might be expected to *interact*: a hot body, for

passed down to Dalton and beyond 'almost without change', with 'insignificant additions'.

[76] See Melissus' argument in Gomperz, pp. 167–168.

[77] Diogenes Laertius, Vol. 2, p. 455, calls the Democritean atom 'impassive and unalterable'. See also Cherniss, p. 102.

[78] Meyerson, p. 93.

[79] *Ibid.*, pp. 91–93.

[80] Kirk and Raven, p. 406; Barnes, Vol. 2, p. 67; Meyerson, p. 234.

[81] *De Caelo*, 1, 7, 275b 31–276a 2.

[82] *De Gen et Corr*, 1, 8, 326a 17.

[83] Hence of course Leibniz's objection that classical Atomism violated his principle of the 'identity of indiscernibles'.

[84] Barnes, Vol. 2, p. 66.

example, seems naturally to impart warmth to others. If atoms interacted, however, they would not be impassive and immutable.[85]

2. If atoms possessed colours, one could not explain changes of colour in terms of their re-arrangement: a body consisting entirely of blue atoms could *only* be blue and could not change to red or green regardless of how those atoms were arranged.

3. All the atoms share a common nature, so if one atom is hot (red, sweet, etc.), so must they all be. But then the contrary qualities (cold, blue, bitter) could not be present in Nature.

This brings us on to one of the most celebrated of all the extant Democritean fragments, i.e. his remark that:

By convention (*nomôi*) is colour, and by convention sweet, and by convention [every] combination (*sunkrisin*), [but in reality (*eteêi*) the void and atoms].[86]

In Diogenes Laertius' summary of the Democritean world-view, the same point is made in different terms:

The first principles of the universe are atoms and empty space; everything else is merely thought to exist.[87]

Colours, heat and cold, smells, tastes, etc., exist only 'by convention' or 'in thought'. They are, according to Democritus, *sensations* (i.e. subjective mental states) which arise in us in virtue of transient modifications of our bodies brought about by the impact of atomic effluvia emitted by objects. If an object emits a stream of atoms which, striking my eye, give me the sensation of redness, it does not follow that anything *resembling* that sensation exists 'out there' in the world. There is no *real quality* of redness, objectively present in the object.[88] (If there were, it would be more difficult – perhaps impossible – to account for the appearance of qualitative change.[89])

---

[85] *Ibid.*, p. 67.

[86] Quoted from Barnes, Vol. 2, p. 68.

[87] Diogenes Laertius, Vol. 2, p. 453.

[88] Barnes, Vol. 2, p. 73, suggests that Democritus' denial that colours exist only makes sense in the context of a denial of the theory of real qualities or quality-stuffs.

[89] We should, in effect, be back with the matter-theory of Anaxagoras, which Democritus severely criticised (see Diogenes Laertius, Vol. 2, p. 445).

To account for Democritus' dark utterances, we have, effectively, attributed to him a distinction between 'primary' and 'secondary' qualities. Primary qualities, let us say, are those conceptually necessary for corporeality: every body as such must be solid and have some determinate size and shape. (Whatever lacks any of these features is not a body.) Hence the atoms, conceived as minute *bodies*, must have those properties. By contrast, heat and cold, colour, smell, taste, etc., are not integral parts of our concept of body: we can conceive (although perhaps not *imagine*[90]) a colourless body.

We can therefore intelligibly assume that atoms *lack* all those qualities not necessary to bare corporeality – this gives us an explanation of their impassivity, a means of explaining qualitative change, and a reductive theory that scores highly on grounds of parsimony of fundamental properties. Since they are corporeal, all atoms are (absolutely) solid; they are differentiated, as Aristotle said, only by differences of size and shape.

What is to be said of the apparent colours, etc., of sensible bodies? Our sensations of red, yellow, etc., may be purely subjective, and give us no warrant for positing corresponding 'real qualities' in bodies, yet it remains the case that we use the manifest colours of things to find our way around the world. The colours of objects are reliably correlated with, and can therefore serve as signs of, more important properties (e.g. the ripeness or otherwise of a fruit). This suggests that colours, although subjective, have objective causal grounds in things. There must be an objective difference between objects we perceive as red and green, respectively, to account both for the difference in our sensations of them and other correlated differences.

The crucial point is one of reductive explanation. If one assumes 'qualityless' atoms and their local motions and spatial arrangements, one can account for the coming-to-be, passing-away and alteration of sensible substances. The qualities of perceived bodies can be explained in terms of atomic arrangements in such a way that a complete specification of the (objective) furniture of the world would make no mention of red, white, hot, cold, sweet, etc. They exist only 'in thought' or 'by convention' – without a perceiver they

---

[90]   To make the primary/secondary quality distinction one needs a sharp distinction between concept and image (mental picture). Hence perhaps Berkeley's attack on Locke's attempt to distinguish primary from secondary qualities.

would have no existence – although they have nonetheless real grounds in objects.

The essential structure of the reductionist programme of the Atomists was well discerned by Aristotle. According to Leucippus and Democritus, he says, differences in 'the elements' (matter* and void) are the causes of all other qualities:

> These differences, they say, are three – shape and order and position. For they say that the real is differentiated only by 'rhythm' and 'intercontact' and 'turning'; and of these rhythm is shape, intercontact is order, and turning is position; for A differs from N in shape, AN from NA in order, ⊥ from H in position.[91]

Interestingly, size and motion are left out of this list of explanatory factors.[92] Democritus does, as we shall see, make use of atomic size as a significant explanatory factor, but little or no use is made of atomic motions in differentiating substances. (For an Atomic theory in which motion plays a much greater rôle, we must turn to Epicurus.)

Let us run briefly through a number of classical Democritean reductions. In the first place, how does he differentiate the 'elements' FAWE? His answers seem to turn on both sizes and shapes of atoms. According to Aristotle, he identified soul with fire, claiming that it

> was a kind of fire or heat. There exist an infinite number of shapes and atoms, and those of the spherical kind are, he said, fire and soul: like the dust-motes in the air called 'atomies' seen in the rays of the sun in doorways . . .[93]

Fire (Soul) atoms are spheres since thus they will be most mobile, 'most adapted to permeate everywhere' (because least angular) and to set other bodies in motion. A, W and E are, by contrast, differentiated 'by the relative size of the atom.[94]

Properties such as odours and tastes are accounted for in terms of atomic shapes[95] – standardly, a smooth, rounded figure will be associated with a sweet, pleasant taste or smell,

---

91 *Metaphysics*, 1, 4, 985b 14–18.

92 See Barnes, Vol. 2, p. 66.

93 *De Anima*, 1, 2, 403b 32–404a 3. See also Diogenes Laertius, Vol. 2, p. 455.

94 *De Caelo*, 3, 4, 303a 16–17. Aristotle objects (*De Caelo*, 3, 5, 303b 30–304a 1) that this means of differentiating A, W and E involves the objectionable notion of absolute sizes.

95 See Bailey, 2, p. 80.

a rough or angular figure with a sharp, bitter or salty[96] one. Thus, as Aristotle says, 'he [Democritus] reduces the sapours to the [atomic] shapes,'[97] a thesis discussed in more detail by Theophrastus.

When we come to discuss their account of colour it is necessary to remember that the Atomists were among the earliest advocates of both the corpuscular theory of light and the intromission theory of vision. Bearing these facts in mind, it is clear that an Atomist explanation of colour must involve (a) the bombardment of the object by light-atoms, (b) the emission of 'eidola' or thin atomic films from bodies and (c) the sensations produced in us by such films impacting on the eye. What it is *in the object* that makes us see it as red or white or black is an *arrangement* of atoms.[98] According to Democritus, whiteness is due to the smoothness of the surface of an object, i.e. to a particular arrangement of its superficial parts. Thus Aristotle:

. . . if to be a white surface is to be a smooth surface, to be white and to be smooth are one and the same.[99]

Smoothness is therefore, for Democritus, the *real essence* of whiteness – it is what whiteness consists in. By focussing on the question of what it is about the object in virtue of which we see it as white, Democritus has now arrived at a more objectivist account of secondary qualities. The sensation of whiteness may be something subjective; its real causal ground in the object is not. Colour as we experience it is not 'real', but smoothness and roughness are. According to Aristotle, Democritus' 'denial of the reality of colour' is 'a corollary from his position: for, according to him, things get coloured by "turning" of the "figures" [atoms]'.[100]

In virtue of their different superficial atomic arrangements, bodies emit different *eidola*; these in turn produce phenomenologically distinct colour-sensations in us.[101] (We see, says Diogenes Laertius in his synopsis of Democritus' views, 'by the impact of [corporeal] images upon our eyes'.[102] This gives

---

96   For Democritus' explanation of salinity, see McDiarmid.

97   *De Sensu*, 442b 11, quoted from Barnes, Vol. 2, p; 71.

98   Bailey, 2, p. 80.

99   *Metaphysics*, 7, 4, 1029b 21–22. See also *De Sensu*, 442b 11, for an account of this Democritean doctrine.

100   *De Gen et Corr*, 1, 2, 316a 2–3.

101   See Barnes, Vol. 2, p. 176 ff.

102   Diogenes Laertius, Vol. 2, p. 455. For Democritus on light and vision, see also Ronchi, pp. 7–8.

Democritus an account which does justice both to the subjectivity of our colour-perceptions *and* to the existence of a real causal ground of them.

If a quality of a body depends on the arrangement of its constituent atoms, their *rearrangement*, clearly, will yield qualitative alteration or *alloiosis*. Thus, says Aristotle, it is possible to explain alteration if one

> 'transposes' *the same* by 'turning' and 'intercontact' and by 'the varieties of figures', as Democritus does.[103]

Alteration occurs by change of order, position, and arrangement of '*the same*' abiding atoms.[104] The same atoms $A_1 \ldots n$ which in one arrangement constitute bread will in another give flesh: this gives Democritus an answer to Anaxagoras' question of how flesh can come from non-flesh.[105]

Another natural consequence of Atomism is the identification of the coming-to-be and ceasing-to-be of substances with the aggregation and dissociation, respectively, of atoms. According to Leucippus, the atoms move in the void, 'and by coming together they produce "coming-to-be"; while by separation they produce "passing-away" '.[106]

How do the atoms combine to give rise to a cohesive body of perceptible magnitude? The answer can be found in a fragment of Aristotle's lost monograph on Democritus, preserved by Simplicius:

> The reason why the substances [atoms] stay together with one another up to a point he [Democritus] finds in the overlappings and interlockings of the bodies; for some of them are scalene, some hooked, some hollow, some convex – and they have innumerable other differences. Thus he thinks that they hold on to one another and stay together for a time, until some stronger necessity comes upon them from their surroundings, shakes them about, and scatters them apart.[107]

In virtue of their irregular shapes, then, the atoms become *entangled*, and thus hang together until some blow shakes

---

103   *De Gen et Corr*, 1, 2, 315b 34–316a 1.

104   See *De Gen et Corr*, 1, 8, 315b 6–9

105   See Paneth, Part 1, p. 11.

106   *De Gen et Corr*, 1, 8, 325a 32–33. See also Diogenes Laertius, Vol. 2, pp. 453–455; Barnes, Vol. 2, p. 132.

107   Quoted from Barnes, Vol. 2, p. 41.

them apart.[108] In the compound body thus formed, however, the atoms *retain* their distinct individual identities: the microscopically acute eyes of *Lynceus*[109] could still discern them. The 'binding' that makes atoms cohere

> does not generate any genuinely single nature whatever out of them; for it is absolutely silly to think that two or more things could ever become one.[110]

Thus also Bailey:

> . . . the atoms never so unite as to merge into one another or lose their identity in a new 'one' which they so form.[111]

Democritean atoms, then, combine to form a mere heap or aggregate: a compound body is conceived as no more than a confused conglomerate of quite distinct elements without any substantial (far less organic) *unity* of its own.[112] Thus Alexander of Aphrodisias:

> According to Democritus there exists no real mixture at all, but what seems to be a mixture is a juxtaposition of bodies that preserve their specific nature which they possessed before mingling. They seem to be mixed because of the failure of our senses to perceive the separate bodies lying side by side for reason of their smallness . . .[113]

Things may appear to be mixed – two substances may seem to be blended into a perfectly homogeneous product – but this appearance is *always* misleading. Since the atoms are, essentially, abiding units, no two atoms can fuse or mingle:[114] the closest they can approach to unity is juxtaposition.[115]

---

[108] See Bailey, 2, pp. 86–87; Bogaard, p. 315.

[109] Lynceus was one of the Argonauts, blessed with particularly acute eyesight. What the eyes of the lynx could nor could not discern in a compound body became in time an issue of great import for matter-theory. Galileo was to become a member of the Academia Lyncei; early microscopists, such as Hooke, were to claim that their instruments at last allowed ordinary men to contend with the lynx-eyed (see the frontispiece to Hooke's *Micrographia*).

[110] From Aristotle's lost monograph on Democritus, quoted from Barnes, Vol. 2, p. 41. Cf. also Kirk and Raven, p. 418.

[111] Bailey, 2, p. 87.

[112] Kirk and Raven, p. 418.

[113] Quoted from Sambursky, 4, p. 27.

[114] See Barnes, Vol. 2, p. 48, and Aristotle, *Metaphysics*, 7, 13, 1039a 7–11, for the impossibility of atomic fusion.

[115] Barnes, Vol. 2, p. 73.

Compound bodies, therefore, never have 'any genuinely single nature' in virtue of which they could be called 'one'. This, Barnes suggests, is 'the metaphysical foundation of the Atomists' view that macroscopic bodies are unreal'.[116] There are in fact no less than *three* convincing reasons why Democritus should place compound bodies, as he does, on the *nomôi* side of the great divide, viz:

1. They are transient, and undergo generation and corruption, while atoms and void are eternal.
2. The so-called 'properties' in terms of which we characterise them (colours, etc.) are not really 'in' the objects, but are subjective.
3. They have no true unity and no cohesiveness other than the accidental mechanical entangling of quite distinct atoms.[117]

While it must be admitted that compound bodies (unlike colours) have objective existence – atoms *do* become entangled and aggregate – they have nevertheless an inferior ontological status to that of the atoms, which are (a) eternal, (b) characterised by primary qualities alone, and (c) true units. Although air and water, plants and animals and men may lie on the *nomôi* side of the divide,[118] this does not mean that they are pure illusions, merely 'thought to exist'. On the contrary, unless in some sense there *are* compound bodies, how can we set out to explain their properties? Deny the existence of compounds, and one loses one's *explanandum* (unless the whole system is to be taken as an 'error-theory', an account of why we tend to hold certain quite *mistaken* views!).

Let us move on now to some *implications* of this reductionist theory. If a 'quality' such as colour has as its real (i.e. objective) causal ground an atomic arrangement, it will follow immediately that it need not be conserved, need not obey PQC (see Part 1). If greenness depends on an arrangement of superficial parts, two green stuffs could mix to produce something of *any* colour: if a new superficial atomic arrangement arises, the resulting colour may be red, orange, or blue – there is no reason why it *must* be green. Moreover, a green body can cease to be green (*pace* the 'material constitution' theory of Empedocles) without gain or loss of

---

116   *Ibid.*, pp. 142–143.

117   See Pancheri, p. 140, for the denial that a Democritean compound can be 'one' in any of the senses allowed by Aristotle in *Metaphysics*, 5, 6. Unity, for Aristotle, is usually conferred by *form*.

118   Barnes, Vol. 2, p. 142.

matter, merely by atomic rearrangement. This point will hold for *all* qualities explained in terms of arrangements, but *not* for those qualities explained in terms of atomic shapes and sizes. The spherical fire (soul) atom and the angular salt-atom will *always* be such, atoms being immutable, and will always contribute their associated 'qualities' to the compound bodies in which they are found. It thus follows that the Atomist should explain all transient, easily-altered qualities in terms of atomic arrangements; and all abiding and apparently conserved qualities in terms of atomic shapes and sizes.[119]

Another corollary of Democritus' Atomism is that every (type of) substance has a distinct *real essence* (R.E.) identifiable with its microstructure (i.e. the sizes, shapes, and arrangement of its constituent atoms). This opens up the prospect of discovering a categorical ground for dispositional properties. Suppose we discover that no substance is soluble both in benzene and in water. This might be left as a positive brute fact, but to the enquiring mind it cries out for further explanation. Perhaps, one might suggest, a particular R.E. both accounts for water-solubility and precludes benzene-solubility. The analogy of the pegboard may be both appropriate and enlightening. A round peg cannot fit (snugly) into a square hole, nor a square peg into a round hole. Now suppose water contains cubic interstitial vacua, while benzene contains spherical ones. Then, if solubility is due to (snug) geometric fit between corpuscle and pore, it will immediately be apparent that water-soluble substances (made of cubic atoms) cannot possibly be benzene-soluble. The necessity of this connection depends on (a) the absolute rigidity of the atoms (they cannot be moulded to fit the pores available) and (b) the principles of elementary geometry. Given these, the necessary connection between solubility in water and insolubility in benzene is *perspicuous*, and suffices to give an intellectually satisfying explanation of what was before obscure. The illustrations may be crude, but the general point made could hardly be more important.

Unfortunately, however, although we know in general what R.E.s are *like* (in what terms they should be characterised), we are entirely in the dark about the R.E.s of particular substances. This accounts for some of Democritus' most sceptical utterances. 'In reality,' he says, 'we know nothing, for truth is in a pit.'[120] According to Aristotle, Democritus

---

[119]   Cf. Gassendi's theory of 'calorific' and 'frigorific' atoms.

[120]   Quoted from Barnes, Vol. 2, p. 257.

believes 'either that there is no truth or that to us at least it is not evident'.[121] What do these dark utterances mean?

Plainly, 'there is no truth' is an incoherent assertion, blatantly paradoxical in nature. 'Truth is not evident to us' hardly seems, *prima facie*, to fare better, but a coherent and plausible doctrine can, I believe, be extracted from it. What 'lies in a pit' and 'is not evident' is *beyond the reach of our senses*. For a Democritean, each body will have a *hidden* R.E., a microstructure that determines its manifest properties and powers but is *imperceptible* to us. Our senses are not so much systematically misleading (although they do tempt us to 'project' our sensations such as colour on to the external world[122]) as *inadequate*, incompetent to plumb Nature's depths.[123] If we were acquainted with the R.E.s of things we could confidently deduce, 'scientifically' (i.e. demonstratively[124]) their other properties and powers; lacking such familiarity, we must either remain mere empirics or make use of an essentially *hypothetical* scientific method. The latter is clearly Democritus' procedure:[125] his lack of confidence in it (and hence in all human knowledge beyond the immediacy of sense-evidence) is well illustrated by his remark that he 'would rather find a single causal explanation than gain the kingdom of Persia'.[126] This dictum is no mere piece of philosophic bravado – a sage's proper contempt for worldly things – it is also an expression of a profoundly pessimistic epistemological attitude.

B. Criticisms of Democritean Reductionism

In this section I shall run briefly through some of the most pertinent objections to the reductionist programme of the

---

121   *Metaphysics*, 4, 5, 1009b 11–12.

122   For this 'projectivist' theory of secondary qualities, see Mackie, Chapter 1.

123   For a denial that Democritus was a thoroughgoing sceptic with regard to the epistemic value of sense-information, see C. C. W. Taylor, pp. 20–21.

124   Cf. of course the scientific methodology outlined in the *Posterior Analytics* of Aristotle.

125   We must therefore regard all the proposed reductive explanations of qualities as *hypothetical* in nature. Lacking immediate intuition of R.E.s, all we can do is construct plausible hypotheses about them, and seek (partially) to confirm those hypotheses by their explanatory power and success.

126   Quoted from Barnes, Vol. 2, p. 110.

early Atomists. Almost without exception, the arguments will be Aristotelian in origin and character. They may be listed as follows:

1.   It is in general 'unsound', says Aristotle, in criticism of Plato's *Timaeus* as well as of Democritus, 'to give a shape to each of the simple bodies'.[127] How, he asks, can proponents of such a view:

> account for the generation of flesh and bone or any other continuous body? The elements alone cannot produce them because their collocation cannot produce a continuum.[128]

The discrete particles of Plato or Democritus would associate to form a compound like a mosaic, or like a mixture of wheat and barley grains, each grain remaining wheat or barley respectively. Democritus, as we have seen, explicitly accepted this consequence of his theory and denied the reality of '*mixis*' (i.e. *blending*) altogether. As Alexander of Aphrodisias says:

> He says that in truth things simply are not mixed, but what is thought to be a mixture is a close juxtaposition of bodies which each preserve its own appropriate nature.[129]

Democritean atoms never so combine as to form any true *unity*[130] whatsoever: a Democritean compound is 'one' only be *accident* and not by *nature*.[131] And yet chemical compounds (as distinct from mechanical aggregates) do seem to have a peculiar nature and unity of their own: they at least appear homogeneous, have highly specific properties and powers, and may be strongly cohesive. They also behave as units: an aggregate can be separated by mechanical processes such as winnowing, whereas a chemical compound, a product of genuine *mixis*,[132] cannot. In some cases (especially of course organic bodies) they also possess a peculiar functional unity in virtue of which they behave, we are tempted to say, as 'organic unities'.

---

[127]   *De Caelo*, 3, 8, 306b 4.

[128]   *Ibid.*, 3, 8, 306b 22–25.

[129]   Quoted from Barnes, Vol. 2, p. 73.

[130]   Kirk and Raven, p. 418.

[131]   In *Metaphysics*, 5, 6, Aristotle discusses the various senses in which something may be called 'one', distinguishing what is 'one by accident' from what is 'one by nature' in one of several senses. Pancheri (p. 140) denies that a Democritean compound is 'one by nature' in any of those senses.

[132]   For Aristotle on '*mixis*', see Joachim.

This unity and cohesiveness of the compound body could not, Aristotle insists, be accounted for by the random agitation and occasional entangling of quite distinct atoms – no true unity could emerge out of such a chaos. In this line of criticism he is followed by many of the ancients, e.g. Galen and Plutarch. Galen objects to the Atomism of Asclepiades and his disciples that on their views 'nature' would not be purposive and craftsman-like,[133] 'all substances being divided and broken up into inharmonious elements and absurd 'molecules' (*ogkoi*)'.[134] Atomism, he insists, cannot account for chemical changes and in particular for the highly specific powers and 'virtues' of medicinal drugs.[135]

Similar objections are found in Plutarch. The atoms of Democritus, he writes,

> receive and inflict blows for all time, and so far are they from being able to produce an animal or mind or natural being that they cannot even produce out of themselves a collective plurality or the unity of a heap in their constant shaking and scattering.[136]

Compound-formation, Plutarch insists,[137] is only possible if the ingredients of the mixture can interact (i.e. undergo mutual action and passion) to produce something new and different from themselves – a new form, an Aristotelian would say. Democritean atoms, however, are hard and quite *impassive*: all that results from their aggregation is 'an uninterrupted series of collisions . . .'. Since 'entanglement' only increases the frequency of atomic collisions, 'generation' produces 'neither mixture nor cohesion, but confusion and conflict'.[138]

Hard and impassive atoms, say these critics, cannot associate to give rise to compounds with the features (unity, apparent homogeneity, cohesiveness, organisation) we detect in perceived bodies.[139] If this objection can be sustained, the

---

133   Galen, p. 45, contrasts his own views on 'Nature' with those of the Atomists. The nature of an organism, he insists, is prior to and the cause of bodily organisation; it is not a mere secondary consequence of the assembly of corpuscles.

134   Galen, pp. 61–63.

135   See Thorndike, Vol. 1, p. 140.

136   Plutarch, Vol. 14, pp. 215–217.

137   *Ibid.*, p. 215.

138   *Ibid.*, p. 219.

139   Bogaard, pp. 315–316.

Atomic Theory of Leucippus and Democritus is false – or at best severely inadequate.

2. Aristotle's next argument turns on the supposed empirical fact of the transmutation of the elements FAWE into one another. If we think of FAWE as *states of matter* (fire, vapour, liquid, solid) rather than as specific chemical substances, Aristotle's insistence that such transformations are part of 'the observed data of sense'[140] becomes readily comprehensible. We *see*, he says, FAWE 'produced from one another, which implies that the same body does not always remain fire or earth . . .'.[141]

When water evaporates, then, it is turned – without residue – into air, a complete transformation of one element into another. This, says Aristotle,[142] refutes the theory of Empedocles that FAWE are not transmuted, but that there can appear to be transmutation when a portion of one element is 'extruded' from a mass of another. If the 'extrusion' theory were correct, Aristotle argues, water could never completely evaporate: it could lose its quota of air but no more. But we see that water can and does evaporate without residue. The elements therefore are generated not by 'extrusion' but 'by changing into one another'.[143]

The Atomists, Aristotle alleges, are 'bound to contradict themselves'[144] on this point:

> For if the elements are atomic, air, earth, and water cannot be differentiated by the relative sizes of their atoms, since then they could not be generated out of one another.[145]

The implied self-contradiction lies in holding the following four propositions:

P1. The elements are atomic – i.e. there are atoms of FAWE.
P2. Atoms are characterised by shapes and sizes only.
P3. Atoms are immutable.
P4. The elements FAWE are transmuted into one another.

To escape this contradiction, the Atomist must either deny P4 (and thus develop a more sophisticated conception of the

---

140   *De Caelo*, 3, 7, 306a 5.
141   *Metaphysics*, 1, 8, 989a 23-24.
142   *De Caelo*, 3, 7, 305a 33-B 26.
143   *De Caelo.*, 3, 7, 305b 29.
144   *Ibid.*, 3, 4, 303a 25.
145   *Ibid.*, 26-28.

elements as ideal, and more remote from sense perception than previously imagined) or reject P1 and generate the elements at a *molecular* level of organisation, thereby permitting transmutation. The Epicureans, as we shall see, preferred the latter course.

3.   The Aristotelian corpus contains a number of passages where it is intimated that Atomism is 'in conflict' with 'the apparent data of sense perception'.[146] What Aristotle generally seems to have in mind is the apparent continuity and homogeneity of compound bodies. Elsewhere, however, he goes further, suggesting that the mere data of sense suffice to refute Democritus' account of alteration:

> We see the same body *liquid* at one time and *solid* at another, without losing its continuity. It has suffered this change not by 'division' nor 'composition' nor yet by 'turning' and 'intercontact' as Democritus asserts; for it has passed from the liquid to the solid state without any change of 'grouping' or 'position' in the constituents of its substance.[147]

Furthermore, he adds, Democritus' theory:

> makes growth and diminution impossible also. For if there is to be an apposition (instead of the growing thing having changed as a whole, either by the admixture of something or by its own transformation), increase of size will not have resulted in any and every part.[148]

The first of these passages seems no more than a question-begging and quite unsupported assertion, but it is illustrative of Aristotle's conviction that 'we see' his views to be correct. Democritus would of course retort that, because of the weakness of our senses, the rearrangement of atoms that occurs on solidification escapes our notice – the fact that we do not perceive any such process occurring does not justify Aristotle's flat denial of its existence. As for the second passage, all Aristotle is entitled to say is that growth as he (Aristotle) conceives it could not occur in a Democritean world. If organic tissue is 'homoeomerous' and grows by increase in size of each and every part (instead of by addition and rearrangement of distinct, fresh parts), then Atomism would preclude organic growth. But of course Democritus

---

[146]   *Ibid.*, 23.

[147]   *De Gen et Corr*, 1, 9, 327a 17–21.

[148]   *Ibid.*, 24–26.

would never accept the Aristotelian theories of homoeo-merous tissues and organic growth in the first place.

4.   If, says Aristotle, there were indeed a uniform matter common to all bodies, like the matter* of the Atomists, and differentiated only by the sizes and shapes of discrete particles, all bodies would share a *common natural motion*:

> The common error of all views which assume a single element is that they allow only one natural movement, which is the same for every body. For it is a matter of observation that a natural body possesses a principle of movement. If then all bodies are one, all will have one movement. With this motion the greater their quantity the more they will move, just as fire, in proportion as its quantity is greater, moves faster with the upward motion which belongs to it. But the fact is that increase in quantity makes many things move the faster downward.[149]

Given the 'unity of matter', all bodies should tend, more or less strongly, in the *same direction*. It would thus be easy, for example, to explain why gold has a greater tendency to fall than iron in terms of its possession of a greater quantity of that common matter. But fire, Aristotle insists, is *absolutely light*[150] – its natural tendency is in the *opposite direction altogether*, viz. upwards (i.e. away from the centre of the universe, which is coincident with the centre of the Earth). This is for Aristotle a weighty – almost conclusive – objection to any variety of monism: the phenomena of contrary natural motions manifest, he feels, the existence of *irreducible qualitative differences* between the 'simple bodies' FAWE.[151] The supposition of a uniform matter* common to all atoms (and hence to all bodies) cannot, he claims, account for such simple natural phenomena as the spontaneous ascent of fire and descent of earth.[152]

5.   According to Democritus, the atom of fire is spherical. But, says Aristotle, if the spherical atom is fiery (i.e. *hot*), what is the contrary shape which would account for coldness? Properly, he says, 'either all or none [of the contrary qualities] should have their distinguishing figures'.[153]

---

[149]   *De Caelo*, 3, 5, 304b 11–20.

[150]   *Ibid.*, 4, 1, 308a 29–30.

[151]   See Cherniss, pp. 7–8.

[152]   *Ibid.*, p. 16.

[153]   *De Caelo*, 3, 8, 307b 10–11.

Aristotle thus seems to think that Democritus thought of the fire-atom as being itself hot, i.e. as possessing the (real) quality of heat. A similar misunderstanding occurs in Theophrastus: Democritus, he remarks, 'says frequently that the shape of the hot is spherical'.[154] But, these peripatetic critics ask, if the sphere corresponds to hotness what contrary figure will correspond to coldness? Is the notion of contrariety properly applicable to shapes at all? If one atom is hot, and another cold, matter is not homogeneous.[155] And, if the atoms have properties such as heat and cold, they will necessarily interact – hot indivisibles will warm cooler ones – and hence not be impassive. Once admit that the spherical atom is hot, and the whole Democritean system collapses.

Such objections fill much of Chapter 8 of Book 1 of *De Generatione*. Throughout, Aristotle imports (quite illegitimately) many of his own concepts, notably his qualitative matter-theory and his doctrine of agent and patient. As Cherniss remarks, the inconsistencies here developed in Democritus' Atomic Theory are 'due to the implications of Aristotle's technical terminology'.[156] The whole critique rests on a gross misconception of Democritus, a failure to realise the radical nature of the Atomist's epistemology and the extent of his departure from common sense. The spherical fire-atom, for Democritus, is *not* itself intrinsically hot (all atoms are *apoioi*); it does not possess the real quality of warmth. Rather, as Cherniss says,

> heat was to be an 'epiphenomenon' arising from the contact of the swiftly moving spherical atoms with other congeries of atoms.[157]

Democritus must deny that Aristotelian 'action' and 'passion' (which presuppose qualitative contrariety) ever actually occur: the atoms, he must emphasise, do not interact in the Aristotelian sense. Mechanical collision of hard qualityless atoms is quite different in kind from the mutual action and passion of Aristotelian substances.

6.   One cannot, Aristotle argues, differentiate the elements by the greatness or smallness of their parts, since:

> This method of distinction makes all judgment relative. There will be no absolute distinction between fire, water,

154   Quoted from Barnes, Vol. 2, p. 68.
155   See Cherniss, p. 99.
156   *Ibid.*, p. 99.
157   *Ibid.*, p. 97.

and air, but one and the same body will be relatively to this
fire, relatively to something else air . . . The principle of
distinction between bodies being quantity, the various sizes
will be in a definite ratio, and whatever bodies are in this
ratio to one another must be air, fire, earth and water
respectively.[158]

At first sight this objection may seem simply wrong-headed
and silly. Could one not, one might ask, decompose matter
into its constituent atoms, then pass those atoms through
successive 'sieves' of mesh-sizes, say, of 4, 2 and 1 units
respectively? If so, one could differentiate atoms of FAWE in
terms of the diameter d as follows:

$$1 > d \quad 2 > d > 1 \quad 4 > d > 2 \quad d > 4$$
$$= \quad F \qquad A \qquad\quad W \qquad\quad E$$

The particles, one is inclined to say, are distinguished by
their absolute sizes, not by their relative ones. But that is
precisely Aristotle's objection. This method of differentiating
FAWE, he is claiming, presupposes the illegitimate notion of
absolute size.

There is for Aristotle nothing absolute about dimension or
quantity: every quantity involves a *relation* between a
magnitude and a measure. A man's height may be six foot
($6 \times 1'$) or seventy-two inches ($72 \times 1''$) but there is nothing
intrinsically six-footish about it.[159] If then the measure
increases simultaneously with, and in the same proportion as,
the thing measured, there will be no discernible change.
(There is no absolute space, in the sense of a reference-frame
quite independent of bodies, in terms of which absolute sizes
can be determined.)

Aristotle is thus giving a definite answer to the question
'What would happen if *everything* doubled in size?' His
answer is 'nothing at all': everything would remain quite
unaltered by such an imaginary *pseudo*-change (it would turn
out to be *no change at all*). But on Democritus' view, he
claims, fire would have become air; air, water; water, earth;
and earth, some other still coarser-grained stuff. This
Aristotelian argument is quite clearly spelt out in Oresme's
Livre du Ciel et du Monde,[160] one of the greatest of the critical
medieval commentaries on *De Caelo*. If all bodies were
enlarged a hundredfold, or a thousandfold, says Oresme,
nothing would really alter one jot.

---

[158] *De Caelo*, 3, 5, 303b 30–304a 6.

[159] Toulmin and Goodfield, 2, p. 102.

[160] Oresme, p. 169.

7.   *De Caelo*, 3, 5, presents a variety of objections to the attempt to differentiate the elements by the *shapes* of their atoms. Suppose that one were to identify the cubic atom with the element earth. Then, unless the cubic earth-atom is mathematically/conceptually indivisible (which Aristotle feels he has sufficiently refuted elsewhere[161]), one will be able to discern other shapes (e.g. pyramids) *within* the cube. And within the pyramids (F atoms, perhaps?) one will find further shapes, and so on *ad infinitum*. There will therefore be elements of elements of elements . . . one is trapped in an infinite regress from which there can be no escape.

This objection need not disturb an Atomist. Suppose we follow Luria and conceive the Democritean atom as itself mathematically/conceptually divisible, composed of a dense mass of minimal parts. It will then follow that within a cubic atom one *can* discern (rough[162]) pyramids, cubes, etc. But that fact simply *does not matter*: the cubic atom is physically indivisible and will never be divided into those parts. Only the shapes of the atoms have physical significance. A spherical part of an atom cannot rotate; a pyramid contained wholly within the cube cannot pierce and cut: the characteristic powers associated with those shapes have in this case no physical reality. They are irrelevant; only separate shapes matter.

Aristotle's mistake seems to have been to assume that on the view he is criticising pyramid-shape was itself sufficient for 'fieriness' and cube-shape for 'earthiness', for example; i.e. that in virtue of a particle's shape, it has a particular qualitative nature. But that *is* an incoherent notion – within any solid we can imagine all manner of different (and overlapping!) shapes without end. Thus every body would have to be *all* of FAWE (and lots of other substances characterised by different particle-shapes) simultaneously!

C. Aristotelian Matter Theory

In the previous section we discussed Aristotle's critique of Democritus' theory of matter; it is now time to examine his own positive views. There are, he says,[163] three 'originative sources' of body, namely prime matter (P.M.), the four primary qualities or contraries Hot, Cold, Dry and Moist

---

161   *De Caelo*, 3, 5, 304a 22-24.

162   Given a theory of mathematical minima (the minimal parts of atoms) these geometrical figures will not be perfect.

163   *De Gen et Corr*, 2, 1, 329a 30-b2.

(HCDM) and the 'simple bodies' or 'so-called elements' Fire, Air, Water and Earth (FAWE). I discuss P.M. at some length in Appendix 2, where it is argued that:

a. No coherent sense can be made of the totality of Aristotle's remarks about matter (*hulē*). To play the rôle of a physical substratum it must be quantified and space-filling; but *qua* ultimate subject of predication it can have no actuality of its own, neither quantity nor quality nor position.

b. The classical doctrine of P.M. as an abiding, indeterminate, common material substratum underlying substantial generation and corruption is essential to Aristotle's physical thought.

c. To develop an intelligible physical account of coming-to-be and passing-away one needs a P.M. which is highly plastic in nature, unlike any sensible stuff, but which is *not* a 'pure potency' devoid of all actuality. A material substratum must be stuff-like (i.e. space-filling, quantified, etc.), and hence of at least some minimal degree of actuality. If then one is to make sense of the matter-theory implicit in the *Physics* and, especially, in *De Generatione* one must ignore some of the darker utterances of the *Metaphysics*.

P.M., then, is featureless and indeterminate stuff, common to all bodies, but never existent as such, devoid of all form. Such a material substratum is necessary to the action and passion of the four basic 'contraries', the 'primary' qualities HCDM. These should not be conceived as themselves substantial (capable of independent existence) – they can only exist, act, and suffer when embodied in an appropriate substrate.[164] They are, says Aristotle, the 'active' (H and C) and 'passive' (D and M.) *powers* constitutive of the natures of material things.[165] The everlasting strife of these pairs of warring opposites[166] is, curiously, the source of the balance and harmony of Nature: fortunately, no one contrary can ever subdue or eradicate its proper antagonist – if this were to happen, the cosmic harmony would be overthrown and chaos let loose.

HCDM are, says Aristotle, the two fundamental pairs of contraries in Nature – from these as principles, other qualities can be derived. 'All other differences,' he says, 'reduce to the

---

164 *Ibid.*, 32–33.

165 *Meteorologica*, 4, 1; *De Gen et Corr*, 2, 2, 329b 25–27.

166 See Kahn, pp. 129–130.

first four, but . . . these admit of no further reduction'.[167] In *De Generatione*, 2, 4, he only derives other tangible qualities (viscosity, brittleness, roughness, smoothness, etc.) from HCDM; elsewhere,[168] however, he makes it plain that he thinks other qualities, such as colours and scents, are also dependent on HCDM ratios.

P.M. combines with the 'powers' HCDM to give the 'simple bodies' FAWE, the simplest things capable of *independent* existence. Although complex in definition (Fire, for example, is P.M. + H + D) and hence not truly *elementary* in nature (Aristotle consistently refers to them as the 'so-called' elements), they are nevertheless the *simplest bodies*. Neither P.M. nor the contraries HCDM can have separate existence. Given (a) that any portion of P.M. must be 'informed' by one or other of each of the fundamental pairs H/C and D/M, and (b) that a pair of contrary qualities cannot co-exist (with full intensity) in the same matter, it follows that there are four of these 'simple bodies', viz.:

P.M. + H + D = Fire
P.M. + H + M = Air
P.M. + C + M = Water
P.M. + C + D = Earth.[169]

These simple bodies are often called 'elements'. What is the meaning of that term? Aristotle offers the following two definitions:

> A body into which other bodies may be analysed, present in them potentially or in actuality (which of these, is still disputable) and not itself divisible into bodies different in form:[170]

> the primary constituent immanent in a thing, and indivisible in kind into other kinds . . . if they *are* divided, their parts are of the same kind, as a part of water is water.[171]

When Aristotle speaks of an element as 'indivisible in kind into other kinds', he clearly has in mind *chemical decomposition* rather than physical analysis. Every homoeomerous stuff (e.g. flesh, bone, wood) is for Aristotle 'indivisible in

167   *De Gen et Corr*, 2, 2, 330a 25–26.

168   *Physics*, 1, 6, 189a 17–19.

169   *De Gen et Corr*, 2, 3.

170   *De Caelo*, 3, 3, 302a 16–18.

171   *Metaphysics*, 5, 3, 1014a 26–31.

kind' in the sense that every part of flesh (or wood) *is* flesh (or wood), but those substances are not elements:

> . . . flesh and wood and all other similar bodies contain potentially fire and earth, since one sees these elements exuded from them; and, on the other hand, neither in potentiality nor in actuality does fire contain flesh and wood, or it would exude them.[172]

Consider a burning branch – a favourite example of later peripatetic chemists.[173] When the wood burns it emits flame (= F) and Smoke (= A); liquor or sap (= W) exudes, hissing from the cut ends; and ashes (= E) are left as a residue. The decomposition of bodies into the elements FAWE was thus for Aristotelians an empirical 'given', a matter of manifest experience.

From FAWE (and their associated 'powers', HCDM) as primitives, Aristotle develops his entire matter-theory. Before going on to discuss compound-formation, let us briefly mention two further points about FAWE. The first concerns their transmutation into one another. Given a common material substratum (P.M.) and the contraries HCDM, it will be straightforward for Aristotle to explain transmutation in terms of the accession and loss of HCDM in given portions of matter – this account is set out in detail in *De Generatione*, 2, 4.

The second point is somewhat more difficult, and concerns the nature and status of the elements. 'In fact,' says Aristotle:

> . . . fire and air and each of the bodies we have mentioned, are not simple, but blended. The 'simple' bodies are indeed similar in nature to them, but not identical with them. Thus the 'simple' body corresponding to fire is 'such-as-fire', not fire: that which corresponds to air is 'such-as-air': and so on with the rest of them.[174]

This passage may be no more than a rather unexciting assertion that the FAWE we perceive around us are never absolutely pure, but always more or less mixed. On the other hand, it does seem to go some way towards a conception of the elements as *ideal*, as substances with (unknown) real essences which we detect merely by their *effects*. On this view,

---

172   *De Caelo*, 3, 3, 302a 21-25.

173   The Aristotelian spokesman, Themistius, uses this example in Boyle's *Sceptical Chemist*. See Boyle, Vol. 1, p. 479.

174   *De Gen et Corr*, 2, 3, 330b 22-25.

the terms 'F' and 'W' would cease to be purely 'observational' and become more heavily 'theory-laden': each would introduce a postulated (but largely unknown) real essence.[175] There is evidence of such developments in Theophrastus and others,[176] but not a lot seems to have come of it – there was after all *insufficient theory* to lend much semantic support to the use of 'F', 'W', etc., as theoretical terms.

This conception of the elements as ideal tended to become associated with a view of them as *quality-bearers*.[177] If, for example, fire is necessary for visibility, earth for tangibility, and water for cohesion, any tangible cohesive and visible body must contain (at least) all three of those elements. *Pure F, A, W or E* now seem to be *impossible in principle*.[178] If each element were actually to contribute a feature conceptually necessary for corporeality, the very notion of a pure element, existing in isolation, would be incoherent. Bodies may on such a view contain more or less of any of the elements, but none can ever be totally lacking. We shall come across a variety of theories of this kind when we come to discuss alchemical matter-theories in Book 2: it is as well that we have at least mentioned them here.

Aristotle draws a sharp distinction between mechanical aggregation (*synthesis*) and chemical combination (*mixis* or *krasis*). *Mixis* proper is, says Joachim, 'the chemical combination of the four elements to form the homoeomers'.[179] The key term here is '*homoeomer*'. A mixture of wheat and barley grains, or of powdered iron and sulphur, is not a *homoeomer*, not a product of *mixis*. In such an aggregate, each discrete particle of one or other of the ingredients retains its distinct nature as such: a sufficient number of physical divisions would separate those particles unaltered. In Aristotelian *mixis*, by contrast, the ingredients undergo mutual action and passion to produce eventually a perfectly homogeneous compound or *mixtum*. Suppose, says Joachim, that the ingredients ABCD of a compound,

---

175   Cf. Paneth's distinction between 'basic substances' and 'simple substances' (Paneth, Part 2, p. 150).

176   See Samburksy, 4, pp. 35–37.

177   See Paneth, Part 2, p. 148.

178   Such a theory is criticised by Plotinus in *Ennead*, 2, 1 'On Heaven' (Vol. 2, pp. 25–27). Plotinus does not, however, attribute this quality-bearer theory of FAWE to any particular source.

179   Joachim, p. 72.

by acting and reacting on one another, produce an alteration in one another's qualities. Suppose further that this reciprocal alteration continues until a resultant, x, emerges, whose qualities are modifications of the components, and yet are different from the qualities of any (and of all) of them. Suppose further that every part of x, however far you subdivide it, retains the character of the whole. And suppose finally that (by appropriate processes of resolution) you can recover (or re-create) from x the original A, B, C and D.[180]

A genuine *mixtum* has therefore the following features:

1. It is homoeomerous – any and every part of x is x.
2. It can be decomposed chemically to release the ingredients of which it was initially formed.
3. It is qualitatively distinct from its ingredients, with a new 'nature' and 'form' of its own.

In *mixis*, the constituents reciprocally alter one another, so the conditions for action and passion must be met. Now for two substances A and B to interact mutually the following conditions must pertain:

1. Both must be material ('the hot' does not act on 'the cold' simpliciter – a hot body warms a cool one).
2. They must be 'contrary' to one another in some respect, i.e. identical in genus but opposed in form or actuality.[181] Only those agents are 'combinable' which involve a contrariety – for these are such as to suffer action reciprocally.[182]

FAWE are therefore ideally suited to play the rôle of elementary ingredients in compound-formation: they are material bodies informed by two pairs of 'contraries', H/C and D/M. These contraries are for Aristotle in a constant state of strife against their respective polar opposites, H against C and D *versus* M. Sometimes the struggle is an uneven one, and one contrary will overcome its antagonist, but, if a measure of *balance* is attained between H and C, and between D and M, no one of these qualities will be sufficiently powerful to subdue its opposite, and 'mixis' (i.e. *blending*) will occur, giving rise to something *intermediate*[183] between the extremes of heat and cold, dryness and moisture. Thus Aristotle:

180   *Ibid.*, p. 75.
181   *De Gen et Corr*, 1, 7, 324a 4–9.
182   *Ibid.*, 1, 10, 328a 32–33.
183   Joachim, p. 82; Bolzan, pp. 140–142.

There are differences of degree of hot and cold. Although, therefore, when either is fully real without qualification, the other will exist potentially; yet, when neither exists in the full completeness of its being, but both by combining destroy one another's excesses so that there exist instead a hot which (for a 'hot') is cold and a cold which (for a 'cold') is hot; then what results from these two contraries will be neither their matter, nor either of them existing in its full reality without qualification. There will result instead an 'intermediate': and this 'intermediate', according as it is potentially more hot than cold or *vice versa*, will possess a power-of-heating that is double or triple its power-of-cooling, or otherwise related thereto in some similar ratio.[184]

Some sort of balance of the contraries HCDM is therefore necessary for *mixis*. If this is attained, the contraries do not annihilate one another and leave bare P.M. - that is impossible - but rather temper or qualify each other, according to a determinate ratio which fixes the specific nature of the final compound. (One HCDM ratio will give flesh, another will give bone, and so on.[185]) The powers HCDM therefore remain present, albeit in an attenuated form, in the *mixtum*.

In what sense, if any, can the elements FAWE be said to be present in a compound C? They cannot for Aristotle be *actually* present in C, since:

1. Physical division of C would reveal only C, not FAWE - the Aristotelian *mixtum* is homogeneous.
2. C has its own specific form in virtue of which it is that particular stuff: there are not a plurality of forms simultaneously present in the same substance.

On the other hand, FAWE are clearly potentially present in C in at least the (weak) sense that they can be extracted from it - C can be chemically decomposed to yield FAWE. Equally clearly, this weak sense of the potential existence of the elements in the compound is insufficient. The associated powers of FAWE (i.e. HCDM) are actually present (albeit in weakened forms) in C, and the actual properties of C are to be accounted for in terms of its specific HCDM ratio. If Aristotle is to say that FAWE have only potential existence in C, he needs some stronger notion of potential presence than mere recoverability. This problem is raised and tackled in *De Generatione*, 1, 10. First, the problem:

[184]   *De Gen et Corr.*, 2, 7, 334b 8–16.
[185]   Joachim, p. 76.

according to some thinkers, it is impossible for one thing to be combined with another. They argue that (i) if *both* the 'combined' constituents persist unaltered, they are no more 'combined' now than they were before, but are in the same condition: while (ii) if *one* has to be destroyed, the constituents have not been 'combined' – on the contrary, one constituent *is* and the other *is not*, whereas 'combination' demands uniformity of condition in both: and on the same principle (iii) even if *both* the combining constituents have been destroyed as a result of their coalescence, *they* cannot have been combined since *they* have no being at all.[186]

Here the familiar problems of identity-through-change are applied to chemical combination with paradoxical results. Aristotle's 'solution' is highly characteristic:

Since, however, some things *are-potentially* while others *are-actually*, the constituents combined in a compound can 'be' in a sense and yet 'not be'. The compound may *be-actually* other than the constituents from which it has resulted; nevertheless each of them may still *be-potentially* what it was before they were combined, and both of them may survive undestroyed . . . The constituents, therefore, neither (a) *persist actually* . . .: nor (b) are they destroyed (either one of them or both), for their 'power of action' is preserved.[187]

The 'power of action' of each constituent element of a compound persists in it – each is present in the compound as an actual '*virtus*' or power. This notion of 'potency' (hereafter, after McMullin, 'V-potency'[188]) is quite distinct from other uses of the term in Aristotle. Something actually hot is for an Aristotelian 'potentially-cold', although coldness is not present in it at all. Clearly, the notion of potency as a present *virtus* is quite distinct from this more orthodox use of the term.

In his paper 'Four Senses of Potency', McMullin distinguishes V-potency from the other senses of 'potency' found in Aristotle, and discusses its influence and significance. The notion was, he suggests, not really developed by Aristotle himself; it was left to scholastic commentators to expound

---

[186] *De Gen et Corr*, 1, 10, 327a 35–b 7.

[187] *Ibid.*, 327b 23–32.

[188] McMullin in McMullin, Ed., 1, p. 306.

and elaborate the conception of potency as a present *virtus*.[189] He also raises the important point that, whereas V-potencies naturally differ in intensity, characterisations of substances in terms of structural properties do not lend themselves to such a pattern of explanation. Either a structure (e.g. an atomic arrangement) is actually present or it is not: it is not the sort of thing that could be present as a V-potency.[190] The two patterns of explanation, one in terms of V-potencies, the other in terms of actual atomic arrangements, do not mesh: hence perhaps the Atomists' persistent disdain for Aristotle's potency/act distinction, and their claim (implicit or explicit) that all the properties and powers of bodies can be accounted for in terms of actual atomic arrangements. The thesis that dispositional properties reduce to categorical ones is one of the most fundamental and characteristic claims of ancient Atomism.

Before leaving the topic of compound-formation, I should like to be permitted one small digression. According to Aristotle, if a single drop of wine falls into a bucket of water, it is simply transformed into water; 'its form is dissolved' and its matter taken over by the form of water.[191] But this is clearly wrong: if many drops of wine fall successively into the bucket, one will eventually discern a difference in its contents – one will have not (pure) water but very dilute wine. The original drop of wine is therefore not lost, not transformed into water – it merely becomes imperceptible due to its extremely low concentration in water.

It was from considerations like these that the Stoics developed their highly idiosyncratic doctrine of *total mixture*. These philosophers retained Aristotle's conception of a true *mixtum* as perfectly homogeneous, and yet dropped his potency/act distinction.[192] They could therefore not avail themselves of Aristotle's solution to the problem of the existence of the elements in the compound. How then did they approach the problem?

In the first place, they classified compound bodies into three types, according as they were produced by:

1.   Mechanical aggregation (e.g. wheat and barley grains).
2.   Fusion – formation of a quite new substance.

189   *Ibid.*, p. 306.

190   *Ibid.*, p. 308.

191   *De Gen et Corr*, 1, 10, 328a 28–29.

192   Sambursky, 3, p. 12.

3. *Krasis* (liquids) or *mixis* (solids): the blending of two or more ingredients to produce a compound body in which each of those ingredients retains its own actual nature (e.g. wine and water).

If the constituents are separable from a mixture they must, say the Stoics, be *actually* present in it. Wherever the ingredients can be extracted from the mixture, then, *mixis* (3) and not *fusion* (2) has taken place. In a *mixtum*, says Sambursky, the ingredients are blended into a homogeneous compound in which they all actually exist: thus

> a complex interpenetration of all the components takes place, and any volume of the mixture, down to the smallest parts, is jointly occupied by all the components in the same proportion, each component preserving its own properties under any circumstances, irrespective of the ratio of its share in the mixture'.[193]

Mixis can, according to the Stoics, take place between a minute body and a massive one: 'There is nothing', says Chrysippus, 'to prevent one drop of wine from mixing with the whole sea'.[194] If one allows a drop of wine to fall into the Mediterranean off Piraeus, and then leaves sufficient time for perfect mixture to take place, a drop of water extracted at Carthage should contain some minute fraction of that drop of wine.[195] This incredible theory of total mixture follows, as Sambursky sees,[196] if one grants:

a. The perfect homogeneity of the *mixtum*.
b. The Stoic conception of the continuum as divisible without end into parts with still smaller parts, and so on *ad infinitum*. (For a Stoic, the continuum has no least or ultimate parts.)
c. The actual existence of the ingredients in the *mixtum* (which, for the Stoic, is entailed by their separability from it).

This Stoic theory was severely criticised in antiquity by Alexander the peripatetic in his *De Mixtione*, and ridiculed by Plutarch in his polemic, 'Against the Stoics on Common Conceptions'. Both allege that the theory involves the interpenetration of dimensions (one body occupying the same place as

---

193   *Ibid.*, p. 13.
194   Quoted from *Ibid.*, p. 13.
195   For a similar example, see Plutarch, Vol. 13, p. 811.
196   Sambursky, 3, pp. 14–15, 121–122.

another) and hence the violation of mathematical truths such as that 2 (Volumes) + 2 (Volumes) = 4 (Volumes).[197] If bodies mutually interpenetrate, Plutarch argues, even such elementary mathematical truths would no longer be inviolate. (Actually, of course, it is a purely synthetic question, an issue of empirical chemistry, whether 2 volumes of water combine with 2 of wine to produce 4 – or more, or less – volumes of *mixtum*.) There is no reason to suppose that the Stoics were unable to answer these questions – Sambursky[198] shows in more detail how this may have been done.

D. The Epicureans on Compound Bodies

One of the greatest differences between the systems of Democritus and Epicurus lay in epistemology. Whereas Democritus, as we have seen, had profound doubts about the epistemic value of the 'bastard' knowledge derived from sense-perception, Epicurus makes sensation the *criterion* of truth. What appears to sense, he insists in his *Canonics*, cannot but be true: there is no superior epistemic authority capable of overriding sensory data.

But what is manifest to sense is a world of hot and cold, red and blue, scented and malodorous bodies: bodies with these qualities must therefore exist not merely 'by convention' (*nomôi*) but 'in reality' (*eteêi*). Sense reveals the existence of such bodies, and reason is powerless to gainsay it. Colours, smells, etc., are therefore objective properties of perceptible bodies.[199]

What then becomes of the distinction between primary and secondary qualities that we discovered in Democritus? Does Epicurus simply reject it? The answer is clearly negative. Although for Epicurus compound bodies have colours, etc., *atoms* have only three inseparable properties, viz. size, shape, and weight. This makes Epicurus a believer in some form of 'emergence' of qualities as one ascends from the atoms to higher levels of organisation. A compound body is for an Epicurean not a mere heap or aggregate but a new entity, almost an organism.[200] Epicurus thus adopts a position similar in some respects to the Aristotelian view that *mixis*

---

[197]   Plutarch, Vol. 13, p. 803. For a more sober discussion of the question of total mixture, see Plotinus, *Ennead*, 2, 7, Vol. 2, p. 192 ff.

[198]   See Sambursky, 3, p. 15.

[199]   See Long, p. 38, for Epicurus' denial of the Democritean thesis of the subjectivity of secondary qualities.

[200]   Bailey, 2, p. 293.

(chemical combination) gives rise to a new and specific *form*. The compound is, says Lucretius, a '*concilium*' different in character from its constituents, with a unity and nature of its own[201] arising out of the harmony of the atomic motions within it.[202] More of this anon.

Epicurus thus believes, as Bailey says, in two equally real 'worlds', each of which is as it seems to us to be. There is the sensory world of colours, smells, tastes, etc.; and there is the 'rationally apprehended' world of the atoms and the void.[203] Enquiry about the nature of this second 'world' must pursue an *analogical* method: we have no choice but to characterise it in terms of concepts derived from the more familiar world of the senses. A judicious and careful use of analogy is therefore essential to Epicurean scientific method: one cannot begin to understand Lucretius in particular until one has grasped the rôle played by analogical reasoning in *De Rerum Natura*.

It is to the 'rationally perceived' world of atoms and vacuum that one turns for causal explanations of the phenomena of the sensory world: one can explain the appearances of heat, cold, colour, etc., in terms of atomic shapes and sizes, motions and arrangements, but not *vice versa*. This is the crucial asymmetry between the two worlds which tempts one to speak of the featureless atoms and vacuum as more real and objective than the multi-coloured compounds. Would Epicurus assent to such a move?

In modern terminology Epicurus would say, I suspect, that colour is *supervenient on* but not *reducible to*[204] atomic arrangement. Bodies really are coloured, just as they appear to be (colour is not a mere sensation caused in us by something itself colourless), but they are so in virtue of the arrangements of the atoms of their superficial layers – the colour of a body could not alter without a corresponding rearrangement of those atoms. Whether any coherent intermediate position is actually possible here between real qualities on the one hand and Democritean reductionism on the other I do not know, but am inclined to doubt it. If pressed with a straight dilemma between those two options, Epicurus would be in a quandary: his epistemology would lead him one way, his matter-theory the other.

201   Pancheri, p. 143.

202   See Bailey, 2, pp. 347–348.

203   *Ibid.*, p. 293.

204   Cf. of course many similar claims in modern philosophy with regard to the dependence of the mental on the physical.

How do compound bodies arise? In Democritus, the explanation is in terms of the mechanical aggregation of discrete atoms which are then locked together by entangling, by hooks and eyes, etc. It is probable that Democritus thought of the resulting structure as *static*,[205] i.e. that the atom was not considered to be in motion in a compound body.

For Epicurus, however, every atom always moves with a speed of q/t, i.e. one space-minimum per time-minimum.[206] No atom, then, is ever at rest: even in a compound, the atoms are in a state of *perpetual agitation*, as Lucretius informs us in one of his most effective images. If, he says, we see flocks of sheep or embattled legions from a sufficient distance, the motions of the individual sheep or men will be imperceptible, and the flocks or legions will be seen as stationary masses of colour. It is similar, we are told, with the motions of the atoms in the compound: we see the resting whole, but not the rapidly agitated parts.[207]

The continual internecine motions of the imperceptible individual parts is then perfectly compatible with the perceived stability of the whole mass, be it a flock of sheep or a compound body. But if the atoms in a body are constantly jostling one another, colliding and recoiling, etc., their 'entanglements' can only be very loose: they cannot be bound fast by hooks or chains. How then is the unity and cohesiveness of the compound body to be explained? Although the Epicureans continue to use the Democritean vocabulary of 'entanglement', the emphasis of their theory has shifted to the motions of the atoms: *kinematic harmony* is more important in conferring unity than mere mechanical entanglement. To see this, let us turn to Lucretius.

It is certain, says the poet, that no rest is allowed to the atoms as they wander in the void. In their unceasing journeys, moreover, they must at times come into collision with one another, after which:

> some when they have dashed together leap back at great intervals apart, others too are thrust but a short way from the blow. And all those which are driven together in a more close-packed union and leap back but a little space apart, entangled by their own close-locking shapes, these make the strong roots of rock and brute bulk of iron and all other

205   Bogaard, pp. 326–327.
206   See of course Chapter 1, part 2, pp. 33–34.
207   Lucretius, Book 2, 310–332, p. 253.

things of their kind. Of the rest which wander through the great void, a few leap apart, and recoil afar with great spaces between; these supply for us thin air and the bright light of the sun.[208]

The various materials of our world can therefore be characterised in terms of the *Mean Free Path* (M.F.P.) of the motion of their constituent atoms, i.e. the distance each atom travels, on average, between successive collisions. (M.F.P. will obviously vary with the amount of inner void a compound contains.) A short M.F.P. corresponds to a densely-packed stuff such as stone or metal; a long M.F.P. gives a rare material like air or fire.[209] Many atoms, however,

> wander on through the great void, which have been cast back from the unions of things, nor have they anywhere else availed to be taken into them and link their movements.[210]

To join a *concilium* or 'union', an atom must be able to 'link its movements' with those of the constituent atoms of that body in such a way as to preserve its particular *kinematic harmony*. (This harmony is not, it must be stressed, a bare uniformity, but a 'harmony in diversity', as Bailey calls it.[211]) It is this harmony or sympathy of motions which then becomes for Lucretius the key factor in explaining the unity and coherence of the compound: to dissolve a *concilium*, for example, it is necessary not merely to disentangle its parts but to *disrupt* its harmony.[212]

A *concilium*, on this view, is still a mechanical aggregate of quite distinct atoms – the atoms do not lose their individual identities and merge together – but it possesses highly specific and peculiar powers and properties of its own.[213] It preserves itself, for example, by effectively 'rejecting' all atoms which cannot combine their motions harmoniously with those of its components. In virtue of this rejection-process, there exists a stock of free atoms which Lucretius likens, in a celebrated passage, to the motes in the sunbeam.[214] The image goes back

---

[208] *Ibid.*, 97-108, pp. 241-243.

[209] See Bailey, 2, p. 341; Bogaard, p. 322.

[210] Lucretius, Book 2, 109-111, p. 243.

[211] Bailey, 2, p. 348.

[212] Bogaard, p. 325. Cf. of course the Epicurean theory of soul and body.

[213] Bailey, 2, pp. 347-348.

[214] Lucretius, Book 2, 112-124, p. 243.

of course to Democritus,[215] but Lucretius uses it with great mastery. It serves a twofold purpose: it is both an analogy or likeness (*simulacrum*) to the motion of the free atoms and an effect of them.[216] The image therefore serves to introduce a concept by analogy ('this is what atomic motions are like') and also serves as part of a causal argument ('one cannot explain this phenomenon without supposing that the dust-particles are being bombarded by still smaller, and hence invisible, corpuscles').

But I digress: the main theme of this section is the formation of compound bodies, not the motion of the free atoms which are kinematically unsuitable for compound-formation.

One point which will have struck the reader is that the difference between iron and stone on the one hand, and air and fire on the other, was described by Lucretius in terms of a difference in atomic arrangements and motions rather than their sizes and shapes. Densely-packed atoms and short M.F.P.s give solids; loosely-associated atoms and long M.F.P.s give rarer, gaseous bodies. (The liquid state presents some more serious difficulties.[217]) This allows the Epicurean to explain the transmutation of the 'elements' FAWE (= states of matter) without doing violence to his Atomic Theory. Thus Lucretius denies that there are fire-atoms, atoms intrinsically fiery in nature. His objection, however, tells also against the Democritean theory that a given atom is a fire-atom simply in virtue of its (spherical) shape. If, says Lucretius, there were such atoms,

> it would be no matter that some should separate and depart, and others be added, and some changed in order; if despite this all retained the nature of heat; for whatever they might create would be in every way fire.[218]

In truth, says Lucretius, 'there are certain bodies, whose meetings, movements, order, position and shapes makes fires';[219] change those meetings, movements, orders, etc., and

---

215   See Aristotle, *De Anima*, 1, 2, 404a 3–4.

216   Lucretius, Book 2, 127–132, p. 243.

217   See Bailey, 2, pp. 340–341. Toulmin and Goodfield, 1, p. 67, argue that the ancient Atomists, by rejecting interparticulate forces, made it impossible for themselves to give an adequate account of the liquid state.

218   Lucretius, Book 1, 680–684, p. 211.

219   *Ibid.*, 685–686, p. 211.

they make some other substance. Although Lucretius does cite shape among the relevant explanatory factors, it is secondary in importance to arrangement and motion. What it is for a body to be F (or W) is for it to have a particular arrangement of atoms and a peculiar harmony of atomic motions. Change this arrangement or those motions, and one no longer has fire.

There are therefore, for the Epicurean, no atoms of FAWE, nor likewise of iron, chalk, cheese, etc. All sensible stuffs exist only at the level of *compounds*; atoms are not, merely in virtue of their size and/or shape, F-atoms or W-atoms. An individual atom may at different times in its eternal career fall into different complexes and become thereby a constituent part of watery or fiery bodies, but it could never truly be described as a W or F atom: it has a 'secret and unseen' nature of its own.[220]

The natures of FAWE, therefore, arise only at a molecular level, i.e. at some level of aggregation above the atomic and, presumably, below the perceptual. Evidence for attribution to Epicurus of a molecular theory rests largely on the following passage in Alexander of Aphrodisias' *De Mixtione*. Epicurus, says Alexander, differed from Democritus over the nature of *mixis*:

> He too belonged to those who regarded a mixture as a juxtaposition of bodies, but he declared that these bodies are not preserved in the process of division but dissolve into elements and atoms. Each of the bodies that consist either of wine or water . . . is somehow composed of these elements and atoms. From the quality of these bodies of which the components consisted the mixture results by synthesis. Thus it is not water and wine that mingle but, so to say, water-producing and wine-producing elements, and the mixture results from a kind of destruction and generation. For the dissolution into the elements is destruction and the synthesis of the components is generation.[221]

There are then no atoms of wine or water but only 'wine-producing' and 'water-producing' atoms. When wine and water are mixed, these *concilia* decompose, releasing those

---

[220] See Lucretius, Book 2, 779, p. 215, for his attribution to the atoms of a 'secret and unseen' nature of their own, quite disanalogous to perceived bodies.

[221] Quoted from Sambursky, 4, p. 27. For the Epicurean molecular theory, see also Masson, p. 45.

atoms: wine and water are therefore corrupted and a new *concilium* (the *mixtum*) generated. The characteristic natures of wine and water are on this account not possessed by their atoms but arise only at some level of aggregation. Lucretius sometimes speaks of 'seeds' (*semina*) of water, fire, etc., as if these are to be conceived as the smallest particles of *particular substances*, i.e. as molecules in our modern sense, but this usage is by no means consistent – elsewhere he uses '*semina*' to refer to the atoms.[222]

Since an individual atom is not, in virtue merely of its shape and/or size, an atom of any particular substance, the transmutation of substances by the disintegration and reassembly of *concilia* is readily conceivable. Even the so-called elements FAWE can be transformed into one another: so long as the elements are not deemed atomic, transmutation is readily accommodated into an Atomist theory of matter.[223]

The *same* atoms, Lucretius insists:

> build up sky, sea, earth, rivers, sun, the same too crops, trees, living creatures, but only when mingled and moving with different things in different ways. Indeed scattered abroad in my verses you see many letters common to many words, and yet you must needs grant that verses and words are unlike both in sense and in the ring of their sound.[224]

The letter-analogy too derives from Democritus,[225] but seems more fitting in Lucretius, where more explanatory weight is placed on the arrangement of atoms and less on their sizes and shapes. (Lucretius does admittedly give a conventional Atomist account of odours and tastes in terms of atomic shapes,[226] but in his discussion of compound bodies in general arrangement and motion are by far the more significant explanatory factors.)

An Epicurean compound or *concilium* has therefore what Bogaard[227] dubs a 'dynamic texture', a combination of *atomic arrangement* and *internecine harmonic motion* in virtue of which it is what it is. (One can still speak meaningfully of an

---

222   See Bailey, 2, pp. 343–344.

223   Many of the great Atomists of the seventeenth century, men such as Newton and Boyle, reconciled their Atomism with their alchemical pursuits in a similar way.

224   Lucretius, Book 1, 820–826, pp. 217–219.

225   See *De Gen et Corr*, 1, 2, 315b 15.

226   See Lucretius, Book 2, 444–477, pp. 259–261.

227   Bogaard, p. 325.

'arrangement' of moving entities if they oscillate about fixed points, e.g. if they have simple harmonic motions.) Aristotle and others had argued that hard and impassive atoms could not combine to form compounds with new and specific natures of their own. Epicurus attempts to evade this objection: his atoms are impassive (they are not intrinsically altered by association) but nevertheless come together to form a new unity with a cohesiveness and individuality of its own, guaranteed by this 'dynamic texture', an Epicurean surrogate for the Aristotelian notion of form. It is of course in virtue of its 'form', not its matter, that a compound body has its own peculiar unity and self-identity.[228]

E. Epicurus and Lucretius on Primary and Secondary Qualities

The atoms, says Epicurus, lack all the properties of observed things other than shape, size and weight.[229] (We shall deal with weight in the next chapter – here I shall concentrate only on size and shape.) They will also, of course, have as 'accidents' such features as motion and position: the precise position and state of motion of an atom at any given time is accidental to it, whereas its size, shape, and weight are inalienable.

What arguments has Epicurus for his claim that the atoms lack colour, smell, taste, temperature, etc.? He answers that:

> . . . every quality changes; but the atoms do not change at all, since there must needs be something which remains solid and indissoluble at the dissolution of compounds, which can cause changes; not changes into the non-existent or from the non-existent, but changes effected by the shifting of position of some particles, and by the addition or departure of others. For this reason it is essential that the bodies which shift their position should not possess the nature of what changes, but parts and configuration of their own. For this much must needs remain constant.[230]

One can detect traces of two distinct arguments here. Epicurus could be saying that, since we can explain secondary

---

228   For an extreme case, consider the Epicurean doctrine of the nature of the gods, which are said to have identity-conditions like those of rivers or waterfalls. It is important to note, however, that the same is true, more or less, of all bodies: gain or loss of matter (atoms) need not jeopardise identity over time; disruption of form ('dynamic texture') does.

229   'Letter to Herodotus', 54, Bailey, 1, p. 31.

230   Ibid., p. 31.

qualities in terms of atomic arrangements, and hence qualitative alteration in terms of the rearrangement of atoms, it is redundant to make the atoms themselves coloured, etc. Worse, it undercuts the point and purpose of the whole reductive explanation: if atoms are red, blue, warm, cold, etc., then an account of the nature and origin of those qualities in terms of atomic arrangements is otiose.

The second point concerns mutability. If atoms are hot and cold, sweet and sour, red and blue, etc., they must undergo change in respect of those qualities: it is presumably taken to be a datum of sense that whatever it is, e.g. coloured, changes colour. But mutability, says the Epicurean, involves perishability: whatever can alter can cease-to-be. (I have yet to find a satisfactory explanation for the widespread acceptance of this highly dubious assumption in Greek philosophy.) Now the atoms are the abiding Eleatic *substratum* behind the coming-to-be and passing-away of substances – as such they must not bear any trace of mortality. Thus, says Lucretius, the atoms cannot be coloured:

> For any colour, whatever it be, changes . . .; but the first-beginnings ought in no wise to do this. For it must needs be that something abides unchangeable, that all things be not utterly brought to nought. For whenever a thing changes and passes out of its own limits, straightway this is the death of that which was before. Therefore take care not to dye with colour the seeds of things, lest you see all things altogether pass away to nought.[231]

In their rôle as the ultimate substratum of physical change, the atoms themselves cannot alter. (Any alteration in an 'atom' could only be explained in terms of the re-arrangement of its parts, but then *these* would be the *true* atoms, and would have to be immutable, on pain of infinite regress.)

Epicurus also presents a further argument for the attribution of shape, but not colour, to his atoms:

> For even in things perceptible to us which change their shape by the withdrawal of matter it is seen that shape remains to them, whereas the qualities do not remain in the changing object in the way shape is left behind, but are lost from the entire body.[232]

This argument seems to founder on the determinable/ determinate distinction. Any perceived body will of course

---

231 Lucretius, Book 2, 750-756, pp. 275-277.
232 'Letter to Herodotus', 55, Bailey, 1, p. 33.

have some determinate value of the determinable quality shape: its determinate shape will vary while its 'shapedness' cannot. But exactly the same can be said of colour: any visible body must have some determinate value of the determinable colour; its determinate colour can alter, but its 'colouredness' cannot. Where, then, lies the difference? Divide a body often enough, and its colour may disappear, but then so will its shape: why then this confidence that the minute and invisible parts into which it has been divided possess shape but lack colour?

Sooner or later, the argument must turn to conceptual questions about the essential qualities of bodies. We are obliged, says the Atomist, to think of the size and shape of a perceived body as made up of the sizes and shapes of its imperceptible parts – we have no way of explaining size and shape as 'emergent' properties of an aggregate of entities lacking those features. The size of a compound body is – loosely speaking – the *sum* of the sizes of its parts. The same cannot be said of colour.

The Epicurean argument can be made to work if phrased as follows. Bodies can only be compounded of bodies: we cannot conceive how a corporeal substance could be made up of non-corporeal elements. Thus the minute atoms of which compound bodies are composed must themselves be bodily in nature, i.e. must possess those properties conceptually necessary to corporeality, which include size and shape. We *must* therefore attribute size and shape to atoms: if the micro-realm is to be accessible to our thought there has to be a positive 'Analogy of Nature' between the sensible and intelligible realms.[233] There is, however, no obligation to attribute colour, temperature, etc., to the atoms – we can conceive of bodies lacking those properties. And if we have an account of how those properties can emerge at a molecular level which does not involve their attribution to the atoms, so much the better – it enables the atoms to play the rôle of a featureless, immutable, Eleatic *substratum*.

Let us once again turn to Lucretius for elucidation and elaboration of themes left obscure in the crabbed and condensed prose of the 'Letter to Herodotus'. If, says Lucretius, the reader imagines that white bodies are composed of white atoms – or indeed of atoms of any colour – he is in error:

---

[233]   For a penetrating study of the nation of an 'Analogy of Nature' between atoms and perceived bodies, see McGuire, 3.

> For the bodies of matter [i.e. atoms] have no colour at all,
> neither like things nor again unlike them. And if by chance
> it seems to you that the mind cannot project itself into these
> bodies, you wander far astray. For since those born blind,
> who have never described the light of the sun, yet know
> bodies by touch, never linked with colour for them from
> the outset of their life, you may know that for our mind
> too, bodies painted with no tint may enter our com-
> prehension. Again, we ourselves feel that whatever we
> touch in blind darkness is not dyed with any colour.[234]

Bodies without colour are therefore perfectly conceivable: the
blind man has a *concept* of body, derived from his sense of
touch, as possessing size, shape, and solidity. Indeed, we
sighted people, as we blunder around in the dark, make use of
a very similar concept of body as tangible extension. There is
therefore no *conceptual* difficulty about the notion of a
colourless body, as there would be about that of a sizeless or
shapeless one. The mind can, says the poet, 'project itself into'
colourless atoms; such 'bodies painted with no tint' may 'enter
our comprehension'. The conceptual point which we
attributed to Epicurus is therefore quite explicit in his
disciple: we *must* attribute size and shape to atoms, but *need
not* attribute to them colours, etc.

Granted that we are under no conceptual obligation to
attribute colour and other secondary qualities to atoms, is
there any reason why we should not nevertheless do so?
Lucretius has several reasons: with colourless atoms, he
begins:

> you can most easily . . . give account, why those things
> which were a little while before of black colour are of a
> sudden to become of a marble whiteness; as the sea . . . is
> turned into white waves of shining marble.[235]

On the other hand, if atoms were coloured, one could give no
intelligible account of colour-changes. Suppose, for example,
that the atoms of the blue sea were themselves blue: then no
mixing or re-arrangement of these (immutable) atoms could
ever produce the white surf. Nor can the atoms which make
up the sea be of all colours, since then we should not see it as
uniformly blue. Explanatory ease and power therefore

---

[234] Lucretius, Book 2, 737-747, p. 275.

[235] *Ibid.*, 763-767, p. 277.

demand that we attribute no colour at all to the 'first bodies'. Moreover, Lucretius continues:

> since colours cannot be without light nor do the first-beginnings of things come out into the light, you may know how they are not clothed in any colour. For what colour can there be in blind darkness?[236]

Colours, for the Epicurean, are 'begotten by a certain stroke of light'. The intensity, angle, and nature of this 'stroke' are of crucial importance to the nature of the hue produced, as the shimmering colours of the peacock's tail bear witness. Atoms, being too small to be susceptible to such blows, could not conceivably undergo the re-arrangement of superficial parts which compound bodies undergo under their influence and which give rise to colours. On this theory colour becomes either identical with or at least supervenient on an *arrangement* of atoms produced in a body by the impact of a stream of light-corpuscles; as such, it can be meaningfully attributed only to compound bodies and not to their constituent simples.

Lucretius' next argument introduces the Epicurean theory of vision, according to which sight results from the transmission of corporeal 'images' (*eidola, simulacra*) from the object to the eye:

> And since the pupil of the eye receives in itself a certain kind of blow, when it is said to perceive white colour, and another again, when it perceives black and the rest, nor does it matter with what colour things you touch may choose to be endowed, but rather with what sort of shape they are fitted, you may know that the first-beginnings have no need of colours, but by their divers forms produce divers kinds of touch.[237]

If fully thought-through, this mechanistic theory of vision should yield as a consequence the thesis of the subjectivity of colours. If the *eidolon* only serves to deliver a *blow* on the eye, and visual sensation arises from this blow, then only such features of the 'image' as its size, shape, motion, and structure are causally relevant to the story. The *eidolon*, then, could itself entirely lack colour: indeed, colour should be a feature *only* of our sensations, a purely subjective effect of an external blow on our sense-organs. This, however, does not

---

[236] *Ibid.*, 795–798, p. 279.
[237] *Ibid.*, 810–816, p. 279.

seem to have been the Epicurean view: in Book 4 Lucretius[238] speaks freely of bodies emitting *simulacra* which resemble them in *colour* as well as in shape. (The 'images', being compound bodies, may themselves intelligibly be said to be coloured.) Objects, he says, emit *simulacra* 'of their own colour'; the *simulacrum* is indeed a 'thin film of surface-colour'.

But if all that matters to the causal story is the mechanics of the blow which the *simulacrum* imparts to the eye, its colour (unless this is simply *identified* with an atomic arrangement) is causally redundant. And if the colour of the 'image' or 'likeness' *is* nothing more than the arrangement of its atoms, we have a full-blown reductive account of colour. Once again we find in Epicureanism an unresolved tension between reductionism on the one hand and naïve realism on the other. But let us move on to Lucretius' next argument, which runs as follows:

> . . . the more each thing is pulled asunder into tiny parts, the more you can perceive colour little by little fading away and being quenched: as comes to pass when purple is plucked apart into small pieces: when it has been unravelled thread by thread, the dark purple or the scarlet, by far the brightest of colours, is utterly destroyed; so that you can know from this that the tiny shreds dissipate all their colour before they are sundered into the seeds of things.[239]

This argument appears to rest on a straightforward extrapolation from sense-experience. As such, it founders on the determinable/determinate distinction: the fading of a particular colour, such as red or purple, is not equivalent to the fading of 'colouredness' as such. (A washed-out pink is as much *a colour* as the brightest scarlet.) The failure of the argument illustrates a weakness in the attempt to make conceptual points by way of perceptual analogies: where the two realms are disanalogous it can lead to misapprehensions like the above. We could never in principle *see* the fading-out of the determinable quality 'colouredness' – all we can do is (a) argue at the conceptual level that colourless bodies are perfectly conceivable, and (b) argue positively *for* colourless

---

238   Lucretius, Book 4, 54–109, pp. 365–367. See also Book 4, 332–335, where the poet speaks of 'seeds of yellow' streaming from the bodies of jaundiced persons and *tinging* the incoming *simulacra* with their colour.

239   Lucretius, Book 2, 826–833, pp. 279–281.

atoms on physical grounds such as explanatory power. Elsewhere, Lucretius handles this mode of argument impeccably; here, he seems to stumble.

Having denied colour to the atoms, Lucretius goes on to deny them heat, cold, sound, taste, and smell.[240] Last but not least, he denies them all sensation – sensation can, *pace* the Platonists, be derived from the non-sentient. If the anti-reductionist objects to deriving compounds with any given property F from atoms lacking that property, Lucretius responds with the following anti-anti-reductionist gibe. If, he jokes, nothing F could arise out of non-F elements, the atoms which make up men would have to laugh and cry and philosophise: 'whatsoever you say speaks and laughs and thinks shall be composed of other particles which do the same things'.[241] But this is 'raving madness', to avoid which one must be prepared to contemplate non-F elements giving rise to F compounds by aggregation.

240   *Ibid.*, 842–846, p. 281.

241   *Ibid.*, 971–990, pp. 287–289. Lucretius uses the same argument against Anaxagoras' anti-reductionism in Book 1, 919–920, p. 223.

THE MECHANICAL PHILOSOPHY: 1
The Ancients

The 'Mechanical Philosophy' I take to be best defined by the following remarks of Thomas Hobbes:

> Whatsoever is at rest, will always be at rest, unless there be some other body besides it, which, by endeavouring to get into its place by motion, suffers it no longer to remain at rest.[1]

No body is caused to move except by 'a body contiguous and moved'.[2] Mechanism can thus be characterised negatively as involving the threefold denial of:

1. Action at a Distance (hereafter, A.D.).
2. The spontaneous initiation of motion.
3. Incorporeal causal agents capable of moving bodies.

There is also a further implication of Hobbes' remarks which takes a little more spelling out. The transmission of motion from one body to another by a mechanical impulse can be aptly characterised as 'blind' causation, devoid of purpose and design. The mechanical philosophy therefore also involved the denial of

4. Final causes not explicable in (reducible to) mechanistic terms.

Hobbes draws this conclusion quite explicitly: the final cause, he says, has no place 'but in such things as have sense and will', and this 'I shall prove . . . to be an efficient cause'.[3] The purposiveness which we experience in ourselves and impute to other men and animals can, he insists, be accommodated within a purely mechanistic explanatory framework. There is no irreducible teleology in Nature – no purposiveness inexplicable in terms of blind mechanical causation.

It may well be that Hobbes was the only major thinker in history to accept wholeheartedly all these theses, and hence the only fully-fledged 'mechanical philosopher' in my sense. I

---

[1] Hobbes, *Works*, Vol. 1, p. 115.

[2] *Ibid*, p. 124.

[3] *Ibid.*, p. 132.

am certainly not going to argue that either Democritus or
Epicurus was such. That the ancient Atomists come close to
mechanism is, however, clear both from their own texts and
those of their critics. Let us take the four major problem-areas
– A.D., spontaneity, incorporeal causes, and teleology – in
turn, starting with A.D.

A. Action at a Distance

Ancient Atomism was from its outset resolutely opposed to
any form of A.D.[4] According to Mary Hesse,[5] the 'primitive
analogy' in terms of which the Atomists conceived all physical
action was that of the labourer. All the labouring activities of
men – pushing, pulling, carrying, etc. – plainly require
contact between agent and patient. In Atomist thought, says
Hesse, this one 'primitive analogy' was developed and
systematised to the exclusion of all others (organic development, attraction, command, craftsmanship) present in the
mythic thought from which Greek philosophy emerged. Why
did this occur – why did the Atomists feel that the denial of
A.D. was essential to their system?

Perhaps the best answer is in terms of their ontology. Only
the atoms ('being') and void ('non-being') exist. Only the
atoms are *real*, i.e. capable of physical action – void is wholly
impassive and inert. The only conceivable form of change is
the local motion of the atoms in the void. A.D. is excluded,
says Hesse, because void is inert and 'no action can be
transmitted through the non-real except by passage of the
real'.[6] A 'force' would have to be something real (causally
efficacious) emanating somehow from atoms and traversing
space without affecting the medium (no field theory is
possible in ancient Atomism) and without any corporeal
bearer. But, says Lucretius, 'nothing can act or suffer without
body': only something tangible – and hence corporeal – can
move a body; whatever is intangible is 'empty void' and thus
causally inert.[7]

The argument seems to run as follows. True A.D. – action
on a body in one place by a body in another, without any
mediation *whatsoever* – is totally unintelligible. When we
speak of such things, then, we tend to imagine some

4   See Dijksterhuis, p. 12.

5   Hesse, 1, p. 33.

6   Hesse, 1, pp. 45–46. For the *a priori* nature of the Atomists' denial of
    A.D., see also Furley, 3, pp. 84–85.

7   Lucretius, Book 1, 433–448, p. 199.

(immaterial) 'influence' emitted by the agent and crossing the intervening space in order to act on the patient. But (a) the featureless void cannot sustain such 'influences', (b) only impact with another body can conceivably move a body.

The upshot of this is that such forces as magnetic and electrical 'attraction' can only be explained in terms of the action of corporeal effluvia emitted by the 'attracting' body. If only the push or pull of another body will impart motion to one at rest, the movement of the iron towards the lodestone or the chaff towards the rubbed amber must be explained in terms of some such effluvia. Let us deal with magnetism – a crucial test-case for mechanical philosophies over the ages – in a little more detail.

The founder of the 'effluvium' theory of magnetism seems to have been Empedocles,[8] although mechanical theories are also found in the *Timaeus*[9] (where Plato attributes it to the effect of the 'circular thrust' caused by the cyclical mutual replacement of bodies) and in Democritus. Since the details of Democritus' account do not survive, we must turn for our primary sources to the views of Epicurus (as reported by Galen) and of Lucretius.

According to Epicurus, the lodestone emits myriads of minute hooked particles, which combine to form *chains* linking the magnet to the iron; having hooked on to the iron, they then draw it back towards the lodestone.[10] The account is crude and schematic, and Galen soon develops some powerful objections, viz:

1.   How can such minute atoms shift a great mass of iron?
2.   How, on this theory, is the 'magnetic power' transmitted from a lodestone to a piece of iron? How, that is, is iron magnetised?
3.   How do corpuscles emitted from the magnet and striking the iron cause it to move towards the magnet rather than, as one might expect, away from it? The mere formation of a 'chain' between two bodies does not suffice to explain how they are drawn together: someone or something must still *pull* on the chain.[11]

In Lucretius,[12] we find a different variation on the same theme. The lodestone, he says, emits 'very many seeds', or 'an

8   Roller, p. 17; Hesse, 1, p. 57.

9   *Timaeus*, 80, p. 108.

10   See Galen, p. 71 ff.

11   Galen, pp. 75-77.

12   Lucretius, Book 2, 991-1064, pp. 567-569.

effluence which with its blows parts asunder all the air which has its place between the stone and the iron', thus producing an empty space between them. Once this void has formed, 'straightway first-beginnings of iron start forward' and 'fall' into it, drawing the whole mass of the iron in their wake – the particles of iron are so entangled that, as soon as a few of them move towards the magnet, the rest must follow.[13] There is also, however, a second and co-operating cause which acts as follows. Since on one side of the iron there is void and, on the other, atmospheric air,

> it straightway comes to pass that all the air which has its place behind, drives . . . and pushes the [iron] ring forward. For the air which is set all around is for ever buffeting things; but it comes to pass that at times like this it pushes the iron forward, because on the other side there is empty space, which receives the ring into itself.[14]

This is Lucretius' ingenious hypothesis,[15] involving no substantial chain between magnet and iron.[16] The motivation behind such hypothetical mechanisms is clearly the *a priori* conviction that, since A.D. is unintelligible, some such mechanism must be in operation. The actual details of the account given are much less important than the *demand* of the intellect for such an explanation. The reader could no doubt devise weighty objections to Lucretius' theory of magnetic action, as Galen did to Epicurus', but the effect of such criticisms could only have been to compel the Epicureans to modify their effluvium-theories: to abandon contact-action and posit A.D. was unthinkable for them.

Another problem-area for the mechanists is the sorting of the 'elements' FAWE such that F goes to F, E to E, etc. The phenomena are plain: their explanation, more difficult. Fire rises, heavy (earthy) bodies fall, water and air 'find their own levels' between these extremes. There seems, as Empedocles would say, to be a sort of 'Love' of like for like, a 'sympathy' between kindred things which draws them together.[17] But such a 'sympathy', if construed as a force acting at a distance, is quite unintelligible to an Atomist like Democritus.

13 *Ibid*, 1003–1008, p. 567.

14 *Ibid.*, 1025–1030, pp. 567–569.

15 See Harré, 3, p. 24.

16 Similar mechanisms were of course also proposed for electrical 'attractions' in antiquity.

17 For the 'primitive analogy' of *attraction*, see Hesse, 1, pp. 31–32; Jammer, 3, p. 27; Roller, p. 16.

The flocking of kindred creatures, 'doves with doves, cranes with cranes', is, replies Democritus, like the separation of grain from chaff in winnowing, or the sorting of pebbles according to size on a shingle beach. The phenomena occur, he admits, 'as if similarity in things possessed some power tending to draw them together.[18] We must take this 'as if' very seriously – for Democritus the appearance of sympathy in Nature is *mechanically explicable*. In the 'whirl' or vortex which constitutes our cosmos, atoms are sorted according to their sizes, producing great conglomerations of similar atoms – the masses of FAWE that make up our world. The purely mechanical operation of the 'whirl' therefore explains why there seems to be an attraction of like for like.[19] Larger atoms have more resistance to the whirling motion of the vortex, as a result of which they have lower angular velocities and fall into the centre of the 'whirl'; smaller atoms rotate more swiftly and fly to its periphery.[20]

There are therefore 'like to like' *tendencies* in Nature, but they are not irreducible examples of sympathy or A.D.: they will turn out to be explicable in terms of mechanical analogues such as the vortex, the winnowing basket,[21] the shingle beach, etc. In the words of Lucretius:

> For from all places all bodies are separated by blows each to its own kind, and they pass on to their own tribes; moisture goes to moisture, with earthy substance earth grows, fires forge fires, and sky sky.[22]

This separation occurs, says the poet, *by blows*, i.e. as a secondary consequence of mechanical causes. Given the conception of the compound body as a *concilium* which owes its unity and coherence to a *harmony* of internecine atomic motions, it will be easy for an Epicurean to render a plausible account of the appearance of 'sympathy' in Nature. Corpuscles of a certain type will only readily combine with certain other types of stuff, so 'seeds' of X will wander randomly around the cosmos until they encounter a mass of X with

---

18   Quoted from Kahn, p. 212. See also Kirk and Raven, p. 411, p. 420.

19   See Hesse, 1, p. 53; Furley, 3, pp. 85–86.

20   For discussions of vortex theories in Presocratic thought, see Ferguson and Tigner.

21   The analogy of the winnowing-basket plays a key rôle in Plato's explanation of the separation of FAWE from the pre-cosmic chaos in *Timaeus* 52–53, pp. 71–72.

22   Lucretius, Book 2, 1112–1115, p. 295.

which to fuse. They are not *attracted* to or by that mass of X but will be 'drawn' to it as it were negatively by indirection: after a sufficient time-period X-seeds will have 'found' masses of X, Y-seeds masses of Y, and so on, looking for all the world as if a like-like force of attraction were in operation. Although the Epicureans rejected the Democritean 'whirl', they could still find some account of the apparent sympathy we discern in the workings of Nature.

Let us now examine the views of the other major schools of ancient philosophy on this topic of A.D. In the *Timaeus*, Plato seems flatly to deny the existence of attraction; the 'puzzling attraction of amber and lodestone' is not, he insists, such:

> Proper investigation shows that there is no void and that circular thrust operates in all these instances; the various bodies part or come together in the course of mutual interchanges of position and what seems like magic is due to the complication of their effects on each other.[23]

The separation of the great masses of inchoate FAWE from the pre-cosmic chaos is also described in mechanical terms. In this state of chaos, says Plato, 'The Receptacle' (Space) was continually agitated:

> And its contents were in constant process of movement and separation, rather like the contents of a winnowing-basket . . . in which the solid or heavy stuff is sifted out and settles on one side, the light and insubstantial on another . . .[24]

FAWE are thus sorted into 'different regions of space' by a mechanical process of grading according to density.[25] But, although this mechanical shaking-process is no longer in evidence, fire still rises to join the mass of fire above, and stones still fall to earth. Why is this so? When, replies Plato, we

> stand on earth and try to weigh earthy substances . . . we lift them into the alien air by force and against their natural tendency, as they cling to the matter kindred to them.[26]

A larger body resists this separation from its 'kin' more strongly than smaller one, and is therefore called 'heavier'. It

23 *Timaeus*, 80, p. 108.
24 *Ibid.*, 52–53, pp. 71–72.
25 Cornford, 1, p. 202.
26 *Timaeus*, 63, p. 88.

is, says Plato, 'the tendency of any body to move towards its kindred aggregate' that makes it heavy.[27] Each element therefore moves 'in accordance with its nature' toward its kin, i.e. the great bulk of that element, wherever it is situated.[28] Plato therefore admits, in the final analysis, at least *one* example of a like-like tendency not explained in mechanical terms. (The Platonic conception of gravity as a matter-matter attraction rather than a tendency of matter to place became of course of crucial importance in the anti-Aristotelian cosmologies of men like Crescas, Oresme, and Copernicus.)

Let us now turn briefly to Aristotle's views on the topic of A.D. Physical action, he would say, divides into natural and constrained. The precise cause of natural motions in Aristotle's philosophy is far from clear, but it is evidently not thought of as involving A.D. The upward motion of a flame and the free fall of a stone are motions directed *toward* the natural places of those bodies, but those places must not be conceived as *causes*. Although place 'exerts a certain influence',[29] it cannot serve as any of the four causes; in particular, it does not 'move existents' as an efficient cause.[30] Whatever be the cause of the free fall of a heavy body, it is not the attraction of its natural place. Nor is it the attraction of similar stuff – if the Earth were removed from its present position and placed in the lunar sphere, earthy bodies would tend not to it but to its present location (the centre of the universe).[31] Such natural motions are therefore not to be conceived as due to attractions (A.D.) either by place or by matter.

When we turn to forced motions, the situation is much simpler. We see it to be universally the case, says Aristotle, that 'that which primarily is moved locally and corporeally must be either in contact with or continuous with that which

---

[27] Or light, according to where one is standing. To an observer in the heavens, fire would be 'heavy' and earth 'light'.

[28] Cornford, 1, pp. 262–264; Solmsen, 4, p. 280.

[29] *Physics*, 4, 1, 208b 9–10.

[30] *Physics*, 4, 1, 209a 17–23. Aristotle's explicit denial that natural place could be the efficient cause of natural motion did not prevent such theories from being developed by the schoolmen. See Grant, 9, pp. 51–52; and, for Aquinas' outright rejection of such a view, Hesse, 1, p. 66. For an explanation of what Aristotle may have meant by saying that place 'exerts a certain influence', see Machamer, p. 378.

[31] *De Caelo*, 4, 3, 310b 2–4.

moves it.[32] All such influences are ultimately reducible to *pushing* and *pulling*, i.e. to purely mechanical processes.[33] We discover the need for such contact 'by induction' (*epagoge*), since we observe it 'in every case'.[34] This leads to the notorious hypothesis of the 'motor-medium': only the medium is in direct contact with the projectile throughout its flight,[35] so only it can provide the required motive force or impulsion.[36]

With regard to forced motions, then, Aristotle is as pure a mechanist as the Atomists: all such motions, he says, involve direct contact between 'mover' and 'moved'. A.D. is ruled out not by *a priori* argument but 'by induction', by taking note of what we observe 'in every case'.[37] It is thus clear that the Stagirite was no advocate of A.D.: neither forced nor natural motions involve, on his view, any such thing.

We turn now to somewhat darker waters. Belief in A.D. was strongly supported by the doctrines of the astrologers, so we must perforce at least dabble in affairs astrological. The *philosophical* roots of astrology lie in Platonic astral theology, in Aristotle's cosmic vision in *Metaphysics* Λ, and, last but not least, in the Stagirite's insistence in *De Generatione*, 2, 10, that the efficient cause of terrestrial generation and corruption is the annual motion of the sun in the plane of the ecliptic.[38] Astrology was, however, essentially an empirical theory; and, as we shall see, one not lacking in evidential support.

The importance of the Sun to terrestrial events could scarcely be more obvious: the cycle of the seasons is sufficient testimony in itself. But the Moon, too, has discernible effects on Earth – the correlation of its phases with our tides was studied in antiquity by the Stoic, Posidonius.[39] (These effects are not due to the Moon's light, since they are unaffected by

32   *Physics*, 7, 1, 242b 24-26.

33   See *Physics*, 7, 2.

34   *Physics*, 7, 2, 244b 3. See also 3, 1, 202a 5.

35   For more details, see Murdoch and Sylla in Lindberg, Ed., p. 210 ff.

36   Cf. of course, one of the anti-vacuist arguments in *Physics*, 4, 8, 215a 13-18.

37   For Aristotle, of course, *epagoge* (usually, but somewhat misleadingly translated as 'induction') may enable the intellect to gain a rational intuition of the essence of a thing and hence to discern the necessary connections in Nature.

38   *De Gen et Corr*, 2, 10.

39   See Jammer, 3, pp. 41-42; Long, p. 221.

clouds.) There is therefore some invisible lunar 'influence' over the tides: we can discern its effects, but know nothing of its nature.

Now, for a believer in a geocentric cosmology, Sun and Moon are only two of the 'planets' or 'wanderers'. If their motions cause noticeable terrestrial events, one should expect those of Mercury, Venus, Mars, Jupiter and Saturn to do likewise. Their 'influences' may be less immediately evident to us, but are nevertheless likely to be in operation. This is the powerful empirical argument at the heart of the astrological tradition. Thus Aristotle:

> This [sublunary] world is tied in some way and in a necessary manner to the local movements of the superior world, in such a way that all the power which resides in our world is governed by these movements.[40]

The authority of Aristotle combines with the plain evidence of the senses to give sanction to astrological theories. Evidence for the fundamental axiom of astrology – the existence of specific 'astral influences' over terrestrial affairs – was considered overwhelming[41] even by those who were sceptical of the exaggerated pretensions of judicial astrologers. The *meteorological* evidence was undeniable: the 'stars', say the astrologers, govern the weather, and hence crops, famines, wars, plagues, etc. This is the physical argument for astrology, and it proved very difficult to refute. Although many thinkers, pagan and Christian alike, were to reject vigorously astral determination of human actions and re-affirm the freedom of the will, few doubted the fundamental correctness of the physico-meteorological argument.

Gradually, astrology developed into a systematic body of doctrine, resting on its unshakeable physical foundation but extending also — and more controversially – into all areas of human life. Its *locus classicus* was the *Tetrabiblios* of Ptolemy: just as his *Almagest* became the authoritative text for subsequent generations of astronomers, so the *Tetrabiblios* became the Bible of the astrologers. As Thorndike says:

> In the *Tetrabiblios* the art of astrology receives sanction and exposition from perhaps the ablest mathematician and closest scientific observer of the day . . . Hence from that

---

40   *Meteorologica*, 1, 4. Quoted from Clagett, 1, p. 122.

41   See Grant in Lindberg, Ed., p. 288, and Thomas, p. 334.

time on astrology was able to take shelter from any criticism under the aegis of his authority.[42]

Ptolemy's defence of judicial astrology proceeds by the route we have outlined: he first sets out the physico-meteorological argument for astral 'influences', and then proceeds to extend and develop the argument to cover human concerns. Although human faculties may be inadequate to attain it, Ptolemy suggests, a predictive science of astrology is in principle possible. His theory is not, however, *fatalistic*: one can read in the stars what is likely to happen, and take steps to *avert* it.[43] (The predictions of the astrologer will therefore at best be only usually true or true 'for the most part' rather than invariably.)

There developed in late antiquity, and particularly among Stoics and Neoplatonists, a doctrine of 'universal sympathy', of a sort of harmony between all the events in and constituents of the cosmos, and in particular between those of the celestial and terrestrial realms. (The doctrine was therefore especially congenial to the claims of the astrologers.) All events, say Neoplatonists such as Plotinus, fall into a *single pattern*: the sage can therefore 'read off' future events from current ones, or terrestrial events from celestial ones:

> just as because of the one principle in a single living thing, by studying one member we can learn something else about a different one.[44]

The Stoics too pursue this organismic analogy: the cosmos, says Chrysippus, 'is a living being, endowed with sense and reason, and having aether for its ruling principle'.[45] A change anywhere will therefore have effects everywhere, just as a cut finger may disturb the whole body.[46]

The doctrine of 'universal sympathy' can either be explained at a *causal* level – as is most natural – or in terms of *correspondences*. Celestial events, says Plotinus, do not directly cause their corresponding terrestrial accompaniments; rather, both are co-caused by the *Anima Mundi* or world-soul operating according to a single pre-established

---

42 Thorndike, Vol. 1, p. 110.

43 *Ibid*, pp. 111–112.

44 Plotinus, *Enneads*, 2, 3, Vol. 2, pp. 69–71.

45 Diogenes Laertius, Vol. 2, p. 243.

46 See Sambursky, 3, p. 9.

program.[47] All things, he says, 'are filled full of signs, and it is a wise man who can learn about one thing from another.[48] Nevertheless, a sign of X is not the cause of X: if celestial events were said to cause terrestrial ones merely because the sage could use them to predict future terrestrial occurrences, it would follow that 'birds would be the causes of what they indicate, and so would everything at which the soothsayers look when they foretell'.[49] The correspondence of a celestial event C.E. with a terrestrial event T.E. does not show that C.E. is the cause of T.E.: both are common effects of a single antecedent state of the *Anima Mundi*. The existence of 'universal sympathy' and the possibility of rational prediction and divination, Plotinus adds, refute the Epicurean pretension that the world is a random aggregation of atoms: in such a disorderly world 'there would be no foretelling or divination'.[50]

It is nevertheless natural to construe the universal harmony as causal. If we forget 'correspondences' and 'signs' and think – as most astrologers did – of celestial events as direct physical causes of terrestrial ones, the question immediately arises: is 'celestial influence' a form of A.D.? Do the seven planets act at a distance on the Earth, or by some mediation?

The Stoic answer is straightforward. All causal influences, including those which sustain 'the sympathy and tension which binds together things in heaven and earth',[51] are transmitted by the *tonos* (tension) of the *pneuma* (spirit), a fine and all-pervading substance compounded of flame and air. The *pneuma* is highly elastic: we may think of it as consisting of fine threads linking the various parts of the cosmos and providing, in Chrysippus' words, a principle of cohesion in things'.[52] If we imagine such threads linking every body with every other, and giving all bodies, in virtue of their *tonos*, a certain coherence, both internally and with one another, we shall have a fair picture of the Stoic cosmos.[53]

47   This theory of 'correspondences' is thus reminiscent of the 'pre-established harmony' of Leibniz.

48   Plotinus, *Enneads*, 2, 3, Vol. 2, pp. 69–71.

49   *Ibid.*, 3, 1, Vol. 3, pp. 23–24. For Seneca's very similar views, see Thorndike, Vol. 1, p. 103.

50   Plotinus, *Enneads*, 3, 1, Vol. 3, p. 17.

51   From Diogenes Laertius, Vol. 2, p. 245.

52   See *Ibid.*, pp. 243–245.

53   The cosmos itself also of course owes its cohesion to the *tonos* of the *pneuma*.

The Stoics therefore deny A.D. but accept a causal interpretation of the 'universal sympathy' principle. Alter any body anywhere and, because of the pneumatic threads linking it to all other bodies, one will produce some effect, however minute, *everywhere* in the cosmos.

In vision, says the Stoic, 'the object seen is reported through stressed air, as if in contact by a stick'.[54] Similarly, says Chrysippus, the soul is present in the body like a spider at the centre of its web: it senses, by the tension in the threads, events occurring in the rest of the web.[55] All physical influences are similarly transmitted through continuous media by the *tonos* of the *pneuma*: there is a 'universal sympathy', e.g. between celestial and terrestrial events, but it is mediated by these quasi-mechanical means – there is no true A.D.[56]

The only major philosophical school to espouse A.D. explicitly were the Neoplatonists. Thus Iamblichus:

> One need not share the view of the Stoics, with whom we will continue to differ, that action takes place by contact and touch. It is much more correct to say that not everything acts by contact and touch, but that action happens according to the appropriateness of the active part with regard to the passive one, and further, that many things are active without any perceptible contact, as we all certainly know.[57]

His examples of alleged A.D. include the resonance of a lyre-string when another, tuned to the same frequency, is plucked; and the inflammability of naphtha at a distance from a fire. One could also of course cite magnetism, the lunar influence over the tides, and 'astral influence' in general.

By late antiquity, the ideas of 'sympathy' and 'antipathy' had been developed into a doctrine of universal applicability. Everything tends towards its kin, as Plato says in the *Timaeus*. One might attempt to postulate some corporeal mediation to explain the workings of 'sympathy', but it is most naturally construed as A.D. Plutarch, for example, explains that heavy bodies fall to the Earth not because it lies at the centre of the universe but because they 'have some affinity and cohesion with her'. For, he says,

---

[54] Quoted from Sambursky, 3, p. 23. The quotation is from Chrysippus.

[55] See Sambursky, 3, p. 24; Long, p. 172.

[56] Sambursky, 3, pp. 29–30.

[57] Quoted from Sambursky, 4, p. 103.

as the sun attracts to itself the parts of which it consists, so the earth too accepts as [her] own the stone that has properly a downward tendency, and consequently every such thing ultimately unites and coheres with her'.[58]

The Sun draws to itself solar matter, the Moon lunar matter, and so on – a doctrine of particular rather than universal gravities.[59]

Similar theories were developed in the medical sphere by the great physician Galen. The phenomena of magnetism should, he insists, 'draw' the mind 'to a belief that there are in all bodies certain faculties by which they attract their own proper qualities'.[60] The nature (*physis*) of each living thing tends, he says, instinctively towards those things which are qualitatively akin to it and hence congenial to its growth and development. 'Nature' attracts like and expels unlike[61] – vomiting is a paradigm of natural antipathy.[62]

Every organ within the body, Galen claims, extracts from the blood material with its own proper qualities: assimilation requires 'a certain community and affinity' of properties.[63] Those who deny 'that there exists in any part of the animal a faculty for attracting its own special quality' are compelled to deny 'obvious facts'.[64]

The processes of nutrition and assimilation can therefore, he argues, only be explained on the basis of a sympathetic attraction of like for like. Likewise, with the operation of drugs: a purgative attracts a particular 'humour' and thus helps to extract it from the body 'by similitude of the whole substance'.[65] Each of the four 'humours' of Hippocratic medicine (blood, phlegm, black bile and yellow bile) can therefore be extracted by a *specific* cathartic drug which is by nature closest in sympathy to it. Galen stresses, however, that this sympathetic A.D. is operative only over short distances.[66]

---

58 Plutarch, Vol. 12, pp. 69–71.

59 Such theories were to become widespread and influential during the Renaissance.

60 Galen, p. 71.

61 *Ibid.*, pp. 45–47.

62 *Ibid.*, pp. 247–249.

63 *Ibid.*, p. 33.

64 *Ibid.*, p. 49. Galen uses the kidneys as a test-case for this theory – they attract, he claims, urinary material from the blood by such a faculty.

65 See Tenkin in Debus, Ed., 7, p. 61.

66 Galen, p. 321.

The notions of 'sympathy' and 'antipathy' are among the commonplaces of Pliny's monumental (but heavily anecdotal) *Natural History*: Pliny clearly believes that even inanimate things have their loves and hatreds, friendships and enmities, etc.[67] All manner of real or imagined natural phenomena are explained on the basis of this supposition, not least those of *sympathetic magic* – the theory of sympathy and antipathy lies at the heart of our magical and occult tradition. The magnet is of course the classical example of sympathetic A.D. (it is easy to show that the lodestone contains iron and is therefore 'akin' to it). As Roller says,

> through the first eleven centuries of our era the lodestone is known principally through medical uses, uses in magic, and as an example of occult powers.[68]

In late antiquity, then, the theories of cosmic harmony and sympathetic attraction supplanted the mechanical-mindedness of earlier Greek thought. In the occult wing of Neoplatonism the doctrines of astral influence, sympathetic medicine and natural magic found philosophical backing.[69] Cases of 'sympathy', and hence – if one ignores the Stoic compromise – of A.D., are multiplied *ad lib*, from the *bona fide* examples of magnetism and the lunar influence over the tides to more dubious 'astral influences', medicinal 'faculties' and 'powers', and the whole occult bandwagon.

B. Spontaneity

The initiation of motion, and hence of new causal chains, was long regarded as a special prerogative of the *psuchê* or 'animator'.[70] Whatever has a 'motor' – i.e. is able to initiate local motion – may therefore legitimately be regarded as *empsuchos* (animate). This conception of the *psuchê* as a motor accounts for Thales' theory of magnetism, as reported by Aristotle:

> Thales too . . . seems to have held soul to be a motive force, since he said that the magnet has a soul in it because it moves the iron.[71]

---

67   Thorndike, Vol. 1, p. 84.

68   Roller, p. 28.

69   See Hansen in Lindberg, Ed., p. 491 ff.

70   See Barnes, Vol. 1, p. 7. To translate *psuchê* as 'soul' is, he insists, misleading, at least without further explanation.

71   *De Anima*, 1, 2, 405a 20–21. See also Diogenes Laertius, Vol. 1, p. 25.

The lodestone clearly does possess a 'motive force' by which it draws iron to itself. But such a power is characteristic only of living things: we experience it in ourselves and witness its operations in other men and animals. Therefore the magnet is itself animate, i.e. possesses a *psuchê* – animistic theories of magnetism remained common throughout the Middle Ages,[72] and even received favourable comment from Gilbert himself.[73]

Only the *psuchê*, then, can initiate motion: inanimate matter can transmit local motion mechanically but cannot act spontaneously. Now if we also think of local motion as 'naturally' decaying – decreasing in 'quantity' in collisions between bodies – we will have all the necessary premisses for an animistic physics. Modern (post seventeenth century) physics is often referred to as positing a 'billiard-ball' model of the universe. In fact, however, the analogy of the billiard table works the other way, i.e. favours animism. One ball may strike another and impart motion to it by impulse, but both balls will sooner or later come to rest, and remain at rest indefinitely unless and until some human *agent* picks up his cue to play another shot. On the table, motion decays – bodies tend to rest – it requires the action of a living being (and hence ultimately of a *psuchê*) to set things going again. Either immediately or mediately, *psuchê* is the efficient cause of all local motions.

This animistic physics is explicitly endorsed by Plato in the *Phaedrus* and the *Laws*.[74] Some things, he says in the latter dialogue, are moved only by external causes; others move spontaneously and may therefore be called 'self-movers'.[75] Self-motion is, he continues, 'ten thousand times superior to all the others'; it is 'the mightiest and most efficient' and should be ranked *first* in the order of things.[76]

Something which is moved only by another cannot, Plato insists, ever *initiate* a causal chain, i.e. be the *first* mover in a sequence. When the self-moved object moves another, however, the second (moved) body can go on to move others,

72 Dijksterhuis, p. 154.

73 'Not without cause,' says Gilbert in *De Magnete*, did Thales hold the magnet to be animate. See especially pp. 208–209.

74 For Plato's animistic physics, see Vlastos, 7, p. 31; Hesse, 1, p. 47; Jammer, 3, p. 29.

75 *Laws*, 894, p. 279.

76 *Ibid.*, 894, p. 279.

to transmit physical action along a causal chain (cf. of course the billiard balls). The beginnings of all motions must therefore lie in 'the change of the self-moving principle':[77] without self-movers (*automata*) there would be no motion at all.

But, the argument continues, the power of self-motion is commonly called *life*. But it is in virtue of its *psuchê* that an animate thing is such; *psuchê* is therefore the *source* of this power of self-motion. Hence, Plato concludes, the proof is complete that 'soul is the first origin and moving power of all that is, or has become, or will be . . .'.[78] If is, he insists, the oldest of all things – it is prior to the body and the ruler of it. At *Phaedrus*, 245C-6A, Plato argues from the conception of *psuchê* as self-mover to its immortality.[79] As the beginning and ultimate source of all things, he claims, soul must be first and hence unbegotten, and, 'else the whole heavens and all creation would collapse and stand still', it must also be incorruptible. (All this argument can show, of course, is that some soul or other always exists, not that any particular soul is immortal.)

In the *Timaeus* too it is asserted that soul is older than and dominant over body, that

> god created the soul before the body and gave it precedence both in time and value, and made it the dominating and controlling partner.[80]

It is important not to think of this theory as peculiarly Platonic. I have used Plato as an example of animism in physics merely because he states its assumptions and arguments so clearly and explicitly. In fact, similar views were commonplace and orthodox until the seventeenth century: matter as such cannot initiate motion; motion in causal chains tends to decay; only a *psuchê* can engender fresh starts of motion – these theses are part of the fabric of Greek and medieval thought.[81]

---

[77] *Ibid.*, 894, p. 280.

[78] *Ibid.*, 896, p. 281.

[79] According to Barnes (Vol. 1, pp. 116–117), the argument is derived from the Pythagorean physician, Alcmeon.

[80] *Timaeus*, 34-35, p. 46.

[81] Although the animism in Aristotelian physics is less blatant than in that of Plato, it is nonetheless present: for Aristotle, too, *psuchê* is the ultimate source of all motions.

How do Leucippus and Democritus come to terms with this universal animism? Every atomic motion, they will say, is the outcome of a prior blow resulting from collision with another atom. The infinite regress of causes here is not vicious and was simply accepted by the first Atomists, according to whom 'the primary bodies are in perpetual movement in the void or infinite'[82] The atoms have always been and will always be moving and colliding with one another.[83]

Now, if the whole universe is not to run down, some sort of conservation principle must be in operation: if, for example, motion tended to be lost in interatomic collisions – as often happens in collisions between perceived bodies – everything would have ground to a halt by now, unless there were some other type of entity capable of spontaneously initiating motion.

But, the Platonist will object, although Democritus can perhaps explain why atom A moves as it does – e.g. in terms of a collision with atom B – he cannot explain why either A or B (or indeed *any* atom) moves *at all*. Without a source or originator of motion, there can be no motion. If, therefore, motion is to be present in the atoms from all eternity, they must be conceived as self-movers (*automata*).

We seem to have several distinct issues tangled up here. In one sense, it is true, the motion of the Democritean atom is *automaton*,[84] innate and intrinsic to it; in another, it is not. The atom may be called a 'self-mover' in the minimal sense that it has been moving from all eternity – there is and can be no explanation of why atoms move at all. They do so, says Bailey, by 'necessity', by 'the law of their own being'.[85] This suggests that motion is natural – perhaps even essential – to a Democritean atom. But that cannot be right: one can surely arrest an atom's motion without destroying its very being? After all, it is just as much a solid corpuscle of matter* when at rest as when in motion.

Suppose one were to restrain a single atom from moving, and then release that check or restraint. Would it immediately 'spring back to life'? This is the crucial question: if answered in the affirmative, it gives a non-mechanical theory involving a measure of spontaneity; if answered in the negative, it yields genuine mechanism. For a mechanist, no body starts to move

82   Aristotle, *De Caelo*, 3, 2, 300b 9.

83   See Barnes, Vol. 2, p. 127.

84   See Partington, 4, p. 250.

85   Bailey, 2, p. 85.

without an external impulse – if it is of the very nature of the atoms to move *unless* hindered or restrained, they are self-movers in something approaching the Platonic sense.[86]

How would Democritus answer the above question? The answer he gives will clearly have momentous consequences for his whole natural philosophy, of which the following three readings R1–R3, may be given:

R1.   *Classical Mechanism*: According to this theory, a law of conservation of motion is in operation. More specifically, the principle of inertia governs all motion *in vacuo*, and all collisions (at the atomic level) are perfectly elastic. There is therefore no need of any fresh starts of motion in Nature.

R2.   *Balme's theory*:[87] The Ancient Atomists, Balme insists, did *not* possess the modern concept of inertial motion.[88] Like their contemporaries, he claims, they thought of forces as being 'spent' or exhausted, of trains of events dying out and coming to an end.[89] But if this is so, some active principle is needed to prevent the universe from grinding to a halt. Motion is sustained, Balme suggests, by the super-elasticity of the atoms, their > 100% coefficient of elasticity.[90] An atom loses speed between collisions, but gains speed due to the 'jostling' that occurs in collisions – without this accelerative effect of 'jostling', the universe would indeed run down. On this theory, a single atom shot off into empty space would gradually decelerate and come to a halt, and would remain at rest until struck by another. No body spontaneously leaps into motion from a state of rest: motion is sustained by the effects of jostling, not by fresh starts *de novo*. The total 'quantity of motion' in the universe is approximately conserved, in the long run, but no conservation *law* holds – motion may be gained and lost.

R3.   *Aristotle's Interpretation*: A further possibility is that Democritus thought of soul (fire) atoms as being *self-moving*, and of all others as being *moved* by them. Spherical atoms, says Aristotle, were identified with soul because:

---

[86]   Hesse, 1, p. 43, likens the atoms to living things.

[87]   Balme, 2, p. 25.

[88]   Aristotle's argument in *Physics*, 4, 8, that motion *in vacuo* would go on *ad infinitum* was meant as a *reductio* of that idea: we may therefore conclude that, at the very least, Democritus had not explicitly assented to that position.

[89]   Balme, 2, p. 24.

[90]   *Ibid.*, p. 25.

atoms of that shape are most adapted to permeate everywhere, and to set all the others moving by being themselves in movement. This implies the view that soul is identical with what produces movement in animals.[91]

Soul-atoms are never at rest (like the motes in the sunbeam) and, by their restless agitation, serve to vivify and enliven – i.e. stir into motion – other, grosser, bodies. This theory would enable Democritus to accommodate animism *within* his Atomic theory: soul-atoms are, like Platonic self-movers, the ultimate source of all motion. Motion in all gross bodies would tend naturally to decay in the absence of the vivifying effect of these minute *automata*. This would involve the denial of the principle of inertia and/or the elasticity of atomic collisions, and an admission of true spontaneity in Nature, i.e. in the motion of the soul-atoms.

Of these three readings, R1 is by far the least plausible: it is quite anachronistic and unhistorical to imagine that the mechanical philosophy could have sprung, fully-formed, from the head of Democritus. As for Balme's theory, R2, it is ingenious but unconvincing: the concept of super-elasticity in particular sounds dubious. (We certainly have no experience of such bodies – our senses teach us that motion tends to be lost rather than gained in collisions.) My preference is therefore for R3: it accommodates a typically Greek (i.e. animistic) dynamics within the framework of the Atomic theory, and establishes a clear positive analogy between the dynamics of the micro- and macro-worlds. Although the paucity of textual evidence licenses only a tentative conclusion, R3 does seem the most probable interpretation.

Does a Democritean atom, alone in empty space, have *weight* (i.e. a tendency to a particular type of motion) as a primary and irreducible quality? If so, it will move – downwards – of its own accord, without need of external blows. An exegetical difficulty arises when we seek to reconcile the following pair of quotations:

Democritus says that each of the indivisible bodies is heavier in proportion to its bulk.[92]

Democritus named two [properties of atoms], size and shape; but Epicurus added a third to these, namely weight . . . – Democritus says that the primary bodies do not

---

[91]   *De Anima*, 1, 2, 404a 5–9. See also Kirk and Raven, p. 417.

[92]   Aristotle, *De Gen et Corr*, 1, 8, 326a 9.

possess weight but move in the infinite as a result of striking one another.[93]

According to Aristotle, the Democritean atom has weight; according to Aetius, it does not. Whom are we to believe?

The atom, say the majority of modern scholars,[94] is in itself and intrinsically without weight. It is only in the cosmic 'whirl' or vortex that an atom acquires weight as a 'secondary' quality, a tendency to fall to Earth. In our world, the weight of an atom is directly proportional to its bulk (matter* is homogeneous); in void space, all atoms are weightless. The bulkier atoms, says Bailey,[95] will *resist* the 'whirling' motion of the vortex more strongly, as a result of which they have lower angular velocities and hence tend to collect in the centre of the vortex.[96] Konstan has shown in more detail how this theory could work: a vortex (e.g. a tornado) can, he insists (in answer to Furley's objections), have a vertical vector – there could therefore be a sort of 'atomic rain' *along* the axis of such a vortex which would account for gravity.[97]

This interpretation is reinforced by some remarks of Aristotle's. Leucippus and Democritus, he alleges in *De Caelo*, say that the atoms are perpetually moving in the void and may therefore properly be asked to explain 'the manner of their motion and the kind of movement which is natural to them'.[98] The Atomists, he objects, attribute no natural motion to their 'primary bodies'. 'The question of movement – whence or how it is to belong to things – these thinkers . . . lazily neglected.'[99] But, if the Democritean atom had possessed weight as a primary and intrinsic quality, it would have had a natural motion, viz. downwards. It is clear that Aristotle regarded weight as being, for Democritus, no more than a product or outcome of the 'whirl', a view which he (Aristotle) resolutely opposes:

> Again, it cannot be the whirl that determines the heavy and the light. Rather that movement caused the pre-existent

---

93 Aetius, 1, 3, 18 (D.K. 68a 47), quoted from Kirk and Raven, p. 414.

94 Kirk and Raven, pp. 415–416; Konstan, 2, pp. 408–410, Hahm, 2, pp. 58–59; Long, p. 35; Cherniss, p. 97; Bailey, 2, pp. 144–146. The only modern dissenter is Furley (3, p. 87), whose objections have been met by Konstan (2, pp. 408–409).

95 Bailey, 2, p. 145. See also Ferguson, p. 106.

96 See Konstan, 2, pp. 408–409; Ferguson p. 114.

97 Konstan, 2, pp. 409–410.

98 *De Caelo*, 3, 1, 300b 9–12.

99 *Metaphysics*, 1, 4, 985b 18–19.

heavy and light things to go to the middle and stay on the surface respectively. Thus, before ever the whirl began, heavy and light existed.[100]

Whether innate or acquired, weight is, for the ancient Atomists, a property of *all* the bodies in our world. Since the atoms are made of a common matter*, every atom has a gravity proportional to its bulk. The gross weight of a body depends on its total volume of matter*; its specific gravity (S.G.) on its matter*/void ratio. All upward motion is therefore due to the 'squeezing out and up' of less dense stuffs by denser ones.[101] There is no absolute 'levity': even fire would fall through some still-rarer medium.

Aristotle objects to this theory – and reaffirms the absolute levity of fire – in *De Caelo*, 4, 2, but his objections miss the point completely. If, as Hahm[102] suggests, Aristotle is in possession of a concept of S.G., it must be granted that he entirely muffs his argumentative use of it[103] in attempting to convict the Atomist account of S.G. of incoherence.

The fullest account of the theory of upthrust, common property to all the Atomists, is to be found in Lucretius:

> Now is the place, I hold, herein to prove this also to you, that no bodily thing can of its own force be carried upwards or move upwards; lest the bodies of flames deceive you herein.[104]

According to Epicurus, weight is a primary quality of the atoms and hence a universal quality of matter – all things therefore, 'as far as in them lies',[105] tend downwards. How then can an Epicurean like Lucretius cope with the apparently natural and spontaneous upward motion of flames and hot air?

First, he reminds us that some things which are unquestionably heavy can be seen on occasion to move in an upward direction. Plants grow, blood spurts from severed arteries, wooden beams immersed in water spring up when released. In

100   *De Caelo*, 2, 13, 295b 4–7.
101   See Bailey, 2, p. 328; Kirk and Raven, p. 415.
102   Hahm, 2, p. 73.
103   See especially *De Caelo*, 4, 2, 309b 9 ff.
104   Lucretius, Book 2, 184–187, pp. 245–247.
105   Lucretius, Book 2, 201, p. 247. For the subsequent history of this phrase '*quantum in se est*', see I. B. Cohen, 2.

like manner, Lucretius reasons, 'flames too must be able when squeezed out to press on upwards through the breezes of air, albeit their weights are fighting, as far as in them lies, to drag them downwards'.[106] Indeed, there are phenomena which show that fiery things naturally fall – meteors, lightning, and the heat of the sun all come down to earth from the heavens.

According to the Atomists, then, all bodies have more or less weight; the upward or downward motions of bodies in various media are determined by relative S.G.s. This theory is naturally congenial to the technical work of Archimedes on the topics of S.G.[107] and upthrust. A solid immersed in a fluid, says Archimedes, experiences an upthrust equal to the weight of the fluid displaced and therefore 'weighs lighter than its true weight by the weight of the fluid displaced'.[108] If the fluid is denser than the solid immersed in it, the net force on that body is upwards – the upthrust overcomes its own weight. Archimedes' work can therefore be seen – at least with the benefit of hindsight – to provide powerful theoretical support for the crude notion of 'squeezing out and up'. For Archimedes, as for the Atomists, weight is a property of all the bodies in our cosmos: those which we call light are not possessed of absolute levity but merely of low S.G.

If we turn now to the writings of Aristotle we can see that he – like his master, Plato – is a believer in natural spontaneity: he believes that there are fresh starts in Nature.[109] The initiation of motion must clearly appertain only to those causes acting 'by nature' rather than 'by constraint'. When an animal starts to move it may be called a 'self-mover' in the sense that it contains in itself the source or spring of its own motion. What of the natural motions of the elements FAWE to their proper places? Whatever is in motion, says Aristotle in the *Physics*, 'must be moved by something'.[110] What then is the mover, i.e. the *efficient cause*, involved in the free fall of a stone? We have already dismissed the idea that this could be the attraction (A.D.) either of a place or a body. Is it then some immanent, intrinsic, incorporeal motor, the thing's 'nature' (*physis*)? Nature, says Aristotle, is

---

[106]   Lucretius, Book 2, 203–205, p. 247.
[107]   See Archimedes, p. 251, for his clear and correct account of S.G.
[108]   Archimedes, p. 258.
[109]   Balme, 1, p. 133.
[110]   *Physics*, 7, 1, 241b 24.

. . . the essence of things which have in themselves, as such, a source of movement . . . nature in this sense is the source of the movement of natural objects.[111]

Every natural body 'has *within itself* a principle of motion and of rest;[112] *physis* may therefore be characterised as 'a source or cause of being moved and of being at rest in that to which it belongs primarily'.[113] The obvious temptation is to identify the *physis* of a thing with its form, and make it the efficient cause of natural motion. This, says Aristotle, is legitimate only for animals, 'for when an animal is in motion its motion is derived from itself'.[114] The efficient cause of the natural motions of FAWE is, however, not evident:

It is impossible to say that their motion is derived from themselves: this is a characteristic of life and peculiar to living things.[115]

To make FAWE self-movers is to capitulate to animism, which is not Aristotle's intention. In such cases, he says, 'the thing does not move itself, but contains within itself the source of motion'[116] – a fine distinction indeed! The true efficient cause of these motions lies elsewhere: the elements in their natural motions are moved:

either by that which brought the thing into existence as such and made it light and heavy, or by that which released what was hindering and preventing it.[117]

We can therefore explain the free fall of a stone in terms of its *generator* (that which made it what it is) and the remover of whatever obstacle had previously prevented it from falling. (There may also be a certain assistance from the medium.[118]) All these factors warrant inclusion in our characterisation of the efficient cause. But this is all we can say:

---

111   *Metaphysics*, 5, 4, 1015 13–18.

112   *Physics*, 2, 1, 192b 15.

113   *Ibid.*, 2, 1, 192b 22.

114   *Ibid.*, 8, 4, 254b 16.

115   *Ibid.*, 8, 4, 255a 5–6.

116   *Ibid.*, 8, 4, 255b 29–30.

117   *Ibid.*, 8, 4, 256a 1–2.

118   *De Caelo*, 3, 2, 301b 30; Machamer, pp. 384–385. Machamer makes a great deal – too much – of Aristotle's isolated remark that natural motions may be 'helped' by the medium.

. . . to ask why fire moves upward and earth downward is the same as to ask why the healable, when moved and changed *qua* healable, attains health and not whiteness.[119]

The answer is simply: 'Because it is its nature (*physis*) to do so', where *physis* is to be conceived not as an internal incorporeal motor but rather as the *law* of a thing's being.[120] Once we have explained how the stone (a) came-to-be and (b) was freed from all restraints, no further question arises as to why it falls.[121] Anyone who asks such a question merely manifests his ignorance of the *physis* of earthy bodies.

Although rejected by Aristotle (except for living things) the conception of *physis* as an internal motor or '*vis*' became widespread in late antiquity. The Stoics, for example, defined *physis* as 'a force moving of itself',[122] a 'power which permeates and pervades the whole universe', a 'guiding force' which acts as the 'organising principle' in all things.[123] Whereas Aristotle carefully distinguished and sharply contrasted 'nature' and 'force',[124] later thinkers are quite happy to characterise *physis* as a particular (i.e. internal) variety of *vis*![125]

The concept of nature as an internal force or power was made popular in later years by the writings of Philoponus, who defines *physis* as:

a kind of force that is diffused through bodies, that is formative of them, and that governs them, it is a principle of motion and rest, and so on.[126]

Philoponus therefore promotes the idea – popular during the Renaissance period – of nature as an immanent force or motive power, the efficient cause of 'natural motions'.[127] This position is, however, as Aristotle saw, implicitly vitalistic – it blurs the distinction between animate and inanimate things.

[119] *De Caelo*, 4, 3, 310b 16–18.

[120] See Randall, 2, p. 174.

[121] Furley, 3, p. 92.

[122] Diogenes Laertius, Vol. 2, p. 253.

[123] Cicero, pp. 134–135.

[124] Jammer, 3, pp. 35–37; Collingwood, p. 45.

[125] This tendency is marked in Galen: according to Jammer (3, p. 35) he lists no less than sixty varieties of *dynamis* (force) which appertain to the human body.

[126] Quoted from Wallace, 5, p. 110.

[127] For the subsequent development of this theory, see Wallace, 5, pp. 110–126.

The natural motions of FAWE are for Aristotle directed toward their respective natural places rather than towards kindred matter. Earth, being absolutely heavy, tends to the centre of the finite geocentric cosmos; fire, being absolutely light, moves to the periphery. Earth moves naturally to the centre of the universe, not merely to the centre of the Earth, although of course the two happen to coincide:

> If one were to remove the earth to where the moon now is, the various fragments of earth would each move not towards it but to the place in which it now is.[128]

There is therefore no tendency of like matter to join like; rather, each element tends to its proper place. When it attains this goal, a body naturally comes to rest: it requires violence to wrest something from its rightful place in the scheme of things. This explains the stability of the Earth at the centre of the cosmos – it rests there because that is its proper natural place.

The movement of each of the elements to its natural place is, says Aristotle, 'motion towards its own form',[129] towards the perfection of its nature. It is therefore a purposive motion: for the peripatetic every thing 'seeks' the full actualisation of its latent potencies, the perfect realisation of its form in the medium of matter.[130] Only at the very centre of the Earth, then, can one find pure E, and only at the periphery of the sublunary world could pure F be discovered.

The free fall of a stone is therefore a true Aristotelian *kinesis*, the actualisation of a potency latent in all earthy bodies.[131] This gives a possible explanation of the acceleration manifest in the free fall of heavy bodies. (That such bodies accelerate is evident[132] – why they do so is a different matter[133]). Given the 'Aristotelian' dynamic formula $v \propto F/R$, an increase in velocity can only result from either an increase in F (weight) or a decrease in R (resistance). Although theories

---

128   *De Caelo*, 4, 3, 310b 2-4.

129   *Ibid*, 4, 3, 310a 35.

130   See Koyré, 7, p. 9.

131   According to Aristotle, *Metaphysics*, 11, 9, 1065b 16-17, motion (*kinesis*) is ' the actuality of the potential as such'.

132   Drop a brick on to a man's head from a height of an inch and it will give him a light blow; let it fall from the rooftops, and it will split his skull. Acceleration was studied empirically in antiquity by Strato of Lampsacus (see Clagett, 2, p. 70).

133   For some of the views held in antiquity on this thorny topic, see Cohen and Drabkin, Eds., p. 209 ff.

of both types were developed, the above remark that in moving to its proper place a body moves also 'towards its own form' suggests the former view, that in free fall an earthy body gains in weight as it approaches the centre of the Earth. The *quality*[134] of weight is intensified as the stone approaches that place where it becomes (in act) most fully what it already (in potency) is.[135]

In Aristotelian physics, therefore, there is true spontaneity or initiation of motion. If removed by violence from its natural place, a body will 'strive' to return to it and, if no obstacle prevents it, will set off in the direction of its 'home' without needing any external push. Although Aristotle is a mechanist about forced motion, in his conception of natural motion he approaches (but seeks to evade) vitalism.

If we turn now to the Epicureans, the situation is different again. In Epicurean physics, every atom always moves one space-minimum per time-minimum.[136] There is therefore no initiation of motion at the atomic level. (This is perfectly compatible with the appearance of such initiation at the level of compound bodies.) The atom can, however, be described as 'self-moving' in virtue of its primary and intrinsic property of weight: Lucretius speaks of the atoms as 'carried downwards straight through the void by their own weight'.[137] It is plausible to think of Epicurus attributing weight (and hence such a natural motion) to his atoms in response to Aristotelian criticisms of Democritus.[138] Aristotle objected that the atom could not have a forced motion without having some natural motion; Epicurus replies by attributing weight to his atoms as a primary quality and thus giving them a natural downward tendency. Without weight as an original source and spring of motion, says Epicurus, the atoms would never move at all[139] – without a prior natural motion there would be no 'blows' and hence no forced motions either. On

---

134 Whether weight is to be classed as a quality or an (additive) quantity was controversial in antiquity. See Sambursky, 4, p. 67, pp. 83–85.

135 Unfortunately, Aristotle also seems to reject this view at *De Caelo*, 1, 8, 276b 22–27, where he flatly denies that distance from its natural place makes any difference to the nature of a thing. It is therefore difficult to attribute any coherent account of acceleration to Aristotle himself: the position outlined in the text is essentially that of the commentator, Alexander.

136 See Chapter 1, Part 2, pp. 32–33.

137 Lucretius, Book 2, 84, p. 241, and 216–217, p. 247.

138 Long, p. 36.

139 Bailey, 2, p. 290.

the other hand, given weight as both (a) an inalienable property of all atoms and (b) a tendency to move, it is easy to explain why the atoms can never halt but go on moving for all eternity.[140]

The weight of the atoms, however, is never an initiator or source of motion. Since every atom always moves at 'atomic speed', no atom ever *starts* to move (from rest). Can the weight of an atom *alter* (e.g. reverse) the direction of its motion? If I fling a heavy body upwards, it soon reverses its course and falls back to Earth. Could this ever happen for an atom? Is the change in direction of the compound body due to 'blows' (e.g. of a downward 'rain' of atoms) or can it change direction by its weight alone? Consider an atom which has been 'squeezed out and up' by some means and is now moving upwards: will it ever, of itself, reverse its path and fall back down again? If the weight of an atom cannot initiate motion itself, can it cause a spontaneous change of direction? If not, an upward-moving atom may move forever with a motion opposed to its very nature!

According to Furley, an Epicurean atom may be displaced upwards for a few instants before its weight 'reasserts itself' and reverses its course.[141] But how, Konstan demands, can this 'reassertion' of weight be conceived, except in terms of the 'blows' of a downward-falling rain of atoms? Why should an upward-moving atom suddenly reverse its motion, without some proper (i.e. mechanical) cause?[142] Konstan's alternative suggestion is that the weight of an atom only asserts itself upon collision with another. All changes of direction, therefore – except the 'swerve' – would result from collisions: in virtue of their weights, atoms would tend to emerge from collisions in some preferred direction, i.e. downwards.[143]

This is ingenious, but perhaps unnecessary, If the Epicureans are prepared to countenance a purely random and uncaused 'swerve', by which atoms occasionally diverge from their original courses, it is surely idle to try to turn them into mechanists. (If an atom can swerve for *no* reason, why not by the 'blow' of its own weight?)

Before leaving the Epicureans, we must at least say something about this curious notion of the 'swerve'. As the atoms fall through the void, says Lucretius,

140   See Konstan, 2, p. 414.

141   *Ibid.*, p. 412.

142   *Ibid.*, p. 413.

143   *Ibid.*, p. 414.

at times quite undetermined and at undetermined spots they push a little from their path: yet only just so much as you could call a change of trend.[144]

This 'swerve' (*clinamen*) explains both world-formation and the power, 'wrested from fate', of free will. But in order to 'save' freedom, the mind or soul atom whose swerve is to produce a free act must deviate from its path not merely at random – mere randomness is as inimical to liberty as hard determinism – but at the *right* time and in the *right* place. But no mechanical explanation of this is possible, and to construe mind-atoms as sentient, and thus as 'knowing' when and where to serve, is futile – as well as contradicting Lucretius' explicit testimony.[145]

The Epicureans therefore tended to retain – without adequate theoretical support – the classical Greek conception of soul as a *free* or *uncaused* cause, capable of giving rise to fresh motions, and with a power of spontaneity not granted to things inanimate. The 'swerve' can, however, account only for indeterminism: it cannot explain the goal-directedness of animal and human behaviour.

The doctrine of the 'swerve' was of course ridiculed in antiquity as a mere subterfuge, an *ad hoc* hypothesis invented to save free will. Epicurus, says Cotta in the *De Natura Deorum*:

> casting about for a way to avoid this determinism, which Democritus had apparently overlooked, . . . said that the atoms, as they fell, just swerved a little![146]

No event, says Plotinus, occurs without a cause: 'we must leave no room for vain "starts" or the sudden movement of bodies which happen without any preceding causation'.[147] The swerve, add the Stoics, would violate the 'cosmic harmony' – the mutual interdependence of all things – and bring about the disintegration of the entire cosmos.[148] No one – Aristotelian or Platonist, Stoic or Sceptic – had a good word to say of the notorious 'swerve'.

---

144   Lucretius, Book 2, 218–220, p. 247.
145   See *ibid.*, 865 ff., p. 281 ff.
146   Cicero, p. 97.
147   Plotinus, *Enneads*, 3, 1, Vol. 3, pp. 9–11.
148   See Sambursky, 3, p. 57.

C. *Incorporeal Causes*

The ancient Atomists were all of course thoroughgoing materialists. They admit into their ontology only matter* (Being) and void (non-Being). Of these, only matter* is real, i.e. capable of causal agency: void, although it exists 'of itself' (*kath auto, per se*), is inert and causally inefficacious. Perhaps the clearest expression of this position can be found in Lucretius. Having argued for the existence of a 'twofold nature' in things, he continues as follows:

> Besides these there is nothing which you could say is parted from all body and sundered from void, which could be discovered as it were a third nature in the list. For whatever shall exist, must needs be something in itself; and if it suffer touch, however small and light, it will swell the sum of body . . . But if it is not to be touched, in as much as it cannot on any side check anything from wandering through it and passing on its way, in truth it will be that which we call empty void. Or again, whatsoever exists by itself, will either act on something or suffer itself while other things act upon it, or it will be such that things may exist and go on in it. But nothing can act or suffer without body, nor afford room again, unless it be void and empty space. And so besides void and bodies no third nature can itself be left in the list of things, which might either at any time fall within the purview of our senses, or be grasped by anyone through reasoning of the mind.[149]

This passage contains a two-pronged argument. What exists, Lucretius begins, is either tangible or intangible. But what is tangible is body, and what is intangible is void. If an opponent (e.g. a Platonist) retorts that there are incorporeal substances (souls) which are intangible and yet distinct from void space, Lucretius counters with the second point: what is intangible cannot act on (or be acted on by) bodies – a body can only be moved by an impulse from another. Incorporeal substances could not therefore be the causes of any of the natural phenomena (i.e. bodily motions) which we experience. Not only are they insensible – this would not worry a Platonist – they are also *inconceivable*: their supposed manner of activity cannot be grasped 'through reasoning of the mind'. This is the crucial point: we cannot conceive how an incorporeal motor could move a body.

---

[149] Lucretius, Book 1, 430–448, pp. 197–199.

This conclusion has obvious implications for psychology. According to common sense, states of the soul (volitions) are the efficient causes of the voluntary motions of men and animals. Hence, say the Atomists, the soul must be material in nature. Thus Epicurus:

> Now it is impossible to conceive the incorporeal as a separate existence, except the void: and the void can neither act nor be acted upon . . . so those who say that the soul is incorporeal are talking idly. For it would not be able to act or be acted on in any respect, if it were of this nature.[150]

And also Lucretius:

> The same reasoning shows that the nature of mind and soul is bodily. For when it is seen to push on the limbs, to pluck the body from sleep, to change the countenance, and to guide and turn the whole man – none of which things we see can come to pass without touch, nor touch in its turn without body – must we not allow that mind and soul are formed of bodily nature? Moreover, you see that our mind suffers along with the body, and shares our feelings together in the body . . . Therefore it must needs be that the nature of the mind is bodily, since it is distressed by the blow of bodily weapons.[151]

Mind acts on body (in volition) and is in turn acted on by it (in sensation). But this causal interaction is only possible if Mind is itself corporeal in nature: given commonsense interactionism and a rationalist conception of causation as an intelligible connection, some form of materialism follows.

The Epicureans[152] thought of the *psuchê* or *anima* as made up of minute, smooth, loosely-connected, fast-moving atoms of 'breath', 'heat', 'air', and some mysterious – and still subtler – fourth element.[153] These soul-atoms are rapidly agitated and, although very small, can move the larger body-atoms by virtue of their great impetuosity. Importantly, they are not themselves sentient: the sentience of a man or animal is due to an organised complex of soul and body atoms, not to any

---

150 'Letter to Herodotus', 67, Bailey, 1, pp. 41–43.

151 Lucretius, Book 3, 161–176, p. 311.

152 See Epicurus' 'Letter to Herodotus', 63–66, Bailey, 1, pp. 39–41, and Lucretius, Book 3.

153 Lucretius, Book 3, 231–250, pp. 313–315.

intrinsic property of the latter. Mind (*animus*) is akin to soul (*anima*) though still finer in nature.

Ranged against this Atomist theory we find Plato's arguments for the *incorporeality* of the *psuchê*. The Platonist, as we have already seen, argues at length for an animistic physics, for a conception of the *psuchê* as the ultimate source of *all* motions. But the power of spontaneity – the ability to initiate motion *de novo* – is not intrinsic to body as such (or else all bodies would have it). Indeed, body as such seems rather to resist being set in motion and to tend to a state of rest.[154] It must therefore be to some non-bodily principle that animate things owe their peculiar powers, i.e. to something incorporeal. The *psuchê* is therefore an incorporeal substance capable of moving bodies as an efficient cause.

The other Platonic argument for the incorporeality of the soul turns on the alleged impossibility of accounting for its powers (sensation, thought, volition) in purely material terms. The attempts of the ancient Atomists to provide such explanations cannot be regarded as success-stories, and Epicurean accounts of the mind were the butt of much ridicule. The reaction of Plotinus is typical:

> And bodies will suffer, compulsorily, when they are struck by atoms, whatever the atoms may bring; but to what movements of atoms will one be able to attribute what soul does and suffers? For by what sort of atomic blow . . . will the soul be engaged in reasonings . . . of a particular kind? And when the soul opposes the affections of the body? By what movements of atoms will one man be compelled to be a geometer, another study arithmetic and astronomy, and another be a philosopher? Our human activity, and our nature as living beings, will be altogether done away with if we are carried about where the [primary] bodies take us, as they push us along like lifeless bodies.[155]

Two points are discernible here. One is a general incredulity about the Atomist approach to psychology, and a tacit denial that such explanations of psychological functions are possible. The questions can be read as rhetorical, meant to bring home the absurdity of the Epicurean position, rather than as genuine requests for further information: Plotinus is querying the legitimacy of the whole project rather than asking for a few more details. The other – and perhaps more important –

---

154   See Jammer, 3, pp. 29–30; Balme, 1, p. 138.
155   Plotinus, *Enneads*, 3, 1, vol. 3, pp. 17–19.

objection is that Atomist takes away the spontaneous activity of the soul and thus denies us the status of *agents*. To 'save' agency – and hence freewill – the Epicureans were forced to postulate the 'swerve', which *ad hoc* hypothesis received short shrift from Plotinus.[156]

For the Platonist, then, soul or *psuchê* has powers and faculties (sensation, thought, volition) irreducible to those of bodies and inexplicable in material terms. The soul must therefore be an incorporeal substance or self-subsistent immaterial entity, capable of causal interaction with bodies.

This Platonic doctrine passed of course into Christian thought: many of the Church fathers were heavily imbued with the spirit of Platonism. For a Christian Neoplatonist such as Augustine, materialism is something worse than a mere philosophical error: those who put matter first (i.e. as a first principle) do so 'because their own minds are subservient to the body'.[157] Belief in the primacy of body is not only error but (a mark or sign of) SIN – it signifies overmuch attachment to the body and the senses. That Epicurus is singled out for special opprobrium is not surprising since he denies all the doctrines – Creation, Providence, Individual Immortality – central to the Christian faith.

If one does posit an incorporeal soul, how can one answer the Epicurean argument that such an entity could not possibly move the body? In the *Laws*, Plato asks this question of the soul and body of the Sun. The soul of the Sun, he replies, may move its body in one of the three following ways:

1. Like the human soul moves it body, i.e. by *vital union*.
2. By providing itself with a subtle body of fire or air, and then setting this in motion, which in turn 'violently propels body by body'.
3. The soul is itself without any body (subtle or gross) but guides the body 'by some extraordinary and wonderful power'.[158]

1 is vital union with a gross body; 2 is vital union with a subtle body, mechanically linked to the gross one; 3 is presence as a *virtus* or power without organic union. Of these, 3 is purely *occult*: no attempt is made to explain how this 'extraordinary and wonderful power' acts. 1, however, is little better: the Epicureans admit the existence of a vital

156   See *Enneads*, 3, 1, Vol. 3, pp. 9–11.

157   *City of God*, Vol. 3, Book 8, v, p. 27.

158   *Laws*, 898, p. 285. See also Hesse, 1, p. 48.

union of soul and body, but claim that very union as evidence of the soul's corporeality.

Theories of type 2 seem – at least superficially – more promising. All manner of *intermediates* can be posited to try to bridge the gap between soul and body, from the 'chariot of the soul' or 'astral body' of the *Phaedrus* to the 'spirits' (natural, vital, and animal) of Galenic physiology. These subtle bodies tended to be conceived as intermediates between soul and body in two quite distinct senses. On the one hand, they are causal intermediaries: all interactions between body and soul require the mediation of some such 'spirit'. So long as these are still thought of a bodily in nature, however, the Epicurean challenge remains: how can an incorporeal substance interact even with a subtle body? Why should a mere difference in density affect the conceptual issues involved?

It was to answer this challenge that 'spirits' became conceived as not merely causally but also *qualitatively* intermediate between soul and body. The conceptual boundaries were blurred so as to allow grades of 'incorporeality' and hence of soul-like-ness. The 'spirits' of Galenic physiology, for example, are still subtle bodies, but they are responsible for many of the lower 'faculties' (e.g. vegetative and sensitive) traditionally attributed to the soul.

The classical compromise of the Middle Ages was thus to ascribe the lower faculties of the soul to the agency of corporeal spirits, and the higher, intellectual, faculties to an incorporeal and divinely-instilled rational soul. The causal problem of *how* the incorporeal soul moves the corporeal *spiritus*[159] was of course never solved, but the compromise did have the merit of licensing a naturalistic approach to plant and animal physiology – so long as it steered clear of matters theological.

For the Stoics, of course, the *pneuma* is both corporeal and active.[160] These thinkers accepted both the Platonists' demand for an 'active principle' and the Atomists' conviction that only something corporeal can be capable of action and passion.[161] Passivity and resistance to motion, they say, characterise only gross matter, not matter as such: subtle matter has active powers different in kind. Matter thus divides into the passive and inert (W, E) and the active (F, A, *pneuma*).[162] The Stoics

159   See D. P. Walker in Debus, Ed., 7.

160   Partington, 6, Vol. 1, p. 151; Dijksterhuis, p. 43.

161   Long, p. 132.

162   *Ibid.*, p. 154; Lapidge, pp. 271–272.

are therefore both materialists and animists: in place of the Platonists' incorporeal *psuchê* they posit their own *pneuma* (= *theos* or 'god' and *logos* or 'reason') as the subtle, active material governing and ruling the cosmos.

Before leaving this topic, two further points must be briefly mentioned. One is the lapse into an occult anthropomorphism by many thinkers – especially in the Neoplatonic school – in late antiquity. Neoplatonism was always congenial to the occult arts, and it gradually became common to explain a variety of natural phenomena in terms of the agency of a *daemon*, a 'spirit' in the supernatural sense of the term. Similar doctrines can be found in the cabbala – a tradition of Jewish religious thought with strong leanings towards angelology and daemonic magic – and in the so-called 'Hermetic corpus'.[163] The Hellenic *phusikos*, seeking to understand natural phenomena in purely naturalistic terms, gives way to the *magus*, seeking to control terrestrial affairs by the invocation of angels or demons.[164]

Platonism demands an incorporeal agency for all natural processes, and hence *a fortiori* for the generation of plants and animals. Now if one is to avoid the occultism of the magical world-view, which might lead one to see angelic or demonic power in every such case,[165] one must seek some more 'natural' explanation. This may perhaps be cast in terms of the action of the *Anima Mundi* or world-soul. But this is diffused throughout the cosmos, no more in once place than in another, and hence cannot provide a complete account of the coming-to-be of particular creatures. To fill this lacuna there was developed the theory of 'seminal reasons' (*rationes seminales*, *spermatikoi logikoi*). These are 'parts' of the *Anima Mundi*, present in particular bodies, which they shape in accordance with some design. Both Stoics and Neoplatonists believed in this immanent formative action of the creative *logos*, and Augustine gives the theory a theistic twist by identifying this *logos* with the *divine* reason.[166] God, he says, is immanent in all His Creation, directing and moving

163 See W. Scott, Ed., Thorndike, Vol. 1, p. 291, and Hopkins, p. 28.

164 For an example of daemonic magic, even among reputable philosophers, see Porphyry's 'Life of Plotinus' in Plotinus, Vol. 1, pp. 33–35.

165 The sixteenth-century *magus*, Henry Cornelius Agrippa, seems actually to have accepted such a view: i.e. that angels 'preside' over' the births of all things and give them their peculiar 'occult qualities'. See Thorndike, Vol. 1, p. 454.

166 See Hesse, 1, p. 78.

all things by means of these 'seminal reasons'. These *semina* (seeds) are conceived by the Stoic as corporeal (though subtle), i.e. as parts of the cosmic *pneuma* or *theos*, and by the Neoplatonist as immaterial 'emanations' from the world-soul (and ultimately from the deity). They are both *information-bearing* and *active*: they serve therefore both as instruction-books and as workmen in the building of an organic body.

D. Teleology

Leucippus and Democritus were, so far as we are aware, the first great anti-teleologists, the first major thinkers to deny that the concept of intelligent design is central to natural philosophy. According to Democritus, all things are formed by chance (*tuche*) and 'necessity' (*ananke*): there is no need for a Mind or Deity to guide the motion of the atoms so as to form and sustain an orderly cosmos.[167] Atoms move quite *randomly* in the void: when a sufficient number aggregate and begin to 'jostle' one another, a 'whirl' or vortex is formed, which in turn sorts the elements FAWE according to particles-sizes and thus produces a world. The constitution of a world depends on the number, sizes, shapes, motions and arrangements of its atoms: other worlds may therefore be quite different from outs – one may lack W, another F, another animals, etc. We have no warrant for supposing our world to be in any way typical: the random processes of world-formation could yield a variety of different results.[168]

What does it mean to say that all things are the result of 'chance' *and* 'necessity'? The meaning of 'necessity' is plain. All things come-to-be as a result of atomic motions, but every atomic motion is the outcome of a prior 'blow'. 'Necessity', therefore, stands for the purely mechanical processes of atomic collisions. At times, Democritus seems to identify it with the 'whirl', as Diogenes Laertius says:

> All things happen by virtue of necessity, the vortex being the cause of the creation of all things, and this he calls necessity.[169]

The basic idea is that it is due to the vortex that the atomic collisions and entanglings which give rise to things occur.

---

167   Kirk and Raven, p. 417. Dijksterhuis, p. 77.
168   Kirk and Raven, pp. 411-412.
169   Diogenes Laertius, Vol. 2, p. 455.

'Necessity' thus stands for a sufficient determining chain of efficient causes, for 'the resistance and movement and blows of matter'.[170] Although it is the ultimate *explanans* of all things, it is not itself an agent – hence *a fortiori* not a purposive agency – but a general name for the fact that all events are causally determined by prior ones.[171]

What of the identification of 'necessity' with 'chance'? This is not so strange as at first sight it may seem. Suppose atoms A and B collide at a place P at a time t_1: we can explain why A was at P at that time in terms of 'necessity' (a determining cause), and likewise why B was there at t_1, but it is purely *coincidental* that *both* were. (The two causal chains may be quite independent.) 'Chance' may therefore be read as 'coincidence', and as involving, essentially, the denial of design or contrivance. If I go to London to meet Smith and happen to meet Jones instead we say the encounter was a 'chance' one: it is not that nothing caused me to go there, but that I did not go for that specific purpose.[172]

'All things come-to-be by chance and necessity' can now be given a clear and coherent sense as involving (a) an assertion of causal determinism and (b) a denial of teleology.[173] In particular, the formation of our cosmos and the orderly motions of the heavenly bodies are not, Democritus insists, due to divine agency – his claim that our world has originated by chance and necessity is clearly intended to exclude theistic cosmologies and cosmogonies.[174] As Aristotle says:

> There are some . . . who ascribe this heavenly sphere and all the worlds to spontaneity. They say that the vortex arose spontaneously, i.e. the motion that separated and arranged in its present order all that exists. This statement might well cause surprise. For they are asserting that chance is not responsible for the existence or generation of animals and plants, nature or mind or something of the kind being the cause of them (for it is not any chance thing that comes from a given seed but an olive from one kind and a man from another): and yet at the same time they assert that the heavenly sphere and the divinest of visible

170   Aetius, 1, 26, 2. Quoted from Kirk and Raven, p. 413.

171   Kirk and Raven, p. 411, p. 413.

172   See Aristotle, *Physics*, 2, 5, 196b 33–197a 4.

173   Barnes, Vol. 2, pp. 122–124.

174   Bailey, 2, p. 121.

things arose spontaneously, having no such cause as is assigned to animals and plants.[175]

This is a curious passage. If indeed Aristotle does have Democritus in mind, it seems that he is attributing to the latter the view that animals and plants arise by 'nature' or 'mind', i.e. through some purposive agency. Yet this is surely *not* the Atomist's view. Elsewhere, Aristotle says that Democritus, 'neglecting the final cause, reduces to necessity all the operations of Nature'.[176] Perhaps the evident fact of the *specificity* of generation (man begets man and not horse) forced Democritus to posit for each living thing a specific nature (*physis*): this could be an atomic arrangement or pattern capable of mechanically stamping its likeness on other bodies. We could then accept that living beings are generated by *physis*, but deny that this involves any non-mechanical and irreducibly teleological mode of causation. I can find no better way of reconciling the two passages quoted above.

Let us turn now to the critics of Atomism and their re-affirmation of teleology. There are two distinct spheres in which *design* might seem especially manifest in non-human Nature: one is the celestial realm, with its orderly and regular motions, the other, the functionally adapted bodies of plants and animals. We begin with the former.

Greek thought, says Collingwood,[177] contains the following pair of deep presuppositions:

Motion ⟶ Soul (Life)
Order ⟶ Mind (Intelligence)

Nowhere are these preconceptions more apparent than in Plato's *Timaeus*. We have already discussed the first (the argument for the *psuchê* as the ultimate source of all motion), and are here concerned primarily with the second. In the *Timaeus*, God (the 'demiurge') is conceived not (as in Presocratic cosmogonies) as the *father* of the cosmos but as its *architect* and *artificer*: the analogy is 'no longer that of reproduction or growth, but of the deliberate constructive activity of a craftsman',[178] imposing form on matter.

The demiurge is needed, for Plato, to account for the intelligibility of our cosmos: its structure, he claims, could only be the product of intelligent contrivance. The cosmos

---

175   *Physics*, 2, 4, 196a 25–35.

176   *De Gen. Animalium*, 5, 8, 789b 2–3.

177   Collingwood, p. 3.

178   See Lee, Introduction to the *Timaeus*, p. 8.

being good – it would be blasphemy to think otherwise – it must have been deliberately made after an ideal model, something 'apprehensible by reason and understanding and eternally unchanging',[179], i.e. a form. Whenever, he adds, 'the maker of anything keeps his eye on the eternally unchanging and uses it as a pattern for the form and function of his product the result must be good'.[180]

The phenomena which most clearly manifest intelligent contrivance are the perfectly regular and orderly motions of the heavenly bodies. Even the apparently irregular planetary motions (e.g. retrogressions) have their proper period, their rightful place in the cosmic dance: the souls of the planets know their steps too well ever to get out of line, to violate the universal harmony. After a 'Great Year', all the planets return to precisely the same places.[181]

The *argument* of the *Timaeus* is wholly implicit. It simply assumes the 'Order ⟶ Mind (Intelligence)' presupposition and develops a cosmology based around it. The implicit claim is that a philosophy such as Democritus', which assumes that the cosmos originated by 'chance and necessity' and not by design, cannot account for the plain facts, notably the perfect regularity of even the apparently disorderly (i.e. irrational) motions of the planets.

The *Timaeus* also contains the rudiments of a teleological approach to biology: the lungs, says Plato, exist to cushion and refresh the heart; the liver stores images and is therefore of use in prophecy and divination.[182] Plato's examples are crude, but in Aristotle and Galen teleological biology attained considerable depth and sophistication.

The all-important thesis is that the structures of the organs of a plant or animal are specially adapted to their functions, to the rôles they play in that organism's way of life. We can *explain* structures in terms of functions, as we can for artefacts. Nature, says Aristotle, 'never fails nor does anything in vain so far as is possible in each case'.[183] Its

---

179   *Timaeus*, 29, p. 41.

180   *Ibid.*, 28, p. 40.

181   See *Timaeus*, 38–39, pp. 51–54. The discovery that the motions of the 'wanderers' (or 'vagabonds') is really orderly and regular was crucial to Plato's astral theology: only perfectly orderly beings can be proper objects of reverence. The solution of the problem of the planets was therefore essential to Plato's theology: no Platonist would worship a mere cosmic vagrant or tramp.

182   *Timaeus*, 70, p. 96.

183   *De Gen. Animalium*, 5, 8, 789a 22.

operations 'are for a final cause and for the sake of what is best in each case'.[184] *Physis* 'belongs to the class of causes which act for the sake of something'.[185]

Suppose Democritus were to object that all the 'natural' processes of organic development may occur merely 'by necessity', like the weather: rain does not fall, for example, *in order* to nourish crops, but as a necessary consequence of prior efficient causes: 'What is drawn up must cool, and what has been cooled must become water and descend, the result of this being that the corn grows'.[186] Why should it not be the same with the parts of animals?:

> e.g. that our teeth should come up of *necessity* – the front teeth sharp, fitted for tearing, the molars broad and useful for grinding down food – since they did not arise for this end, but it was merely a coincident result . . .[187]

In similar manner, Lucretius insists that the eye was not made *for* seeing, nor the leg *for* walking:

> All . . . ideas of this sort, which men proclaim by distorted reasoning set effect for cause, since nothing at all was born in the body that we might be able to use it, but what is born creates its own use.[188]

The organ, Lucretius insists, must always exist *before* its use. How then can the anti-teleologist explain functional adaptation? The answer is as follows:

> Whenever then all the parts came about just what they would have been if they had come to be for an end, such things survived, being organised spontaneously in a fitting way; whereas those which grew otherwise perished.[189]

The anti-teleologist thus invokes a crude and rudimentary notion of 'natural selection' to explain the survival (and hence current existence) of well-adapted organisms without invoking a final cause. According to Lucretius,[190] species of animals have survived through 'craft', 'bravery', 'swiftness' or utility to man – animals lacking such advantages will perish without

---

184   *Ibid.*, 789b 5.

185   *Physics*, 2, 8, 198b 10.

186   *Ibid.*, 198b 19-21.

187   *Ibid.*, 198b 23-26.

188   Lucretius, Book 4, 832-835, p. 405.

189   *Physics*, 2, 8, 198b 28-31.

190   Lucretius, Book 5, 837-877, pp. 477-479.

reproducing their kind, as will any so ill-adapted as to be unable to feed or procreate at all. Many races of living things, says the poet, 'perished and could not beget and propagate their offspring'; nature eventually 'brought their kind to destruction'.

These first 'evolutionary theories' are so primitive as to be of minimal importance in the history of biology – it is no great tribute to Empedocles and Lucretius to call them precursors of Darwin. The greatest biologists of antiquity – men like Aristotle and Galen – hardly took them seriously: random or disorderly motion, says the Stagirite, could not even yield flesh and bone[191] (i.e. FAWE mixed in a correct proportion), far less the complex structures of animal bodies:

> For teeth and all other natural things either invariably or normally come about in a given way; but of not one of the results of chance and spontaneity is this true.[192]

In Nature, ends are achieved 'for the most part' – organisms ingest and assimilate their own proper nutrients, Xs impart their forms to their offspring and thus beget Xs, and so on.[193] But if the development of organisms were a chance affair, we could expect to see all manner of monsters produced in *every generation*: there would only be a minimal chance of organisms breeding true to type. The fact that Xs beget ('for the most part') Xs is evidence that generation occurs by nature and not by chance, and that 'nature is . . . a cause that operates for a purpose'.[194]

Physis, although it acts for an end, must not be conceived as having that end in mind. Plants and animals do not have rational souls: in the development of such lower organisms the realisation of form in matter (the final cause) is unconscious or instinctive. We understand the operations of *physis* by analogy with purposive human action (Art): the best illustration, says Aristotle, is 'a doctor doctoring himself: nature is like that',[195] but it is not aware of what it is doing and why. The explanation trades on the analogy with the activity of the artist, but then adds that this particular 'artist' is unconscious, unaware of what it is making.[196]

191   *De Caelo*, 3, 2, 300b 27.
192   *Physics*, 2, 8, 198b 35-199a 2.
193   See Randall, 2, p. 187.
194   *Physics*, 2, 8, 199b 32-33.
195   *Ibid.*, 2, 8, 199b 30.
196   See Temkin, p. 136.

'Nature', for Aristotle, is not an all-pervasive, quasi-divine formative agency operating in all things: this notion, common in Renaissance thought, is alien to him.[197] When he speaks of *physis*, it is always the nature of a particular thing that he has in mind, i.e. a kind of disposition it has to a certain manner of growth, development, and motion. Plants and animals in particular have natures which are purposive, i.e. act as if intelligent without actually being so.[198]

If we turn from Aristotle to Galen, we find the 'argument from design' taken to its logical conclusion. In the structure of every animal, he argues in *De usu partium*, we can discern the workmanship of a wise and benevolent designer – the adaptation of structure to function is a sure sign not merely of purposive but of intelligent contrivance.[199] Galen nevertheless rejects the Platonic theory that the *Anima Mundi* is the cause of all generation and insists that each creature is what it is in virtue of its own *physis*, which acts 'artistically' (i.e. with foresight) to produce and sustain its body.[200] Structures as complex as animal bodies could not possibly arise from mere chance or the random jostling of 'inharmonious elements'.[201]

Galen therefore accepts the argument from functional adaptation to intelligent design, but denies that God or the *Anima Mundi* is the direct cause of all generation. Is the *physis* of every animal and plant then itself intelligent: does the 'nature' responsible for the development of every new embryo know in advance what it is doing? This would eliminate the distinction between rational and irrational Nature. Perhaps the *physis* of each plant and animal is not itself intelligent, but operates according to a pre-set program ordained, after some manner, by the Deity: this compromise would perhaps make the best sense of Galen's overall position. Without intelligence (Mind, God) there would be no *physis* at all (a purposive program presupposes an intelligent programmer), but the *physis* of a worm or a tree is not itself possessed of Mind: it only acts 'artistically' in the sense that it operates *as if* it were possessed of foresight. The explanatory structure (organs exist *for* some *end*) is the same, the

---

197   Charlton, Ed., Introduction, xxvi–xxvii, and p. 88.
198   *Ibid.*, p. 121.
199   See Thorndike, Vol. 1, p. 149. See also Balbus the Stoic in Cicero, p. 178.
200   Galen, p. 43.
201   Galen, p. 155; Thorndike, Vol. 1, p. 150.

metaphysical implications very different: without the 'as if' clause we find ourselves attributing Mind to all creatures.

This teleological biology was a great success, and gave rise to many valuable insights into the workings of plants and animals. It was also considered an *overwhelmingly powerful* objection to the anti-teleological Atomic theory. (Millennia later, Harvey followed Aristotle and Galen by rejecting mechanistic Atomism on very similar grounds.[202]) In place of the conception of *physis* as craftsman, the Atomist has only his rudimentary – and horribly crude – version of 'natural selection'. Small wonder that biology was considered a massive lacuna in the natural philosophy of ancient Atomism.

The Epicurean conception of the compound body as a *concilium* does, however, offer some hope of an account of organic specificity and (apparent) purposiveness. A *concilium* with a particular 'harmony' of atomic motions will only absorb certain types of molecule – this can explain the specificity of nutrition and assimilation. It will moreover emit 'seeds' (*semina*) of its own nature which in turn will absorb congenial material to engender other organisms of the same type. Horses beget horses and cows cows: this biological conception of natural order (as something quite independent of Mind and hence of Divine control and governance) is central to Lucretius' poem.[203] This explanation of the specificity of such processes as nutrition and reproduction is plausible and persuasive. Each *concilium* has what we have elsewhere characterised as a 'dynamic texture' in virtue of which it (a) absorbs kindred material and (b) emits 'seeds' of that stuff. The appearance of purpose in Nature can be explained in naturalistic terms without having to posit any divine guidance. All things, says Lucretius, 'are produced from fixed seeds', each being is born 'out of that which has in it the substance and first-bodies of each' – men do not arise from the sea, nor fish from the sky. As things grow they 'preserve their own kind'; each is 'fostered out of its own substance'. There is a lot more in the very first argument in *De Rerum Natura* than first meets the eye.[204]

The denial of divine Creation and Providence is of course at the heart of Epicureanism. The creation of matter *ex nihilo* is

---

202   See Toulmin and Goodfield, 1, p. 146.

203   See Solmsen, 1, p. 20 ff., for a fascinating discussion of the rival Platonic (theological) and Epicurean (biological) conceptions of order.

204   See Lucretius, Book 1, 159–214, pp. 185–187. See also, Long, p. 41.

of course quite unthinkable; it is the Platonic theory of the origin of our cosmos by the divine ordering of pre-existing matter that is the butt of Epicurean ridicule. In *De Natura Deorum* the Epicurean, Velleius, begins his oration with a diatribe against the cosmology of the *Timaeus*.[205] Plato's account of the Creation, says Velleius, is a mere figment of the imagination. How did God go about it: 'What tools did he use? What levers? What machines? Who assisted him in so vast an enterprise?'[206]

As for the idea that gods control and govern our world, it can be dismissed out of hand as incompatible with their life of perfect bliss and freedom from all cares. It is, says Epicurus, an unworthy conception of the divine to imagine that they are concerned with the day-to-day running of our cosmos: the Platonic astral deities would be mere drudges, living a life of unending toil!

Some people, adds Lucretius, claim that the gods have ordained all things for the benefit of men, but these dreamers:

> are seen in all things to fall exceeding far away from true reason. For however little I know what the first-beginnings of things are, yet this I would dare to affirm from the very workings of heaven, and to prove from many other things as well, that the nature of the world is by no means made by divine grace for us: so great are the flaws with which it stands beset.[207]

Given the existence of famine, drought, plague, earthquake, volcano, tempest, etc., etc., it does indeed seem perverse to claim that this world was created for us by powerful, wise, and beneficent beings!

The Epicurean opinion was, however, not to become widely accepted: Christian and pagan writers alike re-affirm, against Epicurean 'atheism',[208] the argument from design. The Stoics in particular emphasised the rôle of the divine

---

[205] Cicero, pp. 77–78. Cf. the lengthy attack on Platonic astral theology in Epicurus' 'Letter to Herodotus', 76–77; Bailey, 1, p. 49. According to Solmsen (1 and 2) and Furley, 1, Epicurean cosmology was primarily anti-Platonic in inspiration.

[206] Cicero, p. 77.

[207] Lucretius, Book 2, 174–181, p. 245.

[208] Although the Epicureans were nominally theists, their gods have no control over our world, and care nothing for our fortunes – such deities could hardly become objects of worship for us. From a religious point of view, Epicureanism is an atheistic creed, as Cotta says in the *De Natura Deorum* (Cicero, p. 104, pp. 116–120).

creative *logos* both in the cosmos as a whole and in the formation of organic bodies. For the Stoic, *physis* obeys the dictates of *logos* – the universe is a rational system,[209] the product of *physis*, which is defined as 'an artistically working fire, going on its way to create; which is equivalent to a fiery, creative, or fashioning breath.[210] In Cicero's *De Natura Deorum*, the Stoic, Balbus, appears as a forceful advocate of the argument from design. What, he asks,

> could be more clear and obvious, when we look up to the sky and contemplate the heavens, than that there is some divinity of superior intelligence, by which they are controlled?'[211]

When confronted with such perfectly co-ordinated movements in so many immense bodies, in a pattern 'which has never deviated through all the immeasurable ages', we must speak of design and not of mere chance.[212] The heavenly bodies, he continues, are themselves sentient and aware of their parts in the cosmic dance: 'For nothing can move in a measured and orderly way without the guidance of an intelligence'.[213] The planets are misnamed 'vagabonds' – in fact, their motions too manifest a perfect regularity which cannot be understood

> except as the expression of reason, mind and purpose in the planets themselves, which we must therefore reckon in the number of the gods.[214]

In the heavens, he continues, nothing is accidental, random, or erratic: 'Everywhere is truth, reason, constancy'.[215] Hence, says Balbus,

> My belief is that the universe and everything in it has been created by the providence of the gods and is governed by their providence through all eternity'.[216]

---

209 See Long, p. 120.
210 From Diogenes Laertius, Vol. 2, p. 261. The definition is Zeno's (see Balbus in Cicero, pp. 145–146).
211 Cicero, p. 124.
212 *Ibid.*, p. 129.
213 *Ibid.*, p. 140.
214 *Ibid.*, pp. 144–145.
215 *Ibid.*, p. 145.
216 *Ibid.*, p. 154.

If the organised structures of a ship, a sundial, or a picture all bear witness to an intelligent contriver, how much more so the cosmos?

> Is it not a wonder that anyone can bring himself to believe that a number of solid and separate particles by their chance collisions and moved only by the force of their own weight could bring into being so marvellous and beautiful a world? If anybody thinks that this is possible, I do not see why he should not think that if an infinite number of examples of the twenty-one letters of the alphabet . . . were shaken together and poured out on the ground it would be possible for them to fall so as to spell out, say, the whole text of the *Annals* of Ennius. In fact, I doubt whether chance would permit them to spell out a single verse![217]

Here the Atomists' own alphabet-analogy is, by a cunning twist, turned against them. Granted that different things could be produced by different arrangements of atoms (letters), how probable is it that our organised world-system could have arisen purely at random? The answer must be: minimally – perhaps negligibly – small.

Seneca too lends his voice to the choir: God, he insists, is wholly mind or reason,

> though mortal eyes are so sealed by error that men believe this frame of things to be but a fortuitous concourse of atoms, the sport of chance. And yet than this universe could aught be fairer, more carefully adjusted, more consistent in plan?[218]

In Neoplatonism too the order of the heavens is cited as evidence of Divine Providence. Plotinus writes at length against the gnostics, who decry the sensible (i.e. material) universe and its maker as evil.[219] Surely, he asks, 'what fairer image of the intelligible world could there be?'[220] Given the beautiful order and harmony of the cosmos, 'how should one not call it a clear and noble image of the intelligible gods?'[221] Divine Providence, he insists elsewhere,[222] is all-embracing:

---

[217]   *Ibid.*, pp. 161–162.

[218]   Seneca, p. 7.

[219]   Plotinus, *Enneads*, 2, 9, Vol. 2, p. 220 ff.

[220]   *Ibid.*, p. 239. The influence of the *Timaeus* is obvious.

[221]   *Ibid.*, p. 253.

[222]   *Enneads*, 3, 2 and 3, Vol. 3, p. 38 ff.

all events fall into a single, ideally rational, pattern (of which we humans are only partially aware). Partial evils, from a God's-eye point of view, contribute to the general good, as dark colours and discords may contribute to the beauty of a painting or a piece of music.[223]

This was perhaps the main reason for the unpopularity of Epicureanism in antiquity: few found it remotely credible that our world-system could have arisen from a 'fortuitous concourse' of atoms without any guidance. Religious and philosophical reasons combine to dismiss Atomism as both implicitly atheistic and wildly improbable, unable to account adequately either for the functional adaptation of the parts of animals (cf. Aristotle and Galen and their teleological biology) or for the beauty and harmony of the whole cosmos.

[223] *Ibid.*, Vol. 3, pp. 79–81, p. 107.

INDIVISIBLES: 2

Part 1: The Continuum Revisited

We left the theory of the continuum in a sorry state, torn between two warring factions, neither of whom had a satisfactory account to offer. On the one side we had the Atomists, who denied infinite divisibility and constructed all magnitudes out of fine numbers of discrete *minima*; on the other, the Aristotelians, who accepted the potentially infinite division of every continuous magnitude. The Atomist position was incompatible not only with elementary geometry but with the intuitive thesis (Z1) that every magnitude is divisible at least in thought; the Aristotelian is forced to deny that parts actually exist (Z4) in an extended magnitude, and to claim – counterintuitively – that a part is only actualised by a cut.[1] (The Aristotelian continuum has no ultimate constitution, either of points or minima – Aristotle explicitly denies that any magnitude is composed of indivisibles.)

Arab and medieval discussions of the issue were dominated by a battery of anti-indivisibilist arguments from the *Metaphysics* of Avicenna (ibn Sīnā, 980–1037) which, curiously, became known as 'the arguments of algazel' (al-Ghazzālī, 1058–1111).[2] Avicenna rejects the doctrine that 'a material body is composed of indivisible elements and that a body comes to be from such a composition'[3] for the following reasons, A1–6:

A1.   This is merely a prolix version of Aristotle's dilemma: Do the indivisibles which allegedly make up a continuum touch 'part to part' or 'whole to whole'? Either answer faces insuperable objections. This is how Avicenna puts it:

> For example, if we consider three elements forming a composite, one of these being a median and the other two

---

[1]   For a very frank statement of this counterintuitive peripatetic thesis, see Digby, pp. 12–17. The parts of a body, he insists at p. 12, 'are not really there, till by division they are parcelled out'.

[2]   See Grant, Ed., 4. p. 314 ff.

[3]   Morewedge, Ed., p. 19. I omit, here and in the following quotations, Morewedge's parenthetical transliterations of key Persian terms.

being on the *extrema*, then the median either separates the two *extrema* in such a manner that they do not touch one another, or it does not separate them but allows one to touch the other. If the median separated one from another, then each of the two *extrema* would touch a part of the median which the other *extrema* would not touch. Hence, there would be two positions for the median, and it would therefore have to become divisible. If the median would not connect each element with all other elements (i.e. it would not separate one from another), then any entire element would be indiscernible from any other entire element. Furthermore, the position of both *extrema* would be equal to the position of one median, so that it would stand separately and they would not mingle with each other. As a result, the position of two together would be no more than that of one. Therefore, the distinctness of any two of these combined elements would be no greater than that of one. Likewise, if a third element were combined with them, it would also be combined into one element. Consequently, if thousands upon thousands were combined, they would be equal to one.[4]

Or, as Aristotle would have put it – more briefly but no less cogently – indivisibles cannot touch 'part to part' without, *per impossibile*, being divisible, nor 'whole to whole' without coinciding in the same place, in which case they cannot give rise to anything greater by aggregation.

A2.   Imagine, says Avicenna, the following arrangement of indivisibles:

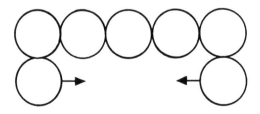

---

4   *Ibid.*, p. 19. See of course Aristotle, *Physics*, 6, 1, 231 b3–6.

The moving atoms will, he continues,

> approach one another with equal motion until they meet. Without a doubt, each cuts a part of the median, causing a part of the median to become one of the end points and another part to become another end point (i.e. the median partakes of the two elements in the *extrema*). If this were not the case, one would have to stand in order that the other could meet it, or both would have to stand and refuse to move at any time. Reason knows, however, that it is possible to bring these two together; where they meet the median is divided into two halves.[5]

A3.   Consider the following situation:

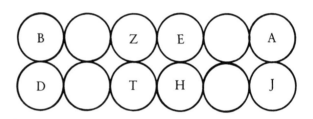

Let an element from A move towards B and another element from D move towards J until they face each other. Then there is no doubt that these elements will be aligned opposite each other at first, and that thereafter they will pass the other. Let us construct this approach in such a manner that their movements take place in equal amounts of time. They will face each other when one half of the total time will have elapsed . . . But the element corresponding to . . . E is H, and that corresponding to Z is T. If the place at which they will meet falls on E and on H, one will have moved three intervals and the other four . . . If one is at E, then another will be on T, or if one is at H, the other will be on Z. Henceforth, they will not be aligned opposite each other. Therefore, they will not pass each other, which is impossible.[6]

---

[5] *Ibid.*, pp. 19–20. The earliest use of this argument with which I am familiar is in Sextus Empiricus, 'Against the Physicists', 142, p. 283.

[6] *Ibid.*, pp. 20–21. The argument is of course strikingly similar to the fourth of Zeno's paradoxes of motion, the 'Moving Rows'. I have claimed that there is no evidence that Zeno used this argument specifically against indivisibles, but Avicenna clearly sees its potential in that regard.

This is perhaps the easiest of the six arguments for the Atomist to rebut: it *assumes* continuity in the course of arguing for it. The advocate of indivisibles must claim that 'X has passed Y' can be true without 'X is opposite Y' ever being true. There is, he will say, simply no instant at which the two moving elements are opposite one another. One moment one is at E and the other at T; the next, the former is at Z and the latter at H.[7]

A4.   Construct a square of indivisibles, thus:

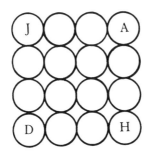

Here, says Avicenna,

> it is . . . evident that diagonally AD is equal to JH. Thus, with respect to the figures constructed from the elements designated by the sixteen . . . points on these four lines, it follows necessarily that AJ is equal to JH, and, likewise, that AH is equal to AD, for, when viewed from any one direction of either longitude, latitude, or the diagonal, no more than four . . . points can be seen.[8]

The indivisibilist position falls foul of the truths of geometry, as is argued in the pseudo-Aristotelian monograph, 'On Indivisible Lines'.

A5.   A vertical pole casts a shadow in the sunlight. What happens, Avicenna asks, to the end-point of the shadow when the sun moves a single indivisible space-unit? If it remains at the same location, 'then a straight line will have to terminate in two branches, which is impossible'. If, on the other hand, it moves, what distance can it move?

---

7   Unfortunately, this raises paradoxes elsewhere, e.g. with regard to the individuation of events. (See Grünbaum in Salmon, Ed., pp. 249–250). Event-status will turn out, on this view, to depend on the relative velocities of moving bodies and hence on (arbitrary) choices of reference-frames.

8   Morewedge, Ed., p. 21.

> If an element is moved whenever the sun moves somewhat, then the movement of that line . . . would be equal to the motion of the sun in the heavens, which is impossible. If the movement of the line were greater than that of the sun, it would be even more of an impossibility. If it moved less, the element would be divided, contradicting thereby the premises of the doctrine.[9]

By identifying a light ray with an Euclidean straight line, geometrical optics can be brought into play. Since the light source moves, the shadow must also move; when the sun moves a single space-unit, the shadow must move *less*, which is impossible.

A6. Let us imagine a mill made of iron or of diamond revolving around itself. The central element of the wheel turns less than the element on the periphery. Hence, every time an element on the periphery is moved by an interval, the element in the centre is moved less than the distance the element on the periphery is moved. This condition necessarily implies the division of an element.[10]

Avicenna concludes that a body is not actually an aggregate of parts, but that parts are only actualised by division:

> The proper doctrine is that a body is not a composite of elements and that in reality (i.e. in its natural state) a body does not essentially have an element until it is made to have one (i.e. until its natural state is altered artificially) since it would otherwise have elements without limit in number and without measure.[11]

This is an explicit acceptance of the Aristotelian doctrine of potentiality: a line, on this view, does not actually have a mid-point unless and until a cut is made there. The Atomist insists, by contrast, on an *actual ground* for every 'potency': the possibility of bisection, he argues, entails the existence of an actual mid-point.

The arguments A1–6 are important, and a little reflection soon shows why. Earlier objections to mathematical minima emphasised their incompatibility with geometry; but it is not obviously an irrational strategy to deny that our rough-hewn material world is isomorphic with the perfect forms of Euclid. Difficulties start to arise, however, with A2, A5 and A6 – all

9   *Ibid.*, pp. 21–22.
10   *Ibid.*, p. 22.
11   *Ibid.*, pp. 22–23.

seem designed to show that one cannot divorce geometry from physics. A2 derives a *physical* paradox from the doctrine of geometrical minima. A5 uses geometrical optics – a flourishing branch of physical science – against mathematical Atomism. Shadows and light-rays are not merely ideal fictions of the mind. Nor, more obviously still, is the *cohesion* of the mill-wheel, which the Atomist is prepared to dissolve with a wave of his hand.

Advocates of mathematical minima are thus driven into physical absurdities: it is this clash with physics which is particularly embarrassing to them. Mathematics just is applicable – and often successfully *applied* – to a vast range of natural phenomena. This is of course one of the leading ideas of the 'geometrisation of Nature' practised by Galileo and Descartes and their disciples in the seventeenth century. In this respect, the Atomist tradition, with its inherent scepticism about the physical truth of geometrical theorems, may have been one of the influences hindering the Scientific Revolution!

For the most forceful advocates of indivisibles among the Arabic philosophical schools, we must turn to the Mutakallimun, whose views are discussed at length by the twelfth-century Jewish scholar, Maimonides. Of the twelve fundamental propositions in terms of which Maimonides sets out this philosophy, the following four are especially relevant here:

P1.   All things are composed of atoms.

P3.   Time is composed of instants or time-atoms.

P11.   The idea of the infinite is equally inadmissible whether it be actual, potential, or accidental.

P12.   The senses mislead, and are in many cases insufficient; their perceptions, therefore, cannot form the basis of any law, or yield data for any proof.[12]

Maimonides goes on to spell out further corollaries of these basic propositions, starting with P1, which gives rise to five sub-theses, 1A–E:

1A.   The universe consists of atoms, which are indivisible because of their *smallness*.

1B.   An individual atom 'has no magnitude; but when several atoms combine, the sum has a magnitude, and thus forms a body'.

1C.   All atoms are identical: there is no respect (e.g. size or shape) in which they can differ. All bodies consist of rows of such atomic units.

12   Maimonides, p. 120.

1D.   Genesis = aggregation, which in turn reduces to 'motion and rest'.

1E.   The atoms can be annihilated and/or recreated at any instant, if it pleases God to do so.[13]

The indivisibility of the atoms is thus guaranteed by their lack of extension. For the Mutakallimun, an extended magnitude must be built up of atomic units: it follows that each of these units cannot itself be extended. (It is not itself compounded of further units, on pain of infinite regress.) But by P11 there can only by a *finite* number of these basic units in any given body. The problem now is to build up extended magnitudes from unextended points. If the points occupy no space at all, their mere juxtaposition will not constitute a body. A finite number of extensionless elements cannot themselves fill any space: the only sort of magnitude to which they can give rise is a 'gappy' one in which the intervals *between* atom-points actually contribute all the volume of the body. On this view, there would be no *continuous* magnitudes at all.

But the atoms, say the Mutakallimun, *do* come into contact:[14] they are arranged in rows, like Epicurean minimal parts, to make up bodies. If then a finite number n of atoms make up a magnitude m, we are surely entitled to assign to each atom the size m/n? What could invalidate such an elementary calculation?

Exactly how unextended atoms could aggregate to form a compound body was left very obscure. A much-debated question concerned the minimum number of atoms sufficient to produce a body – 2, 4, 6, and 8 were all proposed.[15] It seems quite clear, however, that a finite number of extensionless atoms cannot give rise to anything at all: to accommodate the anti-infinitist bias of the Mutakallimun, we must attribute some minimal size to their indivisibles.

Time, according to this school, is composed of time-atoms, indivisible 'on account of their short duration'.[16] This thesis will follow, by a familiar Aristotelian argument, from 1A, the existence of (mathematical) indivisibles of matter and hence of space. The Mutakallimun thus postulate indivisible units of time and motion as well as of space. As in Epicureanism, basic atomic speed is one space-minimum q per time-minimum t: the

13   *Ibid.*, p. 120.

14   Levey, p. 41.

15   *Ibid.*, p. 44.

16   Maimonides, p. 121. For the rôle of atomic time in their metaphysics and theology, see the paper of MacDonald.

only possible velocities are q/t and zero. Differences in the observed velocities of moving bodies are explained not in terms of retrograde motions but in terms of *instants of rest*. The slower a body moves, the more instants it rests. This of course gives the Mutakallimun a weird *dynamics* – no physical cause is given for the atom's stops and starts – but, since they were occasionalists and thought of God as the immediate physical cause of all things, this would probably not disturb them. One moment, they will say, He wills the atom to move; the next, He wills it to stop – that is all that can be said.[17]

The theory of time-atoms cropped up spasmodically in Western thought in the Middle Ages. Encyclopaedists like Isodore of Seville and Bede mention such theories: a commonly accepted figure was 22,560 time-atoms per hour.[18] (This shows that the time-atoms were conceived not as extensionless instants but as minimal periods, intervals than which nothing shorter can exist.) In Bartholomew of England, we can see the breakdown of this figure. An hour, he says, divides into four 'points', each of which consists of ten 'moments'. A 'moment' is in turn made up of twelve 'ounces', and an ounce of forty-seven 'atoms'.[19] $4 \times 10 \times 12 \times 47 = 22{,}560$ time-atoms per hour. Where all these figures come from, and what physical significance – if any – they may have, I do not know.[20]

In the field of geometry says Maimonides,[21] the Mutak-allimun are led to make the following pair of assertions:

M1.   The diagonal of a square is equal to one of its sides.
M2.   The square is not a thing of real existence.

Can we make sense of these apparently paradoxical remarks? I think we can, with a little charity. No one, we presume, would believe M1 to be true of visible (and measurable) squares. What of minute, sub-microscopic ones? By elementary geometry, the same $1{:}\sqrt{2}$ ratio should hold of them too. This is where M2 comes in. No genuinely Euclidean figures exist in Nature, it claims. So, although the diagonals of all visible squares are longer than their sides, the same need not be true of an atomic 'square', which is not a Euclidean figure (and not, strictly speaking, a *square* at all!). For this 'figure' (it is not clear that

---

[17]   See MacDonald, pp. 333–334.

[18]   Boyer, p. 66.

[19]   Thorndike, Vol. 2, p. 419.

[20]   Given the strange figures involved – in particular, 47 – it is perhaps more probable that they had some occult – e.g. numerological – significance.

[21]   Maimonides, p. 122.

the term is appropriate) the 'diagonal' may equal the 'side' (if those terms make sense at all in such a context). This doctrine, as I have attempted to spell it out, effectively denies the applicability not only of geometry but also of our (visual) imagination to the atomic realm. Though it hides in obscurity from outright self-contradiction, it does cohere with other explicitly held doctrines of the Mutakallimun, e.g. with their dismissal of all propositions about irrational numbers and incommensurable lines.[22] No attempt is made of course to rebut Euclid's elegant demonstration of the irrationality of $\sqrt{2}$: the Mutakallimun will simply deny, as Epicurus did, its applicability to the real (i.e. physical) world. Much of geometry, they will say, is physically false – e.g. lines consisting of an odd number of minima cannot be bisected – and what remains can be at best only approximately true.

How can the Mutakallimun answer Avicenna's six arguments, A1–6? Their opponents – men like Avicenna himself and Maimonides – outline a set of possible responses. In answer to A1, says Avicenna:

> adherents to this doctrine do not claim that the median does not separate the *extrema* because they realise that this impossibility [the collapse of all bodies into one indivisible by superposition] would follow, but they do state that the two *extrema* are separated from each other.[23]

Unless the median separates the extremes, any number of indivisibles can collapse into the space of one. Hence, says the mathematical Atomist, the median must separate the extremes: *how* it does so without itself having distinct parts we are not informed.

As regards A2, the Atomists are given an occasionalist answer: the atoms, they claim:

> will not obey an order to move at that place, but they will obey only an order to move up to that place. If is not in the power of God to move these elements to meet in that place in order to divide them.[24]

Even Divine Omnipotence cannot divide the indivisible, cannot command the atoms to move in such a way as to do so. The physics of this seems very curious – does each atom know the position of the other? Is the atom *aware* of the illegitimacy of the order? A preferable alternative is to say – with some Jewish

22   *Ibid.*, p. 122.

23   Morewedge, Ed., p. 19.

24   *Ibid.*, p. 20.

thinkers – that, although God has the *power* to divide the indivisible, He *will* not do so because it would not be 'fitting'. The hypothetical situation envisaged by A2 will therefore never arise: God will never give an order the execution of which would involve division of a minimum.[25]

We have already adumbrated possible Atomist responses to A3 and A4. A3, the defender of indivisibles can assert, presupposes the continuity of motion, while A4 assumes the applicability of elementary geometry to the indivisibles. As for A5, the identification of a light-ray with a Euclidean straight line would be an obvious target for criticism. The greatest difficulty is raised by A6, to evade which one must go to great lengths and resort to truly desperate expedients. Advocates of indivisibles, says Avicenna,

> claim that upon the turning of the mill either all the elements are separated from one another, or the one on the periphery moves, while the one in the middle, which can stand still, remains stationary. The impossibility of this reasoning is obvious, and we shall not stretch the point in order to make it even more apparent.[26]

Maimonides too urges A6 against the Mutakallimun, and imputes to them a similar reply, that the revolving mill-wheel breaks up into its constituent atoms.[27] Of these, those nearest the centre have most instants of rest interrupting their motion, and therefore move more slowly. P12 (the weakness of the senses) explains why we do not perceive this disintegration, any more than we discern the cinematic nature of all local motion: our senses are insufficiently acute to discover the real discontinuities beneath the apparent continuities.[28] Maimonides is, quite rightly, unimpressed: if, he says, the mill-wheel is so hard that it does not break up under hammer-blows, why should it dissolve into its constituent atoms at a mere push?[29]

The balance of these arguments favours, quite clearly, the opponents of mathematical Atomism. Yet, if one retains the intuitive conviction that parts actually exist in the continuum prior to division, and hence that every magnitude has an ultimate constitution, quite independent of whether and how many times it is divided, one will not be satisfied with the

[25] This involves a slight extension of some ideas discussed in Wolfson, 2, pp. 194–195.

[26] Morewedge, Ed., p. 22.

[27] Maimonides, p. 122.

[28] *Ibid.*, p. 133.

[29] *Ibid.*, p. 122.

Aristotelian alternative. But, if neither of the great schools of antiquity has a satisfactory answer, where is one to turn?

One natural possibility is to think of the least parts of a continuum, the true indivisibles, as *unextended mathematical points*. These will be, beyond doubt or question, utterly indivisible, the truly *ultimate* constituents of every magnitude. Since a finite number of them could not give rise to anything, one will require an *actual infinity* of such points to stand a chance of making anything of them. This will give a theory of the continuum compatible both with Z1 (every magnitude is divisible) and hence with geometry, and also with Z4 (parts actually exist prior to division). Such a theory was developed in medieval times by Henry of Harclay (c. 1275-1317). A line, he claims, is made up of an actual infinity of points, immediately-next to one another. These points, he argues, touch 'whole to whole' without superposition: each point has a distinct position adjacent to that of its neighbour.[30] Only if points are actually superimposed, he argues, does addition fail to result in increase in size. (Clearly, Henry is in deep waters here, without a satisfactory response to Aristotle's dilemma.)

Henry gives two main arguments in favour of his conception of the continuum. The first starts out fairly clearly:

> . . . God actually perceives or knows the first inchoative point of a line and any other point capable of designation in the same line. Therefore, God either perceives that a line can fall between this inchoative point and any other point in the same line, or He does not. If not, then He perceives point immediate to point, which is that proposed.[31]

So far, so good. The Aristotelian must opt for the first horn of the dilemma: God does perceive that a line can be drawn between the 'first inchoative point' and any other. But then, says Henry, since points can be placed in this 'mean line', it follows that:

> these mean points would not be perceived by God, which is false. The consequence is evident: For, by hypothesis, a line falls between the first point and any other point of the same line perceived by God, and consequently there exists some mean point between this [first] point and any other point perceived by God; therefore, the mean point is not perceived by God.[32]

---

30   See Grant, Ed., 4, p. 321.

31   *Ibid.*, p. 319.

32   *Ibid.*, p. 319.

This totally opaque argument in fact conceals a transparent quantifier-shift fallacy of the 'Every girl loves a sailor' type. This fallacy was accurately diagnosed by William of Alnwick (c. 1270–1333), who defended Aristotle and orthodoxy against Henry. As Alnwick says, the following pair of propositions, P1 and P2, are quite distinct: P1 is true and P2 false:

P1.   Between the first point of the line and any other, there is a mean line.
P2.   There is some one mean line between the first point and every other point.

Since P1 and P2 are distinct, Alnwick insists, Harclay's *reductio* of P2 does not affect P1.

Henry's second argument raises deeper issues. God can, he claims, annihilate *only* the end-point of a line, leaving the remainder intact. This line will still have a terminus, but between this new end-point and the one corrupted by God there was no line. Therefore, on the original line, the two points were adjacent or immediate to one another.[33]

Alnwick replies that even God cannot annihilate a line point by point:

> For that which only exists positively through the existence of another thing is only annihilated by means of the annihilation of the existence of that [other] thing through which it exists. Hence, as a point possesses existence only through the existence of a line, it is only annihilated by the line's annihilation.[34]

But, replies Henry, consider the following possibility:

$$t_1 \qquad \underline{\qquad \overset{L}{\qquad} \overset{P'}{\qquad} \qquad \overset{R}{\qquad \qquad} }$$

$$^\rightarrow t_2 \qquad \underline{\qquad \qquad \overset{P'}{\qquad} \qquad \qquad \qquad}$$

$$^\rightarrow t_3 \qquad \underline{\qquad \overset{L}{\qquad} \qquad \overset{P'}{\qquad} \overset{R}{\qquad} }$$

If after this transfer of P' from the left-hand line L to the right-hand line R, one were to annihilate R, L would have been shorted by a single point.

Alnwick objects that this deceptively neat argument is in fact question-begging: it assumes that there *is* a successor to P' on L which would become the terminus of the new line. If no such successor-point exists, the process Henry envisages

[33]   *Ibid.*, pp. 319–320.
[34]   *Ibid.*, p. 323.

could not occur.[35] He bolsters this objection with a battery of anti-indivisibilist arguments familiar from Aristotle and Avicenna – I surely do not need to dwell on them.

The conception of the continuum as made up of points strung out next to one another like beads on a line may be a very natural one, but is open to overwhelming objections. Many of the great thinkers of the Middle Ages cut their logical teeth devising sophisticated counter-arguments against any such theory. Atomism, says Roger Bacon, has been one of the greatest of all hindrances to philosophy, yet 'this error is wholly eliminated by the power of geometry'.[36] There follows an argument effectively identical to Avicenna's A4, to the effect that, if magnitudes were composed of indivisibles,

> there will not be more atoms in the diagonal than in the side, and thus they have an aliquot part as a common measure, and the side has just as many parts as the diagonal, both of which conclusions are impossible.[37]

In fact, says Bacon, 'every body and every quantity' is infinitely divisible, 'as all philosophy proclaims':

> Aristotle proves in the sixth book of the *Physics* that there is no division of a quantity into indivisibles, nor is a quantity composed of indivisibles, and therefore there are as many parts in a grain of millet as in the diameter of the world.[38]

Bacon thus explicitly assents to this (Stoic) paradox, which he illustrates with the following geometrical argument:

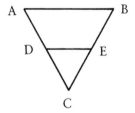

From any point of AB a line can be drawn to C. These lines do not meet before C, hence do not intersect at DE. Therefore, Bacon concludes, DE must contain as many parts as AB, which was the conclusion to be proved.[39]

35   *Ibid.*, p. 323.
36   R. Bacon, Vol. 1, p. 173.
37   *Ibid.*, Vol. 1, p. 173.
38   *Ibid.*, Vol. 2, p. 455.
39   *Ibid.*, Vol. 2, pp. 455–456.

Similar arguments can be found in Duns Scotus, Albert of Saxony, Bradwardine, and numerous other schoolmen.[40] In illustration, I shall cite two elegant examples from the pen of Scotus.[41]

1.   Consider, he says, the following figure. If the circumference of the outer circle is composed of points, take any two allegedly adjacent ones B and C and draw straight lines AB and AC to the common centre A of the two circles.

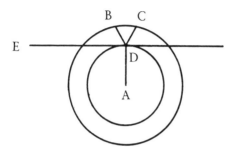

Now if AB and AC cut the smaller circle at different points, then 'there will be as many points in the smaller circle as in the larger one'. But ' it is impossible for two unequal quantities to be composed of parts equal in both magnitude and number.'

This last premiss, as thus formulated, sounds formidably plausible: a truly counterintuitive 'paradox of infinity' is needed to deny it.[42] Scotus considers the possibility that AB and AC cut the inner circle at different points as reduced to absurdity, and thus concludes that they must cut the smaller circle at a single point D.

We could now close the argument with an appeal to geometrical intuition: ADB and ADC cannot both be straight lines. Instead, Scotus draws the line ED, tangential to the inner circle. Then, if ADB and ADC were both straight lines, the angles on them should sum to 180°.

Therefore $A\hat{D}E + E\hat{D}B = 180°$
                      $A\hat{D}E + E\hat{D}C = 180°$
Therefore $E\hat{D}B = E\hat{D}C$

But then, either B = C, contrary to our initial assumption that these were adjacent but distinct, or 'the part' ($E\hat{D}B$) is equal to 'the whole' ($E\hat{D}C$), which is impossible.

[40]   See Boyer, p. 66.

[41]   From Grant, Ed., 4, pp. 317–318.

[42]   Adolf Grünbaum would claim – quite correctly – that only a Cantorean non-denumerable infinity could solve Scotus' problem.

2.   After discussing the incommensurability of the diagonal of a square with its side, Scotus adds the following point. Take a square, mark a diagonal on it, and take two allegedly adjacent points A and B on its side. Lines drawn perpendicular to this side from A and B meet the diagonal at C and D. Are C and D contiguous points?

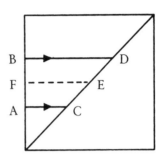

If C and D are contiguous, the diagonal, although longer than the side, consists of the same number of parts (i.e. points) of the same magnitude (zero), which is impossible. So there is a point E between C and D. But then one can draw EF parallel to BD and CD, cutting AB at F. (By definition of parallel lines, EF cannot intersect either BD or AC.) So there is a point F between A and B, in which case A and B are not contiguous points.

These difficulties arise from the notion that every point in a line has a *successor*, a point which is *contiguous* to it. In William of Ockham, we find a conception of the continuum as made up of an (actual) infinity of points which does not make this assumption, i.e. which retains Aristotle's denial that point can be contiguous to point. The theory of Henry of Harclay may be easier to picture (as beads on a string) but is logically incoherent. Ockham explicitly abandons the suspect model: the continuum, he says, is *not* made up of 'things juxtaposed'.[43] Point is never contiguous to point: between any two points there is always another. Although it is true that 'between the first point and any other lies another point' ((x) (Ǝ x) Fx), it is false that 'there is some point between the first point and any other' ((Ǝ x) (X) Fx) – Ockham clearly diagnoses the fallacy involved in this quantifier-shift.[44] Between any two points on a line, he insists, there is an

43   Birch, pp. 494–495.
44   Burns, pp. 510–511.

infinity of others, at any of which the line may be further divided: there are no indivisible minimal magnitudes.[45]

Ockham's conception of the continuum, though strikingly modern, is still riddled with paradoxes. How can a denumerable infinity of points give rise to anything by aggregation? If one imagines placing them at the points $1/2$, $3/4$, $7/8$, etc., of the way along a unit length, they will not fill any space: they would produce only a 'gappy' magnitude rather than a continuous one. Yet how can there be *more* than an infinite number of anything? Can any sense be given to such a concept? It is hardly surprising that Ockham failed to arrive at the solution of Cantor and Dedekind. Yet without their conception of 'super-infinite' sets, his theory of the continuum was unworkable. Indeed, every medieval theory – like every Greek theory – suffered from some major flaw or other.

This becomes clear when we look at Bradwardine's *Tractatus de Continuo*,[46] where he lists and discusses the 'five famous opinions concerning the composition of continua among ancient and modern philosophers'. These options are as follows:

O1.   Aristotle, Averroes, and 'most of the moderns' hold that the continuum is not composed of indivisibles, but of parts divisible without end.
O2.   Leucippus and Democritus hold that it is made up of indivisible bodies or atoms.
O3.   Pythagoras, Plato (?) and Walter Chatton (writing in 1323) say that the continuum divides into a finite number of indivisible points.
O4.   Henry of Harclay claims that an infinite number of points, 'immediately joined', make up an extended magnitude.
O5. Grosseteste (?) argued that the ultimate constitution was of an infinity of points 'mediate to one another'.[47]

We have already seen how the Aristotelian can attack the rival views O2–5. O2 falls foul of geometry, as well as mathematical intuition, and receives its *coup de grâce* from Avicenna. As for O3, if these 'points' are extensionless mathematical points, a finite number of them cannot aggregate to form a magnitude. This applies even to denumerable infinities of points, which will refute O4 and

---

[45]   Birch, p. 495; Burns, pp. 507–508.
[46]   Grant, Ed., 4, p. 314.
[47]   *Ibid.*, p. 314. Question marks indicate dubious or doubtful readings of the thinkers named.

O5. In any event, O4 founders on the notion of point succeeding point, and O5 cannot (yet) cope with continuity: according to Bradwardine, it can yield only a 'gappy' line and not a continuous one. (This is true for a denumerable infinity of points.)

Small wonder then that Bradwardine concludes with a (modified) defence of orthodoxy: continuous magnitudes, he says, although they *include* an infinite number of points, are not *made up out of* those points.[48] The continuum has no ultimate constituents: every continuous magnitude can be divided *ad infinitum* (Aristotle) and thus contains an infinity of points (Ockham), but cannot be said to be composed of those points – neither O4 nor O5 can give an intelligible account of how a continuum could be thus formed.

One of the most fascinating defences of the indivisibilist position can be found in the *Exigit* of Nicholas Autrecourt, where Nicholas engages in a running battle against Aristotelianism.[49] We start with the Stagirite's attempted demonstration of the infinite divisibility of space and time from the existence of differences in velocity. This argument, says Nicholas, is inconclusive:

> One movable thing is faster than another because one rests and the other does not. The fastest movable thing, indeed, is the one which moves without resting at all. The first heaven is like this, I believe. But other movable objects rest to a certain extent, according as they are more or less slow.[50]

So, while the fastest movable travels over three indivisibles abc in successive instants, a body with half its speed:

> will reach a in one instant and rest at a for another instant; similarly, it will reach b in one instant and rest at it for one instant; similarly with c. Thus in traversing the space it will require six instants, three of movement and three of rest.[51]

Nicholas' theory of motion is therefore that of the Mutakallimun – several scholars have noted the likeness and suggested an influence[52] – and not that of the Epicureans.

48 Boyer, p. 67.

49 Nicholas of Autrecourt, pp. 71–87. For commentaries, see Weinberg, 1, pp. 156–159; O'Donnell, pp. 109–111.

50 Nicholas of Autrecourt, p. 72.

51 *Ibid.*, p. 72.

52 See Weinberg, 1, p. 93.

In answer to Aristotle's dilemma, Nicholas claims that points *can* touch 'whole to whole' without superposition, and can by doing so make up an extended magnitude. A point, he insists,

> has its own position and its own mode of being. Therefore, if points are put together and retain their own positions, it is senseless to say that they will not constitute something larger.[53]

If a continuum is made out of points, the peripatetic objects, the quantified is made out of the non-quantified. If, replies, Nicholas, one means by 'the quantified', 'the composite itself or its extension', then a point is not quantified, but if one means 'that which has its own position and a circumscribed being', a point *is* quantified.[54] Aristotle simply assumed, says Nicholas, that addition of an indivisible to a magnitude would make it no larger: this is precisely what advocates of indivisibles must deny.

But if each indivisible adds some increment ϵ to the size of a magnitude, the indivisibles can be said to have that extension ϵ. If so, there can only be a finite number of such minima in any finite continuum. At this stage of the argument, this seems to be Nicholas' position – that of the classical Atomists. Later, as we shall see, he shifts his ground radically.

At this point, Avicenna's A6 (the mill-wheel) is raised. If one marks four points on a small circle around the centre, and 100 such points on the periphery, how far does one of the inner points move when one of the outer ones moves one indivisible space-unit? Clearly, it cannot also move one unit. Equally clearly, 'it cannot be said that it rests, since then there would be discontinuity in the parts, and the parts will be broken up'. Nicholas answers that:

> if in the small circle there are only four points, in the large circumference there will be only four points from which straight lines would end at those four points of the small circle. Lines drawn from other points would meet before they reached the centre.[55]

This is also a tacit answer to Scotus' point about the two concentric circles. Only four straight lines can be drawn, says Nicholas, from particular points on the periphery, through

---

53   Nicholas of Autrecourt, p. 73.

54   *Ibid.*, p. 74.

55   *Ibid.*, p. 74.

the small inner circle, to the centre. When one of these points moves, 'then the point of the small circle moves'; when one of the other peripheral points moves, 'the small circle will not move, but be at rest'.[56]

This, it must be confessed, is pretty feeble. Nicholas seems to visualise the points on the circumference moving *successively*, but in a rigid figure all the parts will move *simultaneously*, in which case Nicholas' expedient collapses.

If, argues the Aristotelian, lines were composed of indivisibles, it would follow that not every line could be perfectly bisected, 'for example a line of an odd number of points, say three or five'.[57] This, replies Nicholas, is not discomforting. The two 'halves' may appear equal, and yet not be exactly so; because 'a point is so small', the difference may be undetectable.

So far Nicholas has been arguing for the classical Atomist position, for a conception of the continuum as made up of a *finite* number of minimal magnitudes. (We have even mentioned lines of 3, 4 and 5 minima.) Each minimum adds some minute increment ϵ, however small, to the size of the whole. This impression is bolstered by the next argument, explicitly directly against the notion of infinite divisibility:

> Whatever has an infinite multitude of quantified parts is infinite in extension, because, since each part is a certain extension, and [the parts] are infinite [in number] it is unintelligible that the extension is not infinite. But this is [how they describe] a continuum.[58]

We can make sense of this argument if we think of an infinity of aliquot or A-parts, generated by a process of A-division.[59] A continuum contains either a finite or an infinite number of distinct (non-overlapping) A-parts: if this number is infinite, the magnitude must be infinite in extent. Hence, says Nicholas, as soon as we have fully grasped the concepts involved (finite, infinite, continuum, contiguum) it becomes clear that 'the concept of this conclusion (that no continuum is divisible into endlessly divisible parts) can be abstracted clearly'.[60]

But, the Aristotelian protests, the continuum is only *potentially* divisible *ad infinitum*; it is not to be thought of as

56   *Ibid.*, p. 74.

57   *Ibid.*, pp. 74–75.

58   *Ibid.*, p. 75.

59   See Chapter 1, Part 1.

60   Nicholas of Autrecourt, p. 76.

made up of an actual infinity of parts. This evasion cuts no ice with Nicholas: potentially infinite divisibility, he insists, presupposes actually infinite dividedness. The parts 'are not potential in such a way as not to be actual': parts of the continuum 'are really distinct if an act of sense power can grasp one and not the other'.[61] If, says Nicholas, parts had no actual existence, one extended magnitude could not be greater than another (it is so in virtue of possessing more A-parts). Division does not bring anything new into existence: our concept of division involves the pre-existence of what is divided:

> as [the parts] are said to have actual being after division, so [they are said to have had it] before'.[62]

Infinite divisibility, therefore, involves actually infinite dividedness, and hence infinite extension. Thus no (finite) continuum is infinitely divisible: every body consists of a finite number of minima or least possible magnitudes. It is *not true*, says Nicholas, 'that for every magnitude a smaller one can be found'.[63]

If, replies the Aristotelian, the continuum were composed of indivisibles, then the diagonal of a square and its side would each be composed of an integral number of discrete minima, and would thus be commensurable with one another, contrary to Book Ten of Euclid.

'A certain reverend master,'[64] Nicholas retorts, claims to have better arguments against incommensurables than Euclid has for them. But, he continues, 'I answer differently':

> The opponents indeed accept as a principle that, if the diagonal were composed of points, it would be composed of a finite number of points . . . Otherwise this argument would be invalid. But they are wrong in accepting this as a principle. For it is not a conclusion from the Master's principles . . . For these stand at the same time: a continuum is composed of points; and a continuum . . . whether sensed or imagined, is composed of infinite points.'[65]

---

[61]   *Ibid.*, p. 78.

[62]   *Ibid.*, p. 78.

[63]   *Ibid.*, p. 79.

[64]   The identity of this 'reverend master' remains dubious.

[65]   Nicholas of Autrecourt, p. 80.

But what of Nicholas' own arguments against infinite divisibility? He has already claimed that each 'point' makes a line larger, i.e. adds some increment ϵ to the length of a line; does it not therefore follow, the Aristotelian retorts, that if these 'points' were infinite in number 'there would be an infinite extension there'?[66] Nicholas now attempts to evade this plain conclusion, but his expedients become ever-more sophistical.

Two conclusions C1 and C2 follow, says Nicholas, from his theory of the continuum:

C1.   The continuum 'is not composed of parts which can always be further divided', but has an ultimate constitution of indivisibles or points.

C2.   A continuum 'is not composed of a finite number of points'. In this way, says Nicholas, 'a departure would be made from the opinion of all those who have posited that a continuum is composed of indivisibles'.[67]

From these conclusions, he continues, it follows that:

C3.   For every magnitude pointed out to sense or imagination, there is a smaller one.

C4.   There is in a thing some magnitude such that a smaller one cannot be found.

These two propositions seem, *prima facie* at least, to be flatly contradictory. In fact, they are not so: C3 claims only that for every sensible (or imaginable) magnitude there is something smaller; C4 denies that this is true of every magnitude *simpliciter*. A single indivisible, Nicholas claims, 'does not come to sense or imagination as a finished being'[68] and is smaller than anything we can either perceive or conceive. One such point, says Nicholas, 'does not constitute the space to be traversed, but several together do'.[69]

We might almost claim that Nicholas was one of the founders of the doctrine of the *infinitesimal*, that curious creature greater than nothing yet less than anything, an infinity of which make up a magnitude. However great its *heuristic* value in the history of mathematics this doctrine is quite incoherent, and the infinitesimal – as was already apparent in Zeno's day – is an impossible entity.

[66]   *Ibid.*, p. 80.

[67]   *Ibid.*, p. 82.

[68]   *Ibid.*, p. 82.

[69]   *Ibid.*, p. 82.

Nicholas' theory of the continuum follows from his acceptance and forceful advocation of Z4, the actual existence of parts prior to division. From this it follows that every continuum possesses an ultimate constitution of true indivisibles of some kind. For a while it seemed as if Nicholas were going to opt for the classical Atomist position that magnitudes are made up of a finite number of minima (cf. Epicurus' 'minimal parts'). This theory is self-consistent, although strongly counterintuitive and implicitly anti-geometric. The geometrical arguments eventually force Nicholas to abandon this position and opt for an actual infinity of 'points' which, though of >0 extension (a point adds something to a body), are smaller than anything we can imagine or conceive, and are indeed, individually, less than fully real. This theory is an incoherent attempt to have the best of both worlds – indivisibles *and* geometry, Z4 *and* Z1. To bring out its essential incoherence, we need only rigorously apply Zeno's dilemma: 'Do these "points" have any (>0) magnitude?' If so, an infinity of them will give an infinite magnitude; if not, a (denumerable) infinity of them will yield nothing at all. The doctrine of infinitesimals is an ultimately untenable attempt to find a *via media*, a third way between the horns of Zeno's dilemma.

If we move on a century to another celebrated Nicholas, Cusanus,[70] we find another variation of the indivisibilist theme, this time with strong Pythagorean overtones. The essential activity of Mind, says Cusanus, is *counting*. To think is to measure: all cogitation involves the determination of numerical ratios. This presupposes the actual existence of measures or quanta in things being cognised: without such a minimum in Nature, all bodies would consist of an actual infinity of parts and would therefore be *unknowable*.[71] (God and infinity cannot be grasped by finite minds.) But ordinary finite magnitudes can be cognised, hence must consist of countable numbers of discrete units.[72] Under mental consideration, says Cusanus,

> that which is continuous becomes divided into the ever divisible, and the multitude of parts progresses to infinity. But by actual division we arrive at an actually indivisible

---

[70] Nicholas of Cusa, p. 77.

[71] The demand for indivisibles, units, or monads as a *precondition* of *intelligibility* is common in Renaissance philosophy. This particular argument is repeated by Bruno, and assumes great importance in the metaphysics of Leibniz.

[72] See Dijksterhuis, pp. 226–227; Michel, p. 145.

part which I call an atom. For an atom is quantity, which on account of its smallness is actually indivisible'.[73]

Although we may think every magnitude divisible, this is in fact not so: in reality, says Cusanus, there are atoms (magnitudes than which nothing can be smaller) and instants of time: motion can be regarded as compounded of serially ordered states of rest.[74] The indivisibilist position was also adopted by Giordano Bruno, perhaps the true disciple of Cusanus. The notion of infinite divisibility, Bruno argues, is the source of all the errors and obscurities of the mathematicians;[75] in opposition to them he develops a very crude theory of mathematical Atomism – Bréhier is not far wrong when he says that Bruno's views seem 'to date from a period prior to Plato, before the discovery of irrationals'.[76] It is far from clear how Bruno would set about answering the multitude of obvious Aristotelian objections – mathematical and physical alike – to his position.[77]

Mathematical Atomism, then, either in its classical form (a finite number of atoms or minima) or its revamped version (an infinity of infinitesimals) was very much a live issue throughout the Middle Ages and Renaissance. If the seventeenth century programme of the 'geometrisation of Nature' is to go ahead, the classical version of mathematical Atomism must be discarded as a backward and antigeometric stance; it was the incoherent but *useful* doctrine of infinitesimals that was to assume great value and importance in the Scientific Revolution.

Part 2: Atoms and Natural Minima

A. The Fate of the Atomic Theory

During the Dark Ages of Western Christendom the Atomic Theory – like so much else – was almost totally forgotten. The only references to Atomism that survive from this period are the garbled and third or fourth hand accounts of the Encyclopaedists, men such as Isodore of Seville (560–636),

---

73   Quoted from Stones, pp. 446–447.

74   For Cusanus' rôle in the development of the notion of an infinitesimal, see Boyer, pp. 91–92.

75   Michel, pp. 145–147.

76   Bréhier, p. 252.

77   He could presumably deny the applicability of geometrical theorems to the concrete physical world, but this would still leave him with physical problems such as those raised by Avicenna's A2, A5 and A6.

Bede (672–735) and Hrabanus Maurus (776–856).[78] There is no trace of a continuous and developing tradition of Atomist thought.

In the Islamic world, the situation was very different. In the attempt to reconcile the doctrines of the Qur'ān with the teachings of Greek philosophy, there sprang up a variety of philosophical theology called *kalām* (literally, 'conversation').[79] Among the schools of the *kalām* a number favoured Atomic theories of matter. We have already encountered the Mutakallimun, the disciples of Ash'ari (876–935) and their Atomic Theory, which involves indivisible minima of space, time and motion, and incorporates Atomism into an occasionalist metaphysic in which Allah is the sole efficient cause, re-creating all things in every instant of time.[80] This Atomic theory is, as Bréhier says, 'all for the glory of Allah', without a trace of the rationalism of Epicurus.[81]

The Mu'tazilite school also accepted the composition of bodies out of indivisible simples. Thus Peters:

> According to the Mu'tazilite view, the first entity perceptible to the senses is the body, which is, however, merely a composite of atoms. Atom was equated with substance. Though the body obviously had qualities, the Mu'tazilites were divided on the question whether or not the atoms themselves had properties. The most common view was that the atoms had no qualities of their own save motion and rest.[82]

To account for the dramatic difference between Greek Atomism (naturalistic and implicitly atheistic) and its Arabic offspring (which espouses occasionalism and continuous creation), some scholars have postulated an Indian influence on the latter – there were similar theories in contemporary Buddhist thought.[83] But I must not stray too far afield . . .

Atomic theories can also be found in much of the Arabic alchemical literature,[84] and in particular in the writings of the

---

78 See Stones, p. 446; Crombie, 1, p. 236.

79 For a detailed account of the origins and nature of the *kalām* see the fascinating study by Peters. See also Levey, p. 40.

80 See Maimonides, p. 120 ff., for the doctrines of the Mutakallimun. On the theory of continuous re-creation, see MacDonald, p. 331 ff.

81 Bréhier, p. 91.

82 Peters, p. 142.

83 MacDonald, pp. 341–343; Peters, p. 143.

84 Atomism, says Levey (p. 40), is 'applied through much of the alchemical literature'.

great physician and chemist al-Rāzī (865–925). Rāzī's Atomic theory owes a considerable debt to Plato's *Timaeus*, as well as to his acknowledged influences, Democritus and the corpuscularian school of Harran.[85] He believed in five eternal principles – Creator, Soul, Matter, Time and Space. Space and Time were considered absolute (of independent existence); Matter was made up of indivisible atoms and void space; the properties of bodies were explained in terms of the amounts of empty space between atoms, and so on.[86]

In Jewish philosophy too – heavily influenced by the *kalām* – the merits of Atomism were vigorously debated. Saadia lists various types of Atomic theory among possible accounts of the Creation;[87] Isaac Israeli denies that a body can be composed of indivisibles. The so-called 'atoms', Isaac argues, are either extended or unextended. If extended, they are bodies, and hence divisible; if unextended, they are true mathematical points, but of these nothing can be made. This distinction, remarks Wolfson,

> actually corresponds to the difference of opinion between the two Mu'tazilite schools of Basra and Baghdad as to whether the atoms are magnitudes or not.[88]

Of the Arab Atomists, then, some thought of atoms as minute but extended bodies; others likened them to unextended mathematical points – an important parting of the ways.

In general, as Wolfson shows, the Jewish philosophers of the period rejected indivisibles, vacuum, and qualityless atoms in favour of a more orthodox (i.e. Aristotelian) position. Nothing extended, say Saadia and Judah Hadassi, can be indivisible; and nothing devoid of form and quality can produce, by aggregation with others of its kind, the forms and qualities that sense reveals.[89]

When learning began to revive in Christendom in the twelfth century, corpuscular theories soon re-emerged. One of the outstanding figures of this era was William of Conches[90] (1080–1145) who, according to Bréhier, 'tried to introduce the corpuscular physics of Constantine the

---

[85]   Peters, p. 171; Wolfson, 2, pp. 156–157.

[86]   Heym, p. 188; Levey, p. 44.

[87]   Wolfson, 2, p. 124 ff.

[88]   *Ibid.*, p. 164.

[89]   *Ibid.*, p. 167, p. 171.

[90]   For William of Conches, see Thorndike, Vol. 2, p. 50 ff.

African',[91] and – perhaps still more importantly – argued for a naturalistic physics free from theology. Although others – e.g. Adelard of Bath and Bartholomew of England – were also to accept corpuscular theories of some form, William remains the most significant figure in this tradition. His matter-theory is a blend of Platonism with Epicureanism, with influences from Greek and Arab Atomism, alchemy, and the medical tradition.[92] In particular, he accepted quality-bearing corpuscles:[93] William was no Democritean. He also attempted to purge Epicurus' Atomic Theory from its associations with atheism:

> For in their assertion that the world is made up of atoms the Epicureans spoke the truth, but in their assertion that these atoms had no beginning and that they flew about separately in a great void, and then coalesced into four great bodies they were telling fairy tales.[94]

In the thirteenth century, the flood of translations of and commentaries on the works of Aristotle overwhelmed this early corpuscular theory, and the authority of Aristotle superseded that of Plato. The attempt of William and others to resurrect the Atomic theory was all but forgotten. Vincent of Beavais (d. c. 1268) shows some familiarity with and sympathy for Atomism,[95] but, by and large, Aristotelianism was totally dominant.

With one notable exception: in the fourteenth century we discover Nicholas of Autrecourt – admittedly a very radical thinker – rejecting the authority of Aristotle and advocating a full-blown Atomic Theory. As Grant says, Nicholas, 'either stands alone in the Middle Ages or belongs to a very small group whose members are not yet known'.[96] Nicholas states quite frankly that he believes in the void and indivisibles, and in many of the other tenets of classical Atomism – e.g. the reduction of all forms of change to local motion.[97] The

---

91 Bréhier, p. 54. Constantine of Carthage (1010–1087) was a physician heavily influenced by Greek and Arabic corpuscular theories.

92 See Hooykaas, p. 69; Dijksterhuis, p. 122.

93 See Meyerson, p. 327; Thorndike, Vol. 2, p. 420.

94 Quoted from Gassendi, p. 397 (*Syntagma*, Physics, Section 1, Book 3, Chapter 8). Like Gassendi many years later, William tried to free the Atomic Theory of matter from its association with atheism and make it acceptable within a Christian world-view.

95 Stones, p. 446.

96 Grant, Ed., 4, p. 352.

97 See Weinberg, 1, p. 154; Bréhier, p. 197.

arguments of Aristotle against the Atomic Theory are, he insists, not conclusive, and the whole fabric of Aristotelian thought unproven and unfounded.

B. *The Aristotelian Theory of Natural Minima*

While the Aristotelians were staunch opponents of the Atomism of Democritus and Epicurus, they nevertheless developed a particulate theory of their own, based largely on Aristotle's *Physics*, 1, 4, where the Stagirite raises four objections to the matter-theory of Anaxagoras. Let us run through numbers two and three:

> Further (2) if the parts of a whole may be of any size in the direction either of greatness or of smallness (by 'parts' I mean components into which a whole can be divided and which are actually present in it), it is necessary that the whole thing itself may be of any size. Clearly, therefore, since it is impossible for an animal or a plant to be infinitely big or small, neither can its parts be such . . . But flesh, bone, and the like are the parts of animals . . . Hence it is obvious that neither flesh, bone, nor any such thing can be of indefinite size in the direction either of the greater or of the less.[98]

The premiss on which this argument rests is that living beings cannot be indefinitely large or small, i.e. that the size z of any organism is bounded by limits of x and y such that $0 < x < z < y < \infty$. Let us assume this to be true. If follows (a) that there will be a minimum organ-size, a smallest heart or eye, and (b) that there will be a least amount of flesh or bone or whatever *in some animal*. (The smallest animal – e.g. the flea – will have the minutest organs and the smallest amounts of tissue.) What does *not* follow is (c) that bone or flesh are made up of such discrete minima, nor (d) that no smaller amount of flesh or bone could exist *at all*. Anaxagoras might admit that no smaller quantity of flesh (than that of the flea) exists in any animal, but claim that still smaller quanta exist in plants, minerals, etc. Aristotle's argument, therefore, gives a plausible case for minimally small organs, or for minimal amounts of flesh, bone, etc., in some organism (no animal exists which has less flesh than a flea) but this tells us nothing about the divisibility of flesh as such.

According to Anaxagoras' matter-theory, Aristotle continues, every stuff is present in, and can therefore be extracted

[98] *Physics*, 1, 4, 187b 14–22.

from, every other, e.g. water from flesh or flesh from water. But (3)

> since every finite body is exhausted by the repeated abstraction of a finite body, it seems obviously to follow that everything *cannot* subsist in everything else. For let flesh be extracted from water and again more flesh be produced from the remainder by repeating the process of separation: then, even though the quantity separated out will continually decrease, still it will not fall below a certain magnitude. If, therefore, the process comes to an end, everything will not be in everything else (for there will be no flesh in the remaining water); if on the other hand it does not, and further extraction is always possible, there will be an infinite multitude of finite equal particles in a finite quantity – which is impossible. Another proof may be added: since every body must diminish in size when something is taken from it, and flesh is quantitatively definite in respect both of greatness and smallness, it is clear that from the minimum quantity of flesh no body can be separated out; for the flesh left would be less than the minimum of flesh.[99]

These arguments explicitly presuppose the existence of a *natural minimum*, a smallest possible amount of flesh. If, as Aristotle asserts, flesh-particles 'will not fall below a certain magnitude', then the infinite process of extraction envisaged by Anaxagoras would indeed yield 'an infinite number of finite equal particles in a finite quantity'. The whole argument rests squarely on this postulation of *natural minima* and thus on the denial of Anaxagoras' dictum: 'There is no smallest, but always a smaller'.[100]

To make sense of this chapter, therefore, we need to attribute to Aristotle a theory of natural minima, at least for living things. Other passages in the Aristotelian corpus suggest similar views. In *Physics*, 6, 10, he links the natural sizes of things to their specific natures: '. . . the limit of increase is to be found in the complete magnitude proper to the peculiar nature of the thing that is increasing . . .'.[101] Each type of thing, therefore, has a proper natural size appropriate to it. In *Metaphysics*, 1, 8, this principle is applied to the elements FAWE:

99 *Ibid.*, 1, 4, 187b 25-188 a2.

100 See Vlastos, 2, p. 459.

101 *Physics*, 6, 10, 241a 32-b3.

... the property of being most elementary of all would seem to belong to the first thing from which they are produced by combination, and *this* property would belong to the most fine-grained and subtle of bodies. For this reason those who make fire the principle would be most in agreement with this argument . . . At least none of those who named one element claimed that earth was the element, evidently because of the coarseness of its grain.[102]

This suggests that each of the elements FAWE has a natural *particle-size*, F the smallest and E the greatest, with W and A intermediate.

Before we go on to attribute a fully-developed doctrine of natural minima to Aristotle, the following reservations are in order:

1.   Aristotle's theory of the continuum explicitly prohibits mathematical minima. If then he is to develop a theory of natural minima, he requires a sharp distinction between mathematical and physical divisibility. Such a distinction can be found in the Greek commentators, but not in Aristotle himself.[103]

2.   The examples in *Physics*, 1, 4, are all drawn from the biological realm: Aristotle nowhere speaks explicitly of minima of non-living matter.[104]

3.   Natural minima occur only in the course of the anti-Anaxagoran thought-experiment. Even there, they are conceived only as *potential* parts, and are not attributed any chemical rôle as discrete units.

4.   They play no rôle at all, so far as I am aware, in Aristotle's most detailed discussion of chemical reaction-mechanisms in *De Generatione et Corruptione*.

The theory of natural minima is, as Van Melsen says, present only in an 'embryonic state' in Aristotle;[105] it is one of many isolated remarks or ideas of 'the philosopher' on which his commentators built towering edifices. In the works of Alexander, Philoponus and Simplicius the doctrine receives more definitive formulation. *Qua* mathematical extension, they say, quantity is (potentially) infinitely divisible; physically, it is not. Each type of substance has its *natural minimum*, beyond which it cannot be further divided:

102   *Metaphysics*, 1, 8, 988b 35–989a 7.
103   See Van Melsen, p. 47.
104   *Ibid.*, p. 43.
105   *Ibid.*, p. 44.

Simplicius speaks of minima of gold and lead (inorganic stuffs) as well as of flesh and bone.[106]

If anyone is tempted to identify this peripatetic theory of natural minima with Democritean or Epicurean Atomism, let me list a few of the more obvious and striking differences:

1.   Natural minima are mathematically divisible *ad infinitum* and perhaps also physically divisible *into other stuffs*: a natural minimum of X is only physically indivisible *into Xs* – for the elementary reason that no smaller unit of X can exist.

2.   The minima still have their own distinctive forms and qualities: minima of FAWE, for example, possess HCDM.

3.   Minima are, in virtue of those forms and qualities, *active* and *passive*, and hence far from immutable. In compound-formation, for example, minima of FAWE interact to produce a homogeneous *mixtum*.

4.   Compounds still have their own characteristic forms: a compound is not a heterogeneous aggregate of discrete particles of FAWE.

5.   The elements FAWE have no actual existence in the compound: their presence is only a 'potential' one.

6.   Minima have no actual, independent, physical existence and identity: they are purely potential parts. One could not trace the life-history of a natural minimum, as one could of an Epicurean atom.

The fundamental matter-theory is still qualitative and Aristotelian, not mechanical and Atomistic. The two theories are utterly distinct and quite incompatible. And yet, by the early seventeenth century, the chemist Daniel Sennert can make the apparently ludicrous claim that the two theories are one and the same: that the theory of natural minima was what Democritus really *meant*.[107] Now Sennert was no fool – even Boyle refers to him as 'the learned Sennertus'. What, then, has happened to the minima theory during the interim period to bring it closer to Atomism?

The doctrine of natural minima entered Western European thought primarily through the writings of the great Aristotelian commentator, Averroes (ibn Rushd, 1126–1198) and his followers. Their basic argument for such minima is an extension and generalisation of that of Aristotle. There are, says the Stagirite, upper and lower bounds set by

---

[106]   *Ibid.*, p. 47.

[107]   M. Boas, 4, p. 428. Albertus Magnus had made a similar identification of atoms and natural minima in the thirteenth century. (See Wallace, 5, p. 13.)

Nature for the size of organisms. The existence of such a lower bound, he argues, entails the existence of minimal quanta of flesh, bone, etc. Following the Greek commentators, Averroes extends this argument to inorganic Nature, and in particular to FAWE:

> When we remove a part of fire and repeat the action again and again we finally reach a quantity which is such that by a further division the fire would perish, because there is a certain minimal quantity of fire.[108]

This raises difficult questions. Can the natural minimum of fire be divided at all? If so, it would have to break down into some other substance. But then one could bring about transmutations – profound chemical changes – by mere mechanical means. It is surely essential to Aristotelian thought that any part of F is F, and that the qualities of a substance depend on its *form* and not on the sizes or shapes of its constituent particles. To preserve his qualitative matter-theory, Averroes must deny the (physical) divisibility of minima:

> A line as a line can be divided infinitely. But such a division is impossible if the line is taken as made of earth.[109]

How is this distinction to work? For Averroes, as for Aristotle, a line is potentially infinitely divisible, but not actually divisible at every point:[110] It can be divided at *any* point but not at *all* its points (simultaneously), just as a scholar can acquire any science but not all (together). We can now apply this analysis to physical bodies composed of minima.

Let the distances AB, MN and YZ represent minimal quanta of this particular substance. This body cannot then be divided between A and B, or between Y and Z, without splitting a minimum. But at any point between B and Y the body is divisible, e.g. it can be cut between M and N. It is not, therefore, made up of distinct and abiding atoms. Only if the body is first cut at M does the magnitude MN become an indivisible. Thus MN, though a natural minimum of this

---

108   Quoted from Van Melsen, p. 59.

109   *Ibid.*, p. 59.

110   See Chapter 1, Part 1, p. 17.

particular substance, does not function like a Democritean atom: the first cut can be made anywhere between B and Y, not just at those points that mark the boundaries of distinct minima.

A body is therefore not to be conceived as necessarily made up of an integral number of actual minima. A substance containing an odd number of (potential) minima could be perfectly bisected. This does not imply the actuality of any half-minimum; the minima are still potential rather than actual.[111]

Some degree of actuality does, however, seem to be attributed to the minima, since Averroes regards them as playing a rôle in chemical reactions, thus:

> The first thing which becomes or perishes in the generation or corruption of a substance is the smallest particle of this substance, for the minimum of all that becomes is of a limited quantity.[112]

This, as Van Melsen remarks, seems to involve a conception of the minima as separate *units* with some independent reality of their own. But this is in tension with the view outlined above. If a body of 4½ minima of a stuff S were to be corrupted one unit at a time, what becomes of the half-minimum that remains? Can it, indeed, exist at all? (It clearly cannot exist as S.) We do seem to have a problem here. If minima are the actual units of chemical reactions, then presumably every body must consist of an integral number of such units. But this contradicts the orthodox Aristotelian premiss that a body is physically divisible at any point.

Having recorded our debt to Averroes, it is time to move on to the Latin scholastics. Aquinas and Albertus Magnus both accept versions of the natural minima theory, but neither contributes greatly to its development.[113] Scotus and Ockham accepted minima only for living things, and denied the existence of natural minima of either FAWE or inorganic compounds. This position is also accepted by Galileo in his youthful *Notebooks*: 'elements and homogeneous compounds,' he writes, 'of themselves have no termini of largeness or smallness'[114] The form of W or F, he continues, permeates all its parts, however small – there is no reason why

---

111   See Levey, p. 52.

112   Quoted from Van Melsen, p. 60.

113   See Wallace, 5, p. 13.

114   Wallace, Ed., 3, p. 218.

it should be corrupted if a very small fragment is cut off from the whole.

But such sceptics were few: most scholastic writers accepted some version of the natural minima theory, often with idiosyncratic modifications. Variants of the theory can be found not only in Averroists such as Siger of Brabant and John of Jandun but in the fourteenth century Paris school of Buridan, Oresme, and Albert of Saxony. The distinction between mathematical and physical divisibility is clearly formulated by Oresme:

> In dividing a log or a stone or another material or destructible object, one can reach a part so small that further division would destroy its substance. But any continuum or magnitude is continually divisible conceptually in the human mind.[115]

What happens, then, to sub-minimal particulars? According to Siger of Brabant, if X is the minimum size for F-particles, fragments of F smaller than X cannot exist independently of other F-stuff but are converted to some other substance.[116] This raises the question of the influence of the environment on the minima. On the view of Albert of Saxony, the size of the natural minimum for a given substance is *environment-dependent*. We could, for example, assemble ten earth-minima, transfer the resulting piece of earth to different surroundings, and then divide it into twelve minima. This shows how far he was from a conception of minima as atom-like, as abiding entities with immutable shapes and sizes. In his own words,

> It is not correct to speak in an absolute sense, and without any indication of environment and conditions, about a minimum below which a substance cannot exist.[117]

The 'actualisation' of minima still has far to go. By the sixteenth century, however, in the Averroistic Paduan Aristotelianism of Nifo and Zabarella, the theory reaches its fruition. Nifo makes extensive use of natural minima in physical explanation. The erosion of rock, he claims, is a discontinuous process, occurring one minimum unit at a time. In qualitative change, one minimum after another undergoes

[115]   Oresme, pp. 45–47.

[116]   See Dijksterhuis, p. 206. Cf. of course Aristotle on water and wine in *De Gen et Corr.*, 1, 10, 328a 28–29.

[117]   Quoted from Van Melsen, p. 63.

alteration, which is therefore the successive change of discrete units.[118] In his commentary on *De Generatione*, Nifo asserts that 'when elements react upon each other they are divided into minima'.[119]

Toletus (1532–1596) develops this theory of the rôle of natural minima in chemical reactions further:

> Concerning the nature of chemical composition, the opinions of authors vary, but they all agree in this: the reagent substances are divided into natural minima. In this division the separated minima of one substance come alongside the minima of the other and act upon each other till a third substance, having the substantial form of the compound, is generated.[120]

Note that the minima here are actual, chemically active particles, capable of separate and independent existence. However, they still have their own characteristic forms and qualities – this remains a qualitative matter-theory – and interact to produce the substantial form of the compound, the *'forma mixti'*. The coming-to-be of a compound is not mere aggregation but aggregation plus interaction.

One of the most important of the natural philosophers of the sixteenth century, Julius Scaliger (1484–1558), applied the theory of natural minima to a wide range of chemical phenomena. The size of the minima, he says, determines the fineness or coarseness of the stuffs they constitute. Fire has the smallest minima, and is therefore very mobile and highly penetrating in nature; Earth has the largest, and is therefore sluggish and inert. Compounds too have minima: a compound is not merely an aggregate of particles of FAWE, but is made up of *'prima mixta'* (molecules) of its own kind. Some of the stuffs we call compounds are, however, mere mechanical aggregates, raising an epistemological difficulty of considerable importance to seventeenth century chemistry. By what criterion do we recognise true *mixis*? When do we know that a new *form* has been produced? Perhaps, the sceptic will suggest, there are no true compounds, in the Aristotelian sense.

According to Aristotle and Aquinas, (the forms of) FAWE do not *actually* persist in compound bodies: the elements are only *potentially* present in the *mixtum*. This led to a vigorous

---

118 See Levey, p. 55.

119 Quoted from Van Melsen, p. 68.

120 *Ibid.*, p. 69.

controversy about the status of the elements in the compound – a subject which we shall discuss more fully elsewhere. Suffice it to say here that both Avicenna and Averroes believed that the forms of FAWE were actually present in some manner in the *mixtum*. According to Avicenna, the forms of the elements are retained but their characteristic qualities lost; while on Averroes' view the forms themselves persist in an attenuated state. Avicenna's theory was adopted by Albertus Magnus in the thirteenth century[121] and the Paduan Averroists in the sixteenth,[122] and squares well with the minima theory. Minima of F, we can say, remain F, even in the compound, although they temporarily assume the characteristic qualities of the *mixtum*. (Aristotelians continued to believe in the perfect homogeneity of true compounds.) When the *mixtum* is dissolved, the F-minima will resume their status as such: their own forms, 'subordinate' for some period to that of the compound, will re-assert themselves.

This gives us a conception of natural minima as *abiding* (eternal?) *actual* particles of FAWE, with their own inalienable natures, aggregating (and interacting) to form compounds. In the *mixtum*, each minimum assumes the qualities of the homogeneous whole, yet *retains* its own specific form in virtue of which it is a minimum of F, A, W or E.[123] This theory is clearly much closer to classical Atomism than anything found in Aristotle – or even in Averroes. Sennert's identification of atoms with natural minima, although mistaken, was not a total and inexplicable blunder.[124]

C. The Rediscovery of Classical Atomism

For many years, Greek Atomism was known only through the critical accounts of men such as Aristotle and Galen: it is ironic that their criticism helped to preserve at least some degree of familiarity with corpuscular theories. In the fifteenth century, however, the situation was changed quite radically by the work of the great humanist scholars. The rediscovery of Diogenes Laertius made available to Western

---

[121]   Hooykaas, p. 68.

[122]   See Dijksterhuis, p. 277.

[123]   *Ibid.*, p. 287.

[124]   Digby too comes close to identifying the two theories. There are, he says (p. 48), atoms of FAWE, 'by which word *Atome*, no body will imagine we intend to express a perfect indivisible, but onely, the least sort of naturall bodies'.

thought the three extant letters of Epicurus, which present an invaluable epitome of much of his system. The great prize, however, was the discovery by Poggio in 1417 of the only surviving complete manuscript of Lucretius. Printed editions appeared at Brescia (1473) and Verona (1486), and soon an Epicurean strain became manifest in the popular anti-Aristotelianism of the Renaissance.[125] By the sixteenth century, Lucretius' atoms and infinite universe were accepted by some radical thinkers, but Epicurean materialism and mechanism were still avoided like the plague. The eventual outcome is a mixture of Atomism and Platonism: Lucretius' material universe is combined with the spirit-world of the Platonists, which *animates* it. This compromise allows Atomism to be combined with (a) Christian-Platonic views of God and the incorporeal realm, and (b) the implicit animism of the occult sciences. Let us briefly mention some of the key figures in this tradition.

In Nicholas of Cusa (1401–1464) we find a mystical strain of Neoplatonism incorporating elements of Atomism due to familiarity with Lucretius.[126] He accepts from classical Atomism indivisibles (atoms), the generation = aggregation thesis, and the principle of the unity of Nature (the denial of Aristotle's sharp distinction between the celestial and terrestrial realms), denying only the mechanism and materialism of Lucretius. In Nicholas' universe, incorporeal agents (God, the *Anima Mundi*, angels, demons, human and other souls) move the atoms and thus provide the ultimate source of all natural processes. *Philosophically*, he is nearer to Plato than to Epicurus: he merely adopts elements of Lucretius into his (Neo) Platonic world-view.

In this tradition we can also locate Fracastoro and Bruno. Fracastoro (1478–1553) studied medicine, mathematics and philosophy, and fell under the influence of Lucretius. He applied the Atomic Theory to medicine, explaining the spread of infectious diseases by reference to 'seeds' (*semina*), minute material particles which act as carriers.[127] All bodies, he says, naturally emit fine *effluvia* of their own specific nature[128] which mediate in their interactions (e.g. sympathy and antipathy) with other things.[129]. This effluvium-theory, as

---

125   See Stones, p. 446.

126   *Ibid.*, pp. 446–447.

127   *Ibid.*, p. 448.

128   Cf. of course the Epicurean theory of *eidola*.

129   See Thorndike, Vol. 5, pp. 495–496.

taken up and developed by men such an Sennert and Boyle, became one of the cornerstones of the 'corpuscular philosophy'.[130]

In Bruno we find the true disciple of Cusanus, and a blend of Lucretius' Atomism[131] with Platonism and the occult sciences. He accepts from Lucretius not only atoms but also the infinite universe, the unity of Nature, and the 'generation = aggregation' thesis. Everything in Bruno's universe, however, is animate (and almost sentient): even the atoms are alive and self-moving. Bruno's Atomism has no trace of the mechanism of Democritus.

By the late sixteenth and early seventeenth century, Atomism was becoming very popular and widespread. I cannot give due attention to all the figures involved, but must at least mention some of them. In England, Nicholas Hill's *Philosophia Epicurea* appeared in 1601,[132] and reflected (perhaps imperfectly) some of the views of the Hariot-Percy circle, a group of savants and wits who gathered around Henry Percy, the 'wizard' Earl of Northumberland, in the earliest years of the seventeenth century.[133] Thomas Hariot (1560–1621) seems to have been the intellectual leader of the group, and a forceful advocate of the Atomic Theory, which, he writes to Kepler, is the key to Natural Philosophy.[134] In his earlier works, Francis Bacon shows sympathy: 'The doctrine of Democritus concerning atoms,' he writes in his *Thoughts on the Nature of Things*, 'is either true or useful for demonstration'.[135] Elsewhere, in the *De Principius Atque Originibus*[136] and the *De Sapientia Veterum*,[137] he discusses the nature and properties of the atom. In later life, however, he seems to have rejected Atomism, explicitly denying the existence of void and indivisibles.[138]

---

130   Thorndike, Vol. 8, p. 174.

131   By his own admission (Bruno, p. 5) he suffered in 1578, in his thirtieth year, an intellectual crisis brought about by the reading of Lucretius.

132   For a discussion of Hill's work, see McColley, 4.

133   See Kargon, 3, and 4, pp. 6–40.

134   See Kargon, 4, p. 24.

135   F. Bacon, *Works*, Vol. 5, p. 419.

136   *Ibid.*, Vol. 5, p. 461 ff.

137   *Ibid.*, Vol. 6, p. 729 ff.

138   *Novum Organum*, Book 2, Aphorism viii, Works, Vol. 4, p. 126. I therefore agree with Kargon (4, p. 43) that Bacon's views about Atomism changed from sympathetic consideration to outright rejection, as against Rees (5, p. 549), who claims that Bacon was never sympathetic to the Democritean world-view.

Other names could be multiplied almost *ad lib*.[139] Galileo speaks favourably of Atomism, as does his French associate, the physician Berigard (1578-1664), who opposes a particulate theory of matter to peripatetic orthodoxy.[140] Emanuel Magnen (1590-1679) attempts, in his *Democritus reviviscens* (1646), to reconstruct the philosophy of Democritus from the surviving fragments.[141] In Holland, Isaac Beeckman (1588-1637) is developing not only an Atomic Theory of matter but also the basic principles of a *Mechanical Philosophy* of Nature. But already the list has become too long.[142]

D. *The Synthesis: Chemical Atomism*

By the year 1600, then, the Classical Atomic theory had been thoroughly revived, and was the subject of heated controversy. Meanwhile, and in parallel to this rebirth of ancient Atomism, the Aristotelian theory of natural minima had been extensively developed by the Paduan Averroists, and was being applied to chemical reactions by men such as Scaliger and Toletus. It was natural for some men – of eclectic and syncretistic leanings – to seek to identify the two doctrines.

The result is a system of chemical atomism, a *Chemical Atomic Theory*.[143] Atoms of the *chemical elements* – usually the peripatetics' FAWE, but sometimes the three Paracelsian 'principles' Mercury, Sulphur and Salt, or any combination from the two lists – are deemed the fundamental building blocks of matter. These 'atoms' are quality-bearing, and have the nature of their particular element – they are themselves FAWE, or whatever. They aggregate (and interact) to form *prima mixta* or molecules, first-order complexes which in turn combine to form second-order compounds, and so on.

In 1624 the Sorbonne – the theological faculty of the University of Paris – condemned as rash and insolent a

---

139 For the revival of Atomism in the early seventeenth century, see Stones, p. 450 ff.; Partington, 6, Vol. 2, p. 382 ff., p. 455 ff.; M. Boas, 4, pp. 422-433; M. Boas Hall, 8, pp. 263-264.

140 Berigard's atoms or corpuscles are still quality-bearing: he is no Democritean. See Meyerson, p. 329.

141 If Partington is right, Magnen's reconstruction is *far* from accurate. (See Partington, 6, vol. 2, pp. 455-458.)

142 I deliberately omit Gassendi from this list: he will be reserved for treatment in Book 3.

143 For a penetrating discussion, see Kuhn, 2, esp. pp. 15-16. According to Hooykaas (pp. 70-71) such a hierarchical theory of matter was widespread in the alchemical tradition centuries earlier.

number of alchemical theses, including such a system of Chemical Atomism.[144] The theses were defended by the chemist and physician Etienne de Clave, who championed a five-element theory (Earth, Water, Mercury, Sulphur and Salt), explained the properties of bodies in terms of the numbers of atoms of each type that they contained, defended the 'generation = aggregation' thesis, etc.[145] Although de Clave's *Cours de Chimie* was in time to become a standard text for seventeenth century chemists, the medical faculty of the University of Paris supported the condemnation of the Sorbonne, and the Parlement ratified it – de Clave was arrested and his theses torn up.[146]

For a leading exponent of this type of theory we turn to Daniel Sennert (1572–1637), who attempted, in eclectic and reconciliatory mood, to synthesise Paracelsian iatrochemistry, Democritean Atomism, and the Aristotelian theory of natural minima.[147] Although his synthesis was of dubious consistency, it was very influential throughout the seventeenth century, and provided a meeting place for a variety of diverse views.

To explain generation and corruption, says Sennert, one must postulate *simple* bodies:

> . . . in considering natural things, subject to generation and corruption, one must necessarily suppose simple bodies of a particular kind, from which the composite bodies arise and into which they are resolved. These simple bodies are physical, not mathematical minima, and are so small that they cannot be perceived by the senses, and these 'minima' are the smallest indivisible particles to which all natural bodies owe their existence.[148]

What are these 'simple bodies' or 'minima'? Are they Democritean atoms or Aristotelian natural minima? The two theories, Sennert replies, are not really distinct:

> Now it appears that the atomic theory must be explained as follows. Since natural things are subject to a perpetual change of generation and corruption, it is necessary that there are certain simple corpuscles of a very special nature

---

144   Thorndike, Vol. 7, p. 185.

145   Thorndike, Vol. 8, p. 122.

146   Thorndike, Vol. 7, pp. 186–187.

147   Walter Charleton (p. 4) lists Sennert among the eclectics.

148   Quoted from Stones, p. 454.

out of which these compound bodies are generated and into which the latter in turn decompose.[149]

These smallest particles, he continued,

> are called minima of nature, atoms, and indivisible corpuscles of bodies. They owe their name to the fact that they cannot be further divided through natural processes, and reversely form the building blocks of all natural bodies.[150]

Are 'atoms' and 'natural minima' no more than different names for the same things? Surely not: even the highly developed minima theory of Nifo and Scaliger is far removed from Democritean Atomism. Although minima may be conceived as actual and abiding particles, they are still quality-bearing and thus capable of mutual action and passion, far from the impassive atoms of Democritus or Epicurus.

Sennert's theory is similar. Since he accepts the Aristotelian elements of FAWE, his 'atoms' are the natural minima of FAWE. These are still qualitatively distinct (fire is hot and dry; water, cold and wet), i.e. with their own inbuilt 'forms'.[151] These minima have distinct sizes, but their *shapes* are hardly mentioned – they are not considered to be of explanatory significance.[152] Partington[153] suggests that Sennert does not think of his 'atoms' as hard and rigid – and hence possessing fixed and determinate shapes – at all. This supposition is confirmed by the explicit testimony of Kenelm Digby, a contemporary advocate of the minima theory, who explicitly denies that minima of FAWE have fixed, determinate, specific shapes. Since these least parts are jumbled, jostled, and crushed, he says:

> It is impossible that the elements should have any other naturall figure in these least parts, than such as chance giveth them.[154]

We are still, then, far from a truly mechanical theory of matter. Sennert, we must note, accepts from Scaliger the

149   Quoted from Van Melsen, p. 80. See also Partington, 6, Vol. 2, p. 274.

150   Quoted from Van Melsen, p. 80.

151   See M. Boas, 4, pp. 428-429.

152   See Gelbart, p. 157, on the differences between natural minima and atoms.

153   Partington, 6, Vol. 2, pp. 271-272.

154   Digby, p. 153.

definition of *mixtio* as 'the motion of the minima towards mutual contact so that union is effected'.[155] But what becomes of the minima of FAWE in the resulting compound C? Do the 'forms' of the elements persist in the compound? This, Sennert admits, is a *very difficult* question, one of 'total darkness'.[156] Juxtaposed minima of F and W must interact in virtue of their opposed qualities: fire, of its very nature, must warm other things. This mutual interaction will produce a uniformity of sensory qualities: C will at least appear to be homogeneous. In a true *mixtum*, this unity will be due to the dominant form of the compound, to which the forms of FAWE are *subordinated*.[157] These forms cannot be *annihilated* since (a) FAWE are the elements of which other things consist, and (b) they are readily extracted from C.

How then do the forms of FAWE persist in C? Here we have the choice of forms without their associated qualities (Avicenna) or forms in an attenuated state (Averroes): Sennert manifests a preference for the former option.[158] Either way, an atom of F will *remain* an F-atom, even in a compound where it does not appear or behave as such. The substantial form of C somehow imposes itself upon all the material constituents of C, organising them into a genuine unity:

> . . . natural things do not come into existence through a casual co-operation of atoms – even if perhaps Democritus thinks so – but through a dominating superior form.[159]

This dominant form assumes explanatory primacy: it is not merely the resultant of the interaction of the elements, but a prior (pre-existent?) principle of organisation, without which the elements would never assemble to form a unified compound body. As Debus says, Sennert's atoms were

> more than purely mechanical in nature; associated with them were formative forces in a sense similar to the archei of Paracelsus.[160]

Within the compound, however, some of the 'atoms' remain minima of F; others, minima of W; and so on. These minima

---

155   See Thorndike, vol. 7, p. 207; Van Melsen, p. 83.

156   Quoted from Van Melsen, p. 83.

157   In his discussion of the theory of subordinate forms in 'The Origin of Forms and Qualities' (*Works*, vol. 3, p. 113 ff.) Boyle cites Sennert as a great champion of that doctrine.

158   See Hooykaas, pp. 74–75.

159   Quoted from Van Melsen, p. 84.

160   Debus, 10, vol. 1, pp. 191–192. See also Partington, 6, Vol. 2, p. 273.

are substantially different, even if to sense – and even to the microscope or the eyes of Lynceus – they *appear* qualitatively identical. An atom of F remains F, even if we could not discern it as such. There is, then, no transmutation of the chemical elements FAWE: the apparent W ⟶ A transmutation in evaporation is merely the formation of *water vapour*.[161]

The atoms – natural minima of FAWE – are the bottom rung of Nature's ladder. They are not, however, the only type of minima:

> Apart from the atoms of elements there is still another kind of atoms which eventually can be called 'prima mista' [first compound principles]. They are the atoms into which other compound bodies can be dissolved as into homogeneous things.[162]

These 'prima mista' are the least particles of compound bodies, of the same form as the compound which they constitute (e.g. lead, gold, milk, flesh, cheese), and indivisible into bodies of the same form.[163] (They may of course be chemically decomposed into FAWE.) *Prima mista* can combine either with similar 'molecules' or with unlike ones, giving rise in the latter case to a still higher level of organisation of matter. The whole hierarchical theory bears of course a marked resemblance to our own molecular theory.

The simplest *prima mista*, says Sennert, are those of the Paracelsian 'principles', Mercury, Sulphur, and Salt. These 'principles' have very powerful forms which dominate those of FAWE and explain the specific powers and virtues of things. Paracelsus' 'principles' may therefore retain their explanatory significance – e.g. in chemical medicine – despite not being truly elementary.[164] Minima of FAWE thus combine to form particles of these '*tria prima*', which in turn give rise to still more complex particles such as those of metals.[165] Thus Sennert can emphatically *deny* the transmutation of his *elements* (FAWE) but *accept* the possibility of alchemical transmutations of metals, which arise only at a fairly complex level of organisation of matter.

---

161 Partington, 6, Vol. 2, pp. 274–275; Kuhn, 2, pp. 22–23.

162 Quoted from Van Melsen, p. 86.

163 Gelbart, p. 157, shows convincingly that the early work of Walter Charleton was heavily influenced by these doctrines.

164 See Boyle, *Works*, Vol. 1, p. 549, and Pagel, 1, pp. 339–340.

165 For the roots of this theory in the alchemical tradition, see Hooykaas, pp. 70–71.

The chemical Atomic Theory of Sennert and de Clave was further developed by a number of seventeenth century thinkers. Gradually, the explanatory power of *arrangement* was well as of material *constitution* became seen. Sebastian Basso, in his *Philosophia Naturalis adversus Aristotelem* (1621), claims that the atoms of FAWE which aggregate to form compounds are immutable and unchanging. But such atoms must *lack* the qualities HCDM: bodies possessing HCDM are naturally suited to the rôles of agent and patient. On this theory, then, the atoms must be conceived as qualityless and the compound as no more than apparently homogeneous.[166] Basso also argues that the FAWE ratio of a compound does not fix (all) its qualities. $F_4A_3W_4E_1$, for example, could correspond to a number of different atomic arrangements (isomers!) and hence to different chemical properties: to explain fully the qualities of a given compound, one must mention not only its material constitution but its corpuscular arrangement.[167]

In Joachim Jungius (1587–1657) too, the ideas of Sennert and de Clave are developed in the direction of classical Atomism.[168] In the compound C, he insists, the constituents are *not* corrupted, and nothing entirely new (no new form) comes-to-be. Instead, says Pagel, he regards C as

> A *similare* which appears to our senses as homogeneous: it is a *mixis pros aisthesin*. In other words, the components remain as such, but are too small to be perceived. The *mistum* is therefore in reality heterogeneous.[169]

The only changes in compound-formation are in the positions and groupings of the atoms, not in their qualities: atoms do not undergo action and passion. *Forms*, says Jungius, are *consequent upon* atomic constitutions (and arrangements): the form of C is not something which can convert atoms, 'according to its own aims', into C-particles.[170] The qualities of C are determined primarily by its atomic constitution, and secondarily by its atomic arrangement: Jungius acknowledges the possibility of qualitative change occurring by atomic

---

166   See M. Boas, 4, p. 427; Partington, 6, Vol. 2, p. 387.

167   See Hooykaas, p. 75.

168   His 'Botanica Democritea' makes plain his purpose: the application of Democritean Atomism to the organic world. See Pagel, 5, p. 100–101.

169   Pagel, 5, p. 102.

170   *Ibid.*, p. 105.

rearrangement alone, but is not entirely sure that such changes occur.[171]

Jungius accepted the existence of chemical elements (limits of analysis) but denied the adequacy of contemporary accounts and lists of them. Any substance immune to chemical decomposition by fire or acid he is prepared to list as elementary. Mercury, Sulphur, Silver, Gold, Salt, Soda, Saltpetre, Talc, and 'a few other bodies' are, he says, the true chemical elements: all other substances can be 'vanquished' chemically.[172] It is perhaps not too far-fetched to see in Jungius — and others of like views - the true precursors of Lavoisier and Dalton.[173]

---

[171] See Kuhn, 2, p. 17.

[172] See Pagel, 5, p. 105; Kuhn, 2, p. 24.

[173] Pagel, 5, pp. 105-106; Kuhn, 2, p. 32. For the influence of Jungius on Boyle - mediated by Hartlib - see O'Brien, Part 1, pp. 13-14.

THE VACUUM: 2

A. *God and the Void, 1: The Annihilation of Bodies*

According to Aristotle, the principles of natural philosophy have the modal status of *necessary* truths: they could not be otherwise. This aspect of Aristotelianism was emphasised by the commentator Averroes and his disciples, the so-called 'Latin Averroists', men like Siger of Brabant and John of Jandun. Now among the allegedly necessary truths of the peripatetic philosophy we find the non-existence of void or vacuum$_2$: if, as Aristotle assumes, three-dimensional (3D) extension is the characteristic mark or essence of a body, then *incorporeal* 3D extension is absurd and impossible. There is therefore no void space, either intracosmic or extracosmic.

Now from the necessary non-existence of extracosmic void there follows, as equally necessary corollaries, the unity and fixity of the cosmos. Without an extracosmic void, not even God could either *move* our cosmos as a whole,[1] or create another one.[2] Furthermore, He cannot on this view annihilate any body without the surrounding bodies rushing together to fill the gap. If void is *per se* impossible, not even divine power can create one.

But all this is *blatantly heretical*. The Christian God is supposed to be omnipotent, able to do as He wills with His Creation: to create or destroy, move or arrest, as He sees fit. The Averroism of Siger of Brabant involves therefore an implicit denial of Divine omnipotence, and was denounced as such. In 1277, Bishop Tempier of Paris issued a condemnation of 219 articles, mostly Averroist in content, thereby reaffirming Christian dogma against the peripatetic doctrine of natural necessity.[3] The spirit of the 219 theses is best captured by Article 147, which states that: 'The impossible cannot be done by God or another agent', where 'impossible is understood according to nature'.[4]

---

[1] This becomes Article 49 of the 219 theses condemned by Tempier in 1277.

[2] Article 34 of the heretical theses.

[3] For the condemnations of 1277, see Grant, 2, p. 271, and 3, pp. 50–51.

[4] Quoted from Grant, 9, p. 108.

After the condemnation of 1277, therefore, it became obligatory for scholastics to distinguish what is naturally possible from what can be achieved by God: the void, it became orthodox to claim, cannot exist by Nature by only by divine (i.e. infinite) power. Thus was a precarious reconciliation attained between Aristotelian philosophy and Christian dogma. On this compromise view, God has the *power* to create a void, move the whole cosmos rectilinearly, create other world, etc., but – miracles apart – such things do not occur.

It now becomes possible to say that God could – if He so chose – annihilate part (or indeed all) of our material world. He could moreover, by His infinite power, prevent bodies rushing into the empty space created. If, for example, He were to annihilate all sublunary matter, the heavenly spheres would strive mightily to collapse into the void thus produced: only omnipotence can overcome this 'abhorrence' of Nature for a vacuum.

Suppose, then, that God has created – and sustained – such a void beneath the sphere of the moon. What can be said about such an empty space? The answer, according to fourteenth century scholastics such as Buridan and Albert of Saxony, is 'nothing at all'. The *interval* or gap has, as such, no positive reality: it cannot even be called 'vacuum' in virtue of its *privation* of *contents*.[5] This remained a standard scholastic view into the sixteenth and seventeenth centuries.[6]

For a little more detail, let us turn to Albert of Saxony. Vacuum, he begins,

> must . . . be imagined in this twofold way: in one way as a separate space in which there is no other body conjointly; in the second way as a body between whose mutually closest sides there is nothing.[7]

In the first sense, then, vacuum is the empty space or interval itself; in the second sense, it is the vessel or container deprived of contents which is properly called the vacuum. Both types, says Albert, are impossible 'by the action of any natural power, but divine agency could produce a vacuum in the

---

[5] This viewpoint derives from Ockham. Pure nothingness, he argues, is unintelligible and inconceivable. Vacuum can therefore only be conceived as a lack or privation appertaining to some positive reality such as a vessel. See Grant, 9, p. 12.

[6] See Grant, 9, pp. 12-13.

[7] From Grant, Ed., 4, p. 324.

second sense (a container empty of all contents).[8] The first conception of vacuum is, however, quite absurd and impossible: on that view, says Albert,

> a vacuum is a separate dimension and this cannot be posited because a separate accident must not be assumed to exist without a subject. Secondly, if such a separate space were posited, it follows that because it is a body (for it would have length, width and depth) an interpenetration of bodies would occur when it received something placed in it, which is impossible.[9]

Albert thus accepts both the principle that 'accidents' such as quantity must inhere in a proper logical subject, and the '3D ⟶ corporeal' premiss. The substance in which length, width and depth inhere must therefore be a body. There cannot be a 3D but incorporeal space: not even omnipotence could create one.

What then would occur if God actually chose to annihilate all the matter beneath the sphere of the Moon? Either the concave surfaces of the heavens remain separate - distant - from one another, or they do not:

> If they are not separated by a distance, they would be conjoined as two leaves of a book or an empty purse; but then it ought not to be conceded that the sky is a vacuum, because in the aforesaid case the sides of the sky would be immediate [in contact]. If, however, it should be stated in this case that the sides of the sky are yet distant . . . there would then be some dimension between them by which they would be distant; hence there would be no vacuum. To this, one might respond that the sides of the sky are not distant by a straight line, although they may well be separated by a curved line. But if it is said that the sides of the sky would be conjoined, I deny this, for the sky would remain spherical, just as now; and its sides would not be in direct contact, even if not separated by a [rectilinear] distance.[10]

We could therefore measure *around*, but not *across*, the gap or interval between opposite surfaces produced, thus:

8   *Ibid.*, p. 325.

9   *Ibid.*, p. 325.

10   *Ibid.*, p. 325.

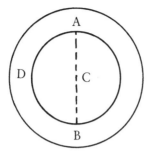

Here ADB is a real distance; ACB is not. ACB is nothing, unreal, or 'imaginary'. But if the lunar sphere is rigid, and undisturbed by the annihilation of its contents, elementary geometry will give us ACB = $2ADB/\pi$, i.e. a *magnitude*. (Exactly the same argument is given by Gassendi centuries later for physically real but incorporeal dimensions in vacuo.[11]) All Albert can say is that this magnitude, because it is not a magnitude *of* anything, is not physically real but merely 'imaginary'.

The denial of genuine magnitude or 3D extension to void space was common in scholastic thought: Aristotle's '3D ⟶ corporeal' premiss was undoubted for centuries. Marsilius of Inghen, for example, denies the possibility of ever measuring or determining dimensions in mere emptiness, and reaffirms the '3D ⟶ corporeal' thesis.[12] Buridan too agrees:

> . . . the space [or distance] between me and you is nothing but the magnitude of the intervening air or of another natural body, if one should intervene.[13]

If no body intervenes, he continues, there is *nothing* between us, in which case we are in contact.[14] One body can be separated from another only by a third. Empty or 'imaginary' space lacks 3D extension.

But, says Henry of Ghent, although void space may not be measurable *per se* it is still measurable *per accidens*. Imagine, for example, that God has annihilated all the contents of a closed calibrated vessel, thus:

---

[11] See Gassendi, p. 387 (*Syntagma*, Physics, Section 2, Chapter 1).

[12] See Grant, 9, p. 123.

[13] *Ibid.*, p. 123.

[14] Cf. of course Descartes' argument that if nothing (void) separates the walls of a vessel they are in contact.

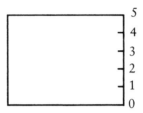

The void within the vessel may in this case be said to be of the same height as the vessel, i.e. five units. Despite its lack of any positive nature or reality of its own, this vacuum$_2$ does nevertheless possess dimensions. (Only a body five units high or less could be placed in the vessel.) Henry thus provides scholastics with an intelligible way of attributing real dimensions to empty spaces.[15]

Another scholastic writer who argued for a similar view was John de Ripa (d. 1370). Imagine, he says, the following arrangement of bodies:

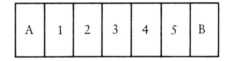

Now suppose that God successively annihilates bodies 1–5. A and B, although at rest in their initial positions, would now come to touch one another, unless they were separated by a real – but incorporeal – distance. Another absurd corollary of the orthodox view is that A and B would be simultaneously at rest (in their original positions) *and* moving towards one another, as 1–5 were annihilated.[16]

These arguments thus implicitly assert the existence of a real, 3D space independent of bodies, a receptacle in which all other things exist. Distances and dimensions must be as real in so-called 'imaginary' space (void) as in 'real' space (plenum).[17] De Ripa is therefore committed to the anti-Aristotelian

15   See Grant, 9, pp. 124–125.

16   For these arguments, and others to the same effect, see Grant, 5, pp. 146–147.

17   Grant, 5, p. 148.

theory of place as 3D 'interval' or vacuum$_1$, and hence to the denial of the '3D $\longrightarrow$ corporeal' premiss.

Such views, however, remained heterodox and unpopular among the schools. Acceptance of Christian dogma demanded assent to at least the *possible* existence of void space, but most scholastics continued to defend Aristotle's '3D $\longrightarrow$ corporeal' premiss and his definition of place as 2D surface of containing body, dismissing such miraculously created vacua as merely 'imaginary' spaces, fictions of the imagination rather than true physical realities.

B. God and the Void, 2: Beyond the Cosmos

According to Aristotle, we inhabit a finite, spherical cosmos, beyond which is 'neither place, nor void, nor time'.[18] Already we have seen grounds for doubt. What are we to say of the fact that the Aristotelian cosmos has itself, as a whole, no (Aristotelian) place, i.e. exists nowhere? What becomes of Lucretius' spear and the Stoic's hand?[19] Can we imagine, or even conceive of, a limit to space itself? Where is God: is He situated in the heavens, or is it more fitting to say that all things are in Him? The difficulties continue to pile up for the Aristotelian.

The crucial problem, however, lies in the theological implications of strict Aristotelianism. If there is no room outside our cosmos, God can neither (a) give the whole world a rectilinear motion, nor (b) create other worlds apart from ours. Even omnipotence requires room to work in. But these Averroist theses were condemned in 1277: it is a denial of God's omnipotence to question His power to move our cosmos or to create others.

Orthodoxy thus involves admission of the existence of some kind of extracosmic space: the motion of our world and the creation of others both presuppose such room. One may dub this space 'imaginary', because actually destitute of bodies – and hence, for the Aristotelian, of 3D extension – but it must have at least some minimal degree of reality.

It follows from God's omnipotence that extracosmic space cannot be bounded or limited in any way: a boundary would involve a limitation to the divine power (e.g. to go on creating

---

18 *De Caelo*, 1, 9, 279a 12-13. Void, says Aristotle, is space devoid of body but possibly occupied by body. But no body can exist beyond the heavens. Therefore there is no void space there.

19 In Bruno's 'On the Infinite Universe and Worlds', the Aristotelian, Burchio, actually claims that they pass out of existence at the outermost perimeter of the cosmos. See Singer, p. 253.

more and more worlds *ad infinitum*). If then God can move our world, or create another, *anywhere* He chooses, the 'imaginary' extracosmic space must be *unbounded*. (It cannot be called infinite in extent since, being dimensionless, it cannot properly be said to be extended at all.)

Another corollary of God's omnipotence is His omnipresence. If God can act at (e.g. create a world at) a place P, He must be actually present at P. But God can act anywhere in the extracosmic void. Therefore, God is actually present everywhere in it. The Scotists attempted to deny this consequence of God's omnipotence, asserting that God can act at a place without being substantially present there – i.e. can be 'present' in a place merely as a *virtus* or power – but most of the schoolmen followed Aquinas in assuming that the deity is substantially (and not merely virtually) omnipresent.[20]

It thus became orthodox scholastic doctrine to conceive God as substantially omnipresent in an unlimited (but dimensionless) extracosmic void or 'imaginary' space. He is present in this space, however, *not* as an extended being with spatially distinct parts, but 'all in every part'. God is completely or totally present in every part of space: like the soul, the deity is perfectly indivisible. The strange and heretical doctrines of More and Spinoza – both of whom attribute extension to God – are still far in the future.

These ideas are of course congenial to the tradition of the divinisation of space which we discussed briefly in Chapter Two.[21] Beyond the outermost sphere, says 'Hermes Trismegistus', is a space filled with the divine spirit. Similar doctrines can be found in the cabbalistic tradition and read into a number of Biblical passages. Although the material universe is limited, beyond its bounds can be found spiritual realities, present in a 'space' that concerns theology rather than physics. (What exactly could be meant by a dimensionless space I still find hard to fathom.)

Let us now attempt to put a little flesh on these bare bones. We shall begin with Bradwardine,[22] and the five corollaries C1-5 which he draws from the premiss of God's immutability:

C1.   God is everywhere in the world.

C2.   And also beyond 'in a place or imaginary infinite void'.

C3.   And can therefore be called immense and unlimited.

---

[20]   See Grant, 9, p. 146, for more details.

[21]   See Chapter 2, pp. 80–82.

[22]   Grant, Ed., 4, p. 555 ff.

C4. Hence we can reply to the question: where was God before the Creation?

C5. Hence void can exist without body but not without God.[23]

What is meant by the expression 'imaginary space'? The answer seems to be along the following lines. If we imagine (i.e. mentally picture) a finite world, we naturally depict it *in* some space. The faculty of *imagination* seems therefore to *demand* the postulation of extracosmic space. Even good peripatetics like Averroes and Aquinas admit that we do picture to ourselves such a space, although they deny it any reality. For a strict Aristotelian, then, 'imaginary' signifies fictitious, unreal, non-existent, etc. According to Averroes, men much influenced in their thinking by the imaginative faculty will tend to believe in such extracosmic space, even though sober reason speaks against it. Aquinas agrees: when we say, he explains, that nothing exists beyond the world, 'beyond' signifies only an imaginary space and not a real one. No reality corresponds to our image of an infinite, 3D, incorporeal vacuum$_1$ or void.[24]

After the condemnation of 1277, however, it became to all intents and purposes obligatory to attribute to the 'imaginary' extracosmic void some measure of actual existence. God could not move our world or create others in a mere fiction or chimera. Imaginary space continues, however, to be conceived of as (a) posited by the faculty of imagination, and (b) lacking genuine 3D extension of its own. What properties then does it have? Let us turn to Bradwardine's C1–5 for enlightenment.

According to C2, the extracosmic void is 'infinite': by His presence in such a void God may be called 'immense and unlimited' (C3). We must presumably take these terms in a metaphorical sense: what is not spatially extended at all cannot be immense in size. We can meaningfully say of such space that it is boundless, but not – at least, literally – that it is infinite in extent.

C4 entails the existence of a pre-Creation space *into which* God created our world, and in which He was present before the Creation.[25] (In Creation, God turns an 'imaginary' space into a 'real' one.) This space is absolute in the sense that its existence is temporally and ontologically prior to the existence of bodies and independent of their existence and arrangement. It can moreover serve as a reference-frame for the determination of

---

23   For discussion, see Grant, 3, p. 44.

24   See Grant, 9, pp. 118–119.

25   *Ibid.*, p. 141.

absolute motions. God can, says Bradwardine – citing the condemnations of 1277 for authority – give our world a rectilinear motion from its current place to another. Now, if this world contains all the matter that exists, this must be an absolute motion (no spatial *relation* between bodies changes), presupposing the existence of an absolute frame of reference for its intelligibility. The identical point can be found in Oresme's *Livre du Ciel et du Monde*:

> . . . beyond this world is an imaginary space infinite and motionless . . . and it is possible to say without contradiction that the whole world moves in this space with a right motion. To say the contrary is to maintain an article condemned in Paris.[26]

But if God can impart to the whole cosmos a rectilinear motion in 'imaginary' space, does this motion have any determinate *velocity*? If so, space must itself have extended magnitude, and the Aristotelian assumption that '3D ⟶ corporeal' collapses. This possible implication of their views seems not to have been seen by Bradwardine and Oresme.

If, Bradwardine continues – in tacit denial of the Scotist view – God can move our world from A to B, or create another world at B, He must already be present at B. Places in this extracosmic void have no existence independently of His presence: space, being itself nothing, is ontologically dependent on God. (Contrast the void space of classical Atomism, which exists *per se* and possesses 3D extension of its own right.)

God, says Bradwardine, is 'infinitely extended without extension and dimension': He is omnipresent in space – real and imaginary alike – but not Himself spatially extended. His presence in space is 'all in every part'. By this presence, He constitutes the being of imaginary space: it has no existence and reality independent of Him. It is, says Oresme, 'the immensity of God', and is as such absolutely dependent on His existence.[27]

If we turn briefly to the writings of Oresme, we find a similar view. 'The human mind,' he says,

> consents naturally . . . to the idea that beyond the heavens and outside the world, which is not infinite, there exists some space whatever it may be, and we cannot easily conceive the contrary.[28]

---

[26]   Oresme, p. 369. Clarke, of course, was to use just this argument against Leibniz in their celebrated correspondence. See H. G. Alexander, Ed., p. 32.

[27]   See Grant, 5, p. 144.

[28]   Oresme, p. 177.

God could, for example, add an extra 'lump' to the heavens as they rotate:

> Thus, outside the heavens, there is an empty incorporeal space quite different from any other plenum or corporeal space.[29]

This space is of real existence, even if destitute of 3D extension.[30] It must moreover be unbounded, unless we are to set limitations on God's power to create other worlds. But this possibility raises acutely the problem of the status of spatial relations within the imaginary void. If God chooses to create another cosmos, will it have a physically real *distance* from ours? If so, do we not have dimensions in void space?

Let us imagine, then, that God has created a body outside our world. Where is this body? The Aristotelian definition of place is inapplicable to it. Buridan answers in terms of the theory of *internal space*, that the place of such a body can be identified with *its own* dimensions.[31] Thus, in answer to the Stoic's hand argument, Buridan replies that:

> it would not be valid to say that he could not place or raise his arm there [simply] because no space exists into which he could extend his hand. For I say that space is nothing but a dimension of body and your space the dimension of your body. And before you raise your arm outside this [last] sphere nothing would be there; but after your arm has been raised, a space would be there, namely the dimensions of your arm.[32]

Similarly, if God were to create a bean in extracosmic space, it would have a place in virtue of its own dimensions. This is the theory of 'internal space' (Toletus) or 'real space' (Suarez), which was in turn to be adopted and championed by Descartes.

This theory of course denies the intelligibility of spatial relations between bodies situated in void space: if God creates another world, it cannot on this view be separated by a distance from ours unless there are *bodies* between. Thus, according to Marsilius of Inghen, if God were to create another world, nothing would separate it from ours, and they must therefore touch one another.[33] No measurable (real)

29 *Ibid.*, p. 177.
30 Grant, 9, p. 120.
31 For this theory of space and place, championed by Buridan, Toletus, Suarez and Descartes, see Grant, 9, pp. 14–15.
32 Quoted from Grant, 9, p. 122.
33 This becomes of course one of Descartes' most celebrated – or notorious – anti-vacuist arguments.

distance could separate them unless God chose to create further bodies in between. Even the infinite might of the deity cannot separate two bodies *in vacuo* without simultaneously creating intervening bodies![34]

Thus Buridan, Marsilius, Albert of Saxony, Bradwardine and others are all adamant that there are no *incorporeal dimensions*. All accept Aristotle's '3D $\longrightarrow$ corporeal' premiss, and deny real magnitude to extracosmic space. But then, either one denies, with Marsilius, that even God could create a second world distant from ours in the extracosmic void, or one accepts that He can do this, but denies that the second world has any spatial relation to the first. The first of these alternatives sounds quite implausible (and possibly heretical to boot); the second founders on elementary geometry and/or kinematics. (What happens if God moves one world towards the other?) A dimensionless space is, in the final analysis, a contradiction in terms.

The theory of 'imaginary' space is therefore an ultimately incoherent attempt to reconcile Aristotle and Scripture. Sooner or later, it had to break down under the weight of internal contradiction and incongruity. Scholastics such as Henry of Ghent and John de Ripa once again pointed the way ahead. If, says Henry, God created another cosmos, not in contact with ours, there would be a *real distance* (measurable *per accidens*) between the two worlds.[35] De Ripa arrived by a very different route at the same conclusion.[36] Unless, he says, there are *positive places* beyond the world, it makes no sense to say that God could move the material cosmos as a whole. Extracosmic space is therefore, in an important sense, real: it is the *ubi definitivum* of spirits, angels, and God Himself.[37] De Ripa fights shy of explicitly attributing 3D extension to this extracosmic void, but this is the conclusion to which his analysis is taking him.

The notion of a God-filled but dimensionless 'imaginary' extracosmic space was adopted and developed by many scholastic thinkers in the fifteenth, sixteenth and seventeenth centuries. (For a much fuller account of these developments,

---

[34] See Grant, 9, p. 122.

[35] *Ibid.*, p. 125.

[36] *Ibid.*, p. 126 ff.

[37] For the distinction between *ubi definitivum* and *ubi circumscriptivum* see Grant, 9, p. 130.

see the writings of Edward Grant.[38]) By and large, however, scholastic writers tended to emphasise the 'imaginary' nature of this space and to deny it 3D extension. Space in itself, says a sixteenth century schoolman, is without dimensions, but its capacity to receive 3D objects (bodies) imparts to it a 'correspondence' to those objects, and hence to dimensionality.[39] (The nature of this 'correspondence' remains obscure.) The motivation behind this is clear: if God by His presence constitutes 'imaginary' space, one cannot construe that space as 3D without risking conceiving God Himself as a 3D and extended Being with distinct spatial parts – i.e., for the scholastics, as a *body*! Such a radical conception of the Deity had to wait for Spinoza for a frank and open advocate. God's omnipresence must, for the scholastic, be taken in a metaphorical or transcendent sense: God is not literally an extended Being. For defenders of incorporeal 3D extension (i.e. the vacuum$_1$ theory of place) one must turn from the predominantly Aristotelian scholastics to the counter-current, the anti-Aristotelian tradition.

C. Anti-Aristotelian Theories of Space

According to Aristotle, place = two-dimensional (2D) surface of a containing body. On this definition, a void or vacuum$_2$ is *a priori* impossible: since there could be no places in such a void, a body situated in it would be *nowhere*. Three-dimensional (3D) extension, says Aristotle, appertains only to body: the '3D ⟶ corporeal' premiss involves as a necessary corollary the non-existence of void space.[40]

This position remained orthodox in scholastic thought although, as we have seen, it had to be modified and amended in order to be reconciled with Christian theology. Among critics and opponents of Aristotle, however, the definition discarded by the Stagirite (place = 3D interval or vacuum$_1$) was often espoused. In Appendix 4 we follow up developments in the Atomist tradition in the Middle Ages: here we trace the influence of Philoponus who, as we have seen, defended the vacuum$_1$ theory of place, denied the '3D ⟶ corporeal' premiss, and claimed that space (vacuum$_1$) has its own incorporeal dimensions, independently of its occupation by matter.[41]

---

38 See especially Grant, 9, p. 148 ff., for later scholastic writings on the topic of extracosmic space.

39 See Grant, 3, p. 52.

40 See Chapter 2, pp. 52–56.

41 See Chapter 2, pp. 74–76.

Although Philoponus' vacuum$_1$ was always occupied by matter – it never became true void or vacuum$_2$ – this was accidental to it: as such, it was not necessarily occupied by any body.

Pride of place in the subsequent anti-Aristotelian tradition must go to Hasdai Crescas (c. 1340–1412), a Spanish Jew who defended the vacuum$_1$ theory of place and launched, in his *Or Adonai*, a vigorous and incisive critique of Aristotle's cosmology.[42] Aristotle's definition of place, Crescas alleges, lands him in severe difficulties. On that definition, for example, the outermost celestial sphere can have no place, except 'accidentally'.[43] Worse still, the sum of the places (= surfaces) of the parts will exceed in size the place of the whole: on the vacuum$_1$ theory, the sum of the places of the parts is equal to that of the whole.[44] The true place of a body, Crescas insists, is not a surrounding surface, but 'the interval between the limits of that which surrounds'.[45] There is, he argues, an infinite, immobile vacuum$_1$ *in* which all things find a place: 'The true place of a body is the void, equal to the body and filled by the body.'[46]

Given this conception of place as vacuum$_1$, as immaterial 3D extension, Crescas must of course reject Aristotle's '3D ⟶ corporeal' premiss. Aristotle, he objects, merely assumed this without argument, but

> . . . he who affirms the existence of an incorporeal infinite magnitude likewise affirms the existence of an incorporeal quantity.[47]

Vacuum$_1$, says Crescas, is measurable but incorporeal, 3D quantity devoid of matter. Mere 3D extension does not, he insists, of itself involve impenetrability:

> . . . the impenetrability of bodies is due not to dimensions existing apart from matter, but rather to dimensions in so far as they are possessed of matter.[48]

Since only dimensions 'possessed of matter' are impenetrable, body and space can interpenetrate but not body and body.[49]

---

42 For a translation of the *Or Adonai*, see Wolfson, 1, p. 131 ff.

43 *Ibid.*, p. 197. See pp. 44–46 for Wolfson's comments.

44 *Ibid.*, p. 199.

45 *Ibid.*, p. 195.

46 *Ibid.*, p. 187.

47 *Ibid.*, p. 179.

48 *Ibid.*, p. 187.

49 *Ibid.*, p. 187. Aristotle's 'cube' argument therefore fails.

Beyond our finite material cosmos, Crescas argues, vacuum$_1$ must extend *ad infinitum*: we cannot conceive of a limit to space itself. Aristotle's cosmology thus gives way to that of the Stoics – a finite material cosmos in an infinite void.[50] But, he continues, Aristotle's proof (from his theory of natural motions), that there can only be one world, is invalid:[51] in the infinite vacuum$_1$, God *may* have created other worlds.[52] (Small wonder that an influence of Crescas on Bruno has been suggested.[53]).

For Crescas, then, space is an infinite, immobile, incorporeal 3D vessel or receptacle *in* which as a container all things have their places, and which is itself indifferent to what goes on in it. A place is simply a determinate part of this all-embracing vacuum$_1$ or absolute space. Within this infinite void, God has created at least one material world: beyond the bounds of our cosmos, He may have created others. We have no knowledge whatsoever of the contents – if any – of the space beyond our world,[54] but it would be both foolhardy and irreverent to deny that God could have created more worlds in that infinite receptacle.[55]

There remains, however, the obvious Aristotelian objection: of what substance are the dimensions of (empty) space accidents? Is space itself an infinite, incorporeal, 3D substance? Surely not.[56] Crescas does not directly raise – or attempt to answer – this question. He does, however, at least hint at a link between God and space. God, he says, is 'figuratively called place' (*māqóm*) in such expressions as 'Blessed be the Place' and 'He is the Place of the World'.[57] This last metaphor, says Crescas, is 'remarkably apt': God, like space, permeates and pervades all the infinite universe, is present everywhere within it, etc. Crescas does not, however, ever take the metaphor so seriously as to claim that God is literally a three-dimensional Being, and that the dimensions of space are therefore parts of

50   *Ibid.*, pp. 60–61, p. 189. See also Grant, 9, p. 332.

51   *Ibid.*, p. 217.

52   *Ibid.*, pp. 117–118.

53   *Ibid.*, p. 35.

54   *Ibid.*, p. 217.

55   Given the Platonic 'Principle of Plenitude', it will follow that God must, by His very nature, have created more worlds, i.e. have actualised further possibilities of creation.

56   For an example of such peripatetic dialectic, see Grant, 5, p. 139.

57   Wolfson, 1, p. 201.

the divine extension.[58] Centuries must pass before anyone dares to draw so radical a conclusion.

The vacuum$_1$ theory of space was vigorously affirmed, in the face of a retreating Aristotelianism, by a number of Renaissance thinkers. Pico della Mirandola drew on Crescas' ideas about vacuum, place and motion, and thus helped to transmit the vacuum$_1$ theory.[59] The writings of Philoponus also became influential in the sixteenth century, especially when they were published in Latin translations.[60] Thus the 'gap' or 'interval' theory of place, rejected by Aristotle but reaffirmed by Philoponus and Crescas, gained new converts. A common move was to accept the place = vacuum$_1$ theory, but then deny the existence of vacuum$_2$ by *filling* one's vacuum$_1$ (absolute space) with some subtle matter. Pico, Telesio, Patrizi and Bruno all opt for such a position.[61] Thus Pico:

> . . . place is space, vacant (vacuum) assuredly of any body, but still never existing as a vacuum alone of itself. It is like the case of matter, which is something other than form; but nevertheless, never without form.[62]

The point, and the analogy, derive alike from Philoponus.[63] A similar position is advocated by Telesio in his *De rerum natura*. Space is, he insists, itself incorporeal and without resistance to the passage of bodies through it, but is – accidentally – filled with bodies: there is no void or vacuum$_2$.

In the writings of Francesco Patrizi (1529–1597) this anti-Aristotelian tradition comes to fruition. Space, Patrizi argues, must be ontologically prior to its contents:

> For all things, whether corporeal or incorporeal, if they are not somewhere, are nowhere; and if they are nowhere they do not even exist.[64]

Space is thus a necessary precondition of any other existence: there can be unoccupied places, but not things without a proper place. Even Creation presupposes space as a necessary

---

58   *Ibid.*, p. 123. Copenhaver (pp. 504–505) agrees with Wolfson that Crescas' approval of the God = *māqôm* formula in fact amounts only to an affirmation of the divine omnipresence, *not* a radical conception of the deity as a 3D and extended Being.

59   Wolfson, 1, pp. 34–36.

60   Grant, 9, p. 19.

61   *Ibid.*, p. 19.

62   Quoted from Grant, 7, p. 558.

63   See Chapter 2, p. 76.

64   Quoted from Grant, 9, p. 203.

prerequisite: God can only create a world *in* a pre-existent space, which is

> everlasting, not admitting destruction, providing a situation for all things that come into being.[65]

What then is space? Is it, as the ancient Atomists said, 'nonbeing' or nothingness? No, replies Patrizi, whatever *is* (has any manner of existence) is something – 'nothing' is no part of the furniture of the world.[66] Space must therefore be something, i.e. have some positive reality. It is, Patrizi suggests, 3D extension itself, without the resistance (*antitypia*) which gives bodies their essential nature.[67] Since bodies are impenetrable only in virtue of this *antitypia*, spatial (i.e. incorporeal) and bodily dimensions can interpenetrate without absurdity:

> Space always remains the same, and always immobile must remain long, wide, and deep in order to release all leaving bodies and to receive all entering bodies. Or else penetration of bodies occurs and this is impossible.[68]

In truth, says Patrizi, 'vacuum', 'spacium', 'plenum', and 'locus' are *all the same thing*.[69] When filled, space is called 'locus' (the place of something) and thus 'plenum'; when empty space is 'vacuum' or void.[70] Whether empty or occupied, space is nevertheless something: its own nature is indifferent to occupation. Space is indeed essentially void and only accidentally filled: void space is ontologically prior to plenum.[71] Thus, says Patrizi – here going beyond Philoponus and Pico – it is an *empirical* matter whether or not there exists perfectly empty or void space. Since vacuum$_1$ *can* exist without body, as empty space or vacuum$_2$, we must turn to experience in order to discover whether in fact it ever does so.[72] (Patrizi, as we shall in time see, gives a positive answer.)

Beyond our cosmos, says Patrizi, there must be infinite extracosmic space. The outermost heaven must have a convex

---

65 Quoted from Henry, p. 554.

66 Patrizi, pp. 225–227.

67 See Henry, p. 562.

68 *Ibid.*, p. 562.

69 Patrizi, p. 231.

70 This is of course Epicurean terminology, as reported by Sextus Empiricus and others. See Chapter 2, pp. 67–69.

71 See Grant, 9, p. 201.

72 Patrizi, p. 232 ff.

surface with a magnitude equal to that of the space with which it has an interface – anything else is unintelligible. The world could, in principle, be moved (by God), or could burn up and expand in a cosmic conflagration (cf. the Stoics' *ecpyrosis*): physically real extracosmic space is a precondition of these possibilities.[73] This extracosmic space, Patrizi affirms, is *actually infinite* in extent.[74] It is moreover of exactly the same nature as the space of our world: 3D, incorporeal, devoid of resistance and *antitypia*, etc. It is entirely accidental and extrinsic to a particular portion of vacuum$_1$, the general space of all things, whether or not it is occupied by any body.[75]

In what category of being, the Aristotelian will demand, does space fall? Is it substance or accident? Space, Patrizi responds, 'is not embraced by any of the categories and is prior to and outside them all'.[76] It is nevertheless closer to the category of substance than to that of accident: it is 'substantial extension . . . subsisting *per se*, inhering in nothing else'.[77] Although of independent existence, however, space is not composed of matter and form, does not participate in action and passion, etc. – we cannot therefore call it a 'substance' in the Aristotelian sense. It is, says Patrizi, *not* a body (because devoid of *antitypia*), but nevertheless not truly incorporeal (because extended); it is therefore 'an incorporeal body and a corporeal non-body', possessing features appropriate to both realms.[78]

Is the extracosmic space of Patrizi a true void, absolutely empty of all substance? No: he first posits an infinite void or vacuum$_2$, then fills it with *light*, which is conceived as a sort of intermediate between corporeal and incorporeal existence, something space-occupying and yet devoid of *antitypia*, physically active and yet 'spiritual' in nature. (Patrizi calls it both 'incorporeal' and 'a body in space' almost within the

73   *Ibid.*, p. 236.

74   See Henry, p. 565; Grant, 9, p. 200.

75   Patrizi, p. 238.

76   *Ibid.*, p. 240. This discussion of the ontological status of space influenced Gassendi and, via Gassendi's disciple Charleton, Newton. See Gassendi, p. 383 ff., Charleton, p. 66 ff., McGuire, 5, and Henry, pp. 567–569.

77   Patrizi, p. 241.

78   *Ibid.*, p. 241. It is worthy of note that Patrizi cites 'Hermes Trismegistus' as one of his authorities for this conception of space. See Chapter 2, pp. 80–82, for the 'Hermetic' view of space.

same sentence.[79]) Like space, light is a 'corporeal incorporeal', extended and yet somehow immaterial.[80] The 'Principle of Plenitude' guarantees that God will leave no part of the universe absolutely empty, devoid of all substance: light is diffused through the whole extent of the 'empty' (i.e. resistanceless) space outside our cosmos. As in the Hermetic writings (which Patrizi translated), the extracosmic space is the domain of some *spiritual* reality: the heavens are filled with the glory (light) of the Lord.

Space is therefore absolute (existing *per se*), actually infinite in extent, physically real, three-dimensional, general, homogeneous, incorporeal (without resistance or *antitypia*), and absolutely *immobile*, a frame of reference within which all things exist and move.[81] Although essentially and in itself it is empty (vacuum$_2$), pure void would be an affront to divine creativity: God's infinite fecundity ensures that all space will be *full* of some substance or other, either material or spiritual in nature.

Another figure who must be considered in this anti-Aristotelian tradition is Giordano Bruno, whose treatise 'On the Infinite Universe and Worlds' contains an almost point by point refutation of Aristotle's arguments in *De Caelo*, and expounds a most radical cosmology, incorporating the infinite universe and worlds of Lucretius. Let us run through a little of this momentous dialogue.

If the universe is finite, asks Philotheus (Bruno's spokesman), *where* is it? According to the Aristotelians, it is 'in itself', but then they also define place as the limits of the confining body, by which definition a finite universe would have no place, and thus be nowhere and non-existent. Aristotelian attempts to evade this conclusion are dismissed as sophistical.[82] The outermost celestial sphere, says Bruno, must itself have an outer surface, which must in turn be bounded by true 3D space: only something with dimensions can *limit* something with dimensions. (Bruno also trots out those old favourites – none the worse for being re-used – Lucretius' spear and the Stoic's hand.)

What, then, lies beyond our cosmos? If you say 'nothing', says Philotheus, I call this *void*. A finite Aristotelian cosmos

---

[79] Grant, 9, p. 203.

[80] See Henry, p. 556. Light, for Patrizi, can also be characterised as the *rarest* of bodies – only God is 100% incorporeal. See Grant, 9, p. 387.

[81] Patrizi quotes 'Hermes' as saying: '. . . a body that is moved is moved through an immovable medium' (see Henry, p. 571).

[82] From Singer, pp. 251–253. See also Michel, p. 166.

thus entails an infinite extracosmic void[83] (cf. of course the Stoic world-view). But in such a homogeneous and undifferentiated void, there is a infinitude of *possibilities* for further creation. If our world is good, the existence of more such worlds must be better. A God who created but a single world and left infinite empty space destitute and void would not be omnipotent and omnibenevolent, i.e. would not be God. The Creator *must*, therefore, by a (moral) necessity of His nature, have *filled* the extracosmic void, and created an actual infinity of worlds in it.[84] God, says Plato in the *Timaeus*, is not 'envious': He grudges existence to no type of being. His infinite goodness spills over into His Creation: only an infinite universe, filled with an infinity of worlds, can truly manifest His greatness.[85]

This then is the thrust of Bruno's argument:

1. Aristotle's cosmos must be somewhere, i.e. in a place.
2. Therefore there must be extracosmic space, with real dimensions.
3. But this space cannot be limited, or the same argument would apply to the supposedly finite whole.
4. The extracosmic space must also be void: Aristotle says that there is *nothing* beyond the heavens.
5. Therefore Aristotle's finite material cosmos must be located in an infinite void.
6. But in an infinite void there is an infinity of opportunities for further creation.
7. Therefore God, by His nature, must have created worlds in those spaces.[86]

The assertion of a finite universe, Philotheus concludes, is an implicit denial of God's omnipotence (or benevolence), and is as such heretical.[87]

Void, says Bruno, is something non-corporeal and without resistance, yet with true physical dimensions. It is not mere nullity or non-being. Aristotle, he alleges, defined void as 'non-being' and disproved its existence under that misleading definition – his arguments are therefore worthless. In fact, however, space can be identified with the infinite 'ethereal'

[83] From Singer, p. 254. This argument was also used by Crescas.
[84] The 'Principle of Plenitude', says Koyré (3, p. 42), dominates Bruno's metaphysical vision.
[85] Singer, p. 255.
[86] *Ibid.*, p. 259.
[87] *Ibid.*, p. 263.

region which contains all things, even perhaps with the 'ether' itself, which, says Bruno, penetrates all bodies and 'becometh inherent in everything'.[88]

Does the 'ether' actually penetrate the solid constituent atoms of bodies, or does it only occupy the interstitial spaces *between* the atoms? On the latter view, favoured by Michel,[89] ether is to be conceived as a subtle fluid medium, *displaced* by the perfectly solid atoms which move through it. As such, it – like all other bodies – would require a *receptacle* or place in which to exist: space or vacuum$_1$ is, says Grant, 'a coextensive container of the ether'.[90]

Unfortunately, Bruno does not seem to be quite as clear about these conceptual niceties as one could wish. He seems in places simply to *identify* 'ether' with 'space', and thus to make it incorporeal, 3D, and all-pervading. On this view, Bruno does not follow Patrizi: instead of arguing for an infinite vacuum$_1$ and then filling it with subtle matter, he identifies the ethereal medium and the 'void' space which it fills. There is, he says, 'a single general space, a single vast immensity which we may freely call void',[91] but it is devoid only of resistance (*antitypia*), not of all substance. Bruno, as Michel notes,[92] seems to use 'void' and 'ether' interchangeably: this all-pervading space is certainly not the non-being or vacuum$_2$ of Democritus. 'Ether', says Michel,[93] is neither body nor void but some distinct type of reality: it is continuous, simple, without resistance to motion, impalpable, and a necessary precondition of local motion.

Of these two conceptions of space, the former is by far the clearer. Grant favours this view, and tries to locate Bruno in the Philoponus-Crescas-Patrizi tradition. On their view, space *itself* is vacuum$_1$, a pure receptacle, accidentally filled by a quite distinct fluid medium, the ether. *Qua* physical medium, Grant insists, the ether itself must be in a *place*, which presupposes the existence of pure space of vacuum$_1$, ontologically prior to its contents.[94] As far as logic goes, says Grant, Bruno's vacuum$_1$ could be truly void (vacuum$_2$). Metaphysics,

---

[88]   *Ibid.*, p. 273.
[89]   Michel, p. 169.
[90]   Grant, 9, p. 189.
[91]   Singer, p. 58.
[92]   Michel, p. 139.
[93]   *Ibid.*, p. 140.
[94]   Grant, 9, p. 189.

however (i.e. the 'Principle of Plenitude'), entails the impossibility of pure void: in the final analysis, says Bruno, 'we shall find the plenum not merely reasonable but inevitable'.[95] On this view, Bruno has *both* an infinite vacuum$_1$ (absolute space, providing a 'receptacle' in which all things exist and move) *and* an infinite ether, a subtle, quasi-material fluid medium *filling* the vacuum$_1$, but without resistance to the passage of bodies.[96]

This is an elegant, and quite possibly sound, reading of Bruno: I shall offer no argument against it. It does, however, involve a generous amount of regimentation, of tidying up the lush growth of Bruno's thought. It also ignores or overlooks Bruno's tendency – oft remarked – to *identify* space with the ether. Perhaps, one ventures to suggest, the Nolan himself was less clear on the conceptual issues involved here than some of his modern commentators.

On either view, space must have the following properties.[97] It is a continuous quantity, existing (ontologically, not temporally) *prior* to the bodies which occupy it and independently of them. It is itself immobile, and quite unaffected by the passage of bodies through it. Although physically real, it is neither matter nor form nor a compound of the two. It is itself homogeneous and isotropic, neither active nor passive, and does not receive forms or qualities. It does not mix with or yield to bodies, since it is a pure receptacle. The dimensions of a place are equal to and interpenetrate with those of the body occupying it. Although neither substance nor accident, it is not merely imaginary but physically real – its existence, indeed, is a precondition of that of all other things. Nothing exists that does not have a place.

D. Motion in the Void, 1: Impetus and Inertia

Whatever is moved, says Aristotle, is moved *by* something: the Greek word even takes the passive voice.[98] For violent motions, the mover is always external to the object moved, and must either be itself in contact with it or operate by means of some 'instrument' which makes such contact – hence the theory of the motor-medium. But in a vacuum$_2$ there is no such corporeal medium, and hence no motive power.

---

95   *Ibid.*, p. 188.

96   *Ibid.*, p. 190.

97   *Ibid.*, p. 189.

98   See Randall, 2, p. 190.

Projectile motion *in vacuo* is, therefore, impossible.[99] The whole argument rests on Aristotle's general conception of motion as a process of change, the traversal of a physical medium,[100] and hence as requiring characterisation in terms of some continuously operating agent. There is in Aristotelian physics a necessary (conceptual) link between local motion and motive power: one cannot adequately or fully conceive motion except in a dynamic context. To evade Aristotle's conclusion of the impossibility of non-natural motion *in vacuo*, two distinct strategies are possible, thus:

S1.   Accept Aristotle's general account of *kinesis*, and postulate some continuously-acting, medium-independent 'motor' to account for projectile motion.

S2.   Deny the need for a continuously operating force to sustain non-natural motion, thereby rejecting what was for Aristotle a conceptual connection, and undermining one of the basic presuppositions of his whole natural philosophy.

Historically, S1 gave rise to the impetus theory, S2 to medieval kinematics and – ultimately – to the modern law of inertia. In the history of physics, S1 must rate as a most valuable and fruitful tradition: the theoretical superiority of S2 only became apparent in the seventeenth century. Before turning to S2, let us follow the tradition associated with S1 through the ages.

We commonly find theories of 'impressed force' or 'impetus' associated with criticisms of the peripatetic theory of the motor-medium. The source of this tradition was Philoponus' highly critical commentary on Aristotle's *Physics*. After dismissing the 'mutual replacement' version of the motor-medium, Philoponus turns to Aristotle's notion that the projector may impart to the medium a special 'power' of moving the projectile. This explanation, says Philoponus, is only superficially more plausible than the other. When one throws a stone, he asks,

> is it by pushing the air behind the stone that one compels the latter to move in a direction contrary to its natural direction? Or does the thrower impart a motive force to the stone, too?[101]

If the motive force is given solely to the air, 'of what advantage is for the bowstring to be in contact with the

---

99   See Chapter 2, pp. 59–60.

100   Moody, 4, pp. 220–221.

101   From Cohen and Drabkin, Eds., p. 222.

notched end of the arrow?'[102] On such a theory, one should be able to propel an arrow simply by agitating a sufficient quantity of air behind it, but

> the fact is that even if you place the arrow or stone upon a line or point . . . and set in motion all the air behind the projectile with all possible force, the projectile will not be moved the distance of a single cubit.[103]

Forced motion, Philoponus concludes, does not require the agency of a motor-medium:

> Rather it is necessary to assume that some incorporeal motive force is imparted by the projector to the projectile, and that the air set in motion contributes either nothing at all or else very little to this motion of the projectile. If, then, forced motion is produced as I have suggested, it is quite evident that if one imparts . . . forced motion to an arrow or a stone the same degree of motion will be produced much more readily in a void than in a plenum. And there will be no need of any agency external to the projector.[104]

The only effect of the medium on motion being to *resist* it, a projectile will move more easily in a void than in a plenum. The causal agency involved, says Philoponus, is not the medium but an 'incorporeal motive force', impressed on the projectile by the projector. Being distinct from the body in which it (temporarily) resides, this incorporeal agent is able to impel it.

Philoponus' views seem to have had a widespread influence among Islamic writers.[105] Thus Avicenna, after a careful scrutiny of various motor-medium theories, reports that

> when we verified the matter we found the most valid opinion to be that of those who hold that the moved receives an inclination (*mail*) from the mover.[106]

Mail qasrī[107] (violent *mail*) is, says Avicenna, capable of persisting indefinitely in the absence of resistance – e.g. in a void:

---

[102]   *Ibid.*, p. 222.

[103]   *Ibid.*, p. 223.

[104]   *Ibid.*, p. 223.

[105]   See Clagett, 4, p. 510 ff.

[106]   *Ibid.*, p. 511.

[107]   For the varieties of *mail* – psychic, natural, and violent – see Grant, 6, p. 49.

If the violent movement of the projectile is produced by a force operating in the void, it ought to persist without annihilation or any kind of interruption.[108]

But, since no abiding non-natural motions are ever found, it is clear that our world does not contain such a 'coacervate' or 'collected' vacuum$_2$. Abū'l-Barakāt (d. c. 1164) modified the *mail qasrī*, making it of self-expending nature.[109] If this *mail* decays naturally with time, there can be non-inertial motion *in vacuo*: a projectile simply comes to a halt when its *mail* runs out.[110]

By the thirteenth century, varieties of the 'impressed force' theory of projectile motion are receiving the critical attention of scholastic authors such as Roger Bacon and Thomas Aquinas. The latter rejects the theory thus:

However, it ought not to be thought that the force of a violent motor impresses in the stone that is moved by violence some force (*virtus*) by means of which it is moved . . . For if so, violent motion would arise from an intrinsic source, which is contrary to the nature (*ratio*) of violent motion. It would also follow that a stone would be altered by being violently moved . . ., which is contrary to sense.[111]

He goes on to defend a more orthodox Aristotelian viewpoint:

For if there were not such a body as air, there would not be violent motion. From which it is evident that air is the necessary instrument of violent motion.[112]

Air is not, for the Aristotelian, itself the *mover* of the projectile: rather, it is the *instrument* by which the true mover (the original projector) moves the missile.[113] For our purposes, however, the distinction carries no great weight: without the instrument, the mover could not act. Forced motion, Aquinas concludes, is not possible *in vacuo*.

Aquinas' two objections to the Philoponus-Avicenna theory are both of importance. The first is that the impressed force

---

108 Quoted from Clagett, 4, p. 513.

109 Grant, 6, p. 49.

110 Precisely parallel moves are made by a number of schoolmen – see Grant, 2, for much relevant material.

111 Quoted from Clagett, 4, p. 517. See also J. F. O'Brien, pp. 355–357.

112 From Clagett, 4, p. 517.

113 See Weisheipl, esp. p. 44, and Wallace, 5, pp. 116–117.

theory, by making the motive force in violent motion intrinsic to the projectile, destroys the distinction between natural and forced motion. The second is that, if impressed force is a new (transient) form, a body should be qualitatively altered by being set in motion, which is not the case.

These objections do serve to highlight the hybrid nature of the 'impressed force' theory, and illustrate some of the conceptual tensions discoverable in it. With the benefit of hindsight we can see it as an attempt to formulate some of the theorems of modern physics in the quite antipathetic terminology of Aristotle's philosophy. Its very incoherence proves instructive. How – unless we revert to pure animism – can a form or quality move a body? And what happens when one moving body strikes another and sets the latter in motion: can a form be transferred from one body to another?

Before following up this line of thought, however, let us turn to the *locus classicus* of the impetus theory, Jean Buridan's 'Questions on the Eight Books of the *Physics* of Aristotle'.[114] In a few brief pages, Buridan provides overwhelmingly powerful objections to the motor-medium theory, and then goes on to sketch his own rival theory of *impetus*.

What, asks Buridan, moves the projectile after it has left the projector? The theory of cyclical mutual replacement fails to impress him: the medium, he insists, resists rather than aids motions through it. In any case, the following 'experiences' (*experientie*) are clearly incompatible with that theory:

1.   A spinning top continues to rotate long after one has ceased to apply any force to it. But it moves without displacing any air: it cannot therefore be claimed that motion of the displaced medium provides the motive force.
2.   A lance with a sharp conical posterior will fly every bit as well as one with a blunt posterior, but 'the air following could not push a sharp end in this way, because the air would easily be divided by the sharpness'.[115]
3.   A ship will continue to move – even upstream – after the motive force has ceased to operate. If it is the air that is propelling it, a sailor on deck should be able to feel such an impulsion, but 'he feels only the air from the front resisting'.[116]

114   Translations of the relevant passages are to be found in Clagett, 4, pp. 532–538, and Grant, Ed., 4, pp. 275–278.
115   Grant, Ed., 4, p. 276.
116   *Ibid.*, p. 276.

What of Aristotle's theory that the projector imparts to the medium a special power of moving the projectile? This account, Buridan insists, fares no better than its rival, since:

1.   It still cannot explain the continued rotation of the spinning top or mill-wheel.
2.   One could cover the back of a moving ship with cloth and then remove the cloth in such a way as to trap all the supposedly 'empowered' air, but the ship would not suddenly come to a halt.
3.   Light objects such as straws are blown to the stern of a moving ship, not to the prow.
4.   Air is fluid and easily divisible: it is not easy to see how it could propel heavy bodies.
5.   One cannot propel a stone or an arrow merely by setting the air behind it in motion.
6.   We can throw stones much further than we can feathers: if the *air* were the motive agency, we should expect to be able to throw feathers considerable distances.

This is a quite devastating set of empirical objections to the motor-medium theory. After the fourteenth century, it survived primarily through the authority of Aristotle rather than by virtue of any intrinsic merits of its own: that a trebuchet can hurl a large rock (which the fiercest of gales will not move an inch) over a hundred yards is surely sufficient to demolish the Aristotelian position.

Having refuted the motor-medium theory, Buridan proceeds to expound his own view:

> Thus we can and ought to say that in the stone or other projectile there is impressed something which is the motive force (*virtus motiva*) of that projectile. And this is evidently better than falling back on the statement that the air continues to move that projectile. For the air appears rather to resist. Therefore, it seems to me that the motor in moving a moving body impresses (*imprimit*) in it a certain impetus (*impetus*) or a certain motive force (*vis motiva*) of the moving body, [which impetus acts] in the direction toward which the mover was moving the moving body, either up or down, or laterally, or circularly. And by the amount the motor moves that moving body more swiftly, by the same amount it will impress in it a stronger impetus. It is by that impetus that the stone is moved after the projector ceases to move. But that impetus is continually decreased (*remitittur*) by the resisting air and by the gravity of the stone, which inclines it in a direction contrary to that

in which the impetus was naturally predisposed to move it. Thus the movement of the stone continually becomes slower, and finally that impetus is so diminished or corrupted that the gravity of the stone wins out over it and moves the stone down to its natural place.[117]

This passage raises a number of important exegetical points:

1.   Buridan explicitly allows a 'circular' impetus as well as a rectilinear one. Some commentators have read this as implying that a stone released from a whirling sling should follow a curved path rather than a tangential one. All Buridan's examples, however (the top, the mill-wheel, the planetary orbs), are of *solid* bodies rotating about an axis, *not* of unattached bodies following circular paths about a central point: the former type of case may be all Buridan had in mind when he admitted 'circular' impetus.[118]

2.   Impetus, says Buridan, is proportional to velocity. We can also see him working in the direction of a concept of mass, and can therefore discern (albeit anachronistically) a formal resemblance between Buridan's 'impetus' and Newton's 'momentum' (= mv).[119] The conceptual differences are, however, profound: impetus, says Buridan, is the *efficient cause* of projectile motion, and must as such be ontologically distinct from the motion it produces.[120]

3.   Impetus is in itself 'permanent' in nature:[121] it decreases because of the resistance of the medium or because of a contrary inclination (e.g. gravity) in the body moved. In a hypothetical void space, such as God might choose to create, there might be neither resistance nor gravity, in which case a projectile might continue to move *ad infinitum*.[122]

After a fascinating passage[123] in which Buridan uses the notion of quantity of matter (*quantitas materiae*) effectively

---

117   *Ibid.*, pp. 276–277.

118   See Drake, 8, pp. 320–321. In time, Benedetti was to explain the rotation of a solid sphere as due to the rectilinear tendencies of all its parts.

119   But see Koyré, 7, p. 14, for some of the conceptual differences between modern physics and the impetus theory.

120   Grant, Ed., 4, p. 278.

121   According to Drake (7, p. 32) Buridan's impetus is 'permanent' only in the sense of that term opposed to 'successive', *not* in the sense opposed to 'temporary'. His arguments for this claim are, however, weak: for a rebuttal of Drake's position, see Franklin.

122   Buridan himself always fights shy of this conclusion. See Grant, 2, pp. 275–280.

123   Grant, Ed., 4, p. 277.

as a mass term, he goes on to make his celebrated remark about the celestial spheres. Since there is no resistance to their motion, and they have no inclination to any other, God could, at the Creation, have given each sphere a suitable (circular) impetus, and left them to rotate indefinitely. As Buridan's pupil Oresme puts it:

> the situation is much like that of a man making a clock and letting it run and continue its own motions by itself.[124]

The heavens may therefore run like clockwork, i.e. without needing the controlling influence of 'intelligences' or angels. Equally radical is the partial dissolution of Aristotle's celestial/terrestrial distinction involved in subordinating the revolutions of the orbs to the laws of terrestrial – or, rather, universal – mechanics.

Impetus, says Buridan, is not something purely successive, as motion is. It is of the nature of motion that one part exists after another, but the impetus of a uniform motion is whole and entire in every part of that motion. This entails, as Boyer[125] sees, that some physical reality must correspond to the idea of instantaneous velocity, viz. the strength of the impetus of a moving body at a particular instant. It also entails the reality of *absolute* motion: there is, for Buridan and Oresme, a real difference between A's moving towards B and B's moving towards A. Impetus theory thus implies the existence of absolute motion and hence also of absolute space: Buridan and Oresme consistently oppose Ockham's relativism.[126] God, Bishop Tempier decreed in 1277, can give the entire cosmos a rectilinear motion. But this makes no sense if space and motion are purely relative.[127] So, Buridan and Oresme conclude, there is such a thing as absolute motion: if A and B are coming together, the impetus really qualifies one or other of them.[128]

Finally, a note on the possible inertial implications of Buridan's theory: impetus, we must remember, although a

---

124   Quoted from Grant's paper in Lindberg, Ed., p. 286. The link with the 'clockwork' universe of the mechanical philosophy should be plain. According to Wallace (5, p. 42), Buridan was anticipated in this idea by Abū'l-Barakat.

125   Boyer, p. 73.

126   See Wallace, 2, p. 21.

127   Cf. Clarke *versus* Leibniz, in H. G. Alexander, Ed., p. 32.

128   See Murdoch and Sylla in Lindberg, Ed., p. 217.

'thing of permanent nature' (*res nature permanentis*),[129] is corrupted not only by the resistance of the medium but by a contrary inclination in the body moved. But for a heavy body within our world there is *always* such a contrary inclination, viz. downwards. Even in a void, then, projectile motion would not be eternal. Only if God annihilated the cosmos altogether, or placed a body in an 'imaginary' space quite free from the influence of gravity, might Buridan's theory yield inertial conclusions for projectile motions[130] – it is hardly surprising that Buridan himself never arrived at such a radical position.[131]

Many advocates of impressed-force theories preferred a self-expending *vis impressiva* to Buridan's permanent impetus. Franciscus de Marchia, a fourteenth century Scotist, became one of the foremost proponents of such a theory.[132] According to Franciscus, the 'force left behind' (*virtus derelicta*) in the projectile by the projector is self-corrupting of its very nature, due to an imperfection in its own being.[133] Even in the absence of resistance, therefore, violent motion will decelerate and decay. Nicholas Bonetus (d. 1343) adopted a similar view, explicitly stating that in violent motion a 'non-permanent and transient form'[134] is impressed in the moving body: motion in a void, he says, is only possible so long as that form endures:

> that force or impressed form continually fails and dimi-
> nishes in moving the mobile, and, as a consequence, moves
> slower. Thus violent motion made in the void has to be
> slower in the end than in the beginning, just as in the
> plenum.[135]

A different theory, but with similar implications, was espoused by Oresme. His impetus was not self-expending of its *own* nature, but was continually remitted by its struggle to

---

129   Grant, Ed., 4, p. 278.

130   This momentous thought-experiment was first propounded – and the inertial conclusion drawn – by Gassendi in his letters 'On Motion'. See Gassendi, p. 139.

131   Buridan, says Grant (9, p. 44), tends to *avoid* discussion of projectile motion *in vacuo*.

132   See Clagett, 4, pp. 526–531.

133   See Wallace, 5, p. 41; Moody, 4, pp. 253–255.

134   Grant, 6, p. 48.

135   Quoted from Grant, 2, p. 274.

overcome the *inclinatio ad quietem* of the body moved.[136] Most theorists who postulated an impetus or impressed force opted *either* for a self-expending *vis* or for such an intrinsic tendency to rest in the body moved, thus deliberately evading any inertial implications.[137] Impetus-physics, as Koyré insists, is most naturally construed so as to be compatible with motion *in vacuo*, but not with inertial motion.[138]

Even Galileo, in his early treatise *De Motu*, favours a self-expending impressed force. After dismissing the motor-medium theory[139] (for reasons which should be familiar by now), he goes on to outline his own view. Hurling a stone upward, he claims, gives it a transient quality of lightness, just as heating a piece of iron gives it a transitory warmth:

> The impressed force gradually diminishes in the projectile when it is no longer in contact with the projector; the heat diminishes in the iron, when the fire is not present. The stone finally comes to rest; the iron similarly returns to its natural coldness.[140]

Likewise, striking a ball temporarily imparts to it a 'sonorous quality' alien to its own nature. Although the stone thrown upwards rises in virtue of its own levity, says Galileo, this motion is still *violent*, since that levity is alien – and even opposed – to the nature of the stone, and is therefore continually corrupted and quickly lost. The stone is only 'preternaturally' light – by force and not by nature.[141] Impressed force, Galileo insists, decreases gradually and continually; otherwise, a violent motion could go on *ad infinitum*, which is 'most absurd'.[142]

The theories which we have been discussing all lie in the S1 tradition: all attempt to *answer* the Aristotelian question: 'What moves the projectile during its flight?' rather than dismissing it as confused or question-begging. Modern physics lies, however, firmly in the tradition of S2. To carry through a programme of explanation based on S2 is a truly revolutionary procedure, involving a complete change in our

---

136   See Shapere, p. 54.
137   See Grant, 2, pp. 265–292, for more details of this story.
138   Koyré, 5, pp. 31–32.
139   G. Galilei, 1, pp. 76–78.
140   *Ibid.*, p. 79.
141   *Ibid.*, pp. 80–81.
142   *Ibid.*, pp. 84–85.

conception of motion. Loosely, motion is for the Aristotelian a *process* (the actualisation of a 'potency'); for the Newtonian, a *state*. Let us attempt to spell out this distinction in a little more detail, and to disentangle some of the many interconnected issues involved:

a. The concept of 'being in motion'

For Aristotle, as we have seen, there is no distinction between 'being in motion' and 'being moved': he had, indeed, only the latter concept available to him. The notion that something could be in motion without being moved was therefore unthinkable for him. Local motion is a variety of change (*kinesis*), a transition from 'potency' to 'act'. This process of actualisation invariably requires some agent, something which brings into actuality the potency of the patient. 'Everything moved is moved by another,' says Aristotle, proposing the principle as a conceptual truth about *kinesis*. Impetus theorists, as we have seen, accepted this peripatetic principle, and tried to reconcile it with the facts of experience by positing a special incorporeal motor for projectile motions.

The wholesale rejection of Aristotle's approach we owe to William of Ockham.[143] By denying the alleged conceptual connection between 'being in motion' and 'being moved', Ockham was able to dispose of both motor-medium and impressed force theories at a stroke. The moving agent, says Ockham, is *indistinguishable* from the object moved. The impetus theory is dismissed as (a) superfluous and (b) incredible:

> For it would be astounding if my hand caused a power in the stone through coming into contact with the stone by local motion.[144]

Motion alone, he insists, is imparted to the stone by the hand, not some strange 'power'.[145] Both the motor-medium and impressed force theories are therefore based on a common – and questionable – assumption. The axiom 'Whatever is moved, is moved by another' is, says Ockham, neither self-evident nor demonstrated.[146]

We can, Ockham insists, conceive of motion without thinking of its cause: there is, *pace* Aristotle, no conceptual

---

143   For Ockham on motion, see Crombie, 2, p. 176 ff., and Leff, p. 585 ff.
144   Quoted from J. F. O'Brien, pp. 361–362.
145   See Boehner's Introduction to Ockham, p. xlviii.
146   See Bréhier, p. 196.

connection between 'being in motion' and 'being moved'. This is of course *not* an assertion of the *actuality* of (non natural) motion in the absence of a propelling force, but only of its conceptual possibility. As the Ockhamist Peter John Olivi says, there is nothing self-contradictory about such an idea. Moody comments that

> The principle of inertia, though not asserted and probably not even glimpsed, is brought within the realm of the possible.[147]

The denial of a conceptual nexus[148] between 'being in motion' and 'being moved' also makes possible the birth of an independent science of kinematics, which may be at least partially attributed to Ockham's influence.[149] Members of the Merton school – e.g. Bradwardine, Heytesbury and Swineshead – distinguish the study of motion *quo ad causam* (dynamics) from that *quo ad effectum* (kinematics).[150] Kinematics, they say, studies the *formal cause* of local motion, which is velocity.[151] The mathematical kinematics of the fourteenth century Paris and Oxford schools reached of course a very high level of sophistication. It is to this body of technical quantitative kinematics that Galileo turns in the *Discorsi*, when after a half-hearted discussion of the rival accounts of the cause of free fall Salviati remarks that

> The present does not seem to be the proper time to investigate the cause of the acceleration of natural motion.[152]

Dismissing contemporary theories of dynamics as mere 'fantasies', Salviati opts for a purely kinematic approach to the nature of free fall.

This development of an independent science of kinetics was historically very important, and does seem at least partly Ockhamist in inspiration. We must not, however, read too much physical significance into Ockham's conceptual point. The fact that we can think about, and quantify, a motion

---

147   Moody, 4, p. 247. See also p. 184.

148   Ockham, pp. 139–140. See also Moody, 4, p. 207.

149   Pederson and Pihl, p. 220.

150   *Ibid.*, p. 220. See also Wallace, 2, p. 20; Clavelin, p. 63.

151   See Wallace, 5, p. 53.

152   G. Galilei, 4, p. 166.

without considering its cause does *not* entail the thesis that uniform motion is force-free – it only makes that thought possible. Ockham and the Mertonians still tend to assume – as a physical truth, not a logical one – that motion requires a mover.[153]

What then is the value of Ockham's discussion of motion? It is twofold: part destructive, part constructive. The negative part is his rejection of the Aristotelians' classification of local motion as a variety of *kinesis*, and hence of the conceptual apparatus with which scholastics customarily tended to investigate the topic of motion. The positive part is just the simpler, kinematic conception of local motion itself. Motion, says Ockham, requires no 'flowing form', no entity of which the parts exist successively (!); indeed, no other inhering form than that of the moving body itself.[154] The moving body remains the same 'in itself' – it gains and loses no forms or qualities – and changes only 'with respect to its surrounding' in such a manner that 'A is at place P' is true at one time t_1 and false at another, t_2.[155] All there is to local motion is 'the successive existence, without intermediate rest, of a continuous identity existing in different places.[156] These different places in the analysans will in turn be reduced to spatial relations between bodies. Thus Ockham:

> If the opponent says that the new effect has some cause; but the local motion is a new effect, etc., I answer: a local motion is not a new effect, neither an absolute nor a relative one. I maintain this because I deny that position (*ubi*) is something.[157]

This is tantamount to denial that locomotion is a variety of *kinesis*, analogous to qualitative change. For Aristotle, just as a white body will not turn black unless some agent acts on it, a body at one place will not move to another except under the influence of some motor. By denying that place is something,

---

153   See Gaukroger, p. 149. Wallace (5, p. 66) construes Ockham as giving an acausal, and even anti-causal, account of motion: if motion is not a new effect, he says, it *needs* no cause. This reading of Ockham must be regarded, to say the least, as questionable. Even a change in spatial relations may require a cause.

154   See Leff, p. 585; Weinberg, 2, p. 58. For a fascinating discussion, clearly inspired by Ockham, of the ontological status of local motion, see Marsilius of Inghen in Clagett, 4, p. 615 ff.

155   See Crombie, 2, p. 177.

156   Quoted from Crombie, 1, p. 249.

157   Ockham, p. 140. See also Crombie, 2, pp. 166–167.

Ockham rejects this analogy. Of course movement involves change of spatial relations between bodies, but that is precisely *all* that motion *is*. And these spatial relations are 'external' ones, not properties of bodies like whiteness and blackness. It is striking that classical Atomism is also committed to this conception of spatial relations as extrinsic to the 'natures' of their terms: immutable atoms are in themselves quite unaffected by their relations to one another. Both Ockhamism and Atomism thus involve the *denial* that locomotion involves Aristotelian *kinesis* – the confluence of these two strains in Nicholas of Autrecourt may now be easier to understand and appreciate.[158]

b. Intensive Magnitudes

Aristotle thought of motion as a flux or flow *in time*: the notion of movement (or, for that matter, rest) in an instant was for him a conceptual absurdity.[159] During medieval times, however, the distinction between extension and intensity was studied by the schoolmen, and applied fruitfully in many fields. The distinction between gross weight and specific gravity is one such application; that between temperature and quantity of heat is another. Both are of fundamental importance to physics.

The intensive/extensive distinction can also profitably be applied to motion. Motion is extended in space and time: considered from the point of view of extension, its existence is successive – one part of a motion succeeds another. But for a uniform motion, its degree or *intensity* is constant throughout. Velocity, say the Mertonians, can be defined as *intensio motus*, intensity of motion.[160] The Mertonians therefore adopted a distinction between 'quantitative' (extensive) and 'qualitative' (intensive) velocity.[161] Quantitative velocity is measured in the normal way, i.e. by distance travelled per unit time; qualitative velocity requires a more sophisticated

---

158   Nicholas of Autrecourt, p. 98 ff. Nicholas objects to the reification of motion, the postulation of a sort of motion-thing, and insists that *all* there is to local motion is change of spatial relations of unchanging things. When we analyse the proposition 'a body moves locally', he says, the following concepts emerge: 'a body which moves; a place or interval between which and the movable object there was previously an interval, but now there is absence of interval there'. See also Weinberg, 1, p. 167 ff.

159   *Physics*, 6, 3, 234a 24–33.

160   See Gaukroger, p. 151.

161   See Clagett, 4, p. 421.

definition. The velocity of a body at a particular instant can, says Heytesbury, be defined in terms of

> the distance which the body would traverse if, in a given interval of time, it was moved uniformly with the velocity it had at the given instant.[162]

To make sense of the concept of instantaneous velocity we must have recourse to a counterfactual: we have not yet attained the sophisticated concept of a *limit*. In Paris, Oresme adopted a definition similar to that of the Mertonians, and developed a system of graphical representation in which degrees of (instantaneous) velocity were represented by lines. By plotting velocity against time, Oresme was able to discover a simple and elegant proof of the Merton Rule:[163] if one bears in mind that the area under such a graph represents the total distance travelled, the Rule follows by elementary geometry.[164]

c. How to apply the Principle of Sufficient Reason?

The crucial concept in this regard is that of a change of state. We tend to assume stability and seek explanations of changes. (This is not merely a psychological fact about us.[165]) For Aristotle, local motion is a change of place, a variety of *kinesis*, a process of attaining some form or perfection. As such, it requires a appropriate causal explanation. But now we have the concept of a state of motion, describable as the quality or intensity of a given motion and measurable (anachronistically) by dy/dt; and we can ask why *this* quality or intensity should alter. Just as a body will not change colour without due cause, neither will it change course or speed. The most explicit statement of this viewpoint comes from Descartes:

> Every reality, in so far as it is simple and undivided, always remains in the same condition so far as it can, and never changes except through external causes. Thus if a piece of matter is square, one readily convinces oneself that it will

---

162   Quoted from Pederson and Pihl, pp. 220–221. See also Clavelin, p. 71.

163   The Merton rule states that a body uniformly accelerating, during a time-period T, from one velocity v_1 to another, v_2, covers as much ground as another body which moves at the average velocity of the first $(v_1 + v_2)/2$ for the same period T.

164   See Clavelin, pp. 80–81. For more on Oresme's 'qualitative geometry', see Molland, 2, pp. 106–119.

165   See Meyerson, p. 147.

remain square for ever, unless something comes along from elsewhere to change its shape. If it is at rest, one thinks it will never begin to move, unless impelled by some cause. Now there is equally no reason to believe that if a body is moving its motion will ever stop, spontaneously that is, and apart from any obstacles. So our conclusion must be: A moving body, so far as it can, goes on moving.[166]

This principle, enunciated at around the same time by Beeckman, Gassendi, Descartes and the pupils of Galileo,[167] was to become enshrined, under the name of Newton's First Law of Motion, at the heart of modern physics. From the standpoint of Aristotle, Descartes' analogy would have seemed the most palpable and childish conceptual error, a veritable howler of a 'category mistake'. In our schools, Newton's first law is often presented as self-evident; it now takes some pains to comprehend Aristotelian physics. Such are the vicissitudes of intellectual history.

The principle of inertia thus derives, as Meyerson[168] saw, from the conception of a motion-state, together with the general principle that what is, *persists*. Given the concept of *intensio motus* (instantaneous velocity), and a novel application of the ancient and time-honoured Principle of Sufficient Reason, the modern conception of a motion-state begins to emerge.

We have now seen both S1 and S2 developed to fruition. S1 gave us the impetus theory, a powerful and sophisticated theory of dynamics, still based on Aristotelian foundations, but mirroring much of modern mechanics. S2 gave us the separation of kinematics from dynamics, the concept of *intensio motus*, and - ultimately - the notion of a motion-state on which the law of inertia depends. Within modern physics - firmly based on S2, of course - we can accommodate many of the theorems of impetus-physics: although the two approaches are conceptually poles apart, they may be indistinguishable in empirical applications.

One thing is clear. Modern dynamics emerged out of those endless scholastic debates on the possibility of motion *in*

---

166 Descartes, *Principles*, 2, XXXVII (Ref 2, p. 216).

167 How close Galileo himself came to the modern principle of inertia remains a matter of controversy.

168 Meyerson, pp. 147-148. All forms of conservation laws have, says Meyerson, high *a priori* probability for us - our minds are naturally disposed to seek such laws.

vacuo. And the final outcome of that controversy is clear: such motion is indeed possible. Either the impetus theory or the modern theory of inertia readily accommodates motion *in vacuo*; the motor-medium was thoroughly discredited by the seventeenth century.

E. Motion in the Void, 2: Resistance and Velocity

According to Aristotle,[169] local motion in a void would have to be instantaneous, which is absurd and contrary to its very nature. Given the $v \propto 1/R$ formula, if $R = 0$, $v = \infty$.[170] Therefore motion *in vacuo* is impossible: any motor, however feeble, would produce an infinite velocity in the body moved. To evade Aristotle's conclusion one must therefore either (a) postulate some source of resistance other than that of the medium, or (b) reject the $v \propto 1/R$ formula outright. In this section, we shall investigate examples of both types of response.

Even in antiquity, 'Aristotelian' dynamics did not go unchallenged. Once again, Philoponus provides both an incisive critique of the peripatetic position and a promising alternative view. We have already encountered Philoponus among the founders of the impetus theory; now we find him rejecting the $v \propto 1/R$ formula:

For if a body moves the distance of a stade through air, and the body is not at the beginning and at the end of the stade at one and the same instant, a definite time will be required, dependent on the particular nature of the body in question, for it to travel from the beginning of the course to the end . . . and this would be true even if the space traversed were a void. But a certain *additional* time is required because of the interference of the medium. For the presence of the medium and the necessity of cutting through it make motion through it more difficult.

Consequently, the thinner we conceive the air to be through which a motion takes place, the less will be the *additional time* consumed in dividing the air. And if we continue indefinitely to make this medium thinner, the additional time will also be reduced indefinitely, since time is indefinitely divisible. But even if the medium be thinned

---

169   *Physics*, 4, 8, 215a 24–b20; Chapter 2, pp. 62–64.

170   Weisheipl (p. 37) insists that this formula should be attributed to Averroes rather than to Aristotle himself, but it is easy to see how a commentator could have extracted it from the text of Aristotle's *Physics*.

out indefinitely in this way, the total time will never be reduced to the time which the body consumes in moving the distance of a stade through a void.[171]

For Aristotle, $v = s/t = kF/R$, so $t = sR/kF$, i.e. $t \propto R$. Double the resistance of the medium, and one doubles the time taken to traverse it. For Philoponus, $t = t' + \Delta t$, where $t' =$ time taken to traverse a given distance *in vacuo*, and $\Delta t =$ *additional time* due to the resistance of the medium. Now $\Delta t \propto R$, so progressive thinning of the medium will make $\Delta t \longrightarrow 0$ and thus $t \longrightarrow t'$. But however thin the medium, if $R > 0$, $t > t'$. Both versions of Aristotle's *reductio* therefore fail: there is no reason to believe either that $t' = 0$ or that for any physical medium $t = t'$. As Philoponus says, progressive rarefaction of the medium will never reduce the additional time Δt to zero. He then puts his finger precisely on the point at issue between Aristotle and himself:

> But it is completely false and contrary to the evidence of experience to argue as follows: 'If a stade is traversed through a plenum in two hours, and through a void in one hour, then if I take a medium half as dense as the first, the same distance will be traversed through this rarer medium in half the time, that is, in one hour: hence the same distance will be traversed through a plenum in the same time as through a void.' For *Aristotle wrongly assumes that the ratio of the times required for motion through various media is equal to the ratio of the densities of the media.*[172]

Aristotle assumes that $t \propto R$, whereas Philoponus has been trying to show that, while $\Delta t \propto R$, t itself is not. In effect, Philoponus replaces the 'Aristotelian' dynamical formula with a rival one, $v = v' - r$, where $v =$ actual velocity, $v' =$ hypothetical velocity *in vacuo*, and $r =$ retardation, where $r \propto R$, the resistance of the medium. The only function of the medium is to *retard* the motion: this dynamical insight links up with Philoponus' embryonic impetus theory to give the germ of a systematic, non-Aristotelian dynamics – a germ that was to lie dormant for centuries.

Among the Arabs, a similar view was advocated by Avempace (Ibn Bajja, d. 1138 or 1139), a Spanish Moslem, whose views became known to the scholastics through the writings of Averroes, who stoutly championed Aristotle's

---

171   From M. R. Cohen and Drabkin, Eds., p. 218.
172   *Ibid.*, p. 219.

viewpoint against Avempace's heterodoxy.[173] Avempace defended Philoponus' $v = v' - r$ formula against the peripatetic $v \propto 1/R$ formula thus:

> The proportion of water to air in density is not the proportion of the motion of the stone in water to its motion in air; but the proportion of the cohesive power of water to that of air is as the proportion of the retardation occurring to the moved body by reason of the medium in which it is moved, namely water, to the retardation occurring to it when it is moved in air.[174]

The greater the 'cohesive power' of the medium, the more difficult it is for a body to move through it, and hence the greater the retardation r. The resistance of the medium is on this view *not* essential to finite velocity. Consider, says Avempace, the heavenly spheres: they move with a uniform and finite velocity, despite the lack of any resistance to their motions. The only quasi-resistance to be overcome in such a case is the extension of space itself.[175] Avempace is the source of the *distantia terminorum* argument of the scholastics, to which we must soon turn our attention.

Averroes' rejection of Avempace's viewpoint is frankly disappointing.[176] He starts by stating that the velocity of motion in a medium 'follows the rarity' of that medium: the rarer the medium, the faster the motion. From this uncontroversial beginning, he abruptly concludes that 'it is obvious that the ratio of the motions varies as the ratio of the media in density and rarity'.[177] This seems simply to confuse a functional dependence (of some kind) of velocity on rarity with a direct proportionality between the two: no argument is given to show that this is in fact the exact function involved. Small wonder that Crescas, in discussing this controversy, dismisses Averroes' reply to Avempace as empty: 'Many words that increase vanity'.[178] Avempace's formula was to be

---

173   See Grant, Ed., 4, pp. 253–262.

174   Quoted from Grant, 2, p. 267.

175   See Moody, 4, p. 240.

176   For some of the deeper metaphysical issues involved in this debate, see Moody, 4, pp. 231–241. Moody characterises the position of Philoponus and Avempace as 'Platonic', since both postulate incorporeal motors within bodies.

177   From Grant, Ed., 4, p. 258.

178   From Wolfson, 1, p. 185.

adopted by vacuists from Crescas[179] to Nicholas of Autrecourt,[180] and used to buttress their position against Aristotelian attacks.

Followers of Philoponus and Avempace thus deny that the external resistance of a medium is essential to motion: resistance, on their view, is not intrinsic to our very conception of local motion. This anti-Aristotelian view was, however, not widely accepted among the schoolmen. Most tended to favour the peripatetic concept of motion – emphasised by Averroes – as involving the successive division or separation of the parts of some fluid medium. If local motion is *essentially* a dynamic process of dividing or 'cleaving' a physical medium, motion in a vacuum$_2$ or void is impossible.[181] The velocity of a motion is, on this view, a measure of how quickly the resistance of the medium is overcome, a concept which cannot have meaningful application in a void.

Attempts were made to accommodate motion *in vacuo* within this general conception of the nature of local motion. Void, say such men as Roger Bacon and Thomas Aquinas, may not need to be cleaved by a moving body, but it consists nevertheless of distinct parts which must be traversed successively. A moving body – even in a void – cannot be present simultaneously both at its *terminus a quo* and its *terminus ad quem*. This argument became known as the *Distantia Terminorum* (D.T.) or *Incompossibilitas Terminorum*.[182] Because of the D.T., says Aquinas, finite velocity is possible even in empty space:

> Just as there is a prior and posterior part in a magnitude traversed by a motion so also we understand that in the motion [itself] there is prior and posterior. From all this it follows that the motion takes place in a definite time.[183]

This conclusion holds for motion in a vacuum$_2$ as well as in a plenum. If a body moves from one place P_1 to another, P_2, it cannot simultaneously be present at both: it must arrive at P_2 *after* leaving P_1. If the void possesses distinct parts, a 'before

179   *Ibid.*, p. 183.

180   See Nicholas of Autrecourt, p. 91.

181   This seems to have been Averroes' view. See Moody, 4, pp. 220–221, p. 239

182   For details, see Grant, 9, p. 27 ff., Moody, 4, p. 166 ff.

183   Quoted from Grant, Ed., 4, p. 334.

and after', there must also be a 'before and after' in motion across it, says Bacon.[184] Even *in vacuo*, motion must be successive and not instantaneous.

It is not clear, however, what Aquinas and Bacon wish to make of the D.T. *Qua* conceptual/kinematic point about motion, it is perfectly sound: our conception of local motion *is* of something essentially successive in nature. Even Aristotle would not have demurred: indeed, that was precisely his point. If $V \propto 1/R$ and $R = 0$, $v = \infty$, but the very idea of an infinite velocity – or an instantaneous motion – is absurd. Hence, says the Aristotelian, there can be no motion *in vacuo*.

Some scholastic writers, faced with this difficulty, tried to use the D.T. as a sort of surrogate for the resistance of a material medium, i.e. turned a valid kinematic point into a piece of spurious dynamics.[185] External resistance, says William of Ware, is inessential to motion: pure motion consists in 'the overcoming of pure distance between positions by a simple motive power'.[186] Siger of Brabant and Albert of Saxony also tried to make the D.T. do what it cannot possibly do – provide a genuine physical resistance to the motion of a body. This confusion of kinematics with dynamics was exposed in a devastating critique by Marsilius of Inghen.[187] Void space, he insists, can provide no substitute for material resistance: mere *distance* cannot exert a retarding influence. Given $R = 0$, all bodies should move through a void with equal – i.e. infinite – speed: the D.T. is powerless to prevent this.

Does the void have 3D extension? The D.T. argument rather seems to imply that it does – that there is a *distance* between the *termini*, even in empty space. But, says the Aristotelian,[188] if void has dimensions, a body moving *in vacuo* must *interpenetrate* with successive parts of space, which is absurd (cf. Aristotle's 'cube' argument). In response to this argument, Walter Burley devised the following ingenious theory.[189] Void, says Burley, is analogous to a very subtle fluid: bodies do not interpenetrate with vacua; rather, vacua 'yield' and move aside to give way to bodies. Thus the

184   See Grant, 9, p. 27.

185   *Ibid.*, p. 28.

186   Quoted from Moody, 4, p. 248.

187   See Grant, 9, p. 37 ff.

188   Roger Bacon and Aegidius of Rome both used this argument. See Grant, 9, pp. 31–32.

189   See Grant, 7, p. 564 ff.; 9, p. 33 ff.

motion of a body 'through' such a 'void' involves the successive displacement of the parts of the 'medium' – this is the source of the resistance of void space. But if the parts of this medium possess tangibility and *antitypia*, the so-called 'void' is merely subtle matter; and, if they do not, they cannot *resist* the passage of a body. Burley is attempting to posit a medium with resistance, but with none of the other features of corporeal substances on which resistance depends: his project, for all its ingenuity, is ultimately incoherent.

Motion in a vacuum$_2$ is therefore significantly disanalogous to motion in a plenum: although it involves the successive occupation of different places, it does *not* involve any cleaving of the parts of a medium. There is therefore no external resistance or 'drag'. If then one continues to accept the v α 1/R formula, finite velocity *in vacuo* must involve some *other variety of resistance*. For a thought-provoking discussion of a variety of possible theories, let us turn to Albert of Saxony.

The starting point for Albert's discussion of motion *in vacuo* is the supposition that God has annihilated all sublunary matter, leaving the heavens a 'vacuum': i.e. a vessel devoid of contents. Interestingly, this annihilation would, on Albert's view, leave the *natural places* of the elements FAWE intact: even in 'imaginary' void space there would be a 'vacuum of fire', a 'vacuum of air', etc.[190] Void is therefore not inert and homogeneous: it is differentiated into distinct places, which (somehow) influence the behaviour of bodies. Even in 'imaginary' space, an airy body tends towards the 'vacuum of air', a fiery body moves naturally to the 'vacuum of fire', and so on.

Albert's variant on Aristotle's dynamical formula can be spelt out as v α F/(R$_E$ + R$_I$), where R$_E$ = *external* resistance and R$_I$ = *internal* resistance. Now imagine a heavy body placed in the 'vacuum of fire'. Its natural motion is towards the 'vacuum of earth', i.e. downwards. Nothing prevents it from falling: the void can offer no resistance to it. Therefore it will fall. But will it not – *per impossibile* – fall instantaneously to its natural place?

Albert's answers are rather bewildering. At one point, he hints that resistance is quite inessential to finite velocity – citing the heavenly spheres in illustration – but elsewhere he seems to *accept* the v α 1/R formula and seek some alternative source of resistance to guarantee finite velocity.

---

[190]   Grant, Ed., 4, p. 337. See also Grant, 9, pp. 51–52.

Like a number of other scholastics, he confuses Aristotelian and anti-Aristotelian ideas, and fails to realise the full implications of the position of Avempace.

All manner of causes may resist a motion, says Albert, other than the external drag of the medium. The free fall of a lump of iron may be retarded by suspending a magnet above it; or the descent of one earthy body retarded by balancing it against another on a pair of scales. The most common source of resistance *in vacuo* would, however, be internal, i.e. due to the intrinsic constitution of the body. All sensible bodies are *compounds* of FAWE: no pure elementary bodies enter our experience. But then, even in empty space, each element strives towards its natural place: the natural motion of a compound body depends on which element is dominant in its make-up. A predominantly earthy body falling freely through the 'vacuum of air' would therefore be retarded by its own quota of fire, which is striving in the opposite direction. In such a vacuum, then, all mixed or compound bodies would have internal resistance and hence finite velocity.[191]

The notion of internal resistance R_I thus enables us to attribute a finite velocity to the natural motions *in vacuo* of all compound bodies. Albert's theory does, however, have curious implications: for example, whenever a compound body crosses the invisible dividing line between natural places, one of its component elements will *instantly* cease to aid its motion and start to oppose it, producing thereby a radical and discontinuous change in the body's motion![192]

But what happens to a piece of *pure earth*, unimpeded by any resistance (external or internal), placed in the 'vacuum of fire'? Since both R_E and R_I equal zero, it should fall with *infinite* velocity, i.e. instantly. Here Albert attempts to draw a distinction between an actually infinite velocity and a potentially infinite one, a velocity 'greater than any speed'.[193] A piece of pure earth, he claims, would fall through the vacua of F, A and W with a staggering, inconceivably fast, but not actually infinite, velocity. (A number of other scholastics – e.g. Roger Bacon and Aegidius of Rome – accepted similar notions, of a motion at once instantaneous *and* successive![194] I still find such a notion plainly incoherent, and can find nothing to say in its favour.)

191   Grant, 9, pp. 49–50.

192   See *Ibid.*, pp. 51–55.

193   Grant, Ed., 4, p. 338.

194   See Grant, 9, pp. 24–25.

One consequence of Albert's position has a much more modern sound. All bodies of the same material constitution, he says, will fall with the same velocity in a void. This is because, while $R_E = 0$ both F (weight) and R_I are proportional to the quantity of matter in the body. But if v α F/R_I, and both F and R_I are directly proportional to the *quantitas materiae*, v will be the same for all bodies of the same make-up. This theory was also accepted by Burley and Bradwardine,[195] and would later be defended by Benedetti and the young Galileo, albeit on different – i.e. Archimedean – grounds.

Aquinas' discussion of motion *in vacuo* is baffling in the extreme. At first sight, he may seem to be a pure Avempacean, but in fact his position is a complex one, involving several distinct elements. 'Several difficulties,' he says, beset the opinion of Aristotle:

> The first of these is that it does not seem to follow that if a motion occurs in a void it would bear no ratio in speed to a motion made in a plenum. Indeed, any motion has a definite velocity [arising] from a ratio of motive power to mobile – even if there should be no resistance. This is obvious by example and reason. An example is that of the celestial bodies, whose motions are not impeded by anything, and yet they have a definite speed in a definite time.'

The appeal to reason is the familiar D.T. argument. It is true, Aquinas admits, that

> in virtue of some impediment [or resisting medium] something could be subtracted from this speed. It is not necessary, therefore, that a ratio of speeds be related as a ratio of resistance to resistance, for then, if there were no resistance, motion would occur instantaneously. But it is necessary that the ratio of retardation to retardation be as the ratio of resistant medium to resistant medium. Thus if motion in a void were assumed, it follows that no retardation would occur beyond the natural velocity, and it would not follow that motion in a void would bear no ratio to motion in a plenum.[196]

Much of this passage is pure Avempace, derived no doubt *via* Averroes. Aquinas explicitly accepts the formula v =

195 *Ibid.*, p. 58. For Bradwardine's acceptance of this position, see Grant, Ed., 4, p. 305.

196 From Grant, Ed., 4, p. 334.

v' − r, where r (retardation) depends on the resistance of the medium – even his example (the heavenly spheres) of finite velocity without resistance derives from Avempace.

Other influences are also present, however. The D.T. guarantees that, *if* there is motion *in vacuo*, it has finite velocity: the very notion of instantaneous velocity is incoherent. Unfortunately, however, the D.T. argument is merely a conceptual/kinematic point: it cannot provide a surrogate for resistance, and Aquinas does not try to use it as such. Given Avempace's dynamical formula instead of Aristotle's, however, do we have any need of a concept of resistance? It seems that we do. A particular force may suffice to move a small body but not a large one. In general, larger bodies are intrinsically more difficult to set in motion: they would be so even in void space. Each body, says Aquinas, has its own '*corpus quantum*', in virtue of which it resists being set in motion (either natural or forced).[197] Every motion, therefore, has a particular finite velocity arising from 'a ratio of motive power to mobile', even in the absence of (external) resistance. Matter as such is passive and inert: it resists the operation of forces which strive to set it in motion.

The motion of a body in a plenum is therefore retarded by two distinct types of resistance. There is, first and foremost, the intrinsic resistance of the *corpus quantum* of that body, its innate reluctance to be set in motion at all. Resistance of this kind is essentially involved in all bodily motions: some variety of resistance thus remains, on Aquinas' account, conceptually necessary to motion.[198] The 'ratio of motive power to mobile', i.e. of F to *corpus quantum*, gives the natural velocity v' for a given motion. There then supervenes, as a secondary and inessential source of further resistance, the external drag of the medium, which gives rise to a retardation r. The natural velocity, less the retardation r, gives the final velocity v. Resistance *of some variety* is essential to local motion; the external drag of the medium is not. There can therefore be (natural) motion, with a finite velocity, *in vacuo* (forced motion requires a motor-medium).[199]

Nicholas Bonetus[200] takes the argument one stage further. The *essential* speed of a motion, he claims, depends only on the nature of the body moved and the strength of the force

---

[197]   See Grant, 9, p. 39.

[198]   See Grant, 2, p. 270.

[199]   Grant, Ed., 4, p. 335.

[200]   See Grant, 2, p. 272.

applied; only the *accidental* speed depends on the resistance of the medium. In a void, therefore, different bodies would move with different, finite, essential velocities. Both natural and violent motions are possible *in vacuo*, violent ones being due to a (self-expending) impetus.

We shall end this section with Galileo's refutation, in his early work *De Motu*, of the 'Aristotelian' argument against finite velocity *in vacuo*. (Galileo's argumentation owes manifest debts both to Avempace and to Archimedes.) The argument that motion *in vacuo* must be instantaneous is, says Galileo, the strongest of the Stagirite's anti-vacuist arguments. Given Aristotle's $v \propto F/R$ premiss, it does indeed follow that $t \propto R$. But then if $R = 0$, $t = 0$, and motion is instantaneous. The inference, says Galileo, is perfectly valid,

> And indeed his conclusions would have been sound and necessary, if he had proved his assumptions, or at least if these assumptions even if unproved, had been true. But he was deceived in this.[201]

Since the $v \propto 1/R$ premiss is quite unfounded, Galileo insists, the whole argument collapses.

Aristotle's claim that there is no *ratio* between zero and any integer holds only, Galileo continues, for *geometrical* ratios, not for *arithmetical* ones. Although there is no non-zero number n such that zero has to n the same geometrical ratio as n has to 2n, zero has the same arithmetical ratio to 2 as 2 does to 4.[201] Galileo, like Avempace, thus formulates his speed-law in arithmetical terms:

> . . . in fact, the ratio of the speeds does depend, in an arithmetical sense, on the relation of the lightness of the first medium to that of the second. For the ratio of the speeds is equal, not to the ratio of the lightness of the first medium to that of the second, but, as has been proved, to the ratio of the excess of the weight of the body over the weight of the first medium to the excess of the weight of the body over the weight of the second medium.[203]

Galileo here neglects entirely such factors as viscosity (the resistance of a medium to being divided) and concentrates solely on *hydrostatic* factors. The crucial determining factor in the rate of free fall is, he says, the different in specific

201   G. Galilei, 1, p. 42.

202   *Ibid.*, p. 43.

203   *Ibid.*, pp. 43–44.

gravity (S.G.) between the falling body b and the medium m, thus:

$$v \propto (S.G._b - S.G._m).$$

Given a body of S.G. = 20 and two media of S.G.s 6 and 12 respectively, we have:

$$v_1 / v_2 = 20 - 12 / 20 - 6 = 8 / 14 \; [204]$$

The same body would fall through a void with a velocity of $20 - 0 = 20$ units. Progressive thinning of the medium would give $v \longrightarrow 20$, which is therefore the *natural* speed of free fall for this particular stuff.

This 'Archimedean' dynamics has the following implications:

1. A heavy body will fall in a void with a finite velocity dependent only on the S.G. of its material.
2. All bodies of the same S.G. will fall with the same velocity: *exactly* the same in a vacuum$_2$ (no aerodynamic effects), approximately the same in a plenum. The scholastic notion of 'internal resistance', such as we find in Albert of Saxony, vanishes into the concept of S.G. Bodies of different S.G.s will, however, still fall *in vacuo* with different speeds.[205]
3. The proper or natural weight of any body is its weight *in vacuo*: in any medium, it will weigh less than its true weight by the Archimedean upthrust of the medium. Thus Galileo:

    . . . if the objects could be weighed in a void, then we surely would find their exact weights, when no weight of the medium would diminish the weight of the object.[206]

The velocity of a body moving *in vacuo* thus depends on (a) the nature or constitution of that body, and (b) the external force (if any) applied to it. The external resistance of a medium is a quite accidental factor. Given the Philoponus-Avempace-Galileo approach to dynamics, there is no reason whatsoever to deny the possibility of finite velocity *in vacuo*; rather the reverse. Motion in a void is the simplest case: a medium merely adds complications.

F. The 'Fuga Vacui'

'Nature', says the schoolmen, 'abhors a vacuum', and does everything possible to prevent vacuum$_2$ formation. A mass of

204   *Ibid.*, pp. 44–45.

205   Contrast, of course, the Galileo of the *Discorsi*.

206   G. Galilei, 1, p. 40.

empirical data can be explained by means of this maxim, notably pneumatic phenomena such as suction and the time-honoured clepsydra experiment.[207] Rather than allow a void to form bodies will forego – or even reverse – their natural motions: in the clepsydra, water does not fall, since to do so would involve leaving a $vacuum_2$; in suction, a fluid can be made to ascend, contrary to its natural inclination.

All this lends experimental support to the principle 'Nature abhors a vacuum', which, although anthropomorphic in formulation, can be rephrased as a sober scientific hypothesis. The *horror vacui* can be replaced by a *fuga vacui* or 'force of a vacuum', a force which draws matter into places where a $vacuum_2$ would otherwise be produced.

It is important to formulate the theory as above: it is not that a $vacuum_2$ actually forms, and then sucks in bodies, but that matter is drawn to *forestall* void-formation. This motion is supposed to be instantaneous: the $vacuum_2$ does not exist even for a minute period of time. But this betokens an *unlimited force*: by the v α F/R formula only an infinite force can overcome a finite resistance to produce an instantaneous motion (assuming for the moment that such a notion even makes sense!). This accounts for the common scholastic belief that *only* God could create a $vacuum_2$ – only omnipotence can overcome Nature's *horror vacui*.

The origins of the *fuga vacui* theory are discussed by Grant,[208] who cites Averroes' commentary on *De Caelo* as a possible source. Averroes, following Alexander of Aphrodisias, explains all manner of pneumatic phenomena in terms of the impossibility of a $vacuum_2$: the elements FAWE will, he says, move in directions contrary to their own natures to prevent formation of a void. But the void is a non-entity, hence not a real efficient cause of anything. 'That a vacuum not exist' cannot be the efficient cause of any phenomenon.[209] Thus Adelard of Bath speaks not of a *horror vacui* but of a 'natural affection' by which air and water, for example, tend to *continuity*. The clepsydra experiment, he claims, is explicable in terms of this positive 'appetite' for continuity on the part of bodies. The four elements FAWE are, he says,

> so united by natural affection that, as no one of them
> desires to exist without another, so no place is or can be

---

[207] For the origin and nature of the clepsydra experiment, see Chapter 2, pp. 76-79.

[208] Grant, 9, p. 67 ff.

[209] See Dijksterhuis, p. 144.

void of them. Therefore immediately one of them leaves its position, another succeeds it without interval, nor can one leave its place unless some other which is especially attached to it can succeed it.[210]

A similar theory was proposed by Grosseteste and Roger Bacon in the next (i.e. thirteenth) century.[211] They note the appearance of a *fuga vacui*, but deny that a mere non-entity can act as an efficient cause. How then can the *fuga vacui* be explained? One must, they reply, assume that all bodies are influenced by two 'natures', *particular* and *universal*. The particular nature of a body gives it its peculiar natural motion, e.g. fire upward and earth downward. Bodies are also, however, influenced by a *natura universalis*, which originates in the celestial realm, whence this celestial force (*virtus caelestis*) is diffused throughout the sublunary world. This *natura universalis* is the vigilant guardian of material continuity, the positive efficient cause of those movements which take place to prevent vacuum$_2$-formation. Wherever this is threatened, 'universal nature' intervenes to forestall it: if necessary, it causes things to act contrary to their particular natures.

This theory was widely accepted among the scholastics, although the precise ontological status of the *natura univeralis* remained a matter of doubt and debate. Is it a universal *property* of all bodies, or a separately existing agent? Bacon and Burley, says Grant,[212] seem to have conceived it as the latter, i.e. as a separate, presumably incorporeal, and purposive (sentient?) agent, a veritable *deus ex machina*.[213]

For a thirteenth century philosopher, void is impossible *simpliciter* – necessarily non-existent. After the condemnation of 1277, however, it became obligatory to grant that God could, if He chose, *overcome* the *fuga vacui* and create a void space. As we shall see, however, this remained no more than a supernatural possibility for most fourteenth century schoolmen: the 'Nature abhors a vacuum' maxim remained unaffected. Thus Albert of Saxony:

> . . . if God should annihilate everything between two walls, those two walls would become rarefied infinitely quickly

[210] Quoted from Thorndike, Vol. 2, p. 38.

[211] See Grant, 9, pp. 69–70.

[212] *Ibid.*, p. 304.

[213] This is precisely the conception of 'Nature' to which Boyle objects in his important monograph, 'A Free Enquiry into the Vulgarly Received Notion of Nature', in *Works*, Vol. 5, p. 158 ff.

[*in infinitum velociter*] in order to avoid a vacuum. This can be proven, because it has already been declared that natural things seek to avoid a vacuum and are inclined toward this action by their very natures. Now, they strive for, or are inclined to, this finitely or infinitely. If the latter, the sides [or walls] would meet infinitely quickly and the intent of this [conclusion] would be had. However, if the striving is only finite, it follows that a finite power (*virtus*) that is greater than the power and inclination of these walls would be sufficient to hold them and prevent their meeting. Hence a finite power could conserve a vacuum and even increase it . . . But this is false, because only God could conserve, increase, and make a vacuum.[214]

One is disappointed with the bald assertion of 'Only God can make a vacuum' to 'prove' the infinite strength of the *fuga vacui*. The assertion of an infinite *fuga vacui* has such obvious empirical implications[215] that one is distressed when Albert appeals to no experimental evidence but concludes that

if God should make a vacuum, natural beings would then strive infinitely to avoid it if their natures could be directed; and this is what is commonly said: 'Nature avoids a vacuum infinitely'. Indeed, before the sky would allow a vacuum to remain, it would descend and fill it.[216]

The word 'before' is crucial here. It is not that a vacuum$_2$ forms, then *sucks* the sky down into itself, but that the sky falls instantaneously as sublunary matter is annihilated, unless God's infinite power prevents it. An infinite *fuga vacui* entails the *natural impossibility* of a vacuum$_2$, a fairly orthodox scholastic belief, compatible both with philosophy (Aristotelianism) and Christian dogma.

In the work of Jean Buridan, a similar position is defended in a more empirical spirit. Universal propositions in natural philosophy, he said, should be established by experimental induction. Now, he continues,

let us show by experience that we cannot separate one body from another unless another body intervenes. Thus if all the holes of a bellows were perfectly stopped up so that no air could enter, we could never separate their surfaces. Not

---

214   From Grant, Ed., 4, p. 339.

215   Even if not conclusively verifiable, it is easily falsifiable – and falsified – by empirical evidence.

216   From Grant, Ed., 4, p. 339.

even twenty horses could do it if ten were to pull on one side and ten on the other; they would never separate the surfaces of the bellows unless something were forced or pierced through and another body could come between the surfaces.[217]

Another experiment involves a hollow reed,

> one end of which you place in wine and the other end in your mouth. By drawing up the air standing in the reed, you [also] draw up the wine by moving it above [even] though it is heavy. This happens because it is necessary that some body always follows immediately after the air which you draw upward in order to prevent the formation of a vacuum.[218]

Buridan concludes that the principle that a vacuum$_2$ is not naturally possible has been given a sound inductive proof. On purely empirical grounds, however, all he has established is the existence of a powerful (but possibly finite) *fuga vacui*: he goes beyond his experimental support when he states that it is (naturally) *impossible* to open the blocked bellows, or that the wine *necessarily* ascends in the straw (to any height).

Curiously, Buridan's thought-experiment with the bellows and the horses bears a striking resemblance to Guericke's actual experiment, three centuries later, with the Magdeburg Hemispheres. But, as Grant remarks,

> . . . where one team of horses laboured mightily to demonstrate nature's abhorrence of a vacuum, the other laboured with equal energy to demonstrate the creation of an artificial vacuum and the powerful pressure of the atmosphere.[219]

We now leave the fourteenth century and jump to the sixteenth, when the rediscovery of ancient texts – notably Diogenes Laertius, Hero, and Lucretius – helped to shake the dominant Aristotelianism of the schoolmen. In the field of pneumatics, the recovery of Hero's classic work was of the greatest importance.[220] According to Hero, there are interstitial vacua between the particles of bodies, and these minute vacua can be 'collected', and a separate vacuum$_2$ formed, by

271   *Ibid.*, p. 326.

218   *Ibid.*, p. 326.

219   *Ibid.*, p. 326.

220   For Hero's transmission and influence, see Boas, 1.

artificial means – by a sufficient (finite) force.[221] This position is, plainly, fundamentally inconsistent with the infinite *fuga vacui* of the schoolmen. Anti-Aristotelian writers like Telesio and Patrizi favoured Hero's position, and used it to attack the authority of the Stagirite. Let us now examine the ensuring controversy.[222]

According to scholastic orthodoxy, no finite force can produce a vacuum$_2$. The blocked bellows, it was argued, would burst asunder rather than open and leave a void. Any vessel can be broken by a sufficiently large, but finite, force: the vessel, says the scholastic Domingo de Soto, will always break *before* a vacuum$_2$ is created.[223] Telesio retorts that a sufficiently strong bellows could be opened without breaking, thereby implicitly affirming the finitude of the *fuga vacui*. The whole debate, however, occurs at the level of thought-experiment: no actual experimental evidence is offered by either side.

Another classical experiment meant to illustrate the workings of the *fuga vacui* is the clepsydra or water-clock. The 'Aristotelian' explanation of this experiment should by now be familiar. This is Patrizi's rival account:

> . . . the clepsydra or garden water-clock, by which many have attempted to establish that a vacuum does not exist, proves the opposite, namely, that empty space can exist in it. For if it has tiny openings in the bottom, and if the hole at the top is closed tightly with a finger, water will not flow through these openings in the bottom. And thus it seems that it does not admit a vacuum. But if one or more openings in the bottom are made large, water will undoubtedly flow through, and before it all flows out there will be a vacuum inside the clock. Experience shows this fact. But indeed, if the water-clock is filled with sand, it will fall through those tiny apertures, being heavier than water. The same is true of honey and oil.[224]

By modifying certain inessential boundary conditions, says Patrizi, the experiment can be made to yield very different results. If, for example, we increase the density of the fluid (how about mercury?), we increase the downward force

221   See Chapter 2, pp. 94–96.

222   The debate between Telesio and Patrizi and their Aristotelian antagonists is well described in Schmitt, 1.

223   Schmitt, 1, p. 357.

224   Patrizi, p. 234.

tending to create a vacuum$_2$: the greater this force, the more likely it is to overcome the (limited) *fuga vacui*.

Patrizi also argues[225] that an empty, sealed pouch *can* be stretched out, creating a vacuum$_2$ inside it.[226] According to the scholastics, the pouch must either break, or leak, or not open at all. In fact, says Schmitt,[227] experience shows the creation of such a vacuum$_2$ to be 'no easy task': once again, Patrizi is probably operating only in the realm of thought-experiment.

Patrizi goes on to argue for interstitial vacua in a manner strikingly reminiscent of Hero. Such vacua are, he claims, needed to account for condensation and rarefaction (not to mention local motion) as 'the wisest of the ancients' saw. Large vacua, he adds, can be made 'by some forces and artificial processes.[228] By sucking the air out of a closed vessel we produce, he says, 'much of a vacuum' inside it. This is proved – *à la* Hero – by the fact that such a vessel will then suck up water, against its natural motion. (The Aristotelian can of course say that the vessel contains 'rarefied air', and attribute to this a force of suction due to its 'desire' to return to its 'natural' state.)

This whole debate, Schmitt stresses, took place in the realm of the thought-experiment.[229] Patrizi and his opponents were largely in agreement about observed cases and disagreed only on the expected outcome of hypothetical experiments. But since these thought-experiments were never made actual – there is no evidence that Telesio or Patrizi actually performed the experiments they suggest – the outcome of the debate could only be inconclusive.[230]

If we are looking for the real, and decisive, experimental refutation of the *infinite fuga vacui*, we must for once quit the academies and go down the mines. Mining engineers of the sixteenth and early seventeenth centuries were already familiar with the fact that water cannot be lifted more than 34′ by a suction pump. A clepsydra more than 34′ high would

---

225 *Ibid.*, p. 234.

226 This argument can be traced back at least as far as an eleventh century Arab writer. See Schwartz, p. 385.

227 Schmitt, 1, p. 356.

228 Patrizi, pp. 232–233.

229 Schmitt, 1, pp. 362–364. Patrizi and his opponents both assumed, for example, that water *contracts* on freezing! A simple *actual* experiment would have sufficed to enlighten them.

230 *Ibid.*, pp. 362–363.

therefore lose water – even if the air hole at the top were blocked. With the much denser fluid, mercury, the critical column height is much less. The allegedly infinite *fuga vacui* will not support even a three foot high column of mercury! Moreover, the vessel does not break under the strain. Nature may 'abhor' a vacuum – the fuga vacui theory can survive this result – but it does *not* do so *infinitely*.

But, as soon as one admits only a finite *fuga vacui* all Hero's other conclusions follow. One can, if one exerts a sufficient force, create a separate vacuum$_2$. There are *actual* interstitial vacua in things. The *fuga vacui* is a finite force of suction, exerted by genuine vacua. (Galileo, for example, explained the *cohesion* of bodies as due to the *fuga vacui* exerted by their interstitial vacua.) The whole scholastic edifice collapses.

G. Conclusion

We have witnessed the disintegration of the world-view of scholastic Aristotelianism. According to scholastic orthodoxy, void or vacuum$_2$ is:

a.   Conceptually absurd: place = 2D surface of container, 3D ⟶ corporeal.
b.   Naturally unattainable: an infinite *fuga vacui* forestalls its formation.
c.   Dynamically impossible: there is no motor-medium to impel projectiles, while for natural motions, if R = 0, v = ∞.

A great rift is made in this beautifully coherent system by the condemnations of 1277, according to which God can, if He chooses, create a void, and even move our world and create others in extracosmic ('imaginary') space. Once one admits 'imaginary' space, however, either within the cosmos or outside it, the question of its *dimensionality* naturally arises. As Henry of Ghent and John de Ripa show, it is difficult – if not impossible – to deny that this 'imaginary' space is truly extended. But, if one admits this, one must deny both Aristotle's definition of place and his '3D ⟶ corporeal' premiss.

While the schoolmen were debating the nature and properties of 'imaginary' space, the anti-Aristotelian theory of place as vacuum$_1$ or 3D incorporeal receptacle was being developed by Crescas, Pico, Patrizi, Bruno and others. Once established, the place = vacuum$_1$ theory could accommodate the scholastics' 'imaginary' space by effectively identifying it with true space, the general or universal space of vacuum$_1$, *in* which all things have their being.

But now if space = vacuum$_1$ or 3D immaterial receptacle, it becomes an *empirical* question whether or not any empty space or vacuum$_2$ exists. The scholastics assumed the existence of an infinite *fuga vacui*, but their evidence did not warrant such a strong conclusion. By the sixteenth century, Telesio and Patrizi are advocating Hero's position, that the *fuga vacui* is finite and can be overcome artificially. The experience of the mining engineers clinched the issue, and set the stage for the next part of the story – the brilliant pneumatical researches of Torricelli, Pascal and Boyle.

Meanwhile, in the realm of dynamics, the impetus theory was supplanting the motor-medium theory of forced motion, and was itself helping to pave the way for the modern theory of inertia. At the same time, the $v = v' - r$ formula of Philoponus and Avempace was gradually replacing the 'Aristotelian' $v \propto F/R$ formula, and thus helping to make sense of a finite velocity in a vacuum$_2$. By around 1600, none of the Aristotelian anti-vacuist arguments remained viable.

MATTER, FORMS, AND QUALITIES: 2

A. The Nature of Matter

There is, as we explained in Appendix 2, a fundamental tension in Aristotle's thinking on the subject of matter. On the one hand, we can think of it as stuff-like, i.e. as actual, space-filling, quantified material. This, according to Sokolowski, was the standpoint of Aristotle himself.[1] 'Matter is extended,' says Sokolowski, does not involve the predication of extension to a non-extended prime matter (P.M.): extension is a precondition of any predication, not something predicated:

> For Aristotle the underlying matter is simply formless, unqualified, space-filling stuff. Its extension is not an attribute, quality or predicate, but is as primary as matter itself.[2]

Opposed to this construal of the doctrine of P.M. we have the rival view, which lays special stress on the rôle of P.M. as the ultimate logical subject of all prediction, and hence, in Aristotle's own words, as

> that which in itself is neither a something, nor a so much, nor a such, nor any of those things by which being is determined.[3]

On this view, P.M. has no actuality at all: matter is actualised by form. But a metaphysical 'principle of potency' cannot serve as the physical substratum underlying generation and corruption.[4] In Aristotle's system, P.M. has to play too many different – and incompatible – rôles: in the final analysis, his account of matter is incoherent.

A number of different responses were possible in the face of this dilemma. Stoics and Atomists insisted alike on the actuality of matter, and ignored or rejected the subtleties of Aristotle's *Metaphysics*. The Neoplatonists, by contrast,

---

1 Sokolowski, p. 278.

2 *Ibid.*, pp. 278–279.

3 *Metaphysics*, 7, 3, 1029a 19–21.

4 The reasons for this claim can be found in Appendix 2.

tended to accept the notion of *hulē* as pure potentiality, somehow intermediate between being and non-being.[5] Thus Plotinus:

> But as for matter, which is said to exist and which we say is all realities potentially, how is it possible to say that it is actually something real? For if it was, it would already have ceased to be potentially all realities. If then, it is nothing real, it necessarily cannot be existent either. How could it, then, be actually something when it is nothing real?[6]

But if matter has no actual existence, how can we meaningfully speak of it? How, indeed, can it be the material component of sensible things? This is because, Plotinus answers,

> it is they potentially. Then, because it is they already potentially, is it therefore just as it is going to be? But its being is no more than an announcement of what it is going to be: it is as if being for it was adjourned to that which it will be. So its potential existence is not being something, but being potentially everything; and since it is nothing in itself – except what it is, matter – it does not exist actually at all. For, if it is to be anything actually, it will be what it is actually and not matter: so it will not be altogether matter but only matter in the way that the bronze is.[7]

A piece of bronze has some positive actuality of its own, prior to being cast into a statue; P.M. by contrast, has none. Matter (*hulē*) is as such 'really unreal', a 'phantasm', even a 'falsity'.[8] As in the *Timaeus*, matter is the receptacle of Forms, the *mirror* in which the true realities (the Forms) are reflected, but Plotinus – even more than Plato – emphasises the illusory nature both of the mirror and the images in it. Matter, he insists, is 'truly non-being; it is a ghostly image of bulk, a tendency towards substantial existence . . .'[9] Its apparent being

> is not real, but a sort of fleeting frivolity; hence the things which seem to come to be in it are frivolities, nothing but

---

[5] For Plotinus on matter, see *Enneads*, 2, 5 (Vol. 2, p. 153 ff.), and 3, 6 (Vol. 3, p. 206 ff.).

[6] Plotinus, *Enneads*, 2, 5, vol. 2, p. 167.

[7] *Ibid.*, pp. 167–169.

[8] *Ibid.*, p. 169.

[9] *Ibid.*, 3, 6, Vol. 3, p. 241.

phantoms in a phantom, like something in a mirror which really exists in one place but is reflected in another, it seems to be filled, and holds nothing; it is all seeming, 'Imitations of real beings pass into and out of it', ghosts into a formless ghost.[10]

Matter is in itself 'impassible', totally unaffected by the 'wraith-like and feeble' images projected on to it.[11] Being devoid of qualities such as HCDM, *hulē* cannot itself be affected by their presence in it. Matter itself, for example, is never set on fire: when a particular *body* burns, a particular portion of *hulē* reflects the form of fire – i.e. the image of fire temporarily appears in it.

One must, then, draw a sharp distinction between matter (*hulē*) and body (*soma*). Matter itself, Plotinus insists, is 'one of the things without body',[12] incorporeal and unextended – its incorporeal nature explains its impassivity, the fact that it is unaffected by bodily things. In conclusion,

> Matter . . . is incorporeal, since body is posterior and a composite, and matter with something else produces body.[13]

What is this 'something else' which, in conjunction with *hulē*, produces bodies? The simplest answer is 'quantity' or 'three-dimensional (3D) extension'. The metaphysical category of pure potentiality can thus combine with the equally abstract principle of 'quantity' to produce a quantified, space-filling stuff, a 'matter signed with quantity', *materia quantitate signata*.[14] Some Neoplatonists followed Proclus in identifying this 'first corporeal form', the source of the 'common corporeity' shared by all bodies, with *light*: the resulting Neoplatonic 'light-metaphysics' is central to the philosophy of Grosseteste and Roger Bacon.[15] Light, says Grosseteste, spontaneously spreads from a point-source to fill a (spherical) space: it is therefore the source of 3D extension, the first of the hierarchy of forms 'informing' P.M. In this theory of matter, forms are 'added' successively to P.M.: the first, the

10   *Ibid.*, p. 243.

11   *Ibid.*, pp. 247–249.

12   *Ibid.*, p. 233.

13   *Ibid.*, p. 241.

14   See Crombie, 1, p. 50; Jammer, 2, pp. 31–32.

15   See Crombie, 2, p. 13.

'corporealising form', makes of it a *body*, something with 'common corporeity', i.e. magnitude and 3D extension;[16] subsequent forms make of it a body of a particular kind.

Neoplatonic influences can be discerned in the discussion of matter by the Aristotelian commentator, Simplicius.[17] He asks the ten thousand dollar question 'Is *hulē* extended?', and has difficulty finding a clear answer in Aristotle, who seems to say both yea and nay. Matter, says the Stagirite, is 'extended body' (*soma*), with 3D extension and magnitude. Yet Aristotle also states that *hulē* lacks quantity: *qua* ultimate subject of predication, it is '*nec qualum, nec quantum, nec quid*', as the schoolmen put it. This contradiction worried Simplicius, and many others in the scholastic tradition.

What Simplicius proposes is the existence of a 'corporeal form' (CF), *prior* to the individual substantial form (SF) of a body, which imposes quantitative and spatial properties on to unextended P.M., giving:

P.M. + C.F. ———→   BODY

This 'body' (*soma*) is 3D stuff, quantified, but still qualitatively indeterminate prior to its 'information' by a particular SF. Is this bare and featureless *soma* capable of separate and independent existence? *Some* of the Aristotelian reasons for denying separate existence to P.M. would no longer apply to it: *soma* could actually exist as such, somewhere, and in some quantity.[18] One still finds it hard, however, to conceive of a qualityless stuff.

In this schema, since P.M. is *always* informed by CF, and their composite (3D body or *soma*) serves as the *substratum* for physical processes, P.M. can be conceived as something very abstract and metaphysical, with little or no physical significance, something 'flowing from being into nonbeing', as Simplicius puts it. The two rôles for *hulē* come apart: metaphysics has its purely potential P.M., devoid of any actuality of its own, while the physicist can treat the 'first actualisation' of matter (P.M. + CF) *as* matter for further actualisation. This new 'corporeal matter' is actual, space-filling, and 3D, i.e. the sort of substratum required by the physicist.[19] It can fulfil many of the functions – quantitative

---

16   *Ibid.*, pp. 106–107. See also Wallace in Lindberg, Ed., p. 95.

17   See Jammer, 2, p. 21 ff.

18   Perhaps, when the alchemists sought a reduction to P.M., what they really meant was P.M. + CF, or 'common corporeity' – this would render their project somewhat less incoherent.

19   See Weisheipl in McMullin, Ed., 1, pp. 151–152.

conservation, individuation, etc. – that a purely potential P.M. could not.

A similar theory occurs in Avicenna, who also postulates a 'corporeal form' which turns P.M. into bodies. A body, he says, is a substance possessing three-dimensional extension,[20] which feature it owes to its 'material form':

> Any receptacle in which form subsists and which is itself other than a form is called simple matter. The material form subsists therefore in the substratum-matter, and from such a form and such a simple matter a body is realised. This, for example, is true of a polo ball realised from wood and from roundness.[21]

A body is an inseparable fusion of P.M. and CF: P.M., Avicenna insists, can have no actuality of its own apart from CF:

> . . . the substratum of a material form is not an actuality without a material form. It is an actual substance due to the material form. In reality, therefore, the material form is the substance. It is not the case that the substratum of matter is in itself an actual thing and that the material form is a necessary accident of it.[22]

The 'material form', says Avicenna (following Aristotle's *Metaphysics*), is a 'this' or substance, not the P.M. P.M. is *actualised* by CF, which imparts to it 3D extension, i.e. makes of it a body:

> Truly, therefore, being a body is due to a form, although there is no doubt that this substratum becomes a body due to the form of materiality.[23]

There is, says Avicenna, a CF common to all bodies as such ('the material form of all bodies is identical . . .'[24]) and distinct from their specific SFs. FAWE, for example, owe their different 'natures' to their own SFs:

> . . . the cause for the position of one body at one place rather than at another must be a natural feature different

---

[20]   Morewedge, Ed., p. 17.

[21]   *Ibid.*, pp. 18-19. See also Afnan, pp. 111-112.

[22]   *Ibid.*, p. 24.

[23]   *Ibid.*, p. 25.

[24]   *Ibid.*, p. 25.

from being a body. Accordingly, the substratum of a body seeks a form other than the form of materiality.[25]

This gives us a plurality of forms in the same matter. Every body has both a CF which imparts corporeality,[26] and a specific SF which gives it its particular nature. Hence all *mādda* (P.M.) is always found 'informed' both by a (general) CF and a (specific) SF. One could not, for example, separate P.M. + CF without a specific SF – this would give a body without any characteristic nature of its own.

In Hasdai Crescas, the fourteenth century Jewish philosopher, the CF theory was given a new twist. In the philosophy of Crescas' day, says Wolfson,[27] matter was not conceived as utterly formless and indeterminate; rather, it was thought of as informed by Simplicius' CF or Plotinus' 'quantity'. The P.M. of Aristotle, it was argued, is unextended and thus incorporeal. But the elements FAWE cannot derive from something incorporeal. So 'between' P.M. and FAWE there must be some intermediate form, endowing *hulē* with extension. This 'first corporeal form' (CF) is *common* to all bodies, as in Avicenna, imparting to them their 'common corporeality' or bodily nature.[28] Crescas does not commit himself to any precise theory of the rôle of CF in conferring 3D extension on matter; instead, he pursues a line of thought originating with Averroes but eventually departing far from the orthodoxy of the above position.

According to Averroes' interpretation of Aristotle,[29] the 'quintessence' Q (the stuff of the heavenly spheres) is without P.M., consisting only of 3D body and its specific form, the former serving as matter to the latter. This lack of P.M. (the *substratum* of generation and corruption) explains, for the peripatetic, the incorruptibility of the heavenly bodies.

---

25   *Ibid.*, p. 25.

26   Exactly *how* CF confers dimensionality on P.M. became a topic of considerable debate among the Arabs. Avicenna, Al-Ghazzali, and Averroes all tackle this thorny problem, and give very different answers. For more details, see Wolfson, 1, p. 101; Jammer, 2, p. 38.

27   Wolfson, 1, p. 100.

28   *Ibid.*, pp. 101–102, p. 261.

29   Textual support for such a reading is not lacking in Aristotle's *Metaphysics*. Perhaps, it is suggested (8, 4, 1044b 6–8) eternal substances 'have no matter', or their matter is liable only to local motion and not to generation and corruption (12, 2, 1069b 25–27). Elsewhere (9, 8, 1050b 16–18) he states that eternal and imperishable things 'exist actually' without any potentiality. But P.M. is the principle of potentiality: a purely actual being must therefore lack P.M.

Sublunary bodies possess P.M. and are forever involved in the cycle of birth and death; superlunary things lack P.M. and are deathless and immortal.

In Crescas, this line of thought is pursued to a radically new conclusion.[30] The spheres, he says, are on this view *immaterial* (without P.M.) but *corporeal* – 3D, impenetrable, etc. Why not, then, apply the same reasoning to terrestrial matter? P.M., on such a view, is quite idle and superfluous: all the essential properties of a body can be possessed by an entity lacking P.M. One can therefore simply abolish P.M. – and with it of course CF – from one's metaphysics, giving a much tidier matter-theory, involving a common, actual matter or 3D body, logically prior both to FAWE and to Q. The whole notion of matter as a metaphysical 'principle of potency' has been dropped: the magnitude of a body is *not* to be predicated of anything in itself unextended. Matter and quantity are *inseparable* and ontologically primary: bodies are not compounded of P.M. plus CF.

The metaphysical shift here is both monumental and momentous. The crucial point is that form (SF) is no longer conceived as the principle of actualisation of a purely potential P.M.; rather, matter exists (partially) informed of its own nature. Form can now be thought of as an accident of 3D body, except of course insofar as a specific SF is still needed to give each substance its particular essence or nature.

So, for Crescas, matter = 3D body, and is a common, actual, extended *stuff*. If we add to this theory of matter his restoration and defence of the vacuum, we can see where this writer is leading us. Although Crescas cannot himself be called an Atomist, there is underlying his arguments, as Wolfson says, 'an attempt to revive Atomism'.[31] The *actuality* or otherwise of matter is, as Wolfson points out, one of the great metaphysical issues dividing Atomists from Aristotelians, and on this issue at least Crescas clearly leans towards the Atomist position.

A similar view gradually emerged among the Christian scholastics. Although Aquinas reaffirmed 'Aristotelian' orthodoxy by denying the actuality of P.M. – not even God, he insists, could give matter actuality independent of form[32] – Scotus and Ockham demurred. Matter, says Aquinas, is not itself substance, 'or else all forms would be accidental, as the

---

30 See Wolfson, 1, pp. 103–104, p. 263.

31 *Ibid.*, pp. 120–121.

32 See Wallace in Lindberg, Ed., p. 100.

early natural philosophers held',[33] and we would lapse into materialism. In itself, says Aquinas, P.M. is purely potential,[34] devoid of actual existence, and hence not substantial. SF is therefore needed as the *principle of actualisation*: it is in virtue of its SF that a particular substance exists at all – this gives a sharp distinction between 'substantial' and 'accidental' forms:

> Accidental form does not merely make a thing *exist*, but makes it exist in some way, i.e. as large or as white, etc. But substantial form merely makes a thing exist, so that accidental form is added to an already existing subject. A substantial form is not added to an actually existing subject, but to a potentially existing subject, namely, prime matter. It is clear, therefore, that one thing may not have many substantial forms. The first form would make a thing merely exist, all others would be added to an already existing subject. So they would be added accidentally.[35]

Aquinas thus affirms the purely potential existence of P.M. and the *unity* of SF in each substance (*versus* the pluralism of Avicenna). Every substance exists, he claims, in virtue of its unitary and unique SF; without that form, it would be no more than 'potency' or P.M. If P.M. were actual, he sees, form would not actualise matter – i.e. there would be no *substantial* forms. The metaphysical thesis of the actuality of matter has radical consequences.

Nevertheless, some scholastics were prepared to take just this step, and to attribute some degree of actuality to matter. According to Scotus, matter has its own idea and hence its own actuality: form is not the principle of actualisation of matter.[36] God could, if He chose, create a piece of actual but formless P.M. In William of Ockham, the argument is taken a stage further: he argues at length for an actual and space-filling P.M., and presents his conclusion in the guise of an exegesis of Aristotle.[37]

SF and P.M. are, says Ockham, mutually interdependent theoretical concepts.[38] The hylomorphic theory of Aristotle,

33   From M. T. Clark, Ed., p. 58.

34   *Ibid.*, p. 67.

35   *Ibid.*, p. 212.

36   See Bréhier, p. 185; Wallace in Lindberg, Ed., p. 108.

37   If Sokolowski is right, then Ockham's reading of Aristotle may be far more authentic than those of most of his contemporaries.

38   See Wolter's paper in McMullin, Ed., 1, p. 129 ff.

though not demonstrable *a priori*, is evident *a posteriori* from the phenomena of generation and corruption. (This presupposes the falsity of the 'Generation = Aggregation' thesis.[39]) P.M. and SF are, he insists, 'two things, really distinct'. But what is really distinct can be separated by God and made to exist on its own. This implies that P.M. has some actuality of its own – it becomes a bare, primitive, space-occupying stuff devoid of forms and qualities.[40]

The basic argument for such a conception of matter is simple. How, one could demand, can a purely potential (i.e. non-existent) P.M. serve as the material principle of things and, crucially, as the underlying substratum for generation and corruption? Aristotle's P.M., as we have said before, just has to do too many incompatible jobs.

Ockham's matter, then, has actual existence as a physical reality, while at the same time being in potency to all forms. P.M. is still devoid of form: it is-actually matter (extended but featureless stuff) and is-potentially FAWE, etc. As Leff says: 'Only in relation to form therefore is it potential; in itself it is always actual'.[41] P.M. is matter without SF, undifferentiated matter. Although not a substance or 'this', it can exist without SF, which is therefore *not – pace* Aquinas – the principle of actualisation of matter. The distinction between matter and substance is not that between 'potency' and 'act', but between different grades of actuality. As Leff says,

> In many ways the difference is a fundamental one; for it helps to endow matter with its own separate identity where previously, both from an Aristotelian point of view, by which it was dependent upon a substantial form, and from an Augustinian point of view, in having its own form of corporeity, matter's actuality was appropriated to a form.[42]

In order to fulfil its rôle of individuating distinct individuals of the same species, matter must consist of numerically distinct parts or quanta. Matter possesses 'quantity', however, not as something over and above its own nature, but as a feature intrinsic to it. Substance and quantity, says Ockham, are not *really distinct*: even Divine power could not

---

[39] Generation = aggregation, says Ockham, *only* for artefacts, not for natural things. See Leff, p. 579.

[40] Several other scholastics favoured similar views. See Wolter, pp. 130–132.

[41] Leff, p. 573.

[42] *Ibid.*, p. 574.

separate them.[43] If God were to separate all the accidents (colour, etc.) from a substance, while leaving the parts of the substance itself unaffected, its quantity (extension) and figure would remain the same.[44] To remove the shape and size from a body would be to destroy the body: matter consists essentially of spatially distinct parts and is therefore inseparable from physical extension. Thus Leff:

> matter without extension is impossible since it must always have extended parts which . . . cannot all occupy the same place. Extension is therefore nothing distinct from quantity.[45]

Ockham's departure from 'Aristotelianism' – if not perhaps from Aristotle himself – is a vast and momentous one. For Ockham, matter does not receive its existence from form – form is not the 'actuality' of matter. Matter exists (actually) prior to receiving form and – in one sense – is not altered by being 'informed', just as wood remains wood when made into tables and chairs. Moreover, P.M. is alike in all bodies – celestial or terrestrial[46] – and only numerically distinct in different individuals. Given such an actual, space-filling, quantified matter, a reductive and materialistic account of forms and qualities becomes once again a real option – if form actualises matter, one cannot explain the nature and origin of forms in terms of material constitution and/or arrangement of material parts.

Ockham himself takes only the first few steps along this path. Only the forms of *artefacts*, he argues, are reducible to the arrangements of their material parts; natural substances such as plants and animals are still deemed to possess true SFs.[47] As for qualities, only shape, size, density and rarity are dependent merely on the spatial arrangement of material parts; other qualities such as colour and temperature are conceived by Ockham as 'real', i.e. separable by divine power. If, says Ockham, a particular quality can be

---

43 If, says Ockham, there is a real distinction between X and Y, God can separate them and sustain them in independent existence. 'Every absolute being,' he insists, 'can exist by divine power after the other absolute is destroyed' (Leff, p. 18). Qualities such as colour, he thinks, pass this test; quantity fails it, and is therefore not to be conceived as a '*res*', an absolute reality in its own right.

44 See Leff, pp. 203–204.

45 *Ibid.*, p. 576.

46 *Ibid.*, p. 576.

47 *Ibid.*, p. 579.

successively affirmed and then denied of a subject merely in virtue of local motion of its constituent parts, that quality is not 'absolute', really distinct from the substance and separable by divine power.[48] Dense, rare, straight, curved, etc., pass this test and are therefore inseparable from substance; red, yellow, hot, cold, etc., do not: their inherence in a given subject is not due to local motion but to a distinct (real) quality.

Although Ockham himself does not really pursue the possibility of a materialistic reduction of forms and qualities very far, he opens the possibility of such an approach to others. Once one grants the *actuality* of matter, the possibility of a materialist account of form emerges. Form, on this view, is an *accident* consequent upon matter.[49] – the position Aquinas attributed to the Presocratic *phusikoi* was to become dominant in the 'corpuscular philosophy' of the seventeenth century.

Let us now turn to a different, but clearly related, topic: the quantification of matter. The notion of quantitative conservation of matter has been axiomatic since the time of Thales: it is one of the most fundamental principles of all Natural Philosophy. For the Aristotelian, however, the axiom has *no clear sense*;[50] if P.M. is 'without quantity', it cannot be conserved by quantity.

Yet Aristotle, as we saw in Appendix 2, *needs* the concept of 'quantity of matter' (Q.M.) if he is to use P.M. to individuate different substances, or to explain why the material of a mouse can form a mushroom but not an oak tree. Physics and common sense alike demand a quantified P.M.

Two obvious, but quite distinct, empirical features of bodies might be used as measures of Q.M., namely, bulk or volume, and weight. On an Atomistic matter-theory, the two measures will in a sense coincide: both volume or bulk of matter*, the stuff of atoms, and gross weight (weight *in vacuo*) measure the total quantity of matter* in a body. Since

---

[48] *Ibid.*, pp. 204–205.

[49] For an interesting intermediary figure, see Bruno's 'Principle Cause and Unity'. According to the ancient Atomists, says Bruno, 'what is not body is nothing'; forms are only 'accidental dispositions of matter' (p. 99). For a while, says Teofilo (Bruno's spokesman), 'I myself . . . adhered to this view' (p. 100), before attempting instead to find a compromise position between the materialism of the Atomists and the immaterialism of Plato and Aristotle.

[50] See Weisheipl in McMullin, Ed., 1, p. 147.

matter* is homogeneous and uniformly possessed of weight, the Q.M. of a body can be identified with the total bulk of its constituent atoms, and can be determined simply by weighing.[51] (Net or measured weight equals gross weight, less the Archimedean upthrust of the ambient medium.) There may be some minor operational difficulties, but the essential conceptual clarity of the account is manifest.

For the Aristotelian, however, all sorts of complexities arise. P.M. itself, the metaphysical 'principle of potency', cannot meaningfully be said to be quantified at all. What about P.M. + CF, the 'first body' or 'first actualisation' of P.M.? This at least possesses 3D extension and magnitude. Unfortunately, however, Aristotelian matter can undergo (intrinsic) condensation and rarefaction: it cannot therefore be quantified according to volume. What is, intuitively, the same Q.M. can fill more or less space (and can do so, for the peripatetic, without leaving pores or interstitial vacua). Weight, however, fares no better as a measure of Q.M. Fire, for a peripatetic, is absolutely light, but nonetheless material: adding fire to a compound would increase its Q.M. but decrease its weight. Weight and lightness were normally construed by Aristotelians as contrary *qualities* of bodies;[52] given such a view, the notion that weight could serve as a measure of Q.M. is as absurd as the supposition that redness or heat could.

The Aristotelians are, therefore, faced with the following quandary. The P.M. of the *Metaphysics* cannot intelligibly be said to be quantified at all. But, even if one thinks of matter as actual, space-filling stuff (P.M. + CF), one still cannot make any clear sense of the concept of Q.M. Our intuitions favour the application of such a concept, but leave us with nothing more than intuition to go on, i.e. with no clear idea of *how* to quantify P.M.

One historically influential attempt to resolve this problem derives from Averroes. P.M., he argued, must be quantified if it is to perform the rôle of individuation. Distinct parts or portions (quanta) of P.M. must correspond to different individuals; two mice must contain (approximately) twice as

---

[51] The Atomists were not alone in antiquity in postulating the conservation of matter by weight. Lucian of Samosta, when asked what smoke weighs, replied: 'Weigh the wood before burning and ashes afterwards – the difference is the weight of the smoke' (from Jammer, 2, p. 26).

[52] One consequence of this conception of weight and lightness as qualities was the denial of the additivity of weights (see Sambursky, 4, p. 67).

much P.M. as one.[53] But Q.M. cannot be identified with the determinate dimensions of a body at some particular time. Rather, as Weisheipl explains,

> there must be another kind of dimensionality prior to every divided form, by which primary matter would be divisible and actually divided from all other segments of itself. This antecedent dimensionality Averroes called *dimensiones indeterminatae*. This allowed him to explain solidity in a given amount of primary matter and variability in actual terminated dimensions.[54]

Thus the determinate (measurable) dimensions of a body will change during condensation or rarefaction, while its 'indeterminate dimensions' – a quantity – remain constant. We have of course no operational method of measuring these so-called 'dimensions', but their *conceptual* rôle as a notion of Q.M. is clear.

Aquinas and Albertus Magnus formulated the classical scholastic theory of matter, with P.M. as pure potency, no common spatialising CF, and each 'part'[55] of P.M. informed – and thereby actualised – by one and only one SF. Quantities of matter can only be conceived as quantities of actual stuffs – chalk, cheese, etc. – not of P.M.[56] This, however, seems counterintuitive. When one volume of water is rarefied to give ten volumes of 'air', which are then condensed to give back the initial quantity of water, our intuitions tell us that the same quantity of matter has been conserved throughout. But, to make sense of that intuition we need to quantify the abiding substratum, not merely the water and air. In rarefaction and condensation, we speak of the same Q.M. as occupying more or less space: this 'quantitative' approach to the nature of dense and rare is a very natural one. Aquinas[57] is inclined to make 'quantity' (= extension) the substratum of condensation and rarefaction. But in such processes it is precisely the extension which changes: the changing accident cannot be identical with the abiding substratum.

Aegidius of Rome takes up this problem from Aquinas, and proposes the following solution. We introduce first another

---

53 See Bobik in McMullin, Ed., 1, p. 283 ff.

54 Weisheipl in McMullin, Ed., 1, p. 155.

55 The sense of this term is not clear in the context of such a theory of matter.

56 Weisheipl in McMullin, Ed., 1, pp. 151-152.

57 See Jammer, 2, pp. 45-46.

concept of quantity, *quantitas materiae*, to serve as sub-stratum. Then this Q.M. is substratum; measurable 3D extension is the changing accident; condensation and rarefaction are changes in ratios between the two quantities.[58] Averroes' 'indeterminate dimensions' thus give way to Aegidius' *'quantitas materiae'*. This Q.M. is distinct from spatial extension: matter is only secondarily extended – its primary unit is the unit of 'mass' or Q.M.[59] This concept is still 'metaphysical' in the sense of being without operational import (I would prefer to call it 'intuitive'), but is none the less fruitful for that.

Marsilius of Inghen arrived at a very similar conclusion – an intuitive but as yet non-operational concept of Q.M. – in his work on condensation and rarefaction.[60] When hot air condenses on cooling, he asks, is there a decrease in quantity? To answer the question one must clearly distinguish two senses of 'quantity', he answers: plainly, the measurable dimensions shrink; equally plainly, the same Q.M. remains present, albeit confined in a smaller space.

Ockham's position on this topic of Q.M. is somewhat paradoxical. By insisting on a conception of P.M. as an actual, space-filling stuff, he might be thought to be sympathetic to the idea of its quantification. And in a way, of course, he is: Ockham's P.M. is stuff-like, extended, capable of individuating different things, etc. It is furthermore, as Leff says, indistinguishable from extension and quantity:

> Nor is it of any determinate measure but varies with the degree of quantity, which itself varies with the quantity of matter whose parts are in turn more or less extended according to the nature of the form to which it is subject: more when the form is fire, less . . . when it is air or water, but always having some distance between its parts.[61]

The 'quantity of matter' is a body is therefore not tied to any determinate set of dimensions, hence not measurable in any straightforward way. This might be thought congenial to the notion of Q.M., characterised perhaps in terms of 'indeterminate dimensions'. Unfortunately, however, Ockham's radical simplification of the concept of quantity seems to prohibit such a notion. 'Continuous permanent quantity,' he says 'is

58   *Ibid.*, pp. 46–48.
59   See Weisheipl in McMullin, Ed., 1, pp. 155–156.
60   In Grant, Ed., 4, pp. 350–352.
61   Leff, p. 576.

nothing but a thing with one part locally different from another part.'[62] To say that something possesses quantity is to say that it consists of non-superimposed parts, parts separated in space. But God could condense all the bodies in the universe to a *point*, thus annihilating 'quantity' (in this sense) altogether. Matter cannot, therefore, be quantified by *any* set of spatial dimensions, determinate or otherwise. Although matter is said to have quantity in virtue of its spatial extension, it could continue to exist without extension (and hence without quantity). The concept of 'quantity of matter' has, therefore, no clear sense.[63]

What this shows is the need for a concept of Q.M. *divorced entirely* from spatial dimensions. If God should condense all the matter of our universe into a single point, in virtue of what physical feature of that singularity (!) would we say that it contained all the matter that once made up the world?[64] Why should we say that God has *condensed* all the matter into that point, rather than that He has *annihilated* everything except it? The answer must lie in the notion of *density* or *compactness*. If, for example, one were to attempt to move this 'world-point' one should expect a quite colossal resistance. If matter as such possesses an *inclinatio ad quietem* (Oresme) or an *'inertia'* (Kepler), i.e. a tendency to rest, the world-point should be all-but impossible to move.

Matter, then, is conserved by mass, i.e. by a *dynamical* measure. This *corpus quantum* (Aquinas) or *massa elementaris* (Swineshead) is conceived of course as a measure of a body's resistance to motion, *not*, as for Newton, as a measure of a body's resistance to change of state of motion or rest. It does, however, play an analogous rôle in Buridan's impetus theory: the more massive a body (i.e. the more matter it contains), he says, the greater the quantity of impetus it can receive, and hence the greater the resistance needed to bring it to rest once it is set in motion.[65]

Q.M. is therefore essentially a measure of *mass*, i.e., for the scholastics, of (a) resistance to being set in motion at all, and (b) receptivity to impetus or motive force. The identification of Q.M. with this primitive notion of mass permits one to quantify matter in terms of some feature quite independent of

---

[62] Ockham, p. 137.

[63] See Weisheipl in McMullin, Ed., 1, p. 161.

[64] The following train of thought is my own: I am not attributing it to any particular scholastic author.

[65] See Grant, Ed., 4, p. 277.

spatial extension, and hence makes sense of the concepts of condensation and rarefaction. Dense bodies have high mass/volume ratios; rare bodies have low mass/volume ratios. Thus, according to Swineshead and the Mertonians, the *massa elementaris* of a body remains constant through all condensations and rarefactions of it. It is the ratio of '*massa*' to volume that determines density and rarity. (As in Newton, mass is the product of one intensive magnitude, density, and one extensive one, volume.[66]) It is clear, says Swineshead,[67] that a cubic foot of earth contains more matter than an equal volume of air – earth is manifestly 'more dense'. What *empirical* criterion did he have in mind here? Was it:

a. Higher Specific Gravity;
b. Greater resistance to passage of bodies through it; or
c. Greater inertia (= resistance of being moved)?

Swineshead himself may have had some combination of these factors in mind. It is, however, the dynamical criterion c which will in time give the notion of Q.M. a clear sense and hence resolve old puzzles about the quantification of matter.

B. The Status of the Elements in Compound Bodies

In what manner do the elements FAWE (or any alternative candidates) persist in a compound body C? Aristotle[68] believed that his matter-theory enabled him to 'save' generation and corruption from Eleatic objections, without falling into the pluralists' equation of generation with aggregation. Substances, for Aristotle, can truly be said to come-to-be and pass away. They are differentiated from one another by their substantial forms (SFs). When FAWE combine and interact to form flesh, the resulting *mixture* is perfectly homogeneous. Every part of flesh is flesh: no process of physical division would yield discrete particles of FAWE. The elements have, therefore, no *actual* existence in the compound. Yet flesh can be decomposed, chemically, to yield FAWE in a characteristic ratio. The elements therefore exist potentially in C in the sense that they can be extracted from it. They also exist in C as 'powers of action' or, in McMullin's phrase, V-potencies:[69] C retains, for example, the power of heating of its constituent

66 Wallace, 5, p. 40.
67 Weisheipl in McMullin, Ed., 1, pp. 164–167.
68 For this aspect of Aristotelian matter-theory, see Chapter 3, pp. 124–126.
69 McMullin in McMullin, Ed., 1, pp. 305–308.

fire, and the power of cooling of its water – both of course in tempered or attenuated forms.

This characteristically Aristotelian position – that FAWE exist potentially (in two distinct senses) in C – was the starting point for a further set of criticisms of the Stagirite. Let us examine the views of some Arabic and Latin writers on this controversial and difficult topic.

According to the Aristotelian, C has only one SF, the form of C. It does not possess also the forms of FAWE. But then, the pluralist will enquire, by what right do we call C a compound of FAWE? Unless the elements persist in C, compound-formation involves the corruption of the elements, not their combination. But the compound C is supposed to possess a particular FAWE ratio, which ratio explains (some of) its qualities – this too surely involves the actual existence in C of its 'constituent' elements? If, moreover, the elementary qualities HCDM are present in C, must not FAWE also be present?

Such considerations led to the development of a 'pluralist' theory, according to which several different forms may co-exist in the same matter. This contradicts the 'monism' of Aristotle and Aquinas, who accept only a single SF in any given portion of matter.[70] (The postulation of a common 'corporeal form' to confer dimensionality on P.M. is already a concession to pluralism.)

The champions of pluralism included both Avicenna and Averroes. According to Avicenna, the SFs of FAWE persist in C, but their associated properties are attenuated or corrupted. The forms thus remain in C in an 'unawakened'[71] sense, i.e. without imposing upon C those qualities with which they are usually conjoined. Instead, the SF of C itself is the *dominant* form, to which all the elementary forms are *subordinated*. On Averroes' view, by contrast, the forms of FAWE themselves exist in the *mixtum* in an attenuated manner. Avicenna thus thinks of SFs as incorporeal entities, ontologically distinct from the qualities to which they normally give rise; Averroes associates forms more closely with their corresponding qualities.

The pluralist thus moves from the facts that the elements (a) constitute C, (b) can be extracted from C, and (c) exist as V-potencies in C, to their actual existence in the compound. On the pluralist view, every mixed body will contain a

---

70 For Aquinas' opposition to pluralism, see Sokolowski, pp. 284–285, and Wallace, 5, p. 12.

71 See Grant, Ed., 4, p. 607.

hierarchy of forms, but will manifest only the form of the compound C itself, which is dominant over the elementary forms. The compound is therefore uniform and homogeneous, despite the presence in it of a plurality of forms.

This theory of subordinate forms has a number of advantages.[72] Suppose we believe, with Aquinas, the following three theses:

a.   SF = principle of actualisation of matter.
b.   There is only *one* SF in any given body.
c.   The SF of a plant or animal is its vegetative or sensitive soul respectively.

When an organism dies, it should on the above view revert to P.M. - or, at least, to its elementary components FAWE. But in fact tissues survive for a while (sometimes, as in the case of bone, for a very long time) after the death of the organism. These tissues may also retain their characteristic (e.g. medicinal) powers or 'virtues' in the corpse. Now, if these were all dependent on the SF of the plant or animal (i.e. its soul), they should all be lost at death. For the pluralist, however, there can be *subordinate* forms - not only of FAWE but of bone, muscle, etc. - which can continue to exist and exert their powers after the death of the organism, thus serving to sustain the natures (and 'virtues') of the parts of the plant or animal. This remained a problem for more orthodox Aristotelians.

It is now time to focus on one representative text, Aquinas' discussion of the status of the elements in the *mixtum*.[73] St. Thomas begins by explaining why many people have favoured the pluralist view, i.e. believe that

> the substantial forms of the elements remain, while the active and passive qualities of the elements are somehow placed, by being altered, in an intermediary state; for if they did not remain, there would seem to be a kind of corruption of the elements, and not a combination.[74]

If FAWE do not survive in C, argues the pluralist, they are not truly the *elements* of things:

> for an element is that out of which something is in the first instance formed, and which remains in it, and is by its

---

[72]   The following is a paraphrase of Sennert's argument for subordinate forms, as outlined in Boyle's 'Origin of Forms and Qualities' (Boyle, *Works*, Vol. 3, p. 113 ff.).

[73]   Larkin, pp. 67–72; Grant, Ed., 4, pp. 603–605.

[74]   Larkin, p. 68.

nature indivisible; for if the substantial forms are with-drawn, the compound will not be formed from the elements in such a way that they remain in it.[75]

There is also of course the obvious point that FAWE can be extracted from C by chemical decomposition, and the fact that the properties of C (HCDM, weight and lightness) result from those of their constituents.

These, then, are the grounds for pluralism. But, Thomas explains, 'it is impossible for the same portion of matter to receive the forms of the different elements'.[76] There being no distinction between 'having the form of F' and 'being F', a piece of matter cannot simultaneously have the forms of F and W without *being* both fire and water, which is absurd and impossible. So if the SFs of FAWE *do* all persist in C, they must do so in different parts of it. The resulting compound would therefore be *heterogeneous* in nature, i.e. not a true Aristotelian *mixtum*. On such a view, says Aquinas, 'there will not be a true mixture, but only an apparent one'.[77]

We now have a very important argument-form. If chemical evidence can convince us that the ingredients which go to make up a given substance actually persist in it, and if we accept Aquinas' assumption that no piece of matter can at the same time have two different SFs, then that substance is not a *homoeomer*, not a true *mixtum* in Aristotle's strict sense. But, as Hooykaas[78] shows, if quantitative conservation of the ingredients is taken as evidence of their persistence in compounds, then one material after another (alloys, solu-tions, salts, etc.) will be erased from the list of true compounds. One might thus begin to suspect that there are no compound substances which satisfy Aristotle's definition of a *mixtum*, and might instead come to favour the following sort of position(s):

a. The actual existence of the elements in all compounds.
b. The 'Generation = Aggregation' thesis.
c. A particulate theory of matter.

The chemical evidence for (a) will be discussed later. Given (a) and Aquinas' denial that the forms of FAWE can co-exist in the same matter, it is easy to derive (b) and (c). The elementary forms must, if they actually exist in C, exist in

75 *Ibid.*, p. 69.
76 *Ibid.*, p. 69.
77 *Ibid.*, p. 69.
78 Hooykaas, pp. 77–78.

different parts of it. Since many compounds at least appear homogeneous, these discrete particles of FAWE must be invisibly small. The aggregation of such particles will then suffice to produce C; their dissociation will cause its corruption. Theses a–c thus constitute an interdependent body of propositions which may form the skeleton of an anti-Aristotelian theory of matter.

Aquinas, however, takes the existence of genuine *homoeomers* as indubitable, and is thus led to deny the actuality of FAWE in C. Some, he continues, have admitted the presence in C of the SFs of FAWE, but claimed that these forms subsist only in an attenuated state in compounds.[79] (It is plainly the theory of Averroes that Thomas has in mind here.) Substance, however, Aquinas insists – and hence also substantial form – does not admit of degrees. A SF is either present or absent: there is no *tertium quid*, no conceivable intermediate state. The notion that FAWE are so lowly in the great hierarchy of things that their forms are only semi-substantial (and semi-accidental) is dismissed by Aquinas as unintelligible. There is and can be, he states, nothing intermediate between substance and accident: the categorial distinction is sharp, and clear.[80]

While SFs do not admit of degrees, qualities and powers do. When FAWE combine to form C, their active and passive powers HCDM interact to produce a substance qualitatively intermediate between the elements. The qualities of each element are tempered by the (opposed) properties of the others, producing eventually the characteristic nature of C, as defined by its chemical formula or FAWE ratio. The elements themselves are present in C only as potencies – both as possibilities of extraction and as present virtues or V-potencies.[81] Aquinas thus concludes his monograph with a blunt reaffirmation of the position of Aristotle: he has not really resolved its attendant difficulties.

The majority of the schoolmen sided on this issue with Aristotle and orthodoxy. The elements, says Oresme,[82] cannot be present in compounds in their proper forms: if they were, there would have to be either (a) distinct SFs in the same portion of matter, which is impossible, or (b) different SFs in distinct portions of matter, in which case every compound

---

[79]   Larkin, p. 70.

[80]   *Ibid.*, pp. 70–71.

[81]   *Ibid.*, pp. 70–71.

[82]   Oresme, p. 617.

body is really heterogeneous. Albert of Saxony[83] too argues that pluralism involves the heterogeneity of the compound: if, he says, the elementary forms persist in C, a compound body is like a syllable such as ab in which the constituent letters a and b remain unaltered. Aristotle, says Albert, is 'absolutely opposed to this':[84] on the peripatetic matter-theory, a *mixtum* is homogeneous throughout. Albert thus defends, against pluralists of all kinds, the three following theses T1–3:

T1. A compound body needs a form distinct from those of its elements.

T2. The SFs of FAWE do not remain in C 'in their intense [i.e. complete] being'.

T3. The SFs of FAWE do not remain in C even in an attenuated state.[85]

Since Albert's arguments in defence of these theses follow closely those of Aquinas, I shall not linger over them. Let us pass instead to the chemical argument for pluralism and its origins.

Avicenna and Albertus Magnus, two of the foremost advocates of pluralism, had argued that FAWE must be actually present in C because:

a. They came together to make it.
b. They can be extracted from it.[86]

This argument is plainly invalid, since it simply assumes that the elements do not undergo corruption and (re)generation in compound formation and dissolution respectively. It is, nevertheless, a mode of argument well suited to the experience of the chemical laboratory. If one thinks one has empirical grounds for the actual existence of the ingredients in a mixture, one will be led by Aquinas' argument to a particulate theory of matter.[87] This step took place among the disciples of Paracelsus, who used fire as an analytical tool, believing that it does not create new substances but only separates the constituents of things. Whatever can be extracted from C by heating, they claim, was an actual constituent of C.[88]

---

[83] See Grant, Ed., 4, pp. 605–613.

[84] *Ibid.*, p. 606.

[85] *Ibid.*, p. 607.

[86] See Hooykaas, p. 70.

[87] *Ibid.*, p. 74.

[88] For the Paracelsian theory of fire-analysis, see Debus, 3, pp. 132–133.

If I mix X and Y to produce Z, and then decompose Z to recover X and Y, by what right do I say that X and Y are *actually* present in Z? After all, Z may possess quite different properties and powers: it need not resemble X or Y in any discernible way. (Just think of the chemical formula of common salt.) If one grants the actuality of the elements in the compound one will – given Aquinas' principle of inference – arrive at a particulate theory of matter. But why should one accept the inference from possibility of extraction to actual presence?[89]

This is the crucial lacuna in Hooykaas' otherwise excellent paper on the empirical roots of the corpuscular theory. Particulate theories of matter, he claims, emerged primarily because of chemical evidence for the actuality of the elements in compounds. Unfortunately, his argumentation is over-simple: the fact that one can recover as much copper from a copper sulphate solution as one put into it does not entail the actual existence of copper, as such, in the 'vitriol'.[90] The blue liquid in the test tube does not evidently contain particles of copper; it does not even manifest the properties of copper as V-potencies. To an Aristotelian, it is a quite different substance, containing copper only 'potentially', in the sense that copper metal can readily be extracted from it.

Two lines of argument will tell against this Aristotelian position and in favour of a corpuscularian view. The first is the general demand[91] for an *actual ground* for every 'potency', the reluctance to admit unexplained and irreducible potentialities of things. There must, we feel, be some *actual* feature of the blue solution in virtue of which we can extract copper (but not lead or iron) from it. The simplest and most intelligible explanation is that actual (though invisibly small) particles of copper persist in the solution.

The second – and more important – point concerns explanatory power. Consider the following experiment as an illustration of what I have in mind. Take one pound of iron and one of tin and dissolve them, together, in a mineral acid, giving a clear solution. If the water is evaporated off, we are left with a mixed salt, from which standard metallurgical techniques for the purification and separation of metals will

[89] Hooykaas, p. 72.

[90] Boyle was of course to urge this point in his *Sceptical Chemist*. See esp. *Works*, Vol. 1, pp. 471–472.

[91] This demand is of course part and parcel of the Atomist tradition in Western thought.

release one pound of tin and one of iron. We could simplify the experiment still further, of course, by considering a simple alloy of the two metals, omitting the intermediate steps.

Does this trivial experiment have deep theoretical implications? Yes and no. Although it does not refute the Aristotelian position, the results of the experiment are far more readily accounted for on the assumption that the solution (or alloy) contains distinct particles of tin and iron. Let us see why this should be so.

For the purposes of this example, I shall assume the alchemists' Mercury-Sulphur (M-S) theory of the constitution of metals. (Exactly the same point could be made in terms of the FAWE theory.) Let us assume, quite arbitrarily, that iron can be represented by the formula M_3S_4 and tin by M_5S_2 (nothing turns on the actual figures used). Their perfect mixture, in equal proportions, would therefore produce an alloy of constitution M_8S_6 (or, dividing by two, M_4S_3). Already a number of minor objections arise. What if M_4S_3 were another metal, say copper? What distinguishes those M-S ratios which constitute the 'seven metals'?

Worse, however, is to come. If the ingredients of the alloy are perfectly mixed, why should it decompose to yield iron and tin rather than, say, M_2S_4 (gold, let us say) and M_6S_2 (silver)? Of course, it always *could* come about – quite mysteriously – that the same ingredients which went in also came out, but there is no good reason, on the theory of total mixture, why they should do so. The most natural and obvious explanation is that our iron/tin alloy contains *actual* particles of iron and tin. Since the 'forms' of iron and tin cannot co-exist in the same matter, they must exist in different parts of the alloy, which is therefore only apparently homogeneous.

This conception of alloys arises naturally out of metallurgical experience. The extraction of tin from bronze would commonly be regarded, I think, as the *separation* of a pre-existent constituent. But the acid solution of our first example is analogous to the alloy: the same substances were extracted from the solution as went into it, and in the same amounts. If abiding particles of copper and tin *explain* the ease of extraction of those metals from bronze, why not from solutions too?

The criterion for the actual existence of Y in Z is therefore, for the 'chemical philosopher', the possibility of extracting Y from Z. But rust comes from iron, as iron from rust. Which is *separated* from which? Is rust actually present in iron as an ingredient? Many good chemists thought so. To answer such a question, one requires a sophisticated hierarchic theory of the structure of matter, based on a theory of chemical Atomism. Atoms of the *chemical elements* (whatever those may turn out to be) exist actually, on such a theory, in all compounds, and combine to form *prima mixta* (molecules), particles of the next level of complexity, and so on. In short, we have the Chemical Atomic Theory of Sennert, de Clave, Jungius and others.[92] Without such a hierarchic matter-theory, the question 'what exists in what?' is unanswerable.

The chemists' insistence on the actuality of the elements in compounds thus naturally combines with the Aristotelian theory of natural minima. Avicenna's pluralism is accepted,[93] but so is Aquinas' proof that this involves the heterogeneity of compound bodies. The Chemical Atomic Theory is well and truly launched.

For a vigorous defence of such a theory, one has only to read Boyle's *Sceptical Chemist*, in which the character of Eleutherius is clearly intended as a portrayal of a 'chemical philosopher'. Eleutherius[94] expresses a clear preference for the Atomists' account of *mixis* as against the peripatetic view. On Aristotle's theory, he says, the elements are

> not so properly said to be altered, as destroyed, since there is no part in the mixed body, how small soever, that can be called either fire, or air, or water, or earth.[95]

Eleutherius favours the theory that corpuscles of the ingredients retain their identities and chemical properties in mixtures, like black and white threads in grey cloth. The eye of the lynx, he suggests, could discern individual black and white threads. After a number of old and very familiar criticisms of the Aristotelian conception of mixis, he launches into a defence of the Atomist view. The separability of the constituents of a 'mixt' involves, he argues, their pre-existence in it:

> ... the *Miscibilia* may be again separated from a mixt body, as is obvious in the chymical resolution of plants and

---

[92] See Chapter 5, Part 2, pp. 225–231.

[93] For the tendency of the chemists to accept Avicenna's pluralism, see Dijksterhuis, p. 209; Hooykaas, p. 74.

[94] See Boyle, *Works*, Vol. 1, p. 502 ff.

[95] *Ibid.*, p. 503.

animals, which could not be, unless they did actually retain their forms in it.[96]

To explain separability we need actual pre-existence, he urges, which in turn entails *heterogeneity*. Although an alloy of gold and silver has new characteristic properties, and appears homogeneous, it is not really so: *aqua fortis* readily dissolves away the silver, leaving the gold unaffected. Both metals have retained their natures in the alloy, despite being mixed 'per minima' with each other. Aristotle's objection that this makes the 'mixt' heterogeneous is simply dismissed as 'no great inconvenience':[97] it may in the final analysis transpire, Eleutherius claims, that Aristotle's definition of *mixis* fails to apply to any actual examples. All compounds may be mechanical aggregates – like a mixture of wheat and barley, but on a smaller scale.

C. Patterns of Explanation

We discovered in Empedocles and Aristotle a method of explaining the qualities of a body in terms of its *material constitution*. According to Empedocles, the FAWE ratio of a substance is constitutive of its nature: bone, for example, is defined as $E_2A_1W_1F_4$.[98] Aristotle adds to this his own account of the 'primary' qualities HCDM: when FAWE combine to form a compound, he explains, HCDM interact to give rise to a *mixtum* with properties intermediate between those of the elements.

What (sorts of) properties are explicable in terms of FAWE (or HCDM) ratios? Aristotle cites colours and odours as examples of 'secondary' qualities,[99] qualities explicable in terms of other ('primary') ones. Generalising further, we might attempt to explain *all* the manifest properties and powers of a given substance in terms of its material constitution. Taken to its logical conclusion, this will yield a purely materialistic conception of form: the form of a given substance will represent, on this view, nothing more than its chemical formula.[100]

---

[96]   *Ibid.*, p. 504. Boyle's spokesman, Carneades, is of course sceptical about the validity of this inference.

[97]   *Ibid.*, p. 505.

[98]   See Chapter 3, p. 93.

[99]   Chapter 3, pp. 98-99.

[100] For such a position, see Digby, p. 150 ff. After explaining at length the FAWE ratios needed to produce various materials, he goes on to claim that *all* the qualities of compound bodies 'arise out of the commixtion of the first qualities and of the elements' (p. 157).

The sheer hopelessness of this programme soon, however, became apparent to many. The only ultimately intelligible explanations that can be given in terms of FAWE (or HCDM) rations are of degrees of the primary qualities themselves: we can understand how H and C interact to yield a moderate warmth or coolness, but not how they produce redness or sweetness, still less the stupendous variety of properties of natural substances. How, for example, should one explain the medicinal virtues of herbs or the 'magical' power of the lodestone? What material constitution could conceivably account for such properties? The highly specific and idiosyncratic powers of natural things seem to defy this pattern of explanation.

The failure of the material constitution theory led to the popular doctrine of *occult qualities*. An occult quality of a body is one that is inexplicable in terms of its elemental constitution (*complexio*) and is as such impossible for us to fathom.[101] Magnetism is commonly cited as an example: no FAWE ratio could be expected, of itself, to confer on a body the highly specific power of attracting iron. We characterise such properties only by their manifest effects (e.g. 'power of drawing iron', 'purgative virtue', etc.): we have no clue of the real grounds of such powers in the objects possessing them.

The occult qualities of a body are therefore (a) not themselves evident to sense (we don't see in the *magnet* anything which would lead us to anticipate its peculiar power) and (b) not rationally deducible from prior principles.[102] They are therefore only discoverable by chance experience, not by rationally-guided experiment. Thus Bonus of Ferrara, a fourteenth century alchemist:

> If you wish to know that pepper is hot and that vinegar is cooling, that colocynth and absinthe are bitter, that honey is sweet and that aconite is poison; that the magnet attracts steel, that arsenic whitens brass, and that tutia turns it of an orange colour, you will, in every one of these cases, have to verify the assertion by experience.[103]

The use of these occult virtues of things to produce spectacular and wonderful effects constitutes natural (i.e. non-demonic) magic.[104] Although the existence of occult

---

[101]   See Dijksterhuis, p. 157; Thorndike, Vol. 2, pp. 854–855.

[102]   See Hansen in Lindberg, Ed., p. 487.

[103]   Quoted from Debus, 11, p. 17.

[104]   See Thorndike, Vol. 2, p. 342 ff., for the defence of natural magic by William of Auvergne, a thirteenth century Bishop of Paris.

qualities was denied by a few sceptics, most scholastics followed Albertus Magnus[105] and Thomas Aquinas[106] in believing it evident from experience that such qualities exist.

Whence are the occult qualities of things derived? The answer is twofold: from the specific form and from astral influences. The combined effects of (a) material constitution, (b) specific form and (c) astral influence together make each object what it is. Roger Bacon was a great advocate of the astral origins of occult qualities. Each herb or stone, he argues, possesses marvellous 'extraneous' virtues, inexplicable in terms of its FAWE ratio.[107] These 'extraneous' (occult) properties are due to the stars, which impart to every creature at birth a 'radical complexion' distinct from its 'common complexion' or material constitution. Thus

> when a child at birth is exposed to a new air, another world as it were, he then receives apexes of celestial pyramids as respects his separate members, and thus receives new impressions, which he never gives up, because what the new jug receives it tastes of when old. And then is formed the radical complexion, which remains to the end of life, although the current complexion may be changed for a whole day. And this radical complexion is followed by inclinations to morals and to sciences . . .[108]

A favourable 'constellation' at birth gives the child a natural tendency to a good life; harmful 'influences' may lead him into all manner of evil.[109] The 'radical complexion' imparted at birth is inalienable and immutable, despite the variability of the 'current complexion' or FAWE ratio.

Most scholastic writers agreed with Bacon in attributing occult qualities to astral influences on the generation of creatures. (Had not Aristotle himself stated that all terrestrial generation and corruption is ultimately due to the annual motion of the sun?) Thus Albertus Magnus admits a stellar influence on the formation of plants,[110] and Peter of Abano

---

[105] *Ibid.*, Vol. 2, pp. 566–567.

[106] *Ibid.*, p. 607.

[107] *Ibid.*, p. 664.

[108] R. Bacon, Vol. 1, p. 159.

[109] Bacon was of course careful to insist that the rational soul of man remains free: the stars impart no more than tendencies to certain types of behaviour. Theses postulating astral determination of human affairs were among the 219 articles condemned at Paris in 1277.

[110] Thorndike, Vol. 2, pp. 565–566.

314   A. J. Pyle

attributes the specific powers of poisons to astral causes.[111] Within such a world-view, astrology becomes the queen of the sciences and the key to terrestrial chemistry and medicine.

One man who attempted to explain the highly specific powers and properties of natural things *without* invoking occult qualities was Nicole Oresme. His notion of 'qualitative configuration' represents an attempt to *demystify* the occult, to render intelligible what was previously mysterious. Every quality, says Oresme, admits of *degrees* of *intensity*, which can be represented graphically by straight lines. (The notion of an intensive magnitude was of course common property by the fourteenth century, and the idea of quantifying degrees of, e.g. HCDM goes back to the Arabic pharmaceutical tradition.[112]) One can now draw the quality-profile for a given body and a given quality, mapping the intensity of that quality (e.g. temperature) over the surface of the body.[113] To explain the precise nature of a body, then, we must know not merely how much H (or C, D or M) it contains, but *how this is distributed*. We must, as Oresme says, know the qualitative configuration of that body. Two objects may, for example, contain the same total amount of heat, but have very different temperature-profiles, thus:

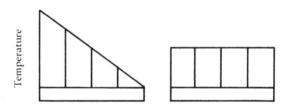

These two bodies, one a 'temperature-triangle' and the other a 'temperature rectangle', contain equal amounts of heat (= area under the graph), but may for all that have very different properties. Oresme's 'qualitative geometry' thus enables him to explain the great variety of highly specific properties of natural things, without admitting any other

[111]   *Ibid.*, p. 906.

[112]   See Thorndike, Vol. 1, pp. 750–751.

[113]   See Molland, 2, p. 111.

primitive qualities than HCDM. The quality-profile of a thing is like a fingerprint, peculiar to the individual – although there may of course be quality-profiles common to the members of a species as well. Qualities previously deemed occult (e.g. the medicinal virtues of gems and herbs) are, says Oresme, really due to the qualitative configurations of things.[114] Granted that we do not know these in any detail, we still find such explanations intelligible: they invoke as primitives only HCDM and spatial arrangement (i.e. basic geometry).[115]

In Oresme, then, the FAWE (or HCDM) ratio does not exhaust a body's material make-up. The *same* HCDM ratio can correspond to many quite different·stuffs, since HCDM do not necessarily mix perfectly, but combine instead in such a manner as to produce specific quality-profiles. Oresme thus effectively denies the homogeneity of compound bodies and stresses instead, like the ancient Atomists (to whose views he likened his position[116]), the explanatory rôle of spatial arrangement and configuration. Although this remains a qualitative matter-theory, Oresme accommodated within it an explanatory factor (configuration) belonging properly to a quite different tradition.

Given such an explanatory programme, Oresme and his disciple, Henry of Hesse,[117] were able to dismiss occult qualities and astral influences from their world-view. The influence of the heavens here below, they insist, is confined to their heat and light: they are not the causes of terrestrial generation and corruption.[118] The curious and multifarious virtues of herbs, gemstones, etc., are not due to the stars but are – in principle – explicable in terms of the twin factors of qualitative constitution and configuration.[119] Even the peculiar power of the lodestone can, Henry claims, be explained in terms of its quality-profile.[120] Since each organ of the body of an animal has its own profile, the whole body will have an enormously complex quality-configuration: this may be far too complex for us ever to discover or characterise in

114 *Ibid.*, pp. 112–113.
115 Thus Thorndike (Vol. 3, p. 490) is quite wrong to dismiss Oresme's explanation of medicinal virtues as no better than that of his opponents: what is at stake is not what a man can in fact know, but what he could in principle understand.
116 See Molland, p. 107.
117 For Henry, see Thorndike, Vol. 3, p. 472 ff.
118 *Ibid.*, p. 414.
119 *Ibid.*, pp. 441–442.
120 *Ibid.*, pp. 481–482.

detail, but it remains in principle something *intelligible* to us, not something totally occult.

The ideas of Oresme were not, however, destined to be influential. A glance at sixteenth century thought reveals the scholastic philosopher, Luiz[121] the physician Fernel,[122] and the Nature-philosopher Cardan[123] all giving conventional defences and illustrations of the doctrine of occult qualities. Some of the properties of a herb, says Fernel, derive from its substantial form (SF) which the plant owes not to its parents but to the stars – and, ultimately, to God. These 'virtues', he insists, may legitimately be called 'occult'; we humans can have no clear conception of how they arise in things and how they produce their effects.

A still more occult conception of these inexplicable virtues can be found in the magician, Henry Cornelius Agrippa, according to whom

> there is work in the world itself for angels who preside over earthly armies, kingdoms, provinces, men, beasts, the nativity and growth of animals, shoots, plants, and other things, giving that virtue which they say is in things from their occult property.[124]

Angels and demons are, for this Renaissance *magus*, the causes of the occult qualities of things: natural philosophy gives way to conjuring, science to superstition. Need one say more?

For those loath to follow this path to its bitter end, there is an alternative account of the origin of 'occult qualities'. It originates from the alchemical technique of distillation. In many distillation processes, the distillate contains, in highly concentrated form, the 'virtues' of the original substance. Wine, for example, will yield 'spirits of wine' or *aqua vitae* (alcohol), and many organic substances will yield their characteristic 'essences' (we still speak, e.g., of 'almond essence', etc.). One can then argue that this small fraction of 'spirit' or 'essence' imparts to the body all its characteristic properties and powers: without its 'essence' the gross matter of a body is dead and chemically insert.[125]

This theory is of considerable antiquity – it was cited by Grosseteste (c. 1175-1253) as the common opinion of 'the

---

121   Thorndike, Vol. 5, p. 551.

122   *Ibid.*, p. 570.

123   *Ibid.*, p. 559.

124   Quoted from Thorndike, Vol. 1, p. 454.

125   See Partington, 6, Vol. 2, p. 141.

doctors of alchemy'[126] – but was especially popularised by the treatise 'The Consideration of the Fifth Essence' by the fourteenth century alchemist, John of Rupescissa.[127] It is concerned primarily with ways of prolonging human life. The source of longevity, says John, must *itself* be incorruptible in nature,[128] and hence of celestial rather than terrestrial origin. The supreme *quintessence* is, he claims, none other than *aqua vitae* (living water) or alcohol, which is a fount of medicinal virtues![129] Being itself spiritual in nature, it serves to extract (by the power of sympathy) the most 'spirituous' parts of other substances, i.e. *their* quintessences: all herbs and simples, aromatics and laxatives are, says John, increased in efficacy by stewing in alcohol. The basic idea is that the peculiar virtue of a given star or planet is present as a quintessence in a particular species of plant, whence it can be extracted by stewing in alcohol and put to medicinal use.

This technique was developed by the so-called 'Lullian' alchemists,[130] although, as we have seen, their true founder was not Ramon Lull but John of Rupescissa. No matter: the 'Lullists' became the great champions of the theory of quintessences. Each body, they claimed, consists largely of dead and inert gross matter, 'enlivened' by a minute proportion of subtle, volatile, chemically active stuff, its quintessence, which is celestial in origin and nature and is causally responsible for all the activity of the body. As Sherwood Taylor says,

> It was clear to the Lullian alchemists that if this active principle could be extracted from a body, it should be a far more active reagent than the body itself.[131]

This belief was borne out by the chemical activity of alcohol itself, and of course of those other distillates, the mineral acids, the discovery of which was to revolutionise chemistry. By distillation, then, the 'spirit' is separated from the 'body':

[126] See Thorndike, Vol. 2, p. 447.

[127] *Ibid.*, Vol. 3, p. 347 ff.

[128] A nice example of what Barnes calls the 'principle of Synonymy'; see Chapter 3, pp. 89–90.

[129] In 1478, the German physician Schrick published a book on distillation in which he praises the wonderful medicinal virtues of brandy. Anyone, he says, 'who drinks half a spoonful of brandy every morning will never be ill.' It can even revive (restore the spirits! of) the dying! See M. Boas Hall, 8, p. 160.

[130] S. Taylor, 2, pp. 111–122.

[131] *Ibid.*, p. 117.

the underlying matter-theory is close to that of the Stoics, with 'spirit' (*pneuma*) as subtle but highly active matter, not as something purely incorporeal.[132]

The theory of quintessence was also adopted by Paracelsus and his followers. The 'life' of each particular thing, Paracelsus insists, is 'none other than a spiritual essence'. Every corporeal thing is endowed by God with its own spirit, 'which spirit it contains after an occult manner within itself', and which 'holds concealed within itself the virtue and power of the thing'.[133] The 'natures' of things are therefore spiritual; indeed, 'nature' itself is

> simply a volatile spirit, fulfilling its office in bodies, and animated by the universal spirit – the divine breath, the central and universal fire, which vivifies all things that exist.[134]

The term 'volatile' has here its normal chemical meaning: a spirit is for Paracelsus a subtle fluid which can be extracted from bodies by distillation. These spirits (*pneumata*) impart to the bodies they 'enliven' their specific powers and virtues, which they (the spirits) possess in a concentrated form – i.e. they are quintessences of things:

> The quintessence, then, is a certain matter extracted from all things which Nature has produced, and from everything which has life corporeally in itself, a matter most subtle purged of all impurities and mortality, and separated from all the elements. From this it is evident that the quintessence is, so to say, a nature, a force, a virtue, and a medicine, once, indeed, shut up within things, but now free from any domicile and from all outward incorporation. The same is also the colour, the life, the properties of things.[135]

It is clear that Paracelsus has in mind the same idea as John of Rupescissa, the idea of an 'essence of X', extractable from X by distillation and purification, and responsible for all the 'virtue' of X. Thus, says Paracelsus,

> Wine contains in itself a great quintessence, by which it has wonderful effects, as is clear. Gall infused into water renders the whole bitter, though the gall is exceeded a

---

132 Pederson and Pihl, p. 164. Toulmin and Goodfield, 1, p. 127.

133 Paracelsus, 1, Vol. 1, p. 135.

134 *Ibid.*, Vol. 1, p. 289.

135 *Ibid.*, Vol. 2, p. 22.

hundredfold in quantity by the water. So the very smallest quantity of saffron tinges a vast body of water.[136]

The rôle of the chemist and physician is to identify – and if possible isolate – the 'spirits' responsible for various powers and properties. Paracelsian chemical medicines are commonly extracts or distillates – in his terminology, quintessences – which leave behind, after extraction, 'a leprous body, wherein is no sweetness or sourness, and no power or virtue remains, save a mixture of the four elements'.[137] FAWE, therefore, remain in this 'leprous body', which can thus retain its corporeal nature. It is of those qualities which the schoolmen labelled 'occult' (i.e. not dependent on FAWE ratio) that the quintessence is the *explanans*. Without its own quintessence, a body lacks any such specific virtues: it has no peculiar and special properties beyond bare corporeality.

We have strayed far from the material constitution theory, to a variant of which we must now return. In his *Meteorologica*,[138] Aristotle postulated the existence of two 'exhalations', emitted by the earth under the influence of the sun's heat. Wet earth emits a moist, vaporous exhalation, dry earth a smoky and combustible one. The dry exhalation explains fiery meteorological phenomena such as comets, meteors, and lightning; the moist exhalation explains mists, dew and rain. The exhalations are also responsible for the formation of minerals in the bowels of the earth: metals in particular result from the condensation of the moist, watery exhalation.

This is the probable source[139] of the alchemical theory of the formation of metals out of Mercury (M) and Sulphur (S),[140] which correspond of course to Aristotle's moist and dry exhalations respectively. This theory was adopted by the great Arabic alchemist, Jābir,[141] and thence passed into the alchemical tradition. The M and S that combine to produce metals, Jābir insisted, are ideal or 'sophic' principles, analogous with but not identical to common quicksilver and brimstone. The M-S theory of the constitution of metals was

---

[136]   *Ibid.*, Vol. 2, p. 23.

[137]   *Ibid.*, Vol. 2, p. 24.

[138]   *Meteorologica*, 1, 4. See also Clagett, 2, p. 124.

[139]   See Partington, 6, Vol. 1, p. 96; S. Taylor, 2, p. 14.

[140]   The theory will not seem so strange if one remembers that many metals are smelted from sulphide ores, and yield on smelting a fluid resembling quicksilver.

[141]   For Jābir, see Holmyard, pp. 74–75.

not conceived as a rival to the FAWE theory: M and S were usually conceived as stable compounds rather than as themselves elementary, thus making the metals themselves tertiary compounds:

FAWE ——————→ M & S ——————→ METALS

FAWE, in this hierarchic matter-theory, combine to form stable secondary particles[142] of M and S, which in turn give rise to the (tertiary) corpuscles of metals: even these may survive all manner of chemical operations (e.g. in alloys, solutions, salts, etc.).

The metals are all, says the alchemist, 'composed of mercury combined and coagulated with sulphur'.[143] Different ratios of M and S (and hence, ultimately, of FAWE) give different metals. Albertus Magnus discusses the M/S ratios characteristic of the different metals,[144] while in a manuscript ascribed to Arnald of Villanova we find 'recipes' for the formation of gold and silver out of FAWE. Gold, says Arnald, can be made out of

1½ lbs. of F   3 lbs. of W
2   lbs of E   3 lbs. of A[145]

Unfortunately, the alchemists also had a tendency to think of M and S as specifically different in the various metals,[146] thus spoiling the clarity of the their material constitution theory.

In Paracelsus, the Mercury-Sulphur theory of the constitution of metals gives way to a Mercury-Sulphur-Salt (MSL) theory of the formation of all material substances.[147] He too stressed the *ideal* nature of his 'three principles' (*tria prima*) – it is not, he insists, of vulgar MSL that he speaks, but of 'the Mercury and Sulphur of the Philosophers'[148] – yet at the same time they (or approximations to them) have to serve as the products of pyrolysis. Fire-analysis, claims Paracelsus, provides empirical confirmation of his matter-theory. Fire is the true key to chemistry:

---

142   For the attribution to the alchemists of a particulate theory of matter, see Hooykaas, pp. 70–71; Dijksterhuis, pp. 207–209.

143   Quoted from Hopkins, p. 121. The quote is attributed to the Latin 'Geber'.

144   Thorndike, Vol. 2, p. 568.

145   *Ibid.*, Vol. 3, p. 59.

146   For such ideas in Geber and in Albertus Magnus, see Hopkins, p. 165, and Grant, Ed., 4, pp. 588–589, respectively.

147   See Debus, 2, p. 27.

148   Paracelsus, 1, Vol. 1, p. 65.

It is this which makes manifest whatever is hidden in anything. In this place, then, we understand by the ultimate matter of everything that into which it is dissolved by fire.[149]

Now a burning branch releases flame (S), smoke (M) and ash (L): since the fire only separates what was previously present in the 'mixt' body, these must have been the material constituents of the tree.

The *tria prima* existed, Paracelsus claims – as 'spirits' or 'sophic' principles – in the primeval chaos from which God created the world: this 'first chaos of the universal matter' was 'constituted of Sulphur, Salt and Mercury'.[150] Of these three 'are composed all the things which are, or are produced, in the four elements'.[151] What of the 'elements' FAWE themselves? They are, says Paracelsus, the 'matrices' (mothers) *in* which things come-to-be. Are they too composed of the *tria prima*? The following passage seems to suggest so:

> Now in water is the primal matter, namely, the three first substances, Fire [= Sulphur], Salt, and Mercury.[152]

The precise relation between FAWE (the mothers or 'matrices' *in* which things come-to-be) and MSL (the 'principles' *from* which they derive their natures) remains very obscure. Paracelsus seems to have favoured the conception of FAWE as posterior to MSL, but many of his followers reaffirmed the traditional conception of M and S (and now L too) as stable second-order compounds of the true elements FAWE.

One noteworthy feature of the Paracelsian theory of the *tria prima* is the extent to which it emphasises an idea already present in the peripatetic FAWE theory, i.e. the conception of the elements as *ideal*, quality-bearing (and imparting) principles rather than as simple material constituents. If each of the elements (FAWE, or MSL, or some other set) contributes a quality necessary for corporeal existence, then the isolation of a perfectly pure element is impossible in principle: every body that falls within the compass of our experience must be composed of all of the elements.

---

149 *Ibid.*, Vol. 1, p. 90. For opposition to the notion of fire-analysis, one
      has only to turn to the chemical writings of Erastus, Jungius, or Boyle.

150 *Ibid.*, Vol. 2, p. 371.

151 *Ibid.*, Vol. 1, p. 204. See also p. 238.

152 *Ibid.*, p. 239.

The antiquity of this idea is evident from Plotinus' critical examination of it in *Ennead*, 2, 1. We must, he says,

> consider whether it is true that earth is not visible without fire and fire is not solid without earth. If this is so, it looks as if no element would ever have its own essential nature by itself, but all are mixed and take their names from the dominant element in each. They say that earth cannot have concrete existence without moisture; the moisture of water is earth's adhesive.[153]

On this view, the elements F, W and E account for properties as follows:

E – Solidity
W – Cohesion
F – Visibility

Thus all the solid, cohesive bodies we see about us must contain at least those three elements (no doubt a similar rôle could be found for air). Plotinus retorts that on this theory there could be no actual particles of pure FAWE; there would, for example, be no E-particles to be 'glued' together by the adhesive power of water. If, he argues, there are discrete particles of pure FAWE, the theory in question is false; if there are no such particles, it is redundant. But if E and W are ideal (incorporeal) principles, rather than material constituents, this objection has no teeth: bodily nature, on such a view, only arises from a confluence and co-operation of two or more 'principles'.

Aristotelian matter-theory remained in a constant state of tension between the rival conceptions of FAWE as either (a) material constituents (capable of separate existence) or (b) ideal principles (incapable in principle of existing independently). No Aristotelian believed, of course, that the FAWE we encounter in everyday experience are perfectly pure and unmixed – Aristotle himself had stated that the true elemental F ought not to be identified with ordinary fire but should rather be called 'such as fire'.[154] The crucial question was whether pure elemental FAWE could ever in principle be discovered or extracted from compounds, Some scholastics suggested that pure F and E may be found at the periphery and geometrical centre of the (sublunary) world[155] – on such a

---

153   Plotinus, *Enneads*, 2, 1, Vol. 2, p. 25.

154   See Chapter 3, pp. 121–122.

155   Albert of Saxony can be cited as an exponent of such a viewpoint. See Grant, Ed., 4, p. 610.

view, pure elements are still in principle discoverable. Others conceded still more to the 'ideal principle' conception of FAWE. Adelard of Bath,[156] for example, argues that no man has ever encountered a piece of true elemental earth. This need not be controversial: the crucial question is the modal one: 'Could there in principle exist such a pure element?' On this question, the schoolmen seem in some doubt – no clear answer is forthcoming. Every mundane object, says Adelard, contains all of FAWE: we deal, in our daily lives, only with compounds. In a plant, for example fire is necessary for upward growth, water and air for 'spreading out', and earth for cohesion – no plant, therefore, can exist without all four elements. In Albert of Saxony, the possession of *sensible qualities* (colour, odour, etc.) is taken as a mark of a *compound* nature: what pure E or F would look (smell, taste) like we have no notion. The elements are therefore not conceived as material constituents but as 'hypothetical abstractions arrived at analytically'[157] – ideal 'principles' necessarily incapable of separate existence.[158]

In the Paracelsian school, it is stressed again and again that the *tria prima* (MSL) are ideal or 'philosophical' principles, identifiable by the properties and powers they impart, rather than isolable material constituents. Sulphur, Mercury, and Salt were commonly associated by the spagyrists with the following properties:

> S   –   Combustibility
> M   –   Metallicity, Volatility
> L   –   Fixity, Resistance to Fire.[159]

The *tria prima* also, however, have other, less well known, virtues:

> On that head, notice this brief information, that all colours proceed from Salt. Salt gives colour, gives balsam and coagulation. Sulphur gives body, substance, and build. Mercury gives virtues, power, and arcana. So these three ought to be combined, nor can one exist without the other.[160]

---

156   See Thorndike, Vol. 2, pp. 34–35.

157   Grant, Ed., 4, p. 610.

158   For a defence of such a matter-theory by a sixteenth century cardinal, see Thorndike, Vol. 5, p. 554.

159   Pagel, 1, p. 84; Dijksterhuis, p. 208; Paneth, Part 2, pp. 144–145.

160   Paracelsus, 1, Vol. 1, p. 245.

All three of the *tria prima* must exist in every material substance: a body lacking any one of them is impossible in principle. If a body is, analytically, something with properties P_1, P_2 and P_3, and one of the *tria prima* is responsible for each of these properties, then only when MSL are combined can a body be generated. The 'principles' are therefore not separable from one another – at least not *as bodies*. (They may perhaps have independent existence as 'spirits'.[161]) In bodies, all three of the *tria prima* are necessarily present, contributing their associated properties and powers.

We know, says the Paracelsian Duchesne, enough about the natures of MSL to be able to estimate their *proportions* in various compound bodies.[162] The *tria prima* theory therefore remains *like* a material constitution theory in that the MSL ratio constitutive of a body explains its specific nature and qualities. Each 'principle' contributes its particular virtues to the compound body: metals are predominantly mercurial; inflammable bodies, sulphureous; and fixed bodies, salty. Given the MSL ratio, the chemist could predict the properties of the resultant compound. (Paracelsus too, however, suggests that each metal has its own characteristic M and S,[163] thus spoiling the generality of the theory.) Although MSL are not material components of bodies but spiritual directives or forces in them, the MSL ratio constitutive of a compound body plays a comparable explanatory rôle to that of the FAWE ratio in peripatetic matter-theory. The balance of these spiritual forces in a body determines its development and hence its ultimate nature.

It was natural for the practical chemist to identify the *tria prima* with the products of fire-analysis, viz. flame, smoke and ash. Since the MSL theory was intended to be confirmed by the results of chemical analysis, Paracelsus could not deny some link between his ideal MSL and these pyrolysis-products. For the chemical philosopher, flame, smoke and ash are materials in which S, M and L, respectively, are the dominant 'spiritual directions'; for the practical chemist, they are (impure forms of) the chemical elements. Theoretically, the distinction is a momentous one, but its immediate practical import is slight. Thus Debus on MSL:

---

161   See Hooykaas, p. 137, who characterises MSL as 'spiritual directives' or 'directive forces' in matter, not as elementary components of it.

162   See Debus, 10, Vol. 1, p. 163.

163   Paracelsus, 1, Vol. 1, p. 266. Gold, says Paracelsus (*Ibid.*, pp. 249–250), is made from the noblest and purest MSL. See also Pagel, 1, p. 103.

For chemical theorists they represented philosophical substances that might never be isolated in reality, whereas for the practical pharmacist they were nothing else but his distillation products.[164]

This notion of quality-bearing chemical 'principles' responsible for specific properties of substances was severely criticised by Robert Boyle.[165] If compounds X, Y and Z all possess a quality Q, Boyle argues, there is not the slightest warrant for the chemist's assumption that they do so in virtue of the presence of a *common* material constituent (or ideal principle). The alchemical assumption 'One Quality ⟶ One Principle', however, survived this critique and reappeared in the phlogiston (= principle of combustibility) theory of the eighteenth century. Even Lavoisier succumbed: what is 'oxygen' after all but the supposed 'principle' of acidity?

We have been concerned so far with attempts to *explain* the qualities of bodies: it is time now to turn to a theory which can only be described as a form of *explanatory surrender* – the theory of *real qualities* (RQs). Some concessions to this doctrine are of course made in any qualitative matter-theory: if the chemical elements (e.g. FAWE) have associated qualities (e.g. HCDM) that they *contribute* to compound bodies, then those qualities may properly be called 'real' in the (weak) sense that they fall under conservation-laws.[166] Other qualities will be *explained* in terms of FAWE ratios, and will therefore not be real even in this weaker sense. No quality is 'real' in the stronger sense of being capable of independent existence: no quality is itself a substance.[167]

Within the alchemical tradition, however, the conception of RQs as quasi-substantial flourished. One source of this tradition was the Stoic conception of qualities as *pneumata* or subtle bodies;[168] another was the metallurgical practice of 'dyeing' metals. Given the notion of RQs, it will follow that, in order to impart the desired qualities (e.g. yellow, malleable, fixed, dense) to a given piece of metal, one needs only to *add* those qualities (=ingredients) to something lacking them. Thus Sherwood Taylor:

---

[164]   Debus, 11, p. 25.

[165]   See Boyle, *Works*, vol. 4, p. 273 ff., for his treatise 'Of the Imperfection of the Chemist's Doctrine of Qualities'.

[166]   See Chapter 3, p. 96.

[167]   This, we argued in Chapter 3 (p. 98), was the position of Aristotle on this topic.

[168]   See Sambursky, 3, p. 16; Reesor, 1.

The principle that lay behind the simplest alchemical processes seems to have been the attempt to introduce the properties which the base metal lacked. It seemed to the alchemist that a metal might be rendered white or yellow by removing the property of yellowness or whiteness from another substance and introducing it into the metal. The colour was a sort of *activity* and therefore a *pneuma* or 'spirit'. They tell us that 'a tinging *pneuma* gives its colour to metals'; that the colour of plants is their *pneuma*.[169]

At its crudest, then, to add the RQ yellow to a piece of lead one should heat it with something yellow (buttercups, saffron, yellow sulphur) in order to extract from that source the desired RQ. If one can add all the requisite RQs, one has gold: all gold is is a cluster of qualities incorporated into matter. To the alchemist, says Hopkins, the glitter was the gold, 'the colour of the metal was the metal'.[170] The theory is of course a natural offshoot of the practices of dyeing and colouring, and, not least, of gilding metal to make it look like gold. One dyes something yellow (by and large) by immersing it in yellow dye: this empirical generalisation has, for the believer in RQs, the status of a necessary truth. Yellowness, he believes, is a RQ inhering in sulphur, saffron, etc., and able to be imparted to other substances.

But of course if one attempts merely to add to lead the qualities of gold, one will encounter resistance and opposition from the qualities already inherent in the lead. One recipe for transmutation – based on a kind of degenerate Aristotelianism – is therefore to reduce the base metal to P.M. and the add, successively, the desired qualities. This would be absurd, of course, if P.M. is construed as the pure potency of the *Metaphysics*, but to produce this does not seem to have been the alchemists' aim. All they require is the simplest body (i.e. P.M. + CF), devoid of qualities. Even this will possess *privations* of qualities: it is usually described as black ('*nigredo*'),[171] i.e. as deprived of colour. The *nigredo* is therefore not wholly qualityless (which would be unintelligible) but rather minimally 'informed', as low on the ladder of being as matter can go while remaining recognisably corporeal. Since this basic matter is common to *all* material substance,[172] one can in principle transmute any material to

---

169   S. Taylor, 2, p. 32. See also Hopkins, pp. 70–71.

170   Hopkins, p. 103.

171   See Hopkins, p. 94; Partington, 6, Vol. 1, p. 109.

172   S. Taylor, 2, pp. 8–9.

any other by first reducing it to this and then adding the desired forms and/or qualities.[173]

Given a common, featureless, indeterminate first matter, and a conception of qualities as thing-like, elementary addition or subtraction of RQs should transform any raw material into any desired product. On this theory, then, the abiding material constitution is insignificant in comparison to the superadded *pneumata* or RQs. These latter are active, and by inhering in bodies such as gold and silver give them their specific properties. Matter is dull and inert stuff which owes all its qualities to the agency of 'spirits'[174]

This is Hopkins' characterisation of early (i.e. Alexandrian) alchemy: the background philosophy clearly owes a debt to Stoicism. The notion of quality-stuffs survived, however, in the very different philosophical climate of Aristotelianism. The great Arabic alchemist Jābir, for example, sought a technique for the preparation of pure HCDM. FAWE, he explains, each possess two of the 'primary' qualities, and are therefore not truly elementary. But suppose one wishes to add coolness to a particular compound without adding *either* D or M – what can one do? Sherwood Taylor takes up the story:

> Jābir's aim is to make 'pure elements' with only one quality, so that he may be able, for example to add coldness to a metal without, at the same time, adding moisture or dryness. Thus he does not want common water .. but one which is cold but no longer moist. To obtain this, he distils common water and redistils it repeatedly, adding substances thought to be very dry, and so capable of removing the moist quality from the water. After hundreds of redistillations, the water, he tells us, becomes white and brilliant and solidifies like salt. This, he says, is a pure element, and is simply the quality of coldness residing in first matter.[175]

Similar processes, says Jābir, will give pure H, D and M. These still inhere in P.M. – they are not absolutely separate and independent of material substance – but may legitimately be called quality-stuffs in a weaker sense. Once in possession of such simples, the art of alchemy is much simplified: transmutation is achieved 'by adding an elixir made of the

---

173   See Holmyard, pp. 26–27; Dijksterhuis, p. 82.

174   Hopkins, p. 18. For the religious background to this idea, see Pagel, 2, p. 128.

175   S. Taylor, 2, p. 83.

pure elements combined in the right proportions to supplement and correct the metal's deficiencies or excesses'.[176]

A similar theory was advocated by Walter of Odington in his *Icocedron*.[177] A precisely weighted mixture of F and E will, he claims, produce pure *dryness*: the H and C cancel each other, leaving a matter that is dry, but neither hot nor cold. Once in possession of such pure HCDM, one could if one wished recreate either FAWE or any desired compound body.

The theory of real qualities also – and most unfortunately – became entangled in the theological mysteries associated with the Eucharist. Although Aquinas denied that a quality such as whiteness could exist without a proper material substance to inhere in,[178] it soon became theologically orthodox to claim that the 'accidents' of water and wine did just that – continued, by divine power, to exist after the annihilation or transformation of their original subjects. God can, decreed Bishop Tempier in 1277, create an accident without a subject, e.g. a quality which inheres in (qualifies) nothing at all. *Qualities*, says Ockham, are *real*,[179] i.e. capable of separate existence, if only by divine power. Quantity, by contrast, is not: God can produce a separate whiteness or warmth, but not a separate length. But *some* qualities (dense and rare, straight and curved, etc.) depend merely on the arrangement of the material parts of bodies, and are then inseparable from material substance. If this idea of explaining the qualities of bodies in terms of the arrangement of their parts is pursued further, and such qualities as colours and odours are explained in such terms, we have (a) a reductive account of so-called RQs as nothing more than arrangements of corporeal particles, and hence as inseparable from material substance, and hence (b) a return to Democritean reductionism. But that still lies far in the future.

Another strategy for explaining the characteristic properties of a thing is to refer to the agency of that object's substantial form (SF). The nature and status of the Aristotelian SF is of course one of the perennial questions of

---

176   *Ibid.*, p. 82.

177   See Thorndike, Vol. 2, p. 54.

178   M. T. Clark, Ed., pp. 78–79.

179   See Leff, pp. 204–205. In the Eucharist, says Ockham, God annihilates the water and wine, after which their accidents persist, although present in no substance.

philosophical disputation. On the one hand, we can identify form with material constitution, and thus obtain a philosophy with a strongly materialistic savour. Enough has been said about this. On the other hand, we can conceive SFs as independent incorporeal substances, separate from the matter they 'inform'. I argue in Appendix 3 that this latter notion is not entirely alien to Aristotle himself, and explain how it came to be read into his *Metaphysics* by many of his commentators.

But surely SF is not *entirely* independent of material constitution? One could not, for example, fabricate an animal or a plant without water (for blood or sap) or fire (vital heat). Too much or too little of any of FAWE will disrupt the vital balance of an organism and thus kill it. A given SF or soul can cope therefore only with a narrow and precise range of FAWE ratios.[180] Socrates, for example, remains identifiably himself when feverish (too hot) or thirsty (too dry), but take either condition to an extreme and he dies.

SF is therefore not entirely independent of material constitution, not an incorporeal entity quite distinct from the body (except perhaps rational souls, which are different again). Rather, it is a formative principle, which can take a particular FAWE ratio (and no other) and make of it a a man, a cat, or an oak tree. The mere material ingredients could not do this spontaneously without the agency of a form: a particular FAWE ratio is necessary but not sufficient for the formation of a specific organism. *How* SF 'moulds' matter into the required form we do not know.

Two more facts about SFs before we proceed: first of all, they are *specific*; in virtue of its SF, each individual is of the particular species that it is, be that man, cat, or oak tree. Secondly, there is a *de re* necessity of kind: an individual with the specific form of a kind K is necessarily a K. The identical individual cannot be at one time an oak and at another time an ash; or at present a cat and at some later time a dog. Of course an oak tree can die, and its material become incorporated into an ash, but this gives no bond of identity between the two trees. Likewise, a cat can devour a mouse and thus assimilate murine tissue into feline, but no one would describe this as a mouse changing into a cat!

We are now in a position to appreciate Avicenna's criticisms of the claims of the alchemists to be able to transmute metals. Avicenna is sceptical:

---

[180]   For a clear account of this idea, see Digby, p. 177.

> . . . it is not in their power to bring about any true change
> of species. They can, however, produce excellent imita-
> tions, dyeing the red [metal] white so that it closely
> resembles silver, or dyeing it yellow so that it closely
> resembles gold . . . Yet in these [dyed metals] the essential
> nature remains unchanged; they are merely so dominated
> by induced qualities that errors may be made concerning
> them . . .[181]

The metals, Avicenna claims, are *distinct species*; hence, just
as a man cannot turn a daisy into a dandelion, or a cow into a
horse, so he cannot transform copper into silver or gold. 'I do
not deny,' he continues, that

> such a degree of accuracy may be reached as to deceive even
> the shrewdest, but the possibility of eliminating or impart-
> ing the specific difference has never been clear to me. On
> the contrary, I regard it is impossible, since there is no way
> of splitting up one combination into another. Those
> properties which are perceived by the senses are probably
> not the differences which separate the metals into species,
> but rather accidents or consequences, the specific differ-
> ences being unknown. And if a thing is unknown, how is it
> possible for anyone to endeavour to produce or destroy
> it?'[182]

Gold cannot be identified with any set of manifest qualities
(yellow, dense, fixed, etc.): its real nature is an unknown
specific form, which is responsible for those properties.
Alchemists may tamper with the superficial appearances of
bodies, but unless they can alter *forms* they cannot genuinely
transmute metals. Alchemical gold, Avicenna insists, is all
fake, and decomposes after a few heatings.

These trenchant criticisms of alchemy by an authoritative
author were often repeated in the Middle Ages.[183] The only
theoretical possibility they leave open to the would-be
alchemist is the arduous route of reduction of the base metal
to P.M., followed by an attempt to induce the desired form.
An alchemical text (falsely) attributed to Avicenna

---

181   From Grant, Ed., 4, p. 572.

182   *Ibid.*, p. 572.

183   Avicenna's arguments are repeated by Lull (see Thorndike, Vol. 4, p. 7)
       and in the 'New Pearl of Great Price', a famous defence of alchemy (in
       Grant, Ed., 4, p. 576).

recommends precisely this (hopeless) course,[184] but most alchemists preferred instead to deny Avicenna's presupposition that the metals are indeed different species. The seven metals, says Albertus Magnus in his defence of the alchemical art, 'differ from one another only in their accidental form, not in their essential form':[185] they can therefore be transmuted one into another without reversion to P.M. The conversion of base metals to gold and silver is often likened to a process of maturation or of cure: lead ⟶ gold is, claims the alchemist, like child ⟶ adult or invalid ⟶ healthy man rather than like oak ⟶ ash or cow ⟶ horse. The alchemists, says Petrus Bonus in his *New Pearl of Great Price*, say that

> there is between metals no specific, but only an accidental difference. They suppose that the base metals are in a diseased condition, while gold and silver exhibit the healthy state of the metallic substance.[186]

The conversion of lead into gold is therefore one of cure, not of transformation. Apart from the 'morbid matter' which they contain, claims Bonus, 'all metals are actually gold – gold is the most mature and healthy metal, the acme of metallic perfection'.[187] This medical analogy derives from Jābir, and his obvious links with the Greek conception of health as a balance (*isonomia*) of the elements FAWE. Gold, says Jābir, has the most perfect harmony or ratio of the elements FAWE: the conversion of base metal to gold is therefore perfectly analogous to the process of cure.[188]

A similar theory can be found in Paracelsus: the conversion of lead to gold is, he claims, like that of ore to metal, i.e. involves only the *separation* and purification of something already present. All the alchemist does is to continue and perfect Nature's work of maturation. 'All sublunary things,' says Paracelsus, 'contain in their inmost centre a precious grain of this elementary gold';[189] it is the task of the alchemist to extract it through the agency of fire, by which 'all that is imperfect is destroyed and taken away'.[190] The artist must

---

184 See Thorndike, Vol. 2, p. 568; Partington, 2, p. 11.

185 From Grant, Ed., 4, p. 588.

186 *Ibid.*, p. 579.

187 *Ibid.*, p. 579.

188 See Crosland, 1, p. 18.

189 Paracelsus, 1, Vol. 1, p. 301.

190 *Ibid.*, Vol. 1, p. 4.

follow and perfect the workings of Nature: Art may *aid* Nature in the production of her marvels, but cannot work against her.

We have discussed in this section a great variety of radically different theories of matter, and have come to the end of it without having arrived at any definite conclusion. We have not even discerned any clear drift or tendency towards consensus: matter-theory remained a chaos of dissenting opinions and conceptual confusion well into the seventeenth century. FAWE, MSL, SFs and RQs, occult qualities and quintessences: all survive to compete with the revival of mechanistic Atomism. I only hope that this section has helped to prepare the way for its successor.

D. Natural Necessity

Within the natural philosophy of Aristotelianism, it is deemed a *necessary* truth that objects impart their SFs to others, that one X tends to make other things X. A similar thesis will be formulable for RQs: an object with a particular RQ will naturally impart that quality to neighbouring things. The theory of forms and qualities is therefore associated with the doctrine of *natural necessity*. Fire, for example, must consume combustible materials: it burns, says Avicenna, 'according to its own nature',[191] i.e. by necessity.

The great Arabic peripatetics, Avicenna and Averroes, explicitly affirm this necessitation of effect by cause. Whatever comes into existence due to a cause, says the former, 'comes into existence by necessity'.[192] Averroes concurs: indeed, 'Averroism' was to become intimately associated with the doctrine of natural necessity.

It was, as might be expected, the Islamic theologians who objected to natural necessity as involving a denial of God's omnipotence and, especially, of His power to work miracles. This theological reaction was given a rational basis by al-Ghazzali, who developed in opposition to Aristotelianism a purely occasionalistic metaphysic with Allah as the only efficient cause. The philosophers, he argues,

> have no other proof [that fire burns cotton] than the observation of the occurrence of the burning, when there is contact with fire, but observation proves only a simultaneity, and in reality there is no other cause but God.[193]

[191]   Morewedge, Ed., p. 43.

[192]   *Ibid.*, p. 47.

[193]   Quoted from Weinberg, 2, p. 118.

Although Allah *can* do anything He pleases, however – e.g. create cold fire, warm ice, or non-magnetic iron – He *chooses*, by and large and of His own free will, to abide by certain *customs*. The regularities we observe in Nature are, Ghazzali insists, nothing more than the habits of the Deity, habits which He could break at any moment. A miracle is simply an infringement of divine custom,[194] a momentary reminder by the deity of the radical *contingency* of all things on His will.

In Ash'ari and his followers, the Mutakallimun, Ghazzali's occasionalism is fused with an Atomic theory of matter. The denial of SFs and RQs is of course implicit in classical Atomism, but in place of Democritean (i.e. mechanistic) natural necessity the Mutakallimun place only the divine will. All the atoms, they claim, are unextended (hence without size and shape) and, in themselves, identical; at any particular moment, however, they possess as *accidents* the properties of the body of which they happen to form part. Thus the atoms making up iron are given – for an instant – the accidents of iron, and so on. Accidents need to be re-created by Allah at every instant, and are purely contingent on His will: if He gives to the atoms which at t_0 made up iron the accidents of copper, then at t_1 we will have copper in place of iron. The body itself has no nature (*physis*) in virtue of which it is necessarily what it is. Divine custom may give us the impression that things have natures, but that impression is seriously misleading. In the final analysis, the deity is the only cause (even of what we think of as our free acts[195]). Since atoms cannot exist without accidents, He can annihilate bodies simply by refusing to renew (re-create) their accidents: He destroys by refraining from creating.[196] Without God's sustaining will, the whole physical universe would pass instantaneously into nothingness.

Let us turn briefly to Maimonides' exposition of the philosophy of the Mutakallimun. Of the twelve propositions into which he condensed this metaphysic, the following are relevant here:

P1.   All things are composed of atoms.
P3.   Time is composed of time-atoms.
P4.   Substance cannot exist without numerous accidents.

---

194   See Wolfson, 2, p. 176.
195   MacDonald, p. 333; Wolfson, 2, p. 200.
196   MacDonald, p. 332.

P5.   Each atom is completely furnished with the accidents and cannot exist without them.

P6.   Accidents do not continue in existence during two time-atoms.

P7.   Both positive and negative properties have a real existence, and are accidents which owe their existence to some *causa efficiens*.

P8.   All existing things, i.e. all creatures, consist of substance and of accidents, and the physical form of a thing is likewise an accident.

P10. The test for the possibility of an imagined object does not consist in its conformity with the existing laws of nature.[197]

The atoms are dimensionless, identical with one another in their own (lack of) nature, but qualified by various 'accidents' which would pass out of existence between successive instants if God did not renew them. God's usual custom is to re-create in each atom the accidents corrupted in it in the previous instant, thus presenting to our senses the misleading appearance of stability and continuity. This is how Wolfson puts it:

> It is the continuousness of God's custom that explains not only the regularity of succession of the newly created things in the world but also why these newly created things appear to us as a continuously existent one thing. Thus when a garment dyed black by its contact with a certain dye seems to us to continue to be black with the same blackness, it is only because it is the custom of God, who has created the blackness in the garment on its coming into contact with that certain dye, to continue to create the same kind of blackness in the garment at every instant.[198]

Of course, God is under no obligation to create *blackness* in the cloth: *a priori*, a black dye could impart a red colour, or a blue one! This can be imagined (mentally pictured) and is therefore possible (see P10 above). Similarly, as Wolfson says,

> fire and water, which by the continuousness of God's custom cause combustion and cooling respectively, may have their actions changed by God so that fire would cause coolness and water would cause combustion.[199]

197   Maimonides, p. 120.

198   Wolfson, 2, pp. 183–184. See also Maimonides, p. 125.

199   *Ibid.*, p. 184.

According to the Aristotelians, fire has a *real essence* (R.E.), in virtue of which it must act as it does: a substance which did not impart heat and (in suitable materials) cause combustion would lack that R.E. and hence not be fire. For the Mutakallimun, by contrast, atoms have only accidents: it is entirely accidental to a given collection of atoms what kind of substance they constitute. God gives them, at one instant t_0, accidents such as 'colour, smell, motion or rest',[200] but only by divine custom will they tend to retain those accidents even till the next instant t_1. And of course no body *ever* imparts its qualities to another: the appearance of such processes is merely another manifestation of divine custom. The Mutakallimun thus, as Maimonides says, explicitly deny the view that 'there exists a natural force from which each body derives its peculiar properties',[201] i.e. they reject the Aristotelian concept of *physis*. There are no 'natures' in things which distinguish, e.g. iron from butter. Even forms such as animality and humanity are relegated to the status of accidents (P8): an atom is accidentally 'humanised', just as it is accidentally 'whitened', as a result of divine whim. A constituent atom of a human body is 'humanised', but this does not touch its nature, since it has none: next instant, it could be part of a totally different substance.

In such a world, everything imaginable, or, more precisely, everything describable without self-contradiction, is possible. The only necessity is logical: not even God can make contradictory propositions simultaneously true. Creatures have no causal powers: 'It should not be said that this is in any respect the cause of that'; 'in the last analysis, God is the sole agent'.[202] There is therefore no natural law and no natural necessity. Objects *need* not impart their forms and qualities to others: there are no Aristotelian SFs and RQs. The principle 'cause resembles effect' may not, even if reformulated in occasionalist vocabulary, be true: such truth as it may possess is only God willing, as a purely contingent matter. On such a theory, as Maimonides sees,[203] the terrestrial globe could turn into the celestial spheres, fire could fall and earth rise, men could have several heads, or fly: whatever is logically possible is naturally possible. It is not for us to say what Allah can or

---

200   Maimonides, p. 123.

201   *Ibid.*, p. 124. See also MacDonald, p. 330.

202   These propositions of the Mutakallimun are quoted from Wolfson, 2, pp. 181–182.

203   Maimonides, p. 128.

cannot do. Fire could indeed be placed next to dry flax, and no combustion ensue, if He so wills it.

Exactly the same issue – the opposition between God's omnipotence and natural necessity – recurred in Christian scholasticism. The main advocates of natural necessity were the so-called 'Latin Averroists', led by Siger of Brabant: by attributing to Aristotelian principles the modal status of necessary truths, they implicitly denied God's capacity to work miracles. If, for example, fire naturally (i.e. necessarily) burns flesh, God could not have delivered Meshach, Shadrach and Abednego from the fiery furnace (Daniel, 3): hence the theological fulminations against Averroism, culminating in the condemnations of 1277.

The Christian equivalent of Al-Ghazzali is William of Ockham, who, like Ghazzali, tried to provide rational justification for what was initially no more than clerical outrage. In Ockham, as in Ghazzali, emphasis on God's freedom and omnipotence led to belief in the contingency of the 'laws' of nature.[204] The first axiom of Ockham's philosophy is that 'all things are possible for God, save such as involve a contradiction'.[205]. Whatever is normally produced by secondary causes, God could produce directly, i.e. by His own power.[206]. As Copleston says, Ockham's main concern as a philosopher was

> to purge Christian theology and philosophy of all traces of Greek necessitarianism, which in his opinion endangered the Christian doctrine of the divine liberty and omnipotence.[207]

What has all this to do with Atomism? The answer of course lies in the denial of Aristotelian SFs and RQs. One cannot accept SFs and RQs and deny natural necessity: they are two sides of the same coin. A *real* quality must be *conserved*; a *substantial* form must tend to imprint its likeness on surrounding things. The Mutakallimun, as we have seen, rejected SFs, RQs and natural necessity wholesale in favour of

---

[204]   And hence, perhaps, to a more empirical theory of knowledge, which may have influenced the rise of the empirical sciences: such a claim sounds plausible and attractive, but I have yet to see it adequately documented.

[205]   Quoted from Boehner's Introduction to his collection of Ockham's writings, p. xix.

[206]   *Ibid.*, pp. xix–xx. See also Leff, pp. 15–17.

[207]   Copleston, 1, pp. 47–48.

atoms, accidents, and the divine will as the sole cause, thus giving them an Atomic theory of matter within an occasionalist metaphysic. In the fourteenth century thinker, Nicholas of Autrecourt, we find another and – if I am right – quite different attempt to combine an Atomic Theory with a denial of natural necessity.

What things, asks Nicholas, can we know with certainty? Only, he answers (a) logical principles dependent on the law of contradiction, and (b) the immediate data of experience.[208] But, he proceeds to argue (along lines later made famous by Berkeley and Hume), logic and sense-data alone do not suffice to give us knowledge of the existence of *substances* and *causes*. From the existence of one thing A we can never justifiably infer that of a distinct reality B:

> From the fact that one thing exists, it cannot be any evidence derived from a first principle be deduced that another thing exists.[209]

Let us ignore substances and concentrate on causes. Cause, says Nicholas, is *logically* independent of effect. Hence, cause and effect are distinct realities, and all distinct realities are, as Ockham said, separable from one another, if only by divine power. Thus, for example, says Nicholas:

> This consequence is not admitted with any evidence deduced from a first principle; 'Fire is brought near to tow and no counteracting cause is present, therefore the tow will be burned'.[210]

If, says Nicholas, we define the term 'fire' in such a way that it is analytic that fire possessed the power to consume tow, we will not be able to tell without trial that a particular substance is truly a fire in this sense; if, on the other hand, we define 'fire' merely by its manifest appearance, it is an inductive (and hence fallible) assumption that all fires will consume tow:

> Whatever conditions we take to be the cause of any effect, we do not evidently know that, those conditions being posited, it follows that the effect must be posited also.[211]

For all we could know, all things could owe their existence to God's direct agency:

208 See Weinberg, 1, p. 7.

209 Quoted from Rashdall, p. 6.

210 *Ibid.*, p. 8.

211 *Ibid.*, p. 10.

> We have no evident knowledge that there can be any cause of any event other than God.

> We do not know that any cause which is not God exercises efficient causality.[212]

Statements like the above led Bréhier[213] to attribute to Nicholas the occasionalism of Ghazzali and the Mutakallimun. But Nicholas has not asserted the *truth* of this doctrine, but merely its *irrefutability*. He uses the irrefutability of occasionalism as part of an anti-Aristotelian *epistemological* argument, not as a positive contribution to metaphysics. As Weinberg insists,[214] Nicholas is concerned not to deny the existence of substances and causes, but to deny that we humans can have certain knowledge of them.

Apart from certain knowledge, however, we also have 'probable' knowledge of things. Nicholas' probabilist metaphysic is one of the most curious parts of his whole philosophy. We assume as a premiss, he begins,[215] that the universe as a whole is absolutely perfect, i.e. that its contents are most perfectly disposed. (This assumption is eventually to be justified by its explanatory power.) From this it follows, Nicholas argues, that nothing ever ceases to be: the death of any reality would impair the universal perfection. And from this in turn there follows the Atomic Theory, with its eternal atoms and its account of generation and corruption as no more than the aggregation and dissociation of these abiding realities.

Among Nicholas' arguments for his Atomic theory, we find the classical Leucippan proof of the existence of the vacuum.[216] This argument clearly turns on the assumption of the absolute *impenetrability* of bodies: without an impenetrable (and incompressible) matter*, Leucippus' argument will not get off the ground.[217] But this gives us a notion of *natural necessity*: if an atom is to move into an occupied place, it *must* displace the previous occupant of that place. Given rigid atoms and elementary geometry, moreover, we can extend

212   *Ibid.*, p. 9.

213   Bréhier, p. 197.

214   Weinberg, 2, p. 66, p. 122.

215   See Nicholas of Autrecourt, p. 37 ff.

216   See Appendix 4.

217   Even the Ash'arites assumed the absolute impenetrability of the atom. Not even divine power, say these Arabic Atomists, could cause the interpenetration of two atoms. See MacDonald, p. 335; Wolfson, 2, pp. 197–198.

this notion of natural necessity in such a way as to explain the manifest properties and powers of bodies as necessary consequences of their intrinsic microstructures. (See the peg-board analogy outlined in Chapter Three.[218])

It now seems that Nicholas is liable to end up replacing the Aristotelian R.E. with the Democritean: instead of character-ising real essences in terms of forms and qualities, he will do so in terms of arrangement of (rigid and impenetrable) atoms. In virtue of its R.E. each substance will (necessarily) possess those powers and properties we discover in it: given its particular microstructure, it could have no others.

But does this not lead us back to an assertion of natural necessity and hence the denial of divine omnipotence? Have we not turned Nicholas, unwittingly and unwillingly, into an Averroist? To answer this question it is first necessary to distinguish what exists from what can be known (or even formulated) by man, i.e. to distinguish Q1 form Q2:

Q1.   Are there necessary connections in Nature?
Q2.   Can we have any knowledge of them?

Aristotelianism gives positive answers to both these questions. Our faculties, says the peripatetic, are adequate to give us some grasp of the R.E.s of things; from these we can 'scientifically' demonstrate many of their other properties.[219] Democritean Atomism tends, by contrast, to give a positive answer to Q1, but a negative one to Q2. While all things possess R.E.s (= microstructures) on which their manifest qualities depend, we humans have no familiarity or acquaintance with those R.E.s. The superficial appearances of things manifest to sense (colours, etc.) give no reliable guide to their R.E.s. Democritus therefore accepts the existence of R.E.s, and hence of natural necessity, which reduces to geometry and mechanics. Given the nature of Democritean atoms, there will be necessary truths about their possible combinations – a round peg cannot fit snugly into a square hole. Although, however, we have some grasp of the general nature of a R.E., the *specific* essences of things are not discoverable either by sense or intellect: hence the enigmatic Democritean remark that 'truth hides in the depths', and his oft-remarked tendency to scepticism.[220]

---

218   Chapter 3, pp. 109–110.

219   Averroes even went so far as to state, in his *Metaphysics*, that there is no question to which the human intellect cannot discover the answer! Nicholas, as might be expected, finds this hard to swallow (see Nicholas of Autrecourt, p. 42).

220   See Chapter 3, p. 110.

Now consider the following propositions:

'The magnet attracts iron.'
'Acid turns blue litmus red.'
'Cheese is nourishing to humans.'

Are these truths necessary or contingent? No doubt, on the Democritean view, there exists a necessary truth corresponding to each of the above. These truths will contain terms referring to (rigidly designating, in the Kripkean sense[221]) the R.E.s of iron, acid, cheese, etc. In our stage of ignorance, however, we can only supply schematic designators. If a particular piece of cheese nourishes me, I may say that anything with *that* R.E. could not fail but nourish any organic body with the same constitution as *mine*, but this is hardly enlightening.

It soon becomes apparent that a radical breach between human and divine (or angelic[222]) knowledge is in the offing. A being with pure intellectual intuition of essences (God) could know *a priori* the outcome of chemical reactions. A being with microscopically acute eyesight (the lynx, Locke's angels) could attain similar knowledge by direct perception. Such beings would use natural kind terms which rigidly designated R.E.s. Let us simply label our natural kind terms with asterisks to produce such putative Kripkean designators. (This procedure may turn out to be question-begging, since it may be that our natural kind terms do not map on to the real natural kinds designated by natural kind* terms.) In the vocabulary of divine (or angelic) chemistry, propositions like 'acid* turns blue litmus* red' and 'cheese* is nourishing to humans*' will be *necessary* truths, dependent on the R.E.s of the kinds in question.

When we examine the human epistemic situation, however, things are quite different. Since we are (or were) utterly ignorant of the R.E.s of things, it is a natural assumption that *our* natural kind terms ('iron', 'acid', 'cheese', etc.) stand only for *nominal essences* (collections of manifest properties). But if that is so, a proposition like 'cheese is nourishing' or 'acid turns blue litmus red' will be *contingent*. There is no necessary connection between the *nominal essence* of cheese (its colour, taste, and smell, let us say) and its nutritive power. The same set of qualities that make up a particular nominal essence may result from very different R.E.s, on

---

221   A 'rigid designator' designates the same object or property in all possible worlds.

222   Cf. Locke's 'angelic chemistry' in the *Essay*.

which very different sets of other properties depend. To allow our natural kind terms to stand for nominal essences is therefore to admit that our system of classification does not cut deep into the heart of Nature, but is a more or less arbitrary superficial imposition.

On this reading, the Atomist admits the existence of natural necessities, but denies that we can formulate them. Divine science is Kripkean; human science must, *faute de mieux*, be Humean. To express the necessary connections that exist in Nature, we would need acquaintance with R.E.s and use of the divine natural kind* predicates to designate them. Even if, as a matter of fact, 'cheese' and 'cheese*' are coextensive, and all actual instances of cheese are, necessarily, nourishing, it will NOT be true to say 'Necessarily, all cheese is nourishing' – our evidence does not extend to the contents of other possible worlds. And, in any case, given our ignorance of R.E.s, how could be ever know that 'cheese' and 'cheese*' were coextensive predicates?

As O'Donnell says,[223] no Science (in the old and honorific sense) can result from Democritean Atomism. There can be no demonstrations of the properties of things by deduction from definitory axioms characterising their R.E.s, since those are unknown to us. We can of course frame conjectures about the hidden R.E.s of things, and can attempt to combine these conjectures into a coherent explanatory system, but such a procedure is always fallible. Atomism, therefore, gives rise to probabilism and scepticism in epistemology, in opposition to Aristotelianism and other 'dogmatic' philosophies.[224]

Classical (i.e. Democritean) Atomism tends therefore to involve an affirmative answer to Q1 and a negative one to Q2. The Atomic theory of the Mutakallimun clearly involved, as we have seen, negative answers to both questions, and an outright rejection of all forms of natural necessity. Where does Nicholas stand on this issue? Bréhier has suggested that he favours the occasionalism of Ghazzali and Ash'ari. While it is clear that he was familiar with such ideas[225] and, furthermore, that he felt such a theory to be irrefutable, this interpretation still seems unsound. In the foregoing discussion we have, I think, discovered the basis for a sounder and juster reading.

---

223   O'Connell, p. 108.

224   Cf. the strain of sceptical empiricism in the Atomist school of the seventeenth century – Gassendi, Charleton, Bernier, and Locke.

225   Nicholas' theory of motion as involving discrete instants of rest was almost certainly derived from the Mutakallimun.

Could God, then, create a cold fire or non-nutritive cheese? *Not* if 'fire' and 'cheese' are defined in terms of their real but unknown essences: cold fire* and non-nourishing cheese* are quite impossible. But if the terms 'fire' and 'cheese', being after all merely terms in a human vocabulary, are taken as standing for their respective nominal essences, it follows that God could create a cold fire or non-nourishing cheese. There could be visible flames that produce no heat, or a substance with all the sensible qualities of cheese but without its nutritive power. Since we only know things from the outside, as it were (i.e. by their appearances to our senses), *our* terms must stand only for nominal essences. We cannot, therefore deny that God could make a cold fire: the necessary truth of a proposition of divine chemistry such as 'all fire* is hot' is of no concern to *us*.

Nicholas' position, if I am right, is essentially that of Locke. But, whereas Locke notices – and condemns – the 'secret supposition' by which we 'tacitly suppose or intend' our natural kind terms to stand for the R.E.s of things,[226] Nicholas simply assumes that our terms can only designate that which is familiar to us, viz., the sensible qualities of things. Both Locke and Nicholas accept the existence of R.E.s capable in principle of being rigidly designated, but reject the Kripkean thesis about the meaning of our natural kind terms. Only some higher beings (God, angels,) can possess a Kripkean vocabulary of rigidly designating natural kind* terms. One can now understand the sheer excitement and enthusiasm of seventeenth century microscopists such as Hooke and Power, who believed that through their labours the R.E.s of things were gradually being made observable, and hence a true Science of Nature made possible. The Democritean depths, they believed, were finally being plumbed, and the real essences (microstructures) of things brought to light.

---

[226]   See Locke, *Essay*, III, 10, xvii–xviii.

THE MECHANICAL PHILOSOPHY: 2

The title of this chapter is somewhat of a misnomer, since there existed, during the period with which it is concerned, no 'mechanical philosophy' worthy of the name. I shall not then be concerned to trace the fortunes of an existing and systematic body of doctrine; rather, I shall be seeking in medieval and Renaissance thought the sources from which sprang the revitalised and world-shaking mechanical philosophy of the seventeenth century. I shall most emphatically *not* be dealing with all the a-mechanical and anti-mechanical ideas current during the period: to explain in detail the workings of the *Anima Mundi* of the Platonists, the 'Universal Nature' of the schoolmen, the *archei* of the Paracelsians, and the angels and demons of the magicians is not a task that appeals to me, even if time and space permitted it. Instead, this chapter is highly selective in content (i.e. it omits a great deal), and concentrates simply on three broad topics or areas of urgent and immediate relevance to mechanism in general, exploring various avenues of thought concerning them. It may help a little to explain the rapid rise of the mechanistic approach to the workings of Nature that typifies much seventeenth century science and philosophy out of the apparently uncongenial background of an animistic (and even occult) world-view. Our three areas of interest are, in order, (a) weight, (b) the theory of 'species', and (c) the nature of spirits.

A. Weight

The nature of gravitation is of obvious relevance to three of the four headings under which we subsumed the content of the 'mechanical philosophy', viz. (a) action at a distance (A.D.), (b) spontaneity, and (c) incorporeal causes. It is natural to think of the free fall of a stone to Earth as involving *either* A.D. (of the Earth on the stone) *or* the spontaneous descent of the stone (of its own volition, as it were, without any external impulse). And either of these causes may imply the operation of some agency other than corporeal in nature: *bodies*, argues the Platonist, act neither at a distance nor spontaneously. The topic of weight is therefore quite central to any putative mechanisation of natural philosophy.

Two great theories of the nature of natural motion dominated medieval and Renaissance thinking on the topic. On the one hand, there was the Aristotelian view that the natural motion of each of the elements FAWE is directed towards its proper place or sphere; on the other hand, there was the (Neo) Platonic conception of a tendency of like stuff to join like: the weight of a stone is for a Platonist merely one instance of the universal force of *sympathy*. This theory took its rise from some remarks in Plato's *Timaeus*,[1] but soon became extended and developed to provide a general account of force. Schematically, then, we have a matter ⟶ place, and a matter ↔ matter theory, Aristotelian and 'Platonist'[2] respectively. As we shall see, the two theories have radically different implications.[3]

On neither view was gravitation normally thought of as involving A.D. Aristotle explicitly denied that natural place is the efficient cause of natural motion: a stone falls, for the peripatetic, in order to attain the perfection of its own form which can be realised only at its natural place.[4] The centre of the universe does not pull the stone, except metaphorically; the stone is moved by its own 'desire' for actualisation of its own latent potencies. For the Platonist, a piece of earth seeks union with other bodies of like kind: the Earth only 'moves' the stone in the sense in which a mouse may be said to be moved by a piece of cheese – the true efficient cause involved is its own desire for the cheese.

For both Platonist and Aristotelian, then, gravity involves spontaneity but not A.D.: the falling stone is moved by an internal principle (analogous to a desire), not by the external pull of either a natural place or kindred material. Some *internal agency* is involved in the stone itself: it is not a mere patient, manipulated by external forces.

Is the stone then a *self-mover*, moved by its own substantial form (SF), or by the motive quality of *gravitas* consequent

---

[1] See Chapter 4, p. 148.

[2] I call this theory 'Platonist', with scare quotes, to indicate that, although developed and advocated by avowed Platonists, their nominal master would not have endorsed the view of his supposed disciples. See Chapter 4, p. 148.

[3] See of course Aristotle's thought-experiment of transferring the Earth to the place of the Moon, discussed in Chapter 4, p. 166.

[4] The movement of a thing towards its natural place is, says Aristotle 'motion toward its own form'. See Chapter 4, pp. 166-167.

upon that form? Aristotle himself was most reluctant to accept such a view: *if*, he says, the form of the falling stone is the efficient cause of its motion, then the distinction between animate and inanimate things collapses – i.e. our philosophy surrenders to animism.[5] Only animals are self-movers. The efficient causes involved in the free fall of a heavy body are, he says, (a) its generator – that which made it what it is – and (b) whatever removes the obstacle or impediment preventing it from falling.

Unfortunately for Aristotle, this evasion will not do. The generator may long since have ceased to exist; the remover of the impediment need impart no downward impulse. Why then does the stone not remain suspended in the air? In peripatetic physics, motion requires for its continuation either an external force or an internal striving. Neither the generator nor the remover of obstacles provides such a force or striving: neither, therefore, is the true motor involved.

In Avicenna and Averroes, the decisive step is taken, and the *motor conjunctus* identified with the SF of the body in question. A freely falling body is, says Avicenna, moved by its own SF as by an indwelling incorporeal motor.[6] Averroes, with some important reservations, concurs: the *gravitas* of a heavy body is, he claims, an accidental form, the instrument by which the SF moves the body.[7] There is also, however, an essential rôle for the medium: the stone, says Averroes, moves the medium, which in turn – by the 'circular replacement' theory – moves it. A freely falling stone is therefore only a self-mover *per accidens*: without the presence of a fluid medium, the natural motions of inanimate bodies would be impossible.[8] In free fall, then, the SF produces the motive quality of *gravitas* in the stone, in virtue of which it can displace the medium beneath it, which will in turn move the stone. The animate/inanimate distinction is preserved: animate beings are self-movers *per se*, inanimate ones only *per accidens*, i.e. *via* the agency of the medium. The form of an inanimate object such as a stone is not, Averroes insists, a distinct entity from the matter it 'informs' – they cannot therefore be thought of as mover and moved respectively. Whereas the soul of a man or animal can be conceived as an

---

5 Chapter 4, pp. 164–165.

6 See Weisheipl, p. 35. For some of the background to and implications of such a point of view, see Wolfson, 1, pp. 88–89.

7 See Wallace, 5, p. 287.

8 Weisheipl, p. 36; Moody, 4, pp. 210–231.

internal incorporeal mover, distinct from the body it inhabits (like a sailor in a ship), the SF of a stone is quite inseparable from it. The stone therefore acts upon the medium, which in turn re-acts upon it to move it: this solution, Averroes feels, avoids the implicit animism of Avicenna.[9] (What actually moves the stone is not its own SF, but an external push from the medium.[10])

For both Avicenna and Averroes, then, the SF of the stone is in some respect the efficient cause of its natural motion, either directly (Avicenna) as an incorporeal motor, or indirectly (Averroes) by the causal mediation of a physical medium. Aquinas flatly denies both these theories: the formal principle of an inanimate body, he insists, cannot act as an efficient cause.[11] Freely falling stones are not self-movers. Aquinas' own position follows closely that of Aristotle: the efficient cause *per se* of such natural motions is, he claims, the generator of the heavy body: the efficient cause *per accidens* is the remover of the obstacle – thereafter, the stone falls by nature (*physis*) without the need for a *motor conjunctus*. (Such natural motions could therefore take place *in vacuo*.) The true mover in such a case is the generator:[12] the falling stone remains an instrument of its progenitor, even if that has long since ceased-to-be. The stone's own SF and *gravitas* are internal principles in virtue of which downward motion is natural to it. But they are not and cannot be the efficient causes of its descent.

These three positions – the implicit animism of Avicenna, the Aristotelianism of Aquinas, and the rather curious compromise view of Averroes – provided the framework for much scholastic disputation. During this dispute, it gradually became clear that peripatetic dynamics demands some *motor conjunctus* for every motion. Even if, as Weisheipl insists,[13] the principle 'Omne quod movetur ab alio movetur' ('whatever is moved, is moved by another') did not initially assert the existence of a *motor conjunctus* for every local motion, it gradually became so interpreted as to mean just that. Scotus, for example, follows Avicenna rather than Aquinas: the nature of a thing is, he states, an internal active

9   See Moody, 4, p. 239.
10 See Wolfson, 1, p. 89.
11 Weisheipl, p. 39; Wallace, 5, p. 288.
12 Weisheipl, p. 40.
13 *Ibid.*, p. 29.

principle, the efficient cause of its natural motion.[14] In late scholasticism, Weisheipl admits, *impetus* and *gravitas* (or *levitas*) were often conceived as internal movers responsible for violent and natural motions respectively.[15]

For a typical late-scholastic discussion of the cause of free fall, let us turn to Vitelleschi, one of the professors at the Jesuit Collegio Romano in the sixteenth century – and, incidentally, one of the teachers of Galileo.[16] What, Vitelleschi asks, is the active principle from which the free fall of a stone proceeds? Is there an internal mover; or can we only, with Aristotle, cite such external features as the generator and the remover of obstacles? Neither of these extrinsic factors, says Vitelleschi, is in contact with the stone throughout its fall: if we seek a *motor conjunctus* to explain natural motion, they will not serve.

External movers, says Vitelleschi, move bodies by the instrumentality of certain types of quality. The magnet, for example, is the efficient cause of the motion of a piece of iron, but its *instrument* is a magnetic quality imposed on the iron. Similarly, one might say that the generator imposes on the stone a quality (*gravitas*) by which it falls: this quality can then be described as the instrument by which the true efficient cause (the generator) moves the stone. Or should we say that the stone's *gravitas* is due to its own SF, which would make that form the efficient cause involved?

Vitelleschi here borrows from Zabarella a distinction between two different kinds of efficient causation. The first type produces its effect on a distinct thing, and is said to be *proprie efficiens*; the second type may produce its effect on itself, and is described as *efficiens per emanationem*.[17] In this second sense, says Vitelleschi – i.e. as emanent efficient cause – the SF of a stone may be described as the efficient cause of its fall.

In natural motion, then, Vitelleschi concludes, the following three potencies are actualised:

1. A potency to the SF itself: this is actuated, of course, by the generator.
2. A potency to the motion that follows from the form: this is actuated *per accidens* by the remover of obstacles.

---

14   See Wallace, 5, p. 288.

15   Weisheipl, p. 44.

16   See Wallace, 5, p. 112 ff.

17   *Ibid.*, p. 113.

3. A potency to a particular terminus or natural place: this is actuated by the motive quality (*gravitas* or *levitas*) acting as an instrument of the SF, from which it flows as an emanent effect.

Every motion will involve, on this view, a *motive quality* in the moved body: an impressed force or impetus for violent motions; *gravitas* or *levitas* for natural. The natural/violent distinction must be made in terms of the source of these qualities: if imposed from without (and opposed by the nature of the body), they are violent; if flowing from the SF of the body itself, they are natural.[18] Even if we continue to speak of these motive qualities as no more than instruments of the true efficient causes (the projector for forced motions; the generator for natural ones), they still function, to all intents and purposes, as *internal motors* capable of impelling the bodies in which they reside.

Although scholastic authors continued to hedge themselves around with scores of subtle distinctions, the general drift of the debate is clear. Despite the authority of Aristotle and the fear of falling into a purely animistic natural philosophy, the tendency was to insist on the need for a *motor conjunctus* for every motion, natural or forced, and to identify this motor in the case of natural motions with the SF or 'nature' (in Philoponus' sense, i.e. *qua* internal and incorporeal causal agency[19]) of the body in question, acting perhaps through the instrumentality of a 'motive quality' such as *gravitas*.

The free fall of a stone is, therefore, due to the motive quality of *gravitas* inherent in it, not to any external agent. Natural motions therefore involve spontaneity rather than A.D.: the animism implicit in Aristotle's physics becomes explicit in Renaissance thought. But does the stone 'desire' to reach its *natural place* in the Aristotelian cosmos, or to attain union with *kindred matter*? Do the phenomena of natural motions betoken a structured and anisotropic *space* made up of distinct natural places, which somehow 'influence' the behaviour of bodies?[20] Or are they to be conceived as yet another manifestation of the universal force of sympathy by

18   This entails the possibility of a neutral motion, neither flowing from nor opposed to the nature of a body – a concept which was to be of great importance in the physics of Galileo. See Wallace, 5, p. 112.

19   See Chapter 4, p. 165; Wallace, 5, p. 110 ff.

20   Aristotle admits that natural place 'exerts a certain influence', but denies that it can function as the efficient cause of natural motion. See Chapter 4, p. 148.

which like things everywhere tend to like? *Within* our cosmos, this might seem a distinction without a difference: since the great bulk of the elements FAWE exist in their respective natural places, the two theories will give the same predictions. The cosmological implications of these two views are, however, utterly different.

For Aristotle, earth moves naturally to the centre of the universe, fire to its periphery. This presupposes a *finite* universe: infinite space can have no centre and no boundaries. Aristotle's account of natural motions is therefore necessarily tied to a conception of the universe as a single, finite set of 'nested' or concentric spheres. If there were two *cosmoi*, he argues, an earthy body would have two opposed natural motions (i.e. to the centre of each world), which is impossible. In any case, the universe as such *can* only have a single and unique centre. A plurality of worlds is, therefore, out of the question.

But, thunders the theologian, God can – by His infinite power – create as many worlds as He pleases. (The proposition that God could not create other worlds was one of the 219 articles condemned in 1277.) Suppose, then, that God creates another cosmos outside ours: what would be the natural motions of the elements in such a world? This question was asked by Oresme in his *Livre du Ciel et du Monde*. His answer is in effect to make gravity a world-relative phenomenon. In this world, he explains, heavy (i.e. earthy) bodies tend naturally to collect at the centre: in another world an exactly analogous process would occur. Earthy bodies there need not (*pace* Aristotle) tend to the centre of our world:

> In their world they would form a single mass possessed of a single place and would be arranged in up and down order . . . just like the mass of heavy bodies in this world. And these two bodies or masses would be of one kind, their natural places would be formally identical, and likewise the two worlds.[21]

The terms 'up' and 'down', argues Oresme, have no sense except with respect to some world or other. Within a particular cosmos, earth tends to the centre, fire to the periphery, and water and air to their intermediate stations – this gives the terms 'up' and 'down' their application within

[21]   Oresme, p. 175.

that world. In extracosmic void space, however, the terms can have no significance at all.[22] There can therefore be two or more formally identical, but numerically distinct, world-centres to which earthy bodies within their respective worlds will tend: the earth in one world (W_1) tending towards the centre of W_1, that in W_2 towards the centre of W_2, and so on.

A piece of earth equidistant between two worlds would find itself, Oresme continues, equally poised between them:

> . . . if it could be separated, one part would go to the centre of one world and the other portion to the centre of the other world. If the portion could not be divided, it would not move at all because of the lack of inclination, being like a piece of iron halfway between two magnets of equal strength. If it were nearer one world than the other, it would move in the direction of the nearer world.[23]

This is most naturally explained along Platonic lines, i.e. as involving a matter-matter force of 'attraction', rather than a tendency of bodies to their proper natural places.[24] Of course one could posit a plurality of Aristotelian natural places for each element, but to do so would be to undermine Aristotle's whole vision of the cosmos. Gravity is, for Oresme, 'the tendency of heavier bodies to go to the centre of spherical masses of matter':[25] it is the distribution of *matter* in the universe which (*pace* Aristotle) determines the natural motion of any given body.

Another beneficiary of the Platonic conception of gravity is of course Copernicus. For Aristotle, earthy bodies tend towards their natural place, the centre of the universe, where they come to rest. The Earth itself is therefore, by its very nature, at rest in this central position: all the heavenly bodies move around it. To suggest that the massive and inert Earth is whirling around the Sun with an annual orbital motion is therefore quite absurd: once at its natural place, an earthly body *rests* there. If, moreover, the Earth is in orbit around the Sun, bodies falling to Earth at different instants are falling to *quite different places*: on such a view, the element earth could have no natural place.

---

22   See Grant, 6, p. 75.

23   Oresme, p. 135.

24   Crombie (1, p. 243) and Kuhn (4, pp. 114–115) attribute just such a theory to Oresme.

25   See Crombie, 1, pp. 255–256.

Copernicus' answer to this line of objection rests heavily on the Platonic theory of gravity as nothing more than a 'natural inclination' of the parts of a body to assemble 'in the form of a sphere':[26] it is, he says,

> a certain natural appetancy implanted in the parts by the divine providence of the universal Artisan, in order that they should unite with one another . . . and come together in the form of a globe.[27]

In Copernicus' physics, the parts of each heavenly body (including our Earth) have a kind of natural sympathy with one another: if separated, they tend (a) to re-unite to form a sphere, and (b) to share in the circular motion proper to such a figure. (The fact that all earthy bodies participate naturally in the diurnal rotation of their mother planet enables Copernicus to meet many of the stock objections to the Earth's rotation.[28])

As our Earth rushes around the Sun in its annual motion, therefore, it is held together by the mutual sympathy of its constituent parts, their 'natural appetancy' for one another. Given such a conception of gravity, there is no need for Aristotle's natural places: space itself could be perfectly homogeneous and isotropic. The Platonic notion of an affinity of like for like, and hence of a 'desire'[29] in the sundered parts of any heavenly body to return to their proper whole, replaces the Aristotelian conception of the cosmos as a structured set of concentric natural places.

For a vision of the universe in which Copernicanism is combined with the doctrine of a plurality of worlds, we have only to turn to Giordano Bruno. God, Oresme had argued, could have created many worlds. But, says Bruno, if God could have created other worlds but refrained from doing so, He is not perfectly benevolent, i.e. He is not God. By a moral necessity of His own nature, therefore, God must have *actually* created an infinite universe and worlds: only thus is His omnipotence and omnibenevolence made manifest in His Creation.[30]

---

26   Quoted from Kuhn, 4, p. 153.

27   From Grant, Ed., 4, p. 515.

28   See Grant, Ed., 4, pp. 513–514; Kuhn, 4, p. 150.

29   For Copernicus' animism, see Zilsel, 1, p. 114.

30   Creation is not, for Bruno, a temporal event. The universe is eternal, but nevertheless a creation, ontologically dependent on its creator.

Bruno's universe is, therefore, actually infinite in extent, containing an infinity of worlds or Copernican solar systems, each consisting of a number of planets orbiting a central Sun (i.e. a star). Within such an infinite universe, Bruno sees, the concepts of centre and periphery have no application: any part of the universe is as much the centre as any other (it is, trivially, the centre of the space surrounding it); no part of the universe serves as a boundary or perimeter.[31] Natural places are mere 'fictions', as are the celestial spheres;[32] space as such is perfectly homogeneous and isotropic – it does not affect the behaviour of bodies one iota.

In such a universe, 'up' and 'down' become relative terms: what appears to us (i.e. from the Earth) to be upward motion may appear elsewhere (e.g. on the Moon) to be a downward one. Only in relation to some given point of reference can the terms be given a clear sense.[33] Weight is nothing more than 'the impulse of the parts toward the whole',[34] by which terrestrial matter tends toward the Earth, solar matter to the Sun, lunar to the Moon, etc. Gravity is therefore *specific* rather than universal: there is no tendency of matter as such to agglomerate into masses. Hence, Bruno argues, worlds themselves are not heavy or light: gravity only appertains to the severed *parts* of a body seeking their proper whole, not to the wholes themselves. There is therefore no tendency of the various heavenly bodies to collapse together into a single mass.

In the fourth dialogue of *De Immenso*,[35] Bruno turns his attention to Aristotle's argument against the plurality of worlds. If, Aristotle had argued, there were another world made up of FAWE, those elements would move naturally towards their natural places in *our* world – the earth in our world and in the other, being specifically alike, would tend towards *numerically* the same point. Being the same stuff, they must share the same natural motion, i.e. towards one and the same natural place. Relative to its own world, then, this otherworldly earth would have to move upward, which is impossible.

Other worlds, Bruno replied, naturally attract *their own proper parts*: just as the parts of one organism are not suited

31   See Singer, p. 280.

32   Michel, p. 186.

33   Singer, p. 281.

34   *Ibid.*, p. 282.

35   *Ibid.*, p. 330 ff.

to another, so it is with worlds. (This organism-world analogy is meant to be taken quite literally and seriously.) Gravity is therefore a quasi-organic function by which the living bodies of suns and 'earths' sustain themselves.[36] Only by constraint could Jovian matter be brought 'down' to Earth: it inclines by its nature to its original source. (Lunar matter must, by contrast, have some degree of affinity with terrestrial: it is the attraction of the Moon which causes our tides.)

The parts of a foreign 'earth' (i.e. another planet like ours) could, says Bruno, be drawn to our Earth, if that other planet is sufficiently similar to our own.[37] It is even possible for a single rock to remain perfectly poised between two earths, like a piece of iron between two magnets. If the two earths are qualitatively identical, and the stone equidistant between them, it will remain unmoved. If, however, one of the two earths is more closely akin in nature to the stone, it will 'fall' towards that earth rather than the other.[38]

The weight of a body, says Bruno, diminishes with its *distance* from its source, i.e. with the number of interjacent layers of air or aether.[39] If, for example, one were to take a piece of earth and move it further and further away from the Earth, its natural appetite for its home planet would weaken. Eventually, the attraction of some other earth may prove stronger than its residual love of the Earth, and it will move instead to join this rival planet. Alternatively, a stone moved sufficiently far from our Earth could entirely lose its desire to return and thus become weightless.[40]

For Bruno then, weight is not a permanent quality inherent in bodies: no body is heavy (or light) by its own nature.[41] The apparent weight of a body is a function of the following factors:

a. Its *mass*, i.e. for Bruno, the number of its constituent atoms.[42]

b. Its *distance* from some given 'earth' or 'sun'.

c. Its degree of *affinity* for that particular heavenly body.

36 *Ibid.*, pp. 330–331.

37 *Ibid.*, p. 333.

38 *Ibid.*, p. 336.

39 Michel, p. 189.

40 *Ibid.*, p. 190.

41 *Ibid.*, p. 232.

42 All Bruno's atoms are identical, so the equation of mass with number of atoms involves no oversimplification.

It is the third factor which sharply differentiates Bruno from Newton. It is not merely that Bruno never attempts to formulate this functional relationship in a quantitative manner, although that is important, but that he could in principle have no such law as that of Newton – factor c eludes such regimentation. In place of universal gravitation, Bruno has the Neoplatonic doctrine of sympathy, the natural affection of like for like, which leads naturally to a doctrine of partial and specific gravities and an anthropomorphic world-view.

Let us now forget for a while the majestic universal vision of Bruno, and concentrate on the behaviour of bodies within our own world. According to the ancient Atomists, all such bodies possess weight, either as a consequence of the cosmic 'whirl' (Democritus), or as an innate and primary quality (Epicurus). Aristotle argued, in opposition to this view, that there is such a quality as *absolute levity* in virtue of which fiery bodies naturally ascend. The Atomists in their turn alleged that the apparently spontaneous upward motion of fire is really due to its being 'squeezed out and up' by denser bodies. All bodies, they argued, have (more or less) weight: they only *seem* 'light' if present in a denser medium which tends to extrude them in an upward direction.

The technical work of Archimedes on the topics of specific gravity (S.G.) and upthrust (= weight of the medium displaced) provided powerful support for the Atomists' position. Given the Archimedean concepts of S.G. and upthrust, it became possible to give a convincing account of all the so-called 'natural motions' of bodies without presupposing the existence of absolute levity. All that is needed is a clear distinction between gross (true) weight and net (measured) weight. The measured weight of a body in a particular medium will equal its gross weight, *less* the upthrust of the displaced medium, thus:

Net Weight = Gross Weight – Upthrust.

If the upthrust exceeds the gross weight, the net weight is negative and the body is 'squeezed out and up', as Lucretius would say – the whole process depending simply on the principle of the balance.

Archimedes' account of weight has therefore no need for the concept of *levitas*: parsimony will demand its exclusion from physics. All bodies, it will follow, have an intrinsic tendency to downward motion. The gross or true weight of each body is what would be measured if the upthrust were zero, i.e. in a hypothetical void. Only *in vacuo*, then, can the true weight of bodies be determined.

These implications were not all seen at once: the impact of Archimedes on medieval science was a gradual one.[43] The Aristotelian concept of absolute weight and lightness survived well into the seventeenth century. But, already by the fourteenth, some radical thinkers were breaking away from the peripatetic position. Crescas, for example, argues[44] that all bodies possess weight in virtue of their internal structures, and that therefore only downward motion can be natural – upward motions are always *forced*, i.e. mechanically explicable in terms of some external impulse. Denser materials strive more strongly to reach the centre of our world, and, as a result, extrude less dense substances lying beneath them: 'lighter' bodies are less heavy, not more light.

If we turn again to Oresme,[45] we find a most brilliant and incisive use of thought-experiment directed against the peripatetic position. It is not, I think, implausible to see shades of Archimedes in the following argument. Air, says Oresme, always falls (naturally) through fire, and rises through water. But now imagine two closed cylindrical tubes A and B, containing fire and water respectively, running vertically from the heavens to the earth – from the sphere of the earth at the bottom to the lowest sphere of the heavens at the top thus:

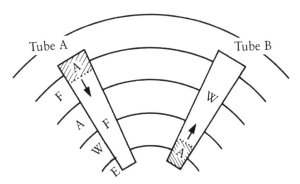

If a little air is admitted into the top of tube A, it will *fall* through the fire until it reaches the bottom of the tube, i.e. the

43   See Clagett, 3, p. 429.

44   Wolfson, 1, p. 58.

45   Oresme, p. 697.

natural place of earth. If, on the other hand, air is admitted to the bottom of tube B, it will ascend until it reaches the very top of that tube, in the heavens. In these two cases, air moves, perfectly naturally – although in opposite directions – *away* from its own supposed natural place and into that of a quite different element! Thus, says Oresme,

> I say it is possible for air to move naturally from the centre up to the heavens and, in the contrary way, from the heavens down to the centre.[46]

The 'natural motion' of a body is therefore *not* directed towards a particular well-defined area of space, its own proper sphere, but depends essentially on the medium in which it is placed. It is natural for air to be above water and below fire, but this does not imply that the four elements have clearly demarcated geometrical zones proper to them. Although the densest stuffs tend to collect at the centre and the less dense stuffs near the periphery, this is for Archimedean rather than Aristotelian reasons:

> The elements do not have the limited and distinct areas stated above, but . . . the entire distance between the centre of the world and the heavens is indeterminate and indifferent to such division, except that the heaviest element is or tends to be beneath the less heavy and lighter elements, or as near to the centre as possible, and likewise with the light or lightest element in relation to the heavens, for such is their natural arrangement and to this end all their natural movements tend.[47]

Although Oresme almost equates 'lighter' with 'less heavy', he does not quite do so: the above passage is most naturally construed as involving the lightest elements (those with most *levitas*) striving most strongly towards the periphery of the cosmos. Nothing in Oresme's discussion demands this, however: his views could easily be reconciled with the assumption that all bodies have weight.

This position is defended by Trevisano (d. 1378), who abandoned Aristotle's absolute weight and lightness in favour of differences of specific gravity (S.G.). Fire, says Trevisano, is not absolutely light but merely the least heavy of the elements: upthrust will explain its supposedly natural upward

---

[46]   *Ibid.*, p. 697.

[47]   *Ibid.*, p. 707.

motion.[48] By positing only differences in S.G. between FAWE, he argues, all the phenomena can be saved: there is, therefore, no reason to posit absolute levity.

By the fifteenth century Nicholas of Cusa is proposing the famous willow-tree experiment (actually performed two centuries later by Van Helmont), which presupposes the conservation of matter *by weight*. By this period, then, the universality of weight (and its quantitative conservation) has become an accepted position among non-Aristotelian natural philosophers.

If we turn now to Galileo's early work, *De Motu*, we find explicitly spelled out all the themes so far implicit in this subsection: the authority of Archimedes has wholly usurped that of Aristotle. Considered in themselves and without reference to any medium, says Galileo, all bodies are heavy:

> not merely water and earth and air have weight, but even fire, and whatever is lighter than fire, and, in short, everything that has size and matter linked with its being.[49]

There is, he argues, no absolute lightness: 'those who came before Aristotle' (i.e. Plato and the Atomists) were right to treat weight and lightness as relative. That all bodies possess (more or less) weight is *obvious*, Galileo claims: to deny it is 'contrary to reason'. Fire cannot be without weight, 'for that is a property of the void':[50] like all other bodies, fire too would fall *in vacuo*.

Given the Archimedean formula for upthrust (= gross weight of the medium displaced), it follows that all bodies weighed in fluid media weigh less than their true weights by that value. Hence only *in vacuo* 'can differences of weight be exactly determined'.[51] The true (gross) weight of a body is its weight as measured in a medium of zero S.G., and hence zero upthrust, i.e. in a void. Once one possesses the concepts of gross (true) weight, net (measured) weight, and, distinct from both, S.G. or gross weight per unit volume, much of Aristotle's argumentation in *De Caelo* – indeed, his whole treatment of the heavy and the light – will be seen to be groundless and futile.

For example, although downward motions may be called 'natural' (i.e. due to some principle intrinsic to heavy bodies –

---

48   See Clagett, 4, pp. 146–147.

49   Galilei, 1, p. 55.

50   *Ibid.*, p. 60.

51   *Ibid.*, p. 61.

i.e. to all bodies), *all* upward motions, even if apparently spontaneous, will turn out to be *forced*, i.e. due to upthrust. If a body of low S.G. is placed in a denser medium (e.g. wood in water), the upthrust of the medium will exceed the gross weight of the wood and thus expel it upwards. Just as a heavy body on one pan of a balance may be forced upwards by placing a greater weight on the other pan, so may a heavy body be extruded from a medium of greater S.G. than itself. Fundamentally the same principle – that of the balance – applies to both cases.[52] When a body ascends, says Galileo, 'it is being raised by the weight of the medium . . .'; it is moved 'by force, and by the extruding action of the medium'. For, he continues,

> when the wood is forcibly submerged, the water violently thrusts it out when, with downward motion, it moves toward its own proper place and is unwilling to permit that which is lighter than itself to remain at rest under it . . . It is therefore clear that this kind of motion may be called forced.[53]

One possible implication of the denial of absolute levity is seen, and explicitly accepted, by Galileo. It is the thesis of the *homogeneity* of matter. According to Aristotle, the existence of opposed natural motions – i.e. of absolute weight and lightness – is sure and certain proof of the existence in Nature of radically distinct stuffs, i.e. of the qualitative heterogeneity of matter.[54] Given the Archimedean account of 'natural motions', however, this argument collapses. All bodies could, as the Atomists claimed, consist of a *common* matter*, intermixed with interstitial vacua. The densest bodies (those with most matter* and least void per unit volume) will then have the highest S.G.s, and will thus tend to displace other, less dense, bodies upwards. In the third edition of *De Motu*, Galileo frankly espouses such a viewpoint:

> But why do we need further arguments? There is a single matter in all bodies, and it is heavy in all of them.[55]

For Galileo, then, gravity is a universal property intrinsic in, and essential to, all bodies as such in virtue of which all

---

52   *Ibid.*, pp. 20–21.

53   *Ibid.*, pp. 22–23.

54   See Chapter 3, p. 115.

55   Galilei, 1, p. 120.

(terrestrial) bodies tend towards a single point, the centre of gravity of our Earth.[56] (There are, presumably, other point-centres of gravitation elsewhere in the universe.) The precise ontological status and mode of operation of 'nature' (the internal mover responsible for 'natural' – i.e. downward – motions) Galileo does not tell us – in later life, he shunned dynamical speculations – but, whatever it is and however it acts, it is non-mechanical. Galileo may well have been influenced, Wallace suggests,[57] by Philoponus' conception of *physis* as an internal *vis* (force), which was widely discussed by contemporary scholastics. In any event, it is clear that Galileo thought of the cause of weight as something non-mechanical, and of downward motion as spontaneous, i.e. as resulting from an internal rather than an external agency.

B. The 'Species' Theory

This theory takes its origin from Aristotle's conception of light as the actualisation of the potential transparency of the medium, i.e. as a qualitative change induced in a physical medium by the action of a luminous object. Light cannot be a corporeal emission, Aristotle argues,[58] since (a) this would involve two bodies or distinct space-filling substances (light itself and the illuminated medium) simultaneously occupying the same place, and (b) light-transmission is instantaneous, but no body can move with infinite velocity.

Does the propagation of light involve A.D.? The answer seems to be that it does not: the light source actualises the potential transparency of the adjacent parts of the medium, which in turn do the same for the next parts, and so on. What may appear to be A.D. involves the successive actualisation of the parts of the medium, each part acting only on its neighbour. But if this process is truly *successive* then, as Roger Bacon saw, light-transmission cannot be instantaneous: the medium near to the light-source must be actualised *before* that further away. Hence, Bacon argues, 'a sensible time is required for vision'.[59] This remained a point of controversy during much of the Middle Ages.

The Aristotelian theory of light became extended and generalised in medieval times, producing eventually the all-

56   See Moody, 4, p. 209.

57   Wallace, 5, pp. 286–299.

58   See *De Anima*, 2, 7.

59   R. Bacon, p. 485. His arguments for this conclusion derive from Alhazen.

embracing species theory of philosophers such as Grosseteste and Bacon. The theory rests firmly on the basic causal axiom: 'Like produces like' – one X tends to make other things X (cf. of course the specific reproduction of living things).[60] The fundamental idea of the species theory is that every object tends to impress its *likeness* ('species') on the ambient medium, which will in turn transmit these 'likenesses' to other parts of the medium, and ultimately to human sense organs. A red object thus engenders in the medium the capacity to transmit red 'species', species which when they reach the eye give rise to a sensation of redness. In perception, we receive the *form* or likeness of an object *without* its matter, just as a piece of wax receives the form of the signet ring but not *its* matter.

One of the great influences on the medieval doctrine of species was al-Kindī's theory of 'rays'. Causal influences such as light, heat, sound and astral virtue are all, he suggested, emitted radially from objects and transmitted as rays.[61] In his *De Radiis Stellatis* Kindī discussed in some detail the influences of stars and planets here below, but he makes it plain that he also thinks of terrestrial things as emitting their proper virtues after a similar manner.[62]

The successors of Kindī were Robert Grosseteste and Roger Bacon,[63] who modelled an entire system of physics (and even of metaphysics!) on the observed nature of light. In Bacon's *Opus Majus*, mathematical optics is hailed as the key to a proper understanding of the powers of Nature.[64] Every efficient cause, it is claimed,

> acts by its own force which it produces on the matter subject to it, as the light of the sun produces its own force in the air, and this force is light diffused through the whole world from the solar light. This force is called likeness, image, species, and by many other names, and it is produced by substance as well as accident and by spiritual

---

60   See Hansen, in Lindberg, Ed., p. 491.

61   The notion of radial emission will yield, by simple geometry, an inverse-square law of the attenuation of 'force' with distance. This inference was made (in the realm of optics) by Kepler in the early years of the seventeenth century.

62   See Jammer, 3, p. 57; Ronchi, p. 44; Thorndike, Vol. 1, pp. 643–644.

63   For the influence of Kindī on Grosseteste and Bacon, see Thorndike, Vol. 2, p. 443. Roger Bacon himself (p. 420) refers to Kindī (with Aristotle and Alhazen) as one of the masters of optical science.

64   See Crombie, 2, p. 13.

substance as well as corporeal. Substance is more productive of it than accident, and spiritual substance than corporeal.[65]

These *species*, says Bacon, act 'on sense, on intellect, and on all the matter in the world for the production of things'.[66] Every substance, in particular, has the power to impart its *form* to the (passive and receptive) ambient medium.[67] If the medium is homogeneous, the species are always propagated in straight lines ('Nature favours the shortest route'), but obstacles or barriers of various sorts can cause reflections, and refraction may occur at the interfaces between different media.[68] The observed behaviour of light thus functions as a model for that of species in general. The same principles of 'multiplication' that govern the propagation of visible species apply equally, Bacon alleges, to *invisible* species such as those responsible for e.g., magnetism and astral influences such as that of the Moon over the tides.[69]

Let us move on to discuss Bacon's theory of light and vision in a little more detail, since his whole world-view is based upon it. The first prerequisite for vision is, he states, the 'species of light'.[70] From every part of a visible object, he explains, 'come forth species in an infinite number, as explained in the laws of multiplications'.[71] After what manner are these species present in the medium? Do species of different colours, for example, mix to produce intermediate hues? Colours of course mix like this:

> But what is true of colours is also true of the species of colours, for the species is of the same nature as that which produceth it.[72]

The species of colours, then, 'mingle at every point of the medium': red and yellow species will, for example, mix to yield orange. Some philosophers, however, maintain that this view leads to insuperable difficulties, and hold instead that:

---

65 R. Bacon, p. 130.

66 *Ibid.*, p. 130.

67 See Hesse, 1, pp. 80–81.

68 R. Bacon, p. 131 ff.

69 *Ibid.*, p. 159.

70 *Ibid.*, p. 450.

71 *Ibid.*, p. 456.

72 *Ibid.*, p. 459.

species have a spiritual existence in the medium and in the sense . . . And because they have a spiritual existence and not a material one they therefore do not observe the laws of material forms, and for this reason do not mix.[73]

This theory, Bacon retorts, contains 'many false and absurd things'.[74] The 'distinction of vision' – the ability of the eye to distinguish the species of different colours – can still be explained, Bacon insists, on the assumption that species are corporeally present in the medium and therefore mix just as colours do. Although a 'mixed species' will come to the eye from that part of the medium where the two differently coloured species mingle, this will only be a 'secondary' or 'accidental' propagation; the 'primary' and 'principal' multiplication of species always proceeds in straight lines from the object perceived.[75] (The resemblance to Huygens' theory of secondary wavelets is quite striking.) Since the eye does not discern the (weaker) secondary species, we do not see, e.g., the orange produced by the mixing of red and yellow species. As for the rival theory, Bacon rejects it utterly:

When they say that species has a spiritual existence in the medium, this use of the word spiritual is not in accord with its proper and primary signification, from spirit, as we say that God and angel and soul are spiritual things; because it is plain that the species of corporeal things are not thus spiritual. Therefore of necessity they will have a corporeal existence, because body and soul are opposed without an intermediate. And if they have a corporeal existence, they also have a material one, and therefore they must obey the laws of material and corporeal things, and therefore they must mix when they are contrary and become one when they are of the same categorical species. And this is again apparent, since species is the similitude of a corporeal thing and not of a spiritual; therefore it will have a corporeal existence. Likewise, it is in a corporeal and material medium . . . Therefore it must have a corporeal existence in a corporeal medium. Moreover, it produces a corporeal result, as, for example, the species of heat warms bodies . . .[76]

---

73   *Ibid.*, p. 459. Bacon cites Aristotle and Averroes as holders of the 'spiritual' conception of species, but it was particularly championed by Avicenna (see Ronchi, p. 60).

74   R. Bacon, p. 459.

75   *Ibid.*, p. 460.

76   *Ibid.*, p. 462.

The study of species belongs as a proper part of *natural* philosophy:

> The species does not . . . differ from the charcoal and the flame, except as the incomplete differs from the complete, the embryo from the child, and the child from the man. But it is agreed that the complete is material; therefore also the incomplete, wheresoever we observe these things, because the incomplete becomes complete.[77]

The species of heat emitted by the burning charcoal are (a) as child is to parent, a similitude or likeness, and (b) capable of kindling fresh flame in other things. Hence, says Bacon, when people say that species have spiritual existence in the medium, 'it is evident that spiritual is not taken from spirit, nor is the word used in its proper sense'.[78]

The presence of an appropriate material medium is absolutely *essential* to the propagation of species, which would therefore be impossible in a vacuum, which has no potencies to actualise. Hence, says Bacon,

> we should not see anything if there were a vacuum. But this would not be due to some nature hindering species, and resisting it, but because of the lack of a nature suitable for the multiplication of species; for species is a natural thing, and therefore needs a natural medium; but in a vacuum nature does not exist.[79]

The propagation of species through a medium requires, moreover, 'a sensible time' – the transmission of light is, *pace* Aristotle, not something instantaneous.[80] According to Alhazen, a light-ray perpendicular to a given surface will reach it more quickly than an oblique ray from the same source, but 'quicker and slower belong only to time'.[81] No limited agent, moreover, can produce an infinite velocity: 'no finite force acts instantaneously'. The basic argument involves, however, the idea of the *successive* 'actualisation' of the parts of the medium. If this process is successive, it cannot also be instantaneous:

[77] *Ibid.*, pp. 462–463.

[78] *Ibid.*, p. 463. Aquinas too argues that this usage must be taken metaphorically rather than literally.

[79] *Ibid.*, p. 485. See of course Aristotle *versus* Democritus in Chapter 2, p. 64.

[80] *Ibid.*, p. 485.

[81] *Ibid.*, p. 487.

since space through which species is carried has a before and after, passage of the ray must have a before and after both in itself and in duration; but before and after exist only in time, since this condition cannot exist in an instant.[82]

Light, heat, and other species, having a corporeal existence in the medium, obey the laws governing corporeal nature: they do not leap instantaneously from one terminus of their path to the other. Only infinite (i.e. divine) power can cause instantaneous motion. It remains, therefore, Bacon concludes, that

> light is multiplied in time, and likewise all species of a visible thing and of vision. But nevertheless the multiplication does not occupy a sensible time and one perceptible by vision, but an imperceptible one.[83]

Light-transmission is so rapid that at dawn the whole sky is illuminated without perceptible delay. This suggests to other scholastic authors, such as Aquinas, that the propagation is instantaneous, but Bacon is adamant. The successive actualisation of a corporeal form in the parts of a fluid medium must occupy some time, however little.[84] But, if light is 'corporeally' present in the medium, and transmitted with finite velocity through it, can we not identify 'species' with a corporeal emission from the light-source? Bacon's answer is flatly negative:

> The species is not a body, nor is it changed as regards itself as a whole, from one place to another, but that which is produced in the first part of the air is not separated from that part, since form cannot be separated from the matter in which it is . . ., but the species forms a likeness to itself in the second position of the air, and so on. Therefore it is not a motion as regards place, but is a propagation multiplied through the different parts of the medium; nor is it a body which is there generated, but a corporeal form, without, however, dimensions *per se*, but it is produced subject to the dimensions of the air; and it is not produced

[82] *Ibid.*, p. 488. The form of the argument is of course identical to that of the *distantia terminorum* argument discussed in Chapter 6.

[83] *Ibid.*, p. 489.

[84] Bacon's conclusion, however impeccably argued, was not popular: most of the schoolmen, including Grosseteste and Aquinas, reaffirmed Aristotle's belief in the instantaneous nature of light-transmission (see Crombie, 2, p. 146).

by a flow from a luminous body, but by a renewing from the potency of the matter of the air.[85]

This is perhaps the clearest statement of the metaphysics of the species theory. A species is not a corporeal emission from the source moving in a medium; it is the successive actualisation in that medium of a potency latent in it. The agent does not *emit* species; it engenders in the medium the capacity to transmit them. (If species *were* corporeal emissions, Bacon alleges, sources of light and other species would all wear away.[86]) The Aristotelian roots of the theory could hardly be more evident.

Although the species theory took its rise – and derived much of its confirmatory evidence – from the field of optics, it was developed by Bacon into a general account of causal influence.[87] All things, he claims, produce species, which all obey the common mathematical principles of 'multiplication'. Geometrical optics is therefore the royal road to an understanding of all Nature's powers. The astral influences of the heavenly bodies work by species; and *'species magnetica'* transmit the power of the lodestone. By means of species the Moon governs the tides, the planets produce their effects, beneficial and otherwise, on the course of human life, and the magnet draws ferrous bodies to itself.

Forces, on this view, do not involve A.D.; instead, the transmission of an 'influence' through a medium involves a chain reaction – one that, in Jammer's works, 'successively energises consecutive parts of the medium or the patient'.[88] The magnet, for example, engenders in its environment a *species magnetica* which spreads radially and is multiplied in the medium until it reaches a piece of iron, which it then endows with a *magnetic quality*. Just as the projector imparts a motive quality, impetus, in the stone, so the magnet imparts (*via* the *species magnetica*) a motive quality in the iron. What appears to be A.D. is therefore explained in terms of the transmission of a causal influence through a physical medium.

How is the species theory relevant to the mechanical philosophy? In its purely peripatetic form, of course, i.e.

---

85   R. Bacon, pp. 489–490.

86   See Hesse, 1, p. 80. The argument derives from Alexander the Peripatetic.

87   See Thorndike, Vol. 2, pp. 667–668.

88   Jammer, 3, p. 60.

conceived as involving the transmission of a form without matter, it is – despite its denial of A.D. – a thoroughly non-mechanistic, and even anti-mechanistic, theory. It tended, however, to become associated with (and sometimes confused with) two other types of theory of a much more mechanistic nature – i.e. wave theories and corpuscular emission theories. Let us start with the former.

Wave-theories of *sound* are of truly ancient lineage, and can be traced back to the earliest Greek thought. It is natural, on reflection, to think of sound as due to the agitation of a fluid medium by, e.g., a vibrating body – so many everyday noises are produced after such a manner. It is no surprise then to find Aristotle saying that for the production of a sound,

> there must be a percussion of solid objects on each other, and on air. This happens when air is confined when struck, and cannot disperse.[89]

If we turn to Vitruvius, we find the key analogy of the wave-theory, the likening of sounds to water-waves. Voice, he says,

> is a flowing breath of air, perceptible to the hearing by contact. It moves in an endless number of circular rounds, like the innumerably increasing circular waves which appear when a stone is thrown into smooth water . . .'.[90]

In Boethius' *De Musica*, the pitch of a note is identified with the *frequency* of these vibrations in the medium, and the mechanical theory of sound is, in all essentials, complete. The wave-analogy is found again in Adelard of Bath,[91] Thomas Aquinas[92] and many other scholastic writers. Sound, says Roger Bacon,[93] is quite different from light: it involves a 'tremor' (i.e. a true local motion) of the medium, not merely the transmission of a similitude or species through it.

Wave-theories of sound were therefore already firmly established in the literature in medieval times – the work of Mersenne and others in this field in the seventeenth century was not an innovation but merely the continuation of an age-old tradition. In Leonardo, however, we find a hint of the radical suggestion that light too may be explicable after a similar manner. After drawing the accepted analogy between

---

89   From *De Anima*, 2, 8.

90   Vitruvius, pp. 138–139.

91   See Thorndike, Vol. 2, p. 32.

92   See Aquinas' commentary on *De Anima*, 2, 8.

93   R. Bacon, p. 490.

sound and water-waves, he proceeds to intimate that light too may be similar:

> Just as the stone thrown into the water becomes the centre and cause of various circles, and the sound made in the air spreads out in circles, so every body placed within the luminous air spreads itself out in circles and fills the surrounding parts with an infinite number of images of itself, and appears all in all and all in each part.[94]

One should be careful not to read too much in this passage: it is far from clear exactly how much positive analogy Leonardo is asserting to exist between sound and light. The positive analogy he has in mind may be no more than that of radial propagation – it would be most unwise to hail Leonardo as the founder of the wave-theory of light on the basis of such a passage (much of which, indeed, sounds like an orthodox statement of the species theory).

Another way of 'mechanising' the species theory would be to identify species with corporeal emissions or effluvia such as Epicurus' *eidola*. According to Epicurus too, each body emits 'similitudes' of itself, but for him these *eidola* are corpuscular emissions with a true local motion through the medium. One might naturally think of odours, for example, as being borne by such effluvia. An odour, says Roger Bacon, involves the emission of a *vapour*, 'which is, in fact, a body diffused in the air'.[95] (He also admits an odoriferous species, but this can now be seen to be redundant.)

In Fracastoro (1478-1553), the re-materialisation of 'species' is well under way. A physician and philosopher of some significance, Fracastoro was much influenced by Lucretius, and applied the Atomic Theory to medicine, explaining the spread of infectious diseases by reference to minute corporeal 'seeds' of contagion.[96] All bodies, he claimed, naturally emit subtle vapours – of which odours are a prime example – which mediate in their interactions with other bodies. Sympathy and antipathy, for example, do not involve pure A.D., but always take place *via* such 'species':

> When two parts of the same whole are separated from each other, each sends toward the other an emanation of its

---

[94] Leonardo, p. 38.

[95] R. Bacon, p. 491.

[96] Stones, p. 448. For the corpuscular theory of the plague, see Thorndike, Vol. 5, p. 496.

substantial form, a species propagated into the intervening space; by the contact of this species each of the parts tends toward the other in order to be united in one single whole; this is the way to explain the mutual attraction of like to like, the sympathy of iron for the magnet being a typical example.[97]

There is no A.D.: all apparent action at a distance will turn out to involve such mediation. Gravity as well as magnetism is explained by Fracastoro as a case of the sympathetic affinity of like for like rather than as a tendency of bodies toward particular places. Not all 'species', however, are conceived as *corporeal*: Fracastoro distinguishes spiritual from material forms, and gross (corporeal) from tenuous (spiritual) species.[98] Magnetic effects, he argues (e.g. those responsible for the operation of the mariner's compass), can operate at such *long range* that they cannot be due to corporeal effluvia: magnetic species are therefore incorporeal in nature.

In Fracastoro, then, we have a *double* species theory: corporeal effluvia or subtle bodies of some kind are responsible for, e.g., odours, the transmission of diseases, etc.; while spiritual species carry, e.g., magnetic influence. A similarly eclectic attitude prevails in Francis Bacon's discussion of sensation in his *Sylva Sylvarum*.[99] Both 'visibles' and 'audibles', he asserts,

in their virtue and working, do not appear to emit any corporeal substance into their mediums, or the orb of their virtue: neither again to raise or stir any evident local motion in their mediums as they pass; but only to carry certain spiritual species.[100]

This looks like a plain statement of the Aristotelian theory of immaterial species or similitudes. One must remember, however, that 'no evident local motion' does not entail 'no local motion' *simpliciter*: not all motion is evident to us as such. Hence Bacon, while retaining a species-theory of light, moves towards a mechanistic account of sound:

The species of visibles seem to be emissions of beams from the object seen; almost like odours; save that they are more

---

97   Quoted from Jammer, 3, p. 73.

98   See Thorndike, vol. 5, pp. 495–496.

99   F. Bacon, *Works*, Vol. 2, p. 429 ff.

100   *Ibid.*, p. 429.

incorporeal: but the species of audibles seem to participate more with local motion, like percussions or impressions made upon the air. So that whereas all bodies do seem to work in two manners; either by the communication of their natures, or by the impressions and signatures of their motions; the diffusion of species visible seemeth to participate more of the former operation, and the species audible of the latter.[101]

The wind, Bacon argues – quite correctly – affects the transmission of sounds but not of sights: 'audibles' are therefore, 'carried more manifestly through the air' than 'visibles', and as such are more dependent on the vicissitudes of that medium.[102] He ends up therefore with a mechanistic theory of sound as involving essentially (consisting of?) the local motion or agitation of the medium, and a true species-theory of light and vision, but – curiously – seems to see the difference as one of *degree* but not of kind. He also accepts an account of odours as due to corporeal emissions,[103] giving him a different type of theory for each sense-modality.

We find therefore in Renaissance thought the following three varieties of 'species' theory, not always sharply differentiated:

S1. The 'Aristotelian' theory – transmission of a form without matter; no true local motion; successive actualisation of a potency latent in the medium, etc.

S2. A mechanical theory of the transmission of 'waves' by the agitation (= local motion of the parts) of a fluid medium.

S3. A corpuscular emission theory, involving the local motion of corporeal effluvia.

These three quite different models of physical action[104] were not always clearly distinguished in Renaissance thought: not until the seventeenth century were the basic principles of the mechanical philosophy formulated with sufficient precision to demarcate sharply the non-mechanical (S1) from the mechanical (S2 and S3).

---

[101] *Ibid.*, p. 430.

[102] *Ibid.*, p. 430.

[103] *Ibid.*, p. 610.

[104] The three models have, for example, very different implications about the transmission of causal influences *in vacuo*: S1 and S2 both attribute an essential rôle to the medium, while S3 does not. The restorers of ancient Atomism must therefore favour S3, at least for light.

Let us now digress a little and examine some medieval and Renaissance theories of magnetism. The following contemporary accounts of magnetic action are discussed by Crescas in his *Or Adonai*:[105]

M1.   The magnet emits an 'effluxion' of particles which move the iron by straightforward contact-action. (The effluvium theory of Epicurus would be familiar, of course, from Galen's critique.)

M2.   The magnet induces a disposition (a transient form or quality) in the iron, which then assists in moving the iron towards the magnet.

M2 is Averroes'[106] explanation of magnetic attraction, and is clearly an attempt to find an account of magnetism compatible with Aristotle's general denial of A.D. The magnet is thus the mover of the iron; the magnetic quality, the instrument by which it does so. But how does the lodestone induce this quality in the iron? If we are to avoid positing A.D., we need something like Roger Bacon's *species magnetica* transmitted through the medium. This then becomes one common scholastic view: the magnet emits such species, which in turn induce in the iron the motive quality in virtue of which it moves towards the lodestone. The appearance of A.D. is thus explained away.

Crescas finds both these accounts implausible, and proposes instead that the iron possesses a 'natural tendency' towards the magnet, like that of a body for its natural place.[107] But, unless *some* 'influence' or other reaches the iron from the magnet, how does the iron 'know' where the lodestone is, and hence in which direction to move? A magnet, after all (unlike an Aristotelian natural place), is mobile, without any permanent resting place.

Another possible account of magnetism is that it involves pure A.D., *by* the magnet *on* the iron. Ockham, for example, argued that there is no legitimate rationale for the rejection of A.D.: the Sun, in his opinion, acts at a distance on the Earth, the lodestone on the iron, and so on.[108] Once we abandon our prejudice against A.D., Ockham argues, we will see that we have no need to posit such intermediates as *species*: the famous razor will dispense with them all as superfluous.

[105]   See Wolfson, 1, p. 255.

[106]   See Crombie, 2, p. 211; Roller, p. 31. Averroes in turn attributes this theory to Alexander.

[107]   Wolfson, 1, p. 255.

[108]   See Crombie, 2, p. 212; Jammer, 3, pp. 64–65; Leff, p. 48.

This theory that the magnet acts directly on the iron is often associated with animism. Thales had attributed a *psuchê* to the lodestone to account for its peculiar power, and his animistic theory of magnetic action did not lack advocates. Mere matter, it was argued, can act, if at all, only by contact: A.D. indicates the operation of some higher cause.[109]

We turn now to that classic work from which the empirical study of electrical and magnetic phenomena takes its rise, the *De Magnete* of William Gilbert. It is difficult for the modern reader to assess the value of Gilbert's achievement, to appreciate the monumental labour involved in clearing away all the myths and old wives' tales, the folklore and fantasy associated for millennia with the topic, and establishing in their place the plain and sober facts of attraction, orientation, declination, etc. It is quite remarkable – and a great tribute to his experimental techniques – that Gilbert erred so rarely in establishing the basic phenomena of his subject-matter.

When we come to the subject of how magnetic and electrical forces operate, however, we find Gilbert very much a child of his time. Simple observation will give data concerning, e.g., angles of declination and deviation; it will not yield insight into the nature of the *causal* processes involved. In fact, Gilbert argues, magnetism and electricity are, despite being lumped together for so long, quite distinct in nature. Magnetic 'coition', the mutual desire of iron and lodestone for union, is 'natural' (due to an intrinsic cause); electrical 'attraction' is forced (due to an external impulse). Let us investigate this distinction in more detail, starting with the *modus operandi* of electrical (i.e. electrostatic) forces.

Solids such as amber, jet, etc., Gilbert claims, emit on rubbing a very tenuous effluvium, a subtle but nevertheless material fluid which acts by *contact* on light bodies such as chaff and straw to draw them to the 'electrick'.[110] The effluvium acts on the chaff or straw to produce

> a singular tendency towards unity, a motion towards its origin and fount, and towards the body emitting the effluvia.[111]

Electrical attraction is therefore (a) unilateral – the 'electric' is not drawn towards the chaff – and (b) mediated by a corporeal effluvium. Can a similar account be given of

---

109   See Roller, p. 32.

110   W. Gilbert, p. 51.

111   *Ibid.*, p. 57.

magnetic phenomena? Gilbert's answer is a decided and emphatic negative. The two types of phenomena, outwardly very similar, are in fact, he claims, totally different:

> In all bodies in the world two causes or principles have been laid down, from which the bodies themselves were produced, matter and form. Electrical motions become strong from matter, but magnetick from form chiefly; and they differ widely from one another and turn out different.[112]

The electrical effluvium is, Gilbert claims, affected by corporeal obstacles: the electrical force of a piece of amber cannot reach through even a thin material barrier such as a piece of paper or linen. The magnetic virtue, by contrast, is unaffected by corporeal barriers – a sure indication of its nobler, more 'formal' origin. Detailed investigation yields, according to Gilbert, a number of other differences between magnetic and electrical action, which can be summarised by the following table:

Electricity	Magnetism
Due to matter	Due to form
Involving a material effluvium	No effluvium
Hindered by corporeal barriers	Unaffected by barriers
Non-specific	Highly specific
Unilateral 'attraction'	Bilateral 'coition'
No orientation	Orientation

One of Gilbert's key presuppositions is made manifest by the following remark:

> Since no action can take place by means of matter except by contact, these electricks are not seen to touch, but, as was necessary, something is sent from the one to the other, something which may touch closely and be the beginning of that incitement.[113]

Matter, Gilbert emphasises, can act only by contact. Electrical attractions are (a) affected by barriers, and (b) non-specific (any light body will be drawn to the rubbed amber), and must therefore be material in origin and nature. (The non-specific nature of electrical attraction shows clearly that it is *not* due to 'sympathy'.) Since there is no direct contact of

---

112   *Ibid.*, p. 52.
113   *Ibid.*, p. 57.

amber and chaff, a corporeal intermediate – i.e. an effluvium – must be posited. The chaff is a mere *patient*, moved by the effluvium: it has no 'appetite' or 'desire' for the amber. Magnetic action is, however – or so Gilbert affirms – quite unaffected by corporeal barriers, and cannot therefore involve any material effluvium. Epicurus' explanation in terms of the formation of chains of linked atoms between magnet and iron cannot be correct, since 'solid and very dense substances interposed, even squared blocks of marble, do not obstruct this power'.[114] All would-be mechanical explanations of magnetism will, he insists, founder on this rock.

After dismissing a number of other rival accounts of the cause of magnetism ('sympathy', 'insensible rays'), Gilbert proceeds to state his own view. Magnetic agency is, he explains, due to form, to the 'prime forces' innate in a particular and unique type of form, which is

the form of the primary and chief spheres and of those parts of them which are homogeneous and not corrupted, a special entity and existence, which we may call a primary and radical and astral form; not the primary form of Aristotle, but that unique form, which preserves and disposes its proper sphere. There is one such in each several globe, in the Sun, the moon and the stars; one also in the earth, which is the true magnetick potency which we call the primary vigour.[115]

This is in effect a flat refusal to explain magnetism in terms of anything more primitive. There is in each of the heavenly bodies, says Gilbert, a 'primary and radical and astral form' by which it draws to itself its own proper parts and thus maintains its cohesion. This magnetic form also helps it to maintain its rightful position and orientation in the Copernican solar system, as Gilbert explains in Book Six.[116] The lodestone is therefore no more than an uncorrupted portion of the planet Earth, possessing its 'primary vigour', i.e. drawing to itself other uncorrupted earthy bodies (viz. iron). Many bodies on or around the surface of the planet have, Gilbert explains, lost the magnetic virtue proper to all homogeneous parts of that great magnet – only iron and the lodestone retain this power to varying degrees.

---

114   *Ibid.*, p. 61.

115   *Ibid.*, p. 65.

116   I have, regrettably, neither space nor time to discuss Gilbert's magnetic cosmology.

In effect then, the magnetic *coitio* is, on Gilbert's view, a movement *natural* to a magnetic body: the 'desire' of iron and magnet for union is a necessary consequence of their very nature or form. Whereas electrical attraction is one-sided and violent (the chaff is moved by force), magnetic 'coition' is mutual and natural to all uncorrupted and homogeneous parts of our Earth.

The opinion of Thales, Gilbert continues, 'was not very absurd': to attribute a soul to the lodestone is far from madness.[117] The magnet's form 'seems to be very like a soul', giving direction and movement to its body. Magnetic force, says Gilbert, 'is animate, or imitates life'.[118] He thus first assimilates magnetic 'coition' to the category of *natural* motions, then conceives these (*pace* Aristotle) as due to the body's own SF as efficient cause, which (as Aristotle saw) involves one in an *animistic* physics. The celestial bodies, including our Earth, are animate, says Gilbert: why not also their 'homogenic' parts?

The strength of a magnetic force decreases with distance.[119] Beyond a certain range, Gilbert claims, its intensity falls to zero, thus giving each magnet an 'orb of virtue' through which its power extends.[120] Does the lodestone induce some abiding qualitative change throughout this *orbis virtutis*? Gilbert's answer is that it does not.[121] Around the magnet, he claims, there exists only a permanent potentiality for attracting iron, not some abiding actuality:

> . . . there exists in nature no orb or permanent or essential virtue spread out through the air. But a lodestone only excites magnetics at convenient distances.[122]

When a ferrous body enters the orb of virtue of a magnet, its form is 'awakened' by the proximity of the magnet and a 'desire' for union with the lodestone excited. The *excitation* in the iron of the magnetic quality is therefore A.D. (by the magnet); thereafter, the 'coition' is due to the mutual desire of lodestone and iron to come together. Only a soul, however, could perform the initial A.D. (matter acts only by contact).

---

117 W. Gilbert, p. 68.

118 *Ibid.*, p. 208.

119 *Ibid.*, pp. 91–92.

120 *Ibid.*, p. 103.

121 See Harré, 3, p. 35.

122 Quoted from Roller, p. 148.

Gilbert's animism is therefore integral to his conception of magnetic action.

There is in Gilbert no sense of tension or strain involved in giving at one moment a mechanistic account of one natural phenomenon (electrical attraction), and then next moment explaining magnetism in purely animistic terms.[123] In 1600, when *De Magnete* was published, the animism/mechanism issue was by no means so clear-cut as it was to have become fifty years later. There was as yet no 'mechanical philosophy', no *a priori demand* that *all* natural phenomena be explained in mechanical terms. A natural philosopher such as Gilbert (or, for that matter, Francis Bacon) can therefore opt for a mechanical explanation of one phenomenon and a non-mechanical account of another. (Bacon, for example, although hailed as a pioneer of the mechanistic world-view, seems to have regarded magnetic action as pure A.D.[124])

Gilbert is, however, quite scathing about the widespread and uncritical use of the concepts of sympathy and antipathy by Renaissance philosophers and (especially) magicians. In fact, Gilbert insists, the idea of a *general* attraction of like for like is totally mistaken:

> neither do similars mutually attract one another, as stone stone, flesh flesh, nor aught else outside the class of magneticks and electricks'.[125]

Of these, electrical attraction is (a) non-specific (and hence non 'sympathetic'[126] and (b) mechanically explicable; while magnetic coition is a *unique* phenomenon, due to the 'radical, primary, and astral' forms of the uncorrupted parts of the celestial bodies. Neither is, then, an instance of a general principle such as 'like attracts like'. Indeed,

> inanimate natural bodies do not attract, and are not attracted by others on the earth, excepting magnetically or electrically.[127]

Imagined examples of A.D. are dismissed either as mere fantasies (e.g. the supposed power of the remora to halt ships) or as more easily explicable without positing A.D.: naphtha is, for example, only said to attract flame because 'it gives off

---

123 See Harré, 3, p. 21.

124 See Hesse, 1, pp. 94-95.

125 W. Gilbert, p. 50.

126 *Ibid.*, p. 48.

127 *Ibid.*, pp. 109-110.

and exhales an inflammable vapour, on which account it is kindled at some distance . . .'[128] Thus, although Gilbert is not committed to the *a priori* denial of A.D. (he is no mechanical philosopher), his empirical spirit leads him to cast doubt on many of the alleged examples of Nature's much-vaunted 'universal sympathy', and to insist in particular on the *unique* and special nature of magnetism.

By 1600, then, quite a variety of natural phenomena – sounds, odours, electricity, etc. – were, or were thought to be, explicable in mechanistic terms. There was as yet, however, no mechanical philosophy (a) to systematise these findings and (b) to demand that other classes of phenomena be explained along the same lines. This programme was set in motion by Mersenne, Descartes, Hobbes, Gassendi, and others: we shall follow its fortunes in Chapter Twelve. The natures of light and of magnetism remained two of the great stumbling-blocks for any such programme.

C. Spirits

It may well be that no term in the history of philosophy has been used in so many quite distinct senses, and in the context of so many totally disparate systems, as the term 'spirit' (*pneuma*, *spiritus*). I have already vowed, however, to eschew all discussion of *incorporeal* agents: I shall therefore omit from this section the *Anima Mundi* of the Platonists, the *Archeus* (inner alchemist) of the Paracelsians, the 'universal Nature' of the schoolmen, and the angels and demons of the magicians, and will concentrate solely on the medical and chemical conception of *spiritus* as a subtle body (usually thought of as a compound of flame and air) of a particularly *active* disposition, causally intermediate between (incorporeal) soul and (gross) body.

Spiritus was often conceived as celestial in nature: its lightness indicates – to a peripatetic, at least, – that its natural place is far above us in the heavenly realm. Even Aristotle was not free from this idea: the fertility of semen, he writes in *De Generatione Animalium*, is due to

> the spirit which is contained in the foamy body of the semen, and the nature which is in the spirit, analogous to the element of the stars.[129]

---

128   *Ibid.*, p. 110.

129   *De Gen. Animalium*, 736b. See also Walker's paper in Debus, Ed., 7, p. 127.

The idea of a subtle *pneuma* causally responsible for the powers and properties of bodies was of course one of the cornerstones of the Stoic philosophy: through Stoicism, the concept of *pneuma* spread far and wide. In Galenic medicine, three types or grades of spirits are admitted: natural, vital and animal.[130] Food, says the Galenist, is transformed in the liver into the four 'humours' (blood, phlegm, black bile, and yellow bile) and *natural* spirits, a fine vapour pervading the venous blood. This blood eventually passes to the lungs, where it reacts with an aerial spirit to produce noble (arterial) blood, infused with *vital* spirits. These in turn pass to the brain, which converts them into *animal* spirits, which provide the causal go-between linking the incorporeal soul to its organic body. Each transformation of the spirits – from natural to vital and finally to animal – represents a further purification and subtilisation of matter: animal spirits represent the highest state matter can reach before passing out of the bodily realm altogether.

The notion of a Great (and continuous) Chain of Being[131] eases the conceptual strain involved in admitting the causal interaction of soul and body. If these are not simply polar opposites, but are linked by a continuous hierarchy of intermediates, then the interaction of the gross and the bodiless may be mediated by something itself intermediate in nature between the two. Animal spirits, although not entirely immaterial, are incredibly tenuous, and possess a power of *activity* (ability to *initiate* causal chains) not normally associated with bodies: they are therefore located midway on the Great Chain between soul and body, and are as such the meeting place of corporeal and incorporeal Nature. The physicians, says Ficino, speak of the *spiritus* as 'the bond of the soul and body',[132] an idea echoed in Donne's *Ecstasy*:

> As our blood labours to beget
>   Spirits, as like souls as it can,
> Because such fingers need to knit
>   That subtle knot, which makes us man.

The 'subtle knot' is of course the union of soul and body which makes a man what he is, a creature straddling two

---

[130]   See Tillyard, pp. 63–64.

[131]   For the history of this idea and its ramifications throughout Western thought I can do no better than to refer the reader to the seminal work of Lovejoy.

[132]   Quoted from Walker, p. 48.

different (and sometimes warring) 'worlds' or realms of being. The purification or subtilisation of spirits – from natural to vital to animal – is seen as part of an upward striving of matter towards the status of soul.[133] Thus Raphael in *Paradise Lost*, explaining to Adam the nature of the Great Chain, states that matter ascends

> Till body up to spirit work, in bounds
> Proportioned to each kind.

Spirits are therefore normally thought of as *both* causally *and* qualitatively intermediate between soul and body. Their causal rôle is as messengers or 'instruments' of the soul: animal spirits flow through the nerves bringing information from the sense organs to the brain (and thence, more mysteriously, to the incorporeal rational soul), and instructions from the brain to the muscles. Both sensation and volition therefore involve the causal mediation of the animal spirits. To fulfil this rôle, it was widely believed, these spirits must themselves have some kinship and affinity with *both* soul and body, i.e. must themselves be intermediate in status and character.[134]

Many Renaissance thinkers followed Ficino in identifying the *spiritus* of the physicians (described by Ficino as 'a certain vapour of the blood, pure, subtle, hot and lucid'[135]) with the Platonic 'chariot of the soul' of the Phaedrus, the 'astral body' of the magicians. This is conceived by the Platonist as quite literally astral in origin: it is received by the soul during its descent to Earth and is, says Ficino, 'not body and almost soul, and not soul and almost body'[136] – a helpful description indeed! In virtue of its origin, the astral body still receives celestial influences (to which it is sympathetically attuned) and may therefore serve to mediate astral influences on humans – even Francis Bacon gives this idea careful attention.[137]

The concept of spirit was not of course the exclusive property of the physicians. It was also, as we have seen,

---

133   See Tillyard, p. 72.

134   Whether the notion of a perfect continuity between soul and body makes sense is a fascinating conceptual question which I must reluctantly decline to pursue.

135   Quoted from Walker, p. 3.

136   From Pagel, 2, p. 128. For Paracelsus on the same topic see also p. 134.

137   F. Bacon, *Works*, Vol. 2, p. 644, and Vol. 4, p. 351. See also Walker, p. 38.

essential to the chemical philosophy of the Renaissance. The specific and characteristic virtues of a substance, say Paracelsus and his disciples, are due not to gross matter but to its contained spirit, i.e. subtle, volatile, and intrinsically active matter.[138] Such 'spirits' (i.e. distillates) as alcohol and the mineral acids seem to bear out this belief in the inherent activity of subtler forms of matter. We discussed in the previous chapter the notion of a particular astral spirit or 'quintessence' as a quality-bearing (and imparting) principle – let it suffice for the moment to have mentioned here the rôle of the chemists in the development of the concept of *spiritus*.

From the combined medical and chemical traditions, therefore – and we must remember that throughout this period chemistry was largely pharmacy, practised as a adjunct to medicine – we derive the concept of *spiritus* as a subtle vapour (usually conceived as a mixture of flame and air), present (more or less) in all bodies, giving them their particular properties and powers. The theologian, says Von Helmont,

> places a *Spirit* in irreconcilable difference to all *corporeal nature*, as an *essence* wholly preternatural. But *Physicians* oppose a *Spirit* against the more *gross compage* or more material and less rarefied substance of a body.[139]

A spirit, says Francis Bacon, is simply:

> a natural body, rarefied to a proportion, and included in the tangible parts of bodies, as in an integument. And they be no less different one from the other than the dense or tangible parts: and they are in all tangible bodies whatsoever, more or less; and they are never (almost) at rest; and from them and their motions proceed . . . most of the effects in nature; . . . for tangible parts in bodies are stupid things; and the spirits do (in effect) all.[140]

Spirits, he insists, are not mysterious detached 'virtues' or 'energies', but *bodies*, 'thin and invisible, and yet having place and dimension, and real'.[141] They are highly active, subtle bodies, compounded of flame and air, causally responsible for, but *not* themselves (*pace* Paracelsus) bearers of, the qualities manifest in gross bodies. The *spiritus* moves, and

---

138 Paracelsus, 1, Vol. 1, p. 135.

139 Quoted from Harré, 2, p. 139.

140 F. Bacon, *Works*, Vol. 2, p. 381.

141 *Ibid.*, p. 381.

acts upon, the tangible parts of a body to make it of a certain nature; it does not itself possess all the qualities its presence imparts. Chemical explanation therefore falls into two distinct phases:

a.   Reduction of the qualities of a body to the configuration and motion of its tangible parts (cf. the 'forms' of heat and whiteness).
b.   Explanation of the motions of these particles in terms of the agency of a contained spirit.

Spirits are therefore the dynamic, motivating agents in Nature. Since tangible matter possesses a 'torpor', a 'natural appetite not to move at all',[142] certain active powers must be operative to prevent all things grinding to a halt. Spirits are therefore a necessary, nay, *vital*, part of Bacon's philosophy of Nature: spirit moves, guides, and directs gross matter, which is dead, inert, and, as Bacon puts it, 'stupid'. The spirits are, therefore, the 'agents and workmen that produce all the effects in the body'.[143]

Let us now digress a little to discuss a problem of a theological nature. Do animals possess incorporeal souls? If they do, will not these souls be immortal? If they do not, how is it that animals have perceptions and desires? To admit that animals possess incorporeal and immortal souls is to deny the unique status of man; to accept that a purely material thing can perceive and will is to risk lapsing into materialism. The orthodox compromise view was to state that man alone possessed an incorporeal and immortal *rational* soul, divinely implanted at birth, and that the psychological functions of animals are all due to the corporeal *spiritus*.[144] But, if the *spiritus* explains sensation, imagination, and appetition in animals, why not also in men? This step is taken by such thinkers as Telesio and Donio in the sixteenth century. Only for *intellectual* activities, one might say, is the incorporeal rational soul necessary. But, since some animals manifest clear signs of intelligence, even this claim will begin to look dubious. A tendency simply to identify soul with (animal) spirits is, says Walker,[145] present in much sixteenth century Natural Philosophy: only fear of heresy must have restrained

---

[142]   *Ibid.*, p. 586. See also the *Novum Organum*, Bk. 2, Aph. xlviii, Vol. 4, p. 232.

[143]   F. Bacon, *Works*, Vol. 5, p. 268.

[144]   See Walker in Debus, Ed., 7, p. 121-122.

[145]   *Ibid.*, p. 125.

a number of people from explicit assertion of such a view. If *all* the psychological functions of animals and *many* of those of man are due to the corporeal *spiritus*, the postulation of an incorporeal soul must seem largely superfluous. In Telesio, Walker explains,[146] the incorporeal soul explains altruistic (non-utilitarian) behaviour; in Donio, his pupil, even that can be attributed to the agency of the *spiritus*. The incorporeal soul can therefore be left to the divines (or dismissed altogether from one's world-view): as far as natural philosophy is concerned, men and animals are driven by subtle but corporeal spirits.

How does a spirit move the body in which it is contained? If we turn again to the writings of Francis Bacon, we shall find the beginnings of an answer. Although Rees argues[147] that the action of spirits is non-mechanical, and that therefore Bacon should not be classed as one of the founders of the mechanical philosophy, the evidence is solidly against him, We do not know, says Bacon, how spirit moves body: we must enquire

> how the compressions, dilations and agitations of the Spiritus (which without doubt is the fountainhead of motion) bend, excite, or drive the corporeal and gross mass of the parts of the body.[148]

It is clear even from this supposed profession of agnosticism what *sort* of explanatory factor Bacon is seeking. It is from spirits and their *motions*[149] that proceed all of Nature's effects: it is by their agitations, compressions, and dilations – all varieties of local motion – that spirits move bodies. One of Bacon's examples of the causal agency of spirit illustrates this contention well. In the explosion of gunpowder in the barrel of a cannon, the force of the explosion is not, Bacon explains, due merely to the expansion of the flame; rather, the nitre emits a 'crude and windy spirit' which dilates very rapidly and thus impels the cannon ball,[150] giving us a clear example of a mechanical action of a spirit on tangible matter.

In the *Novum Organum* too we find hints of a mechanical operation of spirits. The physical qualities of bodies may, says Bacon, by explained in terms of the following factors:

---

[146]   *Ibid.*, p. 124

[147]   Rees, 5, p. 552.

[148]   Quoted from Walker in Debus, Ed., 7, p. 122.

[149]   F. Bacon, *Works*, vol. 2, p. 381.

[150]   *Ibid.*, p. 351.

1. The nature of the contained spirit ('living' spirits, for example, contain more flame and less air than those of non-living things).
2. The texture and arrangement of the tangible parts.
3. The situation of the spirit in the body (in channels, pores, passages, etc.)[151]

It is the third of these factors which is most important to my argument. The distribution of its spirits is for Bacon one of the respects in which an animate body differs from an inanimate one: in a living being, the spirits intercommunicate by veins and channels, and have a central seat where the 'principal spirits' reside; the pneumatic material of a non-living body shows a lower degree of organisation.[152] This suggests some quasi-mechanical interaction of matter and spirit: it makes all the difference between life and death whether spirit is confined (i.e. physically trapped) within a body or allowed to escape, as Bacon explains in the *Historia Vitae et Mortis*.[153]

Another passage from the *Novum Organum* bolsters this case. How, Bacon asks, can a subtle spirit move a massive and inert body? The answer turns on the velocity with which the spirit can move:

> Nor could a small portion of animal Spirit in animals, especially in such vast bodies as those of the Whale and Elephant, have ever bent or directed such a mass of Body, were it not owing to the Velocity of the former, and the slowness of the latter in resisting its motion.[154]

Spirits are therefore enabled by their incredibly rapid motions to move much bulkier bodies – a doctrine reminiscent of the Epicurean theory of the soul. A gross body is inert and torpid unless set in motion by some *mechanical impulse*. Given the nature of the *patient*, we can infer something about the nature of the agent which moves it.

I conclude therefore that Rees is wrong, and that Bacon does think of the action of the *spiritus* in moving gross bodies as mechanical in nature. In Leonardo, the mechanical

---

[151] *Ibid.*, Vol. 4, pp. 125–126; *Novum Organum*, Bk. 2, Aph. vii. See also the longer list of explanatory factors in the *Sylva Sylvarum* (Vol. 2, pp. 618–619). Here too the spatial arrangement of the spirits is relevant to the causal explanation of the properties of a body.

[152] *Ibid.*, vol. 2, pp. 528–530.

[153] See Walker, p. 199.

[154] F. Bacon, *Works*, Vol. 4, p. 212; *Novum Organum*, Bk. 2, Aph. xliv.

operation of the *spiritus* is explicitly stated, and a *hydraulic* model of muscular contraction envisaged:

> The spirit of the sentient animals moves through the limbs of their bodies and when the muscles it has entered are responsive it causes them to swell, and as they swell they shorten and in shortening they pull the tendons that are joined to them. And from this arises the force and movement of human limbs. Consequently material movement springs from spiritual.[155]

The body of a man or animal is therefore a highly complex *machine* moved by the flow of the animal spirits. This illustrates a radically new conception of the respective status of Nature and Art. For the Aristotelian, mechanics is no part of *natural* philosophy: our machines or artefacts classically involve violent motion, i.e. operate *against 'physis'* – their workings cannot therefore be expected to enlighten us about those of Nature's creatures.[156] Leonardo flatly denies this: common mechanical principles, he alleges, apply to natural creatures and artefacts alike. Birds, for example, fly not merely by 'nature' but by exploiting certain physical properties of the air: a heavier-than-air machine could in principle achieve the same result by essentially the same means. Bio-engineering is born:

> The instrumental or mechanical science is the noblest and useful above all others, since by means of it all animated bodies which have movement perform all their actions.[157]

God, being a master-mechanic, has an engineering facility far surpassing ours: His creatures are infinitely more sophisticated and complex than our poor imitations. They rely, however, on the *same* engineering principles – there is nothing occult or in principle incomprehensible about the way a bird flies. In Leonardo we can already discern traces of the mechanistic biology of the seventeenth century. Our mechanical models, though crude, can for him throw real light on the workings of Nature,[158] which, though of incredible complexity, obey the same common set of basic principles (e.g. those of the lever and balance).

---

155 Leonardo, p. 61.

156 See A. R. Hall, 2, p. 18.

157 Leonardo, p. 88.

158 See Rossi in Righini Bonelli and Shea, Eds., pp. 251–253; Marie Boas Hall, 8, p. 211.

D. Conclusion

We are now in a position to understand how the mechanical philosophy of the seventeenth century emerged with such amazing rapidity from the anthropomorphic animism endemic in medieval and Renaissance thought. The key concepts of that anthropomorphic world-view were of course sympathy and antipathy: things, for the animist, are like people in respect of having loves and hatreds, desires and aversions. Like humans, they are 'attracted' by (i.e. move spontaneously towards) those things (usually objects similar to themselves) which they desire, and shun those to which they are averse. Thus, in the thirteenth century 'Book of the marvels of the World', falsely attributed to Albertus Magnus, we read that:

> It was known of Philosophers that all kinds of things move and incline to themselves, because an active . . . virtue is in them . . ., as fire moveth to fire and water to water . . . And generally it is verified . . . by reasons, and diverse experience, that every nature moveth to his kind.[159]

But, in order for a body to move thus (i.e. 'purposively') in the direction of another, it must have some 'perception' of that other – a desire without an associated belief is not sufficient to explain purposive action. The general explanatory framework behind the ideas of sympathy and antipathy is therefore heavily anthropomorphic in nature. One X 'sees' a kindred X, 'desires' union with it, and thus sets off towards it; or, sees an alien object Y, finds it uncongenial, and shuns it. Without perceptions of some kind (perhaps unconscious) the pattern of explanation would not work.

Francis Bacon sees this point quite clearly. Although he often ridicules contemporary accounts of sympathy and antipathy, he does not doubt the existence of the phenomena. A wooden-pointed arrow will, he claims, pass through the wooden hull of a ship, while a metal bullet will not; which fact

> dependeth upon one of the greatest secrets in all nature; which is, that similitude of substance will cause attraction, where the body is wholly freed from the motion of gravity: for it that were taken away, lead would draw lead, and gold would draw gold, and iron would draw iron, without the help of the lodestone.[160]

[159]   Quoted from Hansen in Lindberg, Ed., p. 492.

[160]   F. Bacon, *Works*, Vol. 2, pp. 564–565.

The existence of a general affinity of like for like is explicitly affirmed. Yet this sympathy would be *inexplicable* without some notion of perception applicable even to inanimate things:

> It is certain that all substances whatsoever, though they have no sense, yet they have perception: for when one body is applied to another, there is a kind of election to embrace that which is agreeable, and to exclude or repel that which is ingrate: and whether the body be alterant or altered, evermore a perception precedeth operation.[161]

The flickering flame, says Bacon, perceives the breeze; water feels it way into the pores of a sponge, etc.[162] These perceptions, he continues, are not always dependent on *contact*, but are sometimes 'at distance', e.g. 'when the lodestone draweth iron; or flame fireth naphtha of Babylon, a great distance off'.[163] Even inanimate bodies have, then – in virtue of their attached spirits – 'perception', although they have no 'sense' (conscious awareness). The enclosed spirit in a body perceives the proximity of a 'kindred spirit' and sets the body off in motion towards it. Nothing, it might appear, could be *further* from mechanism: this anthropomorphic world-view seems to involve *either* A.D. (action by the lodestone on the iron) *or* spontaneity (in the motion of the iron), or both. There is moreover the further argument, from A.D. to an incorporeal causal agency, to contend with, and more than a hint of an irreducibly teleological approach to Natural Philosophy.

But now let us introduce one or two themes already discussed in this chapter, notably the two following lines of thought:

a. A mechanistic theory of perception

If all perceptions are caused by mechanical means, i.e. by *either* some corpuscular emission or effluvium, *or* a vibration or agitation of a fluid medium, then one half of the total causal story is mechanised.

b. A mechanistic account of bodily motion

If a gross body such as that of a man can *only* be moved by an *impulse* from another body, it will follow that the *spiritus*

---

161   *Ibid.*, p. 602.
162   See Gregory, p. 93.
163   F. Bacon, *Works*, vol. 2, p. 602.

which moves the body must do so after this manner. The whole body can then be conceived as a complex hydraulically-driven machine – an idea glimpsed by Leonardo and developed extensively in the seventeenth century.

Now a desire must, to be effective, be capable of initiating bodily motion. It must therefore either be itself *identical* with, or the *cause* of, a particular state of motion or agitation of the *spiritus*. But, if the desire is a state of an incorporeal substance, how can it move the corporeal spirits? Only body can, intelligibly, move body. In any case, animals have no incorporeal souls – all their actions are due to the corporeal *spiritus*. In animal psychology, then, the simplest move is to *identify* a given desire-belief complex with state of agitation of the animal spirits. This will enable us to explain (a) how a perception is caused by some mechanical disturbance in, or corpuscular emission through, the medium, and (b) how the desire-belief complex can cause an appropriate bodily response.

What happens, then, is that the anthropomorphic animism of Renaissance thought is *accommodated within* the Mechanical Philosophy. Sympathy and Antipathy are *not* absolutely rejected, but are explained away in mechanical terms – often with great ingenuity. (This process is quite clear to see in Descartes, Gassendi, Charleton, Boyle and other masters of the mechanical philosophy.) What may appear to involve A.D. is explained in terms of some appropriate causal intermediaries; what seems to be spontaneity is no more than the necessary outcome of a particular perception (itself mechanically caused by the impact of corpuscles on the sense organ), and so on. A mechanical process in the medium causes, by impulse, a local motion of the animal spirits, which in turn sets the whole body in motion. We still lack a detailed account of the directionality and appropriateness of the response, but have high hopes that such an explanation will in time be forthcoming.[164]

This is a curious chapter in intellectual history. At first sight, the animism of the Renaissance and the mechanism of the seventeenth century seem utterly opposed. Yet the two coincide without (much) strain in Francis Bacon, and the one emerged within a generation or so from the other by the efforts of Mersenne, Descartes, Hobbes and Gassendi. We are

---

[164] Hobbes attempts to explain these phenomena in terms of the corroboration or otherwise of the vital motions of the heart by the incoming impulse; Hooke favours an account in terms of resonance or sympathetic vibration.

now in a position to see how this process took place. Instead of rejecting wholesale all the ideas of the animists, the mechanical philosophers made use of existing mechanistic accounts of (a) perception and (b) animal motion to provide a framework within which perception and desire, sympathy and antipathy, could be explained *without* invoking A.D., spontaneity, incorporeal causes, or any irreducibly teleological agency (within Nature). In Chapter Twelve we shall study the resulting philosophy in more detail.

INDIVISIBLES: 3

A. *The Labyrinth of the Continuum*

The seventeenth century witnessed some spectacular advances in the field of mathematics, culminating in the discovery of the calculus by Newton and Leibniz. To do justice to this mass of first-rate mathematical thought would require another work and another author: no brief summary is likely to be adequate. I shall therefore in this section omit purely mathematical issues, and concentrate instead on philosophical discussions of the continuum of more immediate relevance to matter-theory. The figures I select for discussion are chosen as representatives of particular well defined points of view: no claim is made, however, for the exhaustiveness of my list. The section should illustrate clearly (a) the sheer diversity of seventeenth century views on the topic, and the lack of consensus achieved, and (b) the abiding relevance of the issues raised by Zeno and the ancient Atomists.

Theories involving indivisibles were by no mean universally accepted: even among the founders of the Mechanical Philosophy there were some who defended the Aristotelian notion of infinite divisibility as something merely *potential*. In Hobbes' *De Corpore*, for example, it is argued that infinite divisibility is not to be so understood,

> as if there might be any infinite or eternal division; but rather to be taken in this sense, *whatever is divided, is divided into such parts as may again be divided*; or thus, *the least divisible thing is not to be given*; or, as geometricians have taken it, *no quantity is so small, but a less may be taken.*[1]

Hobbes therefore follows Aristotle and the Greek geometers in denying the existence of atomic magnitudes. Like the peripatetics, however, he does so at the cost of having no coherent view of the ultimate make-up of the continuum. All that can be *meant* by 'infinitely divisible', on Hobbes' account, is 'divisible as many times as one pleases', i.e.

---

[1] Hobbes, *Works*, Vol. 1, p. 100.

indefinitely.[2] For every natural number n, any continuous magnitude can be divided n times – and then more. Infinite divisibility does not, however, involve the real possibility of infinite *dividedness*. Zeno, we remember, had argued from (1) 'X is infinitely divisible' to (2) 'X consists of an infinite number of aliquot parts' to (3) 'X is infinitely extended'. Hobbes replies that, on his view, (1) does not entail (2), since 'how many parts soever I make, yet their number is finite'.[3] It cannot be true in any other sense than his own, Hobbes claims, that a finite body is infinitely divisible: i.e., he concedes the inference from (2) to (3), and uses his (Aristotelian) conception of infinite divisibility to block the derivation of (2) from (1). Infinite divisibility does *not* involve the possibility of actually infinite dividedness:[4] it functions *only* to proscribe minimal or atomic magnitudes.

A similar view can be discovered in Kenelm Digby's treatise *on Body*, but there the Aristotelianism is still more pronounced. All extended magnitudes, says Digby, are divisible into parts or measures. These parts are, however, not actually present in their respective wholes prior to division:

> Ells, feet, inches, are no more reall Entities in the *whole* that is measured by them . . . than . . . colour, figure, mellownesse, taste, and the like, are several substances in the apple.[5]

These parts, Digby agrees with Aristotle, 'are not really there, till by division they are parcelled out'.[6] Quantity involves divisibility but not dividedness. Before raising the obvious objections, let us follow Digby's argument – essentially, a *reductio* – for this position. If, he argues, parts were actually present in the divisible whole, every quantity would be composed of indivisibles: if a given magnitude

> were divided into all the parts into which it is divisible, it would be divided into indivisibles (for nothing divisible, but not divided, would remain in it) . . .[7]

If parts are actually present in a divisible magnitude, that magnitude is made up of indivisibles: of a finite, or an

---

2 See Laird, p. 106.

3 Hobbes, *Works*, Vol. 1, pp. 63–64.

4 See Boyle, *versus* Hobbes on divisibility in Boyle's *Works*, vol. 1, p. 236.

5 Digby, p. 12.

6 *Ibid.*, p. 12.

7 *Ibid.*, p. 12.

infinite, number? If a *finite* number of indivisible minima constitute an extended magnitude then, clearly, some magnitudes (those composed of an odd number of minima) cannot be bisected. But, says Digby,

> Euclide hath demonstratively proved beyond all cavill . . . that any line whatsoever may be divided into whatsoever number of parts.[8]

There cannot, therefore, be indivisible minima of extension or atomic magnitudes. If a line consists of indivisibles, it must be of extensionless mathematical *points*. But, when point is juxtaposed with point, they become 'drowned in one another': two points, as Aristotle showed, cannot be contiguous without simply *coinciding*.[9] It is therefore impossible to assemble a line out of points: however many such indivisibles are laid down, says Digby, 'they will still drown themselves all in one indivisible point'.[10] Even an infinity of such extensionless points will not suffice to produce a > 0 extension – each in turn will be 'drowned' in the first. Therefore, Digby concludes, he has proved (1), 'That Quantity is not composed of indivisibles (neither finite, nor infinite ones)', and hence (2), 'That parts are not actually in it'.[11]

Sense, Digby admits, may seem to object to this apparently paradoxical conclusion. It may, for example, seem evident that the parts of a man's body are actually distinct even when combined to form the whole. This, he replies, is not something 'given' to sense, but a mistaken abstraction of the understanding. The parts of a body are not genuinely distinct things, but *aspects* of one thing, separable only by an act of abstraction:

> . . . a hand, or eye, or foot, is not a distinct thing by it self; but . . . it is the man, according as he hath a certain virtue or power in him to distinct operations.[12]

Sever an organ from its body, and it can no longer function as such – it is no longer a hand or eye except by equivocation.[13] The hand of a man is, Digby insists, analogous to the

8   *Ibid.*, p. 14.

9   See Chapter 1, p. 15.

10 Digby, p. 14.

11 *Ibid.*, p. 15.

12 *Ibid.*, p. 16.

13 This is of course an orthodox Aristotelian thesis.

whiteness of a wall – an inseparable aspect of a substantial unity. His mistake is clear. It *does not matter* (to this particular issue) if a dissevered hand does not function as such: it remains a distinct, extended, material part. We can *measure* it before and after amputation; which process does not affect its *size*, however much it may destroy its *function*. Digby's error is facilitated by his choice of example: if one thinks of a large lump of lead, it seems quite evident that it actually consists of those parts (smaller lumps of lead) into which it can be divided. Digby's introduction to the concept of function is a red herring, serving only to confuse the issue. The parts of our lump of lead are *not* abstractions but concrete realities: how can the whole lump be a concrete material substance unless its constituent parts are?

In terms of our initial formulation of these issues in Chapter 1, Digby, like Aristotle, denied Z4 (divisibility presupposes the prior existence of distinct parts) and accepted Z1 (every magnitude is divisible). The great difficulty is to accept *both* these apparently intuitive theses without falling foul of Zenonian antinomies. In the matter-theory of Galileo's *Two New Sciences*,[14] an attempt is made to do just that.

Acceptance of Z1 is of course implicit in Galileo's 'geometrisation' of physics – an idea stressed in the writings of Koyré.[15] Nature, says Galileo in the *Assayer*, 'is written in the language of mathematics':[16] the applicability of the theorems of geometry to the concrete realities of the natural world is quite fundamental to Galilean physics. Throughout his works, he never tires of emphasising the *continuity* of all natural processes such as motion and acceleration.[17] If physical space is a Euclidean framework, and all Nature's processes are continuous, the crude mathematical Atomism of the Epicureans must be utterly rejected.

Yet, at the same time, Galileo explicitly accepts Z4: parts are not created by division:

A division and a subdivision which can be carried on indefinitely presupposes that the parts are infinite in number.[18]

14 Galilei, 4, p. 20 ff.

15 See Koyré, 7, p. 35, p. 73, and of course many other passages throughout his *oeuvre*.

16 Galilei, 2, pp. 237–238.

17 See, e.g., Galilei, 4, p. 162 ff.; Clavelin, p. 282.

18 Galilei, 4, p. 34.

Infinite divisibility presupposes infinite dividedness. Every continuum has, therefore, an ultimate or final constitution of (aliquot) parts, parts which cannot conceivably undergo further division. A line is made up of an actual infinity of such 'points': this is Galileo's solution to the age-old problem of the constitution of the continuum. Very crudely, once one grants the existence of an *ultimate* constitution for every extended magnitude, one is left with the following four options, O1–4:

O1.   A finite number of unextended points.
O2.   A finite number of extended minima.
O3.   An infinite number of unextended points.
O4.   An infinite number of extended minima.[19]

Of these, O1 and O4 are obviously hopeless, while O2 is implicitly anti-geometric and contradicts the intuitive Z1.[20] Once we grant Z4, then, and admit the existence of an ultimate constitution of the continuum, we find ourselves left with O3. But how can extensionless points – even an actual infinity of them – make up an extended magnitude? Galileo has no adequate answer. Although $n \times 0 = 0$ for all finite values of n, he claims, such reasoning cannot be applied to infinity, which is by its nature incomprehensible to a finite mind.[21] O3, he insists, is true, even if we can never understand *how* it is true.

This conception of the continuum is developed in the context of Galileo's most extended discussion of matter-theory in Day One of the *Discorsi*. Salviati, Galileo's spokesman, attributes the cohesion of solid bodies to the *fuga vacui* exerted by their interstitial vacua, and suggests that every solid may contain an infinite multitude of such point-vacua:

> . . . let us see if we cannot prove that within a finite extent it is possible to discover an infinite number of vacua.[22]

Bodies, then, are composed of 'infinitely small indivisible particles'[23] or point-atoms, interspersed with these point-vacua. If, says Salviati,

> we imagine the body, by some extreme and final analysis, resolved into its primary elements, infinite in number, then

19   See Clavelin, p. 36.
20   See Shea, p. 25.
21   Galilei, 4, p. 26, p. 30.
22   *Ibid.*, p. 20.
23   *Ibid.*, p. 55.

we shall be able to think of them as indefinitely extended in space, not by the interposition of a finite, but of an infinite number of empty spaces.[24]

Every solid body consists, in the final analysis, of an infinite number of point-atoms, interspersed with an infinitude of point-vacua. (Fluids, by contrast, lack point-vacua and are therefore without cohesiveness.[25]) Do all bodies, then, of whatever magnitude, consist of the *same* number of unextended parts? Galileo toys with the strikingly modern idea of treating one-one correspondence as the criterion of equinumerosity of sets, but eventually concludes that such concepts as 'greater than', 'equal to', and 'less than' are inapplicable to infinite sets.[26] We cannot intelligibly claim either that the greater body contains more point-atoms than the lesser, or that it contains the same number: both propositions lack sense. (It is asking too much of Galileo to imagine that he could anticipate Cantor.) The point-atoms and vacua constitutive of a body are uncountable, without any assignable number.

Galileo thus attempts to develop a theory of the material continuum which reconciles the geometrisation of Nature (Z1) with the demand for an ultimate constitution of matter (Z4).[27] To achieve this synthesis, he requires an actual infinity of point-atoms and vacua. A strikingly similar theory can be found in his English contemporary Hariot. Atomism, Hariot writes to Kepler, is the key to Natural Philosophy.[28] Every continuum, he insists, must be composed of an actual infinity of indivisible elements, which cannot therefore have any assignable magnitude. But even a (denumerable) infinity of extensionless mathematical points could not give rise to a magnitude. (Imaging placing them at points $1/2$, $3/4$, $7/8$, etc., of the way along a line.) If, then, we claim, with Galileo, Hariot, and many others, that every continuum must be composed of indivisible elements, these cannot be identified with mathematical points.

This was the source of the theory of infinitesimals. An infinitesimal is, definitionally, something smaller than any

24   *Ibid.*, p. 25.

25   Galileo even attempted to deny the surface tension of water – his physical intuitions let him down badly here. See Shea, pp. 13-15; Le Grand, p. 203.

26   Galilei, 4, pp. 32-33.

27   See Clavelin, pp. 311-316.

28   See Kargon, 4, pp. 24-26.

assignable size, yet greater than zero. An infinity of these curious entities will constitute a magnitude. Galileo's point-atoms may perhaps best be conceived as infinitesimals rather than as true mathematical points;[29] in his disciple Cavalieri, whose *Geometria Indivisibilibus* (1635) marks the rise of the new analysis, this conception is spelt out quite explicitly.[30] The notion of a continuum as made up of an infinity of infinitesimals became very widespread in the seventeenth century: despite its lack of rigour (and ultimate incoherence) it proved a very fruitful idea, and contributed importantly to the rise of the differential and integral calculi. (Some still speak of the 'infinitesimal calculus', although that concept is now redundant, superseded by the more rigorous notion of a *limit*.)

Strictly, Cavalieri admits, his 'indivisibles' should have no size at all; one can, however, substitute for them, at will, a large but finite number of very minute elements.[31] (The technique derives, ultimately, from Archimedes.) If one is more interested in practical results (the discovery of useful theorems) than in logical rigour, such a method is of considerable value.[32] Many seventeenth century thinkers followed Cavalieri in using the infinitesimal method without bothering overmuch about its logical rigour or coherence. Roberval and Torricelli both made use of such a method, as did Wallis, who actually characterised the size of the infinitesimal as $1/\infty$![33] The infinitesimal value of $1/\infty$ is, he says, a 'non-quantum', an infinitely small magnitude almost, but not quite, vanishingly small. Although smaller than any given magnitude, the infinitesimal is still greater than zero: an infinity of infinitesimals will, while an infinity of mere points will not, give rise to a sizable magnitude.[34]

For a related view, let us turn to Henry More's discussion of the continuum in 'The Immortality of the Soul'. We begin by stating some of the axioms on which his account rests:

> Axiom 11: A Globe touches a Plane in something, though in the least that is conceivable to be reall.

---

[29] Evans, pp. 551–552, interprets Galileo as an advocate of infinitesimals. See also Boyer, p. 112. ff.

[30] See Boyer, p. 111.

[31] See A. R. Hall, 3, p. 91.

[32] Boyer, p. 123.

[33] A. R. Hall, 3, p. 97; Boyer, pp. 170–171.

[34] Boyer, pp. 172–173.

Axiom 12: The least that is conceivable is so little that it cannot be conceived to be discerpible into less.

Axiom 13: As little as this is, the repetition of it will amount to considerable magnitudes.

Axiom 14: Magnitude cannot arise out of mere Non-Magnitudes.[35]

The argument for Axiom 14 is simple. For, says More,

> multiply *Nothing* ten thousand millions of times into nothing, the Product will be still *Nothing*. Besides, if that wherein the Globe touches a Plane were more than *Indiscerpible*, that is, *purely Indivisible*, it is manifest that a *Line* will consist of *Points* Mathematically so called, that is *purely Indivisible*; which is the grandest absurdity that can be admitted in Philosophy, and the most contradictious thing imaginable.[36]

Multiplication of nothings will always yield nothing – a line cannot be made up of unextended mathematical points. Yet every continuous magnitude has an ultimate constitution of indivisible minima, least conceivable units of extension. Of what nature are these? According to More's Axiom 15,

> The same thing by reason of its extreme littleness may be utterly indiscerpible, though intellectually divisible.[37]

Does 'intellectually divisible' imply 'divisible in thought'? Can we *conceive* of the division of such a minimum? Surely not: if so, it would not be a *minimum conceivable* magnitude. Could even divine power divide such an indivisible? Not if our (clear and distinct) ideas mirror the structure of Reality – a rationalist principle endorsed by the Platonist More. Perhaps all that is meant by 'intellectually divisible' is that the minima possess distinct aspects or features which can be abstracted by the understanding – we shall in due course discover support for such a reading.

If, More argues, we admit that matter is divisible *ad infinitum*, then divine power can reduce this potency to act: God can separate a body into as many particles as it is divisible into. But then,

> one of these particles reduced to this perfect Parvitude is then utterly indiscerpible, and yet intellectually divisible,

---

35 H. More, 1, Vol. 1, pp. 26–27.

36 *Ibid.*, p. 27.

37 *Ibid.*, p. 27.

otherwise Magnitude would consist of mere points, which would imply a contradiction.[38]

More evades one paradox here only at the cost of embracing another. If a body is infinitely divisible, it consists of an actual infinity of these 'perfect parvitudes', separable from one another by divine power. But, if each of these minima has > 0 magnitude, the original body must have been infinite in extent. Either the 'perfect parvitudes' are finite in number, in which case not even God can divine a body infinitely, or, if they are infinite in number, they can have no magnitude greater than zero (by however little). Exactly the same dilemma, of course, is applicable to infinitesimals.

If 'perfect parvitudes' are intellectually divisible, says an objector, they must consist of distinct parts, which are therefore separable, if only by divine power. This, More retorts, does not follow:

> Division into parts does not imply any discerpibility, because the parts conceived in one of these *Minima Corporalia* are rather *essential* or *formal* parts than integral, and can no more actually be dissevered, than Sense and Reason from the soul of man.[39]

That *minima corporalia* are indivisible is presented as a conceptual truth accessible to the faculty of Reason. But, if they possess some extension, however little, they surely possess *shapes*, and we can imagine the bisection of any figure. The 'fancy', says More, would have us *picture* the *minima*, e.g. as spheres with void spaces between them, but the evidence of 'fancy' (imagination) is, he insists, 'of no moment in this case, she always representing a Discerpible magnitude in stead of an Indiscerpible one'.[40] Every pictured magnitude consists essentially of distinct parts, divisible in thought or by divine power. The attribution of (any) shape to the *minima corporalia* is therefore to be resisted as an aberration of the imaginative faculty: the minima have (the least possible) size, but are 'without figure'.[41] Just as infinite greatness is without shape, so is infinite smallness. *Any* attribution of shape brings divisibility as an immediate corollary: minimal conceivable units of extension must be

---

38   *Ibid.*, p. 30.

39   *Ibid.*, p. 30.

40   *Ibid.*, p. 30.

41   Quoted by Burtt, p. 128, from More's *Enchiridion Metaphysicum*.

shape*less*. (This doctrine is strongly reminiscent of the Epicurean theory of minimal parts: they too cannot be attributed any shape without absurdity ensuing.[42]) The *minima corporalia* of More cannot, therefore, be *bodies*: there is a radical disanalogy between a body and its ultimate constituent parts.

More never speaks of his 'perfect parvitudes' as infinitesimals, yet, if he is to accept the infinite divisibility of matter (in thought and by divine power) and avoid the crudest form of mathematical Atomism, this must be his position. His 'perfect parvitudes' must, to all intents and purposes, be infinitesimals, magnitudes greater than zero yet less than any assignable value. They cannot of course, be pictured or imagined in any way: More explicitly warns us against this. The difficulties, however, run deeper – conceptual and mathematical problems arise which cannot be dismissed as mere aberrations of the 'fancy'. Between something and nothing there just is no *tertium quid*. There is no least or indivisible magnitude, no smallest unit of extension: if a continuum is composed of indivisibles, it is most certainly not composed of *extended* indivisibles.[43] In the final analysis More's theory, like all theories of infinitesimals, is incoherent.[44]

In Newton and Leibniz, however, the concept of the infinitesimal is gradually phased out, superseded by that of a *limit*. His method, says Newton in *Principia*, does not involve the use of indivisibles *à la* Cavalieri:

> For demonstrations are shorter by the method of indivisibles; but because the hypothesis of indivisibles seems somewhat harsh, and therefore that method is reckoned less geometrical, I chose rather to reduce the demonstrations of the following Propositions to the first and last sums and ratios of nascent and evanescent quantities, that is, to the limits of those sums and ratios, . . . For hereby the same thing is performed as by the method of indivisibles; and now those principles being demonstrated, we may use them with greater safety. Therefore if hereafter I should happen to consider quantities as made up of little particles . . . I would not be understood to mean indivisibles, but

---

[42] See Chapter 1, p. 40.

[43] Were it not for the fact that the young Newton favoured More's theory, I might never have mentioned it. See Westfall, 4, p. 328.

[44] Cf. of course Berkeley's critique of the 'mysteries' of the mathematicians.

evanescent divisible quantities; not the sums and ratios of determinate parts, but always the limits of sums and ratios.[45]

Rigour demands, Newton feels, that we do not consider a line or a time-interval as made up of genuine infinitesimals ds or dt. The value of ds/dt is NOT to be conceived, like $\Delta s/\Delta t$, as the velocity involved in traversing a very small distance in a very short time, but as the ultimate or *limiting* value of the ratio $\Delta s/\Delta t$ as $\Delta t \longrightarrow 0$. This concept makes perfectly good sense, Newton insists, even if infinitesimals are purely fictitious:[46] although we may if we choose continue to use the language of indivisibles, clarity and rigour alike demand that we be prepared to explain away such loose talk in terms of evanescent quantities and limits.

A similar position is attained by Leibniz. Strictly speaking, he says, every magnitude is divisible *ad infinitum*: there are no extended indivisibles. A concept such as that of *conatus* or instantaneous velocity (ds/dt) does not, he insists, depend for its intelligibility on the existence of indivisible minima of space and time.[47] All we need is the idea of taking ever-smaller (but still divisible) quanta Δs and Δt until the value of $\Delta s/\Delta t$ approaches some *limiting* value – the limit of $\Delta s/\Delta t$ as $t \longrightarrow 0$ is the value of the *conatus*. The language of infinitesimals is a mere *façon de parler*, eliminable in favour of a more rigorous account:

> . . . the mathematicians' demand for rigour in their demonstrations will be satisfied if we assume, instead of infinitely small sizes, sizes as small as are needed to show that the error is less than that which any opponent can assign, and consequently that no error can be assigned at all. So, even if the exact infinitesimals which end the decreasing series of assigned sizes were like imaginary roots, this would not at all injure the infinitesimal calculus . . .[48]

The rigorous concept of a limit, although it solved the mathematicians' problems, and helped to put the calculus on

---

[45]   Newton, 1, p. 38. See also Evans, p. 555.

[46]   Newton, 1, p. 39.

[47]   See Costabel, p. 20, and 'The Theory of Abstract Motion' in Leibniz, p. 139 ff.

[48]   From Leibniz's 'Reply to the Thoughts on the System of Pre-established Harmony' (1702) in Leibniz, p. 584.

a sound foundation, did little to solve the philosophers' problems about the ultimate constitution of the continuum. One attitude which gradually became widespread was a sort of agnostic scepticism. *All* theories of the continuum, say men such as Boyle and Locke, are riddled with paradoxes: human minds are simply not competent to penetrate such mysteries. In spite of the time and energy expended on this question, says Boyle in his 'Excellency of Theology',

> yet still . . . the difficulties, till removed, will spread a thick night over the notion of body in general.[49]

Such absurdities and contradictions afflict both sides of the debate, he says, so that some modern thinkers have abandoned the whole problem as *insoluble*. In the 'Discourse of Things Above Reason' he spells out some of these paradoxes more explicitly. Does a line four feet long, he asks, consist of as many ultimate parts as one two feet in length? If so, how could they differ in length? Yet if not, *either* the shorter line is not infinitely divisible, *or* the longer consists of more than an infinity of parts, which is absurd.[50] Whichever way one turns, one faces paradoxes.

A similar attitude prevails in Locke's *Essay*. We cannot, he argues, claim to have any real acquaintance with the essence of material substance, since sooner or later we run up against the problems raised by infinite divisibility, which, 'whether we grant or deny it', yields 'consequences impossible to be explicated or made in our apprehensions clear.[51] Even so fundamental a feature of matter as its extension is, in the final analysis, incomprehensible to us.

One man who drew a radical and quite idiosyncratic conclusion from all these difficulties was Leibniz. In his 'Theory of Abstract Motion' (1671) he asserts plainly that 'there are actually parts in a continuum' and 'these are actually infinite'.[52] With Aristotle and the geometers, he accepts infinite divisibility; against Aristotle, he insists that this involves actually infinite dividedness. From this it follows that every continuous magnitude must contain an infinite number of parts. As Leibniz states in his 'First Truths':

> *There are no atoms*; indeed, there is no body so small that it is not actually subdivided . . . Hence it follows that *every*

---

49   Boyle, *Works*, vol. 4, p. 43.

50   *Ibid.*, p. 409.

51   Locke, *Essay*, II, 23, xxxi, vol. 1, p. 261.

52   Leibniz, p. 139.

small part of the universe contains a world with an infinite number of creatures.[53]

Leibniz is, he insists,

> so much in favour of the actual infinite, that, instead of admitting that nature abhors it, . . . I hold that nature affects it everywhere, in order the better to mark the perfections of its author. So I believe that there is no part of matter which is not, I do not say divisible, but actually divided; and consequently the least particle must be regarded as a world full of an infinity of different creatures.[54]

But, if every continuous magnitude consists thus of an infinity of distinct parts, there can be nothing extended and yet at the same time *simple*. Space, Time and Matter offer no 'principle of unity'. There can be no (mathematical) atoms, no minimal – and hence indivisible – units of extension. Atomism, says Leibniz, is 'contrary to reason'.[55]

What, we might ask, of mathematical points? They are unextended and, after their fashion, truly simple and indivisible. But, Leibniz insists,

> a point is not a certain part of matter, nor would an infinite number of points collected together make an extension.[56]

Here we have all the materials necessary to generate a contradiction. Bodies are extended. Extension involves repetition and hence plurality. But plurality, Leibniz claims, presupposes *units* or *simples* of some kind: this is one of the fundamental premisses of his whole metaphysical system. There must, he asserts,

> be in things indivisible unities, because otherwise there will be in things no true unity, and no reality not borrowed. Which is absurd. For where there is no true unity, there is no true multiplicity. And where there is no reality not borrowed, there will never be any reality.[57]

Every continuum must therefore have an ultimate constitution of simples, parts themselves without further parts. But

---

53   *Ibid.*, pp. 269–270. See also Tymieniecka, pp. 45–46; H. G. Alexander, Ed., p. 43.

54   Quoted from Russell, p. 109.

55   *Ibid.*, p. 104.

56   *Ibid.*, p. 242.

57   *Ibid.*, p. 242. See also Leibniz, p. 643 (*Monadology*, 2).

these units cannot be extended (atoms), nor can the be extensionless (points), and there is no *tertium quid*, no intermediate between extension and non-extension. If all the inferences are sound, it follows that the concept of a continuous magnitude is self-contradictory: it must, and yet cannot, have an ultimate constitution of simples. The continuum is unreal.[58]

Leibniz explicitly assents to this conclusion. Bodies, he says, are no more than *phenomena bene fundata*, appearances of the aggregates of individual substances (monads) that 'constitute' them. Strictly speaking, of course,

> matter is not composed of constitutive unities, but results from them, for matter or extended mass is nothing but a phenomenon founded in things . . . In fact, substantial unities are not parts, but foundations of phenomena.[59]

Matter, Space and Time are *entia rationis*, 'not things but well-founded modes of our consideration.'[60] In Leibniz' philosophy, as Russell says, the discrete is metaphysically real; the continuous, merely phenomenal.[61] The individual substances or monads that provide the real ground for extension must not be thought of as *parts* of extended substances, since they are radically *heterogeneous* with that which they (appear to) constitute.[62] If these monads were of the same nature as the continuous magnitudes which (appear to) result from their aggregation, contradictions would arise. But, if extension in all its varieties is relegated to the status of a phenomenon – an appearance *to* a collection of non-extended realities – the paradoxes fail to bite: on this theory, Leibniz claims:

> there is no extension or composition of the continuum, and all difficulties about points vanish.[63]

Indeed,

> from the very fact that the mathematical body cannot be resolved into its first constituents, we may certainly infer that it is not real, but something mental.[64]

---

[58] See Nason, p. 447; Tymieniecka, pp. 45–46.

[59] Quoted from Russell, p. 243.

[60] From the 'Firth Truths' (Leibniz, p. 270).

[61] Russell, p. 111 ff.

[62] See Costabel, pp. 16–17.

[63] Quoted from Russell, p. 255.

[64] *Ibid*., p. 243.

All extension is continuous, therefore self-contradictory in its very concept, therefore unreal: 'continuous quantity,' says Leibniz, 'is something ideal'.[65] To think of an extended magnitude as something real is to 'entangle ourselves in the labyrinth of the continuum, and in contradictions that cannot be explained'.[66] What more can – or need – be said?

B. Mathematical and Physical Divisibility

What physical implications, if any, can be derived from the mathematicians' conception of the continuum as infinitely divisible? This question, as we shall see, provoked sharply contrasting responses from those two pioneers of the Mechanical Philosophy, Descartes and Gassendi: their respective answers reveal fundamental differences in metaphysics and epistemology.

The essence of matter, says Descartes, is three-dimensional geometrical extension:

> The nature of body consists not in weight, nor in hardness, nor colour and so on, but in extension alone.[67]

Extension, then, *exhausts* the essence of corporeality: all other qualities can be abstracted from a body without affecting its essential nature. We grasp this geometrical essence, says Descartes, *a priori*, by means of the intellect rather than the senses:[68] we possess a clear and distinct *innate* idea of pure extension. Thus 'we know magnitude, figure, etc., quite differently from colour, pain, etc.':[69] we have clear and distinct *ideas* of the former; the latter are mere *sensations*,[70] confused and indistinct.

From this epistemological distinction there follow, for the Cartesian, momentous ontological consequences. Given that God is no deceiver, Descartes argues, our clear and distinct ideas must *mirror* the structure of Reality. 'I clearly and distinctly perceive,' he says in his *Meditations*, that extension, figure, etc., 'exist in objects':[71] no such guarantee is possible

---

65 *Ibid.*, p. 246.

66 From a letter to De Volder in Leibniz, p. 539.

67 *Principles*, 2, 4. From Descartes, 1, Vol. 1, p. 255.

68 Cf. the example from the wax from the *Meditations* (Descartes, 1, Vol. 1, pp. 154–155).

69 *Principles*, 1, 69. Descartes, 1, Vol. 1, p. 248.

70 For the distinction between (representative) ideas and (non-representative) sensations, see Watson, p. 32.

71 Descartes, 1, Vol. 1, p. 164.

for colour or pain. Our ideas of extension, figure, and motion *resemble* corresponding simple natures in things.[72] Nature is therefore 'transparent' to geometrical reasoning:[73] we have a theological guarantee of the intelligibility of the physical universe. Now that I know God, Descartes concludes his fifth Meditation,

> I have the means of acquiring the perfect knowledge of an infinitude of things, not only of those which relate to God himself and other intellectual matters, but also those which pertain to corporeal nature in so far as it is the object of pure mathematics.[74]

Valid deductions from, e.g., our idea of geometrical extension will yield *physical* truths, theorems directly applicable to corporeal Nature.[75] No such assurance holds for our ideas of hard, cold, red, sweet, etc. These are confused and 'adventitious' ideas of sense: their indistinctness threatens their representative status. It is our folly and not God's deceit if we assume that they mirror the structure of Reality.[76]

But our idea of geometrical extension includes that of divisibility: that all bodies are divisible is therefore for Descartes a truth derivable *a priori* from the idea of a body as an extended magnitude.[77] Body is, he affirms, 'by nature always divisible'.[78] The distinction between mathematical and physical divisibility does not arise: geometrical truths are also, and necessarily, physical truths. Infinite mathematical divisibility thus involves infinite physical divisibility. Divisibility is contained in the essence of a body, just as having internal angles that sum to 180° is part of the essence of a triangle. That 'all bodies are divisible' is, says Descartes, as certain as any theorem in Euclid.[79]

From this it follows that the concept of an *indivisible body* is self-contradictory:

> We . . . know that there cannot be any atoms or parts of matter which are indivisible of their own nature .. For,

[72] This is the step to which the sceptic Foucher takes exception. See Watson, pp. 24–27.

[73] See Kemp Smith, p. 324.

[74] Descartes, 1, vol. 1, p. 185.

[75] See Descartes' letter to Mersenne of 16 June 1641 in Kenny, Ed., p. 104.

[76] See *Meditation* Six in Descartes, 1, Vol. 1, p. 192–194.

[77] See Hoenen, p. 356.

[78] From *Meditation* Six (Descartes, 1, vol. 1, p. 196).

[79] Descartes, 1, Vol. 2, p. 54.

however small the parts are supposed to be, yet because they are necessarily extended we are always able in thought to divide any one of them into two or more parts; and thus we know that they are divisible. For there is nothing which we can divide in thought which we do not thereby recognise to be divisible.[80]

Even if a particle of matter were to exist which was so small that no natural force could divide it, God

could not deprive Himself of His power of division, because it is absolutely impossible that He should lessen His own omnipotence.[81]

Every extended magnitude is divisible in thought, in *idea*. But our (clear and distinct) ideas mirror Reality: it therefore follows that every corpuscle of matter, however small, consists of *distinct* parts separable by sufficiently great force. This is how Descartes puts the point in a letter to Gibieuf:

. . . we can say that it implies contradiction that there are atoms (or parts of matter which have extension and yet are indivisible) because one cannot have the idea of an extended thing and not have the idea of half of it or a third of it, and consequently not conceive it as divisible into two or three. For, from the sole fact that I consider the two halves of a part of matter, however small it may be, as two complete substances, and the ideas of them have not been rendered inadequate by an abstraction of the mind, I conclude certainly that they are really divisible. and if one says to me that, notwithstanding that I can conceive them, I do not know for all that whether God has not united or joined them together by a connection so tight that they are entirely inseparable . . . I shall reply that about any connection by which God can have joined them, I am assured that He can likewise disjoin them, so that absolutely speaking, I am right in calling them divisible, since He has given me the faculty of conceiving them as such.[82]

If one thinks of an atom, one thinks of a minute body of some particular size and shape. But this idea involves that of

---

[80]   *Principles*, 2, 20; Descartes, 1, Vol. 1, p. 264.

[81]   *Ibid.*, p. 264.

[82]   From Kenny, Ed., pp. 124–125. See also Weinberg, 2, p. 79.

divisibility: any extended magnitude, however small, consists of *distinct* parts. One can think of half an atom as a substance in its own right: it is therefore certain that such an entity could possess separate existence, if only by divine power. Snow expresses the point clearly and succinctly:

> Conceptually, then, extension implies divisibility. Now to Descartes, that which is conceptually distinct must be real. The modes of extended substance, therefore, being distinctly divisible in thought, must of necessity be divisible in nature.[83]

It is this inference to which Gassendi takes exception. Granted, he says, that we have an *idea* of mathematical extension as infinitely divisible, why should it follow that material Reality corresponds to the idea? However clear and distinct it may be, this does not ensure it an objective reference, an application to the concrete material world. Mathematical extension, he claims, exists only in the realm of ideas, i.e. only in the mind of the geometer. Gassendi is thus a *fictionalist* about geometrical theorems: matter, he says, is not infinitely divisible. Thus Snow again:

> . . . To say that material substance is infinitely divisible because mathematical magnitude is, is simply to imagine a fictitious magnitude which logically admits divisibility, but to identify this logical magnitude with physical matter is pure folly.[84]

A similar spirit prevails in Gassendi's disciple Charleton. The natural philosopher, he claims in his *Physiologia*, does not concern himself with the abstractions of mathematics. The geometer may if he wishes suppose infinite divisibility, and hence an infinity of parts in every continuous magnitude,

> not that he doth, or can really understand it so, but that many convenient conclusions, and no considerable incongruities, follow from the conclusions thereof.[85]

Are the propositions of the geometers true or false? If true, one might claim, they must apply to the physical world; if false, they cannot yield certain and necessary conclusions. Charleton's answer is clear:

[83]   Snow, p. 38.

[84]   *Ibid.*, p. 39.

[85]   Charleton, p. 95.

> Our *expedient* is, that, though we should concede those suppositions to be false, yet may they afford true and necessary conclusions: every novice in logick well knowing how to extract undeniable conclusions out of the most false propositions.[86]

As in astronomy, where fictitious hypotheses may facilitate the discovery of valuable truths, so in mathematics: the geometrical continuum is a fiction, an abstract idea to which no physical reality corresponds, yet one which may be fruitfully invoked by the physicist. This fictionalist philosopher does not, Charleton insists, 'adnihilate the use of geometry':[87] geometrical theorems, even if strictly false, retain considerable utility. Nevertheless, says Charleton, the mathematicians

> well know, that the physiologist is in the right, when he admits no infinity, but only an innumerability of parts in natural continuum.[88]

What, then, is the *real* constitution of the material continuum?[89] It cannot be made of unextended mathematical points: even an infinity of these will not give rise to a magnitude. Point added to point does not produce any increase in quantity. Nor is a continuous magnitude 'a simple entity before division, indistinct', as the Aristotelians would have it: the parts into which a body can be divided have actual existence in it. It remains, therefore, that the material continuum is made up of *atomic magnitudes*.

These atomic magnitudes cannot, however, be simply identified with atoms: the atoms (= physically individual bodies) of Gassendi come in a variety of shapes and sizes, and must therefore be thought of as made up of *parts*. The number of these parts must, however, be finite: an infinitely divisible body must consists of an infinity of (aliquot) parts, and must therefore be infinite in extent. If, moreover, bodies are divisible *ad infinitum*,

> there must be as many parts in a grain of mustard seed as in the whole terrestrial globe.[90]

---

[86]   *Ibid.*, p. 96.

[87]   *Ibid.*, pp. 96–97.

[88]   *Ibid.*, p. 97.

[89]   *Ibid.*, pp. 107–109.

[90]   *Ibid.*, p. 91. For the same argument in Gassendi, see Spink, p. 92. The argument derives of course from Lucretius (see chapter 1, p. 36).

Who, Charleton demands, could seriously entertain the conceit that each grain of sand contains 'a thousand millions of myriads' of parts, each divisible again as many times into parts still further divisible, and so on? Undeluded reason and plain common sense cry out, he feels, against such fantasies.

If one accepts physically indivisible but extended atoms, of a variety of sizes and shapes, but denies the intelligibility of infinite divisibility, one will be left with something like the Epicurean theory of *minimal parts*. This does seem to have been the eventual outcome of the Gassendist position. Thus Gassendi claims that Zeno's paradoxes tell *only* against the infinite divisibility of Space and time, and do not trouble the indivisibilist position,[91] and goes on to outline a discontinuous, cinematic analysis of motion along classical Atomist lines. Moving bodies always move either one space-minimum per time-minimum, or rest: differences in velocity can be explained in terms of different numbers of instants of rest.[92]

The restorers of Epicureanism must therefore postulate, with Epicurus, two different types of minima. There are atoms, physically unsplittable in virtue of their hardness and solidity, but these cannot *also* serve the rôle of minimal magnitudes. Since the atoms vary in size and shape, they must be thought of as compounded of serried ranks of minimal parts, the *smallest physically real entities*. (Mathematical points are, for the Epicurean, merely abstractions, not concrete physical realities.) These minima are so minute that an 'innumerable' (but *not* infinite) number of them make up each invisibly small atom: the theorems of geometry are therefore, though strictly false, at least approximately true of the physical world.

Descartes was not impressed. Some people, he writes, say that

mathematical extension, which I set as the principle of my physics, is nothing but my own thought and that it does not have and cannot have any existence outside my mind, since it is only an abstraction that I make from a physical body; and consequently my entire physics cannot be anything but imaginary or fictitious like all pure mathematics; and that

---

91   Brett, p. 60.

92   *Ibid.* 61. Leibniz argues against this position in his 'Theory of Abstract Motion' (1671), in which he opposes his own notion of continuity to Gassendi's cinematic approach (Leibniz, p. 140).

the real physics about the things that God created must have a real, solid matter, and not an imaginary one.[93]

The challenge is to Descartes' inference from his clear and distinct ideas to the existence of corresponding realities: the ontological status and existential implications of mathematical truth are at stake.[94] On Gassendi's view, geometrical theorems are not true *of* anything; hence cannot, strictly and literally, be called true at all. On such a view, Descartes retorts angrily,

> there is nothing that can be accepted as true except what we cannot understand, conceive, or imagine in the least, that is to say that we must close the door completely on reason and be satisfied with being monkeys or parrots, no longer men . . . [95]

If we cannot built our doctrines on ideas intelligible to us, are we not lost without hope in total darkness?

Let us now quit the difficult topic of mathematical divisibility, and move on to the claim of Gassendi and his followers that matter consists of physically indivisible atoms. These atoms are not, Gassendi insists, mathematical points: to analyse a body into points is to go beyond the point of no return – from points one could not *reconstruct* a body.[96] The atom, says Charleton, is not *partless*, not 'destitute of all magnitude, or no more than a mere mathematical point';[97] though incredibly minute, atoms 'are yet corporeal and possess a determinate extension'.[98] Since they come in a variety of sizes and shapes, they must consist of distinct parts separable at least in thought. Each atom is, however, physically indivisible because, as Charleton puts it, 'there is no force in nature that can divide and resolve it into those parts'.[99]

How do Gassendi and Charleton argue for this thesis? Infinite physical divisibility, Gassendi claims, would violate

---

[93]   From Gassendi, p. 263.

[94]   See Snow, p. 62.

[95]   From Gassendi, p. 263. See also Snow, p. 40.

[96]   See Brett, p. 53.

[97]   Charleton, p. 86.

[98]   *Ibid.*, p. 113. There are, Charleton estimates, 7×10^{17} atoms in a single grain of incense.

[99]   *Ibid.*, p. 441.

the principle 'from nothing comes nothing, and nothing cannot be converted into something'[100] – an argument drawn of course directly from Epicurus.[101] Physical divisibility presupposes inner void: infinite physical divisibility would involve the complete dissolution of a body into vacuum or nothingness. The atom, being a perfectly solid fragment of matter*, without inner void, is *unsplittable*. Atoms, says Charleton, function as the 'Catholique Material Principle' of nature, and must therefore be *incorruptible*: this is, he says, doubted by none, except those who confuse mathematical with physical divisibility and are thereby led to the 'dangerous absurdity' that 'that infinite division of a real continuum is possible'.[102] Only something so solid as to be indivisible by any natural agent can be incorruptible, Charleton argues, and hence fit for the rôle of the universal first matter.

The atom of Gassendi and Charleton is therefore extended in three dimensions, absolutely solid, and physically indivisible by any natural force, though divisible in thought and by divine power. (It is not, however, *infinitely* divisible: not even God can perform something intrinsically self-contradictory.) It is also perfectly *hard* and *rigid*: with hard atoms, unaffected by the buffetings of other bodies, one can explain the constancy of Nature, whereas if the atoms were soft one should expect the natures of things to alter as the atoms were eroded away.[103] (This argument hails from Lucretius, and recurs in Newton's *Opticks*.[104]) The constancy of Nature's operations, says Charleton, entails

> that her materials are certain, constant, and inobnoxious [i.e. not liable] to dissolution, and consequently to mutation: and such are Atoms presumed to be.[105]

With hard atoms and void, moreover, we can explain the formation of soft (compound) bodies; if the *primordia* themselves were soft, one could not account for the existence

---

[100]   See Lindsay, p. 240; Charleton, p. 87.

[101]   See Appendix 1 for an elaboration of this argument.

[102]   Charleton, p. 85. For Gassendi on the rôle of atoms as prime matter, see Syntagma, *Physics*, Section 1, Book 3 (Gassendi, p. 398 ff.).

[103]   Descartes explicitly allows a 'weathering' of the particles of his universal matter: one might therefore expect, in a Cartesian universe, the natures of chemical substances to alter over time.

[104]   See Appendix 1.

[105]   Charleton, p. 89.

in Nature of hard bodies.[106] Gassendi and Charleton thus follow very closely the position of Epicurus and Lucretius: even their arguments are borrowed directly from the texts of those ancient authors. (Gassendi did not, of course, claim to be a thinker of great originality: his physics is, largely, no more than a detailed reconstruction of Epicureanism.)

It soon became apparent that a compromise of sorts was possible between Descartes and Gassendi. One could grant, with the Cartesians, the infinite divisibility of the material continuum as such, but deny that any natural power can actualise this potency. Matter would, on such a view, consist of particles infinitely divisible in thought and by divine power, yet indivisible by any natural agency. Indeed, the physical theory of Atomism will be served by *undivided* corpuscles: the stronger, modal, claim is superfluous except as guarantor and explanans. In Thomas Hobbes, we find such a compromise view in operation; for the understanding, he says in his treatise, *De Corpore*,

> quantity is divisible into divisibles perpetually. And, therefore, if a man could do as much with his hands as he can with his understanding, he would be able to take from any given magnitude a part which should be less than any other magnitude given. But the Omnipotent Creator of the world can actually from a part of anything take another part, as far as we by our understanding can conceive the same to be divisible. Wherefore there is no impossible smallness of bodies.[107]

Matter is divisible *ad infinitum* (a) in thought and (b) by divine power. There are, however, hard 'atoms' (i.e. physically undivided corpuscles): this is one of Hobbes' 'Six Suppositions for saving the Phenomena of Nature'.[108] (In his early works, Hobbes had been a staunch supporter of the Atomic Theory.[109]) Hobbes can therefore accommodate much of the physics of Atomism without any commitment to absolute indivisibles.

Within the Cartesian tradition, a number of thinkers show leanings towards such a compromise view.[110] Thus Cordemoy

---

106   Once again, this argument is Lucretian in origin (see Appendix 1).

107   Hobbes, *Works*, Vol. 1, p. 446.

108   *Ibid.*, p. 416.

109   See Kargon, 4, p. 54.

110   See M. Boas, 4, pp. 455–457.

and Huygens both admitted infinite divisibility in thought and by divine power, but postulated hard and physically unsplittable atoms, indivisible by any natural (i.e. finite) force. Hooke too, though much influenced by Cartesianism, is not convinced of the infinite divisibility of matter. Certainly, he says in his *Micrographia*, 'the quantity or extension of any body may be divisible *in infinitum*, though perhaps not the matter'.[111]

If we turn now to the writings of Robert Boyle we shall discover a sustained attempt to develop a 'corpuscular philosophy' *neutral* between Atomism and Cartesianism: Boyle constantly stresses the similarities between the two philosophies, and tries to play down their differences.[112] His remarks on indivisibles are therefore, as we might expect, guarded and noncommital. There are, he claims, 'great store' of invisibly small corpuscles, each of which is so *solid* that

> though it be mentally, and by divine Omnipotence divisible, yet by reason of its smallness and solidity nature doth scare ever actually divide it; and these may in this sense be called *minima* or *prima naturalia*.[113]

This is a very weak claim. Boyle does not even say that these 'minima' are *always undivided*, far less that they are *indivisible*. Although they 'may almost always escape dissolution' in virtue of their minuteness and the 'strict contact' of their parts, this does not make them genuine indivisibles:

> For to say that it is the nature of every such corpuscle to be indivisible is but to give me cause to demand how that appears; for so important an assertion needs more than a bare affirmation for proof.[114]

No adequate physical explanation has been given, Boyle alleges, for the supposedly unbreakable cohesion of the parts of the atom. Suppose, for example, that three 'atoms', A, B and C, were to come together thus:

---

111 Gunther, Ed., vol. 8, p. 2. For a passage more sympathetic to infinite divisibility, see M. Boas, 4, p. 454.

112 See especially the beginning of the 'Origin of Forms and Qualities' in Boyle's *Works*, vol. 3, p. 7, where Boyle claims to write 'rather for the corpuscularians in general than any part of them' – he attempts a studied neutrality on such potentially divisive issues as indivisibles and vacuum.

113 Boyle, *Works*, Vol. 3, p. 29.

114 *Ibid.*, Vol. 1, p. 412.

A	B	C

Why, Boyle enquires, should not the resulting compound body be *bisected*, i.e. cut into *two* equal parts? If it is divisible between A and B, and between B and C, why should it not be divisible in the middle of B?

> To prove, that the cohesion of the middlemost of the three . . . small dyes with the other two . . . is not so strong as that of the parts of the middlemost corpuscle, notwithstanding that the contact between the two adjoining bodies is supposed to be full . . . to prove this, I say, there must be assigned some better cause of the cohesion of the matter in one part of the proposed body than in the other.[115]

Without an account of the allegedly absolute or perfect cohesion of the parts of the atom, the postulation of physical indivisibles must be greeted with scepticism. (The mere contact, without intervening vacuum, of material parts, will not ensure cohesion; otherwise, the compound body ABC would itself be indivisible, and all bodies would agglomerate into a single super-atom – a criticism first voiced by Aristotle.[116]) Boyle is therefore dubious of the existence of indivisibles: the strongest Atomist thesis we can attribute to him is that of (normally) undivided corpuscles.

Boyle's great contemporary, Newton, was far more sympathetic to the Atomic Theory. In the early notebook entitled *Quaestiones quaedam Philosphiae* (c. 1664–1666) the influence of Charleton is evident, and Atomism is consistently preferred to Cartesianism – an attitude which was to prevail throughout Newton's career.[117] All observable bodies, he asserts in *Principia*, are physically divisible. Are we therefore entitled to assume – by an elementary generalisation – that all bodies whatsoever are so? Newton answers as follows:

> . . . that the divided but contiguous particles of bodies may be separated from one another is a matter of observation; and, in the particles that remain undivided, our minds are able to distinguish yet less parts, as is mathematically

---

115   *Ibid.*, pp. 412–413.
116   See Appendix 1.
117   See Westfall, 2, pp. 172–174.

demonstrated. But whether the parts so distinguished, and not yet divided, may, by the powers of Nature, be actually divided and separated from one another, we cannot certainly determine.[118]

Since every extended magnitude consists of smaller parts, all bodies are divisible at least in thought – Newton is too good a mathematician even to consider any other possibility. There is, however, he claims, a disanalogy between dividing an observable body (which involves the separation of already distinct particles) and dividing one of these single solid corpuscles. The latter process may, though possible for God, be unattainable by any natural (i.e. finite) power. For all we know, God may have created matter in the form of perfectly solid and physically indivisible atoms. Indeed, there must be such absolutely solid particles, if matter is to exist at all. If, says Clarke in the correspondence with Leibniz,

> there be no such perfectly solid atoms, then there is no matter at all in the universe. For, the further the division and subdivision of the parts of any body is carried, before you arrive at parts perfectly solid and without pores; the greater is the proportion of the pores to solid matter in that body. If, therefore, carrying on the division *in infinitum*, you never arrive at parts perfectly solid and without pores; it will follow that all bodies consist of pores only, without any matter at all: which is a manifest absurdity.[119]

Now Clarke's letters, as Princess Caroline wrote to Leibniz, were 'not written without the advice of Chev. Newton'[120] indeed, draft copies exist in Newton's own hand. It is therefore safe to assume Newton's approval of this classical Epicurean argument for the existence of perfectly solid particles of matter*. Hence the following celebrated passage from the *Opticks*:

> It seems probable to me, that God in the Beginning form'd Matter in solid, massy, hard, impenetrable, moveable Particles, of such Sizes and Figures, and with such other Properties, and in such Proportion to Space, as most conduced to the End for which He formed them.[121]

---

[118]   Newton, 1, p. 399.

[119]   From H. G. Alexander, Ed., p. 54.

[120]   *Ibid.*, p. 193.

[121]   Newton, 2, p. 400.

These 'primitive particles' are, Newton continues, perfectly solid plena of homogeneous matter*,

> incomparably harder than any porous Bodies compounded of them; even so very hard, as never to wear or break into pieces; no ordinary power being able to divide what God Himself made one in the first Creation.[122]

What God has joined together, no lesser power can part asunder. The constancy of Nature is certain evidence of this:

> While the Particles continue entire, they may compose Bodies of one and the same Nature and Texture in all Ages: But should they wear away, or break in pieces, the Nature of Things depending on them, would be changed. Water and Earth, composed of old worn Particles and Fragments of Particles, would not be the same Nature and Texture now, with Water and Earth composed of entire Particles in the Beginning.[123]

If atoms could be ground down or eroded away, as the Cartesians believed, Nature would preserve no constancy and regularity through successive generations: as matter was slowly ground ever finer, the natures of all corporeal things would gradually alter – there would be, for example, no such stuff as water, common to us *and* to the ancients. (This argument is, as we have seen, the common property of Atomists from Lucretius to Charleton.) Therefore, Newton continues,

> that Nature may be lasting, the Changes of corporeal Things are to be placed only in the various Separations and new Associations and Motions of these permanent Particles; compound Bodies being apt to break, not in the midst of solid Particles, but where those Particles are laid together, and only touch in a few points.[124]

We find here a hint of a solution to the pressing problem of atomic *cohesion*. Atoms, says Newton, touch one another only at a few 'points' – i.e., presumably, extensionless mathematical points. This enables Newton to evade an objection fatal to Epicurean Atomism. When atoms combine to form compounds, he says in a draft passage intended for the *Principia*, they

---

122   *Ibid.*, p. 400.

123   *Ibid.*, p. 400.

124   *Ibid.*, p. 400.

do not conjoin into solidity but are in mutual contact only at some mathematical point, the remaining spaces being left in each case between them vacant.[125]

The distinction between mathematical points and Epicurean 'minimal parts' is all-important. Two Epicurean atoms could touch, let us suppose, only at a single minimal part – this must be insufficient to cause them to cohere, thus:

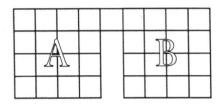

Here atoms A and B touch only at a single minimal part. If the Epicurean theory is to evade Aristotle's objection,[126] they must remain distinct atoms, separable by physical blows. But, if one minimal link can be severed by a finite force, so can any (finite) number of them – the atom itself must therefore be divisible by a sufficiently powerful finite force.

At this point it becomes advantageous to speak of mathematical points rather than of minimal parts. The crucial difference is of course that between the finite and the infinite: a finite number of minimal parts constitute an atom; a finite number of points could not. On the Epicurean theory, no *sharp* distinction can be drawn between the supposedly absolute cohesion of the atom and the cohesiveness of compounds – the difference would be one of degree only. If, however, atoms make contact only at mathematical points, this difficulty disappears. If a finite force is needed to separate two atoms that touch at a mere point, an infinite force (the sum or integral of an infinite number of minute finite forces) would be needed to 'split the atom'. The conceptual distinction between atoms and compounds is restored, and the indivisibility of the atom by any natural (i.e. finite) force accounted for. Only the omnipotent deity can overcome the infinity of forces (each individually minute) which hold each and every atom together. *Why* atoms can only touch at mathematical points, Newton does not tell us: perhaps, we

---

125   Quoted from McGuire, 3, p. 8.

126   See Appendix 1; Konstan, 2.

might speculate, because they are perfect spheres? If two atoms could make contact surface-to-surface they would – as Aristotle alleged – cohere tightly: it is only because this does not happen that bodies do not agglutinate into a single mass.

The distinction between mathematical points and Epicurean natural minima, coupled with an assumption about atomic shapes, will therefore enable Newton to reinstate the classical Atomist argument from lack of inner void to physical indivisibility. The atom lacks inner vacuum: its two halves therefore touch at an infinity of points, and are thus (naturally) inseparable. In a compound, however, the constituent atoms touch only at a finite number of points (between which is vacuum) – the compound is therefore divisible into atoms by a finite force. Compound bodies only appear to be continuous plena: in reality, every compound body contains internal vacuum.

The most acute and penetrating of the many seventeenth century critics of the Atomic theory was undoubtedly Leibniz. Of his many objections to Atomism, I shall concentrate primarily on those belonging properly to the realm of physics, passing over a number of arguments, not without interest, of a metaphysical nature.[127] In his youth, Leibniz writes, 'I also gave in to the notion of vacuum and atoms; but reason brought me into the right way'.[128] The Atomic Theory, he suggests, gives a world-picture that satisfies the imagination, but will not stand prolonged intellectual scrutiny. [129] Above all, there cannot be *material simples*: the least corpuscle, he writes to Clarke, 'is actually subdivided *in infinitum*, and contains a world of other creatures'.[130] How, he asks, could all these hang together to produce a body that is so perfectly and absolutely united as to be immune to any process of division? The absolute cohesiveness of the parts of a Democritean atom is, Leibniz insists in a letter to Hartsoeker of 1711, something inexplicable in natural terms, and hence *miraculous*. Cohesion in general, he says, may be accounted for in terms either of the inner *conatus* (endeavour) of the parts of the body towards one another, or of the external pressure of an ambient fluid. *Neither* of these causes,

---

127   Democritean Atomism, Leibniz claims, violates two of the great principles of metaphysics, the Principle of the Identity of Indiscernibles and the Principle of Plenitude.

128   H. G. Alexander, Ed., p. 43.

129   See Leibniz, p. 454 (from the 'New System' of 1695).

130   From H. G. Alexander, Ed., p. 43.

however, would normally produce an infinite and unshake-able cohesiveness:

> If you allege only the will of God for it, you have recourse to a miracle, and even to a perpetual miracle; for the will of God works through a miracle whenever we are not able to account for that will and its effects from the nature of the objects.[131]

Leibniz is amazed to hear that Huygens favours the Atomic Theory: the absolute 'infrangibility' of the atom is, he writes to his old teacher in 1692, 'a kind of perpetual miracle',[132] totally inexplicable in natural terms. 'I don't,' he writes to Clarke,

> admit in matter, parts perfectly solid, or that are the same throughout, without any variety or particular motion in their parts, as the pretended atoms are imagined to be.[133]

Leibniz thus objects to the supposedly perfect *rigidity* of the atom. Reason demands, he writes to Huygens, 'that there be no atoms or bodies of an infinite firmness at all'.[134] His critique of hard or rigid bodies is one of Leibniz's most significant contributions to physics and, as we shall see, contains his most damaging objections to the classical Atomic theory. In particular, Leibniz writes, hard bodies 'cannot obey the laws of motion':[135] but since those laws hold *universally* there can be no (perfectly) hard bodies in Nature.[136]

In what way might rigid bodies be thought to violate the laws of Nature? In the first place, says Leibniz, they contravene the principle of the conversation of '*vis viva*' (kinetic energy).[137] In a Democritean universe, all things would 'run down', and require *either* non-mechanical ('occult') 'active principles' *or* miraculous divine intervention to keep them running. In particular, Leibniz writes to Huygens,

> the force of two equal bodies which collide directly and with an equal velocity must be lost, since it would seem that it is only elasticity which makes bodies rebound.[138]

---

131 Quoted from Koyré, 4, p. 141.

132 Leibniz, p. 414.

133 From H. G. Alexander, Ed., p. 61.

134 Leibniz, p. 415.

135 *Ibid.*, p. 416.

136 See Snow, p. 44.

137 For Leibniz on the conservation of *vis viva*, see Russell, pp. 89–90.

138 Leibniz, p. 416.

The crucial premiss – common to Leibniz and Newton – is that *deformability* is necessary to the restoration of motion on impact. Perfectly rigid atoms cannot undergo such deformation (the *vis viva* of the whole bodies cannot be transferred momentarily to their constituent parts) and must therefore collide inelastically.[139] Newton explicitly accepts this consequence:

> For Bodies which are either absolutely hard, or so soft as to be void of Elasticity, will not rebound from one another. Impenetrability makes them only stop. If two equal Bodies meet directly *in vacuo*, they will by the Laws of Motion stop where they meet, and lose all their Motion, and remain in rest, unless they be elastick, and receive new Motion from their spring.[140]

When two perfectly hard bodies (such as atoms) collide, writes Clarke, the motion of the wholes

> cannot be dispersed among the parts, because the parts are capable of no tremulous motion for want of elasticity.[141]

Therefore, he continues,

> solid and perfectly hard bodies, void of elasticity, meeting together with equal and contrary forces, lose their whole motion and active force.[142]

In the Newtonian universe, *vis viva* (mv^2) is not conserved. Motion is *lost* in every interatomic collision: the vector quantity momentum $\Sigma\, m\vec{v}$ is conserved, but neither the Cartesian 'quantity of motion' $\Sigma m \mid v \mid$, nor the Leibnizian *vis viva*, mv^2. Motion, says Newton, 'is much more apt to be lost than got, and is always upon the decay':[143] non-mechanical 'active principles' are needed to supplement and sustain the clockwork universe.

This admission of non-mechanical 'active principles' was anathema to Leibniz and his disciples. The conservation of *vis viva* was for Leibniz one of the bulwarks of the whole mechanical world-view: without such a conservation law, natural science lapses into the darkness and obscurity of

---

139   This had already been spotted by Beeckman many years before (see Gabbey, 2, p. 380 ff.).

140   Newton, 2, p. 398.

141   H. G. Alexander, Ed., p. 111.

142   *Ibid.*, p. 112.

143   Newton, 2, p. 398. See also Snow, pp. 156–157.

purely occult causes; far better then to conclude that absolutely rigid bodies do not and cannot exist, and that *vis viva* apparently lost to the wholes in inelastic impacts is really transmitted to their constituent parts,[144] than to abandon the mechanistic vision altogether. Hard bodies, argue Leibniz and his disciple Jean Bernoulli, do not exist in Nature: *all* bodies are (more or less) deformable.[145]

Another principle which would be contravened by the existence of hard bodies is that of continuity, *Natura non facit saltum*.[146] In the collision of two atoms, their velocities would have to change *instantaneously*, by a discrete 'leap', which, says Leibniz, is 'opposed to harmony and order'.[147] In short, he says,

> My axiom, that nature never acts by leaps, . . . is of the greatest use in Physics; it destroys atoms, intervals of rest, globes of the second [Cartesian] element, and other similar chimeras.[148]

That atoms would violate the principle of continuity is due to their absolute rigidity:

> Assuming, then, that there are atoms, that is, bodies of maximum hardness and therefore inflexible, change would obviously occur through a leap or in a moment, for the direct motion becomes retrograde at the very moment of collision, unless we assume that the bodies rest instantaneously after the collision, that is, that they lose their force – a thing which, besides being absurd on other grounds, would still contain a change through a leap, namely, an instantaneous change from motion to rest without passing through intermediate degrees.[149]

If the principle of continuity is true, there are no perfectly rigid bodies:[150] collisions between such bodies would involve velocities 'leaping' instantaneously from $+v$ to $-v$, or from

---

144   See Westfall, 4, p. 295.

145   For a *much* fuller and more detailed account of the ensuing controversy, see Scott, 1, p. 200, and 2, p. 22 ff.

146   For Leibniz's use of this principle, see Russell, p. 222; Westfall, 4, p. 290.

147   From the 'Conversation of Philarète and Ariste' (c. 1711), Leibniz, p. 624.

148   Quoted from Russell, p. 234.

149   From the *Specimen Dynamicum* (1695), Leibniz, p. 446.

150   Or, of course, they never, for one reason or another, come into collision.

+ v to zero. Both possibilities, as well as violating the constraint of continuity, are absurd on other grounds: the former, because only deformable bodies can rebound; the latter, because of the loss of *vis viva* involved. Although the debate was to extend throughout the eighteenth century and into the nineteenth, the defenders of hard bodies were constantly on the defensive – the objections of Leibniz, amplified and developed by his disciples, were eventually to carry the day. Modern physics accepts no perfectly rigid bodies.

Before concluding this section, we must say a little more about the topic of cohesion. This was, from the start, one of the great stumbling-blocks for the Atomic Theory. The cohesiveness of macroscopic, observable bodies can be explained in terms of hooks and eyes, but this leaves the cohesion of the atoms themselves unexplained. Must we, asks Leibniz, 'assume hooks on hooks to infinity?'[151] The Atomists assume, he continues, that physical division must come to an end with the division into atoms, but 'no reason for cohesion and individuality appears within these ultimate corpuscles'.[152] The explanation offered by the ancients was, he claims,

> so inept that their recent followers were ashamed of it, namely, that the parts of atoms cohere because no vacuum comes between them. From this it would follow that all bodies, once they touch each other, ought to cohere inseparably in the manner of atoms, since there can be no intervening vacuum when any two bodies touch. Nothing is more absurd than such perpetual cohesion or more foreign to experience.[153]

Leibniz returns to this problem in his 'Critical Thoughts' on Descartes' *Principles* (1692). After dismissing the Cartesian theory that cohesion is due merely to relative rest,[154] he turns his attention once again to the Atomist account. Some, he begins, 'affirm that perfect unity is itself the cause of firmness', but this, although it may satisfy 'some advocates of atoms', is ultimately either empty or circular – when we ask what makes an atom a perfect unity the answer must be the absolute cohesiveness of its constituent parts.[155] If the Atomist

151   From his writings related to the 'Catholic Demonstrations' (c. 1668–1670), Leibniz, p. 112.
152   *Ibid.*, p. 112.
153   *Ibid.*, p. 112.
154   *Ibid.*, pp. 404–405.
155   *Ibid.*, pp. 405–406.

assumes that mere contact, without intervening vacuum, is sufficient for cohesion we are back with Aristotle's objection – atoms should cohere with one another on contact:

> By a natural progression it would follow . . . that the atoms would continually increase like snowballs rolled through the snow, and the outcome, finally, would be that everything would coalesce into a more than adamantine hardness.[156]

If mere contact were sufficient for cohesion, atoms should gradually coalesce, forming a Parmenidean block, an absolutely solid and immutable mass of matter*. (Leibniz shows no awareness of Newton's attempted solution of this difficulty.) The same argument can be found in the letters to Huygens[157] and in the 1690 monograph entitled 'Demonstration Against Atoms taken from the contact of Atoms'[158] If distinct atoms do not cohere on contact, Leibniz demands, what is the physical difference between the two adjacent cubic atoms A and B, and the atomic parallelepiped C?

Given that they have identical distributions of matter* in space, no void separating A from B, they should be physically indiscernible. (There can be no internecine motion of the parts of a perfectly rigid atom.) Once again, Leibniz argues, the Atomists must have recourse to God (to a miracle), there being no natural difference between AB and C. Atomism, he insists, is 'a perpetual miracle'.

C. The Corpuscular Philosophy versus Chemical Atomism

We discussed in Chapter Five the beginnings of a chemical Atomic theory in the seventeenth century. In figures such as Sennert, de Clave and Jungius we discerned recognisable similarities to the theory of chemical elements associated with the names of Lavoisier, and the chemical Atomic Theory of Dalton. According to these men, there exist atoms (physically indivisible corpuscles) endowed with specific *chemical natures*, atoms *of* the chemical elements, whether the FAWE of the peripatetics, the Mercury, Sulphur and Salt (MSL) of

156   *Ibid.*, p. 406.
157   *Ibid.*, p. 415.
158   See Tymieniecka, p. 68.

the Paracelsians, or some other list. The precise number and nature of these supposed elements was a topic of vigorous controversy in seventeenth century chemical thought: many favoured a five-element theory (Spirit, Oil, Salt, Water and Earth – i.e., roughly MSLWE[159]), while others, such as Jungius, were prepared to accept as elementary any substance resistant to chemical disintegration, and ended up with a lengthier list.[160]

But these, for our purposes, are matters of detail. The important theoretical claim is that atoms (physical indivisibles) instantiate chemical species, and that therefore there are some chemical substances (the elements) immune from transmutation. If there exist, for example, immutable atoms of FAWE, those elements cannot be transmuted into one another. Jungius, for example, denied the possibility of changing W into E: the precipitation of a salt from a solution involves, he argues, the agglutination and settling-out of previously existing particles of salt, not the transmutation of water into a quite different chemical substance.[161]

According to these Chemical Atomists, matter is *heterogeneous*: the primary particles of which it is composed are themselves of radically different natures. There is no 'unity of matter' and no possibility of unrestricted transmutation – compounds may be transformed, but never elements. One of the great opponents of this type of theory was J. B. Van Helmont, whose matter-theory rested on the twin pillars of chemical experimentation and the book of Genesis! Of central importance to his conception of matter was his celebrated willow-tree experiment, in which he proved to his own satisfaction that the gain in weight of a growing plant is derived (almost) entirely from water, not from the earth in which it grows. (He never thought to test the contribution of the air.) But from organic tissue one can separate out, by destructive distillation, various oils, spirits and salts, i.e. the chemists' MSL. We therefore have:

$$W \longrightarrow \text{Plant Tissue} \longrightarrow MSL$$

This experiment thus led Helmont to the radical conclusion that water is the common universal prime matter, the first

---

[159] See the 'improved' chemical matter-theory of Eleutherius in Boyle's 'Sceptical Chemist' (*Works*, vol. 1, p. 544) for an example of such a five-element theory. Boyle derives this theory from the contemporary chemist, Willis (Boyle, *Works*, Vol. 4, p. 503).

[160] See Pagel, 5, p. 105.

[161] *Ibid.*, p. 104.

stuff from which all things (except air, which is something quite different[162]) are made. 'All visible things,' he says, 'are formed of water only'.[163] This opinion is corroborated by the account of the Creation in Genesis (the spirit of the Lord moved across the face of the *waters*) and by the testimony of Thales, one of the *prisca sapientia*, holders of a secret ancient wisdom, known only to an élite few.

Helmont also believed that all substances could be reduced to water by the action of the universal solvent or '*alkahest*'. (The evidential support for this claim was nowhere near so convincing as that given by the willow-tree experiment for the derivability of all stuffs from water.) We have therefore the following schema:

W   Plants→   Other Substances   Alhakest→   W.

In short, we have a single, common, homogeneous matter (water), shaped by a variety of 'seminal' influences into other materials, each of which is reducible to its primary form by the action of the alkahest. Nothing could be further from the chemical Atomic Theory of Helmont's contemporaries. Yet Helmont's views, strange as they may sound, were to have a profound effect on thinkers of the calibre of Robert Boyle and Isaac Newton.

The most penetrating of all the seventeenth century critics of chemical Atomism was Boyle, whose 'Sceptical Chemist' is largely devoted to the overthrow of that theory. There is, he contends, a single, common, homogeneous matter, of which all things are made, which 'universal matter' is 'actually divided into little corpuscles, of several shapes and sizes, variously moved.[164] Of these minute particles, he continues,

> divers of the smallest and neighbouring ones were here and there associated into minute masses or clusters, and did by their coalitions constitute great store of such little primary concretions or masses, as were not easily dissipable into such particles, as composed them.[165]

For example, gold can be alloyed with other metals, dissolved in *aqua regia*, precipitated out of solution as a salt, etc., entire gold-corpuscles remaining unaltered throughout, as is evident on smelting. (The recovery of the gold betokens its *actual* presence in the compound.) It is therefore plausible that

---

[162]   Partington, 1, p. 370.

[163]   *Ibid.*, p. 369.

[164]   Boyle, *Works*, Vol. 1, p. 474.

[165]   *Ibid.*, p. 475.

such little primary masses or clusters . . . may remain
undissipated, notwithstanding their entering into the composition of various concretions, since the corpuscles of gold
and mercury, though they be not primary concretions of
the most minute particles of matter, but confessedly mixed
bodies, are able to concur plentifully to the composition of
several very differing bodies, without losing their own
nature or texture, or having their cohesion violated by the
divorce of their associated parts or ingredients.[166]

Metals such as gold and mercury exist therefore, on Boyle's
account, as semi-stable clusters of atoms – in our terminology, as *molecules*. In this hierarchic matter-theory, the
species of materials arise only at a molecular level – there are
no atoms of gold, but, when atoms aggregate after a certain
manner to produce a cluster with a specific texture, a
corpuscle of gold is produced. These metallic corpuscles are
strongly cohesive, and therefore survive intact through most
chemical reactions, as is evident from chemical experimentation.[167]

If metal corpuscles are clusters of atoms, the transmutation
of one metal into another is in principle possible. All it
requires is an analytical agent sufficiently powerful to
overcome the cohesion of the metal corpuscle, to dissolve,
e.g., a particle of gold into its constituent atoms, which can
then reassemble to form, e.g., silver. Boyle believed himself in
possession of such an 'anti-elixir', capable of degrading gold
into silver,[168] i.e. endowed with

an extraordinary efficacy in reference to gold, not only to
dissolve and otherwise alter it, but to injure the very texture
of that supposedly immutable metal.[169]

Boyle also believed in the possibility of transmuting base
metals into gold, and thought himself very close to the
ultimate achievement of the perennial alchemical quest.[170]
Hence, he concludes,

---

166   *Ibid.*, pp. 475–476.

167   See M. Boas, 7, p. 99.

168   See his 'Strange Chemical Narrative' in *Works*, Vol. 4, p. 371 ff. See
also Ihde for a critical discussion.

169   *Ibid.*, Vol. 3, p. 94.

170   See his 'Experimental Discourse of Quicksilver Growing Hot with Gold'
(*Works*, vol. 4, p. 219 ff.), a work which excited even such men as
Newton and Locke (see L. T. More, 1, pp. 61–64). For more on Boyle's
alchemical techniques, see Dobbs, 4, p. 139.

it seems deducible from what we have delivered, that there may be a real transmutation of one metal into another, even among the perfectest and noblest metals, and that effected by factitious agents . . . after a mechanical manner.[171]

But the transmutation of metals is perfectly compatible with Chemical Atomism, so long as the species of metal are not themselves conceived as elementary. If gold and silver are compounded of the Aristotelians' FAWE, or the Paracelsians' MSL, metallic transmutations will be explicable in terms of the dissociation and aggregation of atoms of these supposed chemical elements. In order to overthrow Chemical Atomism altogether, Boyle needed to show the 'producibleness' of these allegedly elementary substances from one another.

It is at this point that the evidence of Helmont is introduced into the debate. The opinion that 'all mixt bodies spring from one element; and that vegetables, animals . . . minerals, etc., are materially but simple water disguised into these various forms by the plastick or formative vertue of their seeds'[172] is, says Boyle, not an innovation of Helmont, but dates back to Thales and to Moses.[173] In any case, whether or not we believe all the details of Helmont's theory, the willow-tree experiment alone seems to show conclusively that plant tissue derives solely from water.[174] But vegetable matter can be decomposed chemically to yield 'oils' (S), 'spirits' (M), and ashes (L), the *tria prima* of the Paracelsians.[175] These three supposed 'principles' are all, then, producible (albeit indirectly) from water. MSL are not, then, 'primogenial bodies', since

> they are made every day out of plain water by the texture, which the seed or seminal principle of plants puts it into.[176]

Nor need MSL be thought of as *indestructible*, he adds, 'till it be shewn upon what account we are to believe them privileged'.[177] Indeed,

171 Boyle, *Works*, Vol. 3, p. 97.

172 *Ibid.*, Vol. 1, p. 496.

173 *Ibid.*, p. 497. It was of course a common contemporary view that Moses was the true author of the Atomic Theory (see Sailor, 2).

174 *Ibid.*, . 563. For Helmont's influence on Boyle, see Webster, 1; Walton, pp. 11–13; M. Boas, 6, pp. 156–157.

175 Boyle, *Works*, Vol. 1, p. 500.

176 *Ibid.*, p. 509.

177 *Ibid.*, p. 509.

> if it be true, as it is probable, that compounded bodies differ from one another but in the various textures resulting from the bigness, shape, motion, and contrivance of their small parts, it will not be irrational to conceive, that one and the same parcel of the universal matter may, by various alterations and contextures, be brought to deserve the name, sometimes of sulphureous, and sometimes of a terrene, or aqueous body.[178]

The implied theoretical claim – that any chemical substance can be formed from any other – is supported by Boyle with a mass of experimental evidence.[179] If, he argues, MSL and FAWE are producible *de novo*, by a mere change of texture, they cannot legitimately be conceived as chemical elements. Special significance is claimed for the results of one particular experiment, in which the 'essential oil of anniseeds' (which the chemists assume to be largely sulphureous in make-up) yields, on destructive distillation, 'phlegm' (W), a 'volatile salt' (L) and a 'spirit' (M). From this one experiment it may be concluded that

> a substance, that is looked on by chymists as a homogeneous body, and which passes for one of their principles . . . may yet be of such a nature, that, barely by the action of the fire it may be made to afford a very considerable proportion of a substance exceedingly different from that which afforded it.[180]

It 'will scarce be necessary', Boyle concludes,

> to draw so obvious a corollary from what has been already delivered, as this: that it is possible that chymical principles . . . may be made of another.[181]

The results of this experiment are readily explicable in terms of the corpuscular philosophy, so long as one assumes that MSL too arise only at a molecular level. Fire produces an *agitation* of the parts of a body, which agitation may dissolve certain complexes and allow others to form, thus producing a

---

178   *Ibid.*, pp. 493–494.

179   See the Appendix to the 'Sceptical Chemist' on the 'producibleness' *de novo* of all the supposed elements and principles (*Works*, Vol. 1, p. 591 ff.), and the 'Chymical Paradox, grounded upon New Experiments, making it probable, that Chymical Principles are transmutable' (Vol. 4, p. 496 ff.).

180   *Ibid.*, Vol. 4, p. 499.

181   *Ibid.*, p. 500.

change of texture. Transmutation may occur, says Boyle, by the 'transposition' of abiding particles; it does not require the addition or subtraction of matter. The results of the above experiment – transmutation of S to M and L – seem, he concludes, 'very congruous to the mechanical hypothesis, and very unfavourable to that of the chymists'.[182] The chemists' MSL are, then,

> not the first and most simple principles of bodies, but rather primary concretions or corpuscles, or particles more simple than they, as being endowed only with the first, or most radical . . . and most catholick affections of simple bodies, namely, bulk, shape, and motion, or rest; by the different conventions or coalitions of which minutest portions of matter, are made those differing conventions that chemists name salt, sulphur, and mercury.[183]

That MSL can undergo transmutation into one another is certain evidence of their non-elementary status: the corpuscularian can admit them only as molecules, not as atoms.[184] The peripatetic 'elements' FAWE are in the same boat: earth, for example, can, says Boyle, be produced from water. When distilled water is boiled in a glass vessel, a whitish 'earth' is deposited on the walls of the vessel: unless, says Boyle, it can be shown (which, he adds, 'will scarce be pretended'[185]) that the water has dissolved the glass, it will seem probable that

> the earthy powder I obtained from already distilled rain-water, might be a transmutation of some parts of the water into that substance.[186]

The same water can be re-distilled again and again, but will always yield new 'earth', showing that the process is not a mere separation of an earthy fraction in the water, but a true

---

[182] *Ibid.*, p. 501. See also Vol. 3, p. 106.

[183] *Ibid.*, Vol. 1, p. 179.

[184] See M. Boas, p. 99.

[185] Unfortunately, Boyle did not bother to test this possibility (by weighing the glass before and after the experiment). As Marie Boas (7, pp. 224–225) says, he was *looking* for chemical evidence of the transmutation of the supposed elements FAWE – his prior conviction led him astray in this particular instance.

[186] Boyle, *Works*, vol. 3, p. 103. Lavoisier had to re-interpret Boyle's results in defence of his theory of the chemical elements; the white 'earth', Lavoisier claimed (and proceeded to demonstrate) *does* derive from the glass and not the water.

W ⟶ E transmutation, which is, says Boyle, 'no mean confirmation of the corpuscularian principles and hypotheses'.[187] Neither water nor earth, says the corpuscularian, are chemical elements:

> since these bodies are mutually convertible into one another
> . . . they are not either of them ingenerable and incorruptible elements, much less the sole matter of all tangible bodies, but only two of the primordial and most obvious schematisms of that which is indeed the universal matter; which as it comes to have its minute particles associated after this or that manner, may, by a change of their texture and motion, constitute with the same corpuscles sometimes water and sometimes earth.[188]

Exactly the *same* corpuscles make up first the one so-called 'element' and then the other. There are therefore no W-atoms and E-atoms, but undivided corpuscles, themselves *without* a specific chemical nature, which can be arranged in different ways to give water and earth respectively. The alteration involved in the W ⟶ E transmutation is, Boyle remarks, so great that 'it cannot but afford us a considerable instance of what the varied texture of the minute parts may perform in a matter confessedly similar.[189] Striking qualitative alterations may turn merely on a simple rearrangement of abiding parts.

But now if the Paracelsians' MSL and the peripatetics' FAWE fail alike of elementary status, the existence of chemical elements at all may seem dubious. The denial of chemical elements, says Carneades (Boyle's spokesman) in the 'Sceptical Chemist', is 'not absurd'.[190] The corpuscular philosophy, says Boyle elsewhere, has no use for such a hypothesis, which can simply be discarded.[191] A corpuscularian theory of matter is, moreover, much more favourable to transmutations than scholastic Aristotelianism:

> Whereas the schools generally declare the transmutation of one species into another, and particularly that of the baser metals into gold, to be against nature, and physically impossible; the corpuscular doctrine, rejecting the substantial forms of the schools, and making bodies to differ

187   *Ibid.*, p. 105.
188   *Ibid.*, p. 106.
189   *Ibid.*, p. 106.
190   *Ibid.*, Vol. 1, p. 561 ff.
191   *Ibid.*, Vol. 3, pp. 96–97.

but in the magnitude, figure, motion or rest, and situation of their component particles, which may be always infinitely varied, seems much more favourable to the chymical doctrine of the possibility of working wonderful changes, and even transmutations in mixt bodies.[192]

From a single homogeneous matter*, common to all corporeal things, all the incredible variety of Nature may be produced. (For an illustration of this idea, just consider all the many and varied machines that can be made out of *iron*, their differences of form and function depending solely on the *arrangement* of their parts.[193]) The various species of bodies differ from one another only, says Boyle, by their forms or (dynamic) 'textures', i.e. by the arrangements and internecine motions of their constituent parts. Among inanimate bodies, forms are

but peculiar contrivances of the matter, and may by agents, that work but mechanically, that is, by locally moving the parts and changing their size, shape, or texture, be generated and destroyed.[194]

Can *any* material substance by made of *any* other? Is there *universal* transmutability? Boyle comes close to a positive answer:

Though I would not say that any thing can immediately be made of every thing, as a gold ring of a wedge of gold; yet since bodies having but one common matter can be differenced but by accidents, which seem all of them to be the effects and consequences of local motion, I see not why it should not be absurd to think that (at least among inanimate bodies) by the intervention of some very small addition or subtraction of matter (which yet in most cases will scarce by needed) and an orderly series of alterations, disposing by degrees the matter to be transmuted, almost of any thing may at length be made any thing.[195]

In this passage we can detect signs of hedging: Boyle is clearly reluctant to commit himself positively to the strong universal thesis that is in the offing: hence the reference to addition or

192   *Ibid.*, vol. 1, p. 358. For the link between the corpuscular philosophy and the belief in transmutation, see also M. Boas, 7, p. 102.

193   Boyle, *Works*, vol. 1, p. 476. See also Vol. 3, pp. 93–94.

194   *Ibid.*, vol. 3, p. 97.

195   *Ibid.*, p. 35.

subtraction of matter, which is immediately (partially) neutralised by the rider that this will be unnecessary in *most* cases. It is nevertheless plain that Boyle finds the thesis of universal transmutability a plausible and attractive one, readily compatible with his corpuscular philosophy.[196] To make a gold ring out of a wedge of gold is easy; to make oil out of water, though less straightforward, is also possible. One could, for example, convert the water to plant tissue (cf. Helmont's experiment), from which in turn oil may be derived by distillation. By means of a suitable series of intermediates, any substance may, for all we know, be capable of being synthesised from any given set of raw materials. Even if we lack the evidence needed for a definitive proof of the 'anything ↔ anything' thesis, it has a high probability, given the corpuscular constitution of matter, of being true.

Or does it? Boyle's (tentative) acceptance of the 'anything ↔ anything' thesis seems to place the whole explanatory burden on the (dynamic) textures of bodies, rather than on the sizes and shapes of their atomic parts. But a particular texture may *require* particles of a certain shape: one could not, for example, pack spheres so tightly as to leave no interstitial vacuum, or create a smooth surface out of spiky atoms. Only if the following conditions C1-3 are *all* satisfied will the corpuscular philosophy license the 'anything ↔ anything' thesis:

C1.   Chemical species arise at a molecular level, not at the level of the abiding and naturally undivided *minima*.

C2.   The chemical properties of any substance depend on its own specific dynamic texture or 'form'.

C3.   The same dynamic texture can be made from *any* given set of *minima*.

It is C3 which raises difficulties for the advocate of the 'anything ↔ anything' thesis. If the texture T on which the properties of iron depend could only result from the aggregation of angular *minima*, a substance made up entirely of spherical atoms will not be transmutable into iron.[197] Boyle never explicitly considers this possibility – that the sizes and

---

[196]   Kuhn (2, p. 22) and M. Boas (7, p. 103) both attribute the 'anything ↔ anything' thesis to Boyle.

[197]   Unless of course even the *minima* can be broken down – a possibility Boyle does not entirely preclude when he says that they are 'scarce ever' divided in the course of Nature (see Part B of this chapter). Marie Boas (7, p. 169) attributes to Boyle this essentially Cartesian (and anti-Atomist) defence of the 'anything ↔ anything' thesis.

shapes of the *minima* may limit the possibilities of transmutation – although such a thought may lie behind the hedging we detected earlier on the topic of universal transmutability.

In any event, Boyle never devoted any attention to such a reservation: in explaining the properties of a particular substance, it is always the arrangement and motion of its parts that are considered all-important, rather than the sizes and shapes of its (atomic) parts.[198] All chemical species, he clearly believes, arise only at a *molecular* level;[199] free atoms are thus *chemically neutral*: an atom which is a sphere of radius r is not, in virtue of this size and shape, an atom of any particular substance – it can be incorporated into a wide variety of 'dynamic textures', constitutive of different stuffs.[200] All that is required is that it find a suitable rôle in the microstructure constitutive of any particular material.

One of the most significant models used by Boyle in this regard is that of a clock. A timepiece works, he insists, not because its parts are made of brass, or iron, but because of their arrangement in a particular (functional) manner.[201] *Form* (= arrangement or 'texture'), not *matter*, explains qualities. In fact, Boyle admits, the material constitution of a clock is not irrelevant to its function: one could not construct a viable timepiece out of sponge, or water. What Boyle rightly insists on, however, is that, while a certain type of matter may be *necessary* for the manufacture of a given artefact, it can never be *sufficient* – the mere 'brassiness' of the cog-wheels cannot explain the workings of the clock. While material constitution (the stock of available atoms) may constrain the possible forms or textures that can be produced in a given piece of matter, it is still form (not matter) that plays the crucial explanatory rôle.

The chemist is, on this view, simply the mechanic of the micro-realm – chemistry is a form of applied mechanics. By rearranging minute corpuscles, the chemist can form and dissolve the forms or textures responsible for the specific virtues of things. There is no distinction in this regard between artefacts and natural substances,[202] save that the

---

[198] The sizes and shapes of *molecular* parts fall, of course, under the general heading of 'arrangement'.

[199] See Boyle, *Works*, Vol. 1, pp. 506–507.

[200] Free atoms are perhaps, on this view, rarely (if ever) found in Nature: they may be quite unfamiliar to us, and disanalogous to everyday chemical substances.

[201] Boyle, *Works*, Vol. 1, p. 559. See also Kuhn, 2, pp. 19–20.

[202] *Ibid.*, p. 571.

'schematism and texture' of the latter is, as befits the handicraft of the deity, usually vastly more sophisticated and intricate than our poor imitations.

In the writings of Isaac Newton, a strikingly similar matter-theory can be discovered. Newton too believed in a common universal matter, the aether or universal spirit,[203] from which all denser materials have emerged by condensation. The aether condenses first to form *water*, from which in turn other terrestrial materials are derived. (Newton is thus able to accommodate Helmont's matter-theory within his own.[204]) That earth can be derived from water has, Newton claims, been demonstrated by Boyle.[205] In *Principia* we find an account of the origin of all terrestrial substances from celestial 'vapours' (such as the tails of comets), thus:

Vapours ———→ water and 'humid spirits' ———→ Terrestrial matter[206]

In the *Conclusio* (intended as a conclusion to *Principia*) this idea is spelt out more fully:

> . . . water can be transformed by continued fermentation into the more dense substances of animals, vegetables, salts, stones, and various earths. And finally by the very long duration of the operation be coagulated into mineral and metallic substances. For the matter of all things is one and the same, which is transmuted into various forms by the operation of nature . . . [207]

Newton, like Boyle, was impressed by Helmont's apparent proof that vegetable matter is derived entirely from water. All bodies, he suggests, absorb water and/or air, extracting suitable atoms from these ambient fluids and organising them into specific structures. To replenish its stocks of air and water, then, our Earth must attract (by gravity) vapours from, e.g., the tails of comets.[208] All things thus originate from a common universal matter – 'aether' or 'spirit' or whatever – by successive processes of condensation.

203   Newton produced, at different stages in his career, several different (and incompatible) versions of this story. I give here only the barest outline.

204   For Helmont's influence on Newton, see Walton, p. 14 ff.

205   Newton, 2, p. 374. See also Dobbs, 4, p. 201.

206   Newton, 1, p. 542.

207   Quoted from Dobbs, 4, p. 202.

208   Walton, p. 14. For a study of some of the wider implications of this doctrine, see Kubrin.

Newton also developed a sophisticated hierarchic theory of the infrastructure of matter, outlined in the *Opticks*. Bodies, he claims, must be 'much more rare and porous than is commonly believed'[209] – otherwise one could not account for the transparency of such substances as glass and water. Let us consider, he suggests, the relatively complex particles responsible for *colours*:

> Now if we conceive these Particles of Bodies to be so disposed among themselves, that the Intervals or empty Spaces between them may be equal in magnitude to them all; and that these Particles may be composed of other Particles much smaller, which have as much empty Space between them as equals all the Magnitudes of these smaller Particles: And that in like manner these smaller Particles are again composed of others much smaller, all of which together are equal to all the Pores or empty spaces between them: and so on perpetually till you come to solid Particles, such as have no Pores or empty Spaces within them: And if in any gross Body there be, for instance, three such degrees of Particles, the least of which are solid; this body will have seven times more Pores than solid Parts.[210]

if we take perfectly solid atoms (100% matter*) and combine them with an equal volume of void, we obtain corpuscles of the first order of composition (C1s). Combining C1s with equal amounts of vacuum yields C2s, and so on up the scale.[211] If at each stage of aggregation an equal volume of void is intermixed, a corpuscle of the nth grade of composition Cn will contain $1/2^n$ parts of matter * per unit volume.

The smallest corpuscles, says Newton,

> cohere by the strongest Attractions, and compose bigger Particles of weaker Virtue; and many of these may cohere and compose bigger Particles whose Virtue is still weaker, and so on for divers Successions, until the Progression end in the biggest Particles on which the operations in Chymistry, and the Colours of natural Bodies depend, and which by cohering compose Bodies of a sensible magnitude.[212]

[209] Newton, 2, p. 267.

[210] *Ibid.*, pp. 268–269. See also Figala, p. 134.

[211] See Thackray, 2, p. 24.

[212] Newton, 2, p. 398.

Atoms therefore cohere *very strongly*[213] to give C1s; C1s cohere by sightly weaker forces to give C2s; and so on up the scale. (Newton of course incorporates attractive and repulsive *forces* into his Atomic Theory.) Now if chemical species (e.g. the various metals) are not atomic but molecular – i.e. arise at some level Cn where n $\geqslant$ 1, their transmutation is in principle possible. It is, however, difficult, since to achieve it one must overcome the very powerful short-range forces which hold, e.g., C1s and C2s together. If ordinary chemical operations proceed by separating and recombining, say, C4s or C5s, alchemical transmutation will involve separation and recombination of C3s, perhaps even C2s or C1s. (All chemical operations involve the recombination of abiding particles, but different *levels* of analysis are possible.[214]) Whereas acids such as *aqua regis* and *aqua fortis* only dissolve gold and silver, respectively, into particles *of* those metals, a more powerful analytic agent might effect a genuine transmutation:

> If gold could be brought once to ferment and putrifie, it might be turned into any other body whatsoever.[215]

If particles of gold – C4s, say – exist as such in solution in *aqua regia*, a solvent capable of reducing C4s to C3s would make possible the transmutation of gold into some other substance altogether. To achieve such a transmutation, we need an agent capable not merely of separating one gold-particle from another, as *aqua regia* does, but of breaking down their internal structures or 'opening their textures', as Boyle and Newton would put it.[216] The dream of the alchemists – the transmutation of base metals to gold – is not theoretically absurd, but practically very difficult, owing to the strength of the interparticulate attractive forces that need to be overcome.[217]

Newton's matter-theory can also accommodate the alchemical Mercury-Sulphur (M/S) theory of the constitution of metals. If corpuscles of gold and silver are, say, C5s, they may be composed of discrete C4s of M and S in varying

213   Newton is prepared to consider the possibility of forces varying with $1/d^n$ where $n > 2$: such forces could be very strong at short distances and negligible at greater ones.

214   See Thackray, 2, pp. 25–26.

215   Quoted from Thackray, 2, pp. 24–25.

216   See Dobbs, 4, pp. 218–219.

217   Forbes, pp. 34–35.

ratios. 'Opening' a metal will therefore release its *intrinsic* M and S.[218] That each metal should contain its own intrinsic ('philosophical') mercury is therefore readily intelligible in terms of the hierarchic theory – much of Newton's own alchemical work involved the attempt to extract the 'mercuries' of the metals.[219] Thus Newton can find room within his matter-theory even for the quality-bearing 'principles' of the alchemists – so long as they too arise at a molecular level out of *chemically neutral atoms*.[220]

The mater-theory of Boyle and Newton, with its complex hierarchic structure, its rejection of Chemical Atomism, and its implicit scepticism about chemical elements, has been criticised by Kuhn and Thackray as providing, in Kuhn's words, ' a sterile and occasionally adverse intellectual climate for chemistry'.[221] The 'Chemical Revolution' associated with the names of Lavoisier and Dalton marked, says Thackray, a *reversion* to the older tradition of Chemical Atomism, a reaction against the views of Boyle and Newton.[222] For these critics, the influence of Newton and Boyle on the progress of Chemistry was largely *negative*: the science finally attained its modern form by a deliberate reaction against the views of those prestigious authors.[223]

This criticism seems unfair and unwarranted. In the first place, the Newtonian matter-theory is remarkably close to what we now believe to be the truth.[224] Chemical species *do* only arise at a 'molecular' level, i.e. as a result of the aggregation of simpler (and chemical neutral) constituents. The chemical atom of Dalton *is* a highly *complex* structure, made up of neutrons, protons, electrons, etc., held together by powerful interparticulate forces of various kinds. The transmutation of the so-called 'chemical elements' *is* physically possible although, as Newton foresaw, highly difficult

218   See Dobbs, 4, pp. 145–146.

219   *Ibid.*, p. 139ff.; p. 183 ff.

220   *Ibid.*, p. 221.

221   Kuhn, 2, p. 15.

222   Thackray, 2, pp. 196–198.

223   Kuhn, p. 34; Thackray, 2, p. 197. For anti-Newtonian elements in Lavoisier, see Thackray, 2, p. 198.

224   Newtonianism, says Thackray (2, pp. 178–180), was a hindrance to eighteenth century Chemistry, since it entails (a) the remoteness of the *ultimate* particles from our empirical access, (b) the comparative superficiality of chemical operations, and (c) the complexity of all chemical substances. But all these theses can be seen, in the light of modern physics, to be *true*.

owing to the strength of those forces. In the second place, the Boyle-Newton theory can *incorporate* the Lavoisier-Dalton one. Boyle and Newton both emphasised the very considerable *stability* of certain clusters of atoms and their tendency to preserve their chemical identities unchanged through many of the vicissitudes of Nature. One can therefore identify these stable clusters or molecules with the chemical 'atoms' of Dalton, and thus accommodate the (almost) immutable elements of Lavoisier. One must of course continue to admit the *theoretical possibility* of splitting the chemical 'atom', and hence of transmuting the 'elements', but this possibility need not have been felt to be a disturbing one. After all, naturally *undivided* C2s or C3s would suffice for a theory of chemical 'Atomism' – the claim that these particles are absolutely *indivisible* is not essential.

VACUUM: 3

A. Matter, Space, and Deity

The essence of matter, Descartes declares, is three-dimensional (3D) extension, which is not something distinct from particular bodies and can have no existence independent of them, except by a false abstraction of the mind.[1] It is hardly misleading to say that, for Descartes, matter *is* its extension, which feature *exhausts* its essence. Once we grasp our clear and distinct (innate) idea of body, he claims, it becomes evident to us that its nature 'consists not in weight, nor in hardness, nor colour and so on, but in extension alone'.[2]

But if extension alone constitutes body, the very concept of void space is self-contradictory. Wherever X is, there also is the essence of X, and *vice versa*. 3D extension is, for Descartes, both necessary and sufficient for corporeity.[3] The conception of place as an 'interval' or vacuum₁ (a 3D incorporeal receptacle accidentally occupied by a body) is therefore absurd. Place, according to Descartes, divides into 'external' or Aristotelian (the inner surface of the containing body) and 'internal', which can be identified with the dimensions of the object itself. Thus:

> Space or internal place and the corporeal substance which is contained in it are not different otherwise than in the mode in which they are conceived of by us.[4]

The *same* 3D extension constitutes space *and* matter: there is no need to posit separate incorporeal dimensions. Strictly, one should not even speak of place *containing* body – it is the notion of the receptacle as distinct from its contents which Descartes wishes at all costs to avoid.[5]

It follows from this that the concept of a void or vacuum₂ in its strict and philosophical sense – space empty of all

---

[1] Descartes, 1, Vol. 1, p. 57. See also Hoenen, p. 356; Keeling, p. 58.

[2] *Principles*, 2, 4; From Descartes, 1, vol. 1, p. 255.

[3] See Blackwell in McMullin, Ed., 2, p. 59.

[4] *Principles*, 2, 10; Descartes, 1, Vol. 1, p. 259. See also Koyré, 4, p. 164. This conception of 'internal place' was held by a number of scholastic authors, including Buridan and Toletus. See Chapter 6, p. 241.

[5] See Koyré, 3, p. 102.

substance – is self-contradictory: it is, Descartes insists, 'contrary to reason' to affirm its existence:

> As regards a vacuum in the philosophical sense of the word, i.e. a space in which there is no substance, it is evident that such cannot exist, because the extension of space or internal place is not different from that of body. For, from the breadth, and depth, we have reason to conclude that it is a substance, because it is absolutely inconceivable that nothing should possess extension, we ought to conclude also that the same is true of the space which is supposed to be void, i.e. that since there is in it extension, there is necessarily also substance.[6]

But, we might enquire, could not God annihilate the matter contained in a closed vessel, thereby leaving void or vacuum$_2$ inside it? In such a case, Descartes replies,

> the sides of the vessel will . . . come into immediate contiguity with one another. For two bodies must touch when there is nothing between then, because it is manifestly contradictory for these two bodies to be apart from one another, so that there should be a distance between them, and yet that this distance should be nothing; for distance is a mode of extension, and without extended substance it cannot therefore exist.[7]

The following three lines of argument might be given for this conclusion:

1. 3D extension is the essential characteristic of body: wherever there is extension there is body, and *vice versa*. If body is annihilated, so must extension be.
2. The same 3D extension, on the 'internal place' theory, constitutes space and body – the two are only separable by a false abstraction.
3. 3D extension is a property or mode, which must as such inhere in some *substance*. Now there exist, for Descartes, two quite distinct types of substance, spiritual and material respectively. Spirits (God, angels and souls) are, however, not extended in space: 'no incorporeal substance,' says Descartes, 'is in any strict sense extended'.[8] God, he writes to Henry More, 'is not imaginable nor distinguishable into shaped and measurable parts':[9] His omnipresence is *virtual* only. Descartes' God

6   *Principles*, 2, 16; Descartes, 1, Vol. 1, p. 262.
7   *Ibid.*, p. 263. See also Koyré, 4, p. 165.
8   Quoted from McGuire, 5, p. 470.
9   Kenny, Ed., p. 240.

is everywhere in virtue of His power; but in virtue of His essence He has no relation to place at all.[10]

It remains, therefore, that 3D extension inheres in material substance, which is of course its proper and natural logical subject. (How on Descartes' view, one distinguishes the substance from the property is not easy to determine.) As for the suggestion that the dimensions of space may inhere in no substance, that is plainly absurd. As Descartes writes to Arnauld,

> The difficulty in recognising the impossibility of the vacuum seems to arise primarily because we do not sufficiently consider that nothing can have no properties.[11]

Mere nothingness can have no properties, hence no extension: 'Two feet of nothing' is a meaningless phrase. Bodies separated by nothing are in contact.

How can the Cartesian account for the *impenetrability* of bodies? If matter = 3D extension and nothing more, what differentiates a physical body from a geometrical figure? Nothing, it might seem. But geometrical figures can interpenetrate with perfect freedom, whereas bodies cannot.[12] Is this not a refutation of the Cartesian identification of space and matter?

The impenetrability of bodies follows, Descartes replies, from two premisses, (a) the quantitative conservation of matter, and (b) the use of bulk or volume as a measure of quantity of matter. Impenetrability, Descartes explains to More, is not a primary and definitory feature of body, but a mere corollary of extension:

> It is impossible to conceive of one part of extended substance penetrating another equal part without *eo ipso* thinking that half the total extension is taken away or annihilated; but what is annihilated does not penetrate anything else; and so, in my opinion, impenetrability can be shown to belong to the essence of extension and not to that of anything else.[13]

This is a curious account. Instead of locating the impenetrability of bodies in some positive physical feature of them,

10   *Ibid.*, p. 250.

11   *Ibid.*, p. 236. See also Koyré, 3, p. 101.

12   See Keeling, p. 61.

13   Kenny, Ed., pp. 248–249.

he explains it in terms of Nature's observance of a conservation-principle. When two particles come into contact, only the need to sustain the conservation law prevents their mutual interpenetration. In fact, as a number of critics remark, Descartes simply assumes a 'richer' notion of a material particle than his own definition allows[14] – he speaks freely of the 'shock' of an intercorpuscular collision, of the 'impulse' one body gives to another, etc., concepts of questionable sense within his own philosophy. Perhaps the only viable explanation of the physical *enforcement* of the conservation law would be in terms of the occasionalist metaphysic implicit in Cartesianism.[15] On such a view, when two particles meet, God would see that unless He acted the conservation law would be violated, and so would instantly intervene to alter the courses of the particles in such a way as to avoid this eventuality.[16] Weird though this explanation may be, it is perhaps the only account of impenetrability compatible with Cartesian principles.

Turning from Descartes to his great rival Gassendi, we discover a radically different conception of space: in effect, a reaffirmation of the Epicurean position.[17] Space, he claims, = vacuum$_1$, an infinite, 3D, incorporeal receptacle: an occupied part of this universal space is called a *place* (the place *of* some body); an unoccupied part is called *void*. 'Inanity' and 'locality', says Charleton, 'bear one and the same notion essentially':

> The same space, when possessed by a body, is a place, but when left destitute of any corporeal tenant whatever, then it is a vacuum.[18]

The whole universe consists, on this view, of those two mutually exclusive and jointly exhaustive natures, matter*

---

14   Keeling, pp. 62–63; Laing, p. 409.

15   For an occasionalist interpretation of Cartesian 'dynamics', see Hatfield. For two crucial texts, see Descartes, 1, Vol. 2, p. 57, where he claims that 'the present time has no casual dependence on the time immediately preceding it', and Kenny, Ed., p. 257, where he writes (to Henry More) to the effect that the power of moving bodies 'may be the power of God Himself conserving the same amount of translation in matter as He put into it at the first moment of Creation . . .'.

16   It is worthy of note at this point that Descartes derives his 'Laws of Nature' directly from the immutability of God, not from any physical grounds.

17   For the Epicurean conception of space and place, see chapter 2, pp. 68–69.

18   Charleton, p. 62.

and void or vacuum$_2$. The distinction between body and void turns, for the Atomist, on the tangible/intangible dichotomy – body is tangible extension; void, intangible extension. Of itself, 3D extension is ontologically neutral, common to matter* and void alike: only when extension is joined to *antitypia* (impenetrability), the physical ground of tangibility, do we have a body. Bodies can touch and be touched, and can exert and experience physical impulsion on contact; void can do none of these things.[19]

Let us suppose, by way of thought-experiment,[20] that God were to exercise his infinite power and annihilate all sublunary matter. What, Gassendi asks, would ensue? According to Descartes, the surfaces of the surrounding celestial matter would come together, leaving no empty space. On the Atomist position, however, the situation is quite different. Intangible extension – 3D extension without *antitypia* – is perfectly conceivable, hence logically possible. After God has annihilated all sublunary matter, says Gassendi,

> I ask whether or not . . . we do not still conceive the same region between the surface of the lunar sphere that had been there, but now empty of the elements and devoid of every body. That God can preserve this lunar sphere intact and reduce the bodies contained in it to nothing and prevent any other body from taking their place no one would deny, except a man who denies God's power.[21]

Furthermore, he continues,

> since the sphere of the moon is circular, if you take a point on its round surface don't you think that there is a certain interval, or distance, from that point to the one opposite it? And isn't this distance a certain length, namely an incorporeal and indivisible line, which is the diameter of the sphere and on which there is a midpoint that is the centre of the realm or sphere and where once the centre of the earth existed?[22]

Within this empty space we can map out, in imagination, the erstwhile spheres of FAWE: we can construct circles, mark in

---

[19] See Charleton, pp. 16–18.

[20] This thought-experiment was common among scholastic authors – we have previously encountered it in Albert of Saxony (Chapter 6, p. 235).

[21] *Syntagma*, Physics, Section 1, Book 2, Chapter 1; Gassendi, p. 386.

[22] *Ibid.*, p. 387. For exactly the same argument, see Charleton, pp. 63–64.

their radii and diameters, etc. – in short, perform all the operations of elementary geometry. There is therefore a genuinely 3D *space*, with its own incorporeal dimensions, quite independent of those of bodies. The places formerly occupied by bodies survive – and retain their shapes and sizes – after the annihilation of those bodies.[23] (God can annihilate bodies, but not Space itself.[24])

Vacuum $_1$ is therefore absolute space in the ontological sense; it exists *per se*, quite independently of the existence of bodies. Assuming the creation of our physical universe in time, it was created *in* a pre-existent space or vacuum$_1$, which, prior to the Creation, existed as a void or vacuum$_2$. Into this receptacle, God created material things: the corporeal dimensions of bodies interpenetrate with the incorporeal dimensions of the parts of space. To avoid paradoxes, Gassendi insists, it is essential to distinguish sharply between the dimensions of space and those of matter. A place, he explains, is

> the space or interval made up of the three dimensions length, breadth and depth in which it is possible to hold a body, or through which a body may travel. But at the same time it must be said that its dimensions are incorporeal; so place is an interval, or incorporeal space, or incorporeal quantity.[25]

> The length, width, and depth of some water contained in a vase would be corporeal; but the length, width and depth that we would conceive as existing between the walls of the vase if the water and every other body were excluded from it would be spatial.[26]

Two corporeal dimensions (i.e. two bodies) cannot interpenetrate; a corporeal dimension can, however, interpenetrate with a spatial one. Indeed, every body *must* occupy a place equal in magnitude to itself: the idea of a body that does *not* occupy some place is incoherent. 'Wherever,' says Gassendi, 'it is possible to assign corporeal dimensions, we may understand that there are also incorporeal ones corresponding to them'.[27] Incorporeal dimensions are, says Charleton, 'every

---

23 See Grant, 9, p. 208.

24 *Syntagma*, Physics, Section 1, Book 2, Chapter 1; Gassendi, p. 384.

25 *Ibid.*, p. 385.

26 *Ibid.*, p. 385. The originator of this argument for an incorporeal 3D space or vacuum was Philoponus (see Chapter 2, pp. 74–76).

27 Gassendi, p. 388. See also Grant, 9, p. 209.

where coexistent, and compatient, ... with corporeal dimensions, without reciprocal repugnancy.[28] Nature does not 'abhor' the interpenetration of dimensions as such, only of corporeal ones.

But in what *substance*, Cartesians and Aristotelians will demand, do the dimensions of space inhere? 3D extension, Descartes and Aristotle agree, must appertain to something substantial: there can be no quantity of nothing. Here Gassendi follows Patrizi[29] in rejecting the whole structure of the substance-accident metaphysic. Space and Time, he affirms, though undeniably *real* (not fictitious or imaginary[30]) fall under neither category of Being: the substance/accident distinction is therefore not an exhaustive one. Place and Time

> do not come under the division of reality (or being in general) into substance and accident.[31]

Although often conceived as 'corporeal accidents' (accidents *of* bodies), Space and Time in fact exist *per se*, quite independently of bodies: 'even if there were no bodies, there would still be an unchanging place and an evolving time'.[32] They are not, however, incorporeal substances, since they do not partake of action and passion: unlike God and the angels, says Gassendi, Space

> cannot act or suffer anything to happen to it, but merely has the negative quality of allowing other things to occupy it or pass through it.[33]

In short, Space and Time elude the Aristotelian net of categories: they are, Gassendi affirms,

> real things, but of a nature that they are neither substances nor accidents and are not included in that general classification and should rather by considered as fundamental elements of all classification.[34]

Being (*ens*) does not divide neatly into substance and accident, but 'space and time must be added as two members

28 Charleton, p. 68.

29 See Chapter 6, p. 248. Gassendi's debt to Patrizi is freely acknowledged (see Grant, 9, p. 209).

30 Charleton (p. 66) speaks of them as 'realities, things, or not – nothings': empty space, he insists, is *not* a fiction or non-entity.

31 Gassendi, p. 383.

32 *Ibid.*, p. 384.

33 *Ibid.*, p. 389.

34 *Ibid.*, p. 383.

of the classification'.[35] All entities, in this revised ontology, are either substance or accident,

> or *place*, in which all substances and all accidents exist, or *time*, in which all substances and all accidents endure.[36]

All other things exist *in* some place and *for* some period: Space and Time are therefore *necessary preconditions* of their existence. Space and Time, Charleton insists, are categories of Being, independent of and ontologically prior to their contents. (Space and Time could exist without objects and events respectively, but not *vice versa*.)

Between the walls of an empty vessel, then, there is simply space: 3D incorporeal extension – once we have rejected that 'epidemick errour'[37] of scholasticism, the substance/accident metaphysic, this will seem a perfectly reasonable response. This space has magnitude, but to ask 'quantity of what?' is simply to miss the point: it is a quantity inherent *in no substance*. Only the 'tyranny of matter', says Gassendi, leads one to think of quantity as necessarily appertaining to bodies.[38]

Space is therefore an infinite, eternal, three-dimensional, incorporeal (resistanceless) reality, a limitless immobile receptacle within which all substances and accidents have their places or *loci*. (A place or *locus* is no more than a delimited part of this universal vacuum$_1$.) This ontologically absolute space can also function as a *physically* absolute space – i.e. as a reference-frame against which motions and velocities can intelligibly be determined. It *makes sense*, Gassendi affirms (following Oresme and the condemnations of 1277[39]), to suppose that God could move the whole material universe in the infinite extracosmic void. God has created the world in one part of the infinite void; He could, at will, transfer it to another.[40] Without absolute space as a frame of reference, the idea would be senseless.

Space and Time, infinite, eternal, and incorporeal, are, for Gassendi, *uncreated by* and *independent of* the Deity. Space, says Charleton bluntly, 'must be unproduced by, and

---

35   *Ibid.*, p. 384.

36   Quoted from Grant, 9, p. 210. See also Koyré, 4, p. 85.

37   Charleton, pp. 64–65.

38   See Harré, 1, p. 67.

39   See Chapter 6, pp. 239–240.

40   Gassendi, p. 388; Charleton, p. 67.

independent upon, the original of all things, God'.[41] Space
and Time provide the necessarily existent framework within
which He has created our cosmos. Although the postulation
of two infinite, eternal and incorporeal realities independent
of God may sound heretical (even polytheistic), the
alternative – making Space the 'immensity' of the deity
Himself – seems still worse.[42] God cannot be literally
extended in space without having distinct parts, in blatant
contradiction to the 'all in every part' account of the divine
omnipresence and the correlated insistence on the divine
indivisibility. One cannot identify space with God's immensity without making God Himself a spatially extended Being.
Gassendi prefers one heresy (Space an infinite and eternal
reality uncreated by God) to the other (God Himself an
extended – and hence divisible – Being). Although God is of
course omnipresent *in* infinite space, His presence does not
constitute that extension, which exists *per se* and in its own
right.[43]

Of this infinite space or vacuum$_1$, a small portion was
chosen by God to be the place *of* our corporeal world –
Gassendi and Charleton both favour the Stoics' picture of a
finite material cosmos within an infinite extracosmic void.[44]
Given (a) an infinite void and (b) God's omnipotence it
follows, says Charleton,[45] only that God *could* have created
an infinite amount of matter: neither experience nor Holy
Writ tells us that He has actually done so. (Genesis speaks
only of the creation of one world.) Charleton is thus sceptical
of Bruno's use of the Platonic 'Principle of Plenitude' to
provide a theological guarantee of an infinite Creation – we
are not, he asserts, privy to the counsels of the Deity. In all
probability, he concludes, there is but one material cosmos,
surrounded on all sides by a limitless vacuum$_2$.

Within our world, moreover, there must be (interstitial)
vacua$_2$ if there are to be starts of motion: Gassendi reaffirms
the classical Atomist argument from local motion to the

---

41 Charleton, p. 68; Gassendi, p. 389.

42 See Grant, 9, pp. 212-213; Charleton, p. 68. Gassendi attempts to tone
down the heretical sound of this idea by insisting that space is in itself
nihil positivum, not a positive reality in its own right. God therefore
creates all positive things; He need not create space itself. Judged by
medieval standards, however, Gassendi's position still reeks of heresy.

43 Gassendi, p. 396.

44 See Grant, 9, pp. 207-208.

45 Charleton, p. 12 ff.

existence of void space.[46] Given incompressible matter*, immune to intrinsic condensation and rarefaction, only the 'cyclical mutual replacement' theory would serve to evade this conclusion. In a plenum, however, says Gassendi, motion could not *start* and *stop* as it is seen to do in our world; there could be, Charleton agrees, 'no beginning of motion'.[47] There must exist, therefore, interstitial or 'disseminate' vacua, minute void spaces 'betwixt the convened particles of bodies';[48] these may in turn be 'collected' (as in the Toricellian experiment, of which more anon), to produce a 'coacervate' vacuum$_1$ giving in the final count three varieties of void or vacuum$_2$: extracosmic, interstitial and coacervate.

If we turn now to the writings of Henry More, passing over – not without regret – the views of Otto von Guericke,[49] Thomas Hobbes[50] and Isaac Barrow,[51] we discover another, and highly idiosyncratic, conception of space. More accepts both Cartesian dualism and the substance-accident metaphysic, but differs from Descartes on the correct categorisation of the two contrasted types of substance. According to Descartes, body = 3D extension alone; Mind is unextended and characterised by its power of thought: God, angels and souls are present in space only virtually and not substantially. This is the point at which More takes issue with him. A spirit

---

[46]   See Chapter 2, pp. 45–46.

[47]   Dijksterhuis, p. 426; Charleton, p. 22. The argument derives of course from Lucretius (Chapter 2, p. 66): it presupposes the crucial principle that causal priority involves temporal priority.

[48]   Charleton, p. 21. See also Brett, p. 36.

[49]   Guericke's theory of space and the vacuum is strongly influenced by scholastic concepts. Like many late-scholastic authors, he opts for a dimensionless ('imaginary') void which – precisely because it *lacks* real 3D extension – can be identified with the Divine immensity *without* making God a spatially extended Being. (See Grant, 9, p. 215 ff.)

[50]   For Hobbes' intriguing (but ultimately unsatisfactory) distinction between Space in itself (a mere 'phantasm' or internal cinema-screen) and the objective magnitudes of bodies, see *De Corpore* (*Works*, Vol. 1, p. 101 ff.).

[51]   Barrow accepts, *pace* Descartes, a space independent of bodies, but attributes to this space merely a *potential* existence. When one says that an evacuated vessel contains empty space, he claims all this *means* is that a body (of some particular size and shape) can be placed there. Real magnitude involves impenetrability: there are no actual incorporeal dimensions. But surely it is natural to reply, a particular void space can accommodate a body of some specified size and shape only because *it* (i.e. the void) possesses those dimensions – Barrow's attempt at a compromise view is of dubious viability. (See Strong, 4, pp. 160–162; Grant, 9, pp. 233–238; Burtt, pp. 149–150.)

cannot, he argues, act at any place without being present there: there is no *virtus* without substance. But the powers of the human soul are felt throughout the whole organic body; and God's might pervades the entire universe. Spiritual substances are therefore extended in space – this is the inescapable conclusion.[52] The human soul extends throughout its associated body; God is literally omnipresent.[53] Thus More to Descartes:

> You define matter or body in too broad a fashion, for it seems that not only God, but even the angels, and anything which exists by itself, is an extended being . . . Now the reason which makes me believe that God is extended in His fashion is that He is omnipresent, and fills intimately the whole universe and each of its parts.[54]

Every substance, More insists, is extended: 'There is no real entity, but what is in some sense extended.'[55] Extension, however, does not necessarily involve divisibility: if we think of the soul as analogous to a point-source of light, surrounded by a luminous aura, we shall grasp the concept of something extended yet *indivisible* – the 'orb' or 'secondary substance' is extended, yet inseparable from its emanent cause, the indivisible point-source.[56]

If spirits are extended – if, indeed, as More claims, 'there is no Substance but it has in some sort or other the Three dimensions'[57] – in what way is spiritual substance to be distinguished from corporeal? Our idea of a spirit, More replies, involves

> these several powers or properties, viz self-penetration, self-motion, self-contraction and dilation and indivisibility.[58]

How, one might ask, can a spirit contract and expand itself without loss and gain of substance? More's answer runs as follows. If, he says, one folds a sheet of paper, the loss in 2D

---

52 Descartes, says More, is a 'nullibist': he denies that spirits exist *anywhere* in the universe (i.e. in any place).

53 See Čapek, Ed., p. 85; Koyré, 3, p. 111.

54 Quoted from Burtt, p. 137.

55 H. More, 1, Vol. 1, Preface General, p. xii.

56 *Ibid.*, 'The Immortality of the Soul', p. 27, and 'The Antidote against Atheism', p. 150.

57 *Ibid.*, 'The Antidote against Atheism', p. 15.

58 *Ibid.*, p. 15.

surface area is compensated by an increase in depth, conserving the total 3D volume. If, by analogy, we have a loss of 3D bulk or volume, it can be compensated by a gain in the fourth dimension, leaving the total 4D hypervolume the same. In order then to explain the powers of self-contraction and dilation and mutual interpenetration he attributes to his spirits, More finds himself forced to ascribe to them 'essential spissitude' or 4D extension![59]

Matter, by contrast to Spirit, is impenetrable, inert (not self-moving), divisible into distinct parts, and incapable of self-contraction and dilation – this gives us our quite distinct ideas of soul and body. More briefly, the 'immediate' properties of material substance include extension, divisibility, and impenetrability; those of spiritual substance include extension, penetrability, and indivisibility. 'All substance,' says More, 'has Dimensions, that is Length, Breadth, and Depth: but all has not Impenetrability':[60] it follows that 'there is no necessary connexion discernible betwixt Substance, with three dimensions, and impenetrability'.[61] *Pace* Descartes, it is not 3D extension but '*antitypy*' or impenetrability that constitutes something corporeal.[62]

A particular 3D extension or magnitude must, then, given the substance-accident metaphysic,[63] belong to some substance or other; it need not, however, inhere in any *material* substance. Since spirits too are extended, a particular magnitude may qualify such an immaterial substance. Hence there could be a space devoid of all matter (i.e. of resistance and *antitypy*) but nevertheless filled with substance – a physical but *not* an ontological 'void'.

Suppose, then, that God annihilates all the matter inside a particular closed vessel. According to Descartes, 3D extension can only inhere in material substance, so the walls of the vessel must come together. On the rival view of Gassendi, spatial dimensions inhere in no substance: all we can say is that Space remains in the empty vessel. More's answer is quite different. *Pace* Descartes, the walls do not come together. Nevertheless, *pace* Gassendi, quantity must be quantity *of something*. 'I argue,' says More, 'that the divine

[59]   *Ibid.*, 'The Immortality of the Soul', p. 20.

[60]   *Ibid.*, p. 41.

[61]   *Ibid.*, p. 19.

[62]   See Čapek, Ed., p. 86; Koyré, 3, pp. 111–112.

[63]   For More's acceptance of the substance-accident metaphysic, see McGuire, 5, pp. 463–464; Koyré, 3, pp. 146–147.

extension lies between the sides of the vessel'.[64] Spatial dimensions are quantities of God!

Now God is substantially omnipresent throughout the whole limitless universe. If follows that, merely in virtue of His existence, there is an *infinite absolute Space*, a boundless, 3D, immaterial receptacle in which all bodies and created spirits exist. This Space *could* exist devoid of material contents if God so willed, although in point of contingent fact, More insists, there is no such void or vacuum$_2$, since

> the divine creative activity, never idle at any point, has created matter in all places, without leaving the least minute space void.[65]

Only the Platonic 'Principle of Plenitude', then, proscribes the existence of space devoid of all (material) tenants. The existence of such 'empty' space is logically possible; it is merely that, given God's boundless fecundity, no such void will be left in His handiwork.

There is then, says More, an infinite and incorporeal Space, which has the following list of properties. It is, he says,

> Eternal, Complete, Independent, Existing in Itself, Subsisting by Itself, Incorruptible, Necessary, Immense, Uncreated, Uncircumscribed, Incomprehensible, Omnipresent, Incorporeal, All-penetrating, All-embracing, Being by essence, Actual Being, pure act . . .[66]

But all these are also attributes of the deity! Are there then two (or even three, if one includes Time) infinite, incorporeal, necessarily existent Beings? This savours of heresy and polytheism. Perhaps, then Space can be identified with God, in virtue of the above list of common properties? In the secret and esoteric wisdom of the Cabbala, More notes, God is called *māqôm*, the '*place*' of the world.[67] One should not, however, commit oneself unreservedly to the God = Space equation: we do not, after all, usually think of Space as itself active, intelligent, and worthy of our worship. In the *Enchiridion Metaphysicum* (1671) More outlines the following three views:

64 Quoted from Čapek, Ed., p. 87. See also Koyré, 3, p. 112.

65 Quoted from Burtt, p. 138. For this use of the Platonic 'Principle of Plenitude' against vacuism, see Chapter 2, p. 81; Chapter 6, pp. 249–250.

66 From Koyré, 3, p. 148.

67 Copenhaver, p. 519.

V1.   Space = the immensity or omnipresence of the Divine essence.

V2.   Space = the bare possibility of placing a body.

V3.   Space = God Himself.[68]

More finally rejects Barrow's theory (V2), and opts for a hybrid of V1 and V3. Space is, he asserts,

> a certain rather confused and vague representation of the divine essence or essential presence, in so far as it is distinguished from his life and activities.[69]

Our idea of space – encapsulated in the twenty common titles listed above – represents accurately God's immensity and omnipresence, but does not take into account His life and activity. Space is, therefore, an *aspect* of the Deity; His immensity.[70] But, if Space = God's immensity, it is an attribute of God, who is therefore a spatial Being, a substance with spatially distinct parts. This is, as Grant sees,[71] the logical conclusion of More's metaphysics: the radical conception of the Deity as literally extended from infinity to infinity, sustaining by His presence the absolute[72] and incorporeal space within which as a receptacle all things – bodies and spirits alike – are contained.

Newton's early monograph, *De Gravitatione et Aequipondium Fluidorum*,[73] manifests the influences of both More and Gassendi. After a sustained polemic against Descartes' relativistic account of motion, Newton proceeds to state his own positive views. Physics requires, he claims, the concept of Absolute Space as a frame of reference:

> It is necessary that the definition of places, and hence of local motion, be referred to some motionless thing such as extension alone or space in so far as it is seen to be truly distinct from bodies.[74]

---

68   See Burtt, pp. 140–141.

69   Quoted from Burtt, p. 141. See also Power, p. 290.

70   H. More, 1, vol. 1, 'The Antidote against Atheism', p. 163. See also McGuire, 5, p. 480; Grant, 9, p. 223.

71   Grant, 9, p. 227.

72   More argues at length against the relativism of Descartes and in favour of absolute space as a physical *frame of reference*. See H. More, 1, Vol. 1, Preface General, p. xi, and Burtt, pp. 139–140.

73   For an English translation of *De Gravitatione*, see A. R. and M. B. Hall, Eds., 4, p. 122 ff.

74   *Ibid.*, p. 131.

Absolute space *qua* reference-frame depends, then, on ontologically absolute space, space not dependent on matter for its existence and properties. Such a space would, Newton clearly feels, provide a sort of 'grid' with real dimensions and fixed points, within which

> the positions, distances, and local motions of bodies are to be referred to the parts of space.[75]

What is the ontological status of this absolute Space? It is, says Newton, following Gassendi[76] rather than More, neither substance nor accident, but merely Space. It cannot be an accident, since it exists independently of bodies ('we cannot believe that it would perish with the body if God should annihilate a body'[77]), yet it cannot be described as substantial, since

> it is not absolute in itself, but is as it were an emanent effect of God, or a disposition of all being; on the other hand, because it is not among the proper dispositions that denote substance, namely actions . . .[78]

Extension is, says Newton, 'an emanent effect' of God (it stands to God, on More's analogy, as an orb of illumination does to a light-source), and a 'disposition of all being' (whatever exists, exists *somewhere* – i.e. in some place). Space exists actually (*pace* Barrow) as an infinite and immobile incorporeal 'box' or receptacle in which all substances and accidents have their being. The parts of Space must be absolutely immobile: their individual identities are so fixed by their positions *vis-à-vis* one another that it makes no sense to assume that they could move. For a place to move would be to violate its own identity-conditions.[79]

Space, Newton continues, is 'eternal in duration and immutable in nature, and this because it is the emanent effect of an eternal and immutable being'.[80] *Qua* emanent effect of God, space exists necessarily (the existence of an emanent

---

[75]   *Ibid.*, p. 137.

[76]   Newton, says McGuire (5, p. 464), 'is largely a Gassendist on the problems of space, time and existence'. See also Koyré, 4, pp. 85–86; Grant, 9, pp. 241–242.

[77]   A. R. and M. R. Hall, Eds., p. 132.

[78]   *Ibid.*, pp. 131–132.

[79]   *Ibid.*, p. 136.

[80]   *Ibid.*, p. 137.

effect flows simply from that of its cause[81]), and hence eternally. God did not create His own omnipresence: it is co-eternal with Him.

After this lengthy discussion of Space, Newton moves on to investigate the nature of body. All one needs, he claims, to account for the existence of bodies is to suppose (a) that God has endowed certain parts of space with a particular set of powers (impenetrability, inertia, and causal interaction with spirits) and (b) that he allows those powers to be transferred from one part of space to another in accordance with fixed laws. Such a surrogate or pseudo-body will, Newton claims, be indistinguishable from what we think of as a 'real' body, whatever that may be: 'there will be no property of body which this does not possess'.[82] It will be impenetrable, tangible, capable of imparting and receiving impulses, etc. – in short, it will lack nothing necessary to corporeality. It follows, says Newton, that

> if several spaces of this kind should be impervious to bodies and to each other, they would sustain the vicissitudes of corpuscles and exhibit the same phenomena.[83]

Bodies, then, *could* be no more than '*determined quantities of extension which omnipresent God endows with certain conditions*'[84] (mobility, impenetrability, inertia, causal inter-action with spirits). Impenetrability therefore belongs, *pace* Descartes, not to extension as such: it is a divine 'gift' to certain selected parts of space, not a necessary corollary of extended magnitude itself. The link between extension and *antitypia* is a purely contingent one, dependent entirely on the divine Will. Space is necessarily existent, uncreated, eternal, and infinite; matter exists contingently, was created in time, and is limited in extent – to equate matter with space, as Descartes does, is implicitly to assert the necessary and eternal existence of *matter* and thus to lapse into atheism.[85]

The subject of the ontological status of Space recurs in both the *Principia* and the *Opticks*. After defending Absolute Space

---

[81]   See H. More, 1, Vol. 1, 'The Immortality of the Soul', pp. 27–28. An emanent effect is, he says, 'coexistent with the very Substance that is said to be the cause thereof; an emanent cause produces its proper effect 'merely by Being, no other activity or causality interposed . . .'

[82]   A. R. and M. B. Hall, Eds., 4, p. 139.

[83]   *Ibid.*, p. 139.

[84]   *Ibid.*, p. 140.

[85]   *Ibid.*, pp. 142–144. See also Koyré, 4, pp. 93–94.

qua immobile reference-frame in the famous scholium to Book One of *Principia*, Newton moves on in the General Scholium to Book Three to discuss wider philosophical issues, notably God's presence in and dominion over His Creation. God, says Newton,

> is not eternity and infinity, but eternal and infinite; he is not duration and space, but he endures and is present. He endures forever, and is everywhere present; and, by existing always and everywhere, he constitutes duration and space. Since every particle of space is *always*, and every indivisible moment of duration is *everywhere*, certain the Maker and Lord of all things cannot be never and nowhere.[86]

Although Space and Time are ontologically dependent upon God, they cannot be identified with Him. Like More, Newton is aware of the cabbalistic use of *māqôm* ('place') as a divine name, e.g. in the statement that God is the place of the world. But, whereas More tends to identify Space with God in virtue of their common properties, Newton insists that they be distinguished as consequence from condition.[87] The Jews, he claims, do not assert that God = Space in a literal sense: *māqôm* is used as a divine name only as 'figure or illusion', a *metaphor* for the divine omnipresence.[88] Newton is not an advocate of the divinisation of Space of More and Raphson.[89]

Newton does, however, accept from More the *substantial* presence of the Deity in Space. God, he states plainly

> is omnipresent not *virtually* only, but also *substantially*; for virtue cannot subsist without substance.[90]

Thus also Clarke in the correspondence with Leibniz:

> God, being omnipresent, is present to every thing, essentially and substantially. His presence manifests itself indeed by its operation, but it could not operate if it were not there.[91]

[86]   Newton, 1, p. 545.

[87]   Copenhaver, p. 544.

[88]   McGuire, 6, pp. 126-127; Copenhaver, p. 544.

[89]   Raphson takes the God = *māqôm* utterance in 'a genuine and literal sense', and perhaps takes the deification of Space further than anyone else: Absolute Space is, he affirms, 'the very immensity of the first cause'. See Copenhaver, pp. 529-540.

[90]   Newton, 1, p. 545.

[91]   H. G. Alexander, Ed., pp. 33-34.

Or, as Gregory remarks of Newton, 'the plain truth is, that he believes God to be omnipresent in the literal sense'.[92] Absolute space is, he says in the *Opticks*,[93] like God's 'sensorium', the divine equivalent of our sensoria, the mental cinema-screens in which our perceptions occur. (Newton accepts a representative theory of vision.) Leibniz's accusation that this makes God into a sort of pantheistic *Anima Mundi*[94] seems to have more than a little truth.

Newton, then, argues on *physical* grounds[95] for absolute space as a frame of reference (with respect to which God could move the entire material universe![96]); assumes that such a reference-frame must be ontologically independent of bodies, and has doubts and hesitations concerning its exact relation to God. In the *De Gravitatione* it is described as an 'emanative effect' of God, and a 'disposition of all Being', itself neither substance nor accident. In the General Scholium the substantial omnipresence of the deity is asserted, but the God = Space identification denied. In the *Opticks* Space is likened to God's sensorium. In the correspondence with Leibniz Clarke (presumably with Newton's approval) first characterises Space as 'a property, or a consequence of' God,[97] but later abandons 'consequence' and flatly asserts that Space is a *property* of some infinite and incorporeal Being, which can only be God.[98]

The commentators disagree over the proper interpretation of all this material. Grant[99] emphasises Newton's debt to More, and claims that, even if Newtonian Space is not actually to be identified with the deity, it can still be conceived as an *attribute* of Him, His immensity or omnipresence; McGuire[100] stresses Newton's early Gassendism (in *De*

92   Quoted from Thackray, 2, p. 27.

93   Newton, 2, p. 370. The sensorium figure occurs again (without the qualification 'as it were') at p. 403. The *'tanquam'* was also absent in the first Latin edition, a copy of which may have been in the possession of Leibniz (see Koyré and Cohen, 2).

94   See H. G. Alexander, Ed., pp. 11–12, and elsewhere.

95   Cf. of course the bucket and the rotating globes of the Scholium on Space and Time.

96   See H. G. Alexander, Ed., p. 32; Burtt, p. 258.

97   H. G. Alexander, Ed., p. 31.

98   *Ibid.*, p. 121.

99   Grant, 9, pp. 244–245.

100   There is no reason to suppose, says McGuire (6, p. 128), that the Space = God's sensorium idea was meant to be taken literally: this is yet another figure or metaphor.

Gravitatione), his scepticism regarding the substance-accident metaphysic, and his prevailing caution in committing himself to any specific connection between God and Space. Both can accommodate most of the available textual evidence; neither is absolutely satisfying. Our verdict must be *non liquet.*

We turn now from Newton to his arch-rival Leibniz, in whose writings we can discern a radically different conception of space. At first, says Leibniz in his 'New System' (1695), after freeing himself from Aristotelianism, 'I accepted the void and the atoms':[101] traces of this early Atomism are manifest throughout his *oeuvre*. The postulation of a *space-substance*, he claims, rebuts Descartes' anti-vacuist arguments: advocates of the vacuum can simply claim that the dimensions of void space inhere in such a self-subsistent reality.[102] As late as 1672 Leibniz can write that

> in natural philosophy I am perhaps the first to have proved thoroughly . . . that there is a vacuum.[103]

The Cartesian identification of matter with extension is, Leibniz alleges, quite mistaken: from 3D extension alone one cannot legitimately derive all the other features essential to corporeal nature. From such an inadequate reductive basis, one arrives at a weird physics and *no dynamics at all* – hence the lapse into occasionalism. A correct account of the nature of body would have to include not merely extension but also the following factors:

1. Impenetrability or *Antitypia*. This feature of bodies – a feature quite distinct, incidentally, from *hardness* (water is as impenetrable as diamond, though of course much less hard) – does not, Leibniz insists, follow from 3D extension alone:

> Impenetrability is not a consequence of extension; it presupposes something more. Place is extended, but not impenetrable.[104]

Extension alone does not distinguish body from void: the idea of pure extension is that of empty space.[105] What differentiates matter from space is, primarily, *antitypy.*

101   Leibniz, p. 454.

102   See the 'Critical Thoughts' on Descartes' *Principles* (1692) in Leibniz, p. 392.

103   Quoted from Russell, p. 227. Cf. also Northrop, p. 423.

104   *Ibid.*, p. 228.

105   See the 'Theory of Abstract Motion' of 1671 (Leibniz, p. 143), and the 'Critical Thoughts' of '692 (*Ibid.*, p. 390).

2.   *Inertia* or 'passive force'. All bodies resist changes in their state of motion: this resistance is proportional to their bulk or volume (*moles*) of gross matter, the Leibnizian analogue to the Newtonian concept of mass. If, says Leibniz in the *Specimen Dynamicum*, body were nothing more than extension alone, 'it would be no more difficult to move a large body than a small one,[106] which is simply not the case.

3.   '*Active Force*'. To deny to material things all active powers is, says Leibniz, to lapse into pure occasionalism. Yet the matter = extension formula of Descartes does just that: it is impossible for a Cartesian to explain how one body can ever have the power to move another. An adequate theory of matter must incorporate the *vis viva* or 'living force' (α mv^2), possessed by all bodies and conserved throughout the course of Nature. There is in matter, says Leibniz in the *Specimen Dynamicum*, 'something prior to extension, namely, a natural force everywhere implanted by the author of nature', and present 'everywhere in nature', even when not manifest.[107]

Matter, therefore, involves *Extension, Antitypy, Inertia* and *Active Force*: this list gives at least an approximation to its essential nature. But, since extension alone does not of itself involve or imply any of the other three factors, the *concept* of a vacuum$_2$ is perfectly *self*-consistent: it involves no contradiction to divorce extension from *antitypy*, inertia, and force. One must, of course, find a proper subject for the dimensions of a 'void' space to quality – either a space-substance, or God (Leibniz is a staunch supporter of the substance-accident metaphysic) – but neither of these hypotheses is logically deficient in any way. If such theories are to be discarded, it must be on other grounds.

It is only, says Leibniz, through 'higher' (i.e. metaphysical) considerations that the void can be decisively banished from our world-view.[108] A perfect void or vacuum$_2$ must either be itself substantial or be an attribute of some infinite incorporeal Being, e.g. of course the divine immensity. Leibniz has arguments against both views. Although, he says,

I distinguish the notion of extension from that of body, I still believe that there is no vacuum, and even that there is

---

106   From the *Speculum Dynamicum* of 1695 (Leibniz, p. 440). See also Russell, p. 229, who makes the important distinction between mere persistence in a particular state and resistance to a change of state.

107   Leibniz, p. 435. For the significance of Leibniz's dynamics, see Costabel.

108   See the 'Conversation of Philarète and Ariste' (1711) in Leibniz, p. 624.

no substance which can be called space, i.e. that there is no subject having only the attribute of extension.[109]

The parts of such a space-substance would have to be *qualitatively identical* with one another (Space is supposed to be perfectly homogeneous), yet *numerically distinct*. This, says Leibniz, violates his 'Principle of the Identity of Indiscernibles' (P.I.I.). Thus in his 'First Truths' (c. 1680–1684), he writes:

> *There is no vacuum.* For the different parts of empty space would be perfectly similar and congruent with each other and could not by themselves be distinguished. So they would differ in number alone, which is absurd.[110]

Absolute Space is, he claims, 'an idol of some modern Englishmen', but is refuted by sound metaphysical principles. In an Absolute Space, for example, God could have had no *sufficient reason* to create the material universe in one place rather than another: all places are alike. But, without a sufficient reason, God cannot act – i.e. create a world – at all![111]

The postulation of a space-substance independent of the deity is also questionable on theological grounds: to avoid the taint of heresy, says Leibniz,

> some have believed it to be God himself, or one of his attributes, his immensity. But since space consists of parts, it is not a thing which can belong to God.[112]

The God of More (and Newton), being substantially omnipresent, consists of distinct spatial parts: each part of the universe contains a different 'portion' of the deity! But, says Leibniz, such a theory is absurd and impious. For Leibniz, as for Descartes and the schoolmen, God's presence in the world is virtual only: He has no need, for example, of a sensorium in which to perceive what goes on in the universe.[113]

---

109   *Ibid.*, p. 622. Russell (pp. 118–119) is correct to stress Leibniz' concern to refute the existence of an ontologically absolute space or space-substance.

110   Leibniz, p. 269.

111   H. G. Alexander, Ed., pp. 25–27; p. 36. Leibniz' God is like Buridan's ass: He cannot choose arbitrarily between two distinct but equivalent alternatives. Cf. Northrop, p. 427; Ballard, pp. 53–54.

112   H. G. Alexander, Ed., p. 25. See also pp. 67–68.

113   H. G. Alexander, Ed., pp. 11–12. For the virtual omnipresence of Leibniz' God, see Russell, pp. 124–125.

Space and Time, Leibniz concludes, are no more than relations, possible dispositions, types of order: they could have no existence apart from the objects and events that they order. Space is 'an order of coexistences, as time is an order of successions': being 'merely relative' they cannot conceivably exist independently of their proper *relata*.[114] This relational account of Space of Time is, Leibniz argues, perfectly compatible with P.I.I.: if two state-descriptions assign the same spatiotemporal relations to the same set of objects, they are descriptions of one and the same state of affairs. (The absolutist, on the other hand, posits a distinction without a difference.) 'To suppose two things indiscernible,' he says, 'is to suppose the same thing under two names'[115] – the notion that God could move the whole material world as a unit is a supposition of this meaningless kind, 'a change without any change'.[116] (Leibniz still has difficulties, as Clarke is not slow to point out, with accelerations and circular motions, but in a letter to Huygens he explicitly states that the principle of mechanical relativity must apply to these too.[117]) All that can be *meant* by speaking of local motion, he insists, is change of spatial relations with other objects: the notion of space as an absolute frame of reference lacks sense.[118]

We must of course not forget that, for Leibniz, in the final analysis, Space and Time are ideal, not real: they exist only as appearances to the simple substances or 'monads' which truly exist *per se*. The ideality of Space and Time can be proved, Leibniz feels, *either* from P.I.I. (which refutes the existence of a homogeneous space-substance), *or* from the paradoxes of infinity implicit in the conception of Space and Time as *continuous*.[119] Either argument will suffice to show that Space cannot be thought of as something metaphysically real, far less substantial.

114   H. G. Alexander, Ed., p. 25.

115   *Ibid.*, p. 37.

116   *Ibid.*, . 38, pp. 73–74.

117   Leibniz, p. 418. See also Ballard, p. 57.

118   See the relativistic account of the meanings of spatiotemporal terms in H. G. Alexander, Ed., pp. 69–70. Unfortunately, however, Leibniz also admits a real difference between 'an absolute true motion of a body' and 'a mere relative change of its situation with respect to another body' (*Ibid.*, p. 74). The distinction can, he suggests, be made on causal grounds: when the cause of a particular motion is internal to the body in question, he says, 'that body is truly in motion'. It is difficult to see how, if at all, this startling admission can be reconciled with the rest of Leibniz' philosophy (see Russell, pp. 86–87).

119   See of course Chapter 9, Part A.

The existence of empty space or vacuum$_2$ would also, Leibniz feels, violate the Platonic 'Principle of Plenitude'. If, he claims, God had left some space empty, He would not have actualised all His creative possibilities – He would have failed to create some possible creatures. But, says Leibniz (following in the footsteps of Bruno and the whole Platonic tradition[120]), God *must*, in virtue of His (moral) nature, have produced the maximum possible richness and diversity of creatures – leaving a void space would contravene this principle, and would therefore contradict the divine nature.[121] In his 'Paris Notes' of 1676 Leibniz lays it down as an axiom that 'the greatest possible quantity of essence exists': from this it immediately follows, he claims, that 'there is no vacuum in space and time'.[122] In a supposed void, Leibniz asserts, God could always create more creatures (without, crucially, detracting from the perfection of those already in existence[123]); therefore, He must have actually done so.[124] Our universe can only be a plenum.

I conclude this section with an examination of the views of John Locke on the issues in question, as expressed in his famous *Essay*. We begin with his discussion, in Book II, Chapter 4, of the idea of *solidity*. The idea of solidity, Locke begins,

> we receive by touch; and it arises from the resistance which we find in body to the entrance of any other body into the place it possesses, till it has left it. There is no idea that we receive more constantly from sensation than solidity.[125]

Solidity is, he continues, the positive ground of the power of impenetrability, although the two ideas are often confused. It is, moreover, 'the idea most intimately connected with and essential to body'.[126] We discover it in all bodies of sensible

---

120 For an in-depth study of this tradition, see the masterful work of Lovejoy.

121 H. G. Alexander, Ed., p. 16; Koyré, 4, p. 166.

122 Leibniz, p. 157.

123 This premiss is merely stated without argument: a Newtonian such as Clarke is under no pressure to accept it. No doubt, he replies, the present matter/void ratio is the one best suited to God's inscrutable purposes: who are we to deny this? And perhaps the *void itself* contributes to the perfection of the physical universe – e.g. to the stability and longevity of the Newton world-system.

124 H. G. Alexander, Ed., p. 40.

125 Locke, *Essay*, II, 4, i, p. 93.

126 *Ibid.*, p. 94.

magnitude and *impute* it to the invisibly small corpuscles of which those bodies are composed. By its solidity, says Locke, each and every body, however soft, *fills* its allotted portion of space so as to *exclude* other solid substances. Like Leibniz, he distinguishes sharply and clearly between solidity, which is an all-or-nothing concept, and hardness, which admits of degrees.[127]

We can, Locke insists, frame an idea of pure space or extension *without* solidity, i.e. of void or vacuum$_2$:

> For a man may conceive two bodies at a distance so as they may approach one another without touching or displacing any solid thing till their superficies come to rest; whereby, I think, we have a clear *idea* of space without solidity.[128]

The notion of vacuum$_2$ is therefore conceivable: void space is a logical possibility. The idea of a distance or magnitude between the walls of a vessel or container is, he insists, 'equally as clear without, as with the idea of any solid parts within':[129] only confusion can result from the Cartesians' conflation of the distinct ideas of space and body.

We move on now to Chapter 13 of Book II, in which Locke discusses our ideas of Space and Place. He sets out as follows:

> That our *idea* of place is nothing but such a relative position of anything . . . I think is plain and will be easily admitted, when we consider that we can have no *idea* of the place of the universe, though we can of all the parts of it; because beyond that, we have not the *idea* of any fixed, distinct, particular beings, in reference to which we can imagine it to have any relation of distance, but all beyond it is one uniform space or extension, wherein the mind finds no variety, no marks. For to say that the world is somewhere means no more than that it does exist; this, though a phrase borrowed from place, signifying only its existence, not location; and when one can find out and frame in his mind clearly and distinctly the place of the universe, he will be able to tell us whether it moves or stands still in the indistinguishable *inane* or infinite space . . .[130]

127   *Ibid.*, II, 4, ii–iii, p. 94.
128   *Ibid.*, p. 94.
129   *Ibid.*, II, 4, v, p. 96.
130   *Ibid.*, II, 13, x, pp. 136–137.

To understand this passage, we must separate the ontological question ('Does Space exist *per se*, independently of bodies?') from the physical one ('May we use Absolute Space as a frame of reference?'). Failure to make this crucial distinction has led one recent writer to make a terrible mess of his account of Locke, even attributing discordant views and incoherencies to Locke's philosophical masterpiece.[131] It is clear that Locke unhesitatingly accepts ontologically Absolute Space: beyond our finite material world, he says, there extends a limitless void or vacuum$_2$, the existence of which is quite independent of that of matter. (God could annihilate the cosmos and leave only empty space.) The parts of this space are, however, 'indistinguishable', at least to us: it offers no fixed 'marks' in virtue of which it could function as a frame of reference. We have therefore no *ideas* of distinct places in perfectly empty space: the question 'Is the whole material world at rest or in motion?' is therefore a *meaningless* one, at least for *us*. (God's knowledge may perhaps be different in kind from ours.) As far as we are concerned, Locke agrees with Leibniz, the supposition that the world as a whole is in motion (or at rest) is senseless. The universe cannot be said to have any particular *place*: to say that it exists *somewhere* is to say no more than that it exists.

Some people, Locke continues, 'would persuade us that body and extension are the same thing': such people, he retorts, either through confusion or deceit, 'confound very different ideas one with another'.[132] The parts of space are necessarily immobile and without solidity; the parts of matter are solid and mobile.

The Cartesians have, however, another argument up their sleeve. As Locke explains,

> Those who contend that *space and body are the same* bring this dilemma: either this space is something or nothing; if nothing be between two bodies, they must necessarily touch; if it be allowed to be something, they ask: whether it be body or spirit?"[133]

This question provokes Locke's most sustained polemic, both against Cartesian dualism and against the whole substance-

---

[131] See Grant, 9, pp. 238–240. Both relativistic and absolutist notions of space are, says Grant, discernible in the *Essay*. True enough; but the two accounts are, properly understood, perfectly compatible.

[132] Locke, *Essay*, II, 13, xi, p. 137.

[133] *Ibid.*, II, 13, xvi, p. 139.

accident metaphysic. We simply do not know enough about the essences of things, says Locke, to dogmatise about such matters:

> If it be demanded (as usually it is) whether this *space*, void of *body*, be *substance* or *accident*, I shall readily answer, I know not, nor shall be ashamed to own my ignorance, till they that ask me show me a clear distinct idea of substance.[134]

This position is in a way reminiscent of Gassendi, although, while Gassendi accepts the substance-accident metaphysic in general, and questions only its applicability to Space and Time, Locke seems prepared to adopt a very agnostic stance about substance in general. Of substance, he says, 'we have no idea of what it is, but only a confused, obscure one of what it does'.[135] The supposedly terrible dilemma – 'Is space substance or accident?' – loses its terror if the notion of substance itself is so vague as to be all but vacuous.

In any case, says Locke, the argument of the Stoic's hand[136] proves the existence of void space beyond our world. Unless we suppose an infinite amount of matter ('which I think no one will affirm') our material universe must be completely surrounded by limitless void space or vacuum₂.[137] Here Locke moves from a defence of the conceivability of void space to a positive assertion of its *actuality*, at least beyond the bounds of our world. (Since actuality entails possibility, the stronger thesis will of course involve the weaker.)

As well as existing outside our world, void space could be produced in it if God were to annihilate any particular body or bodies. But, says Locke, such recherché examples are hardly necessary:

> not to go so far as beyond the utmost bounds of body in the universe, nor appeal to God's omnipotency to find a *vacuum*, the *motion* of bodies that are in our view and neighbourhood seems to me plainly to evince it.[138]

Two millennia after Leucippus, his argument – from the existence of local motion to that of void space – is still in use.

---

134   *Ibid.*, II, 13, xvii, p. 139.
135   *Ibid.*, II, 13, xix, p. 140.
136   For this argument, see Chapter 2, p. 79.
137   Locke, *Essay*, II, 13, xxi, p. 141.
138   *Ibid.*, II, 13, xxiii, p. 142.

A material corpuscle, says Locke, cannot move unless it has some empty space to move into: 'and let this void space be as little as it will, it destroys the hypothesis of plenitude'.[139]

Locke thus advances from an anti-Cartesian argument for the conceivability of void space to an explicitly and unashamedly Atomistic affirmation of its actuality. The exact status of this empty space is, however, left uncertain:

> But whether anyone will take space to be only a relation resulting from the existence of other beings at a distance; or whether they will think the words of the most knowing King Solomon, *The heaven, and the heaven of heavens, cannot contain thee*, or those more emphatical ones of the inspired philosopher St. Paul, In him we live, and move, and have our being, I leave everyone to consider.[140]

Locke here merely mentions the 'divinisation' of Space associated with the names of More and Raphson; he does not explicitly assent to it.[141] And, even if he were to identify Space with God's immensity, this would not entail abandoning his relativistic account of our idea of space: so long as the divine immensity is featureless and without 'marks' or fixed points, we will never have ideas of absolute places, and our concepts of space and motion must remain relativistic ones. That God's immensity does fill extracosmic space is more plainly asserted in Chapter 15:

> Nor let anyone say that beyond the bounds of body there is nothing at all, unless he will confine GOD within the limits of matter. Solomon, whose understanding was filled and enlarged with wisdom, seems to have other thoughts, when he says, Heaven, and the heaven of heavens, cannot contain thee . . .[142]

After the condemnations of 1277, it became orthodox to posit an infinite, 'imaginary', extracosmic void *in* which God could create other worlds or move our own. By the seventeenth century, this was commonly conceived as a real 3D space, somehow associated with the divine omnipresence. As Locke remarks,

---

139   *Ibid.*, p. 142.

140   *Ibid.*, II, 13, xxvii, p. 144.

141   Grant (9, p. 240) makes Locke far more sympathetic to the God = Space thesis than the text warrants.

142   Locke, *Essay*, II, 15, ii, p. 160.

GOD, everyone easily allows, fills eternity; and it is hard to find a reason why anyone should doubt that he likewise fills immensity.[143]

On the exact relations between Space and God's omnipresence, Locke remains silent – he is not prepared to speculate as to whether Space is God's immensity (More), an 'emanent effect' of God's existence (Newton), or quite independent of the deity (Gassendi). His position on this vexatious issue, it may be safely assumed, was one of agnosticism.

B. The Science of Pneumatics

The flowering of pneumatics in the writings of Torricelli, Pascal, Guericke and Boyle is one of the glories of seventeenth century science, and Boyle's account of the 'Weight and Spring' of the air one of the abiding contributions of the Mechanical Philosophy to our conception of the world. The whole story takes its rise from a single empirical discovery, that of the finite strength of the supposed *horror vacui*, Nature's alleged 'abhorrence' of a vacuum. If this *horror*, or, less anthropomorphically, *fuga vacui*, were of infinite strength, as most scholastics believed, a pump should be able to lift water to any height; in fact, no pump can lift it more than about 34 feet.[144]

Galileo's response was to posit a *fuga vacui* of finite strength, which hypothesis he builds into his theory of matter in an ingenious manner.[145] Matter, he claims, is made up of point-atoms and interstitial point-vacua: the cohesiveness of solid bodies is due to the *fuga vacui* exerted by these actual, but minute, void spaces within bodies. (How a mere *nothing* can exert a 'pull' seems not to have bothered him.) The finite power of this *fuga vacui* explains, on this theory, both (a) the possibility of breaking the cohesion of solid bodies, and (b) the limitations of water-pumps.[146]

But, if the *fuga vacui* is only of finite strength, it can be overcome by artificial means and a 'coacervate' vacuum₂ created. In particular, the *fuga vacui* will not support a column of water over 34 feet in height: above such a column,

---

143   *Ibid.*, II, 15, iii, pp. 160–161.

144   See Chapter 6, pp. 284–285.

145   Galilei, 4, p. 11 ff. For Galileo's awareness of the limitations of water-pumps, see Knowles-Middleton, p. 13.

146   See Dijksterhuis, p. 420.

therefore, one may find a vacuum$_2$. If, for example, one has a 40 foot column of water in a vessel sealed at the top, the fluid will run out of the tube until the height of the column has fallen to about 34 feet, leaving a space above the water apparently empty of all bodies. Such an experiment was performed by a group of *savants* in Rome in the 1630s. As Knowles-Middleton comments, 'Berti and his friends had found a way of producing a fairly good vacuum'.[147]

The argument for conceiving this space to be pure void or vacuum$_2$ was simple. It has been visibly vacated by its previous corporeal tenant, the water, and not – so far as sense was concerned – subsequently occupied by any other body. If, moreover, a tap is opened into this space, air *rushes in* with a hiss (allowing the water in the tube to subside), indicating the lack of a material occupant.[148] A coacervate vacuum$_2$ can be made by 'Art'.

Berti's experiment was performed simply to demonstrate the possibility of artificially creating a coacervate vacuum$_2$; we cannot credit him with the discovery of the barometer. He still thinks in terms of the *fuga vacui*, not of atmospheric pressure. It is to Torricelli that we must turn for the next chapter of our story. Hearing of Berti's results, Torricelli repeated the experiment with mercury instead of water, suspecting that this would give him a column of more manageable proportions. The rest is history. He found that he could produce a mercury-column of c.29–30 inches, about one-fourteenth of the height of Berti's water-column. But the specific gravity (S.G.) of quicksilver is about 14 times that of water. This suggests the theory of *aequipondio*, the idea that the fluid column is *balanced* by the weight of the atmospheric air, pressing down on the surface of the fluid in the lower vessel.[149] (That air possesses weight had been demonstrated by Galileo and others.)

All we need to explain Torricelli's results is, therefore, the principle of the balance. Several miles of atmospheric air counterbalance 34 feet of water or about 29 inches of mercury. 'We live,' Torricelli remarks,

> submerged at the bottom of an ocean of the element air, which by unquestioned experiments is known to have weight.[150]

147 Knowles-Middleton, p. 16.

148 *Ibid.*, p. 14.

149 See A. R. Hall, 3, pp. 250–251; Westfall, 5, p. 44.

150 From M. H. Hall, Ed., 10, p. 185.

The weight of a column of atmospheric air presses down on the mercury in the lower vessel, pushing it up the tube:

> Water also, in a similar but much longer vessel, will rise up to almost eighteen cubits, that is, as much further than the quicksilver rises as quicksilver is heaver than water, in order to come to equilibrium with the same force, which presses alike the one and the other.[151]

The *aequipondio* theory accounts readily for this phenomenon. Hitherto, says Torricelli,

> the force which holds up that quicksilver against its nature to fall down again has been believed . . . to be inside of the vessel . . . and to be due either to vacuum or to that material highly rarefied; but I maintain that it is external and the force comes from without.[152]

If, he argues, there were a force of *suction* exerted either by a vacuum₂ or by highly rarefied matter, one would expect the volume of this 'empty' space to be a causally relevant factor: a larger 'vacuum', one might think, should exert a stronger force of suction. Experiment shows, however, that this is *not* the case: the volume of the 'Torricellian vacuum' is irrelevant to the height of the column.[153]

Given this insight into the nature of the operative causal factor, it is possible to credit Torricelli with the invention of the barometer: he has devised, he says, an instrument ' which would show the changes in the air, which is at times heavier and thicker and at times lighter and more refined'.[154] He admits, however, that his own use of the equipment for that end was less than completely successful.

To show that the 'Torricellian vacuum' is a true void, empty of all matter, it suffices to show that it can 'suck up' water (more strictly, water can be pressed into it) to fill its entire volume. (To posit a subtle matter which condenses to fill no detectable volume, or escapes through the glass walls of the tube, is merely *ad hoc*.) The space above the mercury is therefore empty of all air and, so far as we can tell, absolutely void of matter. 'Many,' says Torricelli,

---

[151] *Ibid.*, p. 186.

[152] *Ibid.*, p. 186.

[153] *Ibid.*, p. 186; Knowles-Middleton, p. 21. The same experiment was subsequently performed – and Torricelli's position confirmed – by Pascal (for whom see Westfall, 5, pp. 45–46).

[154] M. B. Hall, Ed., 10, p. 184; Knowles-Middleton, p. 21, p. 25.

have said that a vacuum cannot be produced, others that it can be produced but with repugnance on the part of nature and with difficulty; so far, I know of no one who has said that it can be produced without effort and without resistance on the part of Nature.[155]

The so-called *fuga vacui* is merely an accidental consequence of the weight of our atmosphere, and of strictly limited power at that.

Torricelli's work raises, as Dijksterhuis[156] sees, two independent questions, (a) Is the so-called 'Torricellian vacuum' really empty of all material substance? and (b) Is the column of mercury supported by the weight of the atmospheric air? Since the two questions are, at least *prima facie*, quite independent, four different responses are possible; and in fact all found advocates, as the following table shows:

	Vacuum	*Column of Air*
Aristotelians	No	No
Cartesians	No	Yes
Roberval	Yes	No
Pascal, Gassendi, Beeckman	Yes	Yes

What grounds were found for the conservative, negative answers to the above questions? Let us take the vacuum first and the 'column of air' later. Apart from conceptual issues about the conceivability of truly empty space or vacuum$_2$, the main anti-vacuist argument turned on the transparency or 'diaphaneity' of the Torricellian vacuum. If light is a medium-dependent phenomenon, either the actualisation of the potential transparency of the medium (Aristotle), or a pressure-wave or pulse transmitted through a subtle fluid (Descartes) then light-transmission *in vacuo* is impossible. This was the main reason why Hobbes abandoned his early belief in (interstitial) vacua and became a staunch advocate of plenism. In his youth, he had accepted a corpuscular emission theory of light and an Atomistic conception of matter and vacuum;[157] in his later years, he turned to a mediumistic

---

155 M. B. Hall, Ed., 10, p. 184.

156 Dijksterhuis, pp. 444–445.

157 For this early phase of Hobbes' career, see Brandt, p. 117. By 1636, says Brandt (p. 145), Hobbes had abandoned the corporeal emission theory of light for a mediumistic one – later still, he came to see that this involves plenism.

theory of light and to the plenum.[158] If we can see clearly through the supposed 'vacuum', he argues, this proves beyond doubt the presence there of a fluid light-transmitting medium, a *materia subtilis*.[159] Similar arguments were brought by the Aristotelians – e.g. by Noël against Pascal[160] and Linus against Boyle. If, says Linus, the supposedly empty space produced by the air-pump were truly void, it 'would appear like a little black pillar, . . . because no visual species could proceed either from it, or through it, unto the eye'.[161] If light is *either* a qualitative *alteration* or a mechanical *agitation* of a medium, it cannot be transmitted *in vacuo*.

As for the *aequipondio* theory, it too had its critics: one or two conceptual confusions needed to be cleared up before it could become widely accepted. Many people, for example, including even the great Galileo,[162] argued (often along Archimedean lines) that a particular substance does not gravitate in itself – air, for example, lacks weight in air. Hence, says Galileo, our atmospheric air, though not without weight, would only exert a downward force in some rarer medium such as a void: air in air is effectively weightless. The *fuga vacui*, he claims, is *irreducible* to any external force. The distinction between net and gross weight dispels this confusion: air has no net weight in air (because gross weight is precisely counterbalanced by upthrust); but the whole column of air nevertheless presses down on whatever is beneath it. For the definitive solution to this problem, one need only turn to Hooke's Cutler Lectures[163] or the 'Hydrostatic Paradoxes' of Robert Boyle.[164] Boyle's first paradox states, clearly, emphatically and correctly, that 'in water, and other fluids, the lower parts are pressed by the upper'.[165]

Another difficulty for the *aequipondio* theory was its exclusive reliance on the principle of the balance: experimenters were misled by Torricelli's emphasis on the weight of

158   As late as 1648, Hobbes can write that he finds the postulation of interstitial vacua plausible; by 1655, when *De Corpore* was released, he is a convinced anti-vacuist (see Brandt, p. 205).

159   See Hobbes' *Problemata De Vacuo* from the 'Seven Philosophical Problems' of 1662 (*Works*, Vol. 7, p. 21); Brandt, p. 206.

160   Koyré, 5, p. 153; Dijksterhuis, pp. 448–449.

161   From Boyle, *Works*, Vol. 1, p. 136.

162   For Galileo's opposition to the *aequipondio* theory, see Knowles-Middleton, p. 11; Dijksterhuis, p. 420.

163   Gunther, Ed., Vol. 8, p. 181 ff.

164   Boyle, *Works*, Vol. 2, p. 738 ff.

165   *Ibid.*, p. 750.

the air and were at a loss to understand why the admission of a small bubble of air into the supposed vacuum should cause the level of quicksilver to drop markedly.[166] (A single bubble of air cannot counterbalance the whole atmosphere.) Weight, moreover, is a *downward* force; yet in some pneumatic experiments it seems to be exerting an upward impulse: how, Torricelli is asked, can this be?[167] To solve these problems, we require a clear distinction between the weight of the air and its *pressure*: the weight of the several miles of air above our heads *compresses* that at ground level, and hence imparts to it a 'desire' to spring back, which manifests itself as an *omnidirectional pressure*. The weight/pressure distinction is hinted at by many thinkers in the first half of the century; it receives its definitive formulation in Boyle's doctrine of the weight and 'spring' (elasticity) of the air.

One of the great protagonists both of vacuism and the 'column of air' theory was Blaise Pascal. Hearing from Mersenne of Torricelli's experiments, he proceeded first to repeat them and then to extend and develop Torricelli's position by a brilliantly designed and executed series of experiments of his own. Let us mention here only the most famous.

On the 'ocean of air' theory, the downward pressure of the atmosphere should be somewhat less at the peaks of mountains than in the troughs of valleys: a shorter column of air should be capable of supporting a lower mercury-column. If, on the other hand, Nature has a *fuga vacui*, there is not the slightest reason for expecting any detectable difference. This is the source of the most celebrated of all Pascal's experiments (carried out, at any rate, under Pascal's directions), his barometric observations on the Puy de Dôme, the results of which provided such convincing confirmation of Torricelli's insight.[168] Air-pressure, as one would expect on the 'ocean of air' theory, falls off with altitude – how the peripatetic can cope with this result is not easy to see.[169] As Pascal demands,

166   See Koyré, 5, p. 128.

167   See Knowles-Middleton, p. 22. For the notion of an omnidirectional air-pressure in Beeckman, Baliani, and Torricelli himself, see also p. 28.

168   See Westfall, 5, p. 47.

169   Boyle, for example, claims in his 'Reply to Linus' that the Puy de Dôme experiment of Pascal provided the *experimentum crucis* which finally settled the issue between the 'column of air' theory and its critics (Boyle, *Works*, vol. 1, p. 151).

Does nature abhor a vacuum more in the highlands than in the lowlands? In damp weather than in fine? Is not its abhorrence the same on a steeple, in an attic, and in the yard?[170]

'Nature', Pascal insists, does *nothing at all* to prevent vacuum-formation: the *horror vacui* is an anthropomorphic myth caused by a total misconstrual of the perfectly natural consequences of the weight of atmospheric air:

The experiment made on mountains has overthrown the universal belief in nature's abhorrence of a vacuum, and given the world the knowledge . . . that nature has no abhorrence of a vacuum, nor does anything to avoid it; and that the weight of the mass of the air is the true cause of all the effects hitherto assigned to that imaginary cause.[171]

Let us now turn from the pneumatic phenomena revealed by the work of Torricelli and Pascal to their explanations in the philosophies of Descartes, Gassendi, and their respective disciples. We begin with the Cartesian school.

Descartes' response to the ideas of Torricelli and Pascal is a predictable one. As one of the great pioneers of the Mechanical Philosophy, we can expect him to favour Torricelli's 'column of air' theory, which substitutes for the occult and anthropomorphic *horror vacui* of the schoolmen a clear, intelligible, and, above all, *mechanical*, cause of pneumatic phenomena.[172] On the other hand, as a staunch *plenist*, we shall expect him to fill the so-called 'vacuum' of Torricelli and Pascal with some kind of subtle matter. Both these expectations are satisfied.

Cartesian matter, though perfectly homogeneous, is differentiated into three 'elements' E1–3 by differences of *particle-size*. E1, the subtlest element, is effectively a perfect fluid,[173] filling the interstices between the larger corpuscles of the other elements. E2 assumes the form of spherical 'boules': the transmission of an impulse along a line of such 'boules' constitutes *light*. Together, E1 and E2 make up a 'subtle

---

170   From Grant, Ed., 4, p. 332.

171   Quoted from A. R. Hall, 3, p. 253.

172   According to J. F. Scott (pp. 11–12) Descartes anticipated Torricelli's work by 'at least a dozen years'; he was also associated with the pneumatical research of Pascal.

173   For Descartes' matter-theory, see M. B. Hall, Ed., 10, p. 302 ff.; Kemp Smith, pp. 117–118. For the genesis of the distinction between E1 and E2, see Love.

matter' which readily pervades all things, passes freely through the pores of solid bodies, etc. This subtle matter is contrasted with gross matter or E3, which consists of the largest corpuscles. Air, for the Cartesian, consists of E3 suspended in a fluid composed of E1 and E2: each particle of E3 is so agitated and whirled around by currents of subtle matter that it 'defends' a space considerably larger than itself against the 'incursions' of other E3-corpuscles. In virtue then of the agitation and interparticulate collisions of these E3 particles, atmospheric air exerts a detectable pressure – Descartes comes close to a purely kinetic explanation of air-pressure.[174]

The elasticity of atmospheric air is due, then, to the presence in it of these bulky E3 corpuscles: subtle matter exerts no detectable pressure. (If one were to attempt to compress pure subtle matter, it would simply flow through the walls of the vessel used as a container.) The 'vacuum' produced by Torricelli is almost pure *materia subtilis*: when the mercury-column descends, E1 and E2 pass freely through the glass walls of the tube to fill the space which would otherwise have been left vacant.[175] The presence of the 'boules' of E2 explains the transmission of light: the absence of E3-particles explains why the 'Torricellian vacuum' exerts no detectable pressure, and hence why the pressure of the external, E3-containing, air can support the weight of a column of mercury.

The Cartesian thus attempts to reap all the benefits of the air-pressure theory, while still avoiding the postulation of a vacuum: to do this, he must assert the permeability of all solid substances by subtle matter. An experiment of Boyle's brings home this point nicely. When a sealed bubble of glass is heated, it *expands*, i.e. comes to occupy a greater volume. Now, if one posits a homogeneous matter* and denies real or intrinsic condensation and rarefaction of matter, one must, to explain this phenomenon, posit *either* a vacuum *or* a subtle matter capable of penetrating glass. Thus, says Boyle, the

---

174   For this Cartesian account of the elasticity of the air, see Boyle, *Works*, vol. 1, p. 12. For a plausible Cartesian explanation of the dependence of pressure on temperature, see p. 180. Descartes' account is, however, not purely kinetic one, since the *shapes* of the E3 particles are a relevant factor in the explanation (they uncoil as the particles are whirled faster). For a full-blooded kinetic theory, in which particle-shapes are irrelevant and motion is all-important, one must turn to Hooke's lecture '*De Potentia Restitutiva*' (Gunther, Ed., vol. 8, p. 333 ff.).

175   For this explanation of the Cartesian position, see Boyle, *Works*, vol. 1, p. 137, and pp. 208–209.

Cartesians, 'the subtilest and wariest champions for a plenum I have yet met with',[176] who 'suppose a subtile matter or aether fine enough to permeate glass',[177] *can* account for this phenomenon; a plenist such as Hobbes, who makes no such assumption, cannot. The expansion caused by heating is due simply, on the Cartesian account, to an ingress of E1: what could be simpler or more logical?

The universe being a homogeneous plenum, says Descartes, it is easy to explain the outcome of the celebrated clepsydra experiment. Instead of positing an occult, unintelligible, and hopelessly anthropomorphic *horror vacui*, he insists, what we ought to say is that

> the wine cannot issue from the cask, because outside it everything is as full as it is capable of being . . . if the wine descended, the part of the air whose place it would then occupy would not be able to find in all the rest of the universe another place into which it could retire – unless and until an opening is made in the upper part of the cask, into which the air, on giving place, may circularly ascend.[178]

This seems flatly inconsistent with what has gone before. If E1 and E2 *can* pass through the walls of the vessel, we can generate a proper cyclical flow, in which case it is possible for the wine to descend. Descartes seems here to suppose that *no* matter can penetrate the walls of the vessel to create the circle or vortex necessary to all movement in a plenum, but this is contradicted by his best pneumatic thought. What he here represents as an all-or-nothing matter is best treated as a question of a balance of opposed forces: the weight of the fluid against the pressure of the atmospheric air. Since the E3 particles responsible for this pressure are themselves too bulky to pass through the walls of the vessel (and thereby equalise the pressure), they will tend to press upwards against the fluid in the vessel, 'striving' to prevent it from escaping. This, then, is how Descartes *should* have explained the results of the famous clepsydra experiment.

For the Gassendist position on these topics, let us turn to the *Physiologia* of Walter Charleton. Gassendi was, not

---

[176]   *Ibid.*, pp. 208–209.

[177]   *Ibid.*, p. 112.

[178]   Quoted from Kemp Smith, p. 110. Exactly the same argument appears in Hobbes' *De Corpore* (*Works*, vol. 1, pp. 414–415), where it is cited as a decisive refutation of the vacuum! Hobbes' pneumatics was not, alas, his best scientific work.

surprisingly, very interested in the experimental work of Torricelli and Pascal, and full of praise for their achievements, which seemed to him to confirm one of the cornerstones of the Atomic world-view, the existence in Nature of void space or vacuum$_2$.[179] His disciple Charleton makes this association between classical Atomism and modern pneumatics very clear. After producing standard Atomist arguments (from motion and from S.G. figures) for the existence of interstitial vacua, he goes on to raise the vexing topic of the so-called *horror vacui*. Experiments with cupping-glasses and water-clocks demonstrate only, he insists, a resistance to the formation of a *coacervate* vacuum, not a 'disseminate' one. In any case, this resistance is best attributed to the 'spring' of the air, its tendency to resume its original state:

> In particular, it is the essential quality of the Aer, that its minute particles preserve their natural contexture, and when forced in rarefaction to a more open order, or in condensation to a more close order, immediately upon the cessation of that expanding, or contracting violence, to reflect or restore themselves to their due and natural contexture.[180]

This may seem innocuous, but is in fact mischievous. Air has no 'natural contexture': in the absence of external pressure it will rarefy *without limit*. (Atmospheric air, says Boyle, is kept in its normal state only by perpetual external 'violence'.[181]) The 'spring' of the air will explain its spontaneous rarefaction after a 'violent' compression; only the external pressure of other bodies will explain its condensation.

Assuming a limited reluctance on the part of 'Nature' to allow the formation of a coacervate vacuum, the question naturally arises: can such a vacuum be made artificially?[182] Hero, Galileo, and Torricelli all affirm this possibility. But perhaps, Charleton suggests, the 'Torricellian vacuum' is not a true void, but is filled with either (a) air or (b) aether. The first of these hypotheses is easily dismissed; the second proves more difficult to refute. There may, Charleton admits, be such a subtle fluid in the so-called 'vacuum', but to posit an aether merely to save plenism is question-begging. The axiom

---

[179]   See W. L. Scott, 2, p. 9; Koyré, 5, pp. 128–129.

[180]   Charleton, p. 34.

[181]   See Boyle, *Works*, Vol. 3, pp. 785–786.

[182]   Charleton, p. 35 ff.

'Nature abhors a vacuum' is, he grants, 'indeed a maxim, and a true one, but not to be understood in any other than a metaphorical sense'[183] – it is a mere figure of speech to attribute desires and aversions to inanimate bodies. Nature's 'reluctancy' to admit the formation of a coacervate vacuum is, moreover, due merely to the 'fluxility of the atomical particles' of which the air is composed, which causes them

> so mutually to compel each other, that no one particle can be removed out of its place, but instantly another succeeds and possesses it; and so there can be no [large] place left empty.[184]

'Nature', then, 'abhors' a vacuum not *per se*, but only *per accidens*, i.e. as an accidental consequence of the 'fluxility' of the air. (Gassendi and Charleton come very close here to an adequate conception of air-pressure.[185])

The Torricellian vacuum, Charleton continues, cannot be 100% void: light and magnetic 'effluvia' pass freely through it. These very subtle bodies do not, however, suffice to make the space a plenum – a considerable proportion of void will no doubt remain.[186] (Since he subscribes to a corpuscular theory of light, he is under no obligation to posit a fluid medium filling the entire space.[187])

As regards the clepsydra, Charleton's position is a somewhat weaker variant on that of Descartes. It is not, on an Atomistic matter-theory, absolutely impossible for the water to flow from the water-clock, since the external matter contains interstitial vacua into which its parts may recede. What happens, however, is that, as the particles of the external air are forcibly compressed into these vacua, the resistance of the air to further condensation increases and the flow of fluid from the clepsydra dries up.[188]

Why, we might ask, will the air sustain a mercury-column of about 29 inches, and no more? This, Charleton replies, is due to

---

183   *Ibid.*, p. 40.

184   *Ibid.*, p. 41.

185   See Koyré, 5, p. 129, for an attribution of the concept of air-pressure to Gassendi. For the influence of Charleton on Boyle, see Kargon, 1.

186   Charleton, pp. 44–45.

187   *Ibid.*, p. 198 ff. Rays of light, says Charleton, are 'certain most tenuous streams of igneous particles . . .'

188   *Ibid.*, pp. 46–47.

the resistance of the parts of the aer, which endures no compression or subingress of its sensible particles, beyond that certain proportion . . .[189]

Why 29 inches? Such, we are told, is the nature of air: it does not befit our lowly human status to enquire further – we are incurably ignorant of the inner secrets and mysteries of Nature. After a brief purple passage of a strongly sceptical tenor, Charleton returns to the issue at hand, and trots out pat the *aequipondio* theory of Torricelli. This theory explains convincingly, he says, why water gives a column of 34 feet and quicksilver of 29 inches: both columns are balanced by the weight of the 'ocean of air'.[190]

Charleton thus possesses a clear grasp of the weight of the atmosphere (for an Atomist it is, of course, axiomatic that *all* matter possesses weight), and of the *aequipondio* theory of Torricelli, and can be seen to be feeling his way towards an adequate conception of the elasticity and effective pressure of the air around us. His true successor in this field is of course Robert Boyle.

Before moving on to discuss Boyle's pneumatics, however, we must at least mention the work of Otto von Guericke, inventor of the air-pump and staunch advocate both of the vacuum and of the concept of atmospheric pressure. His Magdeburg experiment, in which two teams of horses strove to separate two brass hemispheres separated by vacuum$_2$, was a most powerful and dramatic illustration of the sheer strength of the pressure of the air, and one of the milestones of pneumatic research. Hearing of this experiment, Boyle set his assistants (including the ingenious Hooke) to make him an air-pump superior to that of Guericke.[191] Our story follows on from there.

His air-pump is capable, says Boyle, of extracting the air from a closed vessel, leaving a 'vacuum' (known thereafter as the 'Boylian vacuum'),

> by which I here declare once for all, that I understand not a space, wherein there is no body at all, but such as is either altogether, or almost totally devoid of air.[192]

Boyle is less interested in the exact constitution of his eponymous 'vacuum' than in the 'spring' of the atmospheric

189   *Ibid.*, p. 48.
190   *Ibid.*, pp. 52–55.
191   See Boyle, *Works*, Vol. 1, pp. 6–7.
192   *Ibid.*, p. 10.

air revealed by his pneumatic experiments. To explain his results, he claims, one must assume that air either consists of, or abounds with,

> parts of such a nature, that in case they be bent or compressed by the weight of the incumbent part of the atmosphere . . . they do endeavour, as much as in them lieth, to free themselves from that pressure, by bearing against the contiguous bodies that keep them bent; and as soon as those bodies are removed, or reduced to give them way, by presently unbending and stretching out themselves . . . and thereby expanding the whole parcel of air, these elastical bodies compose.[193]

Boyle cites two plausible explanations of the air's elasticity. One is the kinetic theory of Descartes; the other, which he favours, involves the supposition that air consists of particles like little coiled springs, like wool, which

> consists of many slender and flexible hairs; each of which may indeed, like a little spring, be easily bent and rolled up; but will also, like a spring, be still endeavouring to stretch itself out again.[194]

This latter theory Boyle finds 'somewhat more easy' than that of the Cartesians: the picture it gives of our atmosphere as analogous to a spring mattress or a fleece of wool clearly appeals to his penchant for simple mechanical models. Given this model, it is easy to understand how the weight of the atmosphere[195] will compress the low-lying layers of air very considerably:

> A column of air, of many miles in height, leaning upon some springy corpuscles of air here below, may have weight enough to bend their little springs.[196]

Pascal's Puy de Dôme experiment provides, Boyle feels, the decisive confirmation of the 'ocean of air' thesis of Torricelli – no rival theory can cope with the data so plausibly and intelligibly. Air-pressure, says Boyle, is omnidirectional: the weight of the atmosphere compresses the springs, which then 'strive' to expand themselves in any direction. Boyle thus

---

[193]   Ibid., p. 11.

[194]   Ibid., p. 11.

[195]   At p. 81 ff. Boyle estimates the S.G. of atmospheric air to be c. 1/938 that of water.

[196]   Ibid., p. 13.

distinguishes clearly between atmospheric weight and pressure, and correctly diagnoses the causal relation between the two.

Boyle then proceeds to perform a series of ingenious and well designed pneumatic experiments with the aid of his air-pump. The results of the experiments are all, he will claim, readily explicable in terms of the air-pressure theory. In particular, what appears at first sight to be a force of *suction* exerted from *within* the evacuated vessel is best explained as a consequence of the pressure of the external air: bodies are pushed from outside, not pulled from inside, into the 'Boylian vacuum'.[197] Boyle's Aristotelian critic, Linus, insists that we feel our finger *sucked* into the vessel from *within*: since a nothing cannot exert such a pull, he argues, there must be a subtle and strongly elastic substance, a 'funiculus', inside the evacuated vessel which, like a stretched elastic band, *pulls* other bodies inside.[198] Phenomenologically, of course, Linus is absolutely right: we still speak of (and experience?) 'suction' rather than 'pulsion'. Unfortunately Linus' *physics* is hopelessly wrong: the most natural interpretation of our experience is in this instance a misleading one. In a later work, 'The Cause of Attraction by Suction',[199] Boyle supports his position with some crucial experiments. A syringe, for example, cannot raise water inside an evacuated vessel: what we think of as suction is, says Boyle, really due to external pressure. The ascending piston only *makes room* for the water to ascend; the *motive force* involved comes from the pressure of the atmospheric air – hence the failure to raise water *in vacuo*.[200] Similarly, says Boyle, when we suck up water or air: what happens is that in breathing in we increase the capacity of our lungs, which lowers the internal air-pressure – the resulting *difference* of air-pressure produces all the observed effects. This explanation is, he adds, 'very consonant to the mechanical principles', as indeed it is, replacing as it does the occult concept of attraction by the intelligible one of impulsion.[201]

One experiment is of particular theoretical importance: it was indeed, Boyle avers, 'the principal fruit I promised myself

---

[197]   *Ibid.*, pp. 15–16.

[198]   *Ibid.*, p. 127, p. 136.

[199]   *Ibid.*, Vol. 4, p. 128 ff.

[200]   *Ibid.*, pp. 131–132.

[201]   *Ibid.*, pp. 133–134.

from our engine'. It involves the performance of Torricelli's experiment *inside* the exhausted vessel, the production of a 'vacuum in a vacuum'. On the *aequipondium* theory, says Boyle, it follows that

> if this experiment could be tried out of the atmosphere, the quicksilver in the tube would fall down to a level with that in the vessel, since there would be no pressure upon the subjacent, to resist the weight of the incumbent mercury. Whence I inferred . . . that if the experiment could be tried in our engine, the quicksilver should subside below 27 digits, in proportion to the exsuction of air, that should be made out of the receiver.[202]

Performing the Torricellian experiment in an evacuated vessel is therefore as rigorous a test of the 'column of air' theory as performing it in some extracosmic or 'imaginary' void space. (One vital consequence of the development of the air-pump was that many of the thought-experiments of the pioneers of the New Science could be performed for the first time in reality, i.e. transformed into actual experiments.)

While this experiment confirms Torricelli's 'column of air' thesis, it does not support the vacuists' claim that the Boylian vacuum is truly void, empty of all corporeal substance. Although reluctant to enter into the vacuist-plenist controversy, Boyle clearly feels that he must at least take stock of the available evidence. Light rays and magnetic effluvia,[203] says the plenist, both pass freely through the Boylian vacuum: it cannot therefore be 100% void. Boyle agrees: if, he says, vacuum is so defined as to mean a place 'perfectly devoid of all corporeal substance', it may indeed be true that 'there is no such thing in the world'.[204] But on the other hand,

> It may be said, that as for the subtle matter which makes the objects enclosed in our evacuated receiver, visible, and the magnetical effluvia of the earth that may be presumed to pass through it, though we should grant our vessel not to be quite devoid of them, yet with them, as we may suppose, that if they were gathered together into one place without

[202]   *Ibid.*, Vol. 1, p. 33.

[203]   All the classical magnetic phenomena – attraction, orientation, etc. – occur without hindrance in the evacuated vessel (see p. 32). The 'magnetical steams' of the Earth must, therefore, pervade this supposedly empty space.

[204]   *Ibid.*, p. 74.

intervals between them, they would fill but a small part of the whole receiver.[205]

The vacuist can therefore accept the presence in the vessel of a little subtle matter while continuing to affirm that most of that space is void – on the Epicurean theory of light as a corporeal emission, Boyle reckons, light-corpuscles may fill 1% or less of the total volume of the Boylian vacuum.[206]

A further series of experiments has a considerable bearing upon the constitution of the Boylian vacuum. In the evacuated vessel, says Boyle, *sounds* are either quite inaudible, or much diminished in volume,[207] thus confirming the orthodox mechanical account of sound (due to Galileo and Mersenne) as a vibration of the air.[208] Another experiment involves the use of a *pendulum* in the so-called vacuum. He had thought it likely, Boyle admits, that a pendulum would swing faster in the evacuated vessel than in ordinary air, and would continue to oscillate for longer. The result of the experiment was, however, 'other than we expected': no detectable difference was observed, which result does seem to indicate the presence in the vessel of some resisting medium.[209] (Newton certainly drew just such an inference: when the air is exhausted, he says, 'there remains in the glass something much more subtle which damps the motion of the bob'.[210]) On the other hand, operating a bellows inside the evacuated vessel ought to produce aether-currents – if such an aether exists – but this supposed draught is without discernible effects. One can place a feather inside the receiver, pump out the air, and then operate the bellows, without disturbing the feather in the slightest. This is a powerful argument on the side of the vacuists: if, says Boyle, there is some subtle matter in the vessel,

> that aether is such a body as will not be made sensibly to move a light feather by such an impulse as would make the air manifestly move it.[211]

205   *Ibid.*, p. 37.

206   *Ibid.*, p. 136.

207   *Ibid.*, p. 62 ff. See also Vol. 3, pp. 259–261.

208   For more on the Galileo-Mersenne theory of sound, see A. R. Hall, 3, pp. 249–250.

209   Boyle, *Works*, Vol. 1, p. 61.

210   See *De Aere et Aethere* in A. R. and M. B. Hall, Eds., 4, pp. 227–228; Westfall, 4, p. 336.

211   Boyle, *Works*, Vol. 3, p. 256.

If aether-currents cannot even displace a feather, mechanical explanations which invoke them must collapse – unless the subtle matter can exert some impulse on a body, it is physically equivalent to a void.

Another experiment of interest concerns the production of smokes or vapours in the Boylian vacuum. Such vapours descend, and gather, like a pool of liquid, at the bottom of the evacuated vessel, thus confirming Galileo's prediction that all bodies would fall *in vacuo* and refuting the Aristotelian notion of absolute levity.

Another Galilean prediction confirmed experimentally by Boyle was the claim made in the *Discourse*[212] that all bodies should fall equally fast *in vacuo*. In the continuation of his 'New Experiments' Boyle tests this assertion, and discovers that in his evacuated vessel even a light body like a feather falls like a dead weight, not significantly slower than, e.g., a lump of lead.[213] As well as corroborating Galileo's account of free fall (another example of the actualisation of thought-experiments made possible by Boyle's apparatus) this casts further doubt on the postulation of a subtle fluid medium in the vessel. If such an aether exerts no resistance to the fall of a feather, how is it to be distinguished from void?

As for the scholastic *horror vacui*, this is simply an anthropomorphic muddle. It cannot be intelligibly supposed, Boyle insists, that 'hatred or aversion, which is a passion of the soul', can be present 'in water, or such like inanimate body'[214] All we can legitimately say is that the parts of corporeal Nature have been so contrived by their Author as to act *as if* they conspired to prevent vacuum-formation: the supposed 'aversion of nature to a vacuum' is, Boyle emphasises, merely an accidental consequence of the 'fluxility' of atmospheric air.[215] The supposed suction exerted by the Boylian vacuum is, moreover, no more than the 'pulsion' of the external air, which force is finite, measurable, and fairly easily overcome. The imagined *horror vacui* is, therefore:

a.   A misleading anthropomorphic metaphor;
b.   An accidental consequence of air-pressure;
c.   Not 'any such boundless thing, as men have been pleased to imagine'.[216]

---

212   Galilei, 4, p. 72.
213   Boyle, *Works*, vol. 1, p. 256 ff.
214   *Ibid.*, Vol. 1, p. 75. See also Vol. 5, pp. 227–228.
215   Note the terminological resemblance to Charleton.
216   Boyle, *Works*, Vol. 1, p. 76.

After publishing his 'New Experiments' in 1660, Boyle found himself compelled to defend them only two years later against the criticisms of Linus and Hobbes. I shall not dwell on these objections: Boyle has in general no difficulty in rebutting them and defending the essentials of his doctrine – neither Linus' 'funicular hypothesis' nor Hobbes' version of plenism could ever pose a serious threat to Boyle's mastery of the pneumatic domain. It is in the course of his 'Reply to Linus' that he formulates his eponymous 'Law': the same air, he says,

> being brought to a degree of density about twice as great as it had before, obtains a spring twice as strong as formerly.[217]

The $P \propto 1/V$ hypothesis is, says Boyle, empirically well confirmed, at least within the limits of experimental error. The significance of Boyle's Law lies, however, not so much in its accuracy as an isolated theorem (as such, it was superseded long ago) but in its rôle as a great triumph of the Mechanical Philosophy, a successful *extension* of mechanistic concepts and models to the domain of pneumatic (i.e. 'spiritual') bodies. In place of animism (the *horror vacui*) we have mechanism (air-pressure); in place of the *pull* either of a vacuum$_2$, which is absurd, or of a weird subtle matter such as Linus' 'funiculus', we have the impulsion of the atmospheric air, which is merely a more rarefied form of the common, universal matter, subject to the same (mechanical) laws. And when the Mechanical Philosophy:

a.   Takes a simple and intelligible mechanical model (a spring);

b.   Reasons that from that model to a simple mathematical formula;[218] and

c.   Confirms that formula by empirical evidence;

we have a striking illustration of the explanatory power of the overall programme, a convincing demonstration of its ability

---

[217]   *Ibid.*, pp. 156–157. I pass over without comment the question of the true discoverer of 'Boyle's Law': cases have been made for Power, Townley, and Hooke as well as Boyle himself. For the rôles of Power and Townley, see Webster, 2, pp. 158–159; for that of Hooke, see 'Espinasse, p. 46. In his Cutler Lecture, *De Potentia Restitutiva*, Hooke deduces Boyle's law of gases as a corollary of his own law of springs (Gunther, Ed., Vol. 8, p. 348); it is especially noteworthy that Newton attributes the principle to Hooke in *De Aere et Aethere* (A. R. and M. B. Hall, Eds., 4, pp. 223–224).

[218]   This achievement was of course that of Hooke, not of Boyle himself, but for our purposes the distinction does not matter.

to cast light on the darkest subjects: another triumph for the all-conquering Mechanical Philosophy.[219]

In the physics of Newton, the vacuism-plenism issue is given a sharper focus. It is hard, reading Boyle's reports of his pneumatic experiments, to see how a subtle Cartesian aether could ever be distinguished from a $vacuum_2$; after publication of Newton's *Principia* in 1687, the situation was quite different. By giving science a more precise conception of matter, Newton was also responsible for a refinement of the complementary concept of void. Matter, in Newtonian mechanics, is whatever, in virtue of its mass or inertia, is subject to the Laws of Motion: whatever resists changes of state, gives and receives impulses, etc.[220] 'Body,' says Newton,

> I call everything tangible which is resisted by tangible things . . . Vapours and exhalations on account of their rarity lose almost all perceptible resistance, and in the common acceptance often lose even the name of bodies, and are called spirits. And yet they can be called bodies if they are the effluvia of bodies and have a resistance proportional to density. But if the effluvia of bodies were to change them in respect of their forms so that they were to lose all power of resisting and cease to be numbered among the phenomena, these I would no longer call bodies, for I speak with the common people.[221]

Resistance is an immediate consequence of mass or inertia, which is an essential feature of corporeal nature: whatever exerts no resistance has no mass and is therefore properly called immaterial.

But the resistance offered by a fluid to the passage of a body through it does not depend solely on its density: oil may be less dense than water but more resistant. Its extra drag, say the Cartesians, is due to the difficulty in separating its parts, in clearing a path for the passage of the moving body. Let us call this the *viscosity* of the medium. Very small, fast-moving parts are easily separable and therefore constitute fluids of low viscosity; bulkier, slower-moving particles make up more viscous media. The finest Cartesian element E1 is effectively a perfect fluid: its constituent corpuscles are so small and so rapidly agitated that they do not resist separation at all. Such

---

[219] See Westfall, 5, p. 49.

[220] See McMullin, 2, p. 46.

[221] Quoted from McMullin, 4, p. 100.

a fluid medium will therefore, the Cartesian claims, exert no resistance at all.

But, if E1 particles are material (massy), they will *resist* being displaced by the moving body: the mere division of the particles of a fluid into still smaller parts does not detract one iota from the total mass of the fluid.[222] Thus Newton, in Book Two of *Principia*:

> And though air, water, quicksilver and the like fluids, by the division of their parts *in infinitum*, should be subtilised, and become mediums infinitely fluid, nevertheless, the resistance they would make to projected globes would be the same. For the resistance considered in the preceding Propositions arises from the inactivity of the matter; and the inactivity of matter is essential to bodies, and always proportional to the quantity of matter. By the division of the parts of the fluid the resistance arising from the tenacity and friction of the parts may be indeed diminished; but the quantity of matter will not be at all diminished by this division; and if the quantity of matter be the same, its force of inactivity will be the same, as being always proportional to that force.[223]

The resistance experienced by a body projected in a fluid arises form a *combination* of two distinct factors, the *inertia* of the parts of the medium and their 'tenacity and friction', which give rise to the *viscosity* of the fluid. The latter, but *not* the former, is affected by division:

> that part of the Resistance; which arises from the *vis inertiae*, is proportional to the Density of the Matter, and cannot be diminished by dividing the Matter into smaller parts . . .[224]

A perfectly fluid medium, consisting of infinitely tiny parts with stupendously rapid motions, would lack viscosity but would still exert resistance due to its inertia. Resistance arises from a combination of inertia and viscosity, thus:

> The resisting power of fluid Mediums arises partly from the Attrition of the Parts of the Medium, and partly from the *Vis Inertiae* of the Matter.[225]

---

222 For an early statement of this proposition, see A. R. and M. B. Hall, Eds., 4, p. 147.

223 Newton, 1, p. 366.

224 Newton, 2, pp. 365–366.

225 *Ibid.*, p. 365. Cf. also 1, p. 280.

If we label the total resistance R_T, the inertial component R_I, and that due to viscosity, R_v, we obtain the formula $R_T = R_I + R_v$, which, though crude, is not too misleading. A perfect fluid such as Descartes' E1 will have $R_v = 0$ (there will be no difficulty in separating its parts one from another) and hence $R_T = R_I$. For ideal fluids, then, the resistance offered to moving bodies is 'proportional to the densities of the fluids in which they more':[226] quicksilver, which is c.14 times denser than water, should exert about 14 times as much resistance. This prediction can be confirmed experimentally, says Newton, by performing pendulum-experiments in a variety of fluid media.[227] (The $R_T \propto$ density theorem is of course only approximately true of real fluids, since it deliberately omits the R_v factor.)

For a fluid of negligible viscosity, $R_v \approx 0$, and hence $R_T \approx R_I$. A fluid of negligible resistance must therefore possess little matter; a hypothetical fluid of zero resistance is a perfect *void*. When Leibniz argues, therefore (in the correspondence with Clarke), that the reasons given by Torricelli and Guericke to support vacuism are mere 'sophisms', and that the resistance of a fluid is due not to its quantity of matter but to its 'difficulty of giving place',[228] his position has already been undermined in Book Two of *Principia*. Like the Cartesians, Leibniz has hold of a half-truth; correctly identifying the nature and rôle of R_v, he overlooks the inertial factor R_i.[229] And yet Leibniz possessed a clear grasp of the concept of *mass*: it is merely that he fails to attribute it to subtle matter.[230] (In Cartesian physics, of course, mass = quantity of *gross* matter or E3.) As Clarke says, Leibniz' arguments are 'easily answered': he has simply failed to grasp the essential connection between matter, inertia and resistance. However finely one divides the parts of a fluid, this will not detract one jot from the inertial resistance R_I of that medium.

The next step in the argument takes in the crucial empirical evidence. Having analysed R_T into two factors R_I and R_v, and inferred that if $R_T = 0$, $R_I = 0$, Newton now adduces evidence that strongly suggests the absence of resistance in (a)

---

226   Newton, 1, p. 324.

227   Newton, 2, p. 366.

228   H. G. Alexander, Ed., p. 45, pp. 65–66.

229   Ballard, pp. 60–61.

230   Leibniz' aether is, says Koyré (4, p. 138), effectively devoid of mass.

interplanetary space and (b) the Boylian vacuum. Let us deal with them in that order.

The motion of the planets in their orbits is not discernibly resisted by the medium in which they move. This is explained by the Cartesian in terms of a shared motion: the planets are, according to Descartes, carried round in their orbits by a whirling vortex of subtle matter. Since no relative motion is involved (each planet may be said to rest *vis-à-vis* its own part of the vortex), no *drag* is exerted. This, however, fails utterly to account for the motions of the *comets*, which pursue regular Newtonian orbits which intersect the paths of the planets in all manner of eccentric ways. No conceivable vortex theory could cope with these cosmic wayfarers, yet they obey Newton's Laws and follow regular, even predictable, paths, returning after fixed periods within visible distance of our Earth.

As early as his *De Gravitatione*, Newton was aware of this difficulty for the Cartesian position. After all, had not Descartes himself written to Mersenne that

> To suppose a body moving in a non-resistant medium is to suppose that all the parts of the surrounding liquid body are disposed to move at the same speed as the original body in such a way as to leave room for it and take up its room. That is why every kind of liquid resists some movement or other. To imagine matter which resisted none of the different movements of different bodies, you would have to pretend that God or an angel was moving its parts at various speeds to correspond with the speeds of the movement of the body they surround.[231]

We can explain, along Cartesian lines, why the whirling vortices do not resist planetary motions, but for these aether-currents to offer no effective resistance to the eccentric motions of the comets is inexplicable in natural terms. If, says Newton, we set aside resistance, 'we must also reject the corporeal nature': a resistanceless medium, 'I should no longer believe to be subtle matter but a scattered vacuum.' If, therefore,

> there were any aerial or aetherial space of such a kind that it yielded without any resistance to the motions of comets or other projectiles I should believe it was utterly void. For it is impossible that a corporeal fluid should not impede the

---

[231] From Kenny, Ed., p. 62.

motion of bodies passing through it, assuming that . . . it is not disposed to move at the same speed as the body.[232]

Newton thus turns Descartes' own words against him: a medium which does not resist the motions of bodies in *any* direction is not a plenum but a void.

In Book 3 of *Principia*, this argument is elaborated in greater detail. In air, Newton reckons,[233] a planet such as Jupiter would be so decelerated by the drag of the medium that it would lose all its motion inside a year; the fact that the planets pursue their regular courses for millennia without any sensible retardation shows, therefore, that the medium in which they move is whole orders of magnitude rarer than atmospheric air.

The death-blow to the Cartesian vortex-theory comes, however, from observations of comets, which are proved by measurements of parallax to move through interplanetary space. The fact that they pursue a variety of oblique and eccentric paths through this space is, says Newton, as fatal to the Cartesian vortices of subtle matter is it is to the solid orbs of the Aristotelians:

> Hence also it is evident that the celestial spaces are void of resistance; and though comets are carried in oblique paths, and sometimes contrary to the course of the planets, yet they move every way with the greatest freedom, and preserve their motions for an exceeding long time, even where contrary to the course of the planets.[234]

Interplanetary space has, therefore, no resistance to the motions of bodies. But if $R_T = 0$, $R_I = 0$, and the space is *void* of matter. There is no Cartesian aether: far from being necessary to explain celestial motions, it would serve only to impede them, to 'disturb and retard the Motions of those great Bodies, and make the Frame of Nature languish'.[235] Celestial space is 'void of all sensible Resistance, and by consequence of all sensible Matter'[236] – it is, to all intents and purposes, a vast void or vacuum₂, 'utterly void of any corporeal fluid, excepting, perhaps, some extremely rare vapors, and the rays of light'.[237]

232   From A. R. and M. B. Hall, Eds., 4, pp. 146–147.

233   Newton, 1, pp. 418–419.

234   *Ibid.*, p. 497.

235   Newton, 2, p. 368. See also Koyré, 4, p. 108; Snow, p. 70.

236   Newton, 2, p. 365. Cf. also Cotes' preface to *Preface*, p. xxx.

237   Newton, 1, p. 366.

Exactly the same principles apply also the Boylian vacuum, as Newton explains in the General Scholium to Book Three of *Principia*:

> Bodies projected in our air suffer no resistance but from the air. Withdraw the air, as is done in Mr. *Boyle*'s vacuum, and the resistance ceases; for in this void a bit of fine down and a piece of solid gold descend with equal velocity.[238]

And also in the *Opticks*:

> Small Feathers falling in the Open Air meet with great Resistance, but in a tall Glass well emptied of Air, they fall as fast as Lead or Gold, as I have seen several times. Whence the Resistance seems still to decrease in proportion to the density of the Fluid.[239]

The fact that Galileo's law of free fall actually holds in the Boylian vacuum shows that that title is no misnomer, but an entirely appropriate appellation.

In Newtonian thought we have, therefore, the following pattern of argument:

1.	The Laws of Motion.
1→2.	$R_1 \propto$ Density.
3.	Boyle's experiments.
4.	Observations of comets.
3&4→5.	Existence of resistanceless spaces.
2&5→6.	Existence of spaces devoid of mass (inertia).

But mass is essential to matter: it would form part of any adequate definition of corporeality. We may therefore conclude that the spaces inside Boyle's receiver and between the planets are (almost) 100% void of matter. This is the great pro-vacuist argument of *Principia*.

A subtle matter without resistance, says Cotes in his Preface to *Principia*, is both intrinsically absurd and utterly idle. In particular, it would have 'no force to communicate motion with, because it has no inertia'. It follows that it possesses 'no manner of efficacy' and 'does not in the least serve to explain the nature of things'.[240] Given the conception of force (= our concept of impulse) as Δmv common to late seventeenth century physics, and the Third Law of Motion ($+ \Delta mv = - \Delta mv$; $\Sigma m\vec{v} = k$) it follows that only a *massy* object can

---

238 *Ibid.*, p. 543. See also p. 419.

239 Newton, 2, p. 366.

240 See Newton, 1, p. xxxi.

exert a mechanical impulse. The Cartesians require an aether which is at the same time devoid of resistance, yet able to exert a mechanical push on grosser bodies, which is impossible. A fluid without resistance lacks mass, and is therefore without physical efficacy. As Cotes says:

> Those who would have the heavens filled with a fluid matter, but suppose it void of any inertia, do indeed in words deny a vacuum, but allow it in fact. For since a fluid matter of that kind can noways be distinguished from empty space, the dispute is one about the names and not the natures of things.[241]

Another experiment[242] which seems to militate decisively against the postulation of a dense, Cartesian aether is reported by Newton in Book Two of *Principia*. Since, he says,

> it is the opinion of some that there is a certain ethereal medium extremely rare and subtile, which freely pervades the pores of all bodies; and from such a medium so pervading the pores of bodies, some resistance must needs arise; in order to try whether the resistance, which we experience in bodies in motion, be made upon their outward surfaces only, or whether their internal parts meet with any considerable resistance upon their surfaces, I thought of the following experiment.[243]

Take two pendulums of equal length, with two identical wooden boxes for bobs, then fill one of these boxes with metal, leaving the other one empty, and set them to swing. If there is an aether, it should pass easily through the wood, but resist the passage of the much denser metal. (A fluid which exerted no resistance even to the passage of solid metal would be indistinguishable from void.) On the Cartesian theory, then, the first pendulum (with the full box for a bob) should come to rest long before the second. This, however, was not what was observed: the resistance to the first pendulum was only marginally greater than that to the second:

241 *Ibid.*, xxxi. Late seventeenth century Cartesians did indeed posit a resistanceless subtle matter, dynamically indistinguishable from a void (see Aiton, 8, pp. 229–230; 260–261).

242 It is the opinion of Westfall that this experiment was of crucial importance in the development of Newton's thought, effectively converting him (around 1679) *from* the Mechanical Philosophy *to* a quite different world-view involving vacuum and A.D. See especially Westfall, 4, p. 376 ff.

243 Newton, 1, p. 325.

the whole resistance of the box, when full, had not a greater proportion to the resistance of the box, when empty, than 78 to 77.[244]

The resistance to the internal parts is therefore 'either nil or wholly insensible'. The conclusion draws itself: there is no Cartesian aether.

When Newton returned to aether-theories in the 'Queries' of the *Opticks*, these conclusions are *not* abandoned. The existence of a fluid aether is posited to explain a variety of physical phenomena, notably the periodicity or 'fits' associated with the propagation of light[245] and the transmission of heat *in vacuo*,[246] and is soon generalised to provide an explanation of a great variety of natural phenomena, notably those involving *forces* of various kinds. The aether of the *Opticks* is, however, far removed from that of Descartes: it is quite incredibly *rare* and *elastic*. In fact, says Newton,

> the elastick force of this Medium, in proportion to its density, must be above $700,000 \times 700,00$ (that is, above $490,000,000,000$) times greater than the elastick force of Air in proportion to its density.[247]

The aether could therefore be, for example, 700,000 times rarer than air *and* 700,000 times more elastic. It still *possesses* density, and hence mass, but is matter at such a stupendous degree of rarefaction as to be almost totally devoid of resistance. An aetherial medium of this rarity is, Newton claims, perfectly compatible with the stability – for the period of recorded History – of our solar system:

> So small a resistance would scarce make any sensible alteration in the Motions of the Planets in ten thousand Years.[248]

---

244   *Ibid.*, p. 326.

245   Newton consistently opposed the wave theory of light as incompatible with rectilinear propagation (1, p. 367 ff.; 2, pp. 362–363), but had to admit something periodic in nature in order to account for his 'fits'. His final view therefore involves light-corpuscles associated with aether-waves. For the theory of 'fits', see Newton, 2, p. 278 ff.; for a valuable commentary, see Westfall in the *Texas Quarterly*, p. 86 ff.

246   Desaguliers had shown that two thermometers, one placed in a closed vessel in air, and the other in a Boylian vacuum, will keep in step with one another. This entails the transmission of heat across the so-called 'vacuum', which Newton came in his later years to see as evidence of the existence of a subtle aetherial matter. See Home, 3, pp. 2–3; Snow, pp. 105–106; and of course Newton, 2, pp. 348–349.

247   Newton, 2, p. 351.

248   *Ibid.*, pp. 352–353.

The elasticity of the aether is, says Newton, explicable in terms of *interparticulate repulsive forces*. In the unfinished monograph *De Aere et Aethere* he had accounted for the elasticity of the air in terms of such forces, and derived Boyle's Law from a Fα1/d force-law;[249] now he supposes the aether to consist of particles 'which endeavour to recede from one another', and thus constitute a medium 'exceedingly more rare and elastick than Air, and by consequence exceedingly less able to resist the motions of Projectiles, and exceedingly more able to press upon gross bodies by endeavouring to expand it self'.[250]

Newton seems trapped in a cleft stick here: if the aether is so rare as to exert no resistance, how can it exert a mechanical impulse sufficient to move gross bodies? One simply cannot have it both ways. Newton seems to envisage this aether after a mechanical model (its properties are clearly modelled to some extent on those of the air), but its mode of action is not so easily explicable. By simply multiplying the zeros at the end of the velocities of aether-particles, one can empower them to move gross bodies by impulsion, but the expedient has an *ad hoc* ring to it. But, unless the aether suffers a 'reaction' equal and opposite to its action, the Third Law of Motion is violated. (Exactly the same problem emerges if we allow aether-particles to act on grosser bodies by means of forces.) Small wonder then that Newton's aether-theory appeared only in his *Queries*, and was never asserted as a positive doctrine.

Newton's final position, therefore, involves the postulation of an incredibly rare and tenuous aether, incapable of offering any detectable resistance to the passage of bodies through it, yet able, *somehow*, to move grosser bodies. This aether consists of minute and distant particles: its elasticity is due to powerful repulsive forces between those particles. It is therefore composed largely of void space: any particular pair of aether-particles will be separated by a considerable expanse of vacuum$_2$. The postulation of such a subtle medium does not, therefore, affect Newton's overall commitment to *vacuism*. (An aether which *filled* interplanetary space would,

---

[249]   See A. R. and M. B. Hall, Eds., 4, pp. 223-224. For the derivation of Boyle's Law from a F α 1/d force-law, see Newton, 1, p. 300. Such an account has of course the advantage of explaining the sustained elasticity of the air at very low pressures.

[250]   Newton, 2, p. 352.

he claims, be so dense that nothing could ever move through it.) The *Opticks*, like the *Principia*, is an unashamedly pro-vacuist text.

C. Interstitial Vacua and the Constitution of Matter

It is time now to turn our attention to a consequence implicit in the Atomic Theory from the beginning, but of which the full force was not seen until the seventeenth century. It is the thesis that most of the familiar – and apparently solid and space-*filling* – bodies of our everyday experience consist largely of void space. This follows as a corollary from two of the fundamental axioms of the Atomic Theory: the homogeneity of matter* (the stuff of the atoms) and the proposition that matter* has weight proportional to its bulk or volume. From these two premisses there follows the classical Atomist account of *specific gravity* (S.G.) as a measure of the quantity of matter* per unit volume of any given substance. But the S.G. of gold is about 19 times that of water, so even if gold were 100% matter* water would be only about 5% matter* (and 95% void). And, to explain along Atomist lines – i.e. by the penetration of corpuscles – how gold conducts heat, cold, and magnetic 'influence', we must postulate pores even within its apparently dense and compact structure; but, if gold is only, say, 50% matter*, water must be 97.5% void space!

This consequence was seen by a number of critics of the Atomic theory in the early years of the seventeenth century. J. B. Van Helmont, for example, saw that on the Atomist's principles water would be almost entirely empty space, and rejected the theory for that reason. Water, he reminds us, is almost completely *incompressible*, a fact quite inexplicable on the supposition that such a large proportion of void enters into its make-up. (The classical Atomic theories involved, of course, no interatomic *forces*). One must, he claims, conclude that matter is not homogeneous, i.e. that one and the same quantity of matter can *fill* (completely occupy) different volumes at different times, i.e. in different states of condensation.[251]

A similar position is adopted by Kenelm Digby in his treatise *On Body*. The postulation of interstitial vacua would provide, Digby admits, a 'very easie and intelligible' account of condensation and rarefaction; he would, he says, be 'very glad' if it would serve his turn. Unfortunately, however,

---

[251] See Partington, 1, p. 377.

the inconveniences that follow out of the supposition of vacuities, are so great, as it is impossible by any means to slide them over.[252]

Apart from Aristotle's general anti-vacuist arguments, the Atomist position entails that the air we breathe is less than one part in 7,600 solid matter! Galileo, says Digby, has shown that the S.G. of air is about 1/400 that of water; that of water is 1/19 that of gold; gold is not 100% solid matter. Therefore, reasoning along classical Atomist lines, air is less than one part in 7,600 (19 × 400) space-filling matter:

> The aire will by this reckoning appear to be like a net, whose holes and distances, are to the lines and thrids, in the proportion of 7,600 to one.[253]

Those who think to defend *interstitial* vacua along Atomist lines are therefore really committed to a conception of the atmosphere as a great coacervate vacuum, only occasional particles of solid matter interrupting the continuity of the void space. (Lucretius' image of the motes in the sunbeam is far more appropriate than Hero's model of a heap of sand.) Digby, not surprisingly, finds this implication incredible, and opts for a more Aristotelian conception of dense and rare (implicitly, he too denies the homogeneity of matter). A dense body, he says, contains more substance per unit of quantity or volume – such a notion is, he argues, a perfectly intelligible one.[254]

Another important critic of the matter-theory of the Atomists was Francis Bacon. His early 'Thoughts on the Nature of Things' shows considerable sympathy for Atomism ('the doctrine of Democritus concerning atoms is either true or useful for demonstration'[255]) and argues, following Democritus and Hero, for an Atomist account of condensation and rarefaction as involving interstitial vacua,[256] but in

---

252   Digby, p. 25.

253   *Ibid.*, p. 25.

254   *Ibid.*, p. 25.

255   F. Bacon, *Works*, Vol. 5, p. 419.

256   *Ibid.*, p. 420. Condensation, he says, may take place 'either by the exclusion of vacuum in proportion to the contraction; or by the forcing out of some other body previously intermixed; or by some natural (whatever that may be) condensation and rarefaction of bodies'. The third view, that of the peripatetics, he dismisses as unintelligible; the second account, he claims, leads to an infinite regress of subtle matters; we are left, therefore, with the first.

his mature works, and particularly in the *Novum Organum* and the *History of Dense and Rare*, this early attitude is reversed. Bacon's 'pneumatic' matter-theory involves an explicit denial of *both* of the key premises with which this section began, the homogeneity of matter and the universality of weight. Let us deal with the latter first. For tangible or gross matter, we read in the *Novum Organum*, weight is a measure of quantity of matter, but

> . . . Spirit and its Quantity of Matter are not to be computed by Weight, which Spirit rather diminishes than augments.[257]

Spirits or *pneumata* are not merely weight-*less*, they possess positive *levity*: the corpse of an animal, Bacon believes, weighs more than that beast did while alive in virtue of the loss of spirit associated with its demise.[258] S.G. cannot therefore be used as a measure of quantity of matter per unit volume.

Bacon is not even convinced that S.G. measures quantity of tangible or gross matter per unit volume: it may be, he suggests, that close contraction will, 'by reason of the union of force', give rise to a substance of higher S.G. than might have been expected. In such a case, he remarks, the 'calculation' (of quantity of tangible matter from S.G.) 'no doubt fails'.[259] But, even if S.G. did provide a reliable measure of the amount of gross matter per unit volume, this would not imply the existence of void spaces between the tangible parts; and that for two reasons. In the first place, the pores of every body, animate and inanimate alike, are filled with their proper pneumatic bodies or attached spirits:

> . . . no known body in the upper parts of the earth is without a spirit . . . For the cavities of tangible things do not admit of a vacuum, but are filled either with air or the proper spirit of the thing.[260]

And, in the second place, even tangible matter is not homogeneous: Bacon explicitly rejects, in the *Novum Organum*, the Atomist account of condensation and rarefaction he had endorsed in the *Cogitationes*:

257   *Ibid.*, Vol. 4, pp. 197–198; *Novum Organum*, 2, xl.

258   *Ibid.*, Vol. 5, pp. 349–352.

259   *Ibid.*, pp. 343–344.

260   *Ibid.*, Vol. 2, p. 213.

For there is clearly a Folding of Matter, by which it wraps and unwraps itself in space within certain limits, without the intervention of a vacuum.[261]

This 'folding' metaphor is left very obscure, and is in fact totally unenlightening.[262] (When one folds a body, after all, one does not actually alter its total 3D bulk or volume.) It is clear, however, that Bacon is denying the existence of a homogeneous and immutable matter* (the hypothetical stuff of the atoms) in favour of a matter which can undergo *intrinsic* condensation and rarefaction. But, if one accepts such a matter, the Atomist's argument for interstitial vacua collapses: true physical analysis, Bacon insists,

will not bring us to Atoms, which takes for granted the Vacuum, and the immutability of matter (neither of which hypotheses is correct); but to the real particles such as we discover them to be.[263]

In the *History of Dense and Rare* (1624), Bacon's denial of the vacuum is as explicit as one could desire: 'There is,' he flatly states, 'no vacuum in nature, either collected or interspersed'.[264]. The Atomists' argument from condensation and rarefaction to the existence of interstitial vacua is invalid: there is not, Bacon insists, 2,000 times as much vacuum in air as in gold, as would follow from Democritus' principles, 'a fact demonstrated by the very powerful energies of fluids'.[265] If Democritus were right, air would be like a fine dust floating *in vacuo*, but this would lack all continuity and cohesiveness and be quite unable to move gross bodies. The 'virtues of pneumatical bodies' prove, Bacon insists, that air and spirits cannot be composed almost entirely of void space. *All* bodies, for example, share the 'motion of connection', by which they 'delight in mutual contact': this analogue of the scholastic *fuga vacui*[266] would not exist if pneumatic bodies were themselves largely made up of empty space. They too, Bacon concludes, are plena, albeit intrinsically 'thinner' ones.

The cohesiveness of fluids also poses a major problem for the matter-theory of Galileo. Solid bodies, he claims in Day

---

261   *Ibid.*, Vol. 4, p. 231; *Novum Organum*, 2, xlviii.

262   See Reese, 5, p. 561.

263   Bacon, *Works*, vol. 4, p. 126; *Novum Organum*, 2, viii.

264   *Ibid.*, Vol. 5, p. 398.

265   *Ibid.*, Vol. 4, p. 231; *Novum Organum*, 2, xlviii.

266   *Ibid.*, Vol. 5, p. 497.

One of the *Discorsi*,[267] are held together by the *fuga vacui* exerted by their interstitial microvacua; liquids such as water lack such point-vacua and are therefore without cohesion. (The melting of a solid body he explains in terms of the entry of fire-atoms or *igniculi* into the interstitial vacua to neutralise their *fuga vacui*.) Galileo continued to deny, in the face of some powerful evidence, that a fluid such as water possessed *surface tension*[268] – such an admission would have proved fatal to his matter-theory. (It is, looking back, difficult to see how the physics of the *liquid* state could have been solved along classical Atomist lines– i.e. without postulating interparticulate *forces*.[269])

If we turn now to the matter-theory of Descartes, we find a radically different approach to the topics of condensation and rarefaction. Matter, says Descartes, is, essentially, extended substance: it is, as such, to be quantified according to its bulk or volume. To attempt to distinguish 'quantity of matter' from bulk is to be guilty of a vicious abstraction. It is, he insists, quite impossible that any part of matter 'should in any way occupy more space at one time than another'[270] – if one body, X, is larger than another, Y, it simply shows that X contains more matter than Y does. *Volume alone* is the proper measure of quantity of matter, not weight or density: it is, Descartes contends, a childish prejudice to imagine that there is 'more substance or corporeal reality in rocks or metals than in air or water'[271] One cubic foot of air contains exactly the same quantity of matter as an equal volume of gold.

How, then, are we to account for condensation and rarefaction? Rare bodies, Descartes replies, are

> those between whose parts there are many interstices filled with other bodies; and those are called dense bodies, on the other hand, whose parts, by approaching one another, either render those distances less than they were, or remove them altogether, in which case the body is rendered so dense that it cannot be denser. And yet it does not possess less extension than when the parts occupied a greater space, owing to their being further removed from one another.[272]

---

267   Galilei, 4, p. 11 ff.

268   Clavelin, p. 438; see Le Grand in Butts and Pitt (Eds.), p. 200.

269   See Toulmin and Goodfield, 1, p. 67.

270   Descartes, 1, Vol. 1, pp. 263–264. Cf. also Laing, p. 403.

271   Descartes, 1, vol. 1, p. 250.

272   *Ibid.*, pp. 256-257 (*Principles*, 2, 6).

An appropriate analogy is that of a *sponge*, the parts of which do not themselves take up more space when it is immersed in water; the expansion of the whole sponge is due to the fluid absorbed. No other explanation, Descartes asserts, is intelligible:

> . . . it would be undoubtedly contradictory to suppose that any body should be increased by a fresh quantity or a fresh extension, without the addition to it of a new standard substance, i.e. a new body.[273]

The Aristotelian notions of intrinsic condensation and rarefaction being unintelligible, and the concept of void space self-contradictory, it follows that condensation and rarefaction must be identified respectively with the extrusion and ingress of subtle matter or aether. The sponge-particles can be identified with those of Descartes' third element E3, and the water with the subtle matter composed of the two rarer elements E1 and E2. (These are 'rarer', of course, not in the unintelligible sense of containing less matter per unit volume, but only in the sense that they consist of much smaller and more mobile parts.)

Within Cartesian physics, quantity of gross matter or E3 serves as a sort of surrogate concept of mass: the particles of E3 are bulky and sluggish and *resist* all attempts to change their state of motion.[274] 'Dense' and 'rare' are therefore explicable not in terms of quantity of matter, but of E3, per volume-unit: the Cartesian explanation of condensation and rarefaction merely substitutes aether for void in the Atomist's account.

For Descartes, of course, weight is caused by a whirling vortex of subtle matter, which presses the more sluggish corpuscles of E3 towards the centre of the 'whirl'. E3 corpuscles have a slower angular velocity and hence less centrifugal force than the swirling aether, and therefore possess a *net* downward tendency. Only E3 has weight – it cannot meaningfully be attributed to the aether itself. And the gravitation of E3 is no more than an 'accidental' consequence of the existence of the aetherial vortex, not an essential property of matter as such. The Atomists' premiss of the universality of weight is, therefore, expressly denied by Descartes.

---

273   *Ibid.*, p. 257.
274   Meyerson, p. 162; A. R. Hall, 3, p. 120.

One might, nevertheless, be tempted to use S.G. as a measure of the quantity of gross or terrestrial matter, E3, per unit volume.[275] Descartes, however, explicitly *denies* this. The weight of terrestrial bodies, he asserts, 'does not always have the same relation to their matter'; 'gravity alone is not enough to make known how much terrestrial matter [E3] there is in each body'.[276] Quicksilver does not contain nearly fourteen times as much E3 per unit volume as water. The weight of a body depends not merely on its amount of E3, but on the *arrangement* or 'texture' of its gross particles: the surface area of tangible matter exposed to the downward pressure of the aether will clearly be a crucial factor in determining the weight of any given body. One would expect, for example, a substance with texture A to have a higher S.G. than one of texture B, although their E3/volume ratios are identical:

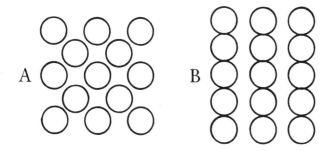

Descartes thus accepts the homogeneity of matter, but rejects the Atomists' assumption of the universality of weight, thereby evading their argument for the conclusion that even solid bodies are largely void space. Given the denial that S.G. is proportional to quantity of E3 per unit volume, one cannot even legitimately conclude, e.g., that water is less than 5% E3, and composed almost entirely of *materia subtilis*. Gold, says Descartes, the S.G. figures notwithstanding, may contain only four or five times as much E3 as water in any given volume.

If we wish then to deny that the apparently solid bodies with which we are familiar in our day-to-day affairs are really almost entirely made up of void space, we must reject one or other of the Atomists' premises, either the homogeneity of matter or the use of S.G. as a measure of quantity of matter

[275] Huygens was to take up just this position (see Dijksterhuis, p. 460).

[276] *Principles*, 4, 25, quoted from Meyerson, pp. 162–163.

per unit volume. For an example of a thinker who soldiers on with *both* these crucial premisses, seemingly unaware of the utter dissolution and disintegration of the fabric of matter that they jointly imply, we have only to turn to the *Physiologia* of Walter Charleton.

Condensation and rarefaction, Charleton argues, can only be explained in terms of a 'disseminate' or interstitial vacuum. When air is compressed, he explains, one of three things must occur. Either (a) the air-corpuscles themselves come to occupy less space, or (b) two or more such particles interpenetrate, or (c) the particles recede into interstitial vacua.[277] (The Cartesian solution is totally ignored.) The first two hypotheses (a) and (b) being absurd, says Charleton, it follows that air must contain a disseminate vacuum to account for its compressibility. But Mersenne has shown that ordinary atmospheric air may be condensed ten or fifteen times,[278] perhaps more: the so-called 'disseminate' vacuum must therefore occupy well over 90% of its total volume. (A still more radical conclusion will of course follow from a cursory glance at a table of S.G.s.[279])

Later in the *Physiologia* Charleton defends the Atomist account of density and rarity against the rival Aristotelian and Cartesian views. A rare body, he insists, contains a little matter in a comparatively large volume: when water is evaporated, for example, its vapour occupies at least 100 times as much space as formerly. It does not, however, *fill* this space – if it did, it could not be condensed back to water. Without interstitial vacua, he claims,

> it must necessarily follow, that in condensation many particles of matter must be reduced into one particle space, which before condensation was adequate onely to one particle of matter: and, on the contrary, in rarefaction, one and the same particle of matter must possess many of space, each whereof, before rarefaction, was in dimensions fully respondent thereto'.[280]

Charleton thus affirms, in opposition to the Aristotelians, the existence of a homogeneous matter*. The peripatetic doctrine of condensation and rarefaction, which allows the

277   Charleton, p. 26.

278   *Ibid.*, p. 26, p. 257.

279   *Ibid.*, p. 33. S.G. figures provide, says Charleton, the third major argument for the existence of interstitial vacua.

280   *Ibid.*, p. 253.

same quantity of matter to occupy at different moments different volumes of space, is dismissed as contrary to reason. The Aristotelians, says Charleton, 'imagine the quantity of a thing to be absolutely distinct from its matter', and suppose that dense and rare arise from different proportions of substance to quantity (volume).[281] There is, however, a 'Law of Nature' to the effect that

> every body in the universe is consigned to its peculiar place, i.e. in such a canton of space, as is exactly respondent to its dimensions . . .[282]

Aristotelian (i.e. intrinsic) condensation and rarefaction are therefore not naturally possible: only God could condense or rarefy matter*.

As for Descartes' explanation of density and rarity in terms simply of particle-size and mobility, it fails to impress Charleton. If, he claims, air were a *plenum*, without interstitial vacua, it would be as dense as gold: it would therefore (a) be quite incompressible, and (b) have the same S.G.[283] This is in effect no more than a frank reaffirmation of the Atomist position: there is no attempt to refute the Cartesian view in detail. A rare body, Charleton insists, simply contains less matter per unit volume than a dense one – given a homogeneous matter*, this can only entail that it contains more vacuum$_2$.

Another phenomenon which indicates the existence in solid bodies such as glass of a sizable proportion of empty space is that of transparency. Given a corpuscular theory of light, a transparent substance must consist of atoms and vacua so arranged as to allow the passage of light-corpuscles through its pores:

> to diaphaneity is required a certain orderly and alternate position of the pores and bodies, or particles.[284]

The transparency of substances tends to vary with their rarity, as one would expect on corpuscularian principles, but is also crucially dependent on atomic arrangement. Paper, though less dense than glass, has its particles in a disorderly state, and is therefore almost totally opaque – it allows only a very diffuse light to penetrate through it. (These brief and

---

[281] *Ibid.*, pp. 262–263. Cf. of course the account of Digby.

[282] *Ibid.*, p. 263.

[283] *Ibid.*, pp. 253–255.

[284] *Ibid.* p. 259.

fragmentary remarks on optical evidence concerning the infrastructure of matter may have stimulated the far more profound and penetrating enquiries of Newton's *Opticks*.)

The peripatetic notion of condensation and rarefaction comes under fire again in Boyle's 'Defence' against Linus. After outlining the three main accounts – Atomist, Cartesian and Aristotelian – of the nature of dense and rare, Boyle goes on to dismiss the peripatetic position as unintelligible, involving as it does the idea that

> the self-same body does not only obtain a greater space in rarefaction, and a lesser in condensation, but adequately and exactly fills it, and so when rarefied acquires larger dimensions without ever leaving any vacuities betwixt its component corpuscles, or admitting between them any new or extraneous substance whatsoever.[285]

Whereas Atomists and Cartesians alike posit a homogeneous matter*, and explain rarefaction in terms of the admixture of void or aether, the Aristotelian rejects matter* and admits true or intrinsic condensation and rarefaction. The absurdity of this position has, Boyle claims, been well brought out by Descartes and Gassendi. In Guericke's Magdeburg experiment, for example, air is rarefied about 2,000 times: does each and every corpuscle of air, then, occupy 2,000 times more space after rarefaction than it did before? For the Cartesians ('who take extension to be but nominally different from body') this supposition will appear self-contradictory; for other mechanists (all of whom regard bodies 'as having a necessary relation to a commensurate space') it will be almost equally objectionable.[286]

The Aristotelian account of condensation is, of course, liable to parallel objections: it would involve, Boyle claims, either the annihilation of matter or the interpenetration of bodies. On Linus' theory,

> two, or perhaps ten thousand, bodies may be crowded into a space, that is adequately filled by one of them apart. And if this be not penetration of dimensions, I desire to be informed what is so.[287]

Boyle is therefore a staunch advocate of a homogeneous matter*. (The so-called 'arguments' of the mechanical

[285]   Boyle, *Works*, Vol. 1, p. 144.

[286]   *Ibid.*, p. 145.

[287]   *Ibid.*, p. 150.

philosophers for this thesis all seem, however, to have an unsatisfactory, even question-begging, character: all that is needed in order to make sense of the intrinsic condensation and rarefaction of the peripatetics is a concept of quantity of matter quite independent of spatial extension. And in the notion of *mass* or *inertia* we have precisely such a concept – the inertial homogeneity of matter is a logically contingent thesis, accepted without sufficient argument by many key figures in seventeenth century thought.) Boyle also favours, as we shall see, the Atomistic account of S.G. and interstitial vacua. In his 'Experiments and Considerations about the Porosity of Bodies'[288] he begins by citing a mass of empirical evidence for the porous nature of apparently solid, even hard, bodies, then moves on to the argument from specific gravities. Even if, he says, as is most unlikely, gold were 100% matter*, a metal such as iron or copper would be less than 50% solid matter, and water as little as 5%.[289]

Boyle thus assumes *both* the homogeneity of matter* *and* its quantification by weight.[290] Once these premises are assumed, however, it will follow that most corporeal substances, however solid they may appear, are almost entirely composed of void space. Boyle, like Charleton, seems blithely unaware of the direction the argument is taking: he argues at length for the existence in bodies of 'pores' or interstitial (!) vacua, apparently blind to the fact that the void is making such inroads into the fabric of bodies that the very existence of solid matter, matter* itself, is coming under threat.

Among seventeenth century thinkers, only Newton and his disciples take the Atomists' premises to their logical conclusion. Even in some of Newton's earliest notebooks, the *Quaestiones Quaedam Philosophicae* (1664–1666), the implications implicit in the Atomic Theory from its beginnings are clearly discerned. Suppose, says the young Newton, that gold consists half of matter*, half of void space. It will follow – assuming that S.G. is a measure of quantity of matter* per unit volume – that water is bout 1 part in 40 of matter*, atmospheric air about 1 in 3,600, and atmospheric air rarefied a thousand times, as in some of Boyle's experiments,

---

288   *Ibid.*, Vol. 4, p. 759 ff.

289   *Ibid.*, p. 778.

290   Fire, he believes, is a very subtle matter, which nevertheless *adds* to the weight of bodies, thus accounting for the gain in weight of calcined metals.

1 in 3,600,000. Yet this extremely rare medium is still *elastic* (it still obeys Boyle's Law), a fact more readily explicable in terms of interparticulate repulsive forces than in terms of mechanical models such as Boyle's little springs.[291]

The phenomenon of transparency is also, Newton notes, powerful evidence of the rarity and tenuity of such materials as glass and water (and hence, of course, of *all* materials of similar S.G.s). Given a corpuscular theory of light, the fact that glass and water will transmit its rays (a) in any direction, and (b) to great distances, will suggest strongly that the constitution of those familiar substances is largely of void space. As early as this, then, Newton is moving in the direction of the matter-theory eventually expounded in the 'Queries' to the *Opticks*.

Before moving on to the *Opticks*, however, we must glance at Book Two of *Principles*, where Newton proves, in opposition to Descartes, the proportionality of mass and weight. His demonstration turns on the isochrony of any two pendulums of the same length, quite regardless of the material constitution of the bob. If weight were *not* directly proportional to mass, a pendulum with a bob of weight/mass ratio 1:1 would swing more quickly than one with a corresponding ratio of 0.9. The fact that this does not happen – that *all* pendulums of length l swing with a periodic time $T = 2\pi\sqrt{l/g}$ shows that mass and weight are always strictly proportional. (The same conclusion will of course follow from Galileo's law of free fall.) As Newton says,

> By experiments made with the greatest accuracy, I have always found the quantity of matter in bodies to be proportional to their weight.[292]

This result simultaneously overthrows the Cartesian theory of gravity and validates the Atomists' use of S.G. as a measure of quantity of matter per unit volume. On the Cartesian theory, weight depends not merely on 'mass' (= quantity of E3) but also on the 'texture' or arrangement of the E3 particles – Newton's experiments show conclusively that this is not the case. The weights of bodies are quite *independent* of their textures, and depend *only* on the quantities of matter they contain. The ancient Atomists were therefore, *pace* Descartes, quite right to use S.G. as a measure of quantity of matter per unit volume.

---

[291]   See Westfall, 2, pp. 180–181.

[292]   Newton, 1, p. 304.

Let us now turn to the matter-theory of the mature Newton, as spelt out in his *Opticks*. The corpuscular theory of light, Newton sees, has momentous implications concerning the structure of matter. From it, he says,

> We may understand that Bodies are much more rare and porous than is commonly believed. Water is nineteen times lighter, and by consequence nineteen times rarer than Gold, and gold is so rare as very readily and without the least opposition to transmit the magnetick Effluvia, and easily to admit Quicksilver into its Pores, and to let Water pass through it . . .[293]

Suppose, then, that gold is less than half solid matter: it will follow that

> . . . Water has above forty times more Pores than [solid] Parts. And he that shall find out an Hypothesis, by which Water may be so rare, and yet not be capable of compression by force, may doubtless by the same Hypothesis make Gold, and Water, and all other Bodies, so much rarer as he pleases; so that Light may find a ready passage through transparent substances.[294]

The argument runs roughly as follows:

1. We cannot account for the observed properties of gold (e.g. solubility in *Aqua Regia*, as well as those mentioned above), unless we assume that it 'abounds with pores', which must constitute *at least half* its total volume.[295]

2. Weight is proportional to mass – a central pillar of the whole architecture of the *Principia*. Hence a particular volume of gold contains about twenty times as much matter as an equal volume of water.

3. There is a homogeneous matter*, corpuscles of which make up all material things. This belief in the *inertial homogeneity* of matter is the crucial premiss for Newton's whole matter-theory: without it, one could simply say that water-particles are intrinsically less dense than those of gold. The assumption of homogeneity is of course a commonplace of the Mechanical Philosophy, but it is for all that a logically

---

[293] Newton, 2, p. 267. For comments, see Snow, p. 88; Thackray, 2, pp. 21–22.

[294] Newton, 2, p. 267.

[295] Elsewhere, in a draft passage intended for *Principia*, Newton estimates the pore/matter ratio in gold to be 5:2. See A. R. and M. H. Hall, Eds., 4, p. 317.

contingent thesis (Newton freely admits its contingency on the Divine Will[296]), a potential Achilles' heel in an otherwise invulnerable argument.[297]

4.   Premises 1 and 2, given the measured S.G.s of water and gold, entail that water contains less than 1/40 of the matter that would be contained in an equal volume of perfectly solid matter*. This, given also 3, entails that it consists of at least 97.5% void space: its particles of matter* occupy less than 2.5% of its total volume.

5.   But water is (all but) *incompressible*: it can be subjected to very great pressure without undergoing significant condensation. To provide a plausible explanation of this fact, we need to postulate *powerful interparticulate repulsive forces*. (As Westfall remarks: 'Populating the universe with forces was the condition for depopulating it with matter'[298])

6.   Now if the impenetrability of a body can be explained without assuming that solid matter fills a significant proportion of its total bulk, the dematerialisation of matter can proceed apace. Gold itself may be one part in ten, or less, of matter*, making water one part in two hundred, or less still. One needs only to posit ever-stronger repulsive forces. The impenetrability of *observed* bodies is now explained *entirely* in terms of these forces, without any need of solid parts. (Newton does, of course, as we saw in Chapter 9, still believe in solid, space-filling atoms – he still thinks of mass, for example, as proportional to *volume* of matter*[299] – but it is the forces which are now doing all the work.)

7.   That this model corresponds to empirical reality is confirmed, Newton claims, by optical evidence. Light, he reminds us,

> is transmitted through pellucid solid Bodies in right lines to very great distances. How Bodies can have a sufficient quantity of Pores for producing these Effects is very difficult to conceive, but perhaps not altogether impossible.[300]

For a stream of light-corpuscles to pass through a transparent body without colliding with any of its solid parts, and hence being 'stifled and lost',

296   Newton, 2, p. 403.

297   See Thackray, 2, p. 16.

298   Westfall, 4, p. 386.

299   See McGuire, 2, p. 161.

300   Newton, 2, p. 268.

one must have recourse to a certain wonderful and exceedingly artificial texture of the particles of bodies by which all bodies, like networks, allow magnetic effluvia and rays of light to pass through them in all directions and offer them a very free passage: and by such a hypothesis the rarity of bodies may be increased at will.[301]

Light passes through transparent bodies from every angle: it is not that the particles are so arranged as to permit only one direction of propagation. To explain this fact consistently with Newton's principles, we must have recourse to an extremely tenuous matter, consisting of far-flung particles of matter* separated by considerably greater void spaces, and held together by interparticulate forces. (To explain *both* cohesion *and* elasticity one will require a net force of attraction at some distances and of repulsion at others.)

Solid matter thus begins gradually to dissolve and disintegrate, to fade away, leaving infinitesimal point-centres of force in an infinite void. Some of Newton's disciples delighted in the immateriality of his eventual model of the physical universe. According to Bentley, the ratio of void to solid matter in our world is 8,575:1 – without providential guidance, he claims, the far-flung atoms could never have assembled to form such an organised system.[302] Clarke, replying to Leibniz's charge of materialism, retorts that Newton's mathematical principles 'prove matter, or body, to be the smallest and most inconsiderable part of the universe'.[303] *All* the solid matter in the whole universe, says Priestley, may occupy as little space as a nut-shell.[304] Take this view to its logical conclusion and we have the matterless universe of Boscovitch, in which atoms are merely point-centres of spheres of force: solid matter has utterly disappeared.

---

301   From a draft passage intended for *Principia* (A. R. and M. B. Hall, Eds., 4, p. 317).

302   See Koyré, 3, p. 182.

303   H. G. Alexander, Ed., p. 12.

304   See Thackray, 1, p. 29.

MATTER, FORMS, AND QUALITIES: 3

A. *Reductionism and Mechanism*

Before proceeding with this chapter, I must first explain my uses of certain key terms, and especially the distinction between the reductionist programme and the Mechanical Philosophy. Reductionism, as I use the term, is a thesis about the origin and nature of qualities, a claim to the effect that (many of) the qualities we discover in bodies are either consequent upon or simply identical with the contrivance and arrangement of their minute parts, i.e. what Boyle and Locke call their 'textures'. If, as seems plausible, we name bodies after their manifest properties, it will follow as an inevitable corollary that generation and corruption can occur merely as a result of the local motion (rearrangement) of these minute corpuscles. If, for example, we call something gold on account of its yellowness, density, malleability, fusibility, fixedness, etc.; and, if by re-ordering its corpuscles these qualities were to be lost, we should say that gold had been corrupted and a new substance generated. (It may, of course, be properly describable as a compound of gold if the dense yellow metal is readily extractable from it.) There is nothing *more* to being gold that the possession of that highly specific texture on which all its manifest properties depend; in particular, no substantial form (S.F.) of gold is required. Forms and qualities are reducible to corpuscular arrangements or *microstructures*.

Robert Boyle sees this point quite clearly. Atomism and Cartesianism, he says, although they differ 'in material points' (e.g. void and indivisibles), 'yet in opposition to Peripatetic and other vulgar doctrines they might be looked upon as one philosophy' – abandoning substantial forms and real qualities (R.Q.s) as unintelligible, 'both the Cartesians and the Atomists explicate . . . phenomena by little bodies variously figured and moved'.[1] Their common ground is just this shared commitment to a particular reductionist programme which, says Boyle, because it explains natural phenomena in terms of corpuscles, may justly be called the *corpuscular philosophy*.

[1] Boyle, *Works*, Vol. 1, p. 355.

Mechanism, as I use the term, is essentially a thesis about the springs of local motion: its gist is best captured by Hobbes' dictum that 'there can be no cause of motion, save in a body contiguous and moved',[2] i.e. that no body can be set in motion except by an impulse from another. As we have interpreted the phrase, then, the 'Mechanical Philosophy' involves, in its purest form, the fourfold denial of (a) Action at a distance, (b) Spontaneity, (c) Incorporeal causes, and (d) Irreducibly teleological explanations.[3]

Unfortunately, my usage is not altogether consistent with that of others: many have tended to accommodate the reductionist programme under the general banner of the Mechanical Philosophy. This is perfectly understandable – after all, the same thinkers are often simultaneously engaged in furthering both programmes, and some commonly used models and metaphors (e.g. the likening of the universe to a giant clock) illustrate *both* the dependence of manifest properties on the arrangement of parts (reductionism) *and* Hobbes' dictum – each cogwheel is moved by another adjacent, moving, one (mechanism). To appreciate the importance of the distinction between reductionism and mechanism, we should at least glance at the variety of thinkers who accepted the former but doubted or denied the latter. This I do in Appendix 5: here I shall only give a skeletal account of the issues involved.

We have already discussed[4] the revival of classical Atomism which occurred around the year 1600: the first generation of the seventeenth century witnessed, we observed, a massive upsurge of interest in and enthusiasm for the particulate matter-theories of Democritus, Epicurus, Plato and Hero. Now among the most attractive and plausible of the doctrines of the ancient Atomists was the 'generation = aggregation' thesis, the idea that substances come-to-be by the coming-together of minute particles and cease-to-be by their dissociation. Even to the layman, there seems something fundamentally sound about such a view. But one can account for the properties of bodies in terms of their material constitution and/or arrangement of parts, and hence explain coming-to-be and ceasing-to-be in terms of the local motion of those parts, without opting for mechanism: a reductive account of qualities is perfectly compatible with the

---

2   Hobbes, *Works*, Vol. p. 124. See also p. 115.

3   See Chapter 4, p. 142.

4   See Chapter 5, p. 222 ff.; M. Boas, 4, p. 426 ff.

admission of non-mechanical agents of various kinds. It is, on the other hand, difficult to accommodate S.F.s and R.Q.s within a mechanistic natural philosophy: it is, as Descartes saw, unintelligible that such entities should 'possess the force adequate to cause motion . . . in bodies'.[5] (Matter and motion, and S.F.s and R.Q.s, are incommensurable causal categories.) We are concerned in this chapter solely with the reductionist programme; its successor tackles the Mechanical Philosophy.

B. Forms and Qualities

The scholastic concepts of 'substantial' forms and 'real' qualities were subjected, in the seventeenth century, to a quite devastating critique. S.F.s and R.Q.s, say such men as Descartes, Hobbes, Boyle, and many others, are (a) unintelligible and (b) redundant, quite superfluous for explaining any of the phenomena of Nature; Boyle adds also a battery of empirical objections. Let us begin, as usual, with the conceptual points.

For a Catholic philosopher such as Descartes expressly to deny R.Q.s was a very dangerous position to adopt, since that doctrine was intimately associated with orthodox accounts of the mystery of the Eucharist.[6] Life was much easier for that heretical Protestant iconoclast, Hobbes, who can dismiss R.Q.s and transubstantiation alike as papistical mumbo-jumbo. In his published works, Descartes never explicitly *attacks* the notion of R.Q.s; rather, he simply omits them from his natural philosophy.[7] In his hypothetical reconstruction of the world, *Le Monde*, for example, he assumes that there existed in the primeval chaos 'none of those forms and qualities so debated in the schools'.[8] In writing to his disciple, Regius, Descartes advises caution and circumspection in attacking scholastic ideas:

> For instance, why did you openly reject substantial forms and real qualities? Do you not remember that . . . I expressly said that I did not at all reject or deny them, but simply found them unnecessary in setting out my explanations.[9]

[5] Descartes, 1, Vol. 1, p. 295.

[6] See Chapter 7, p. 328.

[7] See the letter to Arnauld in Descartes, 1, Vol. 2, p. 116, where he claims – somewhat disingenuously – that he does not actually *deny* the existence of R.Q.s.

[8] Descartes, 1, Vol. 1, p. 107.

[9] From Kenny, Ed., p. 127. See also Kemp Smith, p. 105.

As soon as people realise, he explains, that S.F.s and R.Q.s are unnecessary in Natural Philosophy, that all the phenomena can be accounted for in terms of the sizes, shapes, and motions of the particles of a homogeneous matter*, S.F.s and R.Q.s will be abandoned as redundant without the need for a direct frontal assault.

For Descartes' private views on the topic of S.F.s and R.Q.s, we must turn to his letters, not to his published works. Thus he writes to Mersenne that

> I do not believe there are in nature any real qualities, attached to substances and separable from them by divine power like so many little souls in their bodies.[10]

Whatever can exist separately – even by divine power – is really a substance, not an accident, so the very concept of a 'real' quality is incoherent:

> . . . it is contradictory that real accidents should exist, because whatever is real can exist separately apart from any other subject; but whatever can exist separately is substance not accident.[11]

If R.Q.s can exist apart from matter, they are substances in their own right, i.e. self-subsistent entities. But, *qua* immaterial substances, they must, for Descartes, be soul-like, instances of *res cogitans* rather than *res extensa*. Such a doctrine is, Descartes insists, quite unintelligible, and the supposed R.Q.s totally inconceivable: having no clear and distinct ideas of them, we have no warrant for the assumption that they exist. Matter is differentiated by 'modes' (size, shape and motion), none of which could conceivably possess separate existence, even by divine power.

That the concept of a 'real' quality is self-contradictory was also urged by Digby: philosophers, he says,

> will have them to be reall Entities or Things, distinct from the bodies they accompany: and yet, they deny them a subsistence or self-being . . .,

but

> the first part of their description maketh them compleat substances; which afterwards in words they flatly deny: and it is impossible to reconcile these two meanings.[12]

---

10   Kenny, Ed., p. 135.

11   Descartes, 1, Vol. 2, p. 250.

12   Digby, pp. 48–49.

Advocates of R.Q.s, says Boyle in his *Origin of Forms and Qualities*,

> make then indeed accidents in name, but represent them under such a notion as belongs only to substances; the nature of a substance consisting in this, that it can subsist of itself without being in anything else as a subject of inhesion.[13]

R.Q.s both are, and are not, substances, which doctrine 'is either unintelligible, or manifestly contradictious'.

The critique of S.F.s is similar. In his letter to Regius, Descartes explains that

> . . . the 'substantial form' we are denying means a sort of substance adjoined to matter and constituting along with it a whole that is purely corporeal; something that is not less of a true substance or self-subsistent thing than matter is, but rather more, since it is called 'actuality' and matter 'potentiality'.[14]

Apart from human beings, no other natural creatures consist of an incorporeal substance causally linked to an organic body. (Animals, of course, are automata.) The ontological status of man is unique.[15] The supposition that one needs to posit an S.F. to account for the functional contrivance of the parts of animals and plants is dismissed by Descartes:

> All the arguments to prove substantial forms could be applied to the form of a clock, which nobody says is a substantial form.[16]

It is orthodox scholastic doctrine that only natural creatures possess S.F.s; the forms of artefacts are 'accidental', dependent merely on the arrangement of their material parts.[17] Yet a clock possesses an organised structure in virtue of which it is naturally conceived as a functional unit: why, then, suppose that plants and animals are anything more than sophisticated automata? Economy (Ockham's razor) and intelligibility alike favour the rejection of S.F.s: to posit an incorporeal substance attached to every natural body is extravagance carried to heights of absurdity.

13   Boyle, *Works*, Vol. 3, p. 17.

14   From Kenny, Ed., p. 108.

15   Cf. Boyle, *Works*, Vol. 3, p. 40.

16   Kenny, Ed., p. 129. For the denial of any real distinction between natural and artificial forms, see also p. 61.

17   See the discussion of Ockham in Chapter 7, pp. 296–297.

The spirit of medieval nominalism is clearly manifest in the seventeenth century 'corpuscular philosophy', and nowhere more so than in the rejection of S.F.s and R.Q.s. The postulation of R.Q.s, say men such as Digby, Hobbes and Boyle, results from the mistaken assumption that each different sensation of ours, which we name 'red', 'hot', 'sweet', etc., is produced by a corresponding reality in the object.[18] Humans, Digby explains,[19] are apt to 'confound the true and real natures of things with the conceptions they frame of them in their own minds', which can only lead to error and absurdity. Words, he insists, only represent things 'according to the pictures we make of them in our thoughts'; one and the same *thing* may, then, become known to us through different *aspects*, and thus receive a variety of names. To assume that a distinct reality corresponds to every adjective by which we can truly characterise an object is the merest superstition. Our various senses give us, for example, different 'ideas' (green, sweet, cold, etc.) of an apple: although these ideas are distinct, their causes need not be. Unless one is careful, Digby warns, one may 'give actuall Beings to the quantity, figure, colour, smell, tast, and other accidents of the apple', which is to fall into the 'mistaken subtilities' of the theory of R.Q.s. To assume that the greenness, sweetness, etc., of the apple are distinct and potentially self-subsistent realities is to confuse ideas with things, which is the endemic error of the advocates of R.Q.s.

The scholastics' alleged confusion of names or ideas with things is also rebuked by Hobbes in *De Corpore*[20] and in the *Leviathan*.[21] From the fact that we can consider whiteness, heat, etc., independently of the subjects in which they inhere, and have distinct names for them, it does *not* follow, Hobbes emphasises, that such accidents might exist 'separated from all bodies', i.e. as self-subsistent things. The schoolmen, he alleges, have confused logical with physical analysis, and postulated physical simples to correspond with our simple *ideas* of colours, tastes, smells, etc., which confusion of the real with the ideal (i.e. purely mental) has side-tracked philosophy for centuries.

---

18 For the link between R.Q.s and faith in the senses, see Burtt, p. 5. The relativity and subjectivity of perceived qualities are, as Digby (p. 305) sees, very difficult to reconcile with the theory of R.Q.s.

19 Digby, pp. 2–4.

20 Hobbes, *Works*, Vol. 1, p. 33.

21 *Ibid.*, Vol. 3, p. 24 ff.

A very similar point can be found in Boyle's *Origin of Forms and Qualities*, and again in Locke's *Essay*. Just as, says Boyle, the various rôles which one man can play (husband, father, prince, etc.) are quite clearly not 'so many real and distinct entities in the person so variously denominated', so the given aspects (i.e. sensory qualities) under which an object can appear to us are not genuinely distinct and separate things.[22] The theory of R.Q.s is the result of an unfortunate reificiation: we call our different sensations by different names, and then mistakenly infer a corresponding variety in the objective world:

> We have from our infancy been apt to imagine that these sensible qualities are real beings in the object they denominate.[23]

Locke too lends his voice to the general theme: we call bodies hot or cold, black or white, etc., he says, from the *ideas* (= sensations) they produce in us:

> which qualities are commonly thought to be the same in those bodies that those ideas are in us, the one the perfect resemblance of the other, as they are in a mirror, and it would by most men be judged very extravagant if one should say otherwise.[24]

Aristotelians and ordinary folk alike assume (quite without justification) a resemblance between cause and effect, and 'project' features of our ideas onto external objects. To account for our sensation of red, for example, they posit a corresponding R.Q. of redness: the corpuscular philosophy shows the idleness and absurdity of any such assumption.

If, the argument of the corpuscularians continues, we are obliged to posit S.F.s and R.Q.s, every change that occurs in the course of Nature will involve the (miraculous) creation and annihilation of incorporeal substances. (The idea that a form is somehow 'educed' from the 'potency' of matter is ridiculed by men such as Charleton[25] and Boyle.[26]) Since the creation and annihilation of incorporeal substances is beyond

[22] Boyle, *Works*, Vol. 3, p. 17.

[23] *Ibid.*, p. 23.

[24] Locke, *Essay*, II, 8, xvi, Vol. 1, p. 106. Cf. also P. Alexander, 3, pp. 216–217.

[25] Charleton, p. 423 ff.

[26] Boyle, *Works*, Vol. 3, p. 38. His dismissal of the notion of 'eduction' is derived almost verbatim from Charleton.

the power of mere creatures, Descartes writes, scholastic accounts of generation and corruption (accession and departure of S.F.s) and of alteration (gain and loss of R.Q.s) should, strictly, invoke the aid of the divine power:

> It is inconceivable that a substance should come into existence without a new creation by God; but we see that every day many so-called substantial forms come into existence; and yet the people who think they are substances do not believe that they are created by God; so their view is mistaken.[27]

If we admit S.F.s and R.Q.s, then, we implicitly turn the regular course of Nature into a succession of miracles. A similar point occurs in Cudworth:

> But that in all the Protean transformations of nature . . . there should be real entities thus perpetually produced out of nothing, and reduced to nothing, seemed to be so great a paradox to the ancients, that they could by no means admit of it.[28]

To avoid continually evoking the divine power to create and annihilate substantial realities, one must adopt the Atomists' explanation of forms and qualities as consequent upon the arrangement and motion of insensibly small particles. In Nature, then, no real entities come-to-be or cease-to-be; the same abiding stock of simples are perpetually 'anagrammatised', qualities and forms resulting, like words and sentences, from the arrangement of such entities.

The peripatetics teach, says Boyle, that

> the form is not made of any thing of the matter; nor indeed is it conceivable how a physical agent can turn a material into an immaterial substance, especially matter being, as they themselves confess, as well incorruptible as ingenerable.[29]

Since, he reasons, matter cannot be converted or transformed into an incorporeal substance such as a S.F. is presumed to be, it follows that S.F.s must be created *ex nihilo*. On such a view, every natural body comes-to-be 'partly by generation and partly by creation': the aggregation of its parts must be

---

[27]   Kenny, Ed., p. 129.

[28]   Cudworth, Vol. 1, p. 63.

[29]   Boyle, *Works*, Vol. 3, p. 39.

supplemented by the miraculous creation, *ex nihilo*, of its S.F. All births – and, for that matter, deaths – would be miracles.

The postulation of S.F.s and R.Q.s, the critics continue, is quite devoid of explanatory power. Not only can all the manifest properties of observed bodies be accounted for, along corpuscularian lines, without invoking such pseudo-entities, but to invoke them would leave one none the wiser. As supposed *explanatory* principles, they are otiose. S.F.s are postulated, says Descartes, to explain the 'natures' of things, yet

> no natural action can be explained by these substantial forms, since their defenders admit that they are occult and that they do not understand them themselves. If they say that some action proceeds from a substantial form, it is as if they said that it proceeds from something they do not understand, which explains nothing.[30]

Boyle pursues a similar line of attack in his *Origin of Forms and Qualities*. The wisest of those who admit S.F.,s, he says, 'confess, that they do not well know them',[31] that they are hidden from us, beyond human ken and comprehension. But if knowledge of S.F.s and of the manner of their operation is, by the scholastics' own admission, 'too difficult and abstruse to be attained by them', the postulation of S.F.s can be of no explanatory value: it cannot serve to cast light on any natural phenomenon to be informed that it is produced by a S.F.[32]

Belief in S.F.s is therefore, Boyle concludes:[33]

a.   Unnecessary – natural phenomena can be explained intelligibly (i.e. along corpuscularian lines) without invoking such supposed entities.

b.   Useless – we do not have a single example of an enlightening explanation of a natural phenomenon in terms of the operation of a S.F. If, moreover, S.F.s are and will forever remain beyond the bounds of our comprehension, there could in principle be no such explanations.

c.   Unintelligible – the Aristotelians seem simultaneously to assert and to deny that a S.F. is an incorporeal substance.

---

30   Kenny, Ed., p. 61.

31   Boyle, *Works*, Vol. 3, p. 11.

32   *Ibid.*, p. 38.

33   *Ibid.*, p. 46.

d.   Extravagant – it is in violation of the spirit of Ockham's razor to posit a S.F. for every natural creature.
e.   Absurd – if S.F.s *are* incorporeal substances, the course of nature is a procession of miraculous creations and annihilations.

The postulation of R.Q.s is, Boyle continues, as vacuous and unenlightening as that of S.F.s. To explain the whiteness of a body in terms of the arrangement of its parts is illuminating; to be told that white bodies are such in virtue of the R.Q. of whiteness is not. The schoolmen attribute to the R.Q.s 'a nature distinct from the modification of the matter they belong to, and in some cases separable from all matter whatsoever'.[34] Snow is white, say these scholastics, in virtue of its R.Q. of whiteness: if one asks what this is, says Boyle,

> they will tell you no more in substance, than that it is a real quality which denominates the parcel of matter to which it is joined, white . . .[35]

It is in the *nature* of each R.Q., on this theory, to produce its characteristic effect on our sense-organs, i.e. to give rise to a particular sensation. This doctrine, Boyle avers, is simply *too easy* to be of any value: the putative explanation 'X is white in virtue of the R.Q. of whiteness' is, on scrutiny, quite vacuous.

In opposition to this obscurantism, says Boyle, his chief aim is

> to make it probable to you by experiments . . . that almost all sorts of qualities may be produced mechanically; I mean by such corporeal agents as do not appear . . . to work otherwise than by virtue of the motion, size, figure and contrivance of their own parts (which attributes I call the mechanical affections of matter because to them men willingly refer the various operations of mechanical engines) . . .[36]

Boyle's great contribution to the corpuscular philosophy lies not so much in the *a priori* attack on the intelligibility and explanatory power and value of S.F.s and R.Q.s (in which he was anticipated by Descartes, Gassendi, Hobbes and others), but in the empirical arguments which he brings against those scholastic notions. He was concerned in particular to bring chemical evidence against the theory of R.Q.s: as an

[34]   *Ibid.*, p. 12.
[35]   *Ibid.*, p. 13.
[36]   *Ibid.*, p. 13.

interesting corollary, the same evidence tells also against the 'material constitution' account of qualities. Let us examine one or two of the most significant of his examples.

Combinations of substances, either natural or artificial, may, says Boyle, possess properties quite different from those of their constituents.[37] Thus gunpowder (a mere mixture of nitre, brimstone, and charcoal) is explosive, but none of its ingredients possesses that power. This objection tells significantly against the theory of R.Q.s. If, for example, we have a reaction-mechanism:

$$A + B \longrightarrow C,$$

and C has a particular quality Q^*, then, if Q^* is 'real' (and hence subject to conservation[38]), either A or B must have possessed Q^*. It follows that, if neither A nor B possesses Q^*, that quality is *not* 'real'. Explosiveness, then, is not a R.Q.: the same argument-form will work for many other qualities.

Many *colours*, says Boyle, can be altered by mechanical rearrangement of the parts of a body, e.g. by grinding, hammering, etc. If, for instance,[39] one hammers a piece of ice, one can reduce it into a white powder: place this in the sun and it melts, leaving only a clear fluid. The corpuscularian has an easy explanation of these changes in terms of the rearrangements of the minute parts and the resulting changes of texture; the scholastic, who believes 'that colours are inherent and real qualities',[40] must posit the creation and subsequent disappearance (annihilation?) of the R.Q. of whiteness. In hammering, says Boyle, 'there is nothing really produced by the agent in the patient, save some local motion of its parts or some change of texture consequent upon that motion':[41] if then the quality of whiteness appears in consequence, how can it be conceived as a R.Q.?

Another powerful argument against R.Q.s can be drawn from Helmont's willow-tree experiment,[42] in which water, which is 'fluid, tasteless, inodorous, colourless, volatile, etc.', may, 'by a differing texture of its parts', be transformed into plant tissue, and even into substances (oils, spirits, etc.) with quite different qualities. At the very least, a quasi-substantial

37   *Ibid.*, p. 19.
38   See Chapter 3, pp. 96–97.
39   Boyle, *Works*, Vol. 3, pp. 25–26.
40   *Ibid.*, Vol. 1, p. 693.
41   *Ibid.*, Vol. 3, p. 25.
42   *Ibid.*, p. 69.

R.Q. should fall under a conservation law; Helmont's experiment shows that few, if indeed any, of the sensible properties of bodies are in fact conserved. If, for example, we have:

Water ⟶ Plant Tissue ⟶ Oil, Spirit, Ash, etc.

whence emerge the supposed R.Q.s of the distillation-products? If, for example, we distil from a plant an oil which is inflammable and immiscible in water (i.e. qualitatively contrary to water), how can we account for such a transformation while retaining belief in R.Q.s? The example tells also against the chemists' notion of quality-bearing 'principles': for the Paracelsian, an inflammable oil must be predominantly sulphurous in make-up (sulphur is, after all, the 'principle' of inflammability), but no one would claim, surely, that *water* has a high sulphur content?! And if matter is, as Boyle suspected,[43] universally transmutable, i.e. if the 'anything ⟶ anything' thesis is true, quality-bearing chemical 'principles' must be utterly rejected.

Another example[44] concerns the reaction of camphor with 'oil of vitriol' (concentrated sulphuric acid) to yield a dark red solution lacking the characteristic smell of camphor. Add a little water, and the red colour vanishes, the pungent aroma simultaneously re-emerging. The phenomena are, says Boyle, easily explicable on the corpuscular theory as due to a slight textural change (a loose union of corpuscles) effected by the acid and reversed by the water, but must seem quite mysterious on the supposition that colours and odours are 'real' qualities. On the theory of R.Q.s, we have simply

WHITE + PUNGENT⟶   RED + ODOURLESS⟶
WHITE and PUNGENT,

which must seem quite miraculous and inexplicable. Similar examples can be multiplied *ad lib* from the experience of the chemical laboratory. Consider, for example,

Lead   (SOUR) + Vinegar   (SOUR)⟶'Sugar   of   Lead' (SWEET)

Copper Oxide (BLACK) + Sulphuric Acid (CLEAR)⟶ Copper Sulphate (BLUE)

In any event, Boyle continues,

it is but an ill-grounded hypothesis to suppose that new qualities cannot be introduced into a mixed body, or those

---

[43]   See Chapter 9, pp. 428-431.
[44]   Boyle, *Works*, Vol. 3, p. 76.

that it had before be destroyed, unless by adding or taking away a sensible portion of some one or more of the Aristotelian elements, or chemical principles; for there may be many changes as to quality produced in a body without visibly adding or taking away any ingredient, barely by altering the texture, or the motion of the minute parts it consists of.[45]

For example, the freezing of water (the slowing down of the agitation of its parts) yields ice, while the hammering of silver renders it brittle. Consider also the grinding of glass or almonds, or the whisking of egg white: in all such cases, a change of colour is consequent simply upon a change of texture rather than on the loss or gain of any material substance.[46] These examples refute not only the scholastic doctrine of R.Q.s, but also the Paracelsian theory of quality-bearing principles, which might be regarded as a sort of revamped version of that doctrine. On the 'material constitution' account of qualities, no alteration can occur in a body without a corresponding change in FAWE, or MSL, ratio; Boyle convincingly shows the falsity of that thesis. *Form* (= microstructure), he insists over and over again, not FAWE or MSL ratio, explains qualities. The notion that, if substances X, Y and Z share a common property Q^*, they must owe this to one specific Q^*-imparting 'principle' he describes as 'precarious'. The chemists simply *assume*, he says, that

> where the same quality is to be met with in many bodies, it must belong to them upon the account of some one body whereof they all partake.[47]

Properties such as weight, fluidity, light, etc., cannot, Boyle argues, be plausibly attributed to any of the peripatetic elements or Paracelsian principles: the 'One quality ⟶ One Principle' thesis is unfounded and quite probably *false*. In his monograph, 'On the Imperfection of the Chemist's Doctrine of Qualities',[48] he subjects this thesis to careful scrutiny before decisively rejecting it. Chemical explanations of qualities in terms of FAWE or MSL ratios are, he claims, shallow, unenlightening, and quite frequently erroneous; the corpuscular philosophy is capable (in principle) of giving more profound and perspicuous accounts of the origin of qualities.

---

45   *Ibid.*, p. 295.
46   *Ibid.*, vol. 1, p. 556. For more examples, see Vol. 4, pp. 274–275.
47   *Ibid.*, Vol. 1, p. 551.
48   *Ibid.*, Vol. 4, p. 273 ff.

Let us now turn to the empirical grounds that might be given in favour of the postulation of S.F.s. The scholastics posit S.F.s, says Boyle, in order to account for, e.g., the unity, cohesiveness, and organisation of compound bodies: without S.F.s, they will claim, matter could not give rise to recognisable and stable structures, far less to such functionally unified and organic wholes as vegetable and animal bodies. S.F.s also serve, for the Aristotelian, to explain the spontaneous return of objects to their 'natural' states when driven from them by external violence (e.g. the cooling of hot water) and, last but not least, those ('occult') qualities inexplicable in terms of an object's material constitution.[49]

Much of Boyle's reply is predictable. In the first place, S.F.s, being themselves totally obscure, cannot help us to understand how things work – the postulation of S.F.s is devoid of explanatory power. Furthermore, the phenomena on which the Aristotelian's case depends may be explained without invoking S.F.s: the spontaneous cooling of hot water can be accounted for either in terms of the kinetic theory of Descartes or the fire-atoms theory of Gassendi *without* supposing that water is 'naturally' cold; the contrivance of suitably adjusted parts may explain cohesion and unity without need of recourse to a S.F.; and so on.[50]

Boyle's main argument against S.F.s turns, however, on the properties of synthetic materials and artefacts. Artificial 'vitriol' (ferrous sulphate) can be readily manufactured in the laboratory from iron and sulphuric acid by 'a convenient apposition of the small parts of the saline menstruum to those of the metal'.[51] But this 'factitious' vitriol has exactly the same chemical properties as those of the natural substance (produced by the weathering of iron pyrites): if the juxtaposition and arrangement of corpuscles explains the properties of the synthetic material, why not also of the natural? Yet it is scholastic orthodoxy that artificially manufactured substances *lack* S.F.s . . . the conclusion draws itself. Since gunpowder, glass, alloys, etc., have highly specific chemical properties, and yet are confessedly possessed only of 'accidental' forms, the postulation of substantial forms seems idle and redundant.

---

[49]   *Ibid.*, Vol. 3, p. 40 ff. For the dependence of occult qualities on S.F.s, see also p. 11.

[50]   *Ibid.*, pp. 45–46.

[51]   *Ibid.*, p. 45.

The argument is taken a stage further by the consideration of machines such as clocks and mills. These are, doubtless, without S.F.s, yet by the arrangement of their parts achieve an almost organic functional unity comparable to that of an animal.[52] If we require no S.F. to explain the workings of the former, why must we postulate such a curious entity for the latter? There is, Boyle claims, no absolute distinction between 'Nature' and 'Art': the phenomena which produced the mechanical arts should, he says, 'be looked upon as really belonging to the history of nature in its full and due extent'.[53]

How might we set about showing that any particular quality Q^* is *not* dependent for its existence on a specific S.F.? Boyle mentions four possible strategies:

1. Some qualities (e.g. roughness, smoothness) are so palpably dependent on the arrangement of parts that no one is tempted to posit a special S.F. to account for them.
2. If one can manufacture an artefact with Q^*, that is a sure sign that the presence of that quality in natural things need not be due to an S.F.
3. If Q^* can be induced in a body by rubbing, grinding, hammering, etc., this indicates a mechanical origin for that property.
4. Where the agent producing Q^* works only by local motion, we can expect the quality to be dependent on an arrangement of parts, not on an S.F.[54]

Boyle accepts, in conclusion, the Aristotelian notion of form as an explanatory factor, but rejects the reification involved in the scholastic conception of S.F. Forms, he insists, must not be conceived as incorporeal substances, separable in principle from matter; there is nothing more to a form than an arrangement and motion of parts, a (dynamic) texture or microstructure. A form, he insists bluntly,

> doth . . . consist in such a convention of these . . . mechanical affectations of matter as is necessary to constitute a body of that determinate kind. And so, though I shall for brevity's sake retain the word Form, yet I would be understood to mean by it, not a real substance distinct from matter, but only the matter itself of a natural body, considered with its peculiar manner of existence; which I

52 *Ibid.*, p. 29.
53 *Ibid.*, pp. 435–436.
54 *Ibid.*, Vol. 4, p. 233 ff.

think may not inconveniently be called either its specifical or its demonstrating state, or its essential modification; or, . . . in one word, its stamp.[55]

C. The Argument from Intelligibility

Of our total fund of ideas, says Descartes, some are clear, distinct, and innate; others are 'adventitious', confused products of the Mind-Body union.[56] A permanently disembodied soul would possess[57] all the former ideas but none of the latter: it would be able, for example, to meditate on metaphysics and theology, or to prove theorems in pure mathematics, but would have no sense of the 'bloomin', buzzin' confusion' of colours, sounds, smells, etc., which bombard the embodied minds of ordinary human beings. The clear and distinct ideas of extension, figure, and local motion are grasped, says Descartes, 'by reasoning alone',[58] without any need of sensory input and hence of bodily organs; without ears and eyes, by contrast, we should have no sensations of sound or colour.

The next step is crucial. If, Descartes argues, God had implanted in our souls ideas to which no physical reality corresponded He could not be cleared of the charge of deceit and wilful deception. But God is a morally perfect Being, unable, by His very nature, to be a deceiver.[59] It follows that there must exist in physical reality 'simple natures' corresponding to our clear and distinct ideas. (No such theological guarantee holds for the confused and adventitious ideas of sense.) I clearly and distinctly perceive, says Descartes in his *Meditations*, that extension, figure, motion, etc., exist *in objects*: our clear and distinct ideas mirror the structure of Reality.[60]

Cartesian natural philosophy rests heavily on this theological guarantee of the ultimate intelligibility of Nature. Now that I know God, Descartes concludes Mediation V, I have the means to acquire knowledge of many truths, including those 'which pertain to corporeal nature in so far as it is the

---

55   *Ibid.*, Vol. 3, p. 28.

56   See Kenny, Ed., p. 104.

57   Sensory experience may serve, however, to *elicit* innate ideas, to bring them to our attention.

58   Descartes, 1, Vol. 1, p. 252.

59   Cf. of course Descartes' use of the ontological argument.

60   See Descartes, 1, Vol. 1, p. 164.

object of pure mathematics'[61] – i.e. which follow from our concepts of extension, figure, and motion. The applicability of mathematics to physics is guaranteed *a priori*.

In Nature, then, there exist 'simple natures', which correspond to our clear and distinct ideas. The existence of these *naturae simplices* is avouched in the *Regulae*: we shall, says, Descartes,

> call those only simple, the cognition of which is so clear and so distinct that they cannot be analysed by the mind into others more distinctly known. Such are figure, extension, motion etc.; all others we conceive to be in some way compounded out of these.[62]

Explanation in physics involves, ultimately, a *reduction* to these simples, an *analysis* of a complex phenomenon into its primary (and readily intelligible) elements (cf. of course the method of resolution and composition). In the *Regulae*, for example, Descartes recommends the physicist to concentrate on simple varieties of physical action (e.g. the transmission of an impulse along a stick) before attempting to tackle magnetism or gravity, a methodological recommendation which contains, albeit implicitly, a lot of physics.

Since the *only* clear and distinct ideas we have of the physical world are those of extension, figure, and local motion, the intelligibility of Nature demands that all physical explanations be couched in those terms. As Descartes says,

> I first considered in general all the clear and distinct notions of material things to be found in our understanding, and . . . finding no others except those of shapes, magnitudes, and motions, and the rules whereby these three things can be diversified by each other – which rules are the principles of geometry and mechanics – I judged that all the knowledge men can have of nature must be derived from this source alone:[63]

Or, as Pascal puts it in his aphoristic summary of Cartesian natural philosophy, 'ce qui passe la géométrie nous surpasse'.[64] Strictly, of course, this does not imply that all

---

61   *Ibid.*, p. 185.

62   *Ibid.*, pp. 40–41. For comments, see Kemp Smith, p. 60; Keeling, pp. 53–55.

63   *Principles*, IV, 203, quoted from Kemp Smith, p. 113.

64   Quoted from Cassirer, 3, p. 143.

there *is* to corporeal Nature is extension, figure, and local motion: it merely claims that all *our knowledge* of it must rest on these. (Would God be guilty of deceit – a sin of omission – if He had neglected to give us certain innate ideas necessary for a complete grasp of material Reality?) However that may be, it is not a possibility Descartes seriously considers: he plainly believes that extension, figure, and motion *exhaust* the essence of the material world,[65] which is therefore transparent through and through to our intellects. Nothing in the realm of corporeal Nature is in principle unintelligible to us (though some of her more intricate mechanisms may be difficult for us to discover). Kemp Smith's summary of this whole argument is masterly:

> Descartes is here relying on his fundamental thesis . . . that the possibility of scientific understanding is ultimately conditioned, its range and limits determined, by the 'simples' to which alone the human mind has access.[66]

Cartesian Natural Philosophy involves, therefore, both an *a priori demand* for mechanical explanations of natural phenomena (no other type of explanation is intelligible to us), and an *a priori guarantee*, derived from God's veracity, that such explanations are in principle possible (but *not* that they are necessarily discoverable by weak human intellects: there is all the difference in the world between what is intelligible *per se* and what is discoverable by us). The corporeal world, Descartes affirms, is one vast machine, made up of a perfectly homogeneous plenum of matter*, differentiated into discrete particles only by local motion: 'all the variety in matter, or all the diversity of its forms, depends on motion'.[67] All physical phenomena are subservient to the principles of mechanics, which in turn reduces to geometry and kinematics. (As we shall see in the next chapter, dynamics presents a real problem for the Cartesian.) Once again we turn to Kemp Smith for an apt and perceptive summary:

> Nature, Descartes is . . . asking us to recognise, is non-mysterious; all physical happenings are sheerly mechanical;

---

[65] The essence of body, he states in *Principles*, 2, 4 (Descartes, 1, Vol. 1, p. 255), consists in extension *alone*. See also Hoenen, p. 356.

[66] Kemp Smith, p. 93. The same rationalism, without the theological guarantee, can be found in Hobbes.

[67] *Principles*, 2, 63; Descartes, 1, vol. 1, p. 265. Cf. also Laing, pp. 404–406. Without local motion to produce differentiation and diversification, the material universe of Descartes would be like a Parmenidean *block*.

the only processes operating on the earth or in the heavens are impact and pressure.[68]

Our natural philosophy must, therefore, be mechanistic: Cartesianism serves to 'rule out, as being unintelligible, and for us meaningless, explanation in any other fashion'.[69]

If we turn now to Robert Boyle, we find a somewhat different position. Cartesianism involves, as we have seen, both a demand (for mechanical explanations) and a guarantee (that all natural phenomena are explicable after such a manner). Boyle endorses the demand but has doubts about the guarantee. Is it not impious, he wonders, to claim that the physical universe *must* be intelligible to us? Could not the almighty have created a world wholly opaque to our concepts, the workings of which are totally obscure to us? Brought up in the Calvinist tradition of voluntarist theology, which lays special stress on the power of the divine will, and belittles the cosmic significance of man and his puny facilities, Boyle could hardly be expected to endorse any *a priori* guarantee of Nature's intelligibility to *us*:

> I see no necessity that intelligibility to a human understanding should be necessary to the truth or existence of a thing, any more than that visibility to a human eye should be necessary to the existence of an atom, or a corpuscle of air, or of the effluviums of a lodestone.[70]

If it transpires that much of the physical universe *is* intelligible to us, explicable in terms of concepts which we miserable mortals can comprehend, this is a purely contingent fact, a manifestation of God's benevolence to us, not a matter of *a priori* necessity. There are, after all, genuine mysteries (e.g., Boyle suggests, the union of soul and body) which will forever – barring revelation – surpass the compass of our understanding.[71] It is significant,[72] then, that Boyle's most sceptical critique of the argument from intelligibility occurs in the context of a polemic against those natural philosophers (notably the Epicureans) who 'would exclude the Deity, from intermeddling with matter', i.e. who deny God's providential

---

68   Kemp Smith, p. 94. Cf. *Principles*, 2, 64.

69   Kemp Smith, p. 94.

70   Quoted from Burtt, p. 179. See Boyle, *Works*, Vol. 1, p. 450; Vol. 6, p. 694 ff.

71   See Westfall, 1, Part 2, pp. 108–109.

72   Mandelbaum, pp. 94–95.

governance of His Creation. In some matters, Boyle insists, unintelligibility (to us) does not entail untruth. The Atomists argue that theirs is the only intelligible explanation of some given phenomenon, i.e. that if their account is false that phenomenon is perfectly inexplicable by human reasoning. But, replies Boyle, one can admit this consequence without absurdity:

> for who has demonstrated to us, that men must be able to explicate all nature's phenomena, especially since divers of them are so abstruse, that even the learned atomists scruple not to acknowledge their being unable to give an account of them? And how will it he proved, that the omniscient God, or that admirable contriver, Nature, can exhibit phaenomena by no ways, but such as are explicable by the dim reason of man?'[73]

There may be many things in Nature which are unfathomable by the unaided human intellect, and hence not *explicable* by our philosophy:

> I say explicable, rather than intelligible; because there may be things, which though we might understand well enough, if god, or some more intelligent being than our own, did make it his work to inform us of them, yet we should never of ourselves find out those truths. As an ordinary watchmaker may be able to understand the curiousest contrivance of the skillfullest artificer, if this man take care to explain his engine to him, but would never have understood it, if he had not been taught . . .[74]

Suppose, then, that the precise causes of, e.g., gravity, or magnetism, are such that human ingenuity could never suffice to discover them, and that an angel is sent to explain these mysteries to us. *Will he have to impart to us any radically new concepts or principles?* The Cartesian will reply in the negative: he can admit the possibility of mechanisms of such complexity that human reason cannot bring them to light, but *not* of non-mechanical causes (e.g. A.D.). The sceptic will favour a positive answer: God is not restricted in His creation of the universe to those categories and principles intelligible in principle to us. On the sceptic's view, our angelic tuition may have to include fundamental conceptual revisions; for the Cartesian, it will consist largely of extension and elaboration

[73] Boyle, *Works*, Vol. 2, p. 46.

[74] *Ibid.*, p. 46.

of already familiar ideas, plus perhaps some improvement of our calculatory capacity.

Where does Boyle stand? His theology seems to stand firmly on the side of the sceptic and against the Cartesian. Yet his example of the two watchmakers is pure Cartesianism: Nature's mechanisms may be too complex for mere mortals to fathom, but could be explained to us *without* the need for radically new concepts and principles.

Boyle never seems to face this issue squarely: is he countenancing, in the above passages, merely the possibility of complex and intricate mechanisms (perfectly compatible with Cartesianism), or is he rejecting the argument from intelligibility altogether? The answer, such as it is, seems to lie in his response to the Cartesians' use of clear and distinct ideas as the cornerstone of their system. While agreeing with Descartes in being loath 'to assent to any thing, especially in philosophy, that cannot well be conceived by knowing and considering men',[75] Boyle feels it necessary to make some reservations. Men's powers of conception, he reminds us, differ: the Cartesians, for example, find the existence and causal activity of an incorporeal soul perfectly conceivable, but not a vacuum; the Epicureans can conceive of a vacuum but not of the causal interaction between an incorporeal soul and a gross body.[76]

Where Descartes wanted an absolute, hard-and-fast dichotomy between those ideas which are, and those which are not, clear and distinct, Boyle psychologises and relativises the whole issue, blurring the conceptual boundaries involved. One man can conceive X but not Y, another can conceive Y but not X. For Descartes, one man is right and the other is wrong – there is an *absolute* distinction between what is, and what is not, clearly conceivable. Boyle finds this over-dogmatic: people differ in so many respects – why not also in their powers of conception?[77] Poor old Descartes thus reaps the fruit of trying to build his philosophy on the flimsy psychological foundation of the doctrine of clear and distinct ideas. Unless he can find criteria of clarity and distinctness other than the phenomenological, the whole argument from intelligibility will break down in a welter of subjectivist muddle.

---

75   *Ibid.*, pp. 46–47.

76   *Ibid.*, p. 47.

77   Given an empiricist epistemology, there is no especially favoured list of *innate* concepts: this pre-empts one possible Cartesian response.

We can now invent a plausible Boylian answer to the question we asked earlier, about our angelic tuition in the intricacies of Nature. There being no common, universal stock of clear and distinct ideas, innate in all human minds, there will be no univocal answer to the question: some people will find a particular explanation intelligible without requiring new concepts; others will require a more radical and revolutionary course of tuition – nothing more definite can be said.

Even if however, we have no *a priori* assurance of Nature's intelligibility, no certainty that all her phenomena are explicable in terms of some particular short list of simple concepts, we may still legitimately *demand* intelligibility as a necessary feature of any adequate explanatory hypothesis. An unintelligible explanation is, after all, no explanation at all. Among the requisites of a good hypothesis, then, intelligibility is a factor of the first importance;[78] and Boyle constantly cites the acknowledged clarity and perspicuousness of mechanical explanations as the greatest single advantage of the corpuscular philosophy over its rivals:[79]

> The first thing, that I shall mention to this purpose, is the intelligibleness or clearness of mechanical principles and explications.[80]

Strictly, one might say, all he is entitled to say is that he (Boyle) finds these explanations particularly clear. But, he will reply, *everyone* admits to possession of clear concepts of size, shape and local motion: opponents of the corpuscular hypothesis frequently urge the insufficiency of its principles, never their unintelligibility. If a mechanical explanation of a phenomenon is possible, no right-headed man will prefer any other: even the peripatetics will utilise such explanations where they are readily available.[81]

One can, in sum, accept the Cartesian *demand* for intelligibility in physical hypotheses, acknowledge the special clarity of a particular type of explanatory structure, and thus *seek* to account for more and more natural phenomena after such a fashion, *without* accepting the Cartesian guarantee

---

[78] See the list of requisites of a good hypothesis in Westfall, 1, Part 2, p. 116, and M. B. Hall, 9, p. 135. Intelligibility is listed as requisite number one.

[79] See especially the treatise 'About the Excellency and Grounds of the Mechanical Hypothesis' in Boyle, *Works*, Vol. 4, p. 69 ff.

[80] *Ibid.*, p. 69.

[81] *Ibid.*, p. 232.

that every natural phenomenon will have an intelligible (i.e. mechanical) explanation. Such, I claim, is the position of Boyle. Reason prescribes, as a regulative principle governing scientific enquiry, that we seek for explanations of a certain type, but the ultimate justification for the programme as a whole is empirical. Only if the corpuscular philosophy serves to cast light on many areas of Natural Philosophy have we any reason to believe in its essential correctness. We may have an innate *bias* in favour of a type of explanatory structure particularly congenial to our intellects, but we are not entitled to assume that Nature *must* utilise such mechanisms – to justify corpuscularianism as a whole we must have recourse to experience.[82] Reason tells us what to look for: it does not guarantee that we shall find it, or even that it is there to be found.

D. Transduction and the Analogy of Nature

What properties can we attribute to the sub-microscopically small, forever invisible atomic units, the building-blocks out of which all material things are made up? Which, *if any*, of the qualities of observable bodies also qualify the atoms? This is the problem of transduction.[83] A transductive inference resembles an inductive one, in that it involves a generalisation from the seen to the unseen, but is more radical (and hence more vulnerable to sceptical attacks) in that the inference shifts domains from the perceptible to the imperceptible. Unlike such old favourites as 'all swans are white', or 'all ravens are black', a transductive generalisation is not supported simply by the weight of positive instances (and no negative ones): a transductive inference goes beyond all the evidence we might in principle acquire (at least by direct, i.e. perceptual, means). The crucial question is whether our limits of perceptual (or microscopic) acuity mark any significant boundary in Nature: do the same concepts and principles apply at the sub-microscopic level as in the perceptible realm (the 'Analogy of Nature'), or do quite different ones come into play?

In his discussions of the Atomic Theory, Francis Bacon strongly suggests the existence of a *disanalogy* between the sensible and insensible domains. The atom, he says, is not 'clothed' with properties: 'all compounds are masked and clothed; and there is nothing properly naked except the

---

82   *Ibid.*, p. 234.

83   See McGuire, 3, p. 3 ff.; Mandelbaum, p. 63.

primary particles of things'.[84] Atoms do not, he insists, *resemble* visible bodies: they are of 'a perfectly dark and hidden nature'.[85] They do not even, *pace* Democritus, share a common (downward) motion with compounds: there is 'heterogeneity and exclusion throughout, both in substance and motion'.[86] Atoms are *not* minute particles of FAWE, spirit, aether, or any other visible material: 'Neither is their power and form heavy or light, hot or cold, dense or rare, hard or soft',[87] or, in brief, any of the qualities and powers of sensible stuffs.

Elsewhere, however, Bacon seems to relent somewhat. After all, unless the atom has some positive analogy, however minimal, with sensible things, it is difficult to know how we can characterise it at all. To say that it is of a 'perfectly dark and hidden nature' is scarcely illuminating: one requires something more than such a *via negativa*. The atom, he says, is 'a true being, having matter, form, dimension, place, resistance, appetite, motion, and emanations . . .'.[88] I shall not dwell on this (somewhat strange) list (what, the curious reader will ask, 'emanates' from an atom?); it suffices to make the point that an atom, though disanalogous to sensible stuffs (FAWE, etc.) remains recognisably a body – something with size, shape, and motion, among its more outré attributes.

If we turn from Bacon to Gassendi, we find a somewhat clearer account of the properties of atoms. Compound bodies, says Gassendi, are composed of indivisibles, which have 'no qualities other than size, shape, and weight, or motion . . .'.[89] The 'weight' (*pondus*) of the atom should not be identified with common gravity: it is rather a sort of motive 'urge' instilled by God in each and every atom in virtue of which it may be called (*dei gratia*) a 'self-mover'. As material things, the atoms must have size and shape; to account for the phenomena of Nature, we must also attribute to them such a *pondus*. We cannot, however, attribute to them such qualities as colours: if the atoms of the ocean were

---

84   F. Bacon, *Works*, Vol. 6, p. 731.

85   *Ibid.*, Vol. 5, p. 464.

86   *Ibid.*, p. 465.

87   *Ibid.*, p. 464.

88   *Ibid.*, p. 492. The 'emanations' are no doubt associated with the fact that this Atomic Theory involves A.D. between atoms.

89   *Syntagma*, Physics, Section 1, Book 5, Chapter, Gassendi, p. 424. See also Brett, p. 52.

themselves blue, the agitation produced by the wind could not produce white surf.[90] The simile of the letters is, says Gassendi, 'most apposite': just as varied arrangements of letters give rise to different words and sentences, so 'the same atoms in various transpositions display different qualities, or appearances, to the sense'[91] – the wealth of diversity possible by such elementary means is beautifully illustrated by language.

The *essential* qualities of atoms, says Charleton in his *Physiologia*,[92] are the following:

a.   'Consimilarity of Substance' – all are solid plena of homogeneous matter*.

b.   Magnitude or Quantity – atoms are bodies, not mere points.

c.   Figure or Shape – a necessary adjunct to finite extension.

d.   Weight or '*pondus*' – the principle of atomic motion.

It is not entirely clear whether Gassendi and Charleton are claiming an *a priori* intuition of the essence of matter, or whether the whole body of doctrine is to be confirmed empirically – i.e. by its explanatory power. Given Gassendi's general tendency to a variety of sceptical empiricism, the latter reading seems preferable. Yet Charleton explicitly speaks of the 'essential' qualities of atoms. What are we to make of this?

The answer is simple. There is no inconsistency whatsoever between essentialism and empiricism:[93] we can simply say that it is our hypothesis, empirically grounded, that certain things possess certain essences. We hold a *theory* which attributes essential characteristics XYZ to atoms: within that theory, it will seem a necessary truth that atoms are XYZ, but that such things exist at all (i.e. that our theory mirrors Nature) is a contingent matter, subject to confirmation or refutation by experience.

We cannot conceive how bodies could be constituted out of non-corporeal elements, and are therefore obliged to attribute size and shape to *any* body, however minute, and hence

---

90   *Ibid.*, p. 428. See also Charleton, pp. 130–132. The argument is of course familiar from Lucretius, as is the alphabet-analogy which follows (see chapter 3, p. 138).

91   *Ibid.*, p. 429.

92   Charleton, p. 111.

93   The tension that is often felt between these two famous -isms arises from the difficulty of finding any adequate empirical support for essentialist theses.

ultimately to the atoms. *Pondus*, by contrast, is attributed to the atoms only on grounds of explanatory power: a body lacking a motive 'urge' is perfectly conceivable. Investigating analytically the qualities we find conceptually necessary to corporeality we discover (at least) size and shape: we then argue that atoms with size and shape but no *pondus* could not have given rise to our universe, and thus justify this attribution on empirical grounds. Having arrived at this theoretical framework, we turn to experience for illustration, elaboration, and ultimate confirmation. (To suppose that Nature *must* correspond to our concepts is an unwarranted conceit.)

Although atoms must have some size and shape, we are obliged to turn to experience if we are ever to discover the determinate sizes and shapes of particular atoms. They are, Charleton affirms, fantastically minute: the smallest body discernible in a microscope consists of 'myriads of myriads of thousands of true atoms, which are yet corporeal and possess a determinate extension' – he estimates that there are at least 7×10^{17} atoms in a single grain of incense.[94] As for atomic shapes, we have only analogy to guide us. Salt-crystals, for example, are cubic, and can be seen under the microscope to consist of similar parts: a large cube is no more than 'a meer congeries or assembly of small ones'.[95] It may be, then, that salt consists of cubes made up of cubes right down the ladder of Nature until we arrive at atoms (or molecules) of salt – witness the ingenious use of analogy to yield some limited empirical access to the micro-realm.

Although we must attribute to atoms *some* of the properties of perceptible things (notably, size, shape and motion) we may not also attribute to them colour, smell, temperature, etc.: that the atoms are 'absolutely devoid of all quality'[96] turns out to be essential to an intelligible explanation of qualitative change. If one does allow the atoms to possess colours, etc., one ends up with the philosophy of the ultimate anti-reductionist, Anaxagoras, whose theory of 'seeds' involves a perfect and absolute analogy of Nature rather than merely a partial one. On such a theory, honey is sweet because it contains sweet 'seeds', red bodies contain red 'seeds', etc. – the qualities of bodies are easy (too easy) to explain!

94 Charleton, pp. 113–114.

95 *Ibid.*, p. 119.

96 *Ibid.*, p. 129.

One must, however, remember, says Charleton, that 'all things become so much the more decoloured, by how much the smaller the parts into which they are divided'.[97] As one unravels a piece of cloth, its colour seems to fade – moving right down to the bottom of Nature's ladder, we arrive at colourless atoms. Shape and size, by contrast, cannot vary in intensity, cannot be 'intended' or 'remitted' by degrees: for the atoms to lack either property would therefore require a radical discontinuity in Nature, a violation of the continuous gradation of forms that makes up the Great Chain of Being. Unfortunately, however, this argument founders on the determinate/determinable distinction: colouredness, the determinable property of being coloured, is not remitted any more than shapedness is. The determinate shapes, sizes, and hues of the parts may be visibly altered when the coloured cloth is unravelled; but any visible body has the determinables size, shape and colour. (A pale pink is as much a colour as a deep red.) One cannot, therefore, derive direct ocular evidence of the colourlessness of atoms, even by such a process of extrapolation: to confirm the Epicurean view, one must revert to more indirect means.

In Cartesian Natural Philosophy, of course, there is no problem of transduction:[98] we have an *a priori* guarantee that all material things, regardless of scale, have size, shape and motion or rest, and fall under the same common principles of geometry and kinematics. The micro- and macro-worlds are perfectly analogous: scale is a mere irrelevancy. Since there is no metaphysical possibility of a disanology of Nature,[99] there is no *problem* concerning transduction. Although there are, for Descartes, no indivisible atoms, one can assert *a priori* that any particle of matter, however small, possesses the definitory features of corporeal substances (size, shape, and motion or rest) and falls, in virtue of those properties, under the universal laws of mechanics.

This *apriorism* proves quite unacceptable to an empiricist such as Newton: against Descartes, he insists that

> We do not know the substances of things. We have no idea of them. We gather only their properties from the phenomena, and from the properties what the substances

---

[97]   *Ibid.*, p. 130. This argument too is Lucretian in origin (see chapter 3, p. 140).

[98]   See McGuire, 3, p. 18.

[99]   Laing, p. 403.

may be. That bodies do not penetrate each other we gather from the phenomena alone.[100]

That impenetrability is *essential* to bodies is, he says, something that we cannot know: no reflex act of the mind gives us direct access to the essence of material substance. 'We know nothing of bodies but by sense'.[101] A disanology of Nature is therefore a real possibility: God could have created a universe with a sharp discontinuity between the micro- and macro-world.[102] But, if we cannot say that atoms *must* have properties XYZ because those are essential to or constitutive of matter as such, how can we safely attribute *any* properties to them? Newton's answer turns on his celebrated 'Rule 3', which methodological or heuristic principle also provides the grounds for his argument for universal gravitation. It runs as follows:

> The qualities of bodies, which admit neither intensification nor remission of degrees, and which are found to belong to all bodies within the reach of our experiments, are to be esteemed the universal qualities of all bodies whatsoever.[103]

This gives us two criteria: one rationalistic (liability to intension and remission), the other more empirical. If one were granted an essentialist premise,[104] one could make something of the first criterion. One could argue, for example, that, *if* the essential properties constitutive of corporeal nature were remitted as one passes down the *scala naturae*, then a body could be made up out of, and dissoluble into, non-bodies, which is inconceivable. Without a premise of such a kind, Newton has no argument: the intensity-invariance criterion founders on the determinable/determinate distinction, while the universality criterion is simply not up to the job. All *visible* bodies (even under the microscope) must have some determinate colour, just as they must be of

---

100 From a draft version of the General Scholium, in Hall and Hall, Eds., 4, pp. 360–361.

101 Quoted from McMullin, 4, p. 23.

102 Cf. Newton, 2, pp. 403–404, where Newton affirms the contingency of much of his Natural Philosophy – the inertial homogeneity of matter, the inverse-square law, etc. God, he insists, *could* have made things quite otherwise.

103 Newton 1, p. 398.

104 Occasionally, as McMullin (4, p. 100) sees, Newton seems prepared to countenance such essentialism: whatever lacks resistance (and hence inertia) is, he insists, most definitely *not* a body.

some determinate size and shape. Rule 3 does not therefore, as it stands, warrant any ontological distinction between shape and colour: if 'all bodies whatsoever', and hence *a fortiori* the atoms, are shaped, they must also, by Rule 3, be coloured. Although any particular colour may fade out as one passes down the ladder of Nature to its 'Foot Bottom, or Lowest Round'[105] (i.e. the atoms), one could not possibly have ocular evidence of the fading out of colouredness.[106]

In fact, as McMullin sees,[107] Newton's argument does not really turn on Rule 3 at all. He endorses, for example, Lucretius' argument for the hardness of the atoms: if we posit hard atoms, the poet has claimed, we can explain the softness of some observed bodies in terms of the admixture of atoms and void: if we posit soft atoms, there is no way of accounting for the hardness of flint, diamond, etc. Newton replies in the same way to the charge that not all observed bodies are hard: to account for the phenomena, he argues, we must assume the hardness of the atoms.[108] The whole thing is an 'inference to best explanation', misleadingly presented as a variety of inductive generalisation.

How is the whole argument *supposed* to work? Since, says Newton, 'the qualities of bodies are only known to us by experiments', we must 'hold for universal all such as universally agree with experiments; and such as are not liable to diminution can never be quite taken away'.[109] We discover *by sense alone* that bodies possess extension, hardness, impenetrability, mobility and inertia, realise that these properties are not susceptible to intension and remission, and thus attribute them to atoms:

> The extension, hardness, impenetrability, mobility and inertia of the whole, result from the extension, hardness, impenetrability, mobility, and inertia of the parts; and hence we conclude the least particles of all bodies to be also all extended, and hard and impenetrable, and movable, and endowed with their proper inertia. And this is the foundation of all philosophy.[110]

[105]   McGuire, 3, p. 41.

[106]   See McMullin, 4, pp. 11–12.

[107]   *Ibid.*, p. 22.

[108]   Cf. of course the argument of the *Opticks* – also Lucretian in origin – that the constancy of Nature depends on the hardness of the atoms (Newton, 2, p. 400).

[109]   Newton, 1, p. 398.

[110]   *Ibid.*, p. 399.

Presented as a sort of inductive generalisation, this argument is a non-starter. As McMullin says, it is just not true that all sensible bodies are hard and impenetrable: to present the argument for hard atoms as quasi-inductive is quite far-fetched. As for the intensity-invariance criterion, that fails on the determinable/determinate distinction. The whole argument works much better if we construe it as being concerned primarily with *possibilities* of *explanation*. A property X of compound bodies must also be attributed to the atoms if we cannot conceive how X should arrive out of non-X elements; another quality Y need not be attributed to the atoms if can imagine – or, better still, produce – a plausible reduction of Y to something simpler. Construed thus, we can give Newton an intelligent and intelligible argument for his belief in extended, hard, impenetrable, massive and mobile atoms: it is not easy to conceive how any one of those properties could have arisen from elements devoid of it. The irreducibility of its primitive concepts is stressed by Boyle as one of the great strengths of the corpuscular philosophy:

> Neither can there be any physical principles more simple than matter and motion; neither of them being resoluble into any things, whereof it may truly, or so much as tolerably be said to be compounded.[111]

The truth of the matter seems to be that Newton is really one type of empiricist masquerading as another. His actual epistemological stance is an empiricism sufficiently liberal to accommodate a large body of theory (e.g. classical Atomism) proposed as an explanatory hypothesis subject to indirect justification in terms of its explanatory value. In *Principia*, however, he is pretending to be an extreme empiricist of the naïve inductive type, unwilling to accept any theoretical statement without direct empirical support. His lifelong acceptance of the Atomic Theory of matter and the corpuscular theory of light, and his speculative flirtations with aether-theories of gravity, make it clear to the unprejudiced reader that Newton is no naïve inductivist; in this part of *Principia*, however, he seems intent on claiming that Rule 3 is merely a variety of inductive heuristic of wider-than-usual generality. (One can think of a number of tactical or strategic reasons, mainly associated with the argument for universal

---

111   Boyle, *Works*, Vol. 4, p. 70. For a similar point concerning possibilities of reductive explanation, see P. Alexander, 2, pp. 58–59.

gravitation, why he should have wished to appear in this guise.[112])

Newton's belief in an 'analogy of Nature' rests, then, not so much on Rule 3 (which is in any case incompetent to make the distinction he requires) but on theoretical grounds, notably on his *prior* acceptance of the Atomic Theory.[113] If one accepts classical Atomism, one attributes size, shape, mobility, hardness, impenetrability and inertia to one's atoms: other properties of compound bodies are in principle reducible to these. The alternatives are, on the one hand, to posit a total disanalogy between atoms and sensible things, which makes the micro-realm totally dark and obscure to us; or, on the other, to accept a complete and perfect analogy of Nature, which would bring us back to Anaxagoras. To steer a course between these extremes is not easy, but it must clearly involve the attribution to the atoms of some proper subset of the properties of sensible things. *If* from these primitives one can then account for all natural phenomena, this will indicate that one has not chosen arbitrarily. (Just imagine, by way of contrast, trying to explain the structure of material Reality with, say, red, sweet and hot as one's primitive concepts!)

For what is perhaps the best of the seventeenth century treatments of this problem, we must turn to Robert Boyle's essay 'About the Excellency and Grounds of the Mechanical Hypothesis'. Some philosophers, Boyle begins, accept mechanical explanations of observable, macroscopic phenomena, but

> will not admit, that these principles can be applied to the hidden transactions, that pass among the minute particles of bodies; and therefore think it necessary to refer these to what they call nature, substantial forms, real qualities, and the like unmechanical principles and agents.[114]

Boyle's reply is that

> both the mechanical affections of matter are to be found, and the laws of motion take place, not only in the great masses, and the middle sized lumps, but in the smallest fragments of matter; and a lesser portion of it being as well a body as a greater, must, as necessarily as it, have a determinate bulk and figure: and he, that looks upon sand

---

112   Cf. of course the overt profession of an extreme variety of empiricism in Cotes' *Preface*.

113   Cf. McMullin, 4, p. 22.

114   Boyle, *Works*, Vol. 4, p. 71.

in a good microscope, will easily perceive, that each minute grain of it has as well its own size and shape as a rock or mountain. And when we let fall a great stone or a pebble from the top of a high building, we find not, but that the latter as well as the former moves conformably to the laws of acceleration in heavy bodies descending. And the rules of motion are observed, not only in cannon bullets but in small shot . . .[115]

Although Nature usually works with finer materials than the mechanical arts, the principles of mechanics are scale-independent:

And therefore to say, that though in natural bodies, whose bulk is manifest and their structure visible, the mechanical principles may be usefully admitted, that are not to be extended to such portions of matter, whose parts and texture are invisible; may perhaps look to some, as if a man should allow, that the laws of mechanism may take place in a town clock, but cannot in a pocket watch . . .[116]

This is a nice blend of rationalism and empiricism. We cannot conceive, he agrees with the Cartesian, how the division of a body could ever remove such properties as size and shape: our concept of division does not admit of such a possibility. (The disappearance of colour, scent, etc., on division is, by contrast, perfectly conceivable.) *If* the structure of the material universe is to be intelligible to us, then, properties such as size, shape, and mobility, and the mechanical principles consequent thereon, must be applicable to the atoms. *That* Nature *is* intelligible to us is not itself metaphysically necessary (God could have made things otherwise: we must assent to this thesis, even if we cannot understand it), but functions as a heuristic principle guiding scientific enquiry. We have an *a priori* demand, but no *a priori* guarantee.

Justification of the whole approach is indirect and empirical. Whereas Descartes treats the basic principles of his Natural Philosophy (Book Two of the *Principles*) as demonstrable *a priori*, and seeks empirical confirmation only of the hypothetical physics of Books Three and Four, Boyle construes the *whole* philosophical system as one grand *hypothesis* (witness the title of the monograph) to be validated as a unit by the fruits it brings.

115   *Ibid.*, p. 71.
116   *Ibid.*, p. 71.

Now the thesis that 'Nature is always simple and conformable to herself', as Newton puts it, i.e. that there exists an 'Analogy of Nature' between micro- and macro-realms, is one which has for us a very high *a priori* probability (unless it is true, we are at a loss to understand the micro-world). Fortunately, we also have empirical grounds for this supposition, evidence which strongly confirms our *a priori* demand that the principles of mechanics be applicable to sub-microscopic things. Boyle cites four examples in support of the thesis of a trans-level identity of Nature:

1.   The microscope reveals that minute bodies such as grains of sand possess their own determinate shapes and sizes. (Whatever we can *see*, even under the microscope, *must* have size and shape.) These properties do not 'emerge', in a totally mysterious and unintelligible way, out of sizeless and shapeless elements, but are ubiquitous features of bodies at all levels of organisation accessible to our investigation, from the microscopic to the astronomical.

2.   A boulder and pebble obey alike Galileo's law of free fall – we can vary the scale by several orders of magnitude without producing any appreciable difference.[117] The laws of motion can thus be seen to be scale-independent, at least over a considerable range.

3.   Cannon-balls and small shot likewise obey the same laws, and follow alike the parabolical path described by Galileo in the *Discorsi*.

4.   Town clocks and pocket watches observe the same principles: cogs, wheels, levers, balances, etc., function in the same way on different scales.

It is difficult not to discern an implicit 'and how could it be otherwise?' behind this purportedly empirical argument, a belief that the scale-independence of Nature's mechanical principles is evident *a priori*, or at least so plausible that no rational man would doubt it. Our reading of Boyle accounts *both* for this Cartesian undercurrent *and* for the use of empirical examples to confirm (or merely to illustrate?) a thesis proposed *a priori*. Unless there exists an 'Analogy of Nature' justifying the application of our concepts of size, shape, and motion to sub-microscopic things, the principles of our Natural Philosophy will lack any sound foundation in Reality. This gives that supposition a high *a priori* probability for us, and validates the heuristic principle that we should

---

[117] Cf. of course Boyle's experiments dropping bodies of different weights *in vacuo. Only in vacuo*, of course, will one arrive at the ideal results.

seek explanations of a certain (mechanical) kind. To illustrate the actual existence of an analogy between macrocosm and microcosm, and thus justify the whole explanatory programme, one must turn to experience.

Actually, Boyle's reliance on a hypothetico-deductive method, combined with his belief that a whole body of theory can be assessed as a unit on grounds of explanatory power and success, implies that he (unlike Newton*, the author of Rule 3[118]), has no essential need for the 'Analogy of Nature'. While the naïve empiricist will demand that we only attribute to unobservable entities properties which are a subset of those discoverable in observable things, Boyle's epistemology can dispense with this claim and admit particles with qualities never directly evident to sense.[119] So long as the whole theory works, not every theoretical term need be provided with an observable analogue or illustration. Thus Boyle could, in principle, while Newton* could not, accept the existence of quarks, particles with such recherché (and decidedly non-observable) properties as 'strangeness' and 'charm'. Modern physics has almost entirely abandoned the hypothesis of an 'Analogy of Nature': the sub-atomic particles of quantum mechanics are of a nature quite *disanalogous* to any perceptible things, and obey their own weird and idiosyncratic laws. By relaxing the constraints provided by the 'Analogy of Nature' assumption (i.e. the demand that the properties of unobservable entities be a proper subset of those of observables) modern physics gives us *theories* of great power, but no *picturable* models: the terms and concepts used cannot be understood independently of their rôle in the theory as a whole.[120]

E. The Lonely Corpuscle[121]

We are now in a position to approach the celebrated distinction between primary and secondary qualities, as it appears in the writings of Boyle and Locke. Those properties constitutive of our idea of body (individually necessary and

---

[118]   Newton's actual epistemology, I have suggested, is closer to the position which I here attribute to Boyle. Newton* is a straw man, Newton himself with his naïve empiricist mask on.

[119]   See Mandelbaum, pp. 113–115, and much of the material in P. Alexander, 4.

[120]   Alexander characterises such explanations as 'reductive-holistic' rather than 'purely reductive'.

[121]   I owe this phrase to Alexander, on whose work (notably Alexander, 4) this section is heavily dependent.

jointly sufficient for corporeal existence) must, they will argue, be attributed to the invisibly small corpuscles of which observable bodies are composed. (We cannot conceive how a body could arrive out of non-bodily components.) But these essential qualities of corporeality must be *intrinsic* to each and every body – that one thing is a body does not depend on its relations to other things. It follows that, if all else were annihilated, a single, 'lonely', corpuscle would still *retain* those properties: it could not cease to be a body merely by the elimination of other things.

This argument is plainly set out in Boyle's *Origin of Forms and Qualities*. After establishing those 'two grand and most catholick principles of bodies, matter and motion', he proceeds to enquire into the nature of the minute corpuscles into which the one catholic matter is differentiated by the various local motions of its parts. Each of these 'primitive fragments' must, he says, 'have two attributes, its own magnitude, or rather size, and its own figure or shape'.[122] For,

> being a finite body, its dimensions must be terminated and measurable: and though it may change its figure, yet for the same reason it must necessarily have some figure or other. So that now we have found out, and must admit three essential properties of each intire or undivided, though insensible part of matter; namely magnitude . . ., shape, and either motion or rest.[123]

Whatever might befall such a corpuscle (short of annihilation), it could never, in principle, lose any of these 'moods' (cf. Descartes' 'modes') of matter. It follows that

> if we should conceive that all the rest of the universe were annihilated, except any of these intire and undivided corpuscles, . . . it is hard to say what could be attributed to it, besides matter, motion (or rest), bulk and shape. Whence by the way you may take notice that bulk, though usually taken in a comparative sense, is in our sense an absolute thing, since a body would have it, though there were no other in the world.[124]

The intelligibility of these attributions is discussed in detail by Alexander:[125] only shape or geometrical figure proves

---

122   Boyle, *Works*, Vol. 3, p. 16.

123   *Ibid.*, p. 16.

124   *Ibid.*, p. 22.

125   See Alexander, 4.

unproblematic. To suggest that a lonely corpuscle has an absolute size and motion (or rest) might seem to involve the postulation of absolute space as a frame of reference, which 'wicked' assumption Alexander seeks to avoid. All we can say of the lonely corpuscle, he argues, is that it has the *second-order* properties of being the *sort of entity* to which size and motion could intelligibly be attributed *if* other bodies existed to provide a frame of reference. The lonely corpuscle, has, therefore, the first-order property of being spherical, or cubic, or whatever, but it cannot be assigned any determinate size or state of motion in the absence of a suitable frame of reference. To say that size is 'an absolute thing' is, on this view, to make a *realist* point (however we measure, there *is* something there to be measured: the atom is the sort of thing that is in principle mensurable) rather than to posit absolute space as a reference-frame.

According to Alexander again,[126] the corpuscular philosophy of Boyle is quite central to Locke's *Essay*; the famous (and much criticised) distinction between 'primary' and 'secondary' qualities is taken over ready-made by Locke from Boyle. Can we substantiate this claim by examining the passages in Book Two of Locke's *magnum opus* in which the distinction is first made? We are given, in the *Essay* (II, 8, ix), three purported criteria for a quality's being primary: primary qualities are:

1. Such as are utterly inseparable from the body, in what state soever it be; such as in all the alterations and changes it suffers, all the force that can be used upon it, it constantly keeps.[127]

This may look like an empirical point, but can also be read as a conceptual one. Locke *could* be saying this: if we take a particular body and heat it, hammer it, cool it, etc., it will always retain some determinate size, shape, and state of motion or rest – these are therefore inalienable qualities of bodies. If the argument is meant to go through at this empirical level, it founders on the determinable/determinate distinction. There is, however, another possible reading. If we take 'alterations and changes' to include *division into corpuscles* and 'what state soever' to cover 'loneliness', we can read this passage as making a conceptual point *à la*

126   Alexander 2, p. 55. See also Mandelbaum, p. 3 ff.
127   Locke, *Essay*, II, 8, ix, Vol. 1, p. 104.

Alexander. And the reading is far from implausible: what state *soever* must surely cover a lonely existence, while *all* alterations and changes surely takes in division.

2. And such as sense constantly finds in every particle of matter which has bulk enough to be perceived . . .[128]

This is the only one of Locke's three criteria which is clearly of an empirical nature. If read as any sort of argument, it will come to grief (like the naïve empiricist version of Newton's Rule 3) on the determinate/determinable distinction: it is better conceived as an illustration of the empirical familiarity of the primitive concepts of the corpuscular philosophy. The sorts of qualities attributed by the corpuscularian to his hypothetical particles are not, Locke may be reminding us, occult or recherché; nothing, indeed, could be more familiar from everyday sense-experience than size, shape, and motion or rest.

3. and the mind finds inseparable from every particle of matter, though less than to make itself singly be perceived by our senses.[129]

This rationalistic, almost Cartesian, criterion bears the real weight of the distinction. We can conceive a body without colour, smell etc., but not one devoid of size, shape, or mobility. Our intellectual faculty, working on the heuristic (but not metaphysically necessary) principle of the intelligibility of Nature, thus finds itself obliged to attribute those properties to *every* corpuscle of matter. The 'lonely corpuscle' test is, arguably, implicit here: it is inconceivable that the annihilation of other bodies should destroy or corrupt the corporeal nature of any given corpuscle. Alternatively, one might argue as follows: the number of corpuscles in the universe is a contingent matter; therefore, there might only be one; but this must still retain the properties conceptually necessary to corporeal status.

Locke's third criterion has important implications with regard to the concept of division, as he is not slow to see:

V.G., take a grain of wheat, divide it into two parts, each part has still *solidity*, *extension*, *figure*, and *mobility*; divide it again, and it retains still the same qualities; and so

---

[128]   *Ibid.*
[129]   *Ibid.*

divide it on, till the parts become insensible: they must retain still each of them all those qualities. For division . . . can never take away either solidity, extension, figure, or mobility from any body.[130]

We are asked first of all to envisage the division of a minute but visible body such as a grain of wheat. The bisection of such a body yields two smaller ones, each – perceptibly – extended, figured, solid, and with some determinate state of motion or rest. From observing many cuts we derive our everyday *concept* of division. But, according to that concept, a body can only be divided into bodies, i.e. into parts with all those properties which 'the mind finds inseparable from every particle of matter'. No difference of principle arises when the parts produced by a division-process become invisibly small: the parts must still possess size, shape, solidity, and mobility – no alternative is conceivable.

These qualities which the mind attributes to all bodies, however minute, are, says Locke, the

> original or *primary qualities* of body, which I think we may observe to produce simple *ideas* in us, viz. solidity, extension, figure, motion or rest, and number.[131]

'Number' is certainly a curious addition to this list: the lonely corpuscle does not have number (except in the trivial sense that it is *one* corpuscle), but then Locke's lists of primary qualities do vary considerably from one part of the *Essay* to another, to the confusion and dismay of his commentators.[132] There might also be some doubts about the inclusion of solidity as a quality of bodies: it is plausible to claim that this feature differentiates corporeal from non-corporeal substance but does not differentiate one body from another, since all partake of it equally. If we take seriously the claim that Locke's primary / secondary distinction is derived from Boyle's 'lonely corpuscle' test, we shall obtain Extension (size), Figure (shape), and Mobility (motion or rest) as our basic list of *primary qualities*.

---

130 *Ibid.*

131 *Ibid.*

132 On occasion, for example (e.g. II, 8, x), he lists *texture* as a primary quality, which, on the interpretation favoured here, must be seen as an aberration. Sometimes, it seems, Locke uses the term 'primary qualities' in its strict sense; at other times, all explanatory features of bodies get lumped together under that label.

F. Secondary Qualities

A single corpuscle, whether lonely or not, is composed of solid matter* (its solidity is constitutive of its nature as a material thing) and may be differentiated from other corpuscles (actual or possible) by its particular size, shape, and state of motion or rest. If other corpuscles exist, if will have spatial relations with them in virtue of which it may be said to have its own peculiar location or place. It cannot, however, be said to have any texture, arrangement, or contrivance of parts: only when many corpuscles aggregate to form a 'convention' do such properties become meaningfully applicable. In compound bodies, says Boyle, such features as 'order' and 'posture' emerge, and must be added to our list of possible explanatory factors. In posture, he explains, N differs from Z; in order or arrangement, NA differs from AN:[133]

> And when many corpuscles do so convene together as to compose any distinct body, as a stone or a metal, then from their other accidents (or modes) and from these two last mentioned, there doth emerge a certain disposition or contrivance of the parts in the whole, which we may call the texture of it.[134]

Given differences of size, shape and motion at the level of the individual corpuscles, and of order, arrangement, and texture as those elements associate to form compounds, a staggering wealth of variety is possible in the ultimate microstructures of visibly large (i.e. multiply compounded) bodies. Taking into account all these factors – and all possible combinations of them – one can produce an effectively limitless number of distinct, and often highly specific, corpuscular arrangements or textures. In virtue of this, says Boyle,

> I am apt to look upon those, who think the mechanical principles may serve indeed to give an account of the phaenomena of this or that particular part of natural philosophy, as staticks, hydrostaticks, the theory of the planetary motions, etc., but can never be applied to all the phaenomena of things corporeal; I am apt, I say, to look upon those otherwise learned men, as I would do upon him, that should affirm, that by putting together the letters of the alphabet, one may indeed make up all the words to be found in one book, as in *Euclid*, or *Virgil*; or in one

---

[133] The example, of course, goes right back to Democritus. See chapter 3, p. 104.

[134] Boyle, *Works*, Vol. 3, p. 22.

language, as Latin, or English, but that they can by no means suffice to supply words to all the books of a great library, much less to all the languages in the world.[135]

Our old friend the alphabet-analogy is here used to rebut a possible objection to the corpuscular philosophy, namely, that it cannot account for the infinite variety of nature. As Boyle has no difficulty in showing, the objection simply underestimates the scope, flexibility, and 'comprehensiveness' of corpuscular patterns of explanation.[136]

Each and every compound body has therefore a highly specific texture or microstructure, in virtue of which it is the type of substance it is. Textures may be described as 'secondary' qualities, since (a) they are 'emergent' properties, arising only at some level of organisation of matter (a corpuscle has no texture), and (b) they are *reducible* to the arrangements and motions of simple parts – there is nothing to a texture over and above a particular 'contrivance' of material parts. It is on such textures – secondary qualities of bodies – that, say Boyle and Locke, our sensations of heat, cold, colour, smell, etc., are causally dependent. Secondary Qualities, says Locke, are

> such qualities which in truth are nothing in the objects themselves but powers to produce various sensations in us by their primary qualities, i.e. by the bulk, figure, texture, and motion of their insensible parts, as colours, sounds, states, etc. These I call secondary qualities.[137]

At first glance, it may seem that 'these' refers to 'colours, sounds, tastes, etc.', i.e. that those *are* secondary qualities. On closer reading, however, it is clear that 'these' refers to the 'powers to produce various sensations in us', while the 'colours, sounds, tastes, etc.' are the *sensations produced*. (Strictly, Locke should not have included 'texture' among this list of 'primary qualities' – it positively invites misunderstanding.)

If secondary qualities *are* textures, they are both *objective* and *intrinsic* to bodies. The arrangement of its constituent corpuscles is, Boyle insists, a perfectly objective feature of any body, quite independent of any percipient.[138] What is

---

135 *Ibid.*, vol. 4, pp. 70-71.

136 *Ibid.*, Vol. 3, pp. 297-299.

137 Locke, *Essay*, II, 8, x, Vol. 1, pp. 104-105.

138 Boyle, *Works*, Vol. 3, p. 24.

subjective – dependent on the observer – is the *idea* of a secondary quality, the sensation which it causes in us. If the bodies which we call white produce that particular sensation in us in virtue of a common superficial texture T, they all possess T *objectively*: there is nothing subjective about an arrangement of corpuscles. Ideas, Locke insists again and again, are in minds; qualities (primary, secondary, and tertiary) are in bodies.[139] It is important, he emphasises, to grasp clearly *'the difference between the qualities in bodies, and the ideas produced by them in the mind . . .'*[140] What clearer profession of the objectivity of secondary (and tertiary) qualities could one demand that the following:

> The *qualities*, then, that are in *bodies*, rightly considered, are of *three sorts* . . .[141]

Not only are textures (= secondary qualities) *objective*, they are *intrinsic* to bodies: although they provide the ground for relations and powers, they are not themselves relational or dispositional in nature. In virtue of its structure (the arrangement of its parts) a key K may fit a particular lock L, but K's 'power' to open L is *nothing over and above*[142] its particular structure. (A 'power' not reducible to geometrical structure – i.e. to the pegboard model – would be occult.) The solubility of gold in *aqua regia* is, says Boyle, 'not in the gold anything distinct from its particular texture',[143] and is thus entirely explicable on the key-lock model: that acid possesses just the contrivance of parts necessary to 'open' the texture of that particular metal. When Locke identifies secondary qualities with 'powers to produce various sensations in us', he has this corpuscularian concept of powers (as illustrated by the pegboard, or the lock-key model) in mind. The vocabulary of powers is merely an easy way of talking about (largely unknown) textures.

That the manifest properties of observed bodies are often environment-dependent – a fact emphasised by Boyle in his monograph, 'Of the Systematical or Cosmical Qualities of Things',[144] and reaffirmed in a lengthy passage in Locke's

139   See *Essay*, II, 8, vii; Alexander, 3, p. 204.

140   Locke, *Essay*, II, 8, xxii, Vol. 1, p. 109.

141   *Ibid.*, II, 8, xxiii, Vol. 1, p. 109.

142   Boyle, *Works*, Vol. 3, p. 18.

143   *Ibid.*, p. 18. See also Alexander, 3, p. 210.

144   Boyle, *Works*, Vol. 3, p. 306.

Essay[145] – does not cast doubt on the claim that textures are *intrinsic* to bodies. Most of the observed properties of bodies, says Boyle, are 'relations, upon whose account one body is fitted to act upon others or disposed to be acted on by them and receive impressions from them': of such a kind are, for example, the power of *aqua regia* to dissolve gold and of mercury to amalgamate with it. Many of these powers, Boyle suggests, may be due to 'divers unheeded agents' present, though often unnoticed, in our world-system. In the extracosmic void, perhaps, *aqua regia* would not dissolve gold: in such a state of isolation, bodies 'would retain many of the qualities they are now endowed with, yet they would not have them all'.[146] In our cosmos, then, bodies possess some qualities which,

> because they depend upon some unheeded relations and impressions which these bodies owe to the determinate fabrick of the grand system . . . they are parts of, I have . . . thought fit to name their cosmical or their systematical qualities.[147]

These properties, Boyle suggests, may be due to subterranean or celestial effluvia: take away such slight, yet perhaps vital, influences, he says, and who knows what might happen (cf. the dependence of animal life on an aerial 'quintessence' of some kind). Just as a beast cannot survive without fresh air, nor an engine run without oil, so many of the manifest properties of things may be dependent on unsensed and perhaps unimagined subtle fluids, effluvia, etc.

An identical point occurs in Locke. Most, perhaps all, the qualities we observe in things depend upon relations (often quite unheeded) with other things, and in particular on the subtle fluids which surround and pervade gross bodies. After listing a number of instances – actual and possible – of such dependence, Locke advances the conjecture that the system of Nature is such a tightly interconnected whole that, if a single star were removed from its place, all else would be radically altered. That, he admits, is a mere speculation, but

> This is certain: things, however absolute and entire they seem in themselves, are but retainers of other parts of nature for that which they are most taken notice of by us.

---

145 Locke, *Essay*, IV, 6, xi, Vol. 2, pp. 187–189.

146 Boyle, *Works*, Vol. 3, p. 306.

147 *Ibid.*, p. 306.

Their observable qualities, actions, and powers are owing to something without them; and there is not so complete and perfect a part that we know of nature which does not owe the being it has, and the excellencies of it, to its neighbours; and we must not confine our thoughts within the surface of any body, but look a great deal further, to comprehend perfectly those qualities that are in it.[148]

That the qualities we observe in any given body may, in many cases at least, be dependent on its environment is undeniable: although Locke's presentation of the point may be somewhat exaggerated, he nevertheless captures a significant truth. Unfortunately, however, the passage has misled more than one commentator. Anderson, for example, claims that this passage continues Locke's argument against individual essences (from Book 3, Chapter 6) and involves a denial that the real essence (R.E.) of a thing is *intrinsic* to it.[148] This is surely an error. What the passage does entail is that acquaintance with the intrinsic R.E. of *one* body is not sufficient to allow us to explain its manifest properties – to do that, one requires familiarity with the R.E.s of other neighbouring bodies, effluvia, etc. What we have, then, is an R.E. or microstructure intrinsic to, e.g., gold, in virtue of which it has different observable properties in different environments, thus:

R.E. (gold) in Env. 1 ——→ Qualities $Q_{a \ldots m}$
R.E. (gold) in Env. 2 ——→ Qualities $Q_{f \ldots t}$
R.E. (gold) in Env. 3 ——→ Qualities $Q_{u \ldots z}$, etc.

Given knowledge of the intrinsic R.E. of gold we can know, in advance of experiment, which qualities it will manifest under which conditions – but only so long as we are familiar with the R.E.s of all other relevant environmental factors.

Anderson is wrong, then, to read this as a denial that individual things possess intrinsic R.E.s: it is not their very beings, but their observable powers and properties, that things owe to their neighbours.[150] Anderson's reading of Locke, involving as it does the claim that the R.E.s of all substances are environment-dependent, engenders a holistic epistemology and has as a consequence that only God (who sees all things simultaneously) could be acquainted with the

---

[148]   Locke, *Essay*, IV, 6, ix, Vol. 2, p. 189.

[149]   Anderson, p. 212 ff.

[150]   Since we name things only by their nominal essences, however, an alteration quite extrinsic to a given body (involving no change in its R.E.) may lead *us* to call it by a quite different name.

R.E. of any single body.[151] This violates the *analytical* orientation of the entire epistemological tradition of the Atomist school, of which Locke was very much a member. The basic theme is one of analysis, resolution of the complex into simples, until eventually one arrives at atoms, the properties of which are *entirely independent* of their neighbours. From atoms, one assembles more complex corpuscles, each with a texture dependent *only* on the sizes, shapes, motions, and arrangements of its parts, i.e. on factors *internal* to it. Of course, once one has arrived at particles with highly specific textures, many of the properties and powers of these compounds will be explicable only in terms of *relations* (e.g. 'congruity' and 'incongruity') between those textures. And of course it is legitimate to point out that what may appear to be a diadic relation is in fact a triadic one, dependent on some unguessed third factor such as a subtle fluid or effluvium. It may be that gold only dissolves in *aqua regia* here on Earth because of the presence of some terrestrial effluvium: on Mars, perhaps, things are quite different. If we invoke the spirit of Ockham and his razor, however, we will not postulate any such factor unless we need to – we shall not opt for a wholesale holism far removed from the traditions of Atomism. We can evoke such extraneous factors as and when needed to explain particular properties of things: this admission in no way threatens the supposition that a substance such as gold has a R.E. (a microstructure) intrinsic to it.

Each and every compound body possesses, in summary, a microstructure or (dynamic) 'texture' which is (a) *reducible*, in the sense that it is exhaustively explicable in terms of the sizes, shapes, motions, and spatial arrangement of its constituent simples, (b) *objective*, i.e. independent of any percipient, and (c) *intrinsic* to it, independent of environmental conditions. The next – and all-important – question concerns the relation between this objective 'secondary quality' and the ideas it occasions in our minds.

G. The Mechanical Theory of Perception

The human sense-organs, says Descartes, are purely passive in perception, and receive impressions as a piece of wax does:

> And it should not be thought that all we mean to assert is an analogy between the two. We ought to believe that the

---

151   Anderson, p. 213.

way is entirely the same in which the exterior figure of the sentient body is really modified by the object, as that in which the shape of the surface of the wax is altered by the seal.[152]

Already we have the germ of a mechanical theory of perception. External objects exist and impress 'images' of themselves on the 'wax' of the imagination. We must infer, from the variety of our sense-impressions, a corresponding diversity in their causes. 'although possibly these are not really at all similar to them'.[153] Given that our ideas of colours, etc., are confused ideas, we may not legitimately infer the objective existence of any qualities resembling those sensations – we are not entitled to assume any resemblance between cause and effect in their regard.

We may properly infer, as features of the objective external world, whatever properties of objects are causally relevant to our perceptions of them. But odours are mediated by corporeal effluvia, sounds by the agitation of the air, and light by the transmission of a pressure-pulse along a column of 'boules' of the second element.[154] (The difference between 'wave' and 'corpuscle' theories of light is not germane to this particular argument.) Now the only causally relevant properties invoked in these explanations are geometrical and kinematic (as ever in Cartesianism, dynamic concepts have to be smuggled in to make the account work): these are therefore, Descartes will conclude, the only properties we may legitimately impute to external objects.

In mechanical explanations, the concept of *figure* is all-important: figure, says Descartes, (a) falls under two senses, tough and sight[155] (as well, of course, as being an innate notion), and (b) is 'so common and simple, that it is involved in every object of sense'.[156] A mere variety of figures (= Boyle's 'textures') may, he continues, suffice to account for *colours*: why, for example, may we not

> conceive the diversity existing between white, blue, and red, etc., as being like the differences between the following similar figures?

---

152   Descartes, 1, Vol. 1, p. 36.

153   *Ibid.*, p. 192.

154   *Ibid.*, pp. 292–293.

155   Cf. of course the 'common sensibles' of Aristotle's *De Anima*.

156   Descartes, 1, Vol. 1, p. 37.

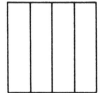 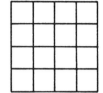 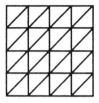

The same argument applies to all cases; for it is certain that the infinitude of figures suffices to express all the differences in sensible things.[157]

The argument will run something like this:

1.   Colours can be accounted for by shapes: there is a sufficiently rich diversity of shapes to account for all the colours of the spectrum.
2.   Shape has privileged epistemological (and hence ontological) status: since we possess a clear and distinct idea of shape, we know that bodies do possess figures. No such guarantee holds for colours.
3.   Therefore intelligibility (explanation in terms of clearly grasped simple ideas) and economy (elimination of the superfluous) alike favour such a reductive account of colour.

Descartes' final account of light and colour is as follows. Light is a pulse of pressure, transmitted instantaneously along a column of 'boules' of the second element. In virtue of its texture, an opaque body may, when it reflects a light-ray, also impart a rotary motion to the 'boules' of which it consists. The different varieties and speeds of spin produce, when the pressure-wave impacts on the retina, different motions of the animal spirits in the optic nerve and, eventually, those various *sensations* which we call colours. Of crucial importance is that nothing red or blue is invoked at any stage in the causal story of how those sensations were produced.

As for the supposition that the cause of a given sensation should *resemble* it (an assumption fundamental to the scholastic theory of sensible species), this does not hold for colours, scents, etc. The notion that only a green body could produce a sensation of greenness is dismissed by Descartes as a mere prejudice, and a palpably false one at that.[158] The

[157]   *Ibid.*, p. 37.
[158]   *Ibid.*, pp. 192–193.

body that causes pain in us by its local motion is not itself painful:

> If a sword moved towards our body cuts it, from this alone pain results which is certainly not less different from the local motion of the sword or the part of the body that is cut, than are colour or sound or smell or taste.[159]

Hobbes is equally emphatic. Consider, he says, the origins of sound:

> The clapper hath not sound in it, but motion, and maketh motion in the internal parts of the bell; so the bell hath motion, and not sound. That imparteth motion to the air; and the air hath motion but not sound. The air imparteth motion to the ear and nerves to the brain; and the brain hath motion but not sound. From the brain it reboundeth back into the nerves outward, and thence it becometh an apparition without, which we call sound.[160]

Light, heat, sound and colour 'are not objects, but phantasms in the sentient':[161] to suppose that our 'phantasms' *resemble* external realities in this respect is to fall into the unintelligible and redundant superstition of real qualities. Mechanical theories of the operation of our five senses involve essentially the attribution to external objects of size, shape and motion (the qualities necessary to corporeal existence) but not of colours, sounds, etc. Granted that our sensations are caused by the bombardment of our sense-organs by minute *bodies* of some kind, we must claim that they are extended, figured, and mobile: to suppose that they are also red, sweet, or loud is superfluous.

The mechanical theory of the operation of our senses has then as a corollary the thesis of the *subjectivity* of colours, smells, odours, etc. This has become widely known as the doctrine of the subjectivity of secondary qualities, a somewhat unfortunate title. In Locke's *Essay*, as we have seen, secondary qualities are to be identified with *textures*, and are perfectly objective properties of compound bodies. For men such as Boyle and Locke, a visible body has a specific texture in virtue of which it produces some 'modification' in the incident light; this 'modified' or 'disturbed' light then produces in us the sensation of a particular colour. Terms

---

159   *Ibid.*, pp. 294-295.
160   Quoted from Brandt, pp. 124-125. For the details of Hobbes' mature theory of sensation, see his *Works*, Vol. 1, p. 389 ff.
161   Hobbes, *Works*, Vol. 1, pp. 391-392. See also Laird, pp. 129-130.

such as 'red', 'blue', and 'green' refer to ideas in the minds of sentient beings, not to properties (primary or secondary) of external objects. Thus, Boyle:

> . . . there is in the body, to which these sensible qualities are attributed, nothing of real and physical but the size, shape, and motion or rest, of its component particles, together with that texture of the whole, which results from their being so contrived as they are; nor is it necessary that they should have in them anything more like to the ideas they occasion in us.[162]

Just as a pin is not itself painful – does not in any way resemble the effect it produces when it punctures our skin – so the cause of a colour-experience need not resemble the experienced colour. (This example is one of the commonplaces of the corpuscular philosophy.)

A similar point occurs in Locke, who insists on more than one occasion on distinguishing ideas (in minds) from their causes (in objects):

> it being one thing to perceive and know the *idea* of white or black, and quite another to examine what kinds of particles they must be and how ranged in the superficies to make any object appear white or black.[163]

The ideas of white and black are familiar to all of us; but the acutest philosophers are still largely in the dark about their proper causes. As Locke says,

> To discover the nature of our *ideas* the better, and to discourse of them intelligibly, it will be convenient to distinguish them as they are *ideas* or perceptions in our minds, and as they are modifications of matter in the bodies that cause such perceptions in us: that we *may not* (as perhaps usually is done) think that they are exactly the images and *resemblances* of something inherent in the subject . . .[164]

The vulgar – and this includes the schoolmen – assume that our sensations of colour, odour, etc., could only be produced by causes with 'real' and objective qualities resembling those of our sensations. Locke replies, with Descartes and Boyle before him, that no such likeness between cause and effect need be postulated: a mere 'modification of matter' (i.e. a texture) in the

---

162 Boyle, *Works*, Vol. 3, p. 23.

163 Locke, *Essay*, II, 8, ii, Vol. 1, p. 102.

164 *Ibid.*, II, 8, vii, Vol. 1, p. 102.

object may cause in me the sensation of yellowness without the least similarity between the secondary quality and the idea it occasions. Black and white, yellow and red, are ideas of secondary qualities (and, as such, like all ideas, mind-dependent), not secondary qualities themselves.[165]

How, Locke asks, do bodies produce sensations in us? It must, he replies, occur 'by impulse, the only way which we can conceive bodies operate in'.[166] To set the animal spirits in our nerves in motion, some impulse or pressure is required in our organs of sense. How then do we perceive distant bodies? Since, Locke replies,

> . . . the extension, figure, number and motion of bodies of an observable bigness may be perceived at a distance by the sight, it is evident some singly imperceptible bodies must come from them to the eyes, and thereby convey to the brain some motion which produces these ideas which we have of them in us.[167]

Qua corporeal things, these 'singly imperceptible bodies' must possess size, shape, and mobility; it is quite superfluous that they should also possess colours,[168] etc.: the corpuscularian account of colour, for example, works as well (or better) without that supposition. In fact, says Locke,

> After the same manner that the ideas of these original [i.e. primary] qualities are produced in us, we may conceive that the ideas of secondary qualities are also produced, viz. by the operation of insensible particles on our senses.[169]

It is evident, he continues, that there exist 'good store of bodies that are so small that we cannot by any of our senses discover either their bulk, figure, or motion'.[170] The corpuscularian's account of the operation of our five senses

---

165   See Alexander, 3, p. 203.

166   Locke, *Essay*, II, 8, xi, Vol. 1, p. 105.

167   *Ibid.*, II, 8, xii, Vol. 1, p. 105.

168   Contrast the Epicurean ideas that the *eidolon* retains the *colour* of the object that emits it. Walter Charleton seems to accept this theory in his *Physiologia* (p. 156): the 'image', he says, 'represents the object from which it was deradiated, in all particulars of the life, i.e. with the same colour, figure, and situation of parts'. This seems to support the claim of Brandt (p. 73 ff.) that colour, in Epicurean Atomism, is not something subjective, but an 'emergent' property of compound bodies. (For Lucretius, see Chapter 3, p. 140.).

169   Locke, *Essay*, II, 8, xiii, Vol. 1, p. 105.

170   *Ibid.*

involves the bombardment of our sense-organs by such minute bodies. But now,

> let us suppose . . . that the different motions and figures, bulk and number, of such particles, affecting the several organs of our senses, produce in us those different sensations which we have from the colours and smells of bodies; viz., that of violet, by the impulse of such insensible parts of matter, of peculiar figures and bulks, and in different degrees and modifications of their motions, causes the *ideas* of the blue colour and sweet scent of that flower to be produced in our minds. It being no more impossible to conceive that God should annex such ideas to such motions, with which they have no similitude, than that he should annex the idea of *pain* to the motion of a piece of steel dividing our flesh, with which that idea hath no resemblance.[171]

We must attribute size, shape, and motion to the insensible corpuscles which cause our perceptions; to attribute to them also colour and scent is to subscribe to the superstition that cause must invariably resemble effect, and thus – ultimately – to fall into the absurdities of real qualities. From the corpuscular theory of perception, says Locke, it follows that

> the ideas of primary qualities of bodies are resemblances of them, and their patterns do really exist in the bodies themselves; but the ideas produced in us by these secondary qualities have no resemblance of them at all. There is nothing like our ideas existing in the bodies themselves. They are, in the bodies we denominate from them, only a power to produce those sensations in us; and what is sweet, blue, or warm in *idea* is but the certain bulk, figure, and motion of the insensible parts in the bodies themselves, which we call so.[172]

Ideas of primary qualities resemble them; ideas of secondary qualities do not. This is presented *not* as an experienced difference in our sensations (far less as the *criterion* by which we are to distinguish primary qualities from secondary), but as a *consequence* of the corpuscular account of perception.[173] The distinction between primary and secondary qualities is, as Alexander says, one of the understanding, not of the senses: 'it concerns conceivable explanations rather than qualitative differences in our sensations'.[174] To explain our

---

[171] *Ibid.*, pp. 105–106.

[172] *Ibid.*, II, 8, xv, Vol. 1, p. 106.

[173] See Alexander, 2, p. 58; Mandelbaum, p. 19.

[174] Alexander, 2, p. 58.

perceptions (ideas) of size and shape, we must posit objects that really possess those features; to account for our ideas of colour, smell, etc., it is idle to postulate corresponding 'real' qualities in objects. The argument is part *a priori* and part *a posteriori*: *a priori*, one can show the conceivability and desirability (intelligibility, economy) of the reductionist programme; *a posteriori*, one can give empirical support for the corpuscularian philosophy as a whole. (Locke refers us, for such evidence, to a number of Boyle's experiments.) The 'resemblance' thesis thus emerges as a corollary of the corpuscular philosophy, not as a phenomenological claim about our sensations. (This renders it immune to the misguided assaults of Berkeley and others.)

It is, we must emphasise, only *qua* determin*ables* that primary qualities need resemble the ideas to which they give rise.[175] Examples of misperceptions of size, shape, and motion are legion: Locke could not conceivably have been unaware of their existence. (Elsewhere in the *Essay*, he refers on a number of occasions to illusions involving primary qualities.) The cause of a triangular image in my visual field need not itself be triangular in shape (although of course it may be so), but it must have some determinate value of the determinable shape.[176] The cause of a yellow idea need not – and, if the corpuscular theory is correct, does not – have any determinate colour at all.

In the previously quoted passage, Locke speaks explicitly of secondary qualities as *causes* of ideas, and of sweetness, blueness, and warmth as the ideas produced, i.e. as subjective. This confirms Alexander's claim that Locke thinks of colours, etc., as ideas of secondary qualities, not as themselves secondary qualities of bodies.[177] The secondary qualities of bodies, he contends, are objective and intrinsic features of them (i.e. textures) in virtue of which they cause in us the *ideas* that are colours, tastes, smells, etc. On this reading, the non-resemblance thesis becomes easy to understand: the ideas as we experience them do not resemble even

---

[175]   See Mandelbaum, pp. 20–24; Curley, pp. 454–458; Alexander, 2, p. 64.

[176]   The one passage in the *Essay* (II, 8, xxi) which *might* be read as a denial of illusions concerning primary qualities is really concerned only, Curley (p. 459) shows, with the sense of *touch*. Even if Locke is right to assert that there are no illusions concerning tactile qualities (which is dubious), the point could bear no theoretical weight – a little rewiring of our nervous system could no doubt produce some.

[177]   Alexander, 3, p. 203.

in type or category their causes in objects. A colour or smell is a different type of thing – in a wholly different category – from a texture.

But surely, one might reply, the same holds for our ideas of primary qualities: Here too a metaphysical divide separates the idea from the quality. Nevertheless, Locke will reply, our ideas of size, shape, and motion are *accurate* representations of their causes (at least *qua* determinables): they represent those primary qualities as they really are (. . . 'a circle or square are the same, whether in idea or existence . . .'[178]). In short, they give us some real insight into the fabric of the physical world. This thesis may be proposed *a priori*, and supported by our Cartesian intuitions, but for empiricists like Boyle and Locke it is the explanatory value of the corpuscular philosophy as a whole (including of course as an essential part its account of sensation) which gives it its ultimate sanction.

We are now in a position to understand a passage which has perhaps provoked more misinterpretations of Locke than any other in the entire *Essay*. It concerns the vexed topic of objectivity, and runs as follows:

> The particular bulk, number, figure, and motion of the parts of fire or snow are really in them, whether anyone's senses perceive them or no; and therefore they may be called real qualities, because they really exist in those bodies. But light, heat, whiteness, or coldness are no more really in them than sickness or pain is in manna. Take away the sensation of them; let not the eyes see light or colours, nor the ears hear sound; let the palate not taste, nor the nose smell; and all colours, tastes, odours and sounds, as they as such particular ideas, vanish and cease, and are reduced to their causes, i.e. bulk, figure, and motion of parts.[179]

If heat, colour, etc., are construed as (secondary) qualities of objects, they are observer-dependent or *subjective* ones. This was the common misinterpretation of Locke, seized upon by Berkeley as a major concession to subjective idealism. If, Berkeley argues, 'secondary qualities' (hot, cold, sweet, etc.) are subjective, and if no phenomenological grounds occur in experience for the distinction between primary and secondary qualities, it is safe to conclude that *all* qualities may be deemed subjective.

---

[178] Locke, *Essay*, II, 8, xviii, Vol. 1, p. 107. See also Alexander, 2, p. 54, p. 65.

[179] Locke, *Essay*, II, 8, xvii, Vol. 1, pp. 106-107.

Our reading of the *Essay* (fundamentally, that of Alexander) evades this tortuous web of distortion and misrepresentation. If, as Alexander has argued, 'hot', 'sweet', 'white', etc., are the names of our *ideas* of secondary qualities, the apparent concession to subjective idealism vanishes. Bulk, figure, motion, etc., may be said to be 'real' qualities simply in the sense that they *are really qualities*, as colour, etc., are not. *All* the qualities of bodies, primary and secondary alike, are objective and observer-independent. Light, heat, etc., 'as they are such particular *ideas*', are subjective: ideas, *qua* states of mind-substances, cannot exist independently of minds. So, for example, the fact that colours do not exist in the dark, or are in some respects relative to percipients, does not entail the subjectivity of secondary qualities.[180] The textures of bodies causally responsible for our perceptions of them are both intrinsic and objective features of those objects.

H. Heat and Cold

In order to illustrate the corpuscular philosophy in action, we must discuss one or two selected examples of reductive explanations. I shall therefore treat the nature of heat and cold in this subsection, reserving colour for its successor. The existence of some association between heat and local motion is the merest commonplace: examples could be multiplied *ad lib* of heat causing motion (e.g. boiling) or motion heat (e.g. friction). But to say that there is nothing to heat over and above a particular variety of local motion is a very radical thesis. In Aristotelian physics, heat and cold are two of the four primitives of a *qualitative* matter-theory: to adopt a kinetic theory of heat is therefore implicitly to abandon that whole world-view. Small wonder then that scholastic philosophers – although noting of course a link between heat and motion – did not espouse a kinetic theory.

One of the first explicit advocates of the kinetic theory of heat was Francis Bacon. (I shall ignore Bacon's claim to have arrived at his account of heat by inductive methods, and treat his conclusion as a hypothesis supported by, but not derived from, the mass of empirical data which he musters.[181]) After

---

180   See Boyle's defence of his account of qualities against the charge of subjectivism in *Works*, Vol. 3, pp. 24–25.

181   The so-called 'inductive' derivation of the 'form' of heat is, says A. R. Hall (3, p. 106), 'nothing more than an instance of how one might compile a justification of a preliminary hypothesis. (As an example of an inductive proof it is worthless.)'

examining his multifarious table of instances, Bacon concludes that

> from the instances taken collectively, as well as singly, the nature whose limit is Heat appears to be Motion.[182]

It is not merely, he insists, that heat causes motion, or *vice versa* but that 'the very essence of Heat, or the Substantial self of Heat, is Motion and nothing else'.[183] The final 'and nothing else' is of crucial importance: it involves the tacit denial of the need for any special fire-particles to generate or impart heat. When one body heats another, says Bacon, no material need be transferred between them, hence, *a fortiori*, no fire-stuff. A cold body can generate frictional heat by rubbing alone.

Nevertheless, heat is not just motion simpliciter; it is motion of a special type, differentiated as follows:

1. It is an expansive motion (whereas cold contracts).
2. It is an upward-tending motion.
3. It is not a uniform expansive motion of the whole heated body, but an agitation of its particles.
4. This agitation is (a) rapid and (b) not of the least parts, but of those of 'some tolerable dimensions'.[184]

The final definition thus runs as follows:

> Heat is an expansive motion restrained, and striving to exert itself in the smaller particles. The Expansion is modified by its tendency to rise though expanding towards the exterior; and the Effort is modified by its not being sluggish, but active and somewhat violent.[185]

Prima facie, then, Bacon has a purely kinetic account of heat. He retains Aristotle's conception of heat and cold as the fundamental 'active powers' – 'nature's two hands', he calls them elsewhere[186] – and produces an account of their natures (as types of motion) which allow them to play that rôle in a mechanistic universe. Before crediting Bacon with a full anticipation of the modern kinetic theory, however, one must make some reservations:

1. On this theory, cold is not merely the privation of heat: it is also a motion, albeit of a different type (contraction rather than expansion).

---

182   F. Bacon, *Works*, Vol. 4, p. 150; *Novum Organum*, 2, xx.
183   *Ibid.*
184   *Ibid.*, p. 154.
185   *Ibid.*, p. 155.
186   F. Bacon, *Works*, Vol. 2, p. 370.

2. The generation of heat in the slaking of lime occurs, says Bacon, through the violent reaction of a pent-up spirit, which reacts like a man enraged by opposition.[187] (Cold, too, is due to its proper spirit.[188]).

3. Bacon accepts as a fundamental part of his whole matter-theory, the existence of a flame-stuff.[189]

4. In the *Sylva Sylvarum*, Bacon seems to countenance a 'species' theory of the transmission of radiant heat: heat and light, he suggests, may be transmitted by the 'communication of natures' rather than by the 'impressions and signatures' of local motion.[190]

On balance, then, it is safer to see Bacon as a somewhat muddled eclectic with a glimmering of the modern kinetic theory than to credit him with a clear-sighted anticipation of later ideas.

For another early seventeenth century discussion of the nature of heat, let us turn to Galileo's *Assayer*. Is it true, Galileo asks, that motion is the cause of heat?[191] It is, he replies, not motion *per se* which produces heat, but the frictional attrition of two hard, rough, surfaces. This produces an agitation of the minute parts of those bodies, with consequent release of minute fire-particles or *igniculi* which are, by virtue of their sharply-pointed shape and especially rapid motion, fit to produce in us that *sensation* which we call heat.

But, it is objected, a hot body suffers no loss of weight: how can this be, if it is emitting copious quantities of *igniculi* in all directions? The loss of weight involved may, Galileo replies, be 'too minute to be perceptible in any balance whatever':[192] since fire must be incredibly subtle and tenuous in order so easily to penetrate and pass through solid bodies, it is no surprise that it is not detectable by weight.

Ignoring some of the details of the controversy between Galileo and his critics, we may characterise his theory of heat as follows. Heat is a sensation (something purely subjective)

---

187   See Gregory, p. 104. Cf. of course the peripatetic theory of *antiperistasis*, ridiculed by Boyle in his *Works*, Vol. 2, p. 659 ff.

188   F. Bacon, *Works*, Vol. 2, p. 371. See also Kargon, 4, pp. 51–52.

189   For the structure of Bacon's 'bi-quaternion' theory of matter, see Rees, 4, p. 113.

190   F. Bacon, *Works*, Vol. 2, p. 429 ff.; Hesse, 1, p. 93.

191   Galilei, p. 277.

192   *Ibid.*, p. 267.

produced in us by the impact on our skins of the sharply-pointed *igniculi* released by the body which we call 'hot'. The mere presence of *igniculi* in a body does not constitute heat: to produce the sensation of warmth in us, the body must be excited in such a way as to make it *emit igniculi* into the ambient medium, and thence to our skins. Quicklime, for example, contains many of these minute fire-particles, but 'locks them up' until they are released on slaking.[193] To produce their characteristic effects, *igniculi* must be freed from their imprisonment in matter and set in motion.

Are the *igniculi* themselves hot? People in general, says Galileo, believe so: the common man thinks that 'heat is a real phenomenon, or property, or quality, which actually resides in the material by which we feel ourselves warmed'.[194] This belief, he remarks, is quite unwarranted. *Qua* material particles, the *igniculi* must have size, shape, and motion or rest, but to attribute to them a 'real' quality of hotness is idle. Heat, he insists, is subjective: there is nothing in the cause which resembles in any way the effect (the sensation) to which it gives rise. Fire-particles are differentiated from those of other stuffs by their size, shape, and motion *alone*: although Galileo accepts a fire-stuff, he does not believe in irreducible qualitative differences[195] – like the ancient Atomists, he accepts the thesis of the homogeneity of matter.

A very similar theory can be found in Gassendi and his disciple Charleton. Heat, says the former, involves the agitation of the parts of a body brought about by the presence and rapid motion of certain minute 'calorific' atoms. These are, Charleton explains,

> certain particles of matter, or atoms, which being essentially endowed with such a determinate magnitude, such a certain figure, and such a particular motion, are comparated to insinuate themselves into concrete bodies, to penetrate them, dissociate their parts and dissolve their contexture.[196]

These 'calorific' atoms are differentiated from those of other substances by their size, shape, and (very rapid) motion, not by any Aristotelian form or quality:[197] although, as Charleton

[193] See Shea, p. 20.

[194] Galilei, 2, p. 274.

[195] *Ibid.*, pp. 277–278.

[196] Charleton, p. 294.

[197] Brett, p. 72.

says, they 'be not hot essentially; yet they do deserve the name *atoms of heat*, or calorific atoms, insomuch as they have a capacity or power to create heat',[198] i.e. to bring about those effects (including of course sensation) that we ascribe to heating. They are, Charleton continues, 'exile in magnitude, spherical in figure, most swift in motion'[199] – these properties account most plausibly for the manifest nature and effects of heat. In virtue of the importance attached by both Gassendi and Charleton to motion as 'the mother of heat' we can legitimately call this a *mixed* 'kinetic' and 'calorific' theory: there are fire-particles but they produce their characteristic effects by means of exciting an agitation of the grosser particles of bodies.[200]

On this theory, a body may be called potentially hot when it merely contains fire-atoms (e.g. unslaked lime), actually hot when it emits them. (All combustible substances must, then, be potentially hot.) If a substance contains such fire-atoms, frictional attrition or agitation of its parts will tend to excite them and thus generate heat, but a substance such as water, which is naturally cold (i.e. contains few fire-atoms) could not be heated by agitation – motion is not *universally* the cause of heat.[201] (Joule would scarcely have been impressed.) The main natural 'seminary' for the calorific atoms is fatty or oily (i.e., for the chemists, sulphureous) matter, the ready inflammability of which testifies to its high concentration of fire-atoms.

Cold, for Gassendi, is *not the privation* of heat, but a positive reality in its own right: the powerful physical effects of cold (freezing, condensation, etc.) indicate the existence of real physical agents.[202] These are 'frigorifick' atoms, bulky, angular (tetrahedral) and sluggish – i.e. well-suited by nature to clog the pores of bodies and damp down atomic motions.[203] ('Calorifick' and 'frigorifick' atoms can easily be accommodated after such a manner *within* a kinetic theory – i.e. as motion-enhancing and motion-inhibiting agents.) Just as unctuous matter is the 'principle' of heat (the main natural source of fire-atoms), so, Gassendi suggests, is *nitre* the

---

198   Charleton, p. 294.

199   *Ibid.*, pp. 294–296.

200   *Ibid.*, pp. 294–296.

201   Charleton, p. 297.

202   *Ibid.*, pp. 306–307.

203   *Ibid.*, p. 308; Kargon, 4, p. 67.

'principle' of cold or *primum frigidum*.[204] In Gassendist physics, then, the qualitative natural philosophy of the peripatetics is given a corpuscular dress. Special atoms are posited to explain different qualities; the 'One Quality ⟶ One Principle' axiom is accepted, etc. – on such a view, 'calorific' atoms emanate from unctuous matter and warm things, 'frigorifick' atoms are emitted by nitre and produce the effects of cooling. Instead of transmitting its qualities to a neighbouring object (cf. the 'species' theory), each body emits instead exhalations of atoms – the effect is much the same.[205]

For a full-blooded and recognisably *kinetic* theory of heat, we must turn to Descartes. Heat, he explains, is nothing more than the agitation of the grosser particles of the third element (E3) by the swirling currents of subtle matter around them; cold is the privation of such agitation (there is no need to posit 'frigorifick atoms'). Here at last we have a true kinetic theory. It is not, Descartes explains to Mersenne, the movement of the subtle matter or aether (E1 and E2) that constitutes heat: that 'quality'

> only consists properly in the agitation of the parts of terrestrial bodies [i.e. E3 corpuscles] because these have more force to move the parts of other bodies and thus burn them.[206]

Agitation of the parts of one body naturally engenders similar motions in the parts of its neighbours: thus is heat transmitted and combustion brought about. If *cold* appears to be conducted from Y to X, this merely indicates that heat (= agitation) is passing from X to Y – cold is a privation, not a positive reality. The tendency of heat to cause expansion and rarefaction is readily explicable on the kinetic theory: as

---

204   The postulation of a *primum frigidum* is subjected to a highly critical examination by Boyle in *Works*, Vol. 2, p. 585 ff. It is, Boyle avers, a very doubtful matter 'whether there be any such thing or no', Three points in particular must be mentioned:

a.   If, as some claim, cold is a privation, there is no need for a *primum frigidum* (p. 585).

b.   The 'One Quality ⟶ One Principle' thesis is 'an unwarrantable conceit' (p. 586).

c.   There exists no plausible candidate for the rôle of *primum frigidum* (p. 586).

205   See Koyré, 5, p. 130, for the primitiveness of this aspect of Gassendi's physics.

206   Quoted from J. F. Scott, p. 220. See also Kemp Smith, p. 105, and of course *Principles*, 4, 29–31.

the particles of E3 vibrate more rapidly and collide more frequently, they will inevitably tend to press outwards and hence, if permitted, to expand into adjacent spaces.[207]

A natural corollary of a true kinetic theory of heat is the non-existence of a special fire-stuff: fire and flame, on such a view, are *states*, not *types*, of matter: the flame produced by burning wood or oil is *materially* nothing other than wood or oil, with its constituent particles in a particularly rapid and violent state of agitation. Such a theory is championed by Hobbes in his *De Corpore*: when, he says,

> a body hath its parts so moved, that it sensibly both heats and shines at the same time, then it is that we say fire is generated.[208]

Flame is not a qualitatively different stuff from wood or oil, not a material distinct from the fuel consumed, but rather 'the matter itself, not simply and always, but then only when it shineth and heateth',[209] i.e. in virtue of its characteristic motions. This is, then, a pure kinetic theory rather than a mixed-kinetic one: it postulates no special fire-stuff to account for heat. Flame is merely 'an aggregate of shining particles';[210] a spark is just a small piece of, e.g., flint or steel with a very rapid rotary motion.

This latter claim was subsequently confirmed by Hooke – a great champion of the kinetic theory – in his *Micrographia*. If, he says, one allows the sparks struck from flint or steel to fall onto a sheet of paper and then examines them under the microscope, one discovers that

> . . . the Spark appearing so bright in the falling, is nothing else but a small piece of the steel or flint, but most commonly of the steel, which by the violence of the stroke is at the same time sever'd and heatt red hot.[211]

These observations, says Hooke, support a purely kinetic theory of heat against all theories of fire-stuffs, calorific atoms, etc.: there is, he insists, nothing to heat over and

---

[207] See M. Boas Hall, 10, p. 310.

[208] Hobbes, *Works*, Vol. 1, p. 451.

[209] *Ibid.*, p. 451.

[210] *Ibid.*, p. 455.

[211] Gunther, Ed., Vol. 13, p. 45. For similar observations, see also Boyle, *Works*, Vol. 5, p. 3.

above 'a brisk and vehement agitation of the parts of a body'.[212]

The importance of the kinetic theory of heat for the corpuscular philosophy was immense, its implications for natural philosophy as a whole momentous. For the peripatetics, heat and cold are two of the four 'primary' qualities or 'powers' of which all material things are constituted, paradigm cases of real qualities if ever there were such.[213] If heat is reducible to local motion, and cold to a mere privation of heat, Aristotelian matter-theory collapses.[214] All the effects of heat and cold, for example, can be set down to *mechanical* causes: if, for example, a change of colour or flavour is produced by heating, that alteration can be set down to local motion. (Boyle makes considerable use of this mode of argument.) When a body is heated, the agitation of its parts may rearrange them and hence alter its *texture*, on which depend its 'forms' and 'qualities'. This, as Boyle sees, refutes the claims of the chemists that fire merely separates bodies into their elements and does not produce anything new. It is in the nature of fire, he says in the *Sceptical Chemist*, 'to set a moving, and thereby to dissociate the parts of bodies, and subdivide them into minute particles',[215] which in turn may re-aggregate to produce a new and quite different specific texture – i.e. a new chemical substance.[216] Fire does not merely separate, it alters things.

Boyle's various discussions of the nature of heat and cold show a certain wavering between the rival 'kinetic' and 'calorific' theories. In his 'New Experiments and Observations touching Cold',[217] he discusses at some length the relative merits of the Cartesian and Gassendist theories of cold, without arriving at any definite conclusion. Although freezing-mix experiments may, he admits, seem favourable 'to their hypothesis, that suppose congelation to be effected by the ingress of frigorifick atoms',[218] the rival kinetic theory is by no means refuted thereby. It is, he eventually concludes, 'a

---

212   Gunther, Ed., Vol. 13, p. 12.

213   HCDM are, I argued in Chapter 3 (pp. 97–99), the only 'real' qualities in Aristotle.

214   Boyle sees this point quite clearly – see his *Works*, Vol. 4, p. 236.

215   *Ibid.*, Vol. 1, p. 488.

216   *Ibid.*, p. 569.

217   *Ibid.*, Vol. 2, p. 462 ff.

218   *Ibid.*, p. 530, p. 748.

work of greater difficulty than every body would imagine', to frame a 'universal and unexceptionable hypothesis of cold': neither Gassendi's theory nor Descartes' copes perfectly with all the available data.[219]

If Boyle is to admit calorific and frigorific atoms, they must be in the rôle of agitation-enhancing and inhibiting agents, i.e. they must find a place *within* a mixed-kinetic theory: Boyle's overall commitment to some variety of kinetic theory is not in doubt. In an aside in the *Origin of Forms and Qualities* he describes heat as simply 'a brisk and confused local motion of the minute parts of a body'[220] – there is no mention of calorific atoms. Similarly, in the *Mechanical Origin or Production of Heat*, we encounter the familiar claim that heat 'seems to consist mainly, if not only, in that mechanical affection of matter we call local motion':[221] more precisely, says Boyle, it is a vehement, random agitation of insensibly small corpuscles. In the dialogue, *Of the Positive or Privative Nature of Cold*,[222] moreover, Boyle skilfully defends the 'privation' theory of cold against the objections of Aristotle and Gassendi: the excitement or inhibition of corpuscular agitation may each have, he suggests, powerful and positive physical effects. A cold body can 'impart' its 'quality' of 'coolness' by damping the agitation of the parts of neighbouring bodies: this may suffice to account for the many and varied physical effects of cooling – including of course sensation.

So far, we have been marshalling evidence for the attribution to Boyle of a pure kinetic theory. If, then, he accepts the general notion of heat as an agitation of minute corpuscles, and of cold as the privation of such agitation, what need has he of calorific and frigorific atoms? Economy surely favours the Cartesian position.[223] To read Boyle as an advocate of a pure kinetic theory would, however, be a grave mistake – it would involve the neglect of his important work on the calcination of metals. The gain in weight that occurs in such a process is, he claims in his *New Experiments to Make*

[219]   *Ibid.*, pp. 478–479. The frigorific atom theory explains, for example, the expansion of water on freezing, while the kinetic theory does not. But most *other* fluids *contract* on freezing . . .

[220]   *Ibid.*, Vol. 3, p. 21.

[221]   *Ibid.*, Vol. 4, p. 244.

[222]   *Ibid.*, Vol. 3, p. 733 ff.

[223]   *Ibid.*, Vol. 2, p. 752.

Fire and Flame Stable and Ponderable,[224] derived from the fire, not from the air.[225] I will show, he claims,

> that flame itself may be, as it were, incorporated with close and solid bodies, so as to increase their bulk and weight.[226]

During calcination, he claims, igneous corpuscles pass through the walls of the vessel used and enter the pores of the metal: since these minute fire-particles 'want not gravity', the effect of this is to increase the weight of the heated metal. Even in hermetically sealed vessels, he claims,[227] heated metals gain in weight. This, *if it were true*, would indeed show that the accession of matter is not from the atmospheric air but from some subtler material. If the gain in weight is not from the air (which cannot penetrate a glass vessel), *nor* from the vessel itself (which loses no weight), whence, Boyle demands,

> can this increase of absolute weight . . . delivered by us in the metals exposed to the mere flame, be deduced, but from some ponderable parts of that flame?[228]

The *Epicurean* theory of fire, Boyle concludes, is seen by these experiments to be superior to the Cartesian.[229] On the kinetic theory of Descartes, all that is transmitted across the glass walls of the vessel is *motion* (agitation): no gross (and hence *weighty*) matter (E3), can be transferred. Yet, says Boyle, the observed gain in weight *must* be attributed to ponderable corpuscles (fire-atoms) passing through the glass and becoming incorporated with the metal – a result easily explicable on the Epicurean theory but not on the Cartesian.

Boyle's final view on the nature of heat is, therefore, a mixed kinetic theory.[230] Heat is a random agitation of the corpuscles of a body, produced (or enhanced) by the action of minute, highly mobile, fire-corpuscles. One can also, he notes, admit into this mixed-kinetic theory the frigorific atoms of Gassendi:

224   *Ibid.*, Vol. 3, p. 706. For seventeenth century ideas on calcination, see also the study of McKie.

225   Such was Boyle's authority in the chemical domain that the actual rôle of the *air* in calcination would not be understood for another century after his death.

226   Boyle, *Works*, Vol. 3, p. 708.

227   *Ibid.*, p. 719 ff.

228   *Ibid.*, pp. 725–726.

229   *Ibid.*, p. 729.

230   See Bentham, pp. 455–457.

For there may be corpuscles of such a nature, as to size, shape, and other attributes, as to be fit to enter the pores, and pierce even into the inward parts of water, . . . so as to repel the calorific corpuscles they chance to meet with, or to clog, or hinder their activity, or . . . considerably to lessen that agitation of the minute parts, by which the fluidity of liquors, and the warmth of other bodies, are maintained.[231]

Even if cold itself is a privation,

yet the cause of it may be a positive agent acting mechanically, by the clogging of the agile calorific particles, or deadening their motion . . .[232]

Let us move now from the physics of heat and cold to their psychology. The sensation of coldness, says Boyle, is caused by the fact that, when a cold body touches our skins, 'we find its particles less vehemently agitated than those of our fingers, or other parts of the organ of touching'. It follows that, 'if the temper of that organ be changed, the object will appear more or less cold to us, though it self continue of one and the same temper'.[233]

Our sensation of cold is, therefore, *relative* to the state of the percipient:

The same lukewarm water will appear hot and cold to the same man's hands, if, when both are plunged into it, one of them shall have been newly held to the fire, and the other benumbed with frost.[234]

The corpuscular explanation sketched above can make sense of this phenomenon, which is a very real difficulty for the advocate of R.Q.s. (Exactly the same example is used to make the identical point in Locke's *Essay*.[235]) If the scholastic accepts both the existence of 'real' qualities of heat and cold, *and* their immediate accessibility to their proper organs of sense, he must conclude that lukewarm water possesses both those contrary and incompatible R.Q.s.

---

[231] Boyle, *Works*, Vol. 2, p. 750.

[232] *Ibid.*, Vol. 4, p. 244.

[233] *Ibid.*, Vol. 2, p. 481.

[234] *Ibid.*, Vol. 4, p. 244.

[235] See Locke, *Essay*, II, 8, xxi, Vol. 1, p. 108. Since Locke's argument adds little or nothing to that of Boyle, there seems little point in repeating it.

We are frequently, Boyle points out, deceived into attributing what are really changes in ourselves to the external world. If the agitation of the particles of my skin falls, I will feel as warm something which previously seemed cool, and may be led to claim, quite erroneously, that a change has taken place in the object. It requires, he warns, 'something of attention and reasoning, if not of philosophy', to become fully aware of the subject-dependence of sensory qualities.[236] (It is possible, he adds, that there exist *specific* heating and cooling agents which, in virtue of their peculiar textures, act only upon particular patients: opium, for example, reduces the body-temperature of humans but has no comparable effect on water.[237])

The distinction between hot and cold is, Boyle suggests, analogous to that between treble and bass: it turns on a difference in *frequency* of vibrations. Cold may, therefore, be a comparative privation of local motion – a sluggish vibration, one of low frequency – and yet still produce a distinct sensation:

> Since it is manifest that bodies in motion are wont to communicate their motion to those more slow bodies they happen to act upon and to lose of their own motion by this communicating of it . . . if, for example, a man take a piece of ice in his hand, the agitation of the particles of the sensory will, in good part, be communicated to the corpuscles of the ice, which, upon that account, will quickly begin to thaw; and the contiguous parts of the hand losing of the motion they thus part with to the ice, there needs nothing else to lessen the agitation they had before. And there needs no more than this slackening or decrement of agitation, to occasion in the mind a new and differing perception, as men have tacitly agreed to refer to coldness.[238]

The kinetic theory here accounts neatly and plausibly *both* for the melting of the ice and for the production in the hand of the sensation of cold. A term such as 'cold', Boyle explains,[239], is *triply* ambiguous between:

---

236   Boyle, *Works*, Vol. 2, pp. 481–482.

237   *Ibid.*, p. 493. For an attempted explanation of this fact, see Vol. 4, pp. 239–240.

238   *Ibid.*, p. 738.

239   *Ibid.*, Vol. 4, p. 244.

1.   A supposedly 'real', tactile quality of things, which soon turns out, paradoxically, to be observer-dependent: 'as it is a tactile quality, in the popular acceptation of it, it is relative to our organs of feeling'.
2.   An objective and in itself insensible state of a body – the frequency of agitation of its parts.
3.   A perception or sensation in the mind.

Conceptual clarity demands caution in using sense 1, and readiness to abandon it in favour of sense 2 in all scientific and explanatory contexts. We have only got by using sense 1 in virtue of certain stabilities in ourselves and our environment: snow will always cool, and steam scald, a living human body. The Aristotelians, says Boyle, have traded on these commonplaces to confuse senses 1 and 3, producing something which is – incredibly – *both* objective *and* observer-dependent: the lukewarm water example poses a genuine dilemma for them. By dividing the universe into the purely objective (sense 2) and the purely subjective (sense 3), and abandoning sense 1 altogether, Boyle effectively undercuts their problem.

I. Colour

By the mid seventeenth century, there existed among the various schools of 'corpuscularian' philosophers a considerable degree of consensus on this issue. Whether one was a disciple of Bacon or Descartes, Gassendi or Hobbes, one could lend one's assent to the following formula. Colour, as it exists in the object (i.e. the causal ground for our sensation) is an arrangement of corpuscles, a disposition of the superficial parts, in virtue of which the object effects a certain 'modification' in the incident light; which, after reflection, produces on impacting on our retinas the sensation that we call red, or blue, or green, etc. Such a formula, carefully phrased so as to be neutral on the very controversial topic of the nature of light, could command very widespread assent among the 'moderns'. For the attempt to put this body of theory onto a sound empirical basis, we must turn to Boyle's *Experiments and Considerations touching Colours*.[240] Although intended primarily as a 'historical' work (an exercise in Baconian data-gathering) Boyle could not forego the opportunity of further argument in favour of his corpuscular philosophy: 'I have added,' he says, 'divers new speculative considerations', which, he hopes, may be of aid to

[240]   *Ibid.*, vol. 2, p. 662 ff.

the theorist in founding 'a solid and comprehensive hypothesis' concerning the nature of colour.[241]

Diversity of colour, Boyle begins, need not indicate any great difference of nature or constitution. (Contrast the view of Aristotelians and chemists.) The different feathers of a multicoloured bird, or parts of a variegated leaf, are very alike in all respects bar colour, which cannot, therefore, be treated as a reliable guide to the natures of things.[242] Colour-changes may, nevertheless, be of considerable utility of indicating the progress of chemical reactions. (Boyle was one of the pioneers of the use of indicators.) In the ripening of a fruit, the heating of a metal, and many of the reactions of the chemical laboratory, colour-changes are indicative of significant alterations in the natures of the substances concerned.

In discussing the nature of colours, Boyle warns, some all-important distinctions must be drawn, lest one lapse into ambiguity and equivocation:

> For colour may be considered, either as it is a quality residing in the body that is said to be coloured, or to modify the light after such or such a manner; or else as the light itself, which so modified, strikes upon the organ of sight, and so causes that sensation which we call colour: and that this latter may be looked upon as the more proper, though not the usual acceptation of the word colour, will be made probable by divers passages in the ensuing part of our discourse.[243]

We have here not two but *three* distinct senses of 'colour', namely:

a.   The superficial texture of a body.
b.   A 'modification' of the incident light.
c.   A sensation produced by the impact on the retina of 'modified' light.

The object may be said to be coloured only in the first of these three senses, i.e. only in so far as it, in virtue of 'a certain disposition of the superficial particles', produces a certain 'modification' in the light reflected from it. We may continue, says Boyle, to speak of objects as being red, green, or blue, etc., so long as the above distinctions are borne in mind.

---

241   *Ibid.*, pp. 662–663.

242   *Ibid.*, p. 669. Cf. of course Bacon's point to the same effect in the *Novum Organum* (F. Bacon, *Works*, Vol. 4, p. 156; *Novum Organum*, 2, xxii).

243   Boyle, *Works*, Vol. 1, p. 671.

Most of his experiments, Boyle continues, concern colour in sense (a), i.e. as it exists in the object. The surfaces of bodies, he notes, although they may appear smooth to the naked eye, are, 'in a strict and rigid sense', not so – the microscope makes evident to us the real roughness of apparently smooth surfaces. Now the sizes, figures, and arrangement of its superficial corpuscles give each visible body that superficial texture which constitutes its colour (in sense [a]). Since colour (sense [a]) is nothing more than an arrangement of parts, a *rearrangement* of those parts should yield a colour-change. This thesis can be supported by a number of *analogies* drawn from the realm of observably large objects: look closely at a cornfield in the wind or the hairs on a dog's back and you will *see* the dependence of colour on texture (this time of macroscopic parts). Varying the angle of illumination may also produce differences of observed colour – an obvious illustration of this would be taffeta.

We are now in a position, Boyle feels, to tackle the question 'Do colours exist in the dark?' The Epicureans, he notes, answer 'no'; most other philosophers favour an affirmative response. The question, Boyle feels, is largely a semantic one about the sense in which we should take the word 'colour':

> . . . if it be taken in the stricter sense, the Epicureans seem to be in the right; for if colour be indeed but light modified, how can we conceive that it can subsist in the dark . . .: but, on the other side, if colour be considered as a certain constant disposition of the superficial parts of the object to trouble the light they reflect after such and such a determinate manner, this constant . . . modifying disposition persevering in the object, whether it be shined upon or no, there seems no just reason to deny, but that in this sense, bodies retain their colour as well in the night as day.[244]

Perhaps what one ought to say, he suggests, is that 'bodies are potentially coloured in the dark, and actually in the light:[245] this locution may help us to evade certain difficulties and paradoxes.

After a lengthy polemic against the scholastics' distinction between 'true' and 'apparent' qualities, and a few critical

---

[244] *Ibid.*, p. 690. Boyle ignores the claim of Lucretius that the superficial texture of a body – the arrangement of its surface-layer of atoms – may be actually produced by a 'stroke' of the light.

[245] Cf. Charleton, p. 186.

remarks about the notion of colour as a real quality, Boyle comes at length to present his own positive views. (On examination, they seem to bear a heavy – and totally unacknowledged – debt to Hobbes' *De Corpore*.[246]) He will attempt, he begins, to account only for whiteness and blackness in terms of 'intelligible and mechanical principles'; the spectral colours represent a far more difficult and conjectural matter. Considered as a quality in the object (sense [a]), whiteness consists in this,

> that the superficies of the body, that is called white, is asperated by almost innumerable small surfaces; which being of an almost specular nature, are also so placed, that some looking this way, and some that way, they yet reflect the rays of light, that fall upon them, not towards one another, but outwards towards the spectator's eye.[247]

A white body has a surface consisting of innumerable little facets, each like a looking-glass, so arranged as to reflect the incident light outwards. White bodies are, therefore, great reflectors of light, as anyone afflicted by snow-blindness will testify.

If this is what whiteness is in the object, Boyle continues, one might expect it to be mechanically producible by rearrangement of the superficial parts of a body: this proves indeed to be the case. Egg white can be whisked to produce meringue; water agitated into froth; glass ground into powder.[248] (This last example is of course particularly apt: ground glass does consist of a multitude of minute mirrors.) The advocate of R.Q.s is in real difficulties here: he must admit *either* (a) the annihilation and/or creation of a R.Q. in every such change, (b) the possibility that a given R.Q. need not always be manifest in its proper nature (i.e. that a body can possess the R.Q. of redness without looking red), or (c) that many of the appearances of colours are not due to R.Q.s at all. None of these options has any real attraction for the scholastic.

If whiteness involves the reflection of incident light, blackness must involve its absorption:

---

246 See Hobbes, *Works*, Vol. 1, p. 456 ff. Boyle's account of whiteness and blackness seems particularly dependent on Hobbes.

247 Boyle, *Works*, Vol. 1, p. 697. The notion that the surface of a white body is like a mass of little mirrors is found in almost identical form in Hobbes.

248 These examples are not original to Boyle: many can be found in Bacon – e.g. *Novum Organum*, II, xxii (*Works*, Vol. 4, p. 156).

That, which makes a body black, is principally a peculiar kind of texture, chiefly of its superficial particles, whereby it does as it were dead the light that falls on it, so that very little is reflected outwards to the eye.[249]

One can explain this property in either of two ways: either

1. The asperities on the surface of the body are so shaped and situated as to reflect the incident light into the depths of the body rather than outwards:[250] Or
2. The superficial particles of the black body are soft and inelastic: they deaden the impact of the light-corpuscle and effectively prevent rebound.

Either theory, says Boyle, will explain why black bodies absorb heat far more readily than white ones – an instance of the great explanatory power of the corpuscular theory as a whole and of its account of colour in particular.[251]

Once again, Boyle serves to cast light on Locke: a number of passages in the *Essay* are illuminated by what has gone before in this section. Consider, for example, Locke's remarks on the colours apparent in porphyry:

Hinder light but from striking on it, and its colours vanish: it no longer produces any such *ideas* in us; upon the return of light it produces these appearances in us again. Can anyone think that any real alterations are made in the *porphyry* by the presence or absence of light; and that those *ideas* of whiteness and redness are really in the porphyry in the light, when it is plain *it has no colour in the dark?* It has, indeed, such a configuration of particles, both night and day, as are apt, by the rays of light rebounding from some parts of that hard stone, to produce in us the *idea* of redness, and from others the *idea* of whiteness; but whiteness or redness are not in it at any time, but such a texture that hath the power to produce such a sensation in us.[252]

This is crystal clear: red and white are *ideas* produced in us when light is reflected from the superficial parts of the porphyry. They are therefore not present in the stone, day or

---

249 Boyle, *Works*, Vol. 1, p. 704.

250 This is Hobbes' explanation: the surface of a black body possesses, he claims, 'little eminent particles erected straight up from the superfices': light striking them is absorbed into the body rather than reflected outwards (Hobbes, *Works*, Vol. 1, p. 464).

251 In this too Boyle was anticipated by Hobbes.

252 Locke, *Essay*, II, 8, xix, Vol. 1, p. 108.

night: it is evident, says Locke, that it has no colour in the dark; reflection convinces us that, since light produces no real alterations in it, it must *also* be without colour in daylight. Whiteness and redness are, therefore, 'not in it at any time' except in so far as it may be said to be white or red in Boyle's sense (a), i.e. in virtue of its particular texture.

The secondary qualities in the object which are causally responsible for its appearance to us *as* coloured are, Locke suggests – following Boyle – producible by mechanical causes. The following is typical of Boyle's mode of argumentation:

> Pound an almond, and the clear white *colour* will be altered into a dirty one, and the sweet *taste* into an oily one. What real alteration can the beating of the pestle make in any body, but an alteration of the *texture* of it?[253]

Pounding and grinding can only, Locke and Boyle agree, produce a mechanical rearrangement of the grosser parts of the almond, yet on this change of texture depend differences in colour and taste. The qualities of bodies on which our perceptions of them as white, sweet, etc., depend are therefore at least supervenient upon, perhaps reducible to, arrangements of corpuscles. Conceptual clarity and economy alike favour the stronger reducibility claim, but that is not without its problems. What we seem to have in Boyle and Locke is a two-phase explanation, in which secondary qualities (i.e. textures) are reducible to primary ones, while colours, tastes, etc., are supervenient upon, but not reducible to, secondary qualities, since the mind-body divide intervenes. We can say that a particular texture T will always produce in the mind an idea I, but we could never in principle explain the nature of I – that is, its experiential or phenomenological nature – in terms of T.

Before leaving this section, we must at least mention Newton's demonstration of the composition of white light, which, he shows, is 'a confused aggregate of Rays indued with all sorts of colours',[254] colours separated, but not created, by a prism. It may seem strange to some readers that this momentous discovery is only mentioned as a sort of afterthought, but the truth is that it does not seriously affect the issues discussed in this section: it supplements and fills out the corpuscularian account of colour rather than requiring

---

253 *Ibid.*, II, 8, xx, Vol. 1, p. 108.
254 Newton, 2, p. 124.

essential modification of it. Instead of saying that light is 'modified' by the reflecting body, we shall simply say that each coloured body absorbs some of the spectral colours and reflects others, in virtue of which it appears to sense as its particular colour. (White bodies reflect the whole of the spectrum; black bodies absorb it all.) For Newton, different hues correspond with different *sizes* of light-corpuscle – one size of light-particle gives a ray of red (or, more strictly,[255] 'rubrifick' or 'red-making') light, another size of light-corpuscle gives rays of blue light; for a wave-theorist, colour will be associated, of course, with *frequency* of vibration.[256]

J. Real Essences and Human Knowledge

Implicit in the Atomic Theory from its Democritean beginnings was the twofold claim, (a) that material substances possess *real essences* (R.E.s) identifiable with their microstructures or (dynamic) 'textures', and (b) that these are inaccessible to human investigation. Given perfectly rigid atoms and the principles of elementary geometry (or some indivisibilist analogue[257]), it will follow that there are *necessary connections* in Nature (cf. the pegboard analogy, or the lock-and-key model). Chemical affinity, solubility, etc., will in principle be explicable in terms of congruity or incongruity of (dynamic) texture. Our senses, however, reveal only the superficial appearances of things, not their hidden R.E.s: sensible qualities such as colours, scents, etc., are not a reliable guide to the mysteries of Nature: hence Democritus' oft-quoted remarks to the effect that 'truth hides in the depths' and that he 'would rather find a single causal explanation than gain the kingdom of Persia'.[258]

The tendency to scepticism associated with the Atomic Theory throughout its lengthy history[259] is clearly manifest in the Gassendist school. Gassendi himself was much influenced by the *crise pyrrhonienne* of early seventeenth century thought, a crisis of confidence in the power of human reason

---

255   *Ibid.*, pp. 124–125. Newton makes it quite plain that he does *not* think that the corpuscles making up a ray of 'red' light are themselves red.

256   Newton himself suggests this fundamental principle of the wave-theory, believed by some of its advocates (e.g. Grimaldi, Malebranche), but by no means all (Huygens was sceptical).

257   For this possibility, see Mau.

258   See Chapter 3, pp. 109–110.

259   An example from the Middle Ages illustrative of this link would be Nicholas of Autrecourt. See Chapter 7, p. 338 ff.

to attain truth brought about partly by the revival of interest in Sextus Empiricus and his demand for a criterion of truth[260] (of vital importance, of course, in the perennial disputes between Protestants and Catholics), and partly by the collapse of confidence in the classical authorities (Aristotle, Ptolemy, Galen), which led to the suspicion that everything is controversial, nothing certain.

In Gassendi's early *Exercises in the Form of Paradoxes in Refutation of the Aristotelians*, this sceptical strain is especially prominent: he seeks in particular to establish the maxim *nihil sciri* ('nothing is known') against the pretensions of the dogmatists.[261] The sceptics, he insists, were right: 'the inner causes of natural events totally elude human investigation'.[262] But without familiarity with R.E.s, the Aristotelian conception of Science as demonstrative knowledge,[263] derived by deduction from axioms ('real' definitions) expressive of the R.E.s of things, is quite inapplicable and useless. Proof, and knowledge, as defined by the peripatetic, simply do not exist.[264] We are acquainted only with *ta phainomena*, the 'appearances' of things: on this scanty and insubstantial basis no towering edifice of demonstrative knowledge can be erected. The wisest men (Socrates, Solomon) have confessed their ignorance of the natures of things: 'no proposition that makes assertions about the nature of a thing according to itself can be affirmed with confidence'.[265]

Gassendi's eventual epistemological position is a 'mitigated' scepticism: we have knowledge of the appearances of things, and may make more or less well-found conjectures as to their causes. The 'inner nature and underground spring' of things, he writes, God has willed to conceal from us: 'when we aspire or presume to know it, we are guilty of immoderation'.[266] The Atomic Theory of Epicurus – accepted as a plausible, albeit

---

260 Logic, says Gassendi in his 'Animadversions', gives no criterion of truth: such a thing 'the wisest of all mortals has never yet been able to attain' (Gassendi, p. 39). For the best study of the scepticism of the period, see Popkin.

261 Gassendi, p. 24.

262 *Ibid.*, p. 19.

263 Cf. of course the account of scientific knowledge outlined in Aristotle's *Posterior Analytics*. Euclidean geometry would serve for many centuries as the paradigm of a Science.

264 Gassendi, p. 68 ff., p. 85 ff.

265 *Ibid.*, pp. 101–102.

266 *Ibid.*, p. 200.

hypothetical, account of natural phenomena – entails that things possess R.E.s, but makes it probable that they will forever elude our enquiries – the true microscopic causes of macroscopic events are unknown to us. We may guess at them, but our hypothetical explanations of heat, cold, colour, etc., have the status of conjectures, not of knowledge. Although sensory evidence may give some degree of confirmation to such hypotheses (e.g. that of the existence of pores in the skin[267]), and the microscope may extend somewhat the range of *ta phainomena*,[268] the whole fabric of our natural philosophy must remain, by and large, conjectural and uncertain. As Charleton says, Nature is 'all depths' – hence the perpetual controversies and uncertainty of our philosophy.[269] We, he says in a memorable passage,

> are *men*, i.e. *moles*; whose weak and narrow opticks are accommodated only to the inspection of the *exterior* and low parts of *Nature*, not perspicuous enough to penetrate and transfix her *interior* and *abstruse* excellencies: nor can we speculate her glorious beauties in the direct and incident line of *essences* and *formal causes*, but in the refracted and reflected one of *effects*; nor that, without so much of obscurity, as leaves a manifest incertitude in our apprehensions, and restrains our ambition of intimate and *apodictical Science*, to the humble and darksome region of mere superficial conjecture.[270]

Had we been born such Lyncei, he continues, as to be able to discern Nature's R.E.s (= microstructures), then we could give the 'express and proper reason' for natural phenomena (e.g. why air-pressure will support only about 29 inches of mercury).[271] Nature has R.E.s (like Gassendi, Charleton is a convinced realist about essences), but they are inaccessible to poor 'moles' like us – to pretend to an adequate understanding of such mysteries is 'one of the highest degrees of madness' for a man.

To understand the position of Descartes on these issues, we must bear in mind the crucial distinction between the *a priori* and demonstrative *Natural Philosophy* of Book 2 of his

---

[267]   *Ibid.*, pp. 333–334.

[268]   *Ibid.*, pp. 334–335. See also M. Boas, 4, p. 433; Brett, p. 80.

[269]   Charleton, p. 5.

[270]   *Ibid.*, pp. 50–51.

[271]   *Ibid.*, p. 51.

Principles and the hypothetico-deductive (H-D) physics of Books 3 and 4. (Exactly the same distinction occurs in Hobbes' *De Corpore*.[272]) Starting from our clear and distinct ideas of matter and motion, and relying throughout on God's veracity for their objective application, he seeks to demonstrate, *a priori*, that corporeal Nature is purely mechanical, that the physical universe is a vast piece of clockwork. The 'matter = 3D extension' thesis, the denial of A.D., the conservation of motion, the rules of impact: all the most general principles of Natural Philosophy are, he claims, deducible *a priori* from the nature of our clear and distinct ideas.

We know, then, that God has created a mechanical universe: what we do not know are the details of the exact mechanism He has selected. We know, *a priori*, that any acceptable physical hypothesis must abide by certain constraints (it may not, for example, involve A.D. or violate the principle of the conservation of motion), but these constraints do not suffice to determine a unique cause for each phenomenon. Different mechanisms may give rise to the same appearances: two clocks, for example, may keep time with one another, yet have quite different arrangements of cogs, wheels, etc. The human physicist is in the position of a man who can directly inspect only the face of the clock and must guess at its inner workings.[273] Thus, says Descartes,

> I have not . . . observed anything which I could not easily explain by the principles which I had discovered. But I must also confess that the power of nature is so ample and so vast, and these principles are so simple and so general, that I observed hardly any particular effect as to which I could not at once recognise that it might be deduced from the principles in many different ways; and my greatest difficulty is usually to discover in which of these ways the effect does depend upon them. As to that, I do not know any other plan but . . . to try to find experiments . . .[274]

---

272 Book 4 of *De Corpore*, says Hobbes (Vol. 1, p. 388), concerns the phenomena of nature, and attempts to explain how they 'may be, I do not say they are, generated'. 'The doctrine of natural causes,' he explains elsewhere (Vol. 7, pp. 3-4), 'hath not infallible and evident principles. For there is no effect which the power of God cannot produce in many different ways'.

273 See Laudan, 1, p. 75. The whole of this excellent paper is relevant to my theme.

274 Descartes, 1, Vol. 1, p. 121.

Whereas Cartesian Natural Philosophy is *a priori* and demonstrative, Cartesian physics is hypothetical and experimental – for Books 3 and 4 of the *Principles* Descartes quite explicitly invokes a H-D method (guided, of course, by certain *a priori* constraints on intelligibility). No doubt, he says, there are an infinite number of different ways in which the omnipotent clockmaker could have fashioned the intricate mechanisms of the corporeal world:

> I believe I have done all that is required of me if the causes I have assigned are such that they correspond to all the phenomena manifested by Nature.[275]

It is, Descartes admits, easy to dream up a cause for any given effect taken singly; far harder to account for *all* the phenomena of a given domain (e.g. magnetism, gravity).[276] And, when the hypotheses invoked to explain phenomena from different realms begin to mesh into a *coherent* body of Natural Philosophy, they receive mutual support from one another. Considered individually, then, mechanical hypotheses might be regarded as *fictions*; taken collectively, their mutual support justifies some confidence on our part that we have hit on the true causes involved in Nature:

> They who observe how many things regarding the magnet, fire, and fabric of the whole world, are here deduced from a very small number of principles, although they considered that I had taken up these principles at random and without good grounds, they will yet acknowledge that it could hardly happen that so much would be coherent if they were false.[277]

Explanatory success, coupled with a coherence-confirmation effect, justifies our 'moral certainty' of the essential correctness of the principles of Descartes' H-D physics.[278] Although odd parts of this physics seem to breathe a fictionalist spirit (e.g. the model used to account for refraction, which is blatantly incompatible with Cartesian principles[279]), Descartes remains, in the final analysis, an *optimist* about the possibility of uncovering Nature's real mechanisms: if our

---

[275]   *Ibid.*, p. 300.

[276]   See Kenny, Ed., p. 58.

[277]   Descartes, 1, Vol. 1, p. 301. See also Sabra, p. 21.

[278]   *Principles*, 4, 206; Descartes, 1, Vol. 1, p. 301.

[279]   See Sabra, p. 18. The *model* involves essentially the assumption that light travels at different speeds in different media, yet in Cartesian optics light-transmission is *instantaneous*.

hypotheses hang together both with one another and with the available data, we may legitimately feel confident that this systematic body of physics gives a true account of the hidden workings of the clockwork universe.

Those desirous of establishing *certain* principles of physics, and loath to settle for anything avowedly hypothetical in nature, will be anxious to establish firmer grounds for our beliefs than are provided by mere coherence or explanatory power. Two attempts were made, during the course of the seventeenth century, to obtain the certain knowledge of Nature's R.E.s (= microstructures) necessary for a demonstrative Science (in the old and honorific sense) of physics. One involves the use of eliminative induction (a method strongly advocated by Bacon); the other turns on microscopy. Both strategies may be discovered in Hooke's *Micrographia*: let us begin with the former.

The microscope, says Hooke, reveals clearly the fine geometrical structures of crystals, and stimulates conjectures as to their possible constitutions: common crystalline figures such as triangles, trapezia, rhombi and parallelograms can all, he remarks, be easily constructed out of regular arrays of spherical globules.[280] Microscopy and geometry may thus aid and assist the empirical investigation of the microstructures of chemical compounds. Given a sufficient body of 'historical' data, says Hooke, we can ask 'by what, and how many means, such and such figures, actions and effects could be produced possibly'. Finally, he adds,

> . . . from all circumstances well weighed, I should have endeavoured to have shown, which of them was most likely, and (if the information by these Enquiries would have born it) to have demonstrated which of them it must be, and was.[281]

The proposed method is to take metals, minerals, stones, etc., dissolve them in 'several menstruums', then crystallise out their salts and examine those crystals under the microscope. If we simultaneously collect a mass of data about the physical and chemical properties of the substance under investigation, we can then attempt a microstructural explanation of the historical data. This will still be hypothetical, but Hooke is clearly optimistic about the prospects for an eliminative induction which would yield a conclusion of the form 'Only

[280] Gunther, Ed., vol. 13, p. 85.
[281] *Ibid.*, pp. 86–87.

microstructure M could account for the properties $P_{1 \ldots n}$'. Such certain knowledge of R.E.s is clearly the goal towards which Hooke is striving.

The basic idea of this 'philosophical algebra'[282] is, therefore, to list in advance all conceivable (i.e. mechanical) explanations of a given body of data, and then to eliminate all bar one, which survivor must be the true explanation, the actual mechanism employed by Nature. Descartes hinted once or twice about a possible use of this method, but its drawbacks are evident.[283] If God's power really is such that He could produce any given phenomenon in an infinite number of ways, the method of eliminative induction will not even get off the ground: one could never, in principle, list all possible hypotheses. And, even if one were to accept only a finite number of possible mechanisms, how could one ever show that one had listed them *all*? Yet, without that assurance, one's conclusion lacks the certainty sought after.

Perhaps, then, we should forget this roundabout method and seek from the microscope *direct ocular acquaintance* with the R.E.s of things. That instrument, says Hooke, may enable ordinary mortals to compete with the legendary *Lynceus*,[284] and to recapture that familiarity with the R.E.s of things lost by Adam through the Fall.[285] It 'seems not improbable', says Hooke, that with the aid of microscopes 'the subtility of the composition of Bodies, the structure of their parts, the various textures of their matter, the instruments and manner of their inward motion. . . may come to be more fully discovered . . .' By such means, he continues,

> . . . we may perhaps be inabled to discern all the secret workings of Nature, almost in the same manner as we do those that are the productions of Art, and are manag'd by Wheels, and Engines, and Springs . . .[286]

282 See Hesse, 3. The precise details of how the 'algebra' was supposed to work remain, regrettably, obscure.

283 See Sabra, p. 41. It is interesting to note that this method gives us one way of accommodating Baconian data-gathering activity within a Cartesian epistemology: one may seek data in order to eliminate possible mechanisms from some prior list.

284 *Non possis oculo quantum contendere Linceus*, we read on the frontispiece of the *Micrographia*. The concept of the Lynx-eyed observer is of great importance for seventeenth century science.

285 This was, of course, a standard Baconian theme. For Adam's familiarity with microstructures, see Gunther, Ed., Vol. 13, p. 154.

286 Gunther, Ed., Vol. 13, *Preface*. Similar claims were being made at around the same time by that other pioneer of microscopy, Power.

The microscope yields observational knowledge of microstructures, which are the R.E.s of things, causally responsible for their manifest appearances. Descartes' clock-metaphor is turned against him by Hooke[287] (himself, incidentally, a notable mechanic and clock-maker). Whereas Descartes had used the clock-metaphor in support of a H-D physics (we can never, he feels, get inside the clock and inspect its actual workings), Hooke claims that the microscope, by giving us *direct* access to Nature's *minutiae*, eliminates all need for a hypothetical approach to physics. (The Royal Society, for which Hooke worked for many years as curator of experiments, favoured a Baconian, anti-hypothetical Natural Philosophy.) With the aid of the microscope, Hooke avers, we can see the R.E.s of things: the truth which, according to Democritus, lay hid in the depths, can now be brought to light.[288]

The reasoning process involved will be something like this:

1.   The microscope reveals that substance S has microstructure M*.
2.   Macroscopic analogues of M* can be observed to have property P.
3.   Therefore, assuming the scale-invariance of the principles of mechanics, S must also have P.

On the assumption of an 'Analogy of Nature', then, we can expect observable mechanical models to cast light on the incomparably subtler (finer-grained) contrivances of Nature. A striking example of this use of macroscopic models to illuminate the microscopic domain is Hooke's investigation of the cellular structure of cork. The observed microstructure of cork is, he says, like that of a mass of bladders or balloons: if we suppose that cork is a partial analogue of such a macroscopic structure we can readily explain its observed properties of low specific gravity, buoyancy (the air-spaces are sealed off from the water by the cell walls) and elasticity. If, Hooke adds, we could as easily discover the 'schematisme and texture' of other substances,

> we might as easily render the true reason of all their Phaenomena; as namely, what were the cause of the springiness, and toughness of some, both as to their flexibility and restitution. What, of the friability and brittleness of some others; and the like; but till such time as

287   Laudan, 1, pp. 101–102.
288   Gunther, Ed., Vol. 13, *Preface*. See also Sprat, p. 79.

our Microscope, or some other means, enable us to discover the true Schematism and Texture of all kinds of bodies, we must grope, as it were, in the dark, and only guess at the true reasons by similitudes and comparison.[289]

Unfortunately, things are not quite so simple. The use of microscopy as a tool for probing into R.E.s suffers from drawbacks both practical and theoretical. The practical limitations are obvious; the theoretical ones rather more profound and illuminating. Suppose, for example, that we are interested in the microstructural grounds of *colours* – can the microscope aid our enquiry? It seems evident that it cannot: one could never, in principle, have direct ocular evidence of the 'emergence' of colour by the arrangement of uncoloured simples. (Whatever we see must have some colour or other.) We could of course see, under the microscope, the spectral colours fading into monochrome; or we could see that what appears one colour to the naked eye appears a different shade at magnification $10 \times$, and perhaps yet another at $100 \times$; we could not see how colouredness itself arose out of uncoloured elements. If, for example, colour (as it is in the object) is a secondary quality or texture, then individual atoms cannot in principle possess colour, and are, therefore, *necessarily invisible*. Newton saw this point quite clearly: the 'more secret and noble works of Nature', he states in his *Opticks*,[290] are invisible for reasons of principle – no improvement in microscopic technique will give us direct ocular acquaintance with them. Locke, unfortunately, seems to have missed the point entirely. 'Had we,' he suggests,

> senses acute enough to discern the minute particles of bodies and the real constitution on which their sensible qualities depend, I doubt not but they would produce quite different ideas in us; and that which is now the yellow colour of gold would then disappear, and instead of it we should see an admirable texture of parts, of a certain size and figure. This microscopes plainly discover to us . . .[291]

But, if we can see these parts, they must themselves be coloured (and hence not elementary). Of course, they need not be *yellow*: bodies which appear one colour to the naked eye may look very different under the microscope (an obvious

289   Gunther, Ed., Vol. 13, p. 112.

290   Newton, 2, p. 262.

291   Locke, *Essay*, II, 23, xi, Vol. 1, p. 250.

example would be blood), but they cannot be entirely colourless. Even the microscope, then, has its limitations: it can help us to survey the 'middle depths' of Nature's ocean, but cannot in principle reach rock bottom.

Neither eliminative induction nor microscopy offering much real prospect of direct acquaintance with R.E.s, we are left with an irredeemably hypothetical physics. This lesson is drawn for us by Robert Boyle. To *see* Nature's microstructures, he says, would require microscopes far superior to our current ones, such as 'I fear are more to be wished than hoped for',[292] while the method of eliminative induction presupposes a finite and exhaustive list of possible mechanisms, and assumes further than we can eliminate all but one of the members of that list, both of which assumptions Boyle finds over-optimistic. Very often, he says, we will be left with several equally possible, yet incompatible, hypothetical explanations of the same phenomenon, with no clear way of deciding between them. Such explanations can only, therefore, be hypothetical in nature: however, confidently naturalists may

> presume to know the true and genuine causes of the things they attempt to explicate, yet very often the utmost that they can attain to, in their explications, is, that the explicated phaenomena may be produced after such a manner as they deliver, but not that they really are so.[293]

Since the same manifest effects can be produced by a variety of different causes, one cannot derive certain knowledge of the cause by observing the effect. If a skilled watchmaker can produce engines with different inner workings, yet producing the same external appearances, how much more variety is possible for the all-powerful architect of Nature? Modesty demands, then, that we claim to assign 'not precisely the true, but possible causes' of natural phenomena[294] – human physics can only be hypothetical in nature:

> For it is one thing to be able to shew it possible, for such and such effects to proceed from the various magnitudes, shapes, motions, and concretions of atoms; and another thing to be able to declare what precise, and determinate figures, sizes, and motions of atoms, will suffice to make

292   Boyle, *Works*, Vol. 1, p. 680.

293   *Ibid.*, Vol. 2, p. 45.

294   *Ibid.*, p. 45.

out the proposed phaenomena, without any incongruity to any others to be met with in nature: as it is one thing for a man ignorant of the mechanicks to make it plausible, that the motions of the famed clock at *Strasburg*, are performed by the means of certain wheels, springs, and weights, etc., and another to be able to ascribe distinctly the magnitude, figures, proportions, motions, and, in short, the whole contrivance either of that admirable engine, or of some other capable to perform the same things.[295]

The analogy, and the moral drawn from it, are both, as Laudan[296] sees, Cartesian in nature and origin, yet appear in Boyle with more than a flavour of the scepticism of Gassendi. Whereas Descartes is confident of finding the true inner mechanisms of things, Boyle is more cautious: rarely, if ever, will he make definite, positive, and unguarded assertions about R.E.s. Instead, he follows Democritus and Gassendi in emphasising the great *difficulty*[297] involved in finding true causal explanations, and recommends modesty and caution in making assertions in the realm of Natural Philosophy. In his *Excellency of Theology*, he flatly denies that the naturalist can attain a degree of certainty unmatched by the divine: the knowledge of Nature accessible to our poor intellects is, he insists, limited in extent, profundity, and certainty.[298] In Boyle, Atomism, Scepticism, and piety combine to form an *attitude* to human knowledge quite different from the Cartesian.

The best 'proof' that we can obtain of any hypothesis – including the corpuscular theory as a whole – is that it *works*, i.e. that it provides convincing and illuminating explanations of natural phenomena:

> For as .. Plato said, that the world was God's epistle written to mankind, and might have added, consonantly to another saying of his, it was written in mathematical letters: so, in the physical explications of the parts and system of the world, methinks, there is somewhat like what happens, when men conjecturally frame several keys to enable us to understand a letter written in cyphers.

---

295   *Ibid.*, Vol. 2, pp. 45–46. The same example is used to make a similar point in Locke's *Essay*, III, 6, ix, Vol. 2, p. 43.

296   See Laudan, 1, pp. 88–89. For the Cartesian influence on Boyle's epistemology, see pp. 92–93.

297   Boyle, *Works*, Vol. 1, p. 307.

298   *Ibid.*, Vol. 4, p. 41 ff.

Although a man cannot demonstrate the truth of any such key *a priori*,

> yet, if due trial being made, the key he proposes, shall be found so agreeable to the characters of the letter, as to enable one to understand them, its suitableness to what it should decypher, is, without either confutations, or extraneous positive proofs, sufficient to be accepted as the right key of that cypher.[299]

We do not need to refute all other possible hypotheses, or give an apodictic proof of that particular one: such demonstrations are not only unattainable, but quite unnecessary, in natural philosophy as in cryptography.

We are now – at last – in a position to understand and appreciate the epistemology of Locke's *Essay*. It lies, quite clearly, in the tradition which we have been examining, though closer to Boyle and Gassendi in spirit that to Descartes. (Locke was a friend of the Gassendist Bernier,[300] and was classed as a member of that school by no less a critic than Leibniz.[301]) The all-important distinction is that between ideal (angelic) and human knowledge. Angels, Locke suggests, may have direct acquaintance with the R.E.s or microstructures of things.[302] This makes possible for them a true *Science* of Nature: i.e. a body of certain and demonstrative knowledge. Angelic physics will begin with axioms, real definitions setting out the R.E.s of things; from these axioms and the laws of geometry and mechanics they will be able to deduce, in advance of experience, all the properties and powers of natural substances. They will be able to discover – without trial – the facts that we must laboriously gather from experience. They will, moreover, not only see that, but understand why, for example, gold dissolves in *aqua regia*: they will be familiar with the *necessary connections* that really exist in Nature, but are beyond our ken.[303] Human science aspires to this ideal, but ever falls short of it:

---

[299] *Ibid.*, p. 77. A similar theme can be found in Descartes, *Principles*, 4, 205 (see Laudan, 1, p. 90). For comments, see also Westfall, 1, Part 1, p. 70.

[300] For the Gassendist influence on Locke, see Mandelbaum, p. 10; O'Connor, p. 17.

[301] See Mandelbaum, p. 5.

[302] Locke, *Essay*, II, 23, xiii, Vol. 1, p. 252.

[303] *Ibid.*, IV, 3, x–xi, Vol. 2, p. 150.

I doubt not but, if we could discover the figure, size, texture, and motion of the minute constituent parts of any two bodies, we should know without trial several of their operations one upon another, as we do now the properties of a square or a triangle. Did we know the mechanical affections of the particles of *rhubarb*, *hemlock*, *opium*, and a *man*, as a watchmaker does those of a watch, whereby it performs its operations, and of a file, which by rubbing on them will alter the figure of any of the wheels, we should be able to tell beforehand that *rhubarb* will purge, *hemlock* kill and *opium* make a man sleep: as well as a watchmaker can that a little piece of paper laid on the balance will keep the watch from going till it be removed; or that some small part of it being rubbed by a file, the machine would quite lose its motion, and the watch go no more. The dissolving of silver in *aqua fortis* and gold in *aqua regia*, and not *vice versa*, would be then perhaps no more difficult to know than it is to a smith to understand why the turning of one key will open a lock and not the turning of another.[304]

This is classical corpuscularian doctrine: the clockwork analogy comes from Descartes; the lock-and-key model derives from Boyle. *If*, says Locke, we were familiar with the R.E.s (= microstructures) of opium, hemlock, etc., we should both know in advance *that* and understand *why* those substances have their wonted effects on us – we should possess scientific knowledge (knowledge through causes) of those facts. This is the type of knowledge the angels possess and human endeavour aspires to.

Unfortunately, however we humans lack – and cannot seriously hope to attain - certain knowledge of R.E.s: the microstructures of things fall far below the reach of our senses:

And therefore I am apt to doubt that, how far soever human industry may advance useful and experimental philosophy in physical things, scientifical will still be out of their reach: because we want perfect and adequate ideas of those very bodies which are nearest to us, and most under our command.[305]

Hence, he concludes, 'certainty and demonstration are things we must not, in these matters, pretend to':[306] we can have no

304   *Ibid.*, IV, 3, xxv, Vol. 2, pp. 160–161. See also Alexander, 2, p. 66.

305   *Ibid.*, IV, 3, xxvi, Vol. 2, p. 161.

306   *Ibid.*, p. 161.

Science (strong sense) of natural philosophy. We have 'experimental' knowledge that rhubarb purges and hemlock kills, but we lack the corresponding 'scientific' (i.e. causal, explanatory) knowledge, to achieve which one would have to discover the precise mechanisms of their operations. Of this, Locke repeats, we are – and are likely to remain – totally in the dark:

> As to a perfect science of natural bodies (not to mention spiritual beings) we are, I think, so far from being capable of any such thing that I conclude it lost labour to seek after it.[307]

Since we are incurably ignorant of the R.E.s of things, and our terms must stand for some *ideas*, our substance-terms such as 'gold' must, Locke argues, stand for *nominal* essences (N.E.s), i.e. for sets of sensible qualities. Our idea of gold, for example, is of a substance yellow, dense, fixed, malleable, fusible, etc. – an element of arbitrariness comes into this list. To imagine that the term 'gold' denotes the unknown R.E. of that metal rather than its familiar N.E. is denounced by Locke as an 'abuse of words'.[308] We sometimes make, he admits, a 'secret supposition' that a natural kind term such as 'gold' designates an unknown R.E, but this 'plain abuse', far from ameliorating the 'imperfection' of our words, serves only to exacerbate it, since it makes that term (and other natural kind terms) stand for something of which we have no *idea* at all.

When we use a term such as 'gold' or '*aqua regia*', then, we must use it to denote some familiar N.E. or other. But the only necessary connections between N.E.s are trivial logico-linguistic ones: if we have defined gold in terms of the N.E. above, only experience will show that it is soluble in *aqua regia* – no necessary connection can be discerned between that power and yellowness, density, etc.[309] By admitting that our natural kind terms designate only N.E.s Locke seems to be making a major concession to such successors as Berkeley and Hume.

Where, then, does the corpuscular philosophy come in? According to a challenging paper by Yost,[310] Locke was sceptical about the point and purpose of sub-microscopic

307   *Ibid.*, IV, 3, xxix, Vol. 2, p. 164.

308   See *Essay*, III, 10, xvii–xviii, Vol. 2, pp. 97–99.

309   *Ibid.*, IV, 3, x, Vol. 2, p. 150.

310   Yost, p. 111 ff.

corpuscular hypotheses, doubtful about the value of their contribution to empirical knowledge. According to Yost, Locke admits the causal dependence of observed phenomena on sub-microscopic causes, grants the potential utility of knowledge of microstructures (cf. angelic physics), but *denies* that unverifiable speculations about sub-microscopic R.E.s are of any value in extending human knowledge of and control over Nature. Yost's Locke is, therefore, a trenchant *critic* of corpuscularianism, eventually advocating a purely Baconian (experimental) natural philosophy in preference to a hypothetical one.

This reading of Locke has been severely – and justly – criticised by Mandelbaum[311] and Laudan.[312] Essentially, they claim, Yost's interpretation ignores or overlooks two vital points. The first is Locke's enthusiastic and lifelong advocation of the (avowedly hypothetical) corpuscular philosophy[313]: the simile of the under-labourer in the Epistle to the Reader (where Locke likens himself to an 'under-labourer', helping to clear the ground for the 'master-builders' Boyle, Huygens, and Newton[314] – all good corpuscularians) is no mere rhetorical flourish, but a sincerely and strongly held belief. (Even the simile itself derives from Boyle![315]) So much of the *Essay* is no more than an exposition, elaboration and development of orthodox corpuscularian doctrine – the number of passages transparently dependent on Boyle is legion – that it is no exaggeration to claim that the work is *primarily* a corpuscularian tract and only secondarily a treatise on epistemology.

The second point missed by Yost is Locke's tendency to use terms such as 'science' and 'knowledge' in their old-fashioned (and very strong) senses.[316] When he sounds most sceptical about the contribution to human knowledge made by the corpuscular philosophy it is often necessary to remember that, on this usage, 'knowledge' means something *certain* or *proven*: even our modern physics and chemistry is by this standard merely 'probable opinion', albeit very plausible and

311   Mandelbaum, p. 11 ff.

312   Laudan, 2.

313   Mandelbaum, p. 3.

314   Locke, *Essay*, 'Epistle to the Reader', p. xxxv.

315   In Boyle's 'Physiological Essays' (*Works*, Vol. 1, p. 307), he likens himself to an 'under-builder', quarrying the materials for future system-builders.

316   Laudan, 2, pp. 213–214; Mandelbaum, pp. 11–12.

well confirmed empirically. For lack of *knowledge*, of which we humans have so very little, we must get by with sound *judgment* – this must suffice in place of proof.[317] Locke's scepticism is of the 'mitigated', Gassendist variety: we have, on such a view, certain knowledge only of appearances, but may form more or less probable *opinions* concerning the causes of the phenomena. These hypotheses, although never attaining the absolute certainty required to give them the status of knowledge, may be so highly probable and well confirmed empirically that no rational man would doubt them. Of such a kind, Locke clearly feels, is the 'corpuscular hypothesis' as a whole, and perhaps also some sub-hypotheses such as the kinetic theory of heat.[318] Such hypotheses serve *both* explanatory *and* knowledge-expanding (e.g. experiment-guiding) functions in any worthwhile natural philosophy. Far from being redundant accessories or appendages to the scientific quest proper, causal hypotheses are – and must remain – an integral part of our physics and chemistry.

---

[317] Laudan, 2, p. 214.
[318] *Ibid.*, pp. 214–215.

THE MECHANICAL PHILOSOPHY: 3

To attempt an in-depth study of the Mechanical Philosophy of the seventeenth century within the course of a single chapter would border on insanity. The subject is a vast one, and at times of quite bewildering complexity. My treatment will therefore be, of necessity, a schematic and selective one: rather than attempt to tell the whole story within the limited compass available, I shall pick up the threads left by Chapter 4, and discuss again the four themes of spontaneity, action at a distance (A.D.), incorporeal causes, and teleology, the joint denial of which, I have claimed, constitutes a complete Mechanical Philosophy. We begin with spontaneity.

A. Spontaneity: The Initiation of Motion

The notion that brute inanimate matter possesses an *inclinatio ad quietem*,[1] an innate tendency (a) to resist being set in motion, and (b) to return spontaneously to a state of rest, is, I suggest, implicit in most Greek and medieval physics.[2] Only on the presumption of a widespread belief in such an innate torpor in bodies can we explain the perennial search for 'motors' of various kinds to account for local motions, natural and non-natural alike. By the late Middle Ages, it had become orthodox to assert that forced motion was due to an *impetus* impressed on the projectile by the mover (the projector); and that natural motions resulted from the motive qualities *gravitas* and *levitas*, themselves 'emanent effects' of the substantial forms of heavy and light bodies respectively, which forms owe their existence to some *generator* or other, which, for an orthodox peripatetic, is the efficient cause *per se* of natural motion.[3] The motive qualities *impetus*, *gravitas*, and *levitas*, although, strictly speaking, only the *instruments* by which the true movers (the projector for forced motion, the generator for natural) work their effects, became widely construed as *motores conjuncti* for forced and natural motions respectively. But, unless brute matter tended to resist

---

[1] The phrase is Oresme's. See Chapter 6, p. 261.

[2] For a similar suggestion, see Koyré, 5, p. 62.

[3] See Chapter 8, pp. 344–348.

even natural motion, there would be no need to posit such continually acting motors. Thus Aquinas, following Avempace, claims that each body, in virtue of its *corpus quantum*, resists being set in motion – even in a motion natural to it. (This explains, as we have seen, how a body can move with a finite velocity in the absence of any external resistance.[4]) For Oresme, impetus, though not self-expending of its own nature, is spent in the attempt to overcome the *inclinatio ad quietem* of the body moved: this explains why no forced motion can be everlasting. Similar views can be found in a wide variety of scholastic thinkers.

The concept of a resistance to motion innate in matter was very common in Renaissance Platonism and survived well into the seventeenth century. Solid bodies, says Francis Bacon, possess a 'torpor', a 'natural appetite not to move at all';[5] in virtue of this 'abhorrence of motion',[6] tangible and earthy bodies tend always to a state of rest. For Kepler, all matter possesses 'inertia', a resistance to motion and 'preference' for rest.[7] (It is curious that the sense of the term 'inertia' should have changed so radically in the period that separated Kepler from Newton.) 'All material bodies,' he says, 'in themselves and by nature, are stationary in whatever place they are at'[8] – to explain motion one must posit either some external force or some internal (and incorporeal) striving. The quantity of matter (*moles*) of a body is, for Kepler, a measure of its innate resistance to being set in motion.

This belief in a 'torpor' or 'inertia' inherent either in all matter (Kepler), or in all gross matter (Bacon) has as a necessary corollary the need for 'active' (motion-initiating and enhancing) principles, unless the whole of Nature is gradually to grind to a halt. If only *gross* matter possesses this resistance to motion, one can posit a subtle but highly active form of matter to fill this rôle: Telesio and Bacon, clearly influenced by the Stoic concept of *pneuma* as well as by contemporary medico-chemical accounts of *spiritus*, pursued this course.[9] Spirits, says Bacon, are 'the most active of bodies',[10] the

4   See Chapter 6, p. 275 ff.

5   F. Bacon, *Works*, Vol. 2, p. 586.

6   *Ibid.*, Vol. 4, p. 230; *Novum Organum*, 2, xlviii.

7   See Jammer, 2, pp. 55–56; Clavelin, p. 255; Dijksterhuis, p. 314.

8   From Koyré, 6, p. 195.

9   For the concept of *spiritus* in Bacon, see Walker in Debus, Ed., 7, Vol. 2, p. 121 ff.

10   F. Bacon, *Works*, vol. 2, pp. 380–381.

'agents and workmen that produce Nature's effects'[11] –
without their vital agency, all motions would soon cease.
Those whose background philosophy was Platonic rather
than Stoic in inspiration tended to posit an 'inertia' in *all*
material things, and thus to insist that a true 'active principle'
could only be *incorporeal*.[12] Neoplatonic philosophers therefore postulated a great hierarchy of incorporeal agents of
various kinds to vivify the corporeal universe: human and
animal souls, angels and demons, the celestial intelligences,
the *Anima Mundi*, the 'seminal reasons' responsible for the
formation of plant and animal bodies, etc. – all served to add
life (and hence motion) to an otherwise dead and inert
corporeal universe.

Belief in an *inclinatio ad quietem* is thus, quite clearly, one
of the cornerstones of an *animistic* physics. It is no surprise,
then, to see this belief challenged by the pioneers of the
Mechanical Philosophy. Matter, Descartes insists, does *not*
resist motion: it has no 'inertia' in Kepler's sense. 'I do not,' he
writes to Mersenne, 'recognise any inertia or natural tardity in
bodies'.[13] In the absence of friction and air-resistance, a
pendulum, for example, would swing for ever. It is, he
claims, 'a serious prejudice' to imagine that more activity is
needed for motion than for rest[14] – motion is a mere
translation of the moving body, not an action. The prejudice
derives from the unreflective experiences of childhood: we see
that moving bodies come to rest once their 'impetus' has
decayed and, generalising over-hastily, come up with a
typically Greek or medieval physics. Mature reflection
teaches, however, that uniform motion is a *state* of a body,
dynamically equivalent to rest (cf. of course the principle of
mechanical relativity), a state that will not be corrupted
without some proper external force. It is valuable to
remember, says Descartes, that 'effort is needed not only to
move external bodies but also, quite often, to arrest their
movement':[15] any change of state requires the operation of an
appropriate cause. Motion and rest are, he insists, 'simply
two different states of a body',[16] and it is the 'First Law of
Nature' in Cartesian physics that 'each thing remains in the

[11]   *Ibid.*, Vol. 5, p. 268.

[12]   See Chapter 4, p. 172.

[13]   Quotes from Koyré, 4, p. 69.

[14]   Descartes, 2, p. 209.

[15]   *Ibid.*, p. 210.

[16]   See Koyré, 4, p. 69.

state in which it is so long as nothing changes it'.[17] Every reality, says Descartes,

> . . . always remains in the same condition so far as it can, and never changes except through external causes. Thus if a piece of matter is square, one readily convinces oneself that it will remain square for ever, unless something comes along from elsewhere to change its shape. If it is at rest, one thinks it will never begin to move, unless impelled by some cause. Now there is equally no reason to believe that if a body is moving its motion will ever stop, spontaneously that is, and apart from any obstacle. So our conclusion must be: A moving body, so far as it can, goes on moving.[18]

A similar point can be found in Hobbes. A resting body, he begins, will remain at rest unless given some external impulse. 'In like manner,' he adds more controversially, 'whatsoever is moved, will always be moved, except there be some other bodies besides it, which causeth it to rest'.[19] There is, therefore, no resistance to motion innate in matter: if there were, a moving body would decelerate and eventually come to rest without need for external resistance.

Among the founders of the Mechanical Philosophy there is, therefore, a widespread denial of the 'torpor' of Bacon or the 'inertia' of Kepler. It remains an obvious empirical truth, however, that it takes more effort to move a large body than a small one, *or* for that matter to arrest it once it has been set in motion. From these commonplaces, together with the new concept of a *motion-state*, there gradually emerged the modern idea of inertia as *resistance to change of state*, proportional to the 'quantity of matter' (or, for the Cartesian, gross matter or E3) in any body. Matter, on this view, is the ultimate conservative: although indifferent to what state of motion it is in, it nevertheless resists all attempts to alter this state in any way. This concept did not spring, fully-formed, from the brain of any single thinker: it represents the fruit of much labour, both empirical and conceptual. Once the insight has been achieved, the basic structure of seventeenth century (and modern) mechanics becomes clear. Rest and uniform motion are alike *states* of bodies; changes of state require the action of forces; bodies resist, with a strength dependent on their quantity of matter, any attempt to compel

17 *Principles*, 2, 37.

18 Descartes, 2, p. 216. Cf. also the letter to Regius in Kenny, Ed., p. 136.

19 Hobbes, *Works*, Vol. 1, p. 115. Cf. also Vol. 3, pp. 3-4.

them to change their state. In virtue of this intrinsic resistance to change of state, a moving body is 'empowered' to impel another upon collision: it must, Descartes warns,

> . . . be carefully observed what it is that constitutes the power of a body to act on another body or resist its action; it is simply the tendency of everything to persist in its present state as far as it can.[20]

While this insight is basically sound, Descartes' attempt to develop it into a body of 'rules of impact' was not a success.[21] For a sure and confident grasp of the modern concept of inertia, we must turn to Newton's *Principia*. The *vis insita*, or 'innate force of matter', is, he says,

> a power of resisting, by which every body, as much as in it lies, continues in its present state, whether it be of rest, or of moving uniformly forwards in a right line.[22]

This innate force may also, he adds, be called *vis inertiae*:

> But a body only exerts this force when another force, impressed upon it, endeavours to change its condition; and the exercise of this force may be considered as both resistance and impulse; it is resistance so far as the body, for maintaining its present state, opposes the force impressed; it is impulse so far as the body, by not easily giving way to the impressed force of another, endeavours to change the state of that other.[23]

It is quite clear from this that *vis inertiae* is *not* (*pace* McMullin[24] and Jammer[25]) a *vis conservans*, a type of impetus sustaining the motion of any body. Newton states quite explicitly that it is only exercised when a force is impressed on a body: uniform rectilinear motion is a state of a body and as such self-explanatory, not in need of explanation in terms of a *vis*.[26] Matter is not active but re-active: it only exerts its *vis*

---

20   Descartes, 2, p. 219. See also Gabbey, 1, p. 24.

21   For the Cartesian rules of impact, see Gabbey, 1, p. 25 ff.

22   Newton, 1, p. 2. For the derivation of Newton's first law from Descartes, see Cohen, 2.

23   Newton, 1, p. 2.

24   McMullin, 4, p. 36.

25   Jammer, 2, p. 65.

26   Herivel, p. 28; Gabbey, 1, p. 35; Shapere in *Texas Quarterly*, p. 205. The concept of *vis inertiae* as a *vis conservans*, Herivel shows quite convincingly, had disappeared from Newton's thought long before publication of *Principia*.

inertiae when a force is impressed upon it.[27] Bodies are inert in the sense that they do not initiate motion: once set in motion, however, they are capable of mechanical action (exerting an impulse) in virtue of their own innate resistance to change.

Uniform motion is, then, a *state* of a body which will tend to persist unless some external force acts upon that body to alter it. There remains, however, one crucial question to be answered: is this uniform, force-free motion circular or rectilinear in nature? Men of the calibre of Galileo, Beeckman, Hobbes and (at times) Gassendi seem quite prepared to countenance the possibility of the former, i.e. of a uniform circular motion. To see why this should be so, we must turn to Galileo, and in particular to his conception of weight.

Consider, says Salviati in Galileo's *Dialogo*,[28] a body at rest on the surface of our Earth. If we give this body an impulse, and set it moving on a flat plane tangential to the circumference of the planet, it will be moving away from the centre of gravity of the Earth, in opposition to its own intrinsic nature, and will therefore decelerate. If, on the other hand, the motion of any body carries it towards the Earth's centre, its original motion is conserved, while its gravity continually impresses new motion on it – i.e., it accelerates. To obtain a *uniform* state of motion (within the bounds of the cosmos) it is therefore necessary that the moving body neither approach nor recede from the point to which its natural 'appetite' tends. Such a 'neutral' motion is provided by a circular path around that point, remaining always at the same fixed distance from it. Being neither accelerated nor retarded by its own weight, a body with such a motion is resisted *only* by external factors such as the drag of the medium: in the ideal case, where this resistance is zero, such a uniform 'neutral' motion could endure forever.[29]

A body with such a circular motion resembles a stone whirled in a sling: it experiences a *centrifugal force*. Such a 'projectile', says Galileo,

---

27 See Westfall, 4, p. 450. *Vis inertiae* may also function, however, as a 'dummy' or pseudo force in the parallelogram principle of combination of forces.

28 Galilei, 3, p. 147. The idea of a 'neutral' circular motion goes back to his early *De Motu* (Galilei, 1, p. 68) and, before that, to contemporary scholastics (see Wallace, 5, p. 112).

29 See Galilei, 2, p. 113.

acquires an impetus to move along the tangent to the arc described by the motion of the projectile at the point of its separation from the thing projecting it.[30]

To explain the centrifugal force experienced by such a body, we must analyse or resolve its actual circular motion into two 'tendencies', a tangential (rectilinear)tendency, restrained by a centrally-directed inclination. Both of these, for Galileo, are *internal* to the body in a state of 'neutral' motion: we impart to it a rectilinear *impeto* when we impel it; its own gravity is intrinsic to it, an 'inherent tendency to move'[31] towards the Earth's centre of gravity. The resulting 'neutral' motion can therefore be called 'natural' in the (attenuated) sense that, *once initiated*, it requires no sustaining external motor. Although brought into being by force (i.e. by some impulse) a neutral motion is, thereafter, self-perpetuating unless impeded from without.

We are now in a position to assess Koyré's claim that Galileo believed in a 'circular inertia'. The claim can be seen to be too strong: a circular inertia, as Drake sees, would not give rise to a centrifugal force.[32] Galileo's analysis of circular motion into a tangential tendency and a centripetal inclination implies that, in the absence of the latter, the circularly moving body would fly off in a tangential (i.e. rectilinear) path, like a stone leaving a sling. If, by analogy, the rotating Earth were annihilated by divine power, bodies resting on it would fly off on tangential paths: they would *not* continue to pursue their regular circular courses. (Galileo's pupils, Cavalieri and Torricelli, make this step quite explicitly.[33])

Where Koyré is right is in his assertion that, for Galileo, gravity is internal to bodies: the weight of any particular body is an inclination or striving *of* that body, *not* an external push or pull.[34] (Wallace suggests that he may have shared Philoponus' concept of *physis* – popular among contemporary scholastics – as an inner incorporeal motor.[35]) This internal motive urge cannot, however, subsist in the absence of other bodies: the 'instinct' of a heavy body is not directed towards a

30   Galilei, 3, pp. 191–192.

31   Galilei, 4, p. 74.

32   Drake, 3, pp. 267–268.

33   See Koyré, 7, pp. 238–241; Clavelin, pp. 376–377; Westfall, 4, pp. 126–127.

34   Koyré, 7, pp. 186–187; Clavelin, p. 356.

35   Wallace, 5, p. 293.

mere mathematical point as such, but towards the centre of gravity of our Earth, i.e. a heavy body.[36] Although intrinsic to a body in the sense that it is an inner striving or urge and not an external compulsion, gravity cannot also be *essential* to body in the sense that no body could conceivably exist without it. (A lonely corpuscle, for example, would be weight*less*.) Weight is *not* listed, in the *Assayer*, among the qualities conceptually necessary to corporeal existence.[37]

Koyré, then, is both right and wrong. Circular motions are favoured by Galileo because *in our world-system* only they can be force-free, uniform, and abiding, contributing to the stability and permanence of the cosmic order.[38] (Any rectilinear motion must be either accelerated or decelerated, and hence non-uniform.) Since both gravity and *impeto* are internal to a body, they may *also* be properly called 'natural': once initiated, they run for ever in the absence of external forces. Nevertheless, weight is *not* essential to matter in the sense of being constitutive of its very nature:[39] if God should annihilate the Earth, leaving only one stone of its matter, this would be weightless but nevertheless material. It would also find itself in a *rectilinear* motion tangential to its original circular path: with the annihilation of the Earth, gravity would be no more. Galileo's (admitted) failure to state the principle of rectilinear inertia stems, then, simply from his inability or unwillingness to 'think away' the cosmos, to make the radical thought-experiment of annihilating the Earth and its gravitational influence. Cavalieri and Torricelli do not, therefore, depart drastically from the principles of their master when they assert that, in the absence of gravity, bodies would have rectilinear inertia: they are, as they themselves saw, merely extracting what was implicit in Galilean physics.[40]

The notion of weight and lightness (*gravitas* and *levitas*) as internal 'motive qualities' is the next target for the mechanists'

---

[36] Galilei, 3, p. 37. Gravity cannot, he insists, be directed towards a mere point *as such*: the centre of the universe, even if such a thing exists, 'is but an imaginary point, a nothing, without any quality'. Similar objections can be found in Bacon, Kepler, and others.

[37] See Galilei, 2, p. 274.

[38] Galilei, 3, pp. 19-20; Clavelin, pp. 212-213.

[39] *Pace* Koyré, 7, p. 187.

[40] Drake even goes so far as to suggest that Galileo's mania for circularity in the *Dialogo* is a mere polemical device; we need not go to this extreme to see the essential correctness of his case against Koyré's 'circular inertia'.

attack. *Levitas*, as we saw in Chapter 8,[41] can be disposed of by the Archimedean theory of upthrust; let us now turn our attention to *gravitas*. For the Aristotelian, this 'motive quality' can be described as an 'emanent effect' of the substantial form (= soul!) of the heavy body. Is this substantial form, then, the *efficient cause* of free fall? Aristotle, as we saw in Chapter 4, denies this: on such a view, he says, a freely falling body could be described as a 'self-mover', which would blur the distinction between animate and inanimate things and thus implicitly lapse into animism.[42] Unfortunately, the demand of peripatetic (and Platonic) physics for a *motor conjunctus* to account for each and every motion led inexorably to that conclusion: despite the tortuous and often sophistical mass of scholastic distinctions woven around this issue,[43] the net result tended to be either mere evasion or a return to the concept of an inner incorporeal motor. Greek and medieval physics is implicitly animistic.

If the substantial form of a heavy body (acting through the instrumentality of its 'emanent effect', the 'motive quality' *gravitas*) is the real efficient cause of its natural motion, we end up positing a 'soul' in every stone or clod of earth. The sheer ontological extravagance of this shocked the mechanists[44] (great wielders of Ockham's razor), while its implicitly pan-psychic, pan-daemonic conception of Nature was found offensive by many on religious grounds. (Men such as Mersenne and Gassendi saw it as a religious duty to purge natural philosophy of the occult animism of much Renaissance thought.) But if the notion of gravity as an inner striving or urge, accepted even by Galileo, involves animism, it follows that to exorcise the ghost from Nature we need to conceive of gravity as a 'passion' in the heavy body rather than an 'action', i.e. as due to an external push or pull rather than to an inner 'desire'.

This step is taken quite explicitly by Kepler. Once a great advocate of animism, he tended in later life to substitute '*vis*' for '*anima*' in his natural philosophy:[45] a small terminological

---

41   Chapter 8, pp. 354–359.

42   See Chapter 4, p. 164.

43   For a small sample of which, see Chapter 8, pp. 344–348.

44   It is absurd, says Hobbes in discussion of the 'internal urge' theory of gravity, to attribute desires to senseless matter (Hobbes, *Works*, Vol. 1, pp. 509–510).

45   See Jammer, 3, p. 90; Dijksterhuis, p. 310.

change with momentous metaphysical implications. Gravity, he states,

> . . . is a mutual affection among related bodies which tends to unite and conjoin them (of which kind also the magnetic faculty is) while the earth attracts the stone rather than the stone tends towards the earth.[46]

Gravity, he insists, is a *passion* in the heavy body:[47] the Earth *pulls* the stone (by means of elastic 'threads') rather than the stone *desiring* the Earth. Explanations of gravity as involving either the external *push* of a subtle matter, or the *pull* of invisible chains, abounded in the seventeenth century: the key idea common to all is the search for an external cause to replace the inner striving of Plato, Aristotle, and Galileo.

This conception of weight as an *external force* of some kind enables us to explain the acceleration evident in free fall. Our concept of a body is – following Descartes – of something *resistant to change of state*. After the addition of an increment of velocity Δv to a body, it will stay in that altered state (i.e. retain its new velocity) until the operation of some force alters it. Suppose then that a freely falling body of mass m experiences a series of discrete pushes, each of which imparts to it an increment of velocity Δv – i.e. a series of impulses each of strength Δmv. If we imagine these impulses ever-fainter, yet separated by shorter and shorter intervals, we obtain an approximation to the operation of a *continuous force*: passing to the limit, we shall obtain the $F = ma$ version of Newton's Second Law and the Newtonian *explanation* of Galileo's $v \propto t$ law of free fall (uniform acceleration is caused by the continuous operation of a constant force).[48]

Kepler's notion of quasi-magnetic elastic 'threads' or 'cables' was adopted by Gassendi in his search for an explanation of gravity. Heavy bodies, he says in his treatise *De Motu*, are 'dragged downward' to Earth by such cables:[49] this is clearly, as Charleton says, 'an imprest motion' produced in the stone by the *pull* of the Earth.[50] But, if gravity is an 'imprest' motion rather than an internal instinct, it

---

[46]   Quoted from Jammer, 3, p. 85.

[47]   Hesse, 1, p. 129; Jammer, 3, p. 82.

[48]   This is not merely a piece of retrospective rational reconstruction: for a very similar pattern of reasoning, see Charleton, p. 450; for comments, see Westfall, 4, p. 108.

[49]   Gassendi, p. 135.

[50]   Charleton, p. 277.

follows that inertial motion is primarily *rectilinear* in nature. The 'neutral' circular motions of Galileo are compounded of rectilinear inertia and a centrally-directed restraining *force*: remove that force, and we have the real possibility of an abiding and uniform rectilinear motion. But now, says Gassendi,

> picture a stone in those imaginary spaces that stretch beyond the world and in which God could create other worlds.[51]

Since no 'magnetic' threads reach from the Earth into such space, a body placed there is weightless, and will naturally *rest* unless moved by some external force. If we give this body an impetus, he continues,

> it is probable that it would move indefinitely in a uniform fashion, slowly or rapidly depending on whether a small or great impetus had been imparted to it.[52]

Here we have a nice example of Atomism assisting in the genesis of a most important and enlightening thought-experiment. In extracosmic space a principle of rectilinear inertia would hold: Gassendi has the honour of having the first published statement of this principle to his name.[53] Having used the thought-experiment (in the best Galilean manner) to clarify the nature of the *ideal* case, Gassendi now comes back down to Earth and shows how the ideal casts light on the actual. In empty space, he says, a body would move forever in a right line:

> And so we deduce from this that absolutely any motion imparted to the stone would be of the same nature taken in itself; so that whatever direction you throw the stone, if you assume that the moment it leaves your hand everything except the stone is reduced to nothing by divine power, the result would be that the stone would be moved eternally by its motion and in the same direction that the hand directed it.[54]

It follows that any deviation from this ideal ('inertial') path must be *explained* in terms of external forces: the drag of the

---

51   Gassendi, p. 136. For comments, see Koyré, 7, p. 247 ff.; Westfall, 4, pp. 101–102.

52   *Ibid.*, p. 139. Cf. also Charleton, p. 466.

53   See Pav, p. 24.

54   Gassendi, p. 141–142.

medium, the *pull* of the Earth, etc. In particular, both acceleration and circular motion must both involve the operation of such forces. (Gassendi, it must be admitted, does not consistently apply this analysis to circular motions: he seems prepared at times to countenance, e.g., the possibility that the circular paths of the heavenly bodies are natural to them.[55])

In Descartes, too, there is an insistence on the *rectilinearity* of inertial motion: the second of his 'Laws of Nature' reads that

> any given piece of matter considered by itself tends to go on moving, not in any oblique path, but only in straight lines.[56]

To explain why inertia is rectilinear rather than circular, says Descartes, we must consider God's re-creation of the universe at every instant of its existence. (For the sake of simplicity, we shall assume that instants can be successive.) Now in His re-creation of the universe at a given instant t_4, says Descartes, God looks only to its predecessor t_3: He *tends* to conserve at t_4 the state of things at t_3. This explains the tangential tendency experienced by bodies in circular motion: only a rectilinear motion can be defined in a single instant. Of all motions, Descartes explains, 'it is the only right-line motion that is entirely simple, and whose nature is comprised in an instant':[57] to define a curvilinear motion one would require two or more past instants. (God would have to look back to t_3, t_2, and perhaps also t_1 in his re-creation of the universe at t_4.)

Every body, then (even those in circular motion), has a rectilinear *tendency*: it *would* move in a straight line if permitted to do so.[58] God will conserve every body in its *present* state of motion (defined in this very instant) *unless* this is impossible (e.g., if it would involve the interpenetration of bodies). A body with a curvilinear motion is being *restrained* (by the presence of other bodies) from pursuing a rectilinear course: the moment the restraint is lifted it will fly off its present path at a tangent.

Any deviation from rectilinearity must, therefore, be explained by the Cartesian in terms of the action of some

55   *Ibid.*, p. 127.

56   *Principles*, 2, 39; Descartes, 2, p. 217.

57   Quoted from Koyré, 4, p. 74.

58   See Prendergast, p. 455; Blackwell, pp. 223–224.

external force on the moving body. Essential to this insight is the conceptualisation of weight as an external force rather than an internal striving: a tendency to gravitate is *not*, Descartes insists again and again, innate in or essential to matter, but is produced in some bodies as a result of the pressure of others. In the absence of external forces, any body is weightless. As Koyré sees,[59] rectilinear inertia and the externality of weight belong together as two sides of the same coin: whereas the Galilean has to imagine the cosmos *annihilated* to arrive at a conception of rectilinear inertia, Gassendi and Descartes can see that principle at work even *within* the cosmos.

To explain uniform circular motion it now becomes necessary to postulate rectilinear inertia and a constant *centripetal* (i.e. centrally-directed) *force*. Attaining this conceptual insight proved difficult to seventeenth century thinkers: it was no doubt highly counterintuitive to think of uniform circular motion as being physically equivalent to free fall, i.e. to uniform acceleration.[60] Robert Hooke's analysis of planetary motion[61] marks the first clear insight into the dynamic structure of the problem; after him, of course, came the *Principia* of Newton, central to the fabric of which is the Moon = projectile analogy and the application of the Galilean laws of *free fall* (e.g. $s \propto t^2$) to the trajectories of the planets. Rectilinear inertia and centripetal force are, arguably, the two most important concepts in the whole great work.

According to the basic principles of mechanism, a body is only moved by an impulse from another: *no body is self-moving*.[62] That a body (i.e. an inert and passive thing) should spontaneously spring into motion is quite unintelligible, hence, for rationalists like Descartes and Hobbes, quite impossible. All bodily motion is, as Cudworth puts it, *heterokinesy* (movement due to another) rather than *autokinesy* (self-movement). Since, he goes on to argue, 'no body could ever move itself', it follows that 'there must be something else in the world besides body, or else there could never have been any motion in it'.[63] To explain why there is motion at all, he argues, we must posit self-movers (*heterokinesy* presupposes *autokinesy*), but these *active* principles must

[59] Koyré, 7, p. 258.

[60] See Westfall, 4, p. 82.

[61] See Gunther, Ed., Vol. 6, pp. 266–267.

[62] For Hobbes' denial of a self-mover, see Brandt, p. 22.

[63] Cudworth, vol. 1, pp. 84–85.

be *incorporeal* in nature. (In essence, the argument goes back to Plato's *Laws*.[64]) Since no body can ever initiate motion *de novo*, material principles alone can never account for its existence in our world. Similar arguments can be found in Cudworth's fellow-Platonist, Henry More,[65] and are even endorsed by Boyle. Lucretius, he says,

> supposes his eternal atoms to have from eternity been their own movers, whereas it is plain, that motion is no way necessary to the essence of matter, which seems to consist principally in extension: for matter is no less matter when it rests than when it is in motion.

> Nor has any man . . . satisfactorily made out how matter can move itself: and indeed, in the bodies, which we here below converse withal, we scarce find, that any thing is moved but by something else.[66]

This argument seems confused. One can surely grant that motion is not *essential* to matter (a body remains such even when at rest), and yet still claim that matter and motion are the co-eternal principles of the physical universe – matter has been in motion from all eternity. Every bodily motion will be explicable in terms of a prior mechanical impulse, and will therefore be *heterokinesy*, in Cudworth's terms, yet there need be no explanation at all of why there is motion: this can be treated as a brute fact. (One will, of course, require some kind of conservation law to explain why the clockwork does not wind down.) After all, any explanatory structure must presuppose something. And, given that the essence of matter is indifferent between rest and motion, why should one demand an explanation of motion rather than of rest? Would a universe of resting matter be any more self-explanatory than one of moving matter? I think not. One can, therefore, assert the co-eternity of matter and motion *without* either making motion essential to matter or construing any body as a self-mover. Cudworth argues against such a vision of the universe – an infinite series of purely mechanical transmissions of motion is, he claims, an impossibility, since it would banish all *action* from Nature[67] – but his argument for this

64   See Chapter 4, p. 156.

65   H. More, *Works*, Vol. 1; 'An Antidote against Atheism', p. 38 ff.; 'The Immortality of the Soul', p. 102.

66   Boyle, *Works*, Vol. 2, p. 42. See also Vol. 1, p. 194, where Boyle's argument against Hobbes involves essentially the same confusion.

67   Cudworth, Vol. 3, pp. 414–416.

conclusion would scarcely have impressed Hobbes (or, for that matter, Democritus).

In the physics of Gassendi, motion is explained in terms of a God-given motive urge or *pondus* instilled in each atom at the Creation. As far as Natural Philosophy is concerned, then, 'matter is not inert, but active':[68] each and every atom can be conceived as a self-mover. Strictly speaking, however, and to avoid making too great a concession to the atheist, one must insist that atoms do not have 'the power to move themselves inherent in their nature': they possess this mobility only *dei gratia*, 'from the power of moving and acting which God instilled in them at their very creation and which functions with his assent'.[69] At the creation of the atoms, says Charleton,

> God invigorated or impregnated them with an internal energy, or faculty motive, which may be conceived the first cause of all natural actions or motions (for they are indistinguishable) performed in the world;
> their internal motive virtue necessitates their perpetual motion among themselves, from the moment of its infusion, to the expiration of nature's lease.[70]

In virtue of its particular *pondus*, each atom strives to move with a certain velocity, to which it will always tend to 'spring back'. Thus Gassendi,

> However much mobility may have been implanted in the atom, it continues to be the same always, so that atoms may indeed be restrained until they do not move, but not to the point where they do not strain and endeavour to disentangle themselves and renew their motion.[71]

When the restraint is lifted, such an atom will spontaneously revert to the motion determined by its particular and inalienable *pondus*. As the critics have said, Gassendi's atoms are like minute *living* things, with a quite unmechanical power of spontaneity. In the micro-realm, as Pav sees, Gassendi's physics is archaic and animistic: the spontaneous return of an atom to its pre-ordained velocity is in flat contradiction to at least two of Newton's Laws.[72]

---

[68]   Quoted from Westfall, 4, p. 103.

[69]   *Syntagma*, Physics, Section 1, Book 3, Chapter 8; Gassendi, p. 399.

[70]   Charleton, p. 126.

[71]   *Syntagma*, Physics, Section 1, Book 4, Chapter 8; Gassendi, p. 417.

[72]   Pav, pp. 30–31; Snow, pp. 41–42; Carré, 2, pp. 116–117.

Some atoms, says Gassendi, are *especially* active (i.e. endowed with a great motive urge): such are the very fine atoms which make up fire and the corporeal souls of animals. (This may, as we saw in Chapter 4, have been the view of Democritus on the question of the springs of motion.[73]) These highly mobile atoms constitute, says Gassendi, the 'bloom' or 'flower' of matter, and are the agents responsible for many natural processes.[74] Matter, then, is moved from *within*, by the *pondus* instilled in each and every atom (though more in some than in others) at the Creation. Since *every* motion results from the inner urge of the atoms, all can be legitimately called 'natural' (due to an internal principle); since all the motions of compounds are consequent upon collisions, they can equally be called 'forced'. The Aristotelian distinction collapses.

This conception of the atoms as self-movers (if only *dei gratia*) must, however, be considered an aberration from the point of view of the Mechanical Philosophy, to which we must now return. The question which now arises is this: given the principle of (rectilinear) inertia, and the concept of mass as resistance to change of state, have we undermined the animist's grand argument for 'active principles'? The answer seems to be that we have not. The principle of inertia gives us a conservation law *only* for unimpeded motions: it remains possible to claim that motion is lost in collisions. (Beeckman, for example – an early advocate of the principle of inertia – believed that motion is lost in the collisions of hard bodies.[75]) But, if motion thus decays, there must be 'active' (i.e. motion-initiating and hence non-mechanical) principles at work in Nature, if the universal clock is not to run down.

To evade this argument, we require a principle of conservation of motion, an essential pillar in the architecture of the Mechanical Philosophy. Now, according to Descartes, God re-creates the physical universe every instant: His immutability is the metaphysical ground for our conservation laws. It follows from the divine perfection, Descartes argues, 'that his operations should occur in a supremely constant and unchangeable manner'. Hence, he continues,

> it is most reasonable to hold that, from the mere fact that God gave pieces of matter various motions at their first

---

73   Chapter 4, pp. 159–160.

74   See Brett, pp. 56–57; Spink, p. 94. See Gassendi, p. 422.

75   Gabbey, 1, p. 18.

creation, and that he now preserves all this matter in being in the same way as he first created it, he must likewise always preserve in it the same quantity of motion.[76]

But if God re-creates, in any instant t_n, the 'quantity of motion' which existed at $t_n - 1$, it follows that this particular amount is eternally invariant, i.e. is conserved for *ever*. This universal 'quantity of motion' is measured by $\Sigma m \mid v \mid$ (note the scalar quantity): collisions between bodies serve as *occasions* for God to transfer 'motion' $(m \mid v \mid)$ from one body to another in accordance with the laws which He has ordained, chief among which is the principle of conservation, $\Sigma m \mid v \mid = k$. (Although this occasionalism is Descartes' considered view,[77] he does at times 'speak with the vulgar' in attributing motive forces to *bodies*:[78] though strictly false, this is permissible in an 'as if' sense – bodies move and 'interact' *as if* they possessed such dynamic features as force and resistance. In reality, however, the only cause is God, forever re-allocating quantities of motion on the occasions of interparticulate 'collisions'.)

Descartes had no need, therefore, for 'active' (motion-initiating) principles in Nature. The total quantity of motion $\Sigma m \mid v \mid$ being conserved, the clockwork universe can run forever without need for intelligences, angels, or an *Anima Mundi*. Cartesian subtle matter may be said to be 'active' only in the much-attenuated, indeed minimal, sense that it is very rapidly agitated and, by its swirling currents, tends to impel grosser bodies: it is most emphatically *not* an 'active principle'; in the Platonic sense.[79]

Unfortunately, however, $\Sigma m \mid v \mid = k$ is simply *false*, as became clear after the experiments on pendulums of Huygens, Wren and Wallis, and was proved beyond cavil by Leibniz in his *Brevis Demonstratio*.[80] (Descartes' 'law', he

---

[76]   Descartes, 2, pp. 215–216.

[77]   For Descartes' occasionalism, see Descartes, 1, vol. 2, p. 56 ('the present time has no causal dependence on the time immediately preceding it') and the letter to More quoted in Kenny, Ed., p. 257 ('. . . the motive force can be that of God Himself, as conserving in matter the same amount of *translatio* as He has set into it in the first moment of creation'). For comments, see Kemp Smith, p. 196, and Hatfield.

[78]   The Cartesian dynamics outlined by Gabbey (1, pp. 8–10) and Westfall (4, p. 61) is, therefore, not an aberration, so long as it is read in an 'as if' sense, i.e. in such a way as to be compatible with an occasionalist metaphysic.

[79]   Westfall regrettably makes this mistake (in Debus, Ed., 7, p. 187).

[80]   See Leibniz, p. 296 ff.

shows, would entail the possibility of a perpetual motion machine, in flat violation of what we now describe as the principle of conservation of energy.) What is conserved in all collisions of bodies is the *vector* quantity $\Sigma m\vec{v}$ (our concept of momentum), but this does not suffice for a mechanical philosopher. Two equal-sized balls of putty colliding with equal and opposite velocities and coming to a dead halt involves no violation of the principle of conservation of momentum; it does, however, seem to imply the possibility of universal decay.

If one accepts the existence of such inelastic collisions, it seems that one is back with the demand for *active principles*. Newton, for example, felt that the postulation of *hard* atoms was necessary to account for the constancy of Nature,[81] yet saw that the collision of such rigid bodies must of necessity be inelastic: 'for Bodies which are either absolutely hard, or so soft as to be void of Elasticity, will not rebound from one another'.[82] In the collision of two such hard bodies momentum ($\Sigma m\vec{v}$ is conserved, but *neither* Descartes' 'quantity of motion' ($\Sigma m \mid v \mid$) nor Leibniz's *vis viva* (Σmv^2). Hence, Newton continues,

> by reason of the Tenacity of Fluids, and Attrition of their Parts, and the weakness of Elasticity in Solids, Motion is much more apt to be lost than got, and is always upon the decay.[83]

> Seeing therefore the variety of Motion which we find in the world is always decreasing, there is a necessity of conserving and recruiting it by active principles.[84]

We identify such agents only by their effects, and characterise them as the *causes* of, e.g. gravity and fermentation. If it were not for these principles, says Newton,

> the Bodies of the Earth, Planets, Comets, Sun, and all things in them, would grow cold and freeze, and become inactive masses; and all Putrefaction, Generation, Vegetation and Life would cease, and the Planets and Comets would not remain in their Orbs.[85]

[81]   Newton, 2, p. 400.

[82]   *Ibid.*, p. 398.

[83]   *Ibid.*, p. 398. See also Snow, pp. 156–157.

[84]   *Ibid.*, pp. 399–400.

[85]   *Ibid.*, p. 400.

As regards the exact ontological status of these agents, Newton is *far* from certain:

> . . . what that principle is, and by means of [what] laws it acts on matter, is a mystery; or how it stands related to matter is . . . difficult to explain.[86]

Never was a truer word spoken: the precise nature and status of Newtonian 'active principles' has baffled commentators as much as, perhaps, it bewildered Newton himself.[87] He seems, at different times in his life, to have proposed a variety of different conjectures: direct divine action, incorporeal 'powers' implanted *in* bodies (though not *of* them), various forms of subtle matter (the 'spirit' of the General Scholium; the 'aether' of the Queries) without ever finding a really satisfactory solution. To invoke God as the cause of all the forces discoverable in Nature is to invite the Leibnizian accusation of 'perpetual miracle'; to posit 'powers' in certain bodies that cannot be explained in terms of their own natures is, in the jargon of the mechanists, to postulate 'occult qualities'. Theories of subtle matter seem to fare little better: it is not easy to explain how the incredibly tenuous aether of the *Opticks* could ever push around such massive bodies as planets and comets. In any event, the ability of the 'active principles' to *initiate* new motion (thus restoring that lost through the collisions of hard bodies) is clearly a power that *transcends the merely mechanical*:[88] in our terminology, Newton is no mechanical philosopher. He retains, importantly, the mechanists' conception of body as something inert (and yet re-active), pushed around by external forces; he diverges from the Mechanical Philosophy on the nature and source of those forces.

For an example of a true mechanical philosopher, resolutely opposed to the admission of active principles within the (phenomenal) realm of physics, we have only to turn to Leibniz. The proper measure of 'motive force', he claims, should be the ability of a moving body to do work, i.e. to raise a weight.[89] Now if work = weight × height (w × h)

---

[86] Quoted from McMullin, 4, p. 46 (from a draft for Query 23 of the *Opticks*).

[87] For an interesting study, see McGuire, 2.

[88] For this issue, see McMullin, 4, p. 2; McGuire, 2, pp. 184–186; Snow, p. 83; Koyré, 4, p. 146, and many others.

[89] See the *Specimen Dynamicum* in Leibniz, pp. 442–443.

and, by Galileo's law of free fall, $v^2 \propto h$, it follows that motive force is proportional to v^2: the proper measure for this *vis viva*, Leibniz insists, is not $m \mid v \mid$ nor $m\vec{v}$ but mv^2. This quantity, he claims (roughly equivalent to our 'kinetic energy'), measures the ability of a moving body to do work, and the total quantity of *vis viva* in the universe, Σmv^2, is *conserved*. In every physical action, *vis viva* is conserved: the cause must always have the same 'force' (energy) as the effect.[90] For the effect to have more *vis viva* than the cause, or less, is impossible: the first alternative gives one 'something for nothing' (a perpetual motion machine), the second gives the decaying, running-down world-machine, which Leibniz finds unintelligible. By construing *vis viva* as something quasi-substantial, constitutive of the nature of bodies, he is enabled to assert its conservation *a priori* with more confidence than the empirical evidence alone would warrant. When Clarke claims, in the course of their celebrated correspondence, that motion is lost in the collision of inelastic bodies, Leibniz responds that

> I answer no. 'Tis true, their wholes lose it with respect to their total motion; but their parts receive it, being shaken [internally] by the force of the course. And therefore that loss of force is only in appearance.[91]

When Clarke replies that *hard* bodies cannot, in virtue of their rigidity, undergo such internal agitation of their parts, Leibniz's reply is, in effect: so much the worse for hard bodies. The existence of perfectly rigid bodies, he claims, would violate two of the great principles of Natural Philosophy, (a) continuity, and (b) conservation of *vis viva*. We may therefore be confident that no such bodies exist. In an apparently inelastic collision such as that of the two lumps of putty mentioned earlier, the motion lost to the wholes is simply transferred to their constituent parts. (One could, in principle, use the kinetic theory of heat to lend empirical support to this claim, but Leibniz misses this point.) All collisions, therefore are really 100% elastic in the sense that no *vis viva* can ever be lost in the course of Nature. (Leibniz berates Newton for making God such an inept clockmaker as to need periodically to 'wind up' His great timepiece.) On the

---

[90]   See Costabel, p. 50, p. 113. The $\Sigma mv^2 = k$ principle was derived by Leibniz from Huygens, but invested by the pupil with a significance it did not have for the teacher.

[91]   H. G. Alexander, Ed., p. 87.

Leibnizian view, *physics* has no need of active principles, but is purely mechanical. (He admits, of course, active principles in metaphysics.) Leibniz's clockwork universe, once wound up (endowed with its proper quantum of *vis viva*), will run forever without requiring 'rewinding' (periodic 'injections' of fresh motion).

B. Action at a Distance

The founders of the Mechanical Philosophy were unanimous in their denial that any body can act upon another at some distance from it. Our concept of a body, they argued, is of a solid, extended, and essentially inert and passive thing: it is quite unintelligible that it should somehow 'reach out' and affect things beyond its bounds. The sympathetic magic and astral influences of the occultists must therefore either be mechanised or denied (for examples of the mechanisation of the occult, see Appendix 6).

Impulsion, says Descartes, is the only form of physical action explicable in terms of our concept of body, hence the only intelligible mode of physical agency. Given (a) that the essence of matter lies in 3D extension, and (b) that the total quantity of matter in the universe is conserved, we can explain the impenetrability of bodies (interpenetration would violate the conservation law). This, he continues, explains the ability of one body to impart an impulse to another on contact,[92] which is the *only* mode of physical action of which we have a clear and distinct idea. As Kemp Smith says, Cartesian natural philosophy serves to rule out 'as being unintelligible, and for us meaningless, explanation in any other fashion';[93] but, since our clear and distinct ideas mirror the structure of Nature, it follows that 'the only causes operating on the Earth or in the heavens are impact and pressure'.[94]

Action at a distance (A.D.), Descartes argues, entails animism: only a soul can know what goes on at some distance from its point of embodiment. When Roberval[95] suggested that gravity is a force of attraction (effectively, A.D.), Descartes replied sarcastically that this would make every

---

[92]   This must, of course, be interpreted in an 'as if' manner compatible with Descartes' occasionalism.

[93]   Kemp Smith, p. 94.

[94]   *Ibid.*, p. 94. Cf. *Principles*, 2, 64.

[95]   For Roberval's attraction theory, see Westfall, 4, p. 265 ff.; Sabra, p. 149 ff.; Koyré, 4, p. 59.

stone and clod of earth animate, indeed, 'vraiment divine': only God knows the positions and masses of all the heavy bodies in the universe.[96] The attraction theory is, therefore, 'most absurd' – it tacitly attributes to bodies powers that can only belong to souls, and must therefore be dismissed *a priori* as unintelligible and hence unreal.

Hobbes is equally blunt: throughout his works, from the 'Little Treatise' of around 1630 to the *De Corpore* of 1655, the denial of A.D. is treated as axiomatic and self-evident. It is indeed stated as an axiom in the 'Little Treatise' that 'That which is in no way touch'd by another, hath nothing added to nor taken from it'[97] by that other: it immediately follows that 'Every Agent that worketh on a distant patient, toucheth it, eyther by the Medium, or by somewhat issueing from it self, which thing so issueing lett be call'd Species'.[98] In the *De Corpore*, likewise, it is asserted flatly that 'there is no cause of motion in any body, except it be contiguous and moved'.[99] A.D. is simply unintelligible, hence, for any good rationalist, unreal.

Gassendi lends his voice to the same refrain: in gravitational or magnetic 'attraction', he asks,

is it not necessary that something act as a medium since no physical action takes place without a physical agent, and no physical agent can act upon a distinct object except through some intervening instrument?[100]

Since all physical action is *impulsion*, some corporeal intermediate must account for those natural forces which seem to involve 'attraction':

When I say impulsion I make no exception of attraction; for to attract is nothing other than to impel toward oneself by means of a curved instrument;[101]

or, as Charleton says, it is a 'law of Nature' that

nothing can act upon a distant subject, or upon such whereunto it is not actually present, either by it self, or by

---

[96] See the letter to Mersenne of 20 April 1646 in Kenny, Ed., p. 191.

[97] From Brandt, p. 13.

[98] *Ibid.*, p. 14.

[99] Hobbes, *Works*, Vol. 1, p. 125. See also Brandt, p. 281.

[100] Gassendi, p. 132.

[101] *Ibid.*, p. 142. See also Koyré, 7, p. 148; Dijksterhius, p. 428; Westfall, 4, p. 101.

some instrument . . . and consequently that no body can move another, but by contact mediate, or immediate.[102]

The sparsity of *argument* for this fundamental thesis testifies to its status, for all these thinkers, as self-evident and indubitable: it is, they feel, a proposition that has only to be understood to be endorsed by the 'natural light' of Reason, something that can be laid down as an unshakeable axiom of any intelligible physics.

What are the *explananda* that forces of attraction might naturally be evoked to explain? Once one accepts that weight is not innate in and intrinsic to a body, an obvious suggestion is that it is an attractive force exerted by the Earth on the heavy body: gravity thus becomes *explanandum* number one. Contrasting with this long-range force, we have a variety of short-range ones, namely electricity and magnetism, capillarity, cohesion, surface tension, and a variety of optical phenomena. These become the crucial *problem-areas* for the Mechanical Philosophy: it is enlightening, as Westfall shows, to see the work of Newton as a lifelong investigation of such problem-areas, a probing, as it were, into the possibilities and limitations of mechanical explanation.[103]

Now that we have our list of *explananda*, a mixed bag of forces of various kinds, we can attempt to account for them. It is important here to keep the term 'force' a theory-neutral concept, non-committal between diverse accounts of the means by which various forces operate. We can speak of electric or magnetic force, measure its strength, and study its attenuation with distance, without knowing (or, if one is an extreme positivist, caring) *how* it works. When I speak of forces, then, I do so in this theory-neutral way: 'attraction', by contrast, is a much more committed term, giving some indication of the mechanism (or rather, lack of it) involved. We have therefore a list of *forces* as *explananda*: it was the aim of the mechanists to account for all those forces *without* invoking any irreducible concept of attraction, i.e. without postulating A.D. To do this made considerable demands on their ingenuity, but they responded heroically to the challenge: by the end of the century there existed a multifarious collection of weird and wonderful hypothetical mechanisms to explain one force or other. Although often far-fetched, the postulation of such mechanisms was far from

---

[102] Charleton, p. 343.

[103] Westfall, 5, p. 141.

irrational: given the (plausible) *a priori* conviction of the impossibility of A.D., some such account of, e.g., magnetism had to be true.

Let us now run through a (strictly limited) selection of these hypothetical mechanisms, starting with some of the simpler models.

a. Elastic threads

Gravity, said Kepler, is caused by quasi-magnetic, elastic 'threads', emitted radially from the attracting body. It is not clear how literally Kepler himself took this model, but in Gassendi's work it is taken very seriously indeed. In his treatise *De Motu* he discusses at some length the nature and rôle of these magnetic 'chains' or 'cables', which hook on to heavy bodies and drag them back to Earth.[104] Some of the implications of this model are very striking. If, for example, the 'threads' are emitted *radially*, it will follow that the strength of the gravitational pull exerted by a given body will diminish with distance from it. (Gassendi sees this implication quite clearly.[105]) One could even, from such a model, derive an inverse-square law for the attenuation of gravity with distance! (In an ingenious series of experiments involving a spring balance and a pendulum clock, Hooke attempted to measure the diminution of weight with altitude, but found it too slight to be detected.[106] Another possible consequence is a crude analogue of the proportionality of mass and weight, giving an explanation of the Galilean thesis that all heavy bodies fall (under ideal conditions) with the same speed. There being, says Charleton, 'a certain commensuration betwixt the force attractive, and the quantity of matter attracted', Galileo's law will follow as a necessary consequence.[107] In fact, however, as Westfall sees,[108] this mechanical model would give at best a crude approximation to Galileo's law: only if (a) all the atoms in a body were identical in mass, (b) each atom has one and only one hook for the attachment of a thread, and (c) the tension in the threads remains invariant, would the 'mass α weight' thesis, and Galileo's law of free fall, be derivable from the model.

---

104 Gassendi, p. 132 ff.

105 *Ibid.*, p. 136. See also Charleton, p. 284.

106 See Gunther, Ed., Vol. 6, pp. 156–159.

107 Charleton, p. 285. See also Gassendi, p. 138.

108 Westfall, 4, p. 110.

Yet another important implication of the 'elastic threads' theory (which must be seen as a most *fruitful model*, for all its shortcomings) is the *mutuality* of, e.g., gravitational and magnetic forces. Kepler saw this point quite clearly: the tension in an elastic thread linking two bodies pulls, he says, equally on each, but (of course) moves the less massive body more than it does the greater one.[109]

The great difficulty of the 'elastic threads' theory lies in the fact that both gravitational and magnetic influences (i.e. those forces which it was principally invoked to explain) pass so freely through such great masses of apparently solid matter. (Gilbert, not implausibly, saw this as a certain proof of the immateriality of magnetic action.[110]) Gassendi explains that the threads are so subtle, and even 'solid' matter so porous, that the threads can pass freely through one body and grasp another one behind it: or, as Charleton says in reply to Gilbert, 'the particles flowing from the earth and loadstone, are of such superlative tenuity, as without impediment to penetrate and permeate the most compact and solid concretions'.[111] One cannot find this altogether satisfying: corporeal chains, one feels, ought to be impeded somewhat by corporeal barriers.

b. 'Tongues'

This theory is really very similar to the first, except that, instead of hooking on and pulling, as the elastic threads do, a 'tongue' of subtle matter curls round behind the attracted body and *pushes* it. This hypothesis clearly owes a debt to Gilbert's effluvium-theory of electrical 'attractions', and was widely used by mechanical philosophers to explain just those phenomena. A good model for electrical attraction, Charleton suggests, would be the tongue of a chameleon;[112] given sufficiently acute eyesight, Gassendi suggests, we should be able to observe the operation of this prehensile 'tongue'.[113]

c. Corporeal 'species'

Each object, on this theory, emits highly specific effluvia or (corporeal) 'species' into the ambient medium, which species

[109]   See Jammer, 3, p. 83, p. 86.

[110]   See Chapter 8, p. 372.

[111]   Charleton, p. 405.

[112]   *Ibid.*, p. 345.

[113]   See Brett, p. 80.

voyage on through that medium until they encounter other bodies. On meeting a 'kindred' body, such a species will excite in the subtler, more 'spirituous' parts of that body a sort of perception or perception-analogue, which in turn 'motivates' the body to return in the direction of the original source of the species. Some such theory, it is clear, must account for animal motion, but the basic pattern of explanation was extended, even by avowed mechanists like Hobbes and Gassendi, to the inanimate realm. Hobbes sketches such a theory in his 'Little Treatise';[114] in Gassendi, we can find such a quasi-animistic account of magnetism.[115] (On such a view, there is a strong positive analogy between the 'attraction' of a magnet for iron and that of cheese for a mouse.) Let Charleton explain:

> As a sensible object, that is convenient and grateful, doth by its species immitted into the sensory of an animal, convert, dispose, and attract the soul of the animal; and its soul being thus converted disposed and attracted toward that object doth by its virtue or power, carry the body, though gross and ponderous, along to the same: exactly so doth the loadstone seem, by its species transfused, to convert, dispose, and attract towards it the (as it were) soul, or spiritual substance of iron; which doth instantly by its power or vertue, move and carry the whole mass.[116]

This presupposes the existence in iron of a subtle matter or 'soul', or perhaps, as Charleton says, something which, 'though it be not perfectly a soul, is yet in some respects analogous to a soul', i.e. partaking of soul-like powers of 'perception' and 'volition'. In virtue of their pleasant and agreeable 'contexture', says Charleton, the magnetic species act upon the subtler parts (the 'soul') of the iron, and 'turn them about toward that part, from whence themselves are derived', i.e. back towards the magnet. While this eliminates A.D., it still has a strongly animistic air, and to produce a satisfactory mechanistic account of the stimulus-response mechanism proved no easy task.

d. Resonance

Each and every body, say men such as Hobbes and Hooke, has a highly specific state (i.e. frequency) of agitation proper to it, in virtue of which it produces vibrations of a particular

114 See Brandt, pp. 30–31.
115 See Brett, pp. 93–94; Spink, p. 94; Snow, pp. 19–21.
116 Charleton, p. 389.

frequency in the ambient medium. (The whole theory rests very heavily on the mechanical theory of *sound* developed by Galileo and Mersenne.) The phenomenon of resonance illustrates quite clearly the powerful physical effects that may be produced by such vibrations: the right note can shatter glass, flatten walls, shake buildings, etc.[117] (We are not just talking about lutestrings). When a train of pulses of frequency f falls upon a body naturally suited to vibrate at that frequency, the pulses will 'corroborate' its own motions and thus intensify them; if, on the other hand, the body is tuned to oscillate at a different frequency f*, the two sets of vibrations will 'cross and jar', and thus damp, one another.

This gives Hobbes his basic model for explaining the stimulus-response mechanism, and hence animal (and human) actions. The impact of corporeal species or pressure-pulses[118] on our sense-organs produces, he explains,[119] agitations in the animal spirits in our nerves. These agitations pass to the heart, where they either 'corroborate' or interfere with its own vital (Harveyan) motions, producing thereby what we call the sensations of pleasure and pain respectively; these, in turn, give rise to the characteristic responses of attraction and aversion. (The animal spirits are pumped into the muscles, and thus bring about a response.)

Like the 'corporeal species' theory, however, this may seem to import too much animism into its explanatory structure: to use such a model to explain gravity, for instance, would involve attributing perception-analogues to every stone and clod of earth. It is tempting, then, to seek a purely external way in which resonance effects could bring kindred ('attuned', 'congruous') bodies together. Hobbes uses for this purpose what is, effectively, merely an updated version of the winnowing-basket image of Plato's *Timaeus*.[120] A body with a 'simple circular motion' (not a rotation but a movement like that involved in using a sieve) tends he says, to produce a periodic agitation or 'fermentation' in the ambient medium, which agitation

---

117   Resonance-effects provide many of Boyle's best examples in his monograph on 'The Great Effects of even languid and unheeded Motion' (Boyle, *Works*, Vol. 5, p. 1 ff.).

118   Hobbes himself abandoned the 'species' theory of the 'Little Treatise' in favour of the mediumistic theories that are to be found in his later works.

119   Hobbes, *Works*, Vol. 1, pp. 406–408.

120   See chapter 4, p. 147.

congregates or gathers into one place such things as naturally float in that medium, if they be homogeneous; and if they be heterogeneous, it separates and dissipates them.[121]

Heterogeneous bodies, he explains,

receive unlike and different motions from the atoms external common movent; and therefore they will not be moved together, that is to say, they will be dissipated. And being dissipated they will necessarily at some time or other meet with bodies like themselves, and be moved alike and together with them; and afterwards meeting with the more bodies like themselves, they will unite and become greater bodies. Wherefore homogeneous bodies are congregated, and heterogeneous dissipated by a simple motion in a medium where they naturally float.[122]

This process of 'fermentation', he claims, explains not only the general tendency of like to associate with like, but also, and more specifically, gravity[123] and magnetism.[124]

For an extension and development of Hobbes' views, we must turn to Hooke's account of 'congruity' and 'incongruity' (mechanistic surrogates for 'sympathy' and 'antipathy') and his conception of music-theory as the key to many of the secrets of Nature.[125] By the terms 'congruity' and 'incongruity', he explains, 'I understand nothing else but an agreement or disagreement of Bodys as to their Magnitudes and motions',[126] and, in particular, as to their proper or native frequencies of vibration. A sensible body is, he says, 'a determinate Space or Extension defended from being penetrated by another, by a power from within',[127] which 'power' is due to the incredibly rapid oscillation of a much smaller body. (One can imagine, he says by way of explanation, that a flat plate one foot square could, by a rapid vibration, 'defend' a cubic foot of space, which would thus constitute a

---

121 Hobbes, *Works*, Vol. 1, p. 323.

122 *Ibid.*, p. 324.

123 *Ibid.*, p. 511 ff. See also Vol. 7, p. 7 ff.

124 *Ibid.*, pp. 527–528.

125 See Gouk, p. 585 ff.

126 Gunther, Ed., Vol. 8, p. 339.

127 *Ibid.*, p. 340.

sensible and impenetrable body of that magnitude.) 'This Vibrative motion,' he continues,

> I do not suppose inherent or inseparable from the Particles of body, but communicated by Impulses given from other bodies in the Universe. This only I suppose, that the Magnitude or bulk of the body doth make it receptive of this or that particular motion that is communicated, and not of any other.[128]

The magnitude of a body, like the thickness and tension of a wire, give it a particular frequency of vibration 'natural' to it. Now 'congruity' of vibrations (unison, or harmony) involves their mutual reinforcement, while 'incongruity' (discord) will produce damping effects. The universal subtle matter transmits these periodic vibrations across space, thus allowing 'congruity' to work at a distance. It is in terms of this concept that Hooke casts his account of gravity. I suppose, he says,

> that there is in the Ball of the Earth such a Motion as I, for distinction sake, will call a Globular Motion, thereby all the Parts thereof have a Vibration towards and fromwards the Center, or of Expansion and Contraction; and that this vibrative Motion is very Short and very quick, as it is in all very hard and very compact Bodies: That this vibrative Motion does communicate or produce a motion in a certain part of the Aether, which is interspersed between these solid vibrating Parts; which communicated Motion does cause this interspersed Fluid to vibrate every way in Orbem, from and towards the Center, in Lines radiating from the same. By which radiating Vibration of this exceeding fluid, and yet exceeding dense Matter, not only all the Parts of the Earth are carried or forced down towards the Centre; but the Motion being continued into the Aether, interspersed between the Air and other kinds of Fluids, it causeth these also to have a tendency towards the Center; and must move any sensible Body whatsoever, that is anywhere placed in the Air, or above it, though at a vast distance.[129]

Centore calls this theory original to Hooke, but it is clearly a variant on the idea discarded by Hobbes, that a light-source has a periodic motion like the systole and diastole of the

---

[128] *Ibid.*, p. 340. See also Vol. 13, p. 15.

[129] Quoted from Centore, p. 104.

heart.[130] As Hooke sees, this model will give an effect that attenuates with distance, and that gives rise to a centrally-directed force (its discrete periodic impulses will give a good approximation to Galileo's v α t law). The only – and crippling – difficulty lies in explaining the *directionality* of the push. It is easy to explain how resonance enhances the innate vibration of a body, but not how it could give rise to a net centripetal force.

The 'resonance' theory of gravity will, as Westfall[131] sees, naturally yield a vision of particular gravities rather than of universal gravitation: the concept of congruity, like that of sympathy, is inherently *specific* rather than general in scope. On such a view, for example, each planet will send out pulses at its own characteristic frequency, which pulses will serve to drive kindred bodies (i.e. those sharing a common, or harmonious, frequency) back to their source. (One can, of course, posit 'harmonies' as and where required, e.g. to account for the Lunar influence over the tides.)

e. Currents of subtle matter

The classical mode of mechanical explanation of forces was, of course, to posit streams of some subtle matter or other. Before giving our attention to the Cartesian theory of vortices, let us briefly mention one or two other examples of the genre. The prize for sheer ingenuity must go to Descartes, for his explanation of the phenomena of magnetism[132] in terms of streams of screw-shaped particles, sufficiently subtle to pass through the pores of most substances without effect, yet so shaped as to work all the variety of magnetic effects on iron. Though undeniably a brilliant *tour de force* of mechanistic thinking, the account cannot escape an air of contrivance and fictionalism: seventeenth as well as twentieth century readers clearly experienced these misgivings. For the members of the early Royal Society, Descartes was a 'wit', a man of undeniable genius yet questionable judgement: his hypotheses *may* be true, but are certainly not well founded.

Another example that should be mentioned is the 'descending aetherial shower' account of gravity of Digby and the young Newton. The sun's heat, said Digby,[133] raises steams, vapours, and subtle matters of various kinds from the Earth,

130   For Hobbes' systole-diastole theory, see Brandt, p. 47, p. 105.

131   Westfall, 3, pp. 246-247.

132   See Descartes' *Principles*, 4, 133-187.

133   See Digby, p. 94 ff.

giving rise effectively to a great convection current: the returning downward stream produces by its pressure on bodies the effect we experience as weight. As an account of gravity, it has to be admitted, this is petty feeble: on such a view, the weight of a body would depend significantly on such factors as climate, time of day or night, etc. Digby attempts to meet some of these obvious objections, but without notable success. The 'descending aetherial shower' theory is, however, independent of Digby's 'convection-current' theory of its origins: in his early Notebooks, the young Newton adopts such a theory and investigates some of its possible implications.[134]

f. The vortex theory

The great advocates of subtle-matter theories were of course the Cartesians. According to Descartes,[135] our Earth is at the centre of a swirling vortex of subtle matter, composed of the two finer 'elements' E1 and E2. The bulkier E3 corpuscles which constitute terrestrial matter share in the diurnal rotation of the Earth, but the whirling subtle matter has a much higher angular velocity ω. Huygens, working from the values of g at different latitudes, estimated it at 17 times greater. The centrifugal force produced at the equator by the Earth's spin subtracts, he discovered, about 1/289 of the weight of a heavy body; but centrifugal force varies as $mr\omega^2$, so if ω were 17 times greater – i.e. if the Earth spun 17 times as fast – gravity at the equator would fall to zero. This provides the definitive answer to one of the classical objections to the Earth's rotation.[136]

In virtue of their low angular velocity ω compared to that of the subtle matter, particles of E3 experience *less* centrifugal force than those of E1 and E2. Assuming then that the whole vortex is bounded (by others), which prevents all its parts flying off on tangential paths, it follows that E3 particles will experience a *net* force towards the centre of the vortex. Although, as Descartes says, all the parts of our Earth, considered singly, 'are not heavy, but light'[137] (i.e. possessed of centrifugal force owing to the Earth's rotation), the greater

134   Westfall, 2, pp. 329–331. According to Rattansi (in Debus, Ed., 7, Vol. 2, p. 176), this theory bears a strong resemblance to that of Digby.

135   *Principles*, 4, 20; Descartes, 1, vol. 1, p. 281.

136   See Westfall, 4, p. 172. For these theorems in *Principia*, see Newton, 1, pp. 425–426.

137   *Principles*, 4, 21; Descartes, 1, Vol. 1, p. 281.

centrifugal tendency of the rapidly whirling aether explains why grosser bodies are pressed towards the centre, 'and thus become heavy'. The weight of a gross body (i.e. one compounded of corpuscles of E3) is, therefore, *not* an intrinsic or essential feature of it, but is due entirely to the external pressure of the aether. (External aether-pressure is also invoked, by many Cartesians, as the only conceivable explanation of the *cohesion* of bodies.[138])

The main difficulties of the vortex theory of gravitation are the following. In order to account for the weight of bodies in buildings, and, worse still, in deep mines, the Cartesian must assume that aether can pass unimpeded through great masses of dense matter. Yet, if it can do this, how can it exert a powerful downward *pressure* on them? – incompatible demands seem to be build into the theory from the outset. (This problem of the activity of the aether was endemic in Cartesian physics: it did not affect only the theory of weight.)

The other great problem concerns the precise structure of the vortex. Gravity, says Descartes quite correctly, 'depresses bodies to the centre of the earth',[139] But it is difficult to envisage a vortex capable of producing such a force. A common 'whirlpool' sends matter to its axis – i.e. to a *line* rather than to a point. In such a vortex, the force of gravity at angle of latitude α would be at that same angle α to the vertical – at the poles, there would be no downward force at all, only a horizontal one. The vortex-theory, its critics allege, would naturally give a cylindrical earth rather than a spherical one, with a force of gravity directed to the *axis* about which the cylinder rotates.[140] Similarly, the swirling aether should cause a horizontal displacement of the falling body as well as the vertical one: it is difficult for the Cartesian to explain why no such displacement is observed. Such problems occupied the defenders of the 'Cartesian' theory, Huygens and Perrault, at the great session of the Académie des Sciences in 1669, specifically devoted to the issue of the nature of

---

138 For a study of this difficult subject, see Millington. One important champion of the aether-pressure theory of cohesion was Hooke: 'All bodies whatsoever would be fluid,' he claims, 'were it not for the external Heterogeneous motion of the ambient' (Gunther, Ed., Vol. 8, p. 344).

139 Descartes, 1, Vol. 1, p. 281.

140 Men as diverse as Robert Hooke and Henry More rejected the Cartesian vortex theory on these grounds: for More, see his 'Antidote against Atheism', p. 39, in Vol. 1, of his *Works*; for Hooke, see Centore, pp. 72–73.

weight.[141] To overcome their difficulties, they found themselves adding greatly to the complexity of their account, multiplying vortices of a bewildering variety of shapes and sizes in the (vain) attempt to account plausibly for the undeniable existence in Nature of a single centrally-directed force. None of the resulting accounts, for all the brain-power invested in them, can be considered wholly successful.

It is time now to turn to the treatment of gravitation in Newton's *Principia*, and the conception of weight implicit in that monumental work. At various places in the text,[142] Newton professes a very positivistic attitude towards his work: he is attempting, he says, only to describe the *mathematical* laws observed by the phenomena, omitting entirely all consideration of their physical causes. When I speak of 'attraction', he says, I do not mean to exclude the possibility that such forces are really due to *impulses* of some kind. The overt professions of a positivist methodology and a complete agnosticism and neutrality concerning causal issues are, however, quite bogus: throughout his life, Newton was obsessed by questions about the causes responsible for the manifest phenomenal laws, and produced a variety of widely differing speculative accounts of the nature and manner of operation of these agents. Reading between the lines in *Principia* itself, we must take note of a number of features of the phenomenal world which (a) constrain possible causal hypotheses, and (b) perhaps even indicate where the cause of gravity *must* lie:

1.   Gravity, says Newton, *may* be due to a mechanical cause of some kind. Yet he also shows, quite clearly, that interplanetary space and the Boylian vacuum are all but 100% void of matter: if there is any matter in such spaces, it is so extremely tenuous as to exert no detectable resistance to the passage of bodies through it.[143] But such an incredibly subtle medium cannot be cited as the cause of gravity: the gravitation of heavy bodies in an evacuated vessel, or of planets in space, cannot readily be explained by such means. (Of course, one can attribute astronomical velocities to aether-particles to compensate for their minute masses, but the whole thing has a very *ad hoc* ring to it – as also, it must be confessed, has the aether-theory of gravity in the *Opticks*.) Once the dense Cartesian aether is banished from the

141   See Aiton, 8, p. 75 ff.

142   Newton, 1, pp. 5–6, p. 192.

143   For this argument, see Chapter 10, pp. 482–491.

universe, there is nothing in interplanetary space capable of imparting a sufficient mechanical impulse to move the massive planets.

2. The weight of any body, Newton shows by means of pendulum-experiments, is directly proportional to its mass: 'by experiments made with the greatest accuracy,' he asserts, 'I have always found the quantity of matter in bodies to be proportional to their weight',[144] factors such as texture and surface area being quite irrelevant. This, he feels, refutes all theories of an aetherial downdraught or pressure (including, of course, his own early view of a descending aether-shower): on any such theory, one would expect the arrangement of the solid parts of a body to make a significant difference to its weight, yet this is simply not the case.[145] The cause of gravity, then, 'knows' exactly how much matter (mass) there is in each body: it is capable of penetrating freely through the most solid of bodies and exerting its pull on each and every constituent atom thereof.

3. To explain the tendency of terrestrial bodies towards the Earth, or of the parts of Jupiter towards that planet, lies within the compass of the vortex theory (each planet, one can say, is situated at the focus of its own vortex of subtle matter): to explain the *universal gravitation*[146] of matter towards matter is quite beyond it. To envisage an arrangement of vortices competent to account of the 'attraction' of every particle of matter in the universe for every other is quite out of the question. Small wonder then that men such as Huygens and Leibniz, still working in a broadly (though much modified) Cartesian tradition, found universal gravitation too much to swallow. It is a 'fantastic assertion'[147] (Huygens), or a 'strange imagination'[148] (Leibniz), to assume that every body in the universe gravitates towards every other – the assertion, they claim (quite justifiably), goes far beyond what is warranted by the available evidence. To accept universal gravitation is tacitly to abandon all hope of a credible vortex theory: while not a refutation in the strictest sense, it is

---

144 Newton, 1, p. 304. Huygens, it is interesting to note, accepted the proportionality of mass and weight, but did not see the damaging implications of this thesis for the vortex theory of weight.

145 See Chapter 10, p. 502.

146 For the grand generalisation by which Newton arrives at the principle of universal gravitation, see *Principia*, p. 549 ff.

147 See Koyré, 4, p. 116 ff.

148 From H. G. Alexander, Ed., p. 66.

nevertheless a death-blow to the whole explanatory programme of the vortex theory.[149]

4.   Newton demands, as Huygens sees, conservation of
momentum ($\Sigma m\vec{v} = k$) among *gross* bodies, e.g. among the
Sun, planets, and moons that make up our solar system. This
is implicitly to deny that those bodies are impelled by currents
of subtle matter: a Cartesian such as Huygens could happily
admit a violation of the conservation law for gross bodies, so
long as the apparent loss or gain of momentum was
compensated for by an equal gain or loss by subtle matter. If
an aether-current gives a planet an impulse $+ \Delta mv$, it must
itself suffer an equal and opposite impulse $- \Delta mv$. (Newton's
third law is thus implicit in the principle of conservation of
momentum, when combined with conception of force as
Δmv: if $\Sigma m\vec{v} = k$, it follows that $+ \Delta mv = - \Delta mv$, i.e. that
'action' and 'reaction' are equal and opposite.) But, if
$\Sigma m\vec{v} = k$ holds for *gross* bodies alone, the natural conclusion
to draw is that the force drawing the Moon towards the Earth
must *also*, and simultaneously, be drawing the Earth towards
the Moon: as Newton himself says, these 'two' forces are
really one and the same.[150] (The elastic threads model is of
value here: if such a thread links two balls, the *same* tension
in the thread pulls equally on both, and produces in each an
acceleration inversely proportional to its mass.) This *mutuality* of gravitational attraction is, as critics and disciples of
Newton alike saw, quite unknown on the vortex theory.[151]

5.   In Newtonian physics, gravitational forces are always (a)
directed towards massive bodies, and (b) proportional in
strength to the masses of those bodies. *Neither* of these
principles need hold in a vortex theory. If the Earth were
annihilated or vaporised, a stone placed in the terrestrial
vortex would still, for a Cartesian, be despatched to the
centre of the vortex, whether or not any gross body were
situated at that spot. What (if anything) occupies the focus is
irrelevant: what counts is the pressure of the swirling aether,
which is quite independent and 'unaware' of what lies at the
centre of the vortex. The $F = GMm/r^2$ formula, then,
according to which the weight of the attracted body depends
on the mass of the attracting one, is thoroughly anti-Cartesian
in its implications – as disciples such as Cotes and critics such

149   See Westfall, 4, p. 465.

150   Newton, 1, p. 569.

151   See Koyré, 4, pp. 278–279.

as Huygens both saw, it does seem to involve a genuine and irreducible concept of *attraction*.[152]

In *Principia*, then, gravity is *not* to be conceived as due to the impulse of a subtle aether: 'attraction' is not eliminable in favour of 'impulse', and the avowed agnosticism about causes is quite disingenuous. Newton was far too intelligent not to see the implications of his principles: at a number of points in the text of *Principia* we find anti-Cartesian asides (often in the form 'let them account for *that* in terms of vortices'[153]), while Book 2 of the *magnum opus* can profitably be read as a sustained attack on the very foundations of Cartesian physics (the aether, the vortex theory, the pulse theory of light).

What, then, were Newton's positive views, at the time of writing of *Principia*, on the subject of the cause of gravitation? Clearly, he believes that a real force of attraction, irreducible to any variety of impulsion, is involved. Equally clearly, he holds the conviction common to all the mechanists that brute matter cannot act at a distance. (Strictly, it cannot act at all: its only minimal degree of activity lies in the power of resistance and re-action inherent in its inertia.) In a letter to Bentley, Newton issues the following warning:

You some times speak of gravity as essential and inherent to matter. Pray, do not ascribe that notion to me; for the cause of gravity is what I do not pretend to know, and therefore would take more time to consider of it.[154]

And, still more strongly,

It is inconceivable, that inanimate brute matter, should, without the mediation of something else, which is not material, operate on and affect other matter without mutual contact . . . That gravity should be innate, inherent, and essential to matter, so that one body may act upon another at a distance through a vacuum, without the mediation of any thing else, by and through which their action and force may be conveyed from one to another, is to me so great an absurdity, that I believe no man, who has in philosophical matters a competent faculty of thinking, can ever fall into it. Gravity must be caused by an agent acting constantly according to certain laws; but whether

---

[152]   *Ibid.*, p. 153.

[153]   See, e.g., Newton, 1, p. 394. The *explanandum* in question is Kepler's third law, $T^2 \propto r^3$.

[154]   Quoted form Cajori's Appendix to Newton, 1, p. 633.

this agent be material or immaterial I have left to the consideration of my readers.[155]

The clear implication of the passage is that the cause of gravity must be immaterial in nature. If weight is *not* produced by the external impulse of any subtle matter, *nor* by an ability to act at a distance innate in matter, it seems that we must seek some non-material principles in order to account for it. The only discussion in *Principia* itself of this issue occurs in the celebrated General Scholium added to the second edition. Much of the scholium is devoted to a rhapsodic eulogy of the 'Lord God PANTOKRATOR' and His absolute dominion over His Creation:[156] it is striking that, with this great hymn of praise to the Deity and His *power* still ringing in our ears, Newton switches abruptly to the question of the cause of gravity, and immediately makes the following point: gravity, he says,

> must proceed from a cause that penetrates to the very centres of the Sun and planets, without suffering the least diminution of its force; that operates not according to the quantity of the surfaces of the particles upon which it acts (as mechanical causes used to do) but according to the quantity of the solid matter which they contain, and propagates its virtue on all sides to immense distances, decreasing always as the inverse square of the distance.[157]

Although Newton may at this point refrain from drawing the obvious conclusion with his famous slogan, 'I feign no hypotheses', the reader is left in no doubt as to the author's real views. Universal gravitation, says Bentley, is

> a direct and positive proof that an immaterial living mind must inform and actuate the dead matter, and support the frame of the world:[158]

> . . . above all mechanism and material causes, and proceeds from a higher principle, a divine energy and impression.[159]

Newton, Fatio de Duillier remarks, 'often seems to incline to think that Gravity had its Foundation only in the arbitrary

155   *Ibid.*, p. 634.
156   *Ibid.*, pp. 544–545.
157   *Ibid.*, p. 546.
158   Quoted from Snow, p. 189.
159   *Ibid.*, p. 190.

Will of God'[160] – the master certainly sanctioned such an interpretation of his views by his pupils,[161] and sought corroboration for such a theory among the writings of the ancient *prisca sapientia*. (The representation of God and matter by the figure of Pan and his pipe shows, he says, that 'it seems to have been an ancient opinion that matter depends upon a Deity for its laws of motion as well as for its existence'.[162]). Only God, it is clear, knows (a) the exact masses, and (b) the precise positions of every particle of matter in the universe: only He, then, is in a position to allocate the forces demanded by the thesis of universal gravitation and the $F = GMm/d^2$ formula.[163] This gives Newton an answer to Descartes' gibe against the attraction-theory of Roberval: a stone is not, but God *is* 'vraiment divine'.

The question that remains is whether gravity is due to direct divine action, or whether, as some Newtonians (e.g. Cheyne) believed, it is an 'original impress'[164] given by God to all particles of matter, a *power* of attracting all other bodies *not* innate or intrinsic to them, or explicable in terms of their nature as corporeal things. The latter position is ridiculed by Leibniz in his *Nouveaux Essais*:

> Thus we can assert that matter will not naturally have the faculty of attraction . . . and will not move by itself in a curved line because it is not possible to conceive how this could take place there, that is, to explain it mechanically; whereas that which is natural must be able to become distinctly conceivable.[165]

What is natural is *intelligible*, says that arch-rationalist. We may not posit in bodies qualities which could not in principle be accounted for in terms of their nature as corporeal things: to posit attraction is therefore to invoke 'occult qualities' (qualities of bodies that are absolutely inexplicable in terms of their natures), and thus 'to renounce Philosophy and Reason' altogether.[166] 'A body,' Leibniz insists, 'is never moved

---

160   Quoted from Cohen, 3, p. 119.

161   Snow, pp. 162–163.

162   Quote from Dobbs, 4, p. 111. On this issue, see especially the paper of McGuire and Rattansi.

163   See Westfall, 4, pp. 397–399.

164   See Koyré. 4, p. 156.

165   *Ibid.*, p. 140.

166   H. G. Alexander, Ed., p. 66.

naturally, except by another body which touches it and pushes it'; attraction is, therefore, 'inexplicable by the natural powers of creatures', and hence supernatural;[167] the Newtonians must suppose attractions

> to be effected by miracles, or else to have recourse to absurdities, this is, to the occult qualities of the schools; which some men begin to revive under the specious name of forces; but they drag us back again into the kingdom of darkness.[168]

The Newtonian is, effectively, offered a choice between occasionalism and occultism: on their view, says Leibniz, gravity is either caused *directly* by the Will of God ('perpetual miracle'[169]), *or* is a quality imparted by God to bodies that is in principle inexplicable and unintelligible to us ('occult qualities'). Faced with this terrible dilemma, Clarke (presumably with his master's approval) opts decisively for the first horn of the dilemma: there is, he says, no absolute distinction between the natural and the miraculous:[170]

> The terms *nature*, and *powers of nature*, and *course of nature*, and the like, are nothing but empty words; and merely signify, that a thing usually or frequently comes to pass.[171]

If the operation of natural law can be identified with the 'ordinary' providence of the Deity,[172] this conclusion seems to follow: some events God brings about regularly and frequently (= the course of Nature); others, rarely and at irregular intervals (= miracles).

Now Newton felt that *Principia* could profitably be used as a model for a prospective science of chemistry:[173] in virtue of the 'Analogy of Nature', the micro-realm with which chemistry is concerned should be significantly analogous to the astronomical domain with which *Principia* primarily deals:

> For if Nature be simple and pretty conformable to herself, causes will operate in the same kind of way in all

[167]   *Ibid.*, p. 66.

[168]   *Ibid.*, p. 92.

[169]   *Ibid.*, p. 94.

[170]   *Ibid.*, p. 24.

[171]   *Ibid.*, p. 114.

[172]   For this identification see McGuire, 4.

[173]   See Hall and Hall, Eds., 4, p. 186.

phenomena, so that the motions of smaller bodies depend upon certain smaller forces just as the motions of larger ones are ruled by the greater force of gravity.[174]

In particular, attractive and repulsive interparticulate forces, subservient to mathematical laws (i.e. varying as various powers of the distance, $1/d^n$), will prove the key to unlock the secrets of the micro-architecture of matter and thus to revolutionise the science of chemistry.[175] The distinction between phenomenal force-laws and causal hypotheses will, of course, still apply, here as in *Principia*, and the crucial question which now arises is this: does Newton feel that the *Principia* can be used only as a mathematical model (an illustration of how to quantify chemistry) or does he *also* feel that the *cause* to which he eventually has to have recourse in order to account for gravity (God) is *also* the direct cause of the short-range forces with which the chemist is concerned? The answer seems to be quite clear: the *Principia* is to be used as a model *only* in the former sense, i.e. as an illustration of the Newtonian method. So far as I am aware, Newton never invokes God to account directly for electricity and magnetism, surface tension, cohesion, capillarity, etc. – to do so would be to lapse into a vicious amalgam of positivism and occasionalism.

Indeed, throughout his working life, Newton seems to have accepted 'subtle matter' explanations of electricity, magnetism, and the whole variety of short-range forces required by chemistry and optics. In the 'Letter to Boyle'[176] and the '*De Aere et Aethere*',[177] he speculates as to aetherial causes of a variety of forces; in the General Scholium to *Principia*[178] he posits an 'electric and elastic' spirit located in the pores of bodies and serving to explain not only electrical forces but also cohesion, sensation and volition, and optical forces; in the Queries to the *Opticks*, these latter functions have been taken over by the all-pervading 'aether'.

In illustration of this lifelong belief in subtle matters, let us discuss briefly Newton's views on the topics of electricity and magnetism. In a series of draft queries intended for inclusion

174 *Ibid.*, p. 307.
175 See Dobbs, 4, pp. 213–216. For the development of this programme, see Thackray, 2.
176 Boyle, *Works*, Vol. 1, pp. cxii–cxvii.
177 Hall and Hall, Eds., 4, p. 221 ff.
178 *Principia*, p. 547.

in the *Opticks*, Newton expands somewhat on the nature of the electric 'spirit':

> Do not bodies by friction emit a subtle exhalation or spirit by which they perform their attractions? And is not this spirit of a very active nature and capable of emitting light by its agitations?[179]

This spirit, he says, 'is agitated in various manners like a wind',[180] a passage which, as Home sees, strongly suggests that the spirit acts 'in a wholly mechanical way'[181] in moving chaff, straws, etc. In another draft query, Newton states flatly that 'electric bodies could not act at a distance without a spirit reaching to that distance':[182] electrostatic forces involve neither A.D. nor direct divine action, but an appropriate material cause.

Newton seems also to have accepted, throughout his life, some sort of effluvium-theory of magnetic action.[183] In *De Aere et Aethere*, this is presented as something evident:

> I believe everyone who sees iron filings arranged into curved lines like meridians by effluvia circulating from pole to pole of the [load] stone will acknowledge that these magnetic effluvia are of this kind.[184]

Newton seems to have retained this belief in magnetic effluvia at least into the 1690s – his pupil Gregory remarks that he believed that magnetism, 'seems to be produced by mechanical means' – and there seems no reason to believe that he ever discarded it.[185]

Yet, after the demonstration of *Principia* that interplanetary space and the Boylian vacuum are all but totally empty of matter, the whole structure of the problem has altered. Subtle matter (the optical aether, magnetic effluvia) capable of operating in such 'vacua' must be very rare indeed: their casual powers must be attributed to their *elastic* nature. The 'aether' of the *Opticks*, Newton reckons, is 700,000 × 700,000 ($= 49 \times 10^{10}$) times more elastic than air in

---

179   Quoted from Home, 3, p. 7.

180   *Ibid.*, p. 8.

181   *Ibid.*, p. 33.

182   *Ibid.*, p. 14.

183   See Home, 2, p. 256.

184   Hall and Hall, Eds., 4, p. 228.

185   Home, 2, pp. 258–259.

proportion to its density.[186] To *explain* this combination of rarity and elasticity, Newton finds himself forced to postulate *interparticulate repulsive forces* between aether-particles[187] (just as, earlier, he had explained Boyle's law in terms of a repulsive force varying as $1/d$ between air-particles[188]). We now seem to be trapped in a regress. We posit subtle matters to avoid postulating A.D.; yet to account for the properties we need to attribute to 'aether' or 'spirit' we find ourselves back with forces; *either* A.D. by matter (metaphysically absurd) *or* inexplicable 'powers' *in* matter but not *of* it ('occult qualities') *or* direct divine action ('perpetual miracle'). There seems no escape.

C. Incorporeal Causes

We witnessed, in Chapter 8, the beginnings of a truly mechanistic biology, centred around an account of the stimulus-response mechanism, which, effectively, conceives of an animal as a sort of hydraulically-driven engine. Corporeal emissions or pressure-waves strike the organs of sense; the blow which they impart produces some sort of motion (e.g. vibration) in the animal spirits in the nerves; these spirits are pumped to the brain and thence to the muscles, where they bring about the muscular contraction that initiates motion. A number of Renaissance thinkers (e.g. Leonardo) came to accept such a hydraulic model, a view obviously associated with the widespread belief in the corporeality of the animal soul. (To attribute to beasts incorporeal – and hence potentially immortal – souls is to threaten the prestigious unique status of man.) Stoic and Epicurean views of the corporeality of the souls of animals were, therefore, theologically acceptable: it is only the rational souls of men, the divines insisted, that must be incorporeal.

The great champions of this mechanistic biology were of course the Cartesian school. According to Descartes, organic bodies, whether human or animal, are mere machines, functioning according to purely mechanical laws. The all-important nervous system operates on hydraulic principles: 'animal spirits' are merely a subtle, swift-moving variety of the universal matter, different only in degree and not in kind from gross matter:

---

186   Newton, 2, p. 351.

187   *Ibid.*, p. 352.

188   See Hall and Hall, Eds., 4, pp. 223–224. For the proof of this thesis, see Book 2 of *Principia* (Newton, 1, p. 300).

> . . . what I here name spirits are nothing but material
> bodies, and their one peculiarity is that they are bodies of
> extreme minuteness and that they move very quickly like
> the particles of the flame which issues from a torch.[189]

Recurrent in Descartes' writing we find the likening of an
organic body to a 'mill' or a 'clock', moved by springs, cogs,
and wheels. The following passage, from the *Traité de
l'homme*, is typical:

> I want you to consider that all these functions in this
> machine follow naturally from the disposition of its organs
> alone, just as the movements of a clock or other automat
> follow from the disposition of its counterweights and
> wheels; so that to explain its functions it is not necessary to
> imagine a vegetative or sensitive soul in the machine, or any
> other principle of movement and life other than its blood
> and spirits agitated by the fire which burns continually in
> its heart . . .[190]

Hobbes is equally blunt: in the Introduction to his *Leviathan*,
we find the following passage:

> For seeing life is but a motion of limbs, the beginning
> whereof is in some principal part within; why may we not
> say, that all *automata* (engines that move themselves by
> springs and wheels as doth a watch) have an artificial life?
> For what is the heart, but a spring; and the nerves, but so
> many strings; and the joints, but so many wheels, giving
> motion to the whole body, such as was intended by the
> artificer?[191]

It was, obviously, only a matter of time before someone
extended the hydraulic model to include man: if, after all,
perception, appetite, and that (not inconsiderable) degree of
intellection sometimes manifest by beasts are all explicable in
purely material terms, it would seem that all we need to
account for the faculties of man is a more complex
mechanical model. This step was taken, quite explicitly, by
Hobbes, one of the great pioneers of philosophical materialism and, unlike Descartes, a *complete* mechanical philosopher. (By admitting incorporeal souls into his system,

---

189   Quoted from Hesse, 1, pp. 111–112. See also Kemp Smith, p. 131.

190   Quoted from Westfall, 5, p. 93.

191   Hobbes, *Works*, Vol. 3, Introduction, p. ix.

Descartes also admits non-mechanical modes of action.) There is, says Hobbes, no such thing as an incorporeal substance: the very concept is self-contradictory. Since 'substance' *means* 'body', 'incorporeal substance', like 'round square', is a contradiction in terms,[192] a senseless notion illegitimately foist upon Scripture by the disciples of those two great pagans, Plato and Aristotle.[193] Moreover, he adds,

> . . . the universe . . . is corporeal, that is to say, body; and hath the dimensions of magnitude, namely, length, breadth, and depth: also every part of body, is likewise body, and hath the like dimensions; and consequently every part of the universe is body, and that which is not body, is no part of the universe: and because the universe is all, that which is no part of it, is *nothing*; and consequently *no where*.[194]

Neither of these arguments, it must be confessed, is worth a great deal. 'Substance' simply does *not* mean 'body': it means 'self-subsistent entity' – philosophers from Plato to Descartes were not guilty of a trivial logico-linguistic error. As for the argument based on the premiss that the universe is body and nothing but, this seems merely question-begging.

Hobbes' real argument for his materialism does not, however, turn on these fallacies. He presents it as a self-evident truth that no body can be moved but by a contiguous, moving, *body*: no other mode of physical agency, he insists, is intelligible.[195] How could something incorporeal push a gross body? Yet, as the Epicureans said,[196] it is evident that the soul moves the body: it follows that the soul must itself be material. From commonsense interactionism, *plus* a conception of cause as intelligible connection, Hobbes derives the inevitable conclusion – a thoroughgoing materialism. This *causal* argument for materialism is much more powerful than the lightweight (and easily rebutted) ones mentioned earlier. Hobbes sees, as others did not, that the Mechanical Philosophy, taken seriously, is implicitly materialistic.

Gassendi comes close to a similar realisation. He too accepts the Epicurean theory of vegetable and animal souls as

---

192   *Ibid.*, p. 27, p. 381. See also Laird, pp. 92–94.

193   Laird, p. 42, p. 94.

194   Hobbes, *Works*, Vol. 3, p. 672.

195   *Ibid.*, Vol. 1, p. 115.

196   For Lucretius' argument to this effect, see Chapter 4, p. 171.

subtle, highly mobile matter (soul-atoms are small and very well endowed with *pondus*), and only accepts the incorporeality of the rational soul of humans on the authority of the Church. (He stoutly opposes Descartes' 'proofs' of the incorporeality of the human soul, advocating instead – for the sake of argument, he says! – a conception of the *anima* as a subtle flame-like or wind-like substance.) Since, he argues, following Lucretius, souls are true causal agents, capable of moving gross bodies, they must be deemed to consist of a 'contexture' of very fine and especially mobile atoms (the 'bloom' of the matter, he says).[197] By their own rapid motions, he says, they are capable of moving and arranging the parts of vegetable, animal, and human bodies, and (for animals and men) of guiding the motions of the whole.

By making the 'active principle' in the physical universe something *material*, he says (the atoms, in virtue of their divinely instilled motive urge, may be described as self-movers), we avoid the difficulties that would he involved in trying to explain how something incorporeal could move bodies. This issue of 'The Initiating Force in Second Causes and the Primary Principle of Action' is discussed in a chapter of the *Syntagma*: what, Gassendi there asks, is the 'internal or root principle of motion in second causes?'[198] Some people (Platonists and Aristotelians) seem to think that it is incorporeal; others (the Stoics) say that it is spirit or subtle matter (*pneuma*). Aristotle, says Gassendi, rebuked the ancient Atomists for their laziness in neglecting to answer this question; but, he replies:

> they did not really omit it; instead they insisted only that the efficient principle was to be distinguished from the material principle as different in thought, but not in fact and substance.[199]

Clearly, he continues,

> they did not consider atoms as inert or motionless, but rather as most active and mobile, so much so that they held them to be the first principle from which things take their motion.[200]

---

[197]   See Brett, p. 57; Spink, p. 97.
[198]   *Syntagma*, Physics, Section 1, Book 4, Chapter 8; Gassendi, p. 409.
[199]   *Ibid.*, p. 411.
[200]   *Ibid.*, pp. 411–412.

After adding a quotation from Lucretius in support of this claim, Gassendi goes on to affirm bluntly that 'the internal principle of action that works in second causes is not some incorporeal substance, but a corporeal one'.[201] This must be so, he claims, since

> it is impossible to conceive how it can bring itself to bear on a body in order to impart an impulse to it if it is not corporeal. Nor is it possible to conceive that it will ever touch the body if it lacks the mass or the sense of touch with which to touch something.[202]

We cannot, in short, conceive how an incorporeal cause could move a body: this is the crux of the whole argument. How, then, are we to account for the causal agency of God, angels and demons, and the incorporeal rational soul of which, the Church assures us, each and every human being is possessed? God, says Gassendi, presents no problem: being omnipotent, there are no restrictions and constraints on His power of action – He can move a body simply by willing it. As for the human intellect or mind, this 'does not stimulate actions except for intellectual, or mental, and incorporeal ones'; in so far as the soul is 'sentient, animate, and endowed with the power of moving bodies', it is corporeal.[203] Yet, unless the intellect can *move* the *corporeal* parts of the soul, intellectual activity is causally idle and inefficacious, an idle wheel that turns nothing. If thought can stimulate volition, and hence lead to action, we are back with our initial problem – Gassendi's evasion solves nothing.

The difficulty, Gassendi admits, is even greater for angels and demons, which are neither all-pervasive and omnipotent, like God, nor 'composed in a certain manner of corporeal and incorporeal parts', like the human soul.[204] How, then, can angels produce physical effects? Many fathers of the Church, says Gassendi, have attributed to them (on Scriptural grounds) *subtle bodies* – this, he feels, would solve the problem. *If*, however, faith demands that we believe angels to be wholly incorporeal, and yet, nevertheless, causally active in the physical realm, this would show that God has given to them causal powers surpassing our powers of comprehension:

---

[201]   *Ibid.*, p. 412.

[202]   *Ibid.*, p. 413.

[203]   *Ibid.*, p. 413.

[204]   *Ibid.*, p. 413.

on such a view, we are committed to something that 'physical reason does not adequately understand', i.e. a mystery of 'sacred theology'.[205]

In the philosophy of Gassendi, then, naturalistic materialism rubs shoulders somewhat uncomfortably with a belief, based largely on faith, in incorporeal causes of various kinds. *Qua* natural philosopher, Gassendi finds incorporeal causes an embarrassment (the manner of their agency is unintelligible to us); *qua* Christian, he finds himself committed to such entities. The resulting tension is never satisfactorily resolved, although the general tendency towards materialism and naturalism is undeniable.

For the Cambridge Platonists, More and Cudworth, the irreducibility of the faculties of the soul to merely material principles is an unshakeable truth. The soul, More insists in his *Antidote against Atheism*, cannot be a 'mere modification of matter': what corporeal thing can possess spontaneous motion, sense, and reason? After a vigorous polemic against the materialism of Hobbes, he re-affirms the traditional (Platonic) conception of soul as an immaterial substance, 'distinct from the body, which uses the animal spirits and brains for instruments'.[206] Cudworth agrees: souls, he claims, 'are plainly real entities distinct from . . . matter and its modifications'.[207] and quite irreducible to material principles.[208] But how, Hobbes and Gassendi will ask, can an incorporeal substance *move* a body? This, More admits, is a problem:

> The greatest difficulty is to fancy how this *Spirit*, being so *Incorporeal*, can be able to move the matter, though it be in it. For it seems so subtile, that it will pass through.[209]

More 'answers' this mystery with another: the cohesiveness of bodies, he says, is just as mysterious, yet undeniably real. This hardly helps: however true it may be that it is difficult to find an adequate physical explanation of cohesion, it scarcely warrants More's conclusion that 'a firm union of *Spirit* and *Matter* is very possible, though we cannot conceive the

205   *Ibid.*, pp. 414–415.

206   H. More, *Works*, Vol. 1, 'The Antidote against Atheism', p. 40.

207   Cudworth, vol. 1, p. 67.

208   *Ibid.*, Vol. 1, p. 89; Vol. 3, p. 114, p. 418.

209   H. More, *Works*, vol. 1, 'The Immortality of the Soul', pp. 80–81. See also 'The Antidote against Atheism', pp. 151–152.

manner thereof'. To cite a second mystery is scarcely to cast any light on the first.

In Cartesian physics, this problem is resolved by the following expedient.[210] For Descartes, the 'quantity of motion' $\Sigma m \mid v \mid$ that is conserved is a *scalar* quantity: a body which reverses its direction while retaining the same speed *retains*, on this view, the same ' motive force'. The transition from $\vec{v}$ to $\overleftarrow{v}$ *not*, therefore, involving any loss or gain of force (no motive force being transferred from one body to another), it follows, for the Cartesian, that no physical *action* is involved.[211] Within such a physics, souls can alter the direction of motion of the animal spirits (and hence of bodies) without contravening any of the principles of physics – incorporeal agents can thus find a place within a mechanistic natural philosophy.

Unfortunately, this will not do. As Huygens, Wren and Wallis were to show, the Cartesian conservation law ($\Sigma m \mid v \mid = k$) is false, and must be replaced by the principle of conservation of momentum, $\Sigma m\vec{v} = k$. But this refutes Descartes' position. Let Leibniz explain:

> . . . the changes which happen in bodies in consequence of modifications of the soul embarrassed him, because they seemed to violate this [conservation] law. He believed, therefore, that he had found an expedient, which is certainly ingenious, by saying that we must distinguish between motion and direction; and that the soul cannot augment or diminish the moving force, but alters the direction, or determination of the course of the animal spirits, and that it is through this that voluntary motions take place.[212]

Unfortunately for Descartes, however, a body tends in fact 'to preserve its determination or direction with its whole force and its whole quantity of motion':[213] the transition from $m\vec{v}$ to $m\overleftarrow{v}$ involves, therefore, a 'force' (impulse) $\Delta mv = 2mv$.[214] From the conservation of the vector quantity $\Sigma m\vec{v}$ (our 'momentum'), it follows, moreover, that $+ \Delta mv = - \Delta mv$,

---

210   This view is attributed to Descartes by Leibniz (see Leibniz, p. 587; Russell, pp. 226-227), but represents a very plausible reading of Cartesianism (see Hatfield, p. 135).

211   See Westfall, 4, p. 67.

212   Quoted from Russell, pp. 226-227. See also Leibniz, p. 587.

213   From the 'Critical Thoughts' on Descartes' *Principles* (Leibniz, p. 396).

214   See Westfall, 4, pp. 291-292.

i.e. that an agent capable of moving a body must itself suffer a *reaction* equal and opposite to the impulse that it imparts. An incorporeal agent (i.e. one without mass) can suffer no such reaction: it follows that the causal agency of incorporeals violates the principle of conservation of momentum and Newton's third law of motion.

The only response available to the advocate of incorporeal agents was either to deny (tacitly or explicitly) the universal applicability of the offending principles of physics, *or* to find some metaphysical expedient that evades the contradiction. The obvious moves were to the occasionalism of Malebranche or the 'pre-established harmony' of Leibniz: both admit the reality of incorporeal souls, but deny that they actually move bodies. If, says Leibniz, Descartes had known of the $\Sigma m\vec{v} = k$ formula, 'he would . . . have come to my system of pre-established harmony'[215] – this, he insists, is less *ad hoc*, and gives a worthier conception of the deity, than the occasionalism of Malebranche.[216]

The Mechanical Philosophy seems therefore, as Hobbes saw, to entail materialism. Combine the principles of the mechanists (cause = impulse, the $\Sigma m\vec{v} = k$ principle, $+ \Delta mv = -\Delta mv$, etc.) with the interactionist view of body and soul implicit in common sense, and we arrive at a materialist conception of the soul. Soul moves body; therefore, must impart an impulse to some bodily part (e.g. animal spirits); therefore, must itself be material. To avoid this conclusion, one must either deny the universal applicability of certain of the laws of physics, *or* lapse into a weird and thoroughly counterintuitive metaphysic such as those of Malebranche and Leibniz. Neither alternative, it must be admitted, it is very attractive.

D. Teleology

Throughout the Middle Ages and Renaissance, the teleological approach to the Natural Philosophy of Plato, Aristotle, and Galen was almost universally accepted. 'Nature', say the scholastics, 'does nothing in vain' (Aristotle) and 'acts like a craftsman' (Galen): such axioms became part and parcel of medieval philosophy. This teleological approach was

---

215   Quoted from Russell, p. 227.

216   See the 'Discourse on Metaphysics' of 1686, in which Leibniz seeks to establish the merits of his own 'System of pre-established harmony' against its rivals, interactionism and occasionalism (Leibniz, p. 324–325).

challenged by Francis Bacon, who claimed that the investigation of final causes had proved deleterious to the study of efficient ones, and preferred the anti-teleological philosophy of Democritus to those of Plato and Aristotle.[217] It was Descartes, however, who launched the great assault on the use of final causes in physics. Mindless creatures, he insists (i.e. vegetables and animals as well as inanimate things), have no purposes of their own: they merely serve those purposes for which God designed them. Yet, he adds, 'we should not take so much upon ourselves as to believe that God should take us into His counsels';[218] 'it seems to me that we cannot without foolhardiness seek to enquire into and profess to discover God's inscrutable ends'.[219] Since Nature is a mere machine, without any immanent purposiveness of its own but blindly carrying out those functions for which it was created by the inscrutable deity, it follows that 'we must not seek into the final, but only into the efficient causes of things'.[220]

This may sound like true piety; the next passage, however, proved a stumbling-block for many pious men. The initial conditions from which our physical universe developed are, Descartes claims, irrelevant to its ultimate course:

> . . . it makes very little difference what we assume in this respect, because it must later be changed according to the laws of nature. Hardly anything can be assumed from which the same effect cannot be derived, though perhaps with greater trouble. For due to these laws *matter takes on, successively, all the forms of which it is capable.* Therefore if we considered these forms in order, we could eventually arrive at that one which is our present world, so that in this respect no false hypothesis can lead us into error.[221]

It follows from this that all things in Nature may have arisen from an initial state of chaos merely by the operation of blind mechanical causes, operating in accordance with the pre-ordained laws of motion. God, on this view, merely creates matter, endows it with its proper 'quantity of motion' $\Sigma m \mid v \mid$, and ordains the laws of motion (conservation, rectilinear inertia, the rules of impact, etc.); thereafter, the machine runs on by itself. (The physicist may legitimately

[217] F. Bacon, *Works*, vol. 4, pp. 363–364.

[218] *Principles*, 1, 28; Descartes, 1, Vol. 1, p. 230.

[219] Quoted from Kemp Smith, p. 176.

[220] Descartes, 1, Vol. 1, p. 230.

[221] *Principles*, 3, 47.

think of Nature in this way; the metaphysician must of course bear in mind the instantaneous divine re-creation of all things.)

This thesis – that our cosmos could have emerged from chaos *without* any contrivance or design[222] – was one of the great grounds for offence preventing the more widespread acceptance of Cartesianism. Descartes' rejection of final causes provoked rebuttals from Gassendi, More and Cudworth, Boyle, Newton, and Leibniz: all explicitly rejected this aspect of Cartesianism and reaffirmed the traditional argument from design. The 'royal road, level and open', to knowledge of God's existence and perfections is, Gassendi writes in his 'Rebuttals' against Descartes, the argument from design, which argues from 'the excellent works of this universe' to the perfections of its Creator.[223] You, he writes to Descartes, choose to ignore the mass of empirical evidence testifying to the existence of a deity, and prefer instead an obscure (and possibly sophistical) *a priori* proof (the ontological argument). But by rejecting final causes,

> it is indeed to be feared that you have rejected the main argument by which God's wisdom, providence, power and even his existence can be proven by the natural light of reason.[224]

Similar claims occur in More and Cudworth: their rejection of final causes, says the latter, gives the Cartesians 'an undiscerned tang of the mechanic Atheism',

> . . . in that their so confident rejecting of all final and intending causality in nature, and admitting of no other causes of things as philosophical, save the material and mechanical only; this being really to banish all mental, and consequently divine causality, quite out of the world; and to make the whole world to be nothing else but a mere heap of dust, fortuitously agitated.[225]

Cudworth eventually comes to see the rejection of final causes as the very spirit of atheism: these 'modern mechanic theists', he claims, are 'cousins' to atheists, 'inspired with a spirit of infidelity, which is the spirit of atheism'.[226] By rejecting the

[222]   See Descartes, 1, vol. 1, p. 109.

[223]   Gassendi, p. 208.

[224]   *Ibid.*, p. 225.

[225]   Cudworth, vol. 1, p. 217.

[226]   *Ibid.*, Vol. 2, p. 620. See also Cragg, p. 12; Sailor, 1, pp. 135–137.

argument from design, one of the best and certainly the most popular and persuasive of arguments for theism, the Cartesians are giving tacit aid to the cause of atheism.[227]

Boyle agrees: the argument from design, he claims, is 'one of the best and most successful arguments' for belief in God, but one which presumes that (*pace* Descartes) we *do* have some knowledge of natural *ends*.[228] That we praise and honour God for His Creation would not make sense unless we presume knowledge that His ends therein have been fulfilled. And *some* of those purposes are so evident to us that it is 'little less than madness' to deny that we have knowledge of them.[229] (This Boyle insists, is not piety but almost wilful blindness and perversity.)

Two realms in particular furnished examples for the advocates of the argument from design. One was the order and harmony of our solar system as a whole; the other, the functional adaptation of the parts of animals. Let us deal with these two examples in that order.

Lucretius supposes, says Boyle, that

> a sufficient number of atoms . . . being granted, there will need nothing but their fortuitous concourse . . . to give a being to all those bodies, that make up the world.[230]

This, Boyle retorts, is quite incredible: that our cosmos should have emerged out of chaos by the random jumbling of atoms is something no rational man could believe. It is, indeed, still more incredible than that

> in a printer's working-house a multitude of small letters, being thrown upon the ground, should fall disposed in such an order, as clearly to exhibit the history of the creation of the world, described in the third or fourth first chapters of Genesis.[231]

Newton is equally forthright: natural philosophy, he claims in the general Scholium to *Principia*,[232] spills over into natural theology – the beauty, order, and variety of our universe are

---

227   *Ibid.*, Vol. 2, pp. 612–613.

228   Boyle, *Works*, vol. 5, pp. 400–402.

229   *Ibid.*, pp. 397–398. Cf. Gassendi, p. 226. *Some* of God's purposes, says Gassendi, He has, as it were, 'put on display', i.e. openly revealed to us.

230   *Ibid.*, Vol. 2, p. 43.

231   *Ibid.*, p. 43. The image is derived from Cicero (see chapter 4, p. 186).

232   Newton, 1, p. 546.

plain evidence of a powerful and intelligent artificer. That the cosmos could have emerged by 'chance and necessity' (as Democritus would have it) is absurd: the 'Book' of Nature reveals clearly the hallmark and signature of its Creator. The laws of motion, he argues in the *Opticks*, may suffice to sustain, but of themselves could never have given rise to, our orderly system of Sun, planets, moons, and comets, the harmony and beauty of which so clearly manifest intelligent design. Hence, says Newton,

> . . . it's unphilosophical to seek for any other Origin of the World, or to pretend that it might arise out of a Chaos by the mere laws of Nature; though being once form'd, it may continue by those laws for many Ages.[233]

The best argument for final causes takes its origin, however, from the functional adaptation of the parts of animals. Here, say Gassendi, More, Cudworth, Newton, Boyle, and Leibniz, the *function* for which a particular organ was designed is often *so* obvious that it would be madness to deny it. The craftsmanship displayed in the parts of animals is clear evidence of design: the eye, says Boyle, is

> so exquisitely adapted for the use of seeing, and that use is so necessary for the welfare of the animal, that it may well be doubted, whether any considering man can really think, that it was not designed for that use.[234]

In the realm of biology, he insists, the use of final causes is permissible and even necessary. It may even be of heuristic value: Harvey's discovery of the circulation of the blood was prompted by asking the question: 'What function is served by the valves in the veins?'[235] Mechanism and teleology are, Boyle claims, perfectly compatible: if one finds a watch, one can ask either 'what is it *for*?' or '*how* does it work?' – the two questions are complementary rather than conflicting in nature.[236].

Newton too cites the parts of animals as clear evidence of intelligent design:

> How came the Bodies of Animals to be contrived with so much Art, and for what ends were their several Parts? Was

---

233   Newton, 2, p. 402.

234   Boyle, *Works*, vol. 5, p. 425.

235   *Ibid.*, p. 427.

236   *Ibid.*, pp. 397–398.

the Eye contrived without knowledge of Opticks, and the Ear without knowledge of Sounds?[237]

Such passages could be multiplied *ad lib*: the reaffirmation, against the extravagancies of Cartesianism, of the argument from design was vigorous and widespread. Yet if we conclude, with More and Cudworth, that mere chance and blind mechanical causes could not account for the organisation and functional adaptation of the parts of animals, how are we to account for them? Are we to assume that God Himself is directly and personally responsible for every flea and mushroom in Nature? Surely not, says Cudworth: in the first place, it is an unfitting conception of the deity to bother Him with all Nature's trivia; in the second place, there could on this account be no monsters, no bungles and botches, in natural things.[238] For, as More says,

> there is no Matter so perverse and stubborn but his Omnipotency could tame; whence there would be no Defects nor Monstrosities in the generation of Animals.[239]

It follows, then, says Cudworth, that there exists a 'plastic nature' (More's 'spirit of nature') created by God which,

> as an inferior and subordinate instrument, doth drudgingly execute that part of his providence, which consists in the regular and orderly motion of matter.[240]

This plastic principle, he continues, is *immanent Art*, 'art itself acting immediately on the matter as an inward principle' – it works not by any external or mechanical means, but 'from within vitally and magically'.[241] Although teleological rather than mechanical in its operations, it is not itself sentient and intelligent: it works by habit (= a prior program) not by reason, by instinct rather than intellect. As More says of his 'spirit of nature', which plays a perfectly analogous rôle in *his* philosophy, it is

> a substance incorporeal, but without Sense and Animadversion, pervading the whole Matter of the Universe, and exercising a Plastical power therein according to the sundry predispositions and occasions in the parts it works upon, raising such Phaenomena in the World, by directing the

237 Newton, 2, pp. 369–370.
238 Cudworth, vol. 1, p. 218.
239 H. More, *Works*, vol. 1, 'The Immortality of the Soul', p. 102.
240 Cudworth, Vol. 1, pp. 223–224.
241 *Ibid.*, pp. 235–236.

parts of Matter and their Motion, as cannot be resolved into mere Mechanical powers.[242]

It is, in short, 'the vicarious Power of God upon this great Automaton, the world'.[243] All the *vital* (i.e. non-mechanical) aspects of Nature, say the Platonists, are due to this principle; without which, says Cudworth, one of two consequences would follow:

> . . . either in the efformation and organisation of the bodies of animals . . . every thing comes to pass fortuitously, and happens to be as it is, without the guidance and direction of any mind or understanding; or else . . . God himself doth all immediately, and, as it were with his own hands, forms the body of every gnat and fly, insect and mite.[244]

Both these alternatives being absurd and impossible, it follows that such natural phenomena as the formation of animal bodies *must* be attributed to the agency of 'plastic nature'.

Cudworth omits however, one important possibility. If one grants the argument from design, and concludes that organisms could not conceivably have arisen by chance and without intelligent contrivance, one still need not posit a purposive semi-deity called 'nature' (an idea much criticised by Boyle both on theological and physical grounds[245]) in order to avoid lapsing into occasionalism. This other option is *preformationism*. At the Creation of the world, we can say, God not only created matter, endowed it with motion, and ordained its laws; He *also* fashioned the 'seeds' or 'rudiments' of living things. These are highly organised, minute particles of matter with complex internal structures in virtue of which they have certain powers – powers to absorb matter and fashion out of it an organic body capable of producing more seeds of the same kind. The operation of these seeds can be purely mechanical: all their special properties and powers, one can say, follow from the complex structures God gave them at the first Creation.

A preformation-theory is hinted at by Gassendi in his account of the generation of plants and animals. Organisms,

---

242   H. More, *Works*, Vol. 1, 'The Immortality of the Soul', p. 193.

243   Quoted from Greene, p. 466.

244   Cudworth, Vol. 1, p. 218.

245   See Boyle's 'Free Enquiry into the Vulgarly Received Notion of Nature' in his *Works*, vol. 5, p. 158 ff., esp. p. 188 ff., where Boyle gives his five main reasons for rejecting such a notion.

he insists, come-to-be out of highly specific 'seeds' or *semina*: in the regular course of Nature, horses beget horses and men beget men. There are, then, as Charleton says, 'certain molecules, small masses, of various figures, which are the seminaries of various productions'.[246] To explain how an organism grows from such a seed, says Gassendi, one must assume one of two things: either

a. 'formative principles', like little artisans, *guide* and direct the motions of atoms, selecting appropriate ones for particular places (like building an artefact); or
b. God originally so placed the atoms that they fell into certain arrangements, forming complex structures with (mechanically explicable) resultant powers.[247]

Gassendi shows a clear preference for the latter (preformationist) option: in the beginning, he says in the *Syntagma*, God made the *seeds* of living things:

> He made the seeds . . . of all things capable of generation . . . from selected atoms he fashioned the first seeds of all things, from which later the propagation of species would occur by generation.[248]

Boyle is another advocate of preformationism. Once one grants, he sees, that animal bodies could not come about without the operation at some point of intelligent contrivance, one is faced with a choice between three options: pure occasionalism, 'plastic nature', and preformationism. Boyle never seriously considers the first of these options, and argues at length against the second: the postulation of a purposive and semi-divine principle called 'Nature' is, he insists, unnecessary, unintelligible, and implicitly heretical, conducive to paganism and Nature-worship.[249] We are left, then, with preformationism: Boyle's God, at the Creation, not only creates matter *ex nihilo*, sets it in motion, and ordains the laws of nature, he also *chooses the initial conditions* – by 'guiding the first motions of the small parts of matter' He brings them 'to convene after the manner requisite to compose the world',[250] including the highly contrived 'seeds' or 'rudiments' that will in time give rise to animal and plant bodies. It is, he insists, greater testimony to God's

---

[246] Charleton, p. 105. See also Brett, p. 90.
[247] See Spink, p. 95.
[248] *Syntagma*, Physics, Section 1, Book 3, Chapter 8; Gassendi, p. 401.
[249] Boyle, *Works*, vol. 5, p. 188 ff.
[250] *Ibid.*, Vol. 3, p. 15. See also p. 48 ff.

wisdom and power if no such thing as 'Nature' (in the sense of a purposive and non-mechanical agent) exists,

> as it more recommends the skill of an engineer to contrive an elaborate engine so, as that there should need nothing to reach his ends in it but the contrivance of parts devoid of understanding, than if it were necessary, that ever and anon a discreet servant should be employed to concur notably to the operations of this or that part, or to hinder the engine from being put out of order; so it more sets off the wisdom of God in the fabric of the universe, that he can make so vast a machine perform all those many things, which he designed it should, by the meer contrivance of brute matter, managed by certain laws of local motion and upheld by his ordinary and general concourse, than if he employed from time to time an intelligent overseer, such as nature is fancied to be, to regulate, assist, and control the motions of the parts.[251]

The schoolmen, he continues, liken the physical universe to a *puppet*; he (Boyle) likens it to a clock:

> Those things, which the school-philosophers ascribe to the agency of nature interposing according to emergencies, I ascribe to the wisdom of God in the first fabric of the universe.[252]

In the mechanism of our physical universe, he continues,

> all things are so skilfully contrived, that the engine being once set a moving, all things proceed according to the artificer's first design . . .[253]

Preformationist biology was raised to great heights by men such as Power[254] and Leibniz, and reached its culmination in the theory of *emboîtement*, i.e. in the idea that every organism has existed since the Creation of the universe as a seed within a seed within a seed, etc. – what is called 'generation' is, on this view, merely the beginning of a process of *development* by which a pre-formed seed will in its turn give rise to a fully-formed organism. The preformationists are *mechanists* (they deny the existence *within* corporeal Nature

251   *Ibid.*, Vol. 5, p. 161.

252   *Ibid.*, p. 163.

253   *Ibid.*, p. 163.

254   For the preformationism of Power, see A. R. Hall, 3, p. 188; Webster, 2, p. 171.

of non-mechanical agents), yet also advocates of the argument from design: they emphatically *deny* the Cartesian claim that the organised bodies of animals could have arisen through blind mechanical causes. It follows that the absolute *generation* of an organism is *not naturally explicable* (i.e. not explicable in terms of the explanatory principles permissible to a mechanist). The inevitable conclusion – drawn so clearly and explicitly by Leibniz – is that generation is *unreal*: what seems to be the absolute beginning of existence of an organism is really a process of growth and maturation, an unfolding of what was latent in a pre-existent *structure*. Thus Leibniz:

> I am of the opinion of Mr. Cudworth, whose excellent work for the most part supports me, that the laws of mechanism by themselves could not form an animal where there is nothing already organised.[255]

Cudworth's immanent purposive agency ('plastic nature') is, however, both *unintelligible* (we cannot conceive how it could bring about its operations) and *superfluous* (transcendent pre-planning eliminates the need for immanent purposiveness). Since organisation can only (intelligibly) arise out of prior organisation, it follows that every organism emerges from an already preformed seed. (Strictly, all the preformationists need to assert is that each seed must possess sufficient information-content for the development of the eventual organism, but most preformationists insisted that this information is stored in *iconic* form – i.e. that the seed or embryo is structurally isomorphic with, though of course much smaller than, the mature organism.) As Leibniz says,

> The organism of animals is a mechanism which supposes a divine preformation. What follows upon it is purely natural, and entirely mechanical.[256]

The microscopic researches of Malpighi, Swammerdam and Leeuwenhoek ('the best observers of our times'), by revealing the complex microstructures of embryos and minute microorganisms, lend, says Leibniz, much empirical support to this theory.[257]

---

255 Leibniz, p. 589.

256 H. G. Alexander, Ed., p. 93.

257 See the 'New System' of 1695 (Leibniz, p. 455). For the microbiology of Malpighi, see Belloni, in Righini Bonelli and Shea, Eds., p. 95. ff.

In Greek and medieval Natural Philosophy, we have remarked elsewhere, there existed the following pair of basic principles or axioms:

Motion $\longrightarrow$ Life (Soul)

Order $\longrightarrow$ Mind (Intellect)

It was characteristic of philosophy up to and including the Renaissance to posit movers and minds *within* the course of Nature, to account for its manifest motion and order. In someone like Boyle, however (a paradigmatic mechanist), *both* inferences are accepted, only in place of immanent teleology in natural things we have a transcendent artificer and a world-machine. God gives corpuscles motion (they do not have it of their own nature), and fashions the 'seeds' which will give rise (by purely mechanical means) to the bodies of organisms. Motion does not occur without life, nor order and organisation without Mind: the error of earlier naturalists, on such a view, lay not in their basic principles but in seeking the sources of motion and order in the *wrong places*, i.e. *within* the frame of corporeal Nature instead of outside it altogether. To posit a quasi-intelligent, purposive 'Nature', men like Boyle and Leibniz agree, is bad theology (conducive to pagan Nature-worship) no less than bad physics (the postulation of unintelligible agencies): Reason and Faith alike favour mechanism and preformationism.

APPENDIX 1
Physical Indivisibility

The classical argument for the physical indivisibility of the atom – an argument that goes back to Democritus and probably to Leucippus – turns on its perfect solidity, its lack of inner void. More precisely, it depends on a premise which we shall entitle 'Axiom A', which states that

> *Axiom A*: A body is susceptible to physical division only if it contains internal vacuum.

If then the Democritean atom is a perfect and absolute plenum of 'Being' or matter*, it will be quite immune to the processes of cutting, shearing, etc. Since it cannot be divided, and cannot simply cease-to-be (matter* is conserved – in Eleatic terminology, 'Being' does not pass away into 'Non-Being') it is an eternal entity.[1]

Before moving on to discuss the views of Epicurus and Lucretius, we must mention one pertinent objection to the above argument. If, says Aristotle, the mere absence of intermediate vacuum is sufficient for cohesion, atoms should *coalesce* on collision. Since all atoms are made of the same fundamental stuff (matter*),

> why, when they come into contact, do they not coalesce into one, as drops of water run together when drop touches drop (for the two cases are precisely parallel)?[2]

The objection is a fundamental one. To evade it, some commentators found themselves forced to ascribe to Democritus the view that the atoms never touched one another, that some void space was always left between them. But this is a truly desperate expedient: without atomic contacts, how can one explain mechanical collision and the formation of compound bodies? Contact is necessary for

---

[1] Kirk and Raven, pp. 407–408; Barnes, Vol. 2, pp. 47–49. There seems to be no reason to doubt that Democritus and Leucippus accepted Axiom A without reservation.

[2] *De Gen et Corr.*, 1, 8, 326a 32–33.

collision; collision, in a universe without forces, is the only form of causal agency.[3]

According to Konstan,[4] the Epicurean school abandoned Axiom A in the face of this Aristotelian objection. In place of Axiom A, says Konstan, the Epicureans proposed a quite distinct Axiom B to the effect that:

Axiom B: The superficial layer of minimal parts of an atom X are *essentially* constituents of X – can only be conceived as parts of X – and hence have no tendency to adhere to a distinct atom Y.

Axiom B, it is clear, would be sufficiently strong to guarantee the physical indivisibility of the atom while allowing the Epicurean an answer to Aristotle. When two atoms X and Y come into contact, the resulting compound body XY is divisible in virtue of the fact that their adjacent superficial layers of minimal parts Sx and Sy are essentially parts of X and Y respectively, thus:

X	Sx	Sy	Y

On this view, Sx is inseparable from X and Sy from Y, but Sx is separable from Sy and hence atom X from atom Y. The compound body XY is therefore physically divisible in spite of the fact that it contains no inner void.

So far, then, we have found two quite different arguments for the physical indivisibility of the atom. There is the simple Democritean argument based on Axiom A, and there is the curious modal argument attributed to the Epicureans by Konstan and resting on Axiom B. Let us turn to the texts of Epicurus and Lucretius in search of an answer to the question: 'Did the Epicureans accept Axiom A?'

What ground does Epicurus propose for his postulation of physical indivisibles? His argument runs as follows:

. . . among bodies some are compounds, and others those of which compounds are formed. And these latter are indivisible and unalterable (if, that is, all things are not to

---

[3]   The no-contact theory was ascribed to Democritus by Philoponus. For discussions, see Barnes, Vol. 2, p. 47, and Konstan, 2, p. 407.

[4]   Konstan, 2, p. 407. Epicurus' lost treatise 'On Contact' (see Diogenes Laertius, p. 557) might, had it survived, have supplied valuable information about these esoteric aspects of Epicurean theory.

be destroyed into the non-existent but something permanent is to remain behind at the dissolution of compounds): they are completely solid in nature, and can by no means be dissolved in any part. So it must needs be that the first-beginnings are indivisible corporeal substances.[5]

Why should Epicurus fear that infinite physical divisibility would 'crush and squander the things that exist into the non-existent'?[6] Bailey suggests that Epicurus does not mean this literally: all that is meant, he alleges, is that if bodies were divisible *ad infinitum* they would in the course of time become ground down into a fine 'cosmic dust' from which nothing could come-to-be. Since the 'forces of destruction' are always more powerful than those of construction, without a limit to division there would be no world as we know it.

If this is indeed Epicurus' meaning, it has been formulated in a most misleading manner; twice in the text Epicurus states quite explicitly that infinite physical divisibility would involve the annihilation of matter, not merely its dissolution into 'cosmic dust'. We should, then, be chary of accepting Bailey's reading if a better one is available.

Now, if we are prepared to attribute to Epicurus acceptance of Axiom A, the argument becomes clear and perspicuous. Given Axiom A, no body can be divided unless it contains at least some minimum quantum of vacuum. But, since every division presupposes the prior existence of a further quantum of void, a body could not be infinitely divisible without being entirely composed of vacuum. An infinitely divisible body would be, quite literally, *'non existent'*: there could be no such *body*. Any given body will only contain a finite number of void spaces, and will only be divisible so many times. Without bodies free from inner void (and hence, by Axiom A, physically indivisible) there would be no matter at all.

The argument can also be approached from another angle. Take two points (or minima) p_1 and p_2 within a compound body. Suppose there to be void space between them. If so, the body can be divided between p_1 and p_2. But now take one of the bodies produced by the cut and ask whether that is divisible (i.e. contains internal void). If so, it is clear, the argument moves on to the products of the next division, and so on, thus:

---

[5] 'Letter to Herodotus', 41; Bailey, 1, p. 23.

[6] 'Letter to Herodotus', 56; Bailey, 1, p. 33.

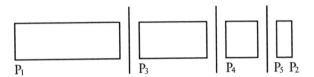

P_1 P_3 P_4 P_5 P_2

Is p_5p_2 divisible? If it is, mark in one more quantum of void and ask the question again about the bodies produced by its division. Without perfectly solid matter* (free from internal vacuum) there would be no matter at all. And if, by Axiom A, perfectly solid matter* is immune from physical division, the case for physical indivisible is complete.

We can therefore attribute to Epicurus a plausible and convincing argument for perfectly solid and hence physically indivisible atoms, bodies which have, literally, 'no *where* or *how* to be divided' – no 'where' because there is no place within the atom for it to be divided; no 'how' because of Axiom A. Without such bodies, he feels, all matter would dissolve into mere non-existence: unless there are solid particles, there can be no matter at all.

To make sense of Epicurus' argument, we have had to attribute to him acceptance of Axiom A, the very principle which, according to Konstan,[7] he must have abandoned in order to rebut Aristotle's objection. This is, to say the least, awkward. If Konstan is right and Epicurus denies Axiom A, the arguments against infinite physical divisibility collapse: if division does not presuppose inner void, one can divide solid bodies without 'squandering' the existent into the non-existent. (One cannot of course divide them infinitely – this would be precluded by the existence of mathematical minima – but the argument for physical indivisibles was intended to be independent of the mathematical arguments[8]). If Axiom A is false, a solid can be divided into solids *without* being weak or sponge-like, capable of being 'crushed' and 'squandered away'. On the other hand, if Epicurus accepts Axiom A, how can he answer Aristotle's objection?

Perhaps, it might be suggested, his argument is meant to work at the level of the cohesiveness or unity of the atom, not

---

7   Konstan, 2, p. 407.
8   In both Epicurus and Lucretius, physically indivisible particles (atoms) are introduced *before* mathematical indivisibles (minimal parts) and independently of them.

merely of its solidity. Perhaps a single atom has a special 'cohesiveness' lacked by two adjacent atoms. How might one attempt to justify such a thesis?

Minimal parts, one might begin, are essentially parts of atoms. They cannot be conceived to exist independently of atoms, in separation from them. They are, one might say, 'nothing' independently of the atoms they constitute. If, then, atoms were divisible, they would be divisible into 'nothing', since no specifiable independently-existing product would survive their dissolution. Minimal parts, Lucretius insists, cannot exist separately – but even if, *per impossibile*, they could, they could not recombine to form atoms and thence compounds.[9]

This sounds promising. The atoms, on this account, are indivisible not simply because of their fullness or solidity but because their constituent parts cannot but constitute them. A compound is divisible not primarily in virtue of containing inner void (although the great majority of them do) but because its constituent atoms can exist separately – atoms are the sorts of things that can have independent existence. We can make sense of the division of a compound because the independent existence of its constituent atoms is perfectly intelligible; by contrast, the division of an atom into minimal parts is unintelligible, since these parts cannot exist (are 'nothing') in isolation.

The whole weight of this curious argument rests on the modal claim that minimal parts are – like surfaces – necessarily parts of bodies. But the Epicureans conceived surfaces as physically real and three-dimensional *parts* of bodies, their outermost 'skins' of atoms. If a surface is thought of as a two-dimensional abstract object, the thesis 'surfaces cannot exist separately from bodies' becomes a plain conceptual truth. If, however, surfaces are thin three-dimensional films, there seems no reason why they should not have independent existence – it is even central to the Epicurean theory of '*eidola*' that they can and do. If then minimal parts are to atoms as surfaces to bodies (on the Epicurean account) there seems to reason why they should not exist separately.

There is, however, a crucial distinction between the minimal parts of atoms and the surfaces of macroscopic bodies. The latter (and hence the *eidola* too) are made up of thin films of atoms, and hence have external surfaces – each atom has an outermost layer of minimal parts. Minimal parts,

[9] Lucretius, Book 2, 628-634, pp. 207-209.

however, cannot – on pain of infinite regress – themselves have surfaces.[10] If they had distinct surfaces, they would not be minimal parts, parts than which nothing smaller can be discriminated by the mind. And there must be such parts if the Zenonian paradoxes of infinite divisibility are to be met.

Minimal parts, then, are somehow *less than bodies*. Being themselves extremities, they cannot without absurdity or regress have extremities, or indeed distinct parts at all. They cannot therefore touch one another, since they lack surfaces. (If they had distinct surfaces, they would be theoretically divisible, and we would be back with the paradoxes of infinite divisibility which the minima were posited precisely in order to avoid.)

Unfortunately, however, this argumentation will not guarantee the existence of physical indivisibles – at least not at the level required. All that the preceding argument has shown is that minimal parts cannot have *separate* existence. Call this Axiom C:

> *Axiom C*: Minimal parts cannot be conceived to exist independently of the atoms they constitute.

It immediately follows from Axiom C that an atom cannot be divided *into its minimal parts*. It does not, however, entail the absolute physical indivisibility of the atom: Axiom C is significantly weaker than Axiom B. It is, for example, perfectly compatible with C, but not with B, for an atom made up of four or more minimal parts to undergo physical division – no minimal part would come to exist in isolation as a result of such a division. (If we think of the majority of atoms as made up of *many* minima, the point becomes still clearer.) Moreover, Axiom C will not bar the following possibility:

Such a separation of two adjacent atoms X and Y does not violate Axiom C (no minima come to have separate existence) but is firmly prohibited by Axiom B.

This yields the following problem-situation. If, following Konstan, we argue that Epicurus denied Axiom A, we need to

10   See Konstan, 2, p. 405.

find an alternative account of the physical indivisibility of the atoms. We have discovered that the Epicureans could plausibly have argued for Axiom C (Lucretius certainly does so). Unfortunately, however, Axiom C does not ensure the absolute physical indivisibility of each and every atom, as the above example shows. To make that guarantee one needs the much stronger Axiom B. But Axiom B seems to involve attributing to each minimum part a sort of 'knowledge' of to which atom it belongs: it is hard to see how one might begin to account for this in terms of the mere distribution of matter* in space. Within the compound body XY, what is *special* about the layers of minima that constitute Sx and Sy? Why should the compound divide between those two layers of minima rather than anywhere else? Does it 'remember' how it came together, and tend to divide in the same way? How then could an Epicurean hope to justify an axiom of the strength of B?

What can we learn from Lucretius on this topic? If we turn to his extended argument for physical indivisibles in Book 1 of *De Rerum Natura* (505-634), we find a series of more or less familiar points, beginning with the almost Eleatic antithesis between (void) space and matter*:

> First, since we have found existing a twofold nature of two things far differing, the nature of body and of space .. it must needs be that each exists alone by itself and unmixed. For whenever space lies empty, which we call the void, body is not there; moreover, wherever body has its station, there is by no means empty void. Therefore the first bodies are solid and free from void.[11]

Matter* and vacuum differ in such a way that, unlike black and white paints, or wine and water, they cannot be mixed. They are *mutually exclusive* by nature. A compound of matter* and vacuum therefore consists of discrete and distinct portions of each. If a body is to exist at all, it must contain some matter*, which can only exist in the form of perfectly solid particles. Matter*, Lucretius continues, delimits and is delimited by vacuum: each nature marks off the boundaries of the other.[12] Moreover,

> if there were nothing which was empty and void, the whole would be solid; unless on the other hand there were bodies determined, to fill all the places that they held, the whole universe would be but empty void space. Body, then, we

---

[11]   Lucretius, Book 1, 505-510, p. 201.

[12]   *Ibid.*, 511-519, p. 203.

may be sure, is marked off from void turn and turn about, since there is neither a world utterly full nor yet quite empty. There are therefore bodies determined, such as can mark off void space from what is full.[13]

Unless, says Lucretius, there were bodies, however small, which completely filled the space they occupied, there would be only empty space. Given a conception of matter* and vacuum as strictly mutually exclusive, there must be solid particles of matter* if there are to be bodies at all. These solid particles, Lucretius continues, must be *eternal*, since they:

> cannot be broken up when hit by blows from without, nor again can they be pierced to the heart and undone, nor by any other way can they be assailed and made to totter . . . For it is clear that nothing could be crushed in without void, or broken or cleft in twain by cutting, nor admit moisture nor likewise spreading cold or piercing flame, whereby all things are brought to their end. And the more each thing keeps void within it, the more it is assailed within by these things and begins to totter. Therefore, if the first bodies are solid and free from void, as I have shown, they must be everlasting.[14]

Bodies can be dissolved either by external 'blows' (cutting, crushing) or by the internal action of heat, cold or moisture. Neither of these types of process, Lucretius insists, can affect a perfectly solid body, since all require the existence of internal vacuum. Thus Lucretius quite explicitly asserts Axiom A and commits himself to the basic Democritean argument for physical indivisibles. What he would say in answer to Aristotle's objection is far from clear. For the moment, however, let us set that difficulty aside and move on:

> Again, if nature had ordained no limit to the breaking of things, by now the bodies of matter would have been so far brought low by the breaking of ages past, that nothing could be conceived out of them within a fixed time, and pass on to the full measure of its life; for we see that anything you will is more easily broken up than renewed. Wherefore what the long limitless age of days, the age of all time that is gone by, had broken ere now, disordering and dissolving, could never be renewed in all time that remains. But as it is, a set limit to breaking has, we may be sure, been appointed, since we see

---

13   *Ibid.*, 520–530, p. 203.
14   *Ibid.*, 530–540, p. 203.

each kind of thing renewed, and at the same fixed seasons ordained for all things after their kind.[15]

An analogy with modern chemistry, although anachronistic, may prove enlightening. Lucretius is claiming in effect that in an association/dissociation reaction the equilibrium constant always favours the dissociation reaction. The nightmare possibility averted by Atomism is therefore the following:

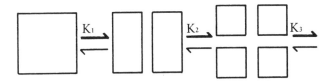

Here the thicker arrows on the left to right (dissociation) steps indicate that these reactions are favoured by K_{1-3}, etc. If there is no limit to this series of reactions, matter will disintegrate into ever-finer 'cosmic dust', into particles too minute ever to generate anything of perceptible size. This, however, does not happen: Nature, as Lucretius never tires of reminding us, maintains a genuine equilibrium of generation and decay. If dissociation is *always* kinetically favoured over association, only a limit to physical division will safeguard this equilibrium. The gestation period for a human foetus was, is, and will for the foreseeable future remain, nine months. If the matter of the universe were disintegrating into ever-finer 'cosmic dust', this period would get progressively longer. Indeed, the generation of men and animals would become ever more difficult generation after generation until eventually the statistical probability of its happening at all became so remote as to be negligible.

Given a limit to division, however, it is easy to ensure equilibrium. If the majority of the atoms in the universe are at any time in a free state, the absolute *rates* of generation and corruption can be equal, despite the fact that the dissociation-reaction has a higher *rate-constant*. (Although the sense of anachronism may be strong in the reader, I do not believe this gloss seriously distorts Lucretius' meaning.)

The next argument is simpler. If, says Lucretius, we posit as first principles bodies which, in virtue of their perfect solidity, are absolutely hard, we can explain the existence of soft (compound) bodies in terms of the admixture of void. If,

[15] *Ibid.*, 530–540, p. 203.

on the other hand, the 'first-beginnings' were soft, it would be impossible to account for the hardness of, e.g., iron and flint.[16] There are then, he concludes,

> bodies that prevail in their solid singleness, by whose more close-packed union all things can be riveted and reveal their stalwart strength.[17]

The eternity of natural kinds and the constancy of Nature's laws provide Lucretius with his next argument for the immutability and eternity of the atoms:

> For if the first beginnings of things could be vanquished in any way and changed, then too would it be doubtful what might come into being, what might not, yea in what way each thing has its power limited and its deep-set boundary stone, nor could the tribes each after their kind so often recall the nature, habits, manner of life, and movements of the parents.[18]

The orderliness and lawfulness of Nature are evidence of the immutability of the atoms. If the 'first-beginnings' were cut up, ground down, or in any way altered by the blows and buffetings they receive, the very course of Nature would be altered. Centuries later, Newton was to make exactly the same point:

> While the particles continue entire, they may compose bodies of one and the same nature and texture in all ages; but should they wear away or break in pieces, the nature of things depending on them would be changed.[19]

Likewise in modern atomic theory all atoms of (the same isotope of) the same element must be regarded as *qualitatively identical*, regardless of their origins and past histories. If, for example, atoms of hydrogen and oxygen could be in any way altered by the various combinations into which they enter during their histories, water would not be a single uniform chemical compound with clearly defined properties. We now have of course a great deal of evidence for what was in Lucretius and Newton primarily a demand, but the basic claim is in essentials the same.

---

16   *Ibid.*, 565–573, p. 205.

17   *Ibid.*, 574–576, p. 205.

18   *Ibid.*, 592–598, p. 207. I have omitted one intervening argument: if it adds anything to the force of this one I do not know what.

19   Newton, 2 (*Opticks*), p. 260.

The last three of Lucretius' arguments for physically indivisible atoms turn on the conception of minimal parts. The first introduces the concept of a 'least', 'partless', or minimal part (like the extremity of an atom) and then argues that, since such a least part 'is itself but a part of another', and hence cannot exist by itself, it follows that atoms, consisting as they do of minimal parts packed 'in close array', must be 'of solid singleness', made up of 'a close, dense mass of least parts'.[20]

This is followed by the mathematical argument against infinite divisibility and for indivisible minima. Since, it concludes, such minima must exist, 'those first-beginnings [atoms] too you must needs own exist, solid and everlasting'.[21] Since infinite divisibility is absurd, there must be minimal parts. But these are, essentially, parts of atoms – they cannot be thought of as having a separate existence. (Nor, if *per impossibile* they were to exist separately, could they give rise to anything by aggregation.[22])

This last trio of arguments seems sufficient evidence of Lucretius' acceptance of Axiom C, the principle that minimal parts cannot exist separately, in isolation from bodies. Unfortunately, as we have already seen, C is insufficiently strong to guarantee the physical indivisibility of the atom – to give that guarantee one needs one or other of the stronger axioms A or B.

The problem-situation is now as follows. Democritus argued for physical indivisibles on the basis of A. Aristotle then deduced from A that atoms should adhere, or even coalesce, on coming into contact. Now in Epicurus and Lucretius we have found strong evidence of acceptance of the classical Atomist Axiom A – both have arguments which quite clearly turn on A. We have also, however, found traces of an argument for C and the supposition (in Lucretius) seems to be that C is sufficient to ensure physical indivisibility. Unfortunately, that supposition is false – to argue for physical indivisibility the stronger principle B would be needed. But there is no trace in either Epicurus or Lucretius of an argument for B; it is moreover hard to see how, consistently with their philosophy of Nature, they could have provided one. How does a given minimal part 'know' to which atom it belongs?

---

20   Lucretius, Book 1, 599–610, p. 207.

21   *Ibid.*, 615–627, p. 207.

22   *Ibid.*, 628–634, pp. 207–209.

I cannot resolve all these mysteries. One suspects, however, that the situation may have been as follows. Epicureanism probably consisted of a popular and exoteric body of doctrine and a more difficult and esoteric research-programme. My suspicion is that the Epicureans, in their popular works at least, went on arguing on the basis of Axiom A even when they knew that this was problematic; and that the developed theory of minimal parts belonged to the more esoteric research work. This research programme clearly led to Axiom C: perhaps the Epicureans thought C sufficient for the physical indivisibility of the atom. If so, the advanced-grade students may have been taught to reject A (for Aristotle's reasons) and argue for physical Atomism on the basis of C. Perhaps they even confused C with B. Who knows?

The above is a purely conjectural piece of rational reconstruction. I find these issues difficult and obscure, and am far from satisfied with my tentative answer. Unfortunately, I know no better account of this thorny problem of Epicurean exegesis.

APPENDIX 2
Aristotle on Prime Matter

When something F becomes non-F, what is it that changes? To what does the term 'something' refer? Clearly, it indicates the existence of some abiding X such that X is F at a time t_0 and is no longer F at a later time t_1. Without such an abiding X, we should have not change but (miraculous) annihilation and *ex nihilo* creation, which, Aristotle agrees with Parmenides, are quite unintelligible. There are, says the Stagirite, three 'principles' of change – the form gained, the form (or privation) lost, and the abiding substratum.[1] This logical analysis of the structure of change has profound implications for Aristotle's view of the physical world.

When a green apple becomes red, or Socrates becomes musical, the question 'What has changed?' is unproblematic; the apple and Socrates are respectively the underlying substrata of the changes mentioned, remaining substantially identical throughout them. For all such cases of accidental change the substratum is the concrete individual substance or 'this' which, perceptibly, remains the same through the change.[2] There are, however, more radical and profound changes in Nature. A mouse dies, its body decays in the earth, and a mushroom springs up on the same spot. No substance or 'this' remains perceptibly self-identical through this process; we speak, therefore, of 'coming-to-be' and 'ceasing-to-be' and not of 'alteration'.[3] The mouse has not become a mushroom: it has simply ceased-to-be, while the mushroom has come-to-be. Is there an underlying substratum here? If so, what is it?

The natural answer turns on identity of material constituents. Democritus, for example, would say that the same Atoms$_1$ . . .n which once stood to one another in the n-adic relation R_1 now stand in a new n-adic relation R_2; the individual atoms, with their characteristic sizes and shapes,

---

[1] *Physics*, 1, 7, esp. 191a 2-7.

[2] See Robinson, p. 171.

[3] See *De Gen et Corr.*, 1, 4, for the distinction between alteration and coming-to-be and ceasing-to-be.

are conserved. (One begins to see the attraction of the 'generation = aggregation' thesis.) This approach is, however, not open to Aristotle: the underlying material substratum which is required for generation and corruption[4] is not to be identified with atoms.

Matter, says Aristotle, 'is to be identified with the *substratum* which is receptive of coming-to-be and passing-away'.[5] But what is this substratum? Perhaps, one might suggest, the material substratum can be identified with the *elements*, Fire, Air, Water and Earth (FAWE). This would give us the position of Empedocles.

Unfortunately, however, Aristotle believed that the 'so-called elements' FAWE could be *transmuted* into one another.[6] What then is the substratum for these changes? When A changes into W, what remains the same?

It is at this point that King[7] takes the heroic step of making the four primary 'powers' Hot, Cold, Dry and Moist (HCDM) the true elements of material things. We have, however, already sufficiently refuted this reading of Aristotle in Chapter 3, Part 1. In the first place, Aristotle's metaphysics commit him to a clear distinction between things and properties;[8] in the second place, he insists, over and over again, that one contrary (e.g. H) does not act on another[9] (e.g. C) but 'both act on a third thing different from both'.[10] Action and passion presuppose a material substrate distinct from 'the contraries' (HCDM) themselves: 'We do not find that the contraries constitute the substance of any thing'.[11]

What then is the substratum which, when combined with the 'powers' HCDM, yields FAWE? At this point, I believe, Aristotle is forced to introduce the concept of *prime matter* (P.M.). If he is to retain both the analysis of change (*Physics*, 1, 7) *and* the transmutation of FAWE (*De Gen et Corr*, 2, 4 and 5), he cannot avoid the postulation of P.M. I therefore

---

4   *Physics*, 1, 7, 190a 34-b4.

5   *De Gen et Corr.*, 1, 4, 320 a 2-3.

6   For a detailed account of the transmutations of FAWE, see *De Gen et Corr.*, 2, Chs. 4 and 5.

7   See King, 2. For criticisms of King, see Solmsen, 3, and Robinson.

8   Robinson, p. 183, insists that HCDM could not exist alone, and are therefore not to be regarded as the elements of things.

9   Sellars in McMullin, Ed., 1, pp. 262-263.

10   *Physics*, 1, 6, 189a 21-27.

11   *Ibid.*, 29-30.

conclude, with Solmsen[12] and Robinson [13] (and against King, Charlton,[14] and Jones,[15]) that the existence of P.M. is essential to Aristotle's natural philosophy. As Solmsen says, the transmutation of FAWE, involving as it does the exchange of HCDM, requires

> a common matter which can exchange, say, the qualities dry and hot against moist and cold and thus make possible the change from fire to water.[16]

Thus also Aristotle:

> The kinds of matter, then, must be as numerous as these bodies [FAWE], i.e. four, but though they are four there must be a common matter of all – particularly if they pass into one another – which in each is in being different.[17]

In its rôle as material substratum P.M. is the *same* in all things; in actuality it is F-like, A-like, etc., in FAWE respectively:

> . . . their matter is in one sense the same, but in another sense different. For that which underlies them, whatever its nature may be *qua* underlying them, is the same: but its actual being is not the same.[18]

Perhaps the most explicit statement of a common matter for FAWE comes from the *Meteorologica*:

> Fire, air, water, earth, we assert, originate from one another, and each of them exists potentially in each, as all things do that can be resolved into a common and ultimate substrate.[19]

I can only presume that Barrington Jones was unfamiliar with this passage when he wrote that *hulē* was for Aristotle *not* a single common stuff *of* which all things are made but instead a mere classificatory label for the multitude of materials

---

12 Solmsen, 3.

13 Robinson, pp. 187-188.

14 Charlton, Ed., p. 129 ff.

15 Jones, p. 474 ff.

16 Solmsen, 3, p. 250.

17 *De Caelo*, 4, 5, 312a 31-34. For the argument from transmutability to a common matter, see also Bolzan, p. 134.

18 *De Gen et Corr.*, 1, 3, 319b 2-4.

19 *Meteorologica*, 1, 3, 339a 36-b1.

(wood, metal, clay, etc.) *from* which things come-to-be.[20]
The above quotations seem clearly to support the classical
conception of P.M. as the one common material substratum
of all bodies.

P.M., of course, never exists *as such*; it will always be
found in some form or other:

> Our . . . doctrine is that although there is a matter of the
> perceptible bodies (a matter out of which the so-called
> 'elements' come to be) it has no separate existence but is
> always bound up with a contrariety.[21]

Even more explicitly:

> We must reckon as an 'originative source' and as 'primary'
> the matter which underlies, though it is inseparable from,
> the contrary qualities: for 'the hot' is not matter for 'the
> cold' nor 'the cold' for 'the hot', but the *substratum* is
> matter for them both. We therefore have to recognise three
> 'originative sources': *firstly* that which is potentially
> perceptible body, *secondly* the contraries (I mean, e.g.,
> heat and cold) and *thirdly* Fire, Water, and the like.[22]

To an adherent of the classical interpretation of Aristotle this
passage is plain sailing: P.M. (= 'potentially perceptible
body') combines with 'the contraries' HCDM to yield the
simple bodies FAWE. What exegetical gymnastics are
required here of the critics of P.M. I cannot imagine.

P.M. is only 'potentially perceptible' because it could not be
actually perceptible without having some form and qualities,
which, in its rôle as the ultimate substratum for coming-to-be
and passing-away, it cannot possess. If any substance can
eventually (i.e. after a sufficient number of intermediate
stages) be transmuted into any other, it follows that P.M.
cannot have a nature *like* any sensible substance. It must be-
potentially all sensible stuffs, and hence *per se* be-actually
none of them.

Unfortunately, Aristotle leaps from 'P.M. is like none of
the sensible stuffs of our experience' to 'P.M. has no nature of
its own at all'. It is, he insists, totally indeterminate in nature,

---

20   Jones, p. 495.

21   *De Gen et Corr.*, 2, 4, 329a 24–27. The long argument at 317b 22–33,
often cited by critics of P.M., may be read as a refutation only of the
actuality of (separate) P.M. A P.M. which always exists 'informed' does
not succumb to the dilemma which Aristotle raises there.

22   *De Gen et Corr.*, 2, 1, 329a 30–36.

devoid of all forms and qualities, a 'pure potency' that is not actually anything at all. This, as we shall see, gives rise to paradoxical consequences.

P.M. is introduced, then, to resolve the identity-problems arising out of substantial coming-to-be and passing-away. But if P.M. is imperceptible and totally indeterminate, does it not become unknowable? Is not the term '*prōtē hulē*' itself just a meaningless sound?[23] How can we say, think, or know anything whatever about it? I can use 'stuff' terms like 'chalk' and 'cheese' in virtue of my ability to point to some things and say 'chalk' and to others and say 'not-chalk'. But any and every material object indifferently is 'potentially P.M.' and 'not actually P.M.' And even the sense of 'potentially' within Aristotle's philosophy has changed here (as it is rather wont to do[24]). To say that X is potentially F is normally to imply the possibility of X becoming F; but P.M. as such can never become perceptible.

How then does Aristotle seek to justify the introduction of this concept? His answer is as follows:

> The underlying nature is an object of scientific knowledge, by an analogy. For as the bronze is to the statue, the wood to the bed, or the matter and the formless before receiving form, so is the underlying nature to substance, i.e. the 'this' or existent.[25]

The 'underlying nature' here is most naturally read as P.M.;[26] indeed it is hard to see what else could only be grasped by such a roundabout analogical method. (Ordinary material stuffs such as wood, metal, etc., are clearly known by more direct means.)

Although this analogy may enable us to attach some minimal significance to the notion of P.M., we shall only appreciate its full import by studying the rôles it plays in Aristotle's system. What, then, are those rôles?

First and foremost, matter, in virtue of its rôle as the substratum for all generation and corruption, must itself be immune to those processes. Matter is precisely what is logically *presupposed* in *genesis* and *phthora*: it is thus

---

23   Charlton, Ed., p. 140.

24   Cf. *Metaphysics*, 9, 6, 1048b 9–17, for the (weak) sense in which the infinite and the void may be said to exist potentially.

25   *Physics*, 1, 7, 191a 7–12.

26   See Owens in McMullin, Ed., 1, p. 91.

incoherent to imagine that it could itself, for example, come-to-be:

> For if it came to be, something must have existed as a primary substratum from which it should come and which should persist in it; but this is its own special nature.[27]

Since matter is logically presupposed as the ultimate substratum of generation and corruption, it cannot itself undergo those processes. It is critically important here that Aristotle defines matter as 'the primary substratum of each thing, from which it comes to be without qualification, and which persists in the result'.[28] The matter of a thing (*pace* Barrington Jones[29]) persists in it. When a product is made of wood, bronze, etc., says Aristotle, 'the first matter is preserved throughout'.[30] Aristotle's *hule* cannot therefore be identified, as Jones is inclined to do, with the pre-existent stuff from which a substance comes-to-be but which itself simultaneously ceases-to-be. To solve the Eleatic problem of identity through change, Aristotle requires something which 'persists in the result'. Since no *actual* material substratum (Democritus' atoms, Empedocles' FAWE, Anaxagoras' quality-stuffs) is acceptable to Aristotle, he is left with P.M. as '*potency*', as a purely potential something.

P.M., therefore, is immune to *genesis* and *phthora*. (It might therefore be thought an appropriate subject for a conservation law, but this, as we shall see, is problematic.) The fact that it is this formless, indeterminate stuff which is conserved shows that the Eleatics were wrong: all sorts of coming-to-be and coming-to-be-F are possible. Without P.M., however, the generation and corruption of substances would be as self-contradictory as Parmenides claimed.[31] To avoid characterising a process of change as 'the F becoming non-F' we need to redescribe it as 'the X which was F becoming non-F'. For accidental change, the 'X' can be replaced by the name of a substance ('Socrates', 'Bucephalus'); but for substantial changes (e.g. mouse to mushroom) no such expression can be found and we need to revert to P.M. One might be inclined to say 'this *portion* of P.M.', and to think of P.M. as stuff-like, space-filling and quantified. (The

---

27 *Physics*, 1, 9, 192a 28–30.
28 *Ibid.*, 31–32.
29 Jones, pp. 488–489.
30 *Metaphysics*, 5, 3, 1014b 30–32.
31 See Fitzgerald in McMullin, Ed., 1, p. 65.

P.M. of a mouse can give rise to a mushroom, but *not* to a pine tree.)

Thus, P.M., in the *physical* works serves as the abiding substratum underlying generation and corruption, guaranteeing (a) the identity of some logical subject, and (b) the conservation of some 'quantity of stuff'. In the *Metaphysics*, however, it serves other (and perhaps incompatible) purposes:

> for the predicates other than substance are predicated of substance, while substance is predicated of matter.[32]

Here P.M. is playing the rôle of the ultimate subject of predication: when all the predicates of the various categories (quality, quantity, etc.) are 'stripped away', says Aristotle, 'evidently nothing but matter remains'.[33] The *apple* is red, sweet, juicy, 3 inches in diameter, etc., but *what* is apple-like? Of what is 'being an apple' predicated? Only P.M. can fill that rôle. Of course P.M. is not *per se* man or horse, but is only accidentally 'humanised' or 'equinised' in particular men and horses:[34] it 'supports' substantial attributes without having them.

We now have two quite distinct definitions of P.M., *viz*:

1. The ultimate substratum of generation and corruption.
2. The ultimate subject of predication.

Lobcowicz[35] asks *why* Aristotle conflated these two quite distinct concepts. The answer becomes clear if we think of a question like '*what* was a mouse and is now a mushroom?' as (a) intelligible (meaningful) and (b) answerable. If such a question *is* intelligible (as Charlton and other critics of P.M. have denied[36]) then the formula 'X is to F at t_0 and non-F at t_1' will be applicable to *all* changes: this seems to have been the assumption which Aristotle (rightly or wrongly) felt he had to make to supply an answer to Parmenides.[37] The place left by the missing substance-term ('X') must, for substantial coming-to-be and ceasing-to-be, be taken by (portions of?) P.M.

32   *Metaphysics*, 7, 3, 1029a 22–23.

33   *Ibid.*, 11.

34   Owens in McMullin, Ed., 1, p. 85.

35   Lobcowicz in McMullin, Ed., 1, p. 96.

36   We should not, says Charlton (p. 140), say that something underlies coming-to-be and ceasing-to-be unless we have a description under which it stays the same throughout. 'Portion of P.M.', presumably, will not do.

37   See McMullin in McMullin, Ed., 1, p. 188.

P.M. also serves a third function, that of the individuation of qualitatively identical substances. If two things are qualitatively identical, Aristotle reasons, they are identical in form and can only be distinguished by their matter. This, however, presupposes quantification of P.M. Because bronze is quantified, one can cast numerically different but specifically similar bronze artefacts. Each new statue requires more bronze: the more one makes, the more bronze is required. Quanta are not created by cutting up the material; rather, the quantification (not necessarily discrete, of course – continuous quanta will suffice) is a necessary condition of the possibility of division.[38] This individuating rôle therefore involves the existence of *bits* or *portions* of P.M. One should be able to speak of 'the P.M. of *this* statue' as distinct from that of another; it should also make sense to say that two such statues contain twice as much P.M. as one.

P.M., however, has only *potential* existence: it is not a substance or 'this' – it is indeed not *actually* anything:

> By matter I mean that which, not being a 'this' actually, is potentially a 'this'.[39]

Not being a substance or 'this', matter cannot have quality, quantity, relation, etc.: all the other categories apply only to substances. Hence matter is in itself:

> neither a particular thing nor of a certain quantity nor assigned to any other of the categories by which being is determined . . . the ultimate substratum is of itself neither a particular thing nor of a particular quantity nor otherwise positively characterised;[40]

> that which is not a substance or a 'this' clearly cannot possess predicates drawn from any of the other Categories either – e.g. we cannot attribute to it a quality, quantity, or position.[41]

P.M., therefore, is devoid of quality, quantity, position, action, passion, etc.: predicates in all these categories are applicable only to substances.[42] I can meaningfully ask 'How much cheese do you want?' or 'Where have you put the cheese?' but questions of that sort do not even make sense for

---

38   See Bobik in McMullin, Ed., 1, p. 283.

39   *Metaphysics*, 8, 1, 1042a 26–27.

40   *Ibid.*, 7, 3, 1029a 20–26.

41   *De Gen et Corr.*, 1, 3, 317b 9–12.

42   See Owens in McMullin, Ed., 1, p. 86.

P.M. P.M. has no amount, no position, no nature – it is-not actually (anything at all). What 'is-potentially', says Aristotle significantly, 'is spoken of both as being and as not-being'.[43]

There is therefore, to put it mildly, a distinct tension in Aristotle's thought. (More bluntly, Aristotle's account of matter is incoherent: no satisfactory and internally consistent reading is possible that takes into account all his scattered remarks on the subject). On the one hand, the physical treatises present us with an account of P.M. as fairly featureless and indeterminate but not verging on 'non-being'; on the other hand, the *Metaphysics* seem effectively to deny that matter is (actually) anything at all.

Several commentators have (rightly, I feel) expressed a preference for the former line of thought. Sokolowski,[44] for example, has argued that P.M. is spatially extended, and that the forms by which P.M. is differentiated do not supply this extension but merely make it manifest. Matter may be present in varying densities in space, but it could never be rarefied to a void, nor compressed to a point. Spatial extension, argues Sokolowski, would not suffice to make P.M. any determinate stuff: it would not therefore jeopardise its rôle as the substratum of *genesis* and *phthora*. Spatiality, he contends, is not itself an attribute but a precondition of attributes.

Similar views were proposed by William of Ockham,[45] who interpreted Aristotle's P.M. as extended, quantified stuff with some minimal degree of *actuality* of its own. (God could, for Ockham, actualise 'uninformed' P.M.) Suppes[46] too opts for this sort of reading: the Aristotelian theory of P.M. is, he thinks, coherent, basically sound, and of potential value for modern physics! 'Energy', after all, is a fairly amorphous, all-embracing concept: we might be tempted to equate 'energy' with formless and plastic P.M. as a material principle. This reading would make P.M. something *actual*, but not actually (like) any sensible stuff, and hence potentially any of them. From the point of view of physics, this seems the best that can be made of Aristotle's position.

Unfortunately, however, this rather attractive physical theory, involving as it does a space-occupying, quantified P.M. with at least some actuality of its own, falls foul of the

---

43   *De Gen et Corr.*, 1, 3, 317b 18.

44   Sokolowski, p. 277.

45   See Wolter's paper on Ockham's reading of Aristotle's doctrine of P.M. in McMullin, Ed., 1, p. 130 ff.

46   Suppes, 2, p. 27.

doctrine of the *Metaphysics*. Whatever has extension has (a) magnitude (quantity) and (b) place. But position and quantity inhere only in substances. P.M. as such can have no such features. Hence McMullin's conclusion that:

> despite the commonsense appeal of a qualityless quantified material that remains invariant through change, it must be said that Aristotle's authority was solidly against it.[47]

Even if P.M. were stuff-like and space-filling, this would not provide a unit for its quantification. Aristotelian matter can (unlike the Atomists' matter*) occupy space more or less intensely: it cannot therefore be quantified by bulk or volume. And, since Aristotle believed that elementary fire was *absolutely light*[48] (i.e. with intrinsic *levity*), quantification of matter by weight is out of the question.

Here we find the Stagirite faced with a terrible dilemma. In order to cope with a variety of physical phenomena he needs to make sense of the idea of the quantification of P.M., its division into distinct portions, and its occupation of space. Unhappily, however, this is precisely what his metaphysical doctrines will not permit.

What reasons have we for the claim that P.M. must be subject to quantification? They are as follows:

1. It is supposed to be the proper subject of a conservation law: no quantum of P.M. can come-to-be or cease-to-be.
2. It functions in the individuation of distinct objects, hence must consist of parts or portions.
3. Aristotle explicitly states that 'Dense differs from rare in containing more matter per unit area'.[49]
4. The rate of transmutation (e.g. $W \longrightarrow A$) would most naturally be characterised in terms of the amount of P.M. transformed in unit time.[50]
5. If one transmutes one volume of W to ten of A, then condenses those 10 volumes of A back to one of W, one feels intuitively that the '*quantity of matter*' has been conserved throughout. (If one regenerated only half the original volume of water, one would feel that some matter had been lost.)

---

[47]   McMullin in McMullin, Ed., 1, p. 20.

[48]   See *De Caelo*, 4, 1.

[49]   *De Caelo*, 3, 1, 299b 8.

[50]   We could of course measure rate of appearance of A and rate of disappearance of W, but this does not capture the idea that the *same* matter is being transformed from one to the other.

6. The matter of a mouse, we said, is-potentially a mushroom, but is not potentially a pine tree.

Natural philosophy seems to demand an actual, space-occupying, quantified substratum for these changes – this seems their most natural explanation. Aristotelian P.M., however, can be quantified neither by weight nor volume. This, however, is a comparatively minor difficulty: we can attribute to P.M. an unseen 'real essence' (disanalogous to those of sensible things) and quantify it in terms of that (cf. quantification of energy in modern physics). The insuperable difficulty arises if one tries to reconcile the physical demand for such a substratum with the explicit statement in the *Metaphysics* that matter has '*nec qualum, nec quantum, nec quid*'.

Aristotle's mistake, as we in retrospect can see, was to *give up too soon* in the search for an abiding and *actual* substratum for *genesis* and *phthora*. Even if one rejects Democritus' atoms and Empedocles' FAWE, there will remain further possibilities. We can provide an abiding Eleatic substrate, with an actuality and nature of its own (and quite *unlike* those of all sensible substances, of course) *without* resorting to the desperate expedient of something purely potential, something that is not actually anything.

The Aristotelian notion of prime matter can therefore only be rescued from utter incoherence if one *plays down* (or simply *rejects*) its metaphysical rôle as something purely potential with no actuality of its own. If in place of 'not actually anything' we put 'not actually any perceptible (type of) thing', we can then allow it its own identity and nature. As such, it *can* serve the physical function of an abiding material substratum of a highly plastic nature.

APPENDIX 3
Aristotle on Substantial Form

Had I written this appendix a year ago, I should have given it the subtitle 'Anatomy of an Error'. The scholastic doctrine of substantial form (S.F.) as itself an immaterial substance, separable (in some sense) from matter and the cause (after some manner) of bodily organisation was, I should have said, clearly a misreading of Aristotle's aims and intentions. I should have devoted my attention to showing where the 'schoolmen' went astray, where they misunderstood and misrepresented the writings of 'the philosopher'.

I am, however, no longer convinced that the scholastic interpretation of Aristotle on S.F. is wrong: if it is mistaken, it is nonetheless a very natural reading of much that is said in the *Metaphysics*, especially in Book Z. I shall therefore present not an 'error-theory' as such (an account of a scholastic misinterpretation) but a deliberately weighted reading of the Stagirite, stressing always those passages which lend credence to the scholastic doctrine of S.F. I wish in effect, while remaining myself agnostic as to Aristotle's own views, to justify labelling the scholastic position 'Aristotelian'; it is, I suspect, a one-sided reading of Aristotle rather than a wrong-headed one. This doctrine of S.F. was of such historical importance that it demands discussion in its own right, quite independently of genetic questions, but it is not beside the point to show how it could have been read into some of the darker passages of the *Metaphysics*.

Let us begin with some examples of forms which are clearly *not* substantial, i.e. not themselves capable of independent existence. Consider first the 'form' of bone: this, says Aristotle, cannot be identified with its matter (FAWE) but is rather to be equated with the specific proportion or ratio in which those elements are combined to make it:

> For even Empedocles says bone exists by virtue of the ratio in it. Now this is the essence and the substance of the thing. But it is similarly necessary that flesh and each of the other tissues should be the ratio of its elements, or that not one of them should; for it is on account of this that both flesh and

bone and everything else will exist, and not on account of the matter, which he names – fire and earth and water and air.[1]

If flesh is $F_3A_2W_1E_1$, the 3:2:1:1 ratio is the essence or substance (*ousia*) of flesh. This arithmetical ratio, although independently intelligible, is 'separable' from the material constituents 'in account' only – no one would feel the slightest temptation to suggest that it could have a fully independent existence of its own.

Another similar example would be the form of a syllable. The syllable ab is, Aristotle insists, something over and above its 'elements' a and b (otherwise ab would be identical with ba). This 'something else' cannot be merely a further element (this would give a 'heap' of *three* elements in need of a principle of order) but is in an important sense the 'cause' and 'substance' of the syllable.[2] Just as the ratio of FAWE is the (formal) cause of flesh, so is the order of its letters the (formal) cause of the syllable, that in virtue of which the syllable is what it is. This form can be 'separately formulated',[3] i.e. captured in a definitory formula, but is not 'separable' (*choriston*) in any stronger sense. It is obvious, says Aristotle, that the forms of some things (e.g. artefacts) cannot exist apart from individual instances of the kinds in question.[4]

A syllable, moreover, can be said to come-to-be by the arrangement of its elements. In this sense, Aristotle admits, the form(= order, arrangement) might be said to come *from* the matter.[5] This is perhaps the closest he comes to an Atomist conception of form as nothing more than arrangement of parts: had Aristotle concentrated his attention on such examples, much of the history of Western philosophy might have been different.[6] On this view, congenial to the Atomist tradition, the material constituents are the primary realities and forms arise from their spatial arrangements.

Form, let us say, is the 'principle of organisation' in structured things. This formula is deliberately ambiguous, the ambiguity depending on how one reads the phrase 'principle

1   *Metaphysics*, 1, 10, 993 a 16–22.

2   *Ibid.*, 7, 17, 1041b 12–34.

3   *Ibid.*, 8, 1, 1042a 28–29.

4   *Ibid.*, 8, 3, 1043b 19–22.

5   *Ibid.*, 5, 24, 1023 35–b3.

6   See Haring, Part 1, pp. 312–313.

of organisation'. One can give it an innocuous sense of making this signify no more than a formal cause, that in virtue of which we call something organised; or one can give it a controversial sense by making form the organising *agency*, i.e. an efficient cause. Now in the case of artefacts – syllables, houses, etc. – the arrangement of parts which constitutes the finished product is worked from without, by an external cause. A human artificer has the form in mind, and seeks to realise that form in some given matter. In the case of natural bodies, however, there is no external artificer: here, the cause of organisation must be put down to 'nature', which, says Aristotle:

> in the primary and strict sense is the essence of things which have in themselves, as such, a source of movement . . . nature in this sense is the source of movement of natural objects . . .[7]

For natural bodies – especially of course organisms – there is some immanent organising agency which, for want of a better word, we call its nature. It is tempting to identify 'nature' with 'form', and thus to treat form as an arrang*er* rather than an arrange*ment*. Thus Farrington:

> The Forms represent the intelligible side of nature, the design in nature. They also represent the active element. Matter is inert, passive. The Forms are active and compel nature to take their shape. The whole activity of nature consists in the bringing of order out of chaos by stamping Form on Matter.[8]

This passage plainly treats Aristotelian forms as the shapers or architects of natural bodies, as the *efficient* causes of their organised structures. Such a view became widespread during the late Middle Ages and Renaissance, but has often been regarded as a scholastic corruption of the actual views of 'the philosopher'. Let us turn to the *Metaphysics* – and in particular to Book Z – in search of the historical Aristotle.

The study of being, Chapter 1 begins, is primarily the study of substance (*ousia*): 'that which "is" primarily is the "what" which indicates the substance of the thing'.[9] All other things exist only as determinants of substance – quantity, quality, relation, etc., exist only in so far as they appertain to some

[7]   *Metaphysics*, 5, 4, 1015a 13–19.

[8]   Farrington, pp. 128–129.

[9]   *Metaphysics*, 7, 1, 1028a 14–15.

ousia or self-subsistent thing. What things, Aristotle asks, are substances? *Candidates* for the part include FAWE, parts of animals and plants, heavenly bodies, mathematical entities, the Platonic Forms, the soul, etc. – these are of course no more than proposals: it is not to be expected that Aristotle will accept them all.

In Chapter 3, Aristotle sets out two tests of substantiality. The first criterion (1) is that substance is 'that which is not predicated of a substratum but of which all else is predicated'.[10] By this criterion, however, matter would be substance, which is impossible, since (2): 'both separability and "thisness" are thought to belong chiefly to substance'.[11] But P.M. is not capable of independent existence, nor is it a 'this' or *particular*. This second criterion (2) for substantiality is decisive and completely overrules (1): there is no further suggestion that matter could be *ousia*.[12] Instead,

> form and the compound of form and matter would be thought to be substance, rather than matter. The substance compounded of both, i.e. of matter and shape, may be dismissed; for it is posterior and its nature is obvious. And matter also is in a sense manifest. But we must enquire into the third kind of substance: for this is the most perplexing.[13]

In his philosophical lexicon, Book Δ, Aristotle lists four senses of 'substance', the last of which is as follows:

> The essence, the formula of which is a definition, is also called the substance of each thing.[14]

The word 'substance' can therefore be properly used to refer to 'that which, being a "this", is also separable – and of this nature is the shape or form of each thing'.[15]

In what sense, however, can a form be called 'separable' (*choriston*)? It is clearly separable *in thought* and prior in *intelligibility* to the concrete thing of which it is the form, but how can it be separable in actuality from the concrete thing, except perhaps as a universal, a Platonic form? When I melt a

10 *Ibid*, 7, 3, 1029a 14-15.

11 *Ibid.*, 28-29.

12 See Chen, p. 50.

13 *Metaphysics*, 7, 3, 1029a 29-33.

14 *Ibid.*, 5, 8, 1017b 22.

15 *Ibid.*, 24-25.

bronze sphere, I might be said to be separating the 'bronze' from the 'sphere'; the bronze (matter) continues to exist, but in what sense does *this* sphere continue to be? Only the universal 'sphericality' survives, but this, as Aristotle insists, is a 'such', not a 'this', and hence by criterion (2) not a *substance*:

> Everything that is common indicates not a 'this' but a 'such', but substance is a 'this'.[16]

The *ousia* of a thing is peculiar to it, whereas a universal is common to many; therefore, no universal can be a substance. If 'humanity' were a substance, all men would have the same *ousia*, i.e. there would only be one man![17] Thus, says Aristotle, 'no universal attribute is a substance', 'no common predicate indicates a "this" but rather a "such"'.[18] The essence or substance of a thing is unique and peculiar to it.

Woods[19] attempts to evade this conclusion by making a distinction between 'being universal' and 'being predicated universally': species names such as 'man' are not, he says, predicted universally of all men. Hence, he suggests, Aristotle can accept that nothing 'predicated universally' is a substance, but nevertheless treat species as *ousiai*. Generic names like 'animal' are, he claims, predicated universally, while species names like 'man' are not. This seems a mere verbal manoeuvre, quite devoid of philosophical justification. 'Humanity' surely *is* common to all men, hence a 'such' and not a 'this'. 'Man' just is the sort of term which for Aristotle stands for a universal, something predicable indifferently of Socrates, Plato, Callias, etc.

If we dismiss Woods' interpretation as ingenious but ultimately empty we are left with what seems the plain message of the text, i.e. with the notion of a form (*eidos*) peculiar to each member of a species.[20] Members of the same species have similar forms but emphatically do not share in or partake of one common form. Each has a numerically distinct instance of the universal form-type.[21]

---

16   *Ibid.*, 3, 6, 1003a 8–9.

17   *Ibid.*, 7, 13, 1038b 7–17.

18   *Ibid.*, 34–36.

19   Woods, pp. 228–229.

20   See Gaukroger, p. 86, and also the important paper by Sellars.

21   Albritton, p. 700. See also Copleston, 2, pp. 93–94, for Aquinas' insistence that each member of a species has a numerically distinct S.F.

Although the form of each member of a species is itself a particular, it 'imparts' universality – it is in virtue of its form that a concrete individual falls under a universal nature.[22] This allows the individual both to be a genuine particular *and* to be knowable in so far as it falls under some universal concept. (Knowledge is always of the universal.[23]) Form, says Haring:

> is not that by which we represent a thing but the very core of the thing, and the ground of the possibility of representation.[24]

Species and genus, though immaterial, are *posterior* to and *abstracted* from concrete things, whereas form is, in an important sense, *prior* to the concrete body. The form of each individual thing makes it what it is; we abstract our concepts of species and genera from concrete things.

A substance (*ousia*) is a 'this' or particular, something determinate. A universal, by contrast, is not a 'this' but a 'such', while matter (*hulē*) is only potentially but not actually a 'this'.[25] A genuine 'this', says Haring, must be 'one, definite, and in some way, primary'.[26] Many of the candidates put forward for the status of *ousia* can now be seen to fail this test:

> Evidently even of the things that are thought to be substances, most are only potencies – both the parts of animals (for none of them exists separately; and when they *are* separated, then too they exist, all of them, merely as matter) and earth and fire and air; for none of them is a unity, but as it were a mere heap, till they are worked up and some unity is made out of them.[27]

FAWE and the parts of animals are not substances: they lack the requisite degree of *unity*.[28] FAWE, although of course they have a 'formal dimension' in so far as they possess the powers HCDM, are nevertheless insufficiently determinate, organised, and unified to be called *ousiai*.[29] They are mere

22 See Haring, Part 3, p. 702.

23 *Metaphysics*, 3, 6, 1003a 14.

24 Haring, Part 3, p. 703.

25 Haring, Part 1, p. 319; Chen, p. 51. See also of course Appendix 2.

26 Haring, Part 1, p. 319.

27 *Metaphysics*, 7, 16, 1040b 5-10.

28 Sokolowski, p. 263 ff.

29 Sokolowski, p. 272.

stuffs, not distinct *things*.[30] The parts of animals too fail the test – separate a finger from a body and it is a mere chunk of stuff, only called a finger by equivocation ('a dead finger,' says Aristotle, 'is a finger only in name'[31]).

Artefacts too will fail the test – they have, says Aristotle, only an apparent or pseudo-unity, achieved by such external means as glue, string, nails, etc.[32] Perhaps, says Aristotle, neither houses, nor utensils,

> not any of the other things which are not formed by nature, are substances at all; for one might say that the nature in natural objects is the only substance to be found in destructible things.[33]

Living things are far more unified and determinate in nature than FAWE, the parts of animals, and artefacts.[34] When Aristotle cites a paradigm *ousia* it is always 'Socrates' or 'that horse', i.e. a man or animal rather than a mountain, lake, hand, or table.[35] Living things manifest a peculiar structural and functional unity which positively thrusts itself on our attention. Other bodies are not substances but only approximate more or less to that status.[36]

To the following types of things or stuffs there therefore correspond the following formal factors:

Organisms (genuine *ousiai*)	– Substantial Forms
Artefacts	– Order, Arrangement, etc.
FAWE, natural stuffs	– Qualities, e.g. HCDM

Only organisms are true substances, hence only they have 'substantial forms' or S.F.s. Artefacts have only 'accidental' forms, imposed on them from without, while FAWE and other natural stuffs have as their 'formal factor' the powers HCDM, which are amorphous, blend readily in *mixis*, etc., and are therefore clearly not *ousiai* or 'thises'. As Haring says,[37] the 'strong implication' of *Metaphysics* Z is that only living beings among perishables and the incorruptible

---

30 According to Aquinas, FAWE have forms of the lowest possible grade. See M. T. Clark, Ed., p. 43.

31 *Metaphysics*, 7, 10, 1035b 25.

32 *Ibid.*, 5, 6, 1016a 4–5.

33 *Ibid.*, 8, 3, 1043b 22–23.

34 Haring, Part 3, pp. 709–710.

35 Sokolowski, pp. 282–293.

36 See Albritton, p. 706.

37 Haring, Part 1, p. 311.

heavenly bodies are material *ousiai* – only they have the requisite unity, determinacy, and completeness.

These substances derive their unity – and hence their substantiality – not from their matter (bone, flesh, sinew, etc.) which is only potentially *ousia*, but from their forms:[38]

> For the form, or the thing as having form, should be said to be the thing, but the material element by itself must never be said to be so.[39]

True substances can therefore be said to possess S.F.s, immaterial *ousiai* prior[40] (in some important sense) to the bodies which they are responsible for organising. For living things, says Aristotle, S.F. can be identified with *soul*:

> . . . since the soul of animals (for this is the substance of a living thing) is their substance according to the formula, i.e. the form and essence of a body of a certain kind . . . so that the parts of soul are prior, either all or some of them, to the concrete animal.[41]

The S.F. (soul) of an animal is an immaterial particular or 'this', 'separable' in some respect from its body, and the cause of its existence as that particular creature. In what sense or senses can the S.F. be called 'separable' (*choriston*) from the body and the cause of bodily organisation? If S.F. is an independent incorporeal substance that can act as an efficient cause, what becomes of the sophisticated *functional* theory of the soul outlined in *De Anima*? Has Aristotle not capitulated to Platonism?

My personal suspicion is that he may well have done just that, and that too much may have been made of the alleged functionalism of *De Anima*. If we turn to that treatise, what do we find? The soul, says Aristotle,

> must be a substance in the sense of the form of a natural body having life potentially within it. But substance is actuality, and thus soul is the actuality of a body as above characterised.[42]

Since the soul is the form or actuality of the body,

---

[38] A concrete thing derives its 'separability' and its 'thisness' from its *form*! See Chen, p. 52.

[39] *Metaphysics*, 7, 10, 1035a 7–9.

[40] See *Ibid.*, 12–14.

[41] *Ibid.*, 14–19.

[42] *De Anima*, 2, 1, 412a 20–22.

it is unnecessary to enquire whether the soul and body be *one*, any more than whether the wax and an impression made in it are one; or in general the matter of anything whatever and that of which it is the matter. For, while *one* and *being* are predicated in many ways, what is properly so is actuality.[43]

Many scholars have – quite naturally – read this as an assertion of the essential unity of soul and body, and of the impossibility of the former existing independently of the latter. (Clearly, the impression cannot exist without the wax.) Yet this does not seem to be Aristotle's final and definitive view. In the *Metaphysics*[44] he leaves it very much an open question whether (some) forms can exist apart, and in *De Anima* itself he denies that the intellectual faculty of a man is the 'act' of any part of his body, which suggests that it may perhaps exist separately. Those 'parts' (or faculties) of soul which are the actualities of particular bodily organs (as perception is of the organs of sense) are by contrast more intimately associated with the body:

> Therefore it is evident that the soul is inseparable from the body – or certain parts of it, if it naturally has parts; or it is of certain bodily parts that it is the act. But with respect to certain of its parts there is nothing to prevent its being separated, because these are acts of nothing bodily. Furthermore it is not clear that the soul is not the 'act' of the body in the way that a sailor is of his ship.[45]

But a sailor is a *quite distinct substance*: he 'actuates' the ship in the sense of making it function as such, but is not the *actuality* of the ship. It is not in virtue of his presence that the ship is such – *unless* of course 'ship' is defined in terms of the actual performance of its function, i.e., in scholastic jargon, in 'second act', in which case an unmanned vessel is like a severed hand, i.e. a ship (hand) only in name and not in fact. If we pursue this functional analogy, the soul is separate from the body but not *vice versa*: the sailor can exist (though not, perhaps, *qua* sailor) without any ship, but a ship cannot exist as an actually functioning unit without a crew.

We have therefore uncovered a deep ambiguity in the formula 'X is the actuality of Y'. It would naturally be thought to indicate no more than Y's proper functioning as

43   *Ibid.*, 412b 6–9.

44   *Metaphysics*, 8, 3, 1043b 18–13.

45   *De Anima*, 2, 1, 413a 4–9.

such 'if the eye were an animal,' says Aristotle, 'sight would it its soul', i.e. its 'act',[46] or it can indicate a separate substance which causes Y to function as such (the sailor in the ship). Now the former type of 'act' is clearly inseparable from that of which it is the 'act'; whereas the latter (the sailor) can exist (at least *qua* man, if not *qua* sailor) independently of any ship. If then the soul is in the body like a sailor in a ship, it can exist separately, although its functional characterisation will have to be altered.

The soul is, says Aristotle, the 'cause and principle' of the living body:

> Now these words can be used in many ways. The soul, however, is a cause in three established senses: for it is whence comes movement, that 'for the same of which' and as the essence of living bodies.[47]

The soul is therefore (a) the efficient cause of bodily movement (b) the final cause of the body's existence and (c) the formal cause of organic structure. The first of these senses is the most important for our purposes: the soul, Aristotle intimates, is the *efficient* cause not only of locomotion but of sensation, growth, and other organic processes.[48] The fact that soul can function as an efficient cause refutes, he alleges, the Pythagorean theory of the soul as a 'harmony'.[49] It also strongly suggests the 'sailor in ship' conception of soul as a separate (incorporeal) substance.

Can S.F.s (souls) exist apart from matter? In *Metaphysics* H Aristotle states that 'of substances completely expressible in a formula [viz. forms] some are separable and some are not'.[50] Whether the *ousiai* of perishable things can exist apart from their matter is, he admits, 'not clear', except that it is plainly impossible for some things (e.g. artefacts) to do so.[51] In *De Anima*, the implicit answer seems to be that animal faculties associated with the 'act' of particular bodily organs are probably not separable from those organs:

> . . . we must examine whether any form also survives afterwards. For in some cases there is nothing to prevent

---

[46] *Ibid.*, 412b 19.

[47] *Ibid.*, 2, 4, 415b 8–11.

[48] *Ibid.*, 22–27.

[49] See *De Anima*, 1, 4.

[50] *Metaphysics*, 8, 1, 1042a 32–33.

[51] *Ibid.*, 8, 2, 1043b 28–20 and 12, 3, 1070a 13–18.

this; e.g. the soul may be of this sort – not all soul but the reason; for presumably it is impossible that all soul should survive.[52]

It is considered of great significance that the intellect is not the 'act' of any organ; 'Mind', says the philosopher,

> cannot reasonably be regarded as blended with the body: if so, it would acquire some quality, e.g. warmth or cold, or even have an organ like the sensitive faculty: as it is, it has none.[53]

If the intellect can exist without a bodily organ, it can exist separately: the *rational* soul at least may be independent of the body. Such disembodied 'intelligences' are posted in Book Λ of the *Metaphysics* as the movers of the heavenly spheres.[54] Books Z and H may therefore be seen as preparing the ground for the cosmic vision of Λ,[55] the theological apogee of the *Metaphysics*.[56] The intelligences and god are, says Aristotle, pure 'act' without any 'potency'[57] (and hence without any matter).

We are now approaching a possible reading of Aristotle on S.F. Only animate things have true S.F.s – FAWE, other natural stuffs, and artefacts are insufficiently unified and cohesive to count as *ousiai*. For S.F.s, the general definition 'form = the actuality of the informed thing' is widely applicable, but cloaks a deep ambiguity. For the lower faculties of the soul (vegetative, sensitive, appetitive) these 'forms' are intimately associated with – and hence inseparable from – parts of the body: there would be no sight without eyes, no animal motion without nerves and muscles, etc. The highest grade of soul, however (viz. intellect.[58]), is the 'act' of the body in the sense of an actuator (like a sailor in a ship) and

---

52   *Ibid.*, 12, 3, 1070a 24–27.

53   *De Anima*, 3, 4, 429a 24–27.

54   The intelligences are most certainly *particulars* (see Albritton, pp. 700–701).

55   See Haring, Part 3, p. 713.

56   Ryan has attempted to deny that Aristotle thought of the 'intelligences' as *pure forms*, but his arguments are weak and purely defensive in character.

57   See *Metaphysics*, 12, 6.

58   I deliberately ignore here the difficult distinction between the 'active' and the 'passive' intellects. Neither have I anything to say as regards the Averroist doctrine that the 'active intellect' is common to all men.

can therefore be thought of as an independent reality. Of this kind are god and the intelligences.

All S.F.s therefore are separable (*choriston*) in thought from and prior in intelligibility to bodies, but only the highest S.F.s (the rational soul, the intelligences, god) can exist in a disembodied state. It is therefore, the 'vegetative' and 'sensitive' souls that provide the real metaphysical headache for the peripatetic: these are S.F.s, hence *ousiai*, but, *qua* 'actualities' of bodily organs, inseparable from matter. The student of Nature, says Aristotle in the *Physics*,

> is concerned only with things whose forms are separable indeed, but do not exist apart from matter.[59]

Are these S.F.s *choriston* only in thought? Or are they separ*able* in principle from matter but never in fact separat*ed*? The latter suggestion may seem attractive, but is in fact untenable. It is a thesis of Aristotle's embryonic modal logic that whatever is possible becomes at some time actual: the notion of a possibility which was never actualised was unintelligible to him. If then a particular type of S.F. is *always* associated with matter, it is *necessarily* associated with matter.

We are left, then, with on the one hand the denial that FAWE, artefacts, etc., *have* S.F.s at all, and, on the other hand, the substantivalist theory of 'intelligences'. In between, there seems to be a great and unbridgeable gulf: perhaps the most baffling thing about Aristotle's treatment of the whole problem is his apparent tendency to treat what seems to us a difference in *kind* as it if were a difference in *degree*. Somewhere in this gulf fall the 'souls' of animals and plants: they are clearly *ousiai*, but never exist – and hence cannot exist – independently of matter. Their exact metaphysical status is therefore, to say the least, obscure.

The schoolmen were of course obliged to lend assent to the Platonic-Christian doctrine of the soul as a distinct incorporeal substance contingently connected to the body. They therefore tended to read such a doctrine into Aristotle – and, as we have seen, such a reading of the Stagirite on the rational soul is perfectly permissible. The doctrine of S.F.s has therefore a tendency to 'slide' into animism: there is a strong temptation to think of other S.F.s as incorporeal substances existing quite independently of matter, shaping and organising it.

---

[59] *Physics*, 2, 2, 194b 12–13.

Many continued to insist, at the same time, that the only S.F. actually separable from all bodies is the human soul: other S.F.s (e.g. plant and animal souls) may be ontologically distinct from – and prior in explanatory order to – bodies, but are nevertheless not capable of totally separate existence. The existence of these lower forms is, says Aquinas, 'not without matter'.[60] Even the human soul, he insists (rather puzzlingly), is naturally embodied and only supernaturally separated from the body by divine power[61] (hence of course the resurrection of the body).

Is S.F. the efficient cause of the organisation of a plant or animal body? Such a move is a natural extension of peripatetic doctrine: it was by the time of the Renaissance commonplace to speak of the soul as 'the architect of its mansion'. S.F.s came therefore be conceived as immaterial substances, efficient causes of the movements and structural organisation of bodies. This is the 'scholastic' doctrine of S.F. to which so many seventeenth century thinkers objected. Although it is a one-sided reading of the scholastics' one-sided reading of Aristotle, a certain *continuity* of view is nevertheless apparent. The scholastics read their Aristotle very closely: it is simply a mistake to suppose that they foisted on him a theory antipathetic in spirit to the letter of his text. There is plenty of scope, as we have seen, in the dark utterances of the *Metaphysics*, for the full scholastic doctrine of S.F.

---

[60]   M. T. Clark, Ed., p. 43. See also p. 249.

[61]   See Copleston, 2, p. 90, p. 99, p. 168.

APPENDIX 4
Nicholas of Autrecourt on the Vacuum

The existence or otherwise of void space was an issue of sustained controversy among Arabic philosophers. Thinkers such as Avicenna and Averroes continued to reject the void – for familiar Aristotelian reasons[1] – while the Atomistic schools of the kalām accepted it. Al-Rāzī postulated the existence of absolute space or vacuum$_1$, which could exist independently of all matter and was in itself 'void' (khalā),[2] only accidentally occupied by bodies. This space, he insists, extends beyond our cosmos: since we cannot conceive the contrary, reason and not merely imagination attests to the existence of extracosmic space.[3]

The Mu'tazilite and Mutakallimun schools also embraced the void. For the latter, we have the testimony of Maimonides:

> The original Mutakallimun also believe that there is a vacuum, i.e. one space, or several spaces which contain nothing, which are not occupied by anything whatsoever, and which are devoid of all substance.[4]

Since these philosophers accept the classical Atomists' identification of coming-to-be and ceasing-to-be with the aggregation and separation of atoms, they

> are compelled to assume a vacuum, in order that the atoms may combine, separate, and move in that vacuum which does not contain any thing or any atom.[5]

This is Leucippus' argument from the existence of local motion to that of empty space or vacuum$_2$. The same contention can also be found in Ibn Mattawayh (d. 1076), a

---

[1] For Avicenna, see Afnan, p. 217. Avicenna, like Aristotle, found the thought of *nothing* having either position or magnitude unintelligible.

[2] Peters, p. 170.

[3] See Grant, 9, p. 332.

[4] Maimonides, p. 121.

[5] *Ibid.*, p. 121.

Mu'tazilite of the Basra school, and his teacher Abd al Jabbar (d. 1025). 'We believe,' says the former, 'that the world is not full, and that there are vacant places in it'.[6] He too sketches Leucippus' argument, although he also adds some evidence from empirical pneumatics.

This extensive discussion of the vacuum in Islamic philosophy provides the background for the position of Nicholas of Autrecourt, perhaps the only true Atomist among the schoolmen. The vacuum, he says, following Aristotle, can be defined as 'that in which there is no body, but in which there can be a body',[7] i.e. a place accidentally bereft of material occupants. It seems, says Nicholas,[8] that such void space must exist:

> because otherwise it would follow that there could be no local motion in a straight line, either because two bodies would coincide, or because all things would have to move and change place in a single motion. We can be left with nothing so probable as [that there is] a vacuum; therefore, one must posit a vacuum.[9]

When a body quits one place for another, the place into which it moves was previously either (a) empty – in which case the argument is conceded – or (b) filled by some other body. But in the latter case,

> either the body remains and so the two bodies will coincide, which is one of the unsuitable [alternatives] or it withdraws and passes to another position, and then I ask my former question.[10]

Local motion in a plenum thus involves either (a) the interpenetration of bodies, or (b) a progression *ad infinitum*. (This need not in fact involve *all* bodies in simultaneous motion: one could surely have a closed loop or vortex?)

But, replies the Aristotelian antagonist, 'finally there will be a halt because a compacting will be made', i.e. because some portion of matter becomes *intrinsically* denser. Nicholas retorts that 'we cannot make this assertion so as to deny a

6   Quoted from Schwartz, p. 385.

7   Nicholas of Autrecourt, p. 87.

8   For Nicholas' vacuism, see Grant, Ed., 4, p. 352 ff.; Weinberg, 1, p. 160 ff.

9   Nicholas of Autrecourt, p. 87.

10   *Ibid.*, p. 88.

vacuum, although perhaps Aristotle could on the basis of his principles'.[11] For, he explains:

> we do not say that an object would be compact because of the generating of some new quantity which did not exist before, but that it would be compact only through the withdrawal of bodies, as in the case of wool, or because the parts come together, that is, they are closer than before. And a body will be rarefied only because its parts are more widely separated than before; and so it is compact or rarefied only through local movement of its parts.[12]

Nicholas thus reaffirms the Atomist account of condensation and rarefaction as involving only the local motion of qualitatively unchanging parts, and thus tacitly commits himself to the notion of matter*, the homogeneous and incompressible material of the atoms. Granted such matter*, the notion of 'compacting' will not save the peripatetic: indeed, the possibility of condensation will turn on the existence of void into which the particles of the condensed body can withdraw.

This leaves only the theory of mutual cyclical replacement. On that view, says Nicholas, if a replaces b and b replaces a simultaneously, one cannot say which object *moves* which, i.e. which is agent and which patient.[13] (This is the argument which I have attributed to Lucretius.[14]) Cause, Nicholas insists, must precede effect in time as well as in nature.

We now move on to a more detailed discussion of condensation and rarefaction. Wine expands on fermentation, Nicholas admits, but this

> is not because of the generation of a new quality .. or because of the addition of new bodies. It only seems to be, because the combining parts . . . would now separate and be further apart than before.[15]

Rarefaction and condensation are, he continues (following Ockham), no more than the separation and coming-together of parts. The 'usual definition' of 'compact', he says, is 'a thing whose parts lie closer together':

---

[11]   *Ibid.*, p. 88.

[12]   *Ibid.*, p. 88.

[13]   *Ibid.*, pp. 88–89.

[14]   See Chapter 2, p. 66.

[15]   Nicholas of Autrecourt, p. 89.

And so in a rarefied thing the parts somehow are separated, so that a vacuum intervenes; that is, the parts could be closer to one another; there is something where body is not, yet could be.[16]

Ockham's conception of dense and rare thus involves, Nicholas claims, the existence of interstitial vacua. If, he adds, 'compact' is defined as 'less yielding to the touch', the enquiring mind naturally asks *why* one body yields less than another, and the notion of close-packed parts provides the natural answer to that question.

What of Aristotle's use of the $v \propto 1/R$ formula to show that motion *in vacuo* would have to be instantaneous rather than successive in nature? This argument, replies Nicholas, is ineffective

against the way of positing a vacuum that I have used. For I do not posit a separate, pre-existing vacuum through which movement would take place, but a vacuum among the parts of the body.[17]

Nicholas is arguing only for interstitial vacua, not for a separate or 'collected' vacuum₂. Nevertheless, he adds, Aristotle's argument is still invalid. The *Distantia Terminorum* (D.T.)[18] guarantees the successive nature of local motion even *in vacuo*:

in motion there is a certain essential characteristic so that, just as one part of space is before another, so it is traversed first; and that essential characteristic cannot be removed.[19]

Even in a void, then, the moving body will traverse the parts of space successively, and can therefore have only finite velocity. Slowness, Nicholas reminds us, is due to intervals of rest during the course of a body's motion.[20] The movable object, he continues, rests either because of (a) 'the determination of its nature', or (b) 'the resistance of the medium'. In a vacuum₂, (b) does not apply, but from this it does not follow that the moving body does not rest but that 'the resting is caused by a determination of the moving object itself'.[21] The

---

16 *Ibid.*, p. 89.
17 *Ibid.*, p. 90.
18 See Chapter 6, pp. 271–272.
19 Nicholas of Autrecourt, p. 91.
20 See Chapter 5, p. 204.
21 Nicholas of Autrecourt, p. 91.

whole theory is so clearly Avempacean that it is no surprise to hear Nicholas say that

> that theory which I have set down about succession, that it depends on the order of parts in space, was the theory of Avempeche, as it seems.[22]

Nicholas has thus reaffirmed the traditional Leucippan argument from local motion to the existence of void space, and denied that motion in such a vacuum₂ would have to be instantaneous. He closes his discussion with a few remarks on the 'natural experiments' which seem to testify to Nature's abhorrence of a vacuum. Notable among these is the clepsydra:

> If a finger is held at the top, the water will not come out, though its natural tendency is to fall. If the finger is removed, the water falls. Its not falling in the first case seems only because of the repugnance to a vacuum, because there would be a vacuum if [the water] came out.[23]

Nevertheless, Nicholas replies,

> this argument is not opposed to the conclusion which we hold, because we have not yet posited that it would be possible for such a separate vacuum to exist as would be the case after the removal of the water.[24]

The existence of a *fugi vacui*, a force 'striving' to prevent formation of a separate of 'collected' vacuum₂, is perfectly compatible with the actual existence of microvacua between the particles of bodies.[25] Alternatively, says Nicholas, there is a 'subtle' answer to the question why the water cannot leave the clepsydra. This is, he conjectures, because 'the whole universe has its proper measure of fullness': the water cannot flow out without causing a preternatural condensation. (This 'subtle answer' is viable for a true plenist such as Descartes, but is hardly a real option for Nicholas, given his admission of interstitial vacua.)

Nicholas is therefore deterred by none of the traditional objections to the vacuum, conceptual, dynamical, or pneumatical: his defence of the void faces them squarely.

---

[22]   *Ibid.*, p. 91. For Avempace's rejection of the peripatetic v α 1/R formula, see Chapter 6, pp. 269–271.

[23]   *Ibid.*, pp. 91–92.

[24]   *Ibid.*, p. 92.

[25]   Cf. of course the matter-theory of Galileo.

Although Nicholas' physics was probably not historically influential, he does show that even a fourteenth century scholastic could escape from the spell of Aristotle and advocate a truly Atomist world-view.

APPENDIX 5
Reductionism and Mechanism

This appendix arises out of my introductory remarks to Chapter 11: its aim is to illustrate the essential independence of the reductionist theses implicit in most particulate theories of matter from the mechanistic doctrines of men such as Descartes and Hobbes. As I said in that place, a reductionist account of qualities is perfectly compatible with the postulation of non-mechanical agents or various kinds. The following examples should help to clarify this point.

In Giordano Bruno we find explicit acceptance of Atomism, a homogeneous matter* constitutive of the atoms, the 'generation = aggregation' thesis, etc.,[1] but not a trace of mechanism. The ultimate constituents of the material universe (the atoms) are, says Bruno, immortal, illustrating the Hermetic thesis that 'nothing dies': by their coming-together and separating compound bodies arise and are dissolved. Every change involves only the local motion of something abiding. The efficient cause of this 'cosmic metabolism' is, however, the World-Soul (*Anima Mundi*), as diversified and differentiated in natural bodies. Every material thing, down to the least atom, participates in the *life* of the universe and thus has a 'share' in the world-soul.[2] Bruno's philosophy is, therefore, animistic and pan-psychic rather than mechanistic: even his atoms are alive, moved by an internal vital principle analogous to a soul. (Even in the later – and much more mechanistic – Atomic Theory of Gassendi, each atom has its own internal motive principle, and can thus be said – *dei gratia* of course – to be a 'self-mover'.) Brunonian matter is *intrinsically* animate: no body is totally deficient in soul, which 'exists permanently together with matter' in indissoluble union.[3] The order and harmony of our universe is due *not* to a fortuitous concourse of randomly agitated atoms – Democritean mechanism is dismissed as 'impious'[4] –

---

1 See Michel, p. 131, p. 142.

2 Bruno, p. 25; Michel, pp. 251–252.

3 Bruno, p. 26, p. 49.

4 See Michel, p. 256.

but to the purposive agency of that 'universal physical efficient cause', the universal intellect, which, says Bruno, is 'the first and principal faculty of the world-soul'.[5]

Although he accepts the actuality of matter and the 'generation = aggregation' thesis, we cannot call Bruno a reductionist without careful qualifications. Not only, he claims, is each *atom* a centre of life and vital force; any viable compound must also have its own characteristic 'form' (= soul[6]) to sustain its unity, cohesion and organisation. The burden of explanation thus falls on the 'form' of the compound: a mere aggregation of atoms is necessary, but not sufficient, for generation proper. Each and every compound body, however lowly on Nature's ladder, must have its own vital agency, an inner force inseparable from matter yet irreducible to material terms, organising and directing the motions of its constituent atoms. These 'forms' too emanate from the universal intellect or 'inner craftsman', so-called because it shapes matter *from within*.[7] It is this universal intellect, immanent in matter, which is the true cause of all Nature's processes, the 'formative and organising force which operates from the depths of nature . . .'[8]

Another thinker who must be mentioned in this connection is Galileo, who, although consistently sympathetic to particulate theories of matter, could not be described as a fully-fledged mechanical philosopher. While accepting the homogeneity of matter, the subjectivity of such sensations as those of heat and colour, an idiosyncratic variety of Atomic Theory, and the 'generation = aggregation' thesis, he admits a non-mechanical 'nature' as the cause of the downward ('natural') motions of heavy bodies. The case for seeing Galileo as a reductionist is a very powerful one. There is, he says in a revised version of *De Motu*, 'a single matter in all bodies', differentiated only by the sizes and shapes and motions of the discrete particles of which it consists.[9] Colour, taste, smell, heat and cold are all purely subjective: they are, Galileo insists in a famous passage in the *Assayer*,

> no more than names so far as the object in which we place them is concerned. They reside only in the consciousness.[10]

5   Bruno, p. 81.

6   *All* forms, says Bruno (p. 85), = souls.

7   *Ibid.*, pp. 81–82.

8   *Ibid.*, p. 24.

9   Galilei, 1, p. 120.

10   Galilei, 2, p. 274.

We have, then, one common matter, diversified only by the sizes, shapes, and motions of its particles, *not* by 'real' qualities of colour, heat, etc. This universal matter makes up celestial and terrestrial stuffs alike: Aristotle's sharp distinction between the heavens and the Earth is unfounded.[11] By the motion of the corpuscles of this common matter, all things come-to-be and cease-to-be:

> If I fancy to myself a body under one aspect, and then under another quite different, I do not think it impossible for transformation to occur by a simple transposition of parts, without any corruption or the generation of anything new.[12]

On the question of the springs of motion, however, Galileo departs from mechanism. *All* bodies, he says, possess a tendency to move towards, and gather at, the centre of some particular world. What causes this tendency, which we of course call gravitation? Galileo never tells us: in later life, he shunned the whole field of dynamics. It is clear, however, that he thinks of weight as something *intrinsic* to matter, and hence of gravitation as due to an inner striving or *conatus* rather than to any external impulse. Wallace[13] speculates that Galileo may have accepted Philoponus' notion of *physis* as an internal *vis*, an incorporeal *motor conjunctus*. (Such a doctrine was widespread among contemporary scholastics.) While this is purely speculative – Galileo himself never explicitly endorses such a view – the suggestion is not wholly implausible. But such a reading of *physis* makes it, quite clearly, a non-mechanical agency, an internal striving implanted by God in all heavy bodies (i.e., for Galileo, *all* bodies), and constitutive of their nature as such. While Galileo undeniably endorses a reductionist approach to the explanation of qualities, he is not a mechanical philosopher in the sense that Hobbes and Descartes are.[14]

For a further example to the same effect, we need only turn to the philosophy of the Cambridge Platonists, Henry More and Ralph Cudworth. (I shall concentrate on the views of the latter, but their overall philosophical positions are closely

---

11   For an extended polemic against the celestial/terrestrial distinction, see of course Day One of the *Dialogo*.

12   Galilei, 3, p. 40.

13   See Wallace, 5, p. 286 ff., esp. p. 293.

14   The explanation of gravitation (*natural* motion) in terms of an *external* impulse seems to be the crucial test that distinguishes the true mechanist.

akin.) According to the Atomic Theory, says Cudworth in his monumental *True Intellectual System*,

> there is no need of anything else besides the simple elements of magnitude, figure, site and motion . . . to solve the corporeal phenomena by; and therefore, not of any substantial forms distinct from the matter; nor of any other qualities really existing in the bodies without, besides the results or aggregates of those simple elements, and the disposition of the insensible parts of bodies in respect of figure, site, and motion; nor of any intentional species or shows, propagated from the objects of our senses; nor, lastly, of any other kind of motion or action really distinct from local motion (such as generation and alteration) they being neither intelligible as modes of extended substance, nor any ways necessary.[15]

This is a lucid and fundamentally accurate resumé of the essentials of the Atomic theory (albeit presented in a very Cartesian guise): the forms and qualities of bodies arise out of the arrangement and motion of simples, as words and sentences are composed out of letters; generation, corruption and alteration are all reducible to the local motion of atoms, etc. Heat, cold, colour, etc., are – as we experience them – purely subjective things or sensations:

> . . . those sensible ideas of light and colours, heat and cold, sweet and bitter, as they are distinct things from the figure, site, and motion of the insensible parts of bodies, seem plainly to be nothing else but our own fancies, passions, and sensations, however they be vulgarly mistaken for qualities in the bodies without us.[16]

Cudworth accepts, for example, a kinetic theory of heat: there is, he says, 'nothing in fire and flame, or a kindled body, different from other bodies, but only the motion or mechanism, and fancy of it'.[17] In the body, heat is a type of local motion, an agitation of its constituent particles in virtue of which it produces a particular sensation in our minds. Strictly, bodies are not red or blue, sweet or bitter, hot or cold: those terms denote only sensations produced in us by the impact of various subtle bodies on our sense-organs.

---

15   Cudworth, Vol. 1, p. 12.

16   *Ibid.*, p. 13.

17   *Ibid.*, p. 83.

Cudworth thus accepts without question the reductionism of Descartes and the Atomists: like them, he rejects the substantial forms (S.F.s) and real qualities (R.Q.s) of the schools[18] and accepts instead a homogeneous and qualityless matter* differentiated into discrete particles only by size, shape, and motion. But, he argues, this impoverished conception of matter indicates the need for incorporeal agents of various kinds to supplement our Natural Philosophy.[19] Souls, for example,

> are plainly real entities distinct from . . . matter and its modifications, and men and brutes are not mere machines, neither can life and cogitation, sense and consciousness, reason and understanding, appetite and will, ever result from magnitudes, figures, sites, and motions.[20]

Even the lower psychological functions (perception, imagination, appetition) are, Cudworth insists, irreducible to material causes.[21] No *body*, for example, can possess the faculty of imagination ('fancy'): the type of reductive explanation that works for heat or colour fails palpably here. One can account for colour as a mere appearance, *to* a sentient being, *of* something itself uncoloured, but this explanation presupposes the existence of a conscious *subject*.[22] Awareness itself cannot be reduced in a similar manner to a mere appearance, since the appearing would appear both as explanandum and as explanans. To say 'colour is but fancy' is to attempt a genuine reduction; to say 'fancy is but fancy' is empty or viciously circular, a prime example of a failed reduction – it presupposes what it sets out to explain.[23]

Cudworth thus incorporates into his Natural Philosophy a variety of incorporeal causal agents, not only human and animal souls, but a universal 'plastick nature' responsible for, e.g., the unity and cohesiveness of the cosmos as a whole and, at a smaller scale, the 'vegetative' functions of Nature. Without such an agent, he argues, all things would disintegrate due to the random agitation of atoms: to prevent cosmic

---

[18]   *Ibid.*, p. 63–64.

[19]   *Ibid.*, p. 87.

[20]   *Ibid.*, p. 67. See also Vol. 3, p. 114.

[21]   *Ibid.*, Vol. 1, p. 80, p. 84.

[22]   *Ibid.*, p. 85.

[23]   *Ibid.*, Vol. 3, p. 418. The same point has been urged in modern philosophy by Thomas Nagel.

dissolution, God would have to intervene in person and effect directly all the minutiae of Nature.[24] To escape this eventuality, says Cudworth, He created 'plastick nature' as a kind of drudge, operating according to a divinely-ordained programme, and executing 'that part of his providence which consists in the regular and orderly motion of matter'.[25] The random agitation of atoms could never, Cudworth argues, have given rise to such beautiful and orderly structures as those of our cosmos as a whole, and of plants and animals in particular: some 'immanent Art' is plainly at work within corporeal Nature, moving the atoms to form certain contrived and functionally unified complexes.

The Cambridge Platonists, then, accept Atomism as a matter-theory, and endorse a reductionist account of qualities such as heat and cold, colours, etc., but insist on the *irreducibility* of the properties of soul to material causes and the need for incorporeal (and hence non-mechanical) *agents* of various kinds to guide and direct the motions of bodies. They are, in my terminology, reductionists but anti-mechanists, believers in incorporeal causes and an irreducible teleology within the physical universe. For such philosophers, 'nature' is *not* 'blind': there will be a *reason*, not merely a mechanical cause, for any given atomic motion: it may be, for example, that it is moved by 'Nature' (i.e., as far as we can see, spontaneously) into its most fitting or appropriate place in the grand scheme of things.

If we turn now to the writings of Francis Bacon, we discover a somewhat different position. First and foremost, he insists, with the Atomists, on the *actuality* of matter: matter, he asserts bluntly, is not a mere 'principle of potency', actualised by form. A purely potential *materia prima* is not actually anything, hence actually nothing, hence non-existent.[26] Matter, like Proteus,[27] may be indefinitely plastic, capable of assuming a great variety of forms; it remains, however, something with a positive reality of its own, an extended, space-occupying stuff, not an abstract 'principle'. As far as *natural* philosophy is concerned, matter and its

24   *Ibid.*, Vol. 1, p. 218.

25   *Ibid.*, pp. 223–224. For More's analogous notion of a 'Spirit of nature', see his *Works*, Vol. 1, 'The Immortality of the Soul', p. 169, p. 193. For an interesting controversy between More and Boyle on the need to posit a 'Spirit of Nature', see Greene.

26   F. Bacon, *Works*, vol. 5, p. 492.

27   *Ibid.*, Vol. 6, p. 725.

immanent activity must be treated as primitives: we must not attempt to derive either from any prior principles.[28] (The attempt to derive matter and its activity from logic, mathematics, or 'common notions' involves turning aside from the study of Nature to the realm of mere abstractions.)

This insistence on the actuality of matter makes possible a reductive account of the origin of forms and qualities.[29] The 'form' of a given 'nature' or quality is, for Bacon, nothing more than an arrangement and motion of its (tangible) parts. It is to such forms – inseparable, of course, from matter – that we turn for *explanations*: in virtue of their forms, individuals fall under general laws.[30] Inductive methods may, Bacon claims, enable us to discover the (hidden) forms of such manifest qualities as whiteness and heat. For example,

> all bodies or parts of bodies which are unequal equally, that is in a simple proportion, do represent whiteness.[31]

His discovery of the 'form' of whiteness, he adds, confirms the Democritean view that colours depend merely on the superficial textures of things: colour, he explains in the *Novum Organum*,

> has not much to do with the intrinsic Natures of any body, but depends only on the coarser and as it were mechanical arrangement of the parts.[32]

Obvious examples of colour-differences without any associated difference of inner nature include, he says, the leaves of variegated plants and the veins in marble – one could add further examples from one's own experience. The fundamental claim is that the 'form' of a quality such as colour is a particular microstructure or texture, that the quality we see depends on the sizes, shapes and arrangement of insensibly small building-blocks.[33] Grind glass, or agitate water, he reminds us, and one can produce whiteness. A

---

28   *Ibid.*, Vol. 5. p. 462. The Presocratics, says Bacon (p. 467), were not guilty of this error; unfortunately, their philosophy was superseded by the abstractions of Plato and Aristotle.

29   Cf. of course our discussion of Ockhamism in chapter 7, pp. 295–297.

30   F. Bacon, *Works*, Vol. 3, p. 355.

31   *Ibid.*, Vol. 3, p. 237.

32   *Ibid.*, Vol. 4, p. 156; *Novum Organum*, 2, xxii.

33   *Ibid.*, Vol. 2, p. 619; Vol. 4, pp. 125–126; *Novum Organum*, 2, vii.

further example to the same effect is of course the celebrated definition of *heat* as

> an expansive motion restrained, and striving to exert itself in the smaller particles. The Expansion is modified by its tendency to rise though expanding towards the exterior; and the Effort is modified by its not being sluggish, but active and somewhat violent.[34]

It is not merely, Bacon insists, that heat *causes* motion, or *vice versa* (these are mere commonplaces), but that 'the very essence of Heat, or the Substantial self of Heat, is Motion and nothing else'.[35] There is nothing to heat but a particular kind of agitation of the parts of the body.

But if we name bodies after their manifest qualities, and those turn out to be reducible after this fashion to arrangements and motions of parts, it follows, as Bacon sees, that what men call generation and corruption may be due to local motion alone. Coming-to-be and ceasing-to-be are, he insists, 'nothing else but the works and effects of motion'[36] – in particular, they do not involve the accession or loss of an Aristotelian S.F. They are, he claims, 'the sums or products of simple motions, rather than primitive motions'.[37]

Bacon is, then, quite clearly a materialist (he believes in the actuality of matter) and a reductionist. Is he also a mechanist? Yes and no. The tangible parts of gross bodies are, he claims, moved, agitated, and arranged by the action of *spirits* or *pneumata*: the powers and properties exhibited by a given body are, then, due to the presence and activity of its particular confined spirit, 'for tangible parts in bodies are stupid things; and the spirits do (in effect) all'.[38] Explanation of the properties of bodies falls, therefore, into two phases:

1. Reduction: explanation of a given quality in terms of the arrangement and (especially) motion of its tangible parts.
2. Explanation of these internecine motions in terms of the agency of a spirit.

Spirits are active, purposive, almost sentient things: they can initiate causal chains and guide the motions of tangible parts in an apparently intentional manner. (Spirits have

[34] *Ibid.*, Vol. 4, p. 155; *Novum Organum*, 2, xx.

[35] *Ibid.*, Vol. 4, p. 150; *Novum Organum*, 2, xx.

[36] *Ibid.*, Vol. 5, pp. 425–426.

[37] *Ibid.*, Vol. 4, pp. 355–356.

[38] *Ibid.*, Vol. 2, p. 381.

'perceptions', though no conscious awareness, of the external world.[39]) Although I have argued elsewhere[40] that they move gross and inert bodies by means of mechanical impulses, Bacon has no mechanical explanation to offer either of their *activity* (ability to initiate motion *de novo*[41]) or of their purposive, almost intelligent, manner of operation. *If* we are to read Bacon as a mechanical philosopher, he is a very incomplete one: animistic concepts remain in his thought, unassimilated into the emergent world-view of the Mechanical Philosophy.

---

[39] *Ibid.*, Vol. 2, p. 602.

[40] Chapter 8, pp. 381–383.

[41] In Vol. 3, pp. 243–244, of his *Works*, Bacon recommends a detailed and intensive study of the springs of motion in Nature. Since he believes that gross matter has a natural 'torpor', a tendency to resist being set in motion at all, it will follow that his Natural Philosophy requires, nay demands, the existence of 'active principles' capable of initiating and sustaining motion. Spirits, it seems, play just this vital rôle.

APPENDIX 6
Mechanism and the Occult

At first sight, one would expect the champions of the Mechanical Philosophy to be highly sceptical about occult qualities, sympathy and antipathy, magical cures, etc. – the whole stock-in-trade of the natural magician of the Renaissance. And of course in a sense they were: men such as Mersenne and Gassendi, Descartes and Hobbes, take their places, quite rightly, in histories of philosophy as the men who exorcised the 'ghost' from the 'machine' of Nature, the founders of the vision of a clockwork universe, utterly subservient to the universal laws of mechanics.

In another sense, however, this picture is highly misleading. While the mechanists were anxious to overthrow the philosophy of occultism (angels and demons, sympathy and antipathy, occult qualities, etc.), they were often content to accommodate the (supposed) *phenomena* ('magical' cures, etc.) within their own world-view. Instead of arguing: 'Nature is purely mechanical; there can be no plausible or convincing mechanical explanation of X; therefore, X is merely fabulous', they argue: 'Nature is purely mechanical; X is a real phenomenon; therefore, some mechanical explanation of X (however far-fetched) must be true'.

It is not difficult to account for this strategy on the part of the founders of the Mechanical Philosophy. In the first place, their doctrines were highly controversial: it was therefore good policy to show how their principles could *explain* the empirical 'facts' from which the occultists claimed to derive support for their curious hypotheses. Furthermore, *some* (in the final analysis, perhaps, only a small proportion) of these 'facts' were facts rather than fantasies: two obvious examples would be magnetism[1] and the 'influence' of the Moon over the tides.[2] If we must posit complex and ingenious mechanistic hypotheses to account for, e.g., magnetic attraction and

1 For Descartes' elaborate and ingenious explanation of magnetic phenomena in terms of streams of minute screw-shaped particles, see the lengthy account in his *Principles*, 4, 133-187.

2 For the Cartesian theory of the tides, see *Principles*, 4, 49; for commentary, see Aiton, 1. The claim that tidal motions are due to

orientation, why not also postulate similar mechanisms for the fabulous powers of the basilisk or the remora? Where should we draw the line?

In Descartes' *Principles*, magnetism is seized upon as an occult quality *par excellence*: it was of course forever being cited, by natural philosophers and magicians alike, as an *undeniable* instance of something truly occult and inexplicable, a power that simply cannot be fathomed by the human intellect. After elaborating his own ingenious explanation of magnetic phenomena in terms of streams of minute corkscrew-like particles, Descartes comes to the moral of the story. 'From what has been said,' he concludes,

> we see what may be the causes of all other remarkable results which are usually referred to occult qualities.[3]

Magnetism becomes a test-case for the Mechanical Philosophy as a whole. If all the curious and obscure phenomena of magnetic action (attraction, orientation, declination, etc.) can be accounted for in terms of currents of subtle matter, there seems no reason to doubt that similar *highly complex but in principle intelligible* mechanisms may account for other phenomena allegedly due to occult causes. That the wounds of a murdered man should begin to bleed on the approach of his killer; or that blood, milk, flesh, stones, or animals should fall from the sky did not seem utterly incredible to Descartes: one can account intelligibly (i.e. mechanically), he insists, for all such phenomena.[4] Rather than play the sceptic and deny the reality of the supposed phenomena, Descartes is prepared to dream up weird and fantastic mechanical explanations of them; *every* natural phenomenon, he insists, has a perfectly intelligible explanation in terms of this principles:

> No qualities are known which are so occult and no effects of sympathy and antipathy so marvellous and strange, and finally nothing else in nature so rare (provided it proceeds entirely from purely material causes lacking in thought or free will) for which the reason cannot be given by means of these same principles.[5]

For a particularly striking account of the assimilation of the occult within the framework of the Mechanical Philosophy

differential pressures of a subtle matter was effectively refuted by barometric experiments made by Wren and Boyle.

[3] *Principles*, 4, 187; Descartes, 1, Vol. 1, p. 288.

[4] See Westfall, 4, p. 88.

[5] *Principles*, 4, 187; quoted from Westfall, 4, p. 86.

we have only to turn to Kenelm Digby's idiosyncratic account of the operation of sympathetic medicine. Digby was one of the second-rank proponents of the Mechanical Philosophy: although, as Dobbs says, he 'presented a systematic corpuscular philosophy to the seventeenth century',[6] he never really made the crucial break with Aristotle. An adequate account of the nature and limitations of corporeal things is necessary, he asserts in the Preface of his treatise 'On Body', to any valid proof of the incorporeality of the human soul. On the theory of occult qualities, he argues, one could simply say that sensation, intellection, and volition were occult (inexplicable) features of *bodies*, thus subverting the argument for the soul's immateriality. A mechanical philosophy of bodies is therefore of great value to metaphysics and theology: once we know what body is, we see immediately that the powers of the soul can only appertain to an incorporeal substance.

All the phenomena of material Nature, he insists, 'may be effected by an exact disposition and ordering (though intricate) of quantitative and corporeal parts', without any need for 'hidden and unexplicable qualities'.[7] It is a mere refuge for ignorance to 'explain' natural phenomena in terms of occult qualities, and far worse to invoke angelic or demonic powers:[8] even the most elaborate workings of the 'wreathy labyrinth of nature', though perhaps not readily unravelled by the human mind, are *in principle* explicable by us. The contrivances of animal and plant bodies, in particular, are of this kind:

> . . . the causes of them are palpably material, and the admirable artifice of them consisteth only in the Dedelean and wonderfull ingenious ordering and ranging them with one another.[9]

Bodies act on one another, he claims, by their local motion alone: all the operations of bodies 'are either locall motion or such as follow out of locall motion'.[10] Once we understand this, he says, we may be in a position to give a general account of how those properties usually deemed occult or attributed

---

6   Dobbs, 1, p. 1.

7   Digby, Preface.

8   Digby, pp. 252–253. Agrippa of Netteshiem and other magicians believed that the occult qualities of things were due to demonic agencies.

9   *Ibid.*, p. 253.

10   *Ibid.*, p. 44.

to daemonic agency may be explained in terms of material principles (organisation of parts, emission of effluvia, streams of subtle matter, etc.). His particular explanations, he admits, are both schematic and hypothetical: they serve merely to illustrate the potential of a particular reductive programme. What may appear to be sympathetic action (A.D.) is really, he avers, due either to 'congruity of temper' (texture, local motion) between two bodies, or to the passage of effluvia between them: in this way one might hope to explain all alleged instances of sympathy and antipathy without invoking A.D. or incorporeal agencies.[11]

We can now focus our attention on Digby's own peculiar version of the weapon-salve therapy. The notion that there existed some bond of sympathy between the blood on the weapon responsible for a given wound (or an old bandage used to bind it) and that in the wound itself was an aspect of folk-medicine adopted by the Paracelsians and defended by J. B. Van Helmont in his treatise on 'The Magnetical Cure of Wounds'. If, these sympathists argue, there exists some affinity or attunement between the blood on the weapon and that still in the patient's body, one should be able to work therapeutic effects on the latter by performing various operations on the former. Apply a healing balm to the blood-stained weapon, for example, and its curative virtues are transmitted 'sympathetically, i.e. after some occult and magical manner, to the wound.[12]

The weapon-salve theory was a great source of controversy in seventeenth century medicine: some physicians attacked it as mere magic and mumbo-jumbo; others defended its therapeutic record. (Since the recommended method of treatment involved cleansing the wound with a sterile fluid such as fresh urine, it is in retrospect hardly surprising that it produced positive results.) It is therefore rather odd to find in Digby – an early champion of the Mechanical Philosophy – a staunch advocate of the weapon-salve technique. Indeed, much of Digby's claim to fame rests on his notorious 'sympathetic power' and its alleged medicinal virtues.

Digby's powder was, in fact, nothing more remarkable than anhydrous ferrous sulphate or 'vitriol', already familiar to contemporary pharmacists. If this 'sympathetic powder' is added to weapon or bandage, he claims, a remarkable cure may be effected. *Prima facie*, this seems occult and magical;

[11]   *Ibid.*, p. 403 ff.

[12]   See Dobbs, 1, p. 9.

Digby insists, however, that the cure is perfectly natural and explicable without invoking occult qualities, demons, etc.[13] He demanded, in fact, a genuinely mechanical explanation, involving only matter and motion, of how the effect is produced. And, against all the odds, he managed to come up with such an account.

A fire, Digby explains, is responsible for a variety of attraction: as flame, smoke and hot air ascend from it, it draws in cooler air from its surroundings – in more modern terminology, it sets up a convection current. Other hot bodies, he adds, set up similar currents: they too 'draw' ambient air in to themselves. This explanation of (apparent) 'attraction' may, Digby suggests, prove of *general* application:[14] in particular, it may explain certain magical practices often deemed diabolical (because of the appearance of A.D. involved). But an open wound, says Digby, is a hot body: by its emission of 'abundance of hot fiery spirits',[15] it, like a fire, establishes a convection current drawing air in to itself.

The next general principle to remember is that all bodies emit 'steams' or 'vapours' of some kind or other.[16] Of such a kind is the subtle effluvium emitted by the anointed weapon or bandage, and consisting of 'atomes' of the more 'spirituous' parts of the blood and vitriol, tightly bound together. If these effluvia are emitted in the vicinity of the wound, they are drawn in to it by the mini-vortex which it has set up: the air-current being 'sucked' into the gash 'will come to incorporate at last the atomes and Spirits of the Blood and Vitriol'.[17] Blood-atoms, finding their natural 'home' in the wound, are readily re-absorbed into the patient's tissue, and thus draw with them also the attached particles of vitriol - these thus become embedded in the wound and there effect the cure. (This method is superior to that of directly applying the vitriol to the wound, says Digby, in that it involves a separation and subtilisation of the vitriol – on his method, only the volatile and 'basalmic' spirits of the salt, not its fixed and 'caustic' parts, ever make contact with the wound.)

This explanation seems so far-fetched as to verge on the grotesque: its account of the *directionality* involved in the

13   *Ibid.*, p. 11.

14   Digby, p. 203.

15   See Dobbs, 1, p. 12.

16   Digby, p. 205.

17   From Dobbs, 1, p. 12.

cure is quite hopeless. In the first place, an open wound produces no especially intense heat sufficient to form its own mini-vortex; in the second place, a multitude of other bodies serve as heat-sources. (If the patient were kept in a heated chamber, the 'atomes' of vitriol would surely be drawn into the fire rather than to the wound.) The fantastic and incredible nature of Digby's account seems evident. Yet is it *so much* more absurd than, for example, the Cartesian explanation of magnetism? And, if one wishes to account for the therapeutic success of the technique without violating one's mechanistic principles, is one not obliged to come up with some such account?

Fundamentally the same dilemma – which of the supposed 'facts' of the occultists to deny, which to endorse and 'mechanise' – is evident in Gassendi and his disciple, Charleton. Like the other pioneers of the Mechanical Philosophy, Gassendi takes several of the *examples* of occult powers quite seriously: he soon finds himself trying to account, along corpuscularian lines, for the evil eye, and even – amazingly – for the power of incantations![18] Although he denies the therapeutic value of the weapon-salve[19] (cures, he alleges, are due to unaided nature), he is prepared to lend corpuscular dress to many of the stock illustrations of occult virtues.

Chapter 15 of Book 3 of Charleton's *Physiologia* is entitled 'Occult Qualities made Manifest', and discusses in some detail many of Nature's supposedly magical phenomena. The schools, Charleton begins, have made the following pair of assumptions:

1.   That observed sensible qualities depend on familiar causes, and are apprehended by known faculties, and are therefore properly called 'manifest'.
2.   That all properties not directly cognisable by one proper sense derive from undiscoverable and obscure causes and are therefore aptly characterised as occult.

This distinction, Charleton claims, is a spurious one: 'to ourselves all the operations of nature are meer secrets'.[20] The fact that we perceive sensible qualities such as redness and sweetness immediately, by their proper senses, does not entail

---

18   Brett, p. 81. Yet, as Aquinas saw clearly, the physical efficacy of *incantations* seems a sure sign of daemonic action: only an intelligent being can understand the *meanings* of words, act on commands, etc.

19   Thorndike, vol. 7, p. 455.

20   Charleton, p. 341.

that their *causes* are familiar to us. Neither does it follow from the fact that we cannot directly perceive the manner of operation of a magnet that this must be *unintelligible* to us. We can construct plausible ('verisimilious') explanations both of 'manifest' and of 'occult' qualities, although our state of ignorance is, regrettably, such that we shall never know whether those accounts are in fact true.

The Aristotelians tend, Charleton continues, to classify all properties not directly explicable in terms of HCDM ratios as occult and inexplicable, but this is a mere refuge for ignorance. As for sympathy and antipathy, these concepts too are 'no less a refuge for the idle and ignorant'[21] – all the *bona fide* examples of (apparent) attraction in Nature must observe the following 'general laws of Nature':

1. That every effect must have its cause.
2. That no cause can act but by motion.
3. That nothing can act upon a distant subject, or upon such whereunto it is not actually present, either by it self, or by some instrument . . . and consequently that no body can move another, but by contact mediate, or immediate.[22]

It follows that all the operations of 'sympathy' and 'antipathy' are performed

> by the same wayes and means, whereby we observe one body to attract and hold fast another, or one body to repell and avoid conjunction with another, in all sensible and mechanique operations.[23]

Although these means may be insensible, the basic laws governing them will be the same as those of pushing and pulling: the mere fact that we do not perceive such subtle mechanisms does not entail that they do not exist.[24] (Electrical attractions, Charleton suggests, may be caused by prehensile 'tongues' like that of the chameleon.)

The existence in Nature of a general like ↔ like tendency is granted by Charleton, who seeks to explain this 'universal sympathy' in terms of (a) congruity of texture, and (b) the

---

21   *Ibid.*, p. 343.

22   *Ibid.*, p. 343.

23   *Ibid.*, p. 343.

24   *Ibid.*, pp. 344-347.

emission of specific effluvia. A body with a particular texture T will, he explains, tend to emit T-bearing effluvia: when these come into contact with a 'congruous' body (i.e. another body with texture T) they will produce in its subtler ('spirituous') parts a state of agitation analogous to a psychological state (a perception-desire complex), in virtue of which the second body sets off towards the first. A.D. is eliminated, but the explanation of the directionality of the response turns on the importation of elements of *animism*: there is no genuine attempt to give a purely mechanistic account of 'perception' and 'desire'.

We now move on to a more detailed discussion of occult qualities, dividing them, after Fracastoro, into 'general and 'special'. On the subject of 'general' occult qualities, Charleton first dismisses the supposed *horror vacui*, then turns to the vexatious topic of celestial 'influences' here below. After rebuking the pride and ostentation of the astrologers, and denying the astral determination of human lives, he proceeds also to query most of the so-called 'influences' of the heavenly bodies on our planet, including even that of the moon over the tides. Such celestial influences as do occur – e.g. the heliotrope 'following' the sun – are, he claims, naturally explicable and hardly marvellous.[25]

What of 'special' occult qualities? Dealing first with the inorganic realm, then with the organic, Charleton proceeds methodically either to cast doubt on the alleged phenomena (often merely fabulous) *or* to give mechanical explanations of those which are undeniably real, which include:[26]

a. The 'amity' of gold and mercury – explicable in terms of 'congruity' of textures.
b. The 'attraction' of a greater fire by a less, which is no more than 'the extension of a greater flame to the fewel of a less'.
c. The 'attraction' of flames by naphtha, due to the emission from it of an inflammable vapour
d. Capillarity.[27]

---

[25] *Ibid.*, pp. 349–353.

[26] *Ibid.*, pp. 354–356.

[27] Charleton has no plausible mechanical account of capillarity. For the best of the seventeenth century attempts to 'mechanise' this phenomenon, see Hooke's *Micrographia* (Gunther, Ed., vol. 13, p. 11 ff.), in which capillary phenomena are attributed to a differential air-pressure. This explanation satisfied Newton for many years, until he discovered the existence of capillary phenomena in the Boylian vacuum.

e.   The resonance of two lutestrings tuned to the same frequency – explicable in terms of the Galileo-Mersenne (mechanical) theory of sound.[28]

Other alleged examples are dismissed as fabulous. Moving on to the vegetable world, Charleton finds the 'sympathy' and 'antipathy' of different species of plants easy to account for: if, he says, plant X extracts from the soil a substance noxious to plant Y, X's presence will benefit Y; if, on the other hand, they find themselves in competition in any respect, they will be mutually harmful. In other cases, substantial effluvia may be responsible for the observed phenomena.[29]

In the animal world, too, a similarly mixed strategy of scepticism and explaining-away is manifest. The supposed aversion of a lion for a cock, or an elephant for a pig, are simply denied. Charleton is also sceptical of the alleged powers of the remora and the basilisk, although he admits that the latter could perhaps produce its fatal effect by some sort of corpuscular emission.[30]

In sharp contrast with this scepticism, we have some examples of sheer credulity. The wolf, for example, when it sees a nice juicy sheep,

> darts forth from his brain certain streams of subtle effluvias, which being part of those spirits, whereof his newly formed idea of diliniating and devouring the sheep, is composed, serve as forerunners or messengers of destruction to the sheep.[31]

Here, quite astonishingly, we find a seventeenth century mechanist taking quite seriously, and even seeking to mechanise, the sort of power attributed to the faculty of imagination in Elizabethan poetry! That the corpse of a murdered man should open his eyes in the presence of his killer may, Charleton suggests, be explicable after a similar manner.[32] Natural fascination and the 'evil eye' may, he adds, be explicable in terms of corporeal emissions from the eye of the witch.[33]

---

28   The Galileo-Mersenne theory of sound as a periodic vibration of the air was generalised by men such as Hobbes and Hooke in an attempt to provide mechanical analogues for sympathy and antipathy. For an interesting study, see Gouk.

29   Charleton, pp. 358–361.

30   *Ibid.*, pp. 365–366.

31   *Ibid.*, p. 363. The same idea can be found in Bacon's *Sylva Sylvarum*.

32   *Ibid.*, pp. 364–365.

33   *Ibid.*, p. 375.

We come now at last to the weapon-salve, a crucial test-case for seventeenth century thinkers. In his earlier works (heavily influenced by Van Helmont,[34]), Charleton had shared Digby's view: the salve works, he had claimed, but does not involve A.D. – the cure is effected by 'a semi-immaterial thread of Atomes' passing from weapon to wound.[35] In his later works, however, including the Physiologia, Charleton turned from the Helmontian weapon-salve theory altogether: any therapeutic effects, he argued (following Gassendi), are due to 'nature' alone, the salve being quite inefficacious.[36]

I close this discussion with an examination of some of the less well known works of Robert Boyle, notably his essay on 'The Great Effects of even Languid and Unheeded Motion',[37] the 'Notes, etc., about the Atmospheres of consistent Bodies here below',[38] the 'Essays on Effluviums',[39] and the 'Experiments and Considerations about the Porosity of Bodies'.[40] These minor works, taken together, serve to illustrate Boyle's general contention that

> he had often looked upon these three doctrines of effluvia, of pores and figures, and of unheeded motions, as the principal keys to the philosophy of occult qualities.[41]

The essay on 'The Great Effects of even languid and unheeded Motion' illustrates the immense importance which attaches to local motion (the universal *explanans*) in Boyle's philosophy. Since, he says, matter and motion are the twin principles of the mechanistic world-view, the study of local motion – treated so superficially by the scholastics – lies at the very heart of natural philosophy: many of the effects usually attributed to occult qualities are, he suggests, really due to 'unheeded' motions.[42] The essay consists of a series of 'observations' in support of this contention, each illustrated

---

[34] For the story of Charleton's change of allegiance *from* Helmont to the Mechanical Philosophy, see Gelbart.

[35] Gelbart, p. 155.

[36] Charleton, pp. 381–382.

[37] Boyle, *Works*, vol. 5, p. 1 ff.

[38] *Ibid.*, Vol. 3, p. 277 ff.

[39] *Ibid.*, Vol. 3, p. 659 ff.

[40] *Ibid.*, Vol. 4, p. 759 ff.

[41] Quoted from Thorndike, Vol. 8, p. 184.

[42] Boyle, *Works*, Vol. 5, p. 2.

by appropriate examples. The local motions of invisibly small corpuscles, or of tenuous fluids such as air, are responsible, Boyle has no difficulty in showing, for a great variety of natural phenomena: acoustics provides many of his most effective examples. The transmission of sounds through various media; resonance effects; the destruction brought about by loud noises – all serve to illustrate Boyle's general thesis. So also do the specific responses of organisms to subtle but swift-moving effluvia, the physical effects of heat (an agitation of the constituent corpuscles of a body) and many other examples. In *all* bodies, Boyle suggests, there is more or less internecine motion: it may be that no body in the universe is absolutely quiescent, without even any agitation of its parts.[43]

In the 'Notes, etc., about the Atmospheres of consistent Bodies here below', Boyle argues that not only soft and fluid bodies but also hard and solid ones emit effluvia, and hence possess their own specific 'atmospheres'. These corporeal effluxions are, he intimates, responsible for many of those mysteries of Nature usually referred to occult qualities or to sympathy and antipathy. Although imperceptibly subtle, effluvia operate, he insists, after a perfectly intelligible (i.e. mechanical) manner: their precise mechanisms may be difficult to fathom, but are not in any sense occult.[44]

Experience abounds, Boyle teaches, with examples of effluvia (or 'vapours', 'steams', or 'exhalations' – the terminological niceties are of no great moment). Odorous bodies such as spices, 'magneticks' and 'electricks', bodies subjected to heat, all tend to emit their more volatile or 'spirituous' parts. Even gemstones may emit effluvia, which would account for their alleged medicinal powers.[45] To account for the many and varied physical effects which the corpuscularian must attribute to the agency of effluvia, we must bear in mind three factors about them, namely:

1. Their great 'subtility', in virtue of which they can freely pervade the parts of gross bodies and work their effects from within. (The physical efficacy of effluvia is, for Boyle, a good indication of the existence of *pores* even in apparently solid bodies.[46])

---

43   See *Works*, Vol. 1, p. 444 ff.

44   *Ibid.*, Vol. 3, p. 278.

45   *Ibid.*, p. 512 ff.

46   *Ibid.*, Vol. 4, p. 759.

2. Their great causal 'efficacy', which Boyle explains as being due to the great number and velocity of their particles, their penetrative nature, and their ability to excite internecine motions within the body they act upon.

3. Their highly specific and determinate natures (= textures or contrivances of parts). Given the acceptably mechanistic concept of 'congruity' (a mechanical analogue, of course, of sympathy), and the lock-and-key model of physical action outlined in the *Origin of Forms and Qualities*,[47] explanation of the operation of effluvia on gross bodies will be straightforward.

Our atmosphere, Boyle deduces, is not, as many have imagined, composed solely of the homogeneous element air, but contains 'a confused aggregate of effluviums' from all manner of different sources: there is, he claims, 'scarce a more heterogeneous body in the world'.[48] Among this 'store of effluviums', most will no doubt be of a terrestrial origin (such are the 'steams' responsible for magnetic phenomena); others, Boyle conjectures, may be astral in origin: corporeal emanations from the heavenly bodies may perhaps

reach to our air, and mingle with those of our globe in that great receptacle or rendesvous of celestial and terrestrial effluviums, the atmosphere.[49]

Here the Neoplatonic/Hermetic conception of the atmosphere as a universal pool of 'spirit', a receptacle for celestial and terrestrial 'virtues', is given a corpuscularian dress. In place of spirits or virtues we have corporeal effluvia; otherwise, the account goes ahead as before. Celestial influences on terrestrial events may, says Boyle, occur through the mediation of such effluvia: the vanities and pretences of the astrologers *do not*, he insists, 'null or take away the possibility of the thing simply'.[50] A sane science of astrology is a real possibility.

The fact that a constant supply of fresh air is necessary for the sustenance of a flame leads our author to suspect that

there may be dispersed through the rest of the atmosphere some odd substance, either of a solar, or astral, or some

---

[47] *Ibid.*, Vol. 3, p. 18 ff.

[48] *Ibid.*, Vol. 4, p. 85.

[49] *Ibid.*, p. 85.

[50] *Ibid.*, Vol. 5, p. 639.

other exotic nature, on whose account the air is so necessary to the sustenance of flame.[51]

This subtle celestial effluvium may perhaps be the same as that minute fraction of the air necessary to animal life, that

vital quintessence (if I may so call it) which serves to the refreshment and restauration of our vital spirits, for which use the grosser and incomparably greater part of the air being unserviceable.[52]

If the physiologists' notion of a *flamma vitalis* is a sound one, and the life of an animal is indeed sustained by a sort of vital heat, it might be expected that an animal and a flame would require the same celestial 'quintessence', without which they both perish. Why this should be thought of as astral, even specifically *solar* in nature and origin, Boyle does not tell us. All he has done is to accommodate two of the commonplaces of Renaissance chemistry – quintessences, and the life-giving rôle of the Sun[53] – within his corpuscularian world-view. The views of authors such as Sendivogius and d'Espagnet are here simply given corpuscularian dress. (In Newton's *Principia*, the life-sustaining quintessence is derived from the vaporous tails of comets.[54])

The *textures* of bodies, Boyle continues, determine their fitness or otherwise to be acted upon by particular, often highly specific, effluvia. (A mechanical congruity of figures, and/or motions accounts for effects traditionally attributed to sympathy.) In virtue of their textures, then, some familiar substances

may be receptacles, if not also attractives, of the sidereal, and other exotic effluviums that rove up and down in our air.[55]

This is classical alchemical doctrine![56] For John of Rupescissa and the so-called 'Lullian' alchemists, particular terrestrial

---

[51]   *Ibid.*, Vol. 4, p. 90.

[52]   *Ibid.*, Vol. 1, p. 107.

[53]   For quintescences, see Chapter 7, pp. 316–319. That the Sun was a fount of life-giving spirit was a commonplace of Renaissance Neoplatonism.

[54]   Newton, 1, p. 530. The tails of comets, he says, may provide us with that 'spirit' which 'is indeed the smallest but the most subtle and useful part of our air, and so much required to sustain the life of all things with us'.

[55]   Boyle, *Works*, Vol. 4, p. 95.

[56]   See Chapter 7, p. 317.

substances (usually plants), act as chemical 'magnets', drawing specific celestial 'virtues' from the atmosphere: violets, for example, might be receptive to the influence of Jupiter, while celandines have a particular affinity for Saturn – the 'quintessences' of Jovian and Saturnine 'virtue' would then be extractable by distillation from violets and celandines respectively. Here again, then, we find in Boyle a corpuscularian version of an old alchemical doctrine: a particular terrestrial substance may have such a texture that it readily absorbs celestial effluvia of some particular kind.

Although Boyle accepts the existence of chemical 'magnets' (from writers such as Sendivogius and d'Espagnet, i.e. authors on the alchemical and Hermetic wing of the Neoplatonic tradition[57]), he denies that they actually *attract* effluvia from the air. (Contrast Newton, who seems prepared to countenance just that.[58]) So long as the air is in a state of perpetual agitation, and the 'magnet' absorbs all those 'congruous particles that fall upon it', it will act *as if* it genuinely attracted those particles from the ambient medium. A deliquescent salt, Boyle argues, does not actually draw water-vapour from the air: it merely absorbs those water-corpuscles that happen to fall upon it.[59]

Although difficult, Boyle admits, it is not in principle impossible to devise, discover, or invent 'magnets' for all manner of exotic subterranean or sidereal effluvia. These may be of great value to the physician: if a particular vapour or effluvium has a healing virtue, a 'magnet' capable of extracting, concentrating and storing that effluvium may be of considerable utility to medicine.

In this curious treatise, then, we find the mechanist and corpuscularian, Boyle, coming to terms with a variety of Neoplatonic, Hermetic, and alchemical ideas. The Neoplatonist's 'aire', a universal pool of 'spirit', has become an aggregate of effluvia; celestial 'virtues' or 'influences' are now carried by discrete particles; the vital spirit emanating from the Sun and enlivening all things has become an 'odd substance' of an 'astral', perhaps of a 'solar', nature, and so on. The doctrines of Platonists and Hermetics, astrologers and alchemists are re-cast, *with surprisingly little loss*, in the terminology of the corpuscular philosophy.

---

57 For the theory of chemical magnets, see Dobbs, 4, p. 152 ff.

58 Antimony was considered for some reason as a chemical magnet *par excellence*, capable of extracting the 'philosophical mercuries' of the metals. See again Dobbs, 4, p. 152 ff.

59 Boyle, *Works*, Vol. 4, p. 96.

BIBLIOGRAPHY

Abbreviations of Names of Journals
A.S.   Annals of Science
B.J.H.S.   British Journal for the History of Science
H.S.   History of Science
J.H.I.   Journal of the History of Ideas
J.H.P.   Journal for the History of Philosophy
N.R.R.S.L. Notes and Records of the Royal Society of London
P.R.   Philosophical Review
S.H.P.S.   Studies in the History and Philosophy of Science

W. E. Abraham 'The Nature of Zeno's Argument against Plurality, in D.K. 29 B1', *Phronesis*, 17, 1972, pp. 40-52.

S. M. Afnan *Avicenna: His Life and Works*, Allen & Unwin, London, 1958.

E. J. Aiton 1. 'Descartes' Theory of the Tides' *A.S.*, 11, 1955, pp. 337-348.

—— 2. 'The Vortex Theory of Planetary Motions', *A.S.*, 13, 1957, pp. 249-264.

—— 3. 'The Cartesian Theory of Gravity', *A.S.*, 15, 1959, pp. 27-49.

—— 4. 'The Celestial Mechanics of Leibniz in the light of Newtonian Criticism', *A.S.*, 18, 1962, pp. 31-41.

—— 5. 'The Celestial Mechanics of Leibniz: A New Interpretation', *A.S.*, 20, 1964, pp. 111-123.

—— 6. 'Kepler's Second Law of Planetary Motion', *Isis*, 60, 1969, pp. 75-90.

—— 7. 'Newton's Aether-Stream Hypothesis', *A.S.*, 25, 1969, pp. 255-260.

—— 8. *The Vortex Theory of Planetary Motions*, Mac-Donald, London, 1972.

R. Albritton 'Forms of Particular Substances in Aristotle's Metaphysics', *Journal of Philosophy*, 54, 1957, pp. 699-708.

H. G. Alexander (ed.) *The Leibniz-Clarke Correspondence*, University of Manchester Press, Manchester, 1956.

P. Alexander 1. 'Curley on Locke and Boyle', *P.R.*, 83, 1974, pp. 229-237.

—— 2. 'Boyle and Locke on Primary and Secondary Qualities', *Ratio*, 16, 1974, pp. 51–67.

—— 3. 'The Names of Secondary Qualities', *Proceedings of the Aristotelian Society*, 77, 1977, pp. 203–220.

—— 4. 'The Case of the Lonely Corpuscle: Reductive Explanation and Primitive Expressions' in R. Healey (ed.), 'Reduction, Time and Reality, CUP, Cambridge, 1981, pp. 17–35.

P. Anderson *The Philosophy of Francis Bacon*, University of Chicago Press, Chicago, 1948.

R. F. Anderson 'Locke on the Knowledge of Material Things', *J.H.P.*, 3, 1965, pp. 205–215.

J. P. Anton & G. L. Kustas (eds.) *Essays in Ancient Greek Philosophy*, University of New York Press, Albany, 1971.

W. Applebaum 'Boyle and Hobbes: A Reconsideration', *J.H.I.*, 25, 1964, pp. 117–119.

Archimedes *Works*, Ed. and Trans. T. L. Heath, Dover Books, New York, 1897.

Aristotle *Works*, 12 Volumes, General Editor W. D. Ross, Clarendon, Oxford, 1908–1963.

A. Armitage 1. 'The Cosmology of Giordano Bruno', *A.S.*, 6, 1948, pp. 24–31.

—— 2. ' "Borell's Hypothesis" and the rise of Celestial Mechanics', *A.S.*, 6, 1950, pp. 268–282.

W. H. G. Armytage 'The Early Utopists and Science in England', *A.S.*, 12, 1956, p. 247–254.

Augustine *The City of God*, Vol. 3, Tr. D. S. Wiesen, Heinemann, London, 1968.

F. Bacon *Works*, 14 Volumes, Ed. J. Spedding, R. L. Ellis & D. D. Heath, Longmans, London, 1862–1874.

R. Bacon *Opus Majus*, Trans. R. B. Burke, 2 Vols., University of Pennsylvania Press, Philadelphia, 1928.

C. Bailey (ed.) 1. *Epicurus*: The Extant Remains, Clarendon, Oxford, 1926.

—— 2. *The Greek Atomists and Epicurus*, Russell & Russell, New York, 1964.

K. E. Ballard 'Leibniz's Theory of Space and Time', *J.H.I.*, 21, 1960, pp. 49–65.

D. M. Balme 1. 'Greek Science and Mechanism 1: Aristotle on Nature and Chance', *Classical Quarterly*, 33, 1939, pp. 129–138.

—— 2. 'Greek Science and Mechanism 2: The Atomists' *Classical Quarterly*, 35, 1941, pp. 23–28.

J. Barnes *The Presocratic Philosophers*, 2 Vols. RKP, London, 1979.

A. C. Bell *Christian Huygens and the Development of Science in the Seventeenth Century*, Arnold, London, 1947.

M. Bentham 'Some Seventeenth-Century Views Concerning the Nature of Heat and Cold', *A.S.*, 2, 1937, pp. 431–450.

H. Bett *Nicholas of Cusa*, Methuen, London, 1932.

P. J. Bicknell 'Parmenides' Refutation of Motion and an Implication', *Phronesis*, 12, 1967, pp. 1–5.

T. B. Birch 'The Theory of Continuity of William of Ockham', *Philosophy of Science*, 3, 1936, pp. 494–505.

M. Black *Problems of Analysis*, RKP, London, 1954.

R. J. Blackwell 'Descartes' Laws of Motion', *Isis*, 57, 1966, pp. 220–234.

M. Boas 1. 'Hero's Pneumatica: A study of its Transmission and Influence', *Isis*, 40, 1949, pp. 38–48.

—— 2. 'Boyle as a Theoretical Scientist', *Isis*, 41, 1950, pp. 261–268.

—— 3. 'Bacon and Gilbert', *J.H.I.*, 12, 1951, pp. 466–467.

—— 4. 'The Establishment of the Mechanical Philosophy', *Osiris*, 10, 1952, pp. 412–541.

—— 5. 'Newton and the Theory of Chemical Solution', *Isis*, 43, 1952, p. 123.

—— 6. 'An Early Version of Boyle's Sceptical Chemist', *Isis*, 45, 1954, pp. 153–168.

—— 7. *Robert Boyle and Seventeenth Century Chemistry*, CUP, Cambridge, 1958.

M. Boas (Hall) 8. *The Scientific Revolution, 1450–1630*, Collins, London, 1962.

—— 9. *Robert Boyle on Natural Philosophy*, Indiana University Press, Bloomington, Indiana, 1966.

—— 10. (ed.) *Nature and Nature's Laws*, Macmillan, London, 1970.

P. A. Bogaard 'The Status of Complex Bodies in Epicurean Atomism', *S.H.P.S.*, 6, 1975, pp. 315–329.

J. E. Bolzan 'Chemical Combination according to Aristotle', *Ambix*, 23, 1976, pp. 134–144.

N. B. Booth '. 'Were Zeno's Arguments a Reply to Attacks upon Parmenides?', *Phronesis*, 2, 1957, pp. 1–9.

—— 2. 'Were Zeno's Arguments directed against the Pythagoreans?', *Phronesis*, 2, 1957, pp. 90–103.

C. B. Boyer *The History of the Calculus and its Conceptual Development*, Dover, New York, 1949.

R. Boyle *Works*, 6 Volumes, Ed. T. Birch, London, 1772.

F. Brandt *Thomas Hobbes' Mechanical Conception of Nature*, Tr. V. Maxwell & A. I. Fausbøll, Levin & Munksgaard, Copenhagen, 1928.

E. Bréhier *The History of Philosophy*, Vol. 3, University of Chicago Press, Chicago, 1965.

G. S. Brett *The Philosophy of Gassendi*, London, 1908.

A. Browne 'J. B. Van Helmont's Attack on Aristotle', *A.S.*, 36, 1979, pp. 575-591.

G. Bruno *Cause, Principle and Unity*, Tr. & Ed. J. Lindsay, Greenwood Press, Westport, Connecticut, 1976.

J. Burnet *Early Greek Philosophy*, 3rd Ed., A. & C. Black, London, 1920

F. D. Burnham 'The More-Vaughan Controversy: The Revolt against Philosophical Enthusiasm', *J.H.I.*, 35, 1974, pp. 33-49.

C. D. Burns 'William of Ockham on Continuity', *Mind*, 25, 1916, pp. 506-512.

E. A. Burtt *The Metaphysical Foundations of Modern Physical Science*, Kegan Paul, Trench, Trubner & Co., London, 1932.

H. Butterfield *The Origins of Modern Science*, Bell, London, 1965.

R. E. Butts & J. W. Davis (eds.) *The Methodological Heritage of Newton*, Blackwell, Oxford, 1970.

R. E. Butts & J. C. Pitt (eds.) *New Perspectives on Galileo*, Reidel, Dordrecht, 1978.

J. Campbell 'Locke on Qualities', *Canadian Journal of Philosophy*, 10, 1980, pp. 567-585.

M. Čapek (ed.) *The Concepts of Space and Time*, Reidel, Dordrecht, 1976.

M. H. Carré 1. 'Ralph Cudworth', *Philosophical Quarterly*, 3, 1953, pp. 342-351.

—— 2. 'Pierre Gassendi and the New Philosophy', *Philosophy*, 33, 1958, pp. 112-120.

E. Cassirer 1. 'Giovanni Pico della Mirandola', 2 Parts, *J.H.I.*, 3, 1942, pp. 123-144, 319-346.

—— 2.'Newton and Leibniz, *Philosophical Review*, 52, 1943, pp. 366-391.

—— 3. *The Platonic Renaissance in England*, Tr. J. P. Pettegrove, Nelson, Edinburgh, 1953.

F. F. Centore *Robert Hooke's Contributions to Mechanics*, Martinus Nijhoff, The Hague, 1970.

W. Charleton *Physiologia Epicuro-Gassendo-Charletoniana*, London, 1654.

W. Charlton (ed.) *Aristotle's Physics, Books 1 & 2*, Clarendon, Oxford, 1970.

C. H. Chen 'Aristotle's Concept of Primary Substance in Books Z and H of the Metaphysics', *Phronesis*, 2, 1957, pp. 46–59.

H. Cherniss *Aristotle's Criticism of Presocratic Philosophy*, Octagon Books, New York, 1971.

Cicero *On the Nature of the Gods*, Tr. H. C. P. MacGregor, Penguin, Harmondsworth, 1972.

M. Clagett 1. 'Some General Aspects of Physics in the Middle Ages', *Isis*, 39, 1948, pp. 29–44.

—— 2. *Greek Science in Antiquity*, Abelard-Schuman, London, 1957.

—— 3. 'The Impact of Archimedes on Medieval Science', *Isis*, 50, 1959, pp. 419–429.

—— 4. *The Science of Mechanics in the Middle Ages*, University of Wisconsin Press, Madison, 1959.

M. T. Clark (ed.) *An Aquinas Reader: Selections from the Writings of Thomas Aquinas*, Hodder & Stoughton, London, 1974.

J. T. Clark 'Pierre Gassendi and the Physics of Galileo', *Isis*, 54, 1963, pp. 352–370.

D. Clarke 'Physics and Metaphysics in Descartes' Principles', *S.H.P.S.*, 10, 1970, pp. 89–112.

M. Clavelin *The Natural Philosophy of Galileo*, Tr. A. J. Pomerans, M.I.T. Press, Cambridge, Mass., 1974.

I. B. Cohen 1. 'Newton in the Light of Recent Scholarship', *Isis*, 51, 1960, pp., 489–514.

—— 2. ' "Quantum in se est": Newton's Concept of Inertia in Relation to Descartes and Lucretius', *N.R.R.S.L.*, 19, 1964, pp. 131–155.

—— 3. *The Newtonian Revolution*, CUP, Cambridge, 1980.

L. D. Cohen 'Descartes and Henry More on the Beast-Machine', *A.S.*, 1, 1936, pp. 48–61.

M. R. Cohen & I. E. Dradkin (eds.) *A Source Book in Greek Science*, Harvard University Press, 1966.

R. G. Collingwood *The Idea of Nature*, OUP, Oxford, 1960.

B. P. Copenhaver 'Jewish Theologies of Space in the Scientific Revolution: Henry More, Joseph Raphson, Isaac Newton and their Predecessors', *A.S.*, 37, 1980, pp. 489–548.

F. Coppleston 1. *A History of Philosophy, Vol. 3, Ockham to Suarez*, Burns Oates & Washbourne, London, 1953.

—— 2. *Aquinas*, Penguin, Harmondsworth, 1955.

F. M. Cornford 1. *Plato's Cosmology* (translation of and commentary on the *Timaeus*), RKP, London, 1937.

—— 2. *Plato and Parmenides*, RKP, London, 1939.

—— 3. 'The Invention of Space' in Capek (ed.), pp. 3–26.

P. Costabel *Leibniz and Dynamics*, Tr. R. E. W. Maddison, Hermann, Paris, 1973.

G. R. Cragg (ed.) *The Cambridge Platonists*, OUP, Oxford, 1968.

A. C. Crombie 1. *Augustine to Galileo*, Falcon, London, 1952.

—— 2. *Robert Grosseteste and the Origins of Experimental Science*, Clarendon, Oxford, 1953.

—— 3. 'Quantification in Medieval Physics', *Isis*, 52, 1961, pp. 143–160.

M. P. Crossland 1. *Historical Studies in the Language of Chemistry*, Heinemann, London, 1962.

M. P. Crossland (ed.) 2. *The Science of Matter*, Penguin, Harmondsworth, 1971.

R. Cudworth *The True Intellectual System of the Universe*, 3 Vols. Ed. J. Harrison, Tegg, London, 1845.

E. M. Curley 'Locke, Boyle, and the Distinction between Primary and Secondary Qualities', *P.R.*, 81, 1972, pp. 438–464.

T. L. Davis 'Boyle's Conception of the Elements', *Isis*, 16, 1931, pp. 82–91.

A. G. Debus 1. 'The Paracelsian Aerial Niter', *Isis*, 55, 1964, pp. 43–61.

—— 2. *The English Paracelsians*, Oldbourne, London, 1965.

—— 3. 'Fire Analysis and the Elements in the Sixteenth and the Seventeenth Centuries', *A.S.*, 23, 1967, pp. 127–147.

—— 4. 'Alchemy and the Historian of Science', *H.S.*, 6, 1967, pp. 128–137.

—— 5. 'Renaissance Chemistry and the work of Robert Fludd', *Ambix*, 14, 1967, pp. 42–59.

—— 6. 'Mathematics and Nature in the Chemical Texts of the Renaissance', *Ambix*, 15, 1968, pp. 1–28.

—— (ed.) 7. *Science, Medicine and Society in the Renaissance: Essays to honour Walter Pagel*, Science History Publications, New York, 1972.

—— 8. 'Motion in Renaissance Chemistry', *Isis*, 64, 1973, pp. 5–17.

—— —— 9. 'The Chemical Philosophers: Chemical Medicine from Paracelsus to Van Helmont', *H.S.*, 12, 1974, pp. 235–259.

—— 10. *The Chemical Philosophy*, 2 Vols., Neale Watson, New York, 1977.

—— 11. *Man and Nature in the Renaissance*, CUP, Cambridge, 1978.

W. C. De Pauley *The Candle of the Lord: Studies in the Cambridge Platonists*, Macmillan, New York, 1937.

R. Descartes, 1. *Philosophical Works*, 2 Vols., Tr. & Ed. E. S. Haldane & G. R. T. Ross, CUP, Cambridge, 1911 and 1912.

—— 2. *Philosophical Writings*, Tr. & Ed. G. E. M. Anscombe & P. T. Geach, Nelson, London, 1970.

K. Digby *Two Treatises: Of Bodies, and of Man's Soul*, London, 1645.

E. J. Dijksterhuis *The Mechanisation of the World Picture*, Tr. C. Dikshoorn, Clarendon, Oxford, 1961.

Diogenes Laertius *Lives of Eminent Philosophers*, 2 Vols., Tr. R. D. Hicks, Heinemann, London, 1925.

B. J. T. Dobbs 1. 'Studies in the Natural Philosophy of Sir Kenelm Digby, Part 1', *Ambix*, 18, 1971, pp. 1-25.

—— 2. 'Studies in the Natural Philosophy of Sir Kenelm Digby, Part 2', *Ambix*, 20, 1973, pp. 143-163.

—— 3. 'Studies in the Natural Philosophy of Sir Kenelm Digby, Part 3', *Ambix*, 21, 1974, pp. 1-28.

—— 4. *The Foundations of Newton's Alchemy*, CUP, Cambridge, 1975.

I. E. Drabkin 'Notes on the Laws of Motion in Aristotle', *American Journal of Philology*, 59, 1938, pp. 60-84.

S. Drake 1. Galileo and the Law of Inertia', *American Journal of Physics*, 32, 1964, pp. 601-608.

—— 2. 'Galileo's 1604 Fragment on Falling Bodies', *B.J.H.S.*, 4, 1969, pp. 340-358.

—— 3. *Galileo Studies*, University of Michigan Press, Michigan, 1970.

—— 4. 'Medieval Ratio Theory or Compound Medicines in the Origins of Bradwardine's Rule', *Isis*, 64, 1973, pp. 67-77.

—— 5. 'Galileo's Experimental Confirmation of Horizontal Inertia', *Isis*, 64, 1973, pp. 291-305.

—— 6. 'Free Fall from Albert of Saxony to Honoré Fabri', *S.H.P.S.*, 5, 1975, pp. 347-366.

—— 7. 'Impetus Theory Reappraised', *J.H.I.*, 36, 1975, pp. 27-46.

—— 8. 'A Further Reappraisal of Impetus Theory', *S.H.P.S.*, 7, 1976, pp. 319-336.

D. B. Durand 'Nicole Oresme and the Medieval Origins of Modern Sciences', *Speculum*, 16, 1941, pp. 167-185.

A. Edel *Aristotle's Theory of the Infinite*, Ann Arbor, Michigan and London, 1979.

B. D. Ellis 'Newton's Concept of Motive Force', *J.H.I.*, 23, 1962, pp. 273–278.

M. 'Espinasse *Robert Hooke*, University of California Press, Berkeley and Los Angeles, 1962.

M. G. Evans 'Aristotle, Newton, and the Theory of Continuous Magnitude', *J.H.I.*, 16, 1955, pp. 548–557.

B. Farrington *Greek Science: Its Meaning for Us*, Pelican, Harmondsworth, 1953.

J. Ferguson 'Dinos', *Phronesis*, 16, 1971, pp. 97–115.

M. Ficino 'Extracts from the *Platonic Theology*', Trans. J. L. Burroughs, *J.H.I.*, 5, 1944, pp. 227–242.

K. Figala 'Newton as Alchemist', *H.S.*, 15, 1977, pp. 102–137.

R. J. Forbes 'Was Newton an Alchemist?', *Chymia*, 2, 1949, pp. 27–36.

A. Franklin 'Stillman Drake's "Impetus Theory Reappraised" ', *J.H.I.*, 38, 1977, pp. 307–314.

P. J. French *John Dee: The World of an Elizabethan Magus*, RKP, London, 1972.

J. F. Fulton 'Robert Boyle and his Influence on Thought in the Seventeenth Century', *Isis*, 18, 1932, pp. 77–102.

D. J. Furley 1. 'Lucretius and the Stoics', *Institute of Classical Studies Bulletin*, 13, 1966, pp. 13–33.

—— 2. *Two Studies in the Greek Atomists*, Princeton University Press, Princeton, New Jersey, 1967.

—— 3. 'Aristotle and the Atomists on Motion in a Void' in Machamer & Turnbull (eds.), pp. 83–100.

A. Gabbey 1. 'Force and Inertia in Seventeenth Century Dynamics', *S.H.P.S.*, 2, 1971, pp. 1–67.

—— 2. 'Essay Review of W. L. Scott, "The Conflict between Atomism and Conservation Theory 1644-1869" ', *S.H.P.S.*, 3, 1973, pp. 373–385.

Galen *On the Natural Faculties*, Tr. A. J. Brock, Heinemann, London, 1956.

G. Galilei 1. *On Motion and On Mechanics*, Tr. & Ed. I. E. Drabkin & S. Drake, University of Wisconsin Press, 1960.

—— 2. *Discoveries and Opinions*, Tr. & Ed. S. Drake, Doubleday Anchor, New York, 1957.

—— 3. *Dialogue Concerning the Two Chief World Systems*, Tr. S. Drake, University of California Press, Berkeley and Los Angeles, 1967.

—— 4. *Two New Sciences*, Macmillan, New Work, 1914.

M. Galston 'A Re-examination of al-Fārābī's Neoplatonism', *J.H.P.*, 15, 1977, pp. 13–32.

P. Gassendi *Selected Works*, Trans. C. B. Brush, Johnson Reprint Corporation, New York, 1972.

S. Gaukroger *Explanatory Structures*, Harvester, Sussex, 1978.

R. K. Gaye 'On Aristotle's Physics Z9 239b 33–240a 18', *Journal of Philology*, 31, 1907, pp. 95–116.

N. R. Gelbart 'The Intellectual Development of Walter Charleton', *Ambix*, 18, 1971, pp. 149–168.

D. Geoghegan 'Some Indications of Newton's Attitude towards Alchemy', *Ambix* 6, 1957, pp. 102–106.

A. Gewirtz 'Experience and the non-Mathematical in the Cartesian Method', *J.H.I.*, 2, 1941, pp. 183–210.

N. W. Gilbert 'Galileo and the School of Padua', *J.H.P.*, 1, 1963, pp. 223–231.

W. Gilbert *On the Magnet*, Basic Books, New York, 1958.

T. Gomperz *The Greek Thinkers*, Vol. 1, Tr. L. Magnus, 7th Imp., Murray, London, 1964.

P. Gouk 'The Role of Acoustics and Music Theory in the Scientific Work of Robert Hooke', *A.S.*, 37, 1980, pp. 573–605.

E. Grant 1. 'Late Medieval Thought, Copernicus, and the Scientific Revolution', *J.H.I.*, 23, 1962, pp. 197–220.

—— 2. 'Motion in the Void and the Principle of Inertia in the Middle Ages', *Isis*, 55, 1964, p. 265–292.

—— 3. 'Medieval and Seventeenth Century Conceptions of an Infinite Void Space beyond the Cosmos', *Isis*, 60, 1969, pp. 39–60.

—— (ed.)4. *A Source Book in Medieval Science*, Harvard University Press, Cambridge, Mass., 1974.

—— 5. 'Place and Space in Medieval Physical Theory' in Machamer & Turnbull (eds.), p. 137–167.

—— 6. *Physical Science in the Middle Ages*, CUP, Cambridge, 1977.

—— 7. 'The Principle of the Impenetrability of Bodies in the History of Concepts of Separate Space from the Middle Ages to the Seventeenth Century', *Isis*, 69, 1978, pp. 551–571.

—— 8. 'Aristotelianism and the Longevity of the Medieval World View', *H.S.*, 16, 1978, pp. 93–106.

—— 9. *Much Ado about Nothing: Theories of Space and Vacuum from the Middle Ages to the Scientific Revolution*, CUP, Cambridge, 1981.

R. A. Greene 'Henry More and Robert Boyle on the Spirit of Nature', *J.H.I.*, 23, 1962, pp. 451–474.

J. C. Gregory 'Chemistry and Alchemy in the Natural Philosophy of Sir Francis Bacon', *Ambix*, 2, 1938, pp. 93–111.

T. R. Girill 'Galileo and Platonism', *J.H.I.*, 31, 1970, pp. 501–520.

W. E. Gross 'Relativity of Motion: From Occam to Galileo', *A.S.*, 31, 1974, pp. 529–545.

A. Grünbaum 1. 'A Consistent Conception of the Extended Linear Continuum as an Aggregate of Unextended Elements', *Philosophy of Science*, 19, 1952, pp. 288–306.

—— 2. 'Zeno's Metrical Paradox of Extension' in W. Salmon (ed.), pp. 176–199.

J. Grundy 'Descartes and Atomism', *Nature*, 173, 4393, 1954, p. 89.

H. Guerlac 1. 'Newton's Optical Aether', *N.R.R.S.L.*, 22, 1967, pp. 45–57.

—— 2. 'Copernicus and Aristotle's Cosmos', *J.H.I.*, 29, 1968, pp. 109–113.

R. T. Gunther (ed.) *Early Science in Oxford*, 15 Vols., OUP, 1923–1967 (for Hooke, see Vols, 6, 7, 8, 10 & 13).

W. K. C. Guthrie *A History of Greek Philosophy*, CUP, Cambridge, Vol. 1 (1962) and Vol. 2 (1965).

D. E. Hahm 1. 'Chrysippus' Solution to the Democritean Dilemma of the Cone', *Isis*, 63, 1972, pp. 205–220.

—— 2. 'Weight and Lightness in Aristotle and his Predecessors' in Machamer & Turnbull (eds.), pp. 56–82.

A. R. Hall 1. 'Two Unpublished Lectures of Robert Hooke', *Isis*, 42, 1951, pp. 219–230.

—— 2. *The Scientific Revolution, 1500–1800*, Longmans Green & Co., London, 1954.

—— 3. *From Galileo to Newton, 1635–1720*, Collins, London, 1963.

—— 4. 'Galileo and the Science of Motion', *B.J.H.S.*, 2, 1964, pp. 185–199.

—— 5. 'Mechanics and the Royal Society, 1668–1670', *B.J.H.S.*, 3, 1966, pp. 24–38.

A. R. & M. B. Hall 1. 'Newton's Mechanical Principles', *J.H.I.*, 20, 1959, pp. 167–178.

—— 2. 'Newton's Electric Spirit – 4 oddities', *Isis*, 50, 1959, pp. 473–476.

—— 3. 'Newton's Theory of Matter', *Isis*, 51, 1960, pp. 131–144.

—— (eds.) 4. *Unpublished Scientific Papers of Isaac Newton*, CUP, Cambridge, 1962.

T. L. Hankins 'The Influence of Malebranche on the Science of Mechanics during the Eighteenth Century', *J.H.I.*, 28, 1967, pp. 193–210.

N. R. Hanson 1. 'Waves, Particles, and Newton's "Fits" ', *J.H.I.*, 21, 1960, pp. 370–391.

—— 2. 'The Copernican Disturbance and the Keplerian Revolution', *J.H.I.*, 22, 1961, pp. 169–184.

E. S. Haring 'Substantial form in Aristotle's Metaphysics Z', 3 parts, *Review of Metaphysics*, 10, 1957, pp. 308–332, 482–501, and 698–713.

R. Harré 1. *Matter and Method*, Macmillan, London, 1964

—— 2. *Early Seventeenth Century Scientists*, Pergamon Press, Oxford, 1965.

—— 3. *The Method of Science, 8*, Wykeham Publications, London & Winchester, 1970.

I. B. Hart *The Mechanical Investigations of Leonardo da Vinci*, 2nd Ed., University of California Press, Berkeley and Los Angeles, 1963.

G. C. Hatfield 'Force (God) in Descartes' Physics', *S.H.P.S.*, 10, 1979, pp. 113–140.

J. L. Hawes 1. 'Newton and Electrical Attraction Unexcited', *A.S.*, 24, 1968, pp. 121–130.

—— 2. 'Newton's Two Electricities', *A.S.*, 27, 1971, pp. 95–103.

T. Heath 1. *A History of Greek Mathematics*, Vol. 1, Clarendon, Oxford, 1921.

—— 2. *Mathematics in Aristotle*, Clarendon, Oxford, 1949.

W. A. Heidel 1. *The Heroic Age of Science*, Williams & Wilkins, Baltimore, 1933.

—— 2. 'Qualitative Change in Presocratic Philosophy' in Mourelatos (ed.), pp. 86–95.

J. Henry 'Francesco Patrizi da Cherso's Concept of Space and its later Influence', *A.S.*, 36, 1979, pp. 549–573.

J. Herivel *The Background of Newton's Principia*, Clarendon, Oxford, 1965.

Hero *Pneumatics*, Tr. & Ed. B. Woodcroft, Taylor Walton & Maberley, London, 1851.

M. Hesse 1. *Forces and Fields*, Nelson, London, 1961.

—— 2. 'Changing Views of Matter', *H.S.*, 3, 1964, pp. 79–84.

—— 3. 'Hooke's Philosophical Algebra', *Isis*, 57, 1966, pp. 67–83.

—— 4. 'Hooke's Vibration Theory and the Isochrony of Springs', *Isis*, 57, 1966, pp. 433–441.

G. Heym 'Al Rāzī and Alchemy', *Ambix*, 1, 1938, pp. 184–191.

W. L. Hine 'Mersenne and Copernicanism', *Isis*, 64, 1973, pp. 18–32.

T. Hobbes *Works*, 11 Vols., Ed. W. Molesworth, Bohn, London, 1839.

P. H. J. Hoenen 'Descartes' Mechanism' in W. Doney (ed.), *Descartes: A Collection of Critical Essays*, Macmillan, London, 1967, pp. 353–368.

E. J. Holmyard *Alchemy*, Pelican, Harmondsworth, 1957.

R. W. Home 1. 'The Third Law in Newton's Mechanics', *B.J.H.S.*, 4, 1968, pp. 39–51.

—— 2. 'Newtonianism and the Theory of the Magnet', *H.S.*, 15, 1977, pp. 252–266.

—— 3. 'Newton on Electricity and the Aether', unpublished.

R. Hooke *Philosophical Experiments and Observations*, Ed. W. Derham, Innys, London, 1726.

R. Hooykaas 'The Experimental Origin of Chemical Atomic and Molecular Theory before Boyle', *Chymia*, 2, 1949, pp. 65–80.

A. J. Hopkins *Alchemy, Child of Greek Philosophy*, Columbia University Press, New York, 1934.

R. A. Horne 'Aristotelian Chemistry', *Chymia*, 11, 1966, pp. 21–27.

M. A. Hoskin & A. G. Molland 'Swineshead on Falling Bodies: An Example of Fourteenth-Century Physics', *B.J.H.S.*, 3, 1966, pp. 150–182.

H. M. Howe 'A Root of Van Helmont's Tree', *Isis*, 56, 1965, pp. 408–419.

W. C. Humphreys 'Galileo, Falling Bodies and Inclined Planes', *B.J.H.S.*, 3, 1967, pp. 225–244.

A. J. Ihde 'Alchemy in Reverse: Robert Boyle on the Degradation of Gold', *Chymia*, 9, 1964, pp. 47–57.

C. Iltis 1. 'Leibniz and the Vis Viva Controversy', *Isis*, 62, 1971, pp. 21–35.

—— 2. 'The Decline of Cartesianism in Mechanics: The Leibnizian-Cartesian debates', *Isis*, 64, 1973, pp. 356–373.

J. A. Irving 'Leibniz's Theory of Matter', *Philosophy of Science*, 3, 1936, pp. 208–214.

M. Jammer 1. *Concepts of Space*, Harvard University Press, Cambridge, Mass., 1954.

—— 2. *Concepts of Mass*, Harvard University Press, Cambridge, Mass., 1961.

—— 3. *Concepts of Force*, Harper Torchbooks, New York, 1962.

H. Joachim 'Aristotle's Conception of Chemical Combination', *Journal of Philology*, 29, 1904, pp. 72–86.

H. J. Johnson 'Three Ancient Meanings of Matter: Democritus, Plato and Aristotle', *J.H.I.*, 28, 1967, pp. 3–16.

B. Jones 'Aristotle's Introduction of Matter', *P.R.*, 83, 1974, pp. 474–500.

R. F. Jones *Ancients and Moderns*, 2nd Ed., University of California Press, Berkeley and Los Angeles, 1965.

C. H. Josten 'A Translation of John Dee's "*Monas Hieroglyphica*" (Antwerp, 1564) with an introduction and notes', *Ambix*, 12, 1964, pp. 84–221.

C. Kahn *Anaximander and the Origins of Greek Cosmology*, Columbia University Press, New York, 1960.

R. Kargon 1. 'Walter Charleton, Robert Boyle and the Acceptance of Epicurean Atomism in England', *Isis*, 55, 1964, pp. 184–192.

—— 2. 'William Petty's Mechanical Philosophy', *Isis*, 56, 1965, pp. 63–66.

—— 3. 'Thomas Hariot, the Northumberland Circle and Early Atomism in England', *J.H.I.*, 27, 1966, pp. 128–136.

—— 4. *Atomism from Hariot to Newton*, OUP, Oxford, 1966.

S. V. Keeling 'Cartesian Mechanism', *Philosophy*, 9, 1934, pp. 51–66.

N. Kemp Smith *New Studies in the Philosophy of Descartes*, Macmillan, London, 1966.

A. Kenny (ed.) *Descartes' Philosophical Letters*, Clarendon, Oxford, 1970.

G. B. Kerferd 'Anaxagoras and the Concept of Matter before Aristotle' in Mourelatos (ed.), pp. 489–503.

H. R. King 1. 'Aristotle and the Paradoxes of Zeno', *Journal of Philosophy*, 46, 1949, pp. 657–670.

—— 2. 'Aristotle without *Prima Materia*', *J.H.I.*, 17, 1956, pp. 370–389.

G. S. Kirk & J. S. Raven *The Presocratic Philosophers*, CUP, Cambridge, 1957.

G. S. Kirk & M. C. Stokes 'Parmenides' Refutation of Motion', *Phronesis*, 5, 1960, pp. 1–4.

W. E. Knowles-Middleton 'The Place of Torricelli in the History of the Barometer', *Isis*, 54, 1963, pp. 11–28.

D. Konstan 1. 'Epicurus on "Up" and "Down" ', *Phronesis*, 17, 1972, pp. 269–278.

—— 2. 'Problems in Epicurean Physics', *Isis*, 70, 1979, pp. 394–418.

A. Koyré 1. 'Galileo and Plato', *J.H.I.*, 4, 1943, pp. 400–428.

—— 2. 'An Unpublished Letter from Robert Hooke to Isaac Newton', *Isis*, 43, 1952, pp. 312–337.

—— 3. *From the Closed World to the Infinite Universe*, Johns Hopkins Press, Baltimore, 1957.

—— 4. *Newtonian Studies*, Chapman & Hall, London, 1965.

—— 5. *Metaphysics and Measurement: Essays in the Scientific Revolution*, Chapman & Hall, London, 1968.

—— 6. *The Astronomical Revolution: Copernicus, Kepler, Borelli*, Tr. R. E. W. Maddison, Ithaca, New York, 1963.

—— 7. *Galileo Studies*, Tr. J. Mepham, Harvester, Sussex, 1978.

A. Koyré & I. B. Cohen 1. 'Newton's "Electric and Elastic Spirit" ', *Isis*, 51, 1960, p. 337.

—— 2. 'The Case of the Missing Tanquam: Newton, Leibniz, and Clarke', *Isis*, 52, 1961, pp. 555–566.

P. O. Kristeller 'Ficino and Pomponazzi on the Place of Man in the Universe', *J.H.I.*, 5, 1944, pp. 220–226.

D. Kubrin 'Newton and the Cyclical Cosmos: Providence and the Mechanical Philosophy', *J.H.I.*, 28, 1967, pp. 325–346.

T. Kuhn 1. 'Newton's "31st Query" and the Degradation of Gold', *Isis*, 42, 1951, pp. 296–298.

—— 2. 'Robert Boyle and Structural Chemistry', *Isis*, 43, 1952, pp. 12–36.

—— 3. 'The Independence of Density and Pore-Size in Newton's Theory of Matter', *Isis*, 43, 1952, pp. 364–365.

—— 4. *The Copernican Revolution*, Harvard University Press, Cambridge, Mass., 1957.

A. R. Lacey 'The Eleatics and Aristotle on some problems of Change', *J.H.I.*, 26, 1965, pp. 451–468.

T. Lai 'Nicholas of Cusa and the Finite Universe', *J.H.P.*, 11, 1973, pp. 161–167.

B. M. Laing 'Descartes on Material things', *Philosophy*, 16, 1941, pp. 398–411.

J. Laird *Hobbes*, Benn, London, 1934.

M. Lapidge, 'A Problem in Stoic Cosmology', *Phronesis*, 18, 1973, pp. 240–278.

V. R. Larkin (ed.) 'Aquinas on the Combining of the Elements', *Isis*, 51, 1960, pp. 68–72.

R. E. Larsen 'The Aristotelianism of Bacon's Novum Organum', *J.H.I.*, 23, 1962, pp. 435–450.

L. Laudan 1. 'The Clock Metaphor and Probabilism: The Impact of Descartes on English Methodological Thought, 1650–1665', *A.S.*, 22, 1966, pp. 73–104.

—— 2. 'The Nature and Sources of Locke's Views on Hypothesis', *J.H.I.*, 28, 1967, pp. 211–223.

H. D. P. Lee *Zeno of Elea*, CUP, Cambridge, 1936.

G. Leff *William of Ockham*, Manchester University Press, Manchester, 1975.

G. W. Leibniz *Philosophical Papers and Letters*, Ed. L. E. Loemker, 2nd Ed., Reidel, Dordrecht, 1969.

Leonardo da Vinci *Notebooks*, Ed. I. A. Richter, OUP, Oxford, 1980.

M. Levey 'Studies in the Development of the Atomic Theory', *Chymia*, 7, 1961, pp. 40–56.

D. C. Lindberg (ed.) *Science in the Middle Ages*, University of Chicago Press, Chicago, 1978.

S. J. Linden 'Francis Bacon and Alchemy: The Reformation of Vulcan', *J.H.I.*, 35, 1974, pp. 547–560.

R. Lindsay 'Pierre Gassendi and the Revival of Atomism in the Renaissance', *American Journal of Physics*, 13, 1945, pp. 235–242.

G. E. R. Lloyd 'The Hot and the Cold, The Dry and the Wet in Greek Philosophy', *Journal of Hellenic Studies*, 84, 1964, pp. 92–106.

J. Locke 1. *An Essay Concerning Human Understanding*, 2 Vols., Everyman, London, 1961.

—— 2. 'Elements of Natural Philosophy' in *Philosophical Works*, Vol. 2, Ed. J. A. St. John, Bohn, London, 1854, pp. 472–496.

L. E. Loemker 1. 'Boyle and Leibniz', *J.H.I.*, 16, 1955, pp. 22–43.

—— 2. *Struggle for Synthesis*, Harvard University Press, Cambridge, Mass., 1972.

A. A. Long *Hellenistic Philosophy*, Duckworth, London, 1974.

J. Losee *A Historical Introduction to the Philosophy of Science*, OUP, Oxford, 1972.

R. Love 'Revisions of Descartes' Matter-Theory in *Le Monde*', *B.J.H.S.*, 8, 1975, pp. 127–137.

A. O. Lovejoy *The Great Chain of Being*, Harvard University Press, Cambridge, Mass., 1936.

Lucretius *On the Nature of Things*, Text, Translation and Commentary in 3 Volumes, Ed. C. Bailey, Clarendon, Oxford, 1947.

D. J. Lysaght 'Hooke's Theory of Combustion', *Ambix*, 1, 1937, pp. 93-108.

D. B. MacDonald 'Continuous Creation and Atomic Time in Muslim Scholastic Theology', *Isis*, 9, 1927, pp. 326-344.

P. K. Machamer 'Aristotle on Natural Place and Natural Notion', *Isis*, 69, 1978, pp. 377-387.

P. K. Machamer & R. G. Turnbull (eds.) *Motion and Time, Space and Matter*, Ohio State University Press, 1976.

J. Mackie *Problems from Locke*, Clarendon, Oxford, 1976.

M. Maimonides *Guide for the Perplexed*, Tr. M. Friedländer, Routledge, London, 1919.

M. Mandelbaum *Philosophy, Science and Sense Perception*, Johns Hopkins Press, Baltimore, 1964.

R. S. Marx 'A Thirteenth Century Theory of Heat as a Form of Motion', *Isis*, 22, 1934, pp. 19-20.

D. Massa 'Giordano Bruno and the Top-Sail Experiment', *A.S.*, 30, 1973, pp. 201-211.

J. Masson *The Atomic Theory of Lucretius*, Bell, London, 1884.

J. Mau 'Was there a special Epicurean Mathematics?' in *Phronesis*, Supp. Vol. 1, 1973, Eds. E. N. Lee, A. P. D. Mourelatos & R. M. Rorty.

G. McColley 1. 'The Theory of the Diurnal Rotation of the Earth', *Isis*, 26, 1936, pp. 392-404.

—— 2. 'The Seventeenth Century Doctrine of a Plurality of Worlds', *A.S.*, 1, 1936, pp. 385-430.

—— 3. 'Gilbert and Bruno', *A.S.*, 2, 1937, pp. 353-354.

—— 4. 'Nicholas Hill and the Philosophia Epicurea', *A.S.*, 4, 1940, pp. 390-405.

J. B. McDiarmid 'Theophrastus de Sensibus 66: Democritus' Explanation of Salinity', *American Journal of Philology*, 80, 1959, pp. 56-66.

D. McGibbon 'The Atomists and Melissus', *Mnemosyne*, 4 ser 17, 1964, pp. 248-255.

J. E. McGuire 1. 'Transmutation and Immutability: Newton's Doctrine of Physical Qualities', *Ambix*, 14, 1967, pp. 69-95.

—— 2. 'Force, Active Principles, and Newton's Invisible Realm', *Ambix*, 15, 1968, pp. 154-208.

—— 3. 'Atoms and the "Analogy of Nature" ', *S.H.P.S.*, 1, 1970, pp. 3–58.

—— 4. 'Boyle's Conception of Nature', *J.H.I.*, 33, 1972, pp. 523–542.

—— 5. 'Existence, Actuality and Necessity: Newton on Space and Time', *A.S.*, 35, 1978, pp. 463–508.

—— 6. 'Newton on Place, Time, and God: An Unpublished Source', *B.J.H.S.*, 11, 1978, pp. 114–129.

J. E. McGuire & P. M. Rattansi 'Newton and the Pipes of Pan', *N.R.R.S.L.*, 21, 1966, pp. 108–143.

D. McKie 'Some Early Work on Combustion, Respiration, and Calcination', *Ambix*, 1, 1938, pp. 143–165.

E. McMullin (ed.) 1. *The Concept of Matter in Greek and Medieval Philosophy*, University of Notre Dame Press, Notre Dame, Indiana, 1963.

—— 2. *The Concept of Matter in Modern Philosophy*, University of Notre Dame Press, Notre Dame, Indiana, 1963.

—— 3. *Galileo: Man of Science*, Basic Books, New York, 1967.

—— 4. *Matter and Activity in Newton*, University of Notre Dame Press, Notre Dame, Indiana, 1978.

C. Merchant 'The Vitalism of Francis Mercury van Helmont: Its Influence on Leibniz', *Ambix*, 26, 1979, pp. 170–183.

E. Meyerson *Identity and Reality*, Tr. K. Loewenberg, Dover, New York, 1962.

P.H. Michel *The Cosmology of Giordano Bruno*, Tr. R. E. W. Maddison, Methuen, London, 1973.

W. Miles 'Sir Kenelm Digby: Alchemist, Scholar, Courtier and Man of Adventure', *Chymia*, 2, 1949, pp. 119–128.

E. C. Millington 'Theories of Cohesion in the Seventeenth Century', *A.S.*, 5, 1945, pp. 253–269.

S. I. Mintz '. 'Galileo, Hobbes, and the Circle of Perfection', *Isis*, 43, 1952, pp. 98–100.

—— 2. *The Hunting of Leviathan*, CUP, Cambridge, 1962.

J. Mittelstrasss 'The Galilean Revolution', *S.H.P.S.*, 2, 1972, pp. 297–328.

A. G. Molland 1. 'The Geometrical Background to the Merton School', *B.J.H.S.*, 4, 1968, pp. 108–125.

—— 2. 'Oresme Redivivus', *H.S.*, 8, 1969, pp. 106–119.

E. A. Moody 1. 'Ockham and Aegidius of Rome', *Franciscan Studies*, 9, 1949, pp. 417–442.

—— 2. 'Galileo and Avempace: the Dynamics of the Leaning Tower Experiment', 2 parts, *J.H.I.*, 12, 1951, pp. 163–193, 375–422.

—— 3. 'The Laws of Motion in Medieval Physics', *The Scientific Monthly*, 72, 1951, pp. 18–23.

—— 4. *Studies in Medieval Philosophy, Science, and Logic*, University of California Press, Berkeley & Los Angeles, 1975.

E. A. Moody & M. Clagett (eds.) *The Medieval Science of Weights*, University of Wisconsin Press, Madison, 1952.

H. More 1. *A Collection of Several Philosophical Writings*, 2 Vols., London, 1662, reprinted Garland, New York, 1978.

—— 2. *Philosophical Writings*, Ed. F. I. MacKinnon, OUP, Oxford, 1925.

L. T. More 1. 'Boyle as Alchemist', *J.H.I.*, 2, 1941, pp. 61–76.

—— 2. *The Life and Works of the Honourable Robert Boyle*, OUP, Oxford, 1944.

P. Morewedge (ed.) *The Metaphysics of Avicenna*, Tr. plus notes and commentary, RKP, London, 1973.

A. P. D. Mourelatos (ed.) *The Presocratics*, Anchor Doubleday, New York, 1964.

N. W. Nason 'Leibniz's Attack on the Cartesian Doctrine of Extension', *J.H.I.*, 7, 1946, pp. 447–483.

R. H. Naylor 'Galileo and the Problem of Free Fall', *B.J.H.S.*, 7, 1974, pp. 105–134.

I. Newton 1. *Mathematical Principles of Natural Philosophy*, 2 Vols., Tr. A. Motte, 1729; revised translation F. Cajori, University of California Press, Berkeley, 1934.

I. Newton 2. *Opticks*, 4th London Edition, 1730; Dover, New York, 1952.

Nicholas of Autrecourt *Universal Treatise*, trans. L. A. Kennedy, R. E. Arnold & A. E. Millward with an introduction by L. A. Kennedy, Marquette University Press, Milwaukee, Wisconsin, 1971.

Nicholas of Cusa *Of Learned Ignorance*, Tr. G. Heron, RKP, London, 1954.

A. T. Nicol 'Indivisible Lines', *Classical Quarterly*, 30, 1936, pp. 120–126.

F. C. S. Northrop 'Liebniz's Theory of Space', *J.H.I.*, 7, 1946, pp. 422-446.

J. F. O'Brien 'Some Medieval Anticipations of Inertia', *New Scholasticism*, 44, 1970, pp. 345–371.

J. J. O'Brien 'Samuel Hartlib's Influence on Robert Boyle's Scientific Development', 2 parts, *A.S.*, 21, 1965, pp. 1–14, 257–276.

William of Ockham *Philosophical Writings*, Tr. P. Boehner, St. Bonaventure, New York, 1957.

D. J. O'Connor *John Locke*, Dover, New York, 1967.

J. R. O'Donnell 'The Philosophy of Nicholas of Autrecourt and his Appraisal of aristotle', *Medieval Studies*, 4, 1942, pp. 97–125.

N. Oresme *Le Livre du Ciel et du Monde*, Ed. A. D. Menut & A. J. Denomy, Tr. A. D. Menut, University of Wisconsin Press, Madison, Milwaukee and London, 1968.

M. J. Osler 1. 'John Locke and the Changing Ideal of Scientific Knowledge', *J.H.R.*, 31, 1970, pp. 3–16.

—— 2. 'Galileo, Motion, and Essences', *Isis*, 64, 1973, pp. 504–509.

—— 3. 'Descartes and Charleton on Nature and God', *J.H.I.*, 40, 1979, pp. 445–456.

G. E. L. Owen 'Zeno and the Mathematicians', *Proceedings of the Aristotelian Society*, 58, 1957–1958, pp. 199–222.

W. Pagel 1. *Paracelsus*, Karger, Basle, 1958.

—— 2. 'Paracelsus and the Neoplatonic and Gnostic Tradition', *Ambix*, 8, 1960, pp. 125–166.

—— 3. 'The Prime Matter of Paracelsus', *Ambix*, 9, 1961, pp. 117–135.

—— 4. 'The "Wild Spirit" of J. B. Van Helmont and Paracelsus', *Ambix*, 10, 1962, pp. 1–13.

—— 5. 'Chemistry at the Cross-Roads: The Ideas of Joachim Jungius', *Ambix*, 16, 1969, pp. 100–108.

W. Pagel & M. Winder. 'The Higher Elements and Prime Matter in Renaissance Naturalism and in Paracelsus', *Ambix*, 21, 1974, pp. 93–127.

L. U. Pancheri 'Greek Atomism and the One and the Many', *J.H.P.*, 13, 1975, pp. 139–144.

F. A. Paneth 'The Epistemological Status of the Chemical Concept of Element', 2 parts, *British Journal for the Philosophy of Science*, 13, 1962, pp. 1–14, 144–160.

Paracelsus 1. *Hermetic and Alchemical Writings*, 2 Vols., Ed. & Tr. A. E. White, University Books, New York, 1967.

—— 2. *Selected Writings*, Ed. J. Jacobi, RKP, London, 1951.

J. R. Partington 1. 'Joan Baptista Van Helmont', *A.S.*, 1, 1936, pp. 359–384.

—— 2. 'Albertus Magnus on Alchemy', *Ambix*, 1, 1937, pp. 3–20.

—— 3. 'The Chemistry of Rāsī *Ambix*, 1, 1938, pp. 192–196.

—— 4. 'The Origins of the Atomic Theory', *A.S.*, 4, 1939, pp. 245–282.

—— 5. 'The Life and Work of John Mayow', 2 parts, *Isis*, 47, 1956, pp. 217–230, 405–417.

—— 6. *A History of Chemistry*, 4 Vols., Macmillan, London, 1961–1970.

J. A. Passmore *Ralph Cudworth*, CUP, Cambridge, 1951.

C. A. Patrides 1. 'The Numerological Approach to Cosmic Order during the English Renaissance', *Isis*, 49, 1958, pp. 390–397.

—— (ed.) 2. *The Cambridge Platonists*, Arnold, London, 1969.

F. Patrizi 'On Physical Space', Tr. B. Brickman, *J.H.I.*, 4, 1943, pp. 224–245.

L. D. Patterson 'Hooke's Gravitation Theory', *Isis*, 40, 1949, pp. 327–341.

P. A. Pav 'Gassendi's Statement of the Principle of Inertia', *Isis*, 57, 1966, pp. 24–34.

O. Pederson & M. Pihl *Early Physics and Astronomy*, MacDonald & Janes, London, 1974.

M. E. Perl 'Physics and Metaphysics in Newton, Leibniz and Clarke', *J.H.I.*, 30, 1969, pp. 507–526.

Plato 1. *Timaeus*, Tr. D. Lee, Penguin, Harmondsworth, 1965.

—— 2. *Laws* in *The Dialogues of Plato*, Tr. & Ed. B. Jowett, Vol. 5, 3rd Ed., OUP, Oxford, 1892.

Plotinus *Enneads 1–3*, Tr. A. H. Armstrong, Heinemann, London, 1966–1967.

Plutarch *Moralia*, 16 Vols., Heinemann, London, 1927–1969.

W. Pohle 'The Mathematical Foundations of Plato's Atomic Physics', *Isis*, 62, 1971, pp. 36–46.

R. H. Popkin *The History of Scepticism from Erasmus to Descartes*, Van Gorcum, Assen, 1964.

J. E. Power 'Henry More and Isaac Newton on Absolute Space', *J.H.I.*, 31, 1970, pp. 289–296.

T. L. Prendergast 'Motion, Action and Tendency in Descartes' Physics', *J.H.P.*, 13, 1975, pp. 453–462.

D. B. Quinn & J. W. Shirley 'A Contemporary List of Hariot References', *Renaissance Quarterly*, 22, 1969, p. 9–25.

J. H. Randall 1. 'The Development of Scientific Method in the School of Padua', *J.H.I.*, 1, 1940, pp. 177–206.

—— 2. *Aristotle*, Columbia University Press, New York, 1960.

H. Rashdall 'Nicholas de Ultricuria, a medieval Hume', *Proceedings of the Aristotelian Society*, 7, 1907, pp. 1–27.

P. M. Rattansi 'Paracelsus and the Puritan Revolution', *Ambix*, 11, 1963, pp. 24–32.

G. Rees 1. 'Francis Bacon's Semi-Paracelsian Cosmology', *Ambix*, 22, 1975, pp. 81–101.

—— 2. 'Francis Bacon's Semi-Paracelsian Cosmology and the Great Instauration', *Ambix*, 22, 1975, pp. 161–173.

—— 3. 'The Fate of Bacon's Cosmology in the Seventeenth Century', *Ambix*, 24, 1977, pp. 27–38.

—— 4. 'Matter-Theory: A Unifying Factor in Bacon's Natural Philosophy?', *Ambix*, 24, 1977, pp. 110–125.

—— 5. 'Atomism and "Subtlety" in Francis Bacon's Philosophy', *A.S.*, 37, 1980, pp. 549–571.

M. Reesor 1. 'The Stoic Concept of Quality', *American Journal of Philology*, 75, 1954, pp. 40–58.

—— 2. 'The Problem of Anaxagoras' in Anton & Kustas (eds.), pp. 81–87.

M. L. Righini Bonelli & W. R. Shea (eds.) *Reason, Experiment and Mysticism in the Scientific Revolution*, Macmillan, London, 1975.

H. M. Robinson 'Prime Matter in Aristotle', *Phronesis*, 19, 1974, pp. 168–188.

G. A. J. Rogers 'Boyle, Locke, and Reason', *J.H.I.*, 27, 1966, pp. 205–216.

D. H. D. Roller *The De Magnete of William Gilbert*, Menno Hertzberger, Amsterdam, 1959.

P. Romanell 'Some Medico-Philosophical Excerpts from the Mellon Collection of Locke papers', *J.H.I.*, 25, 1964, pp. 107–116.

V. Ronchi *The Nature of Light*, Tr. V. Barocas, Heinemann, London, 1970.

E. Rosen 1. 'Galileo's Misstatements about Copernicus', *Isis*, 49, 1958, pp. 319–330.

—— (ed.) 2. *Three Copernican Treatises*, Dover, New York, 1959.

P. Rossi *Francis Bacon: From Magic to Science*, Tr. S. Rabinovitch, RKP, London, 1968.

B. Russell *The Philosophy of Leibniz*, 2nd Ed., Allen & Unwin, London, 1937.

E. E. Ryan 'Pure Form in Aristotle', *Phronesis*, 18, 1973, pp. 209–224.

A. I. Sabra *Theories of Light from Descartes to Newton*, Oldbourne, London, 1967.

D. B. Sailor 'Cudworth and Descartes', *J.H.I.*, 23, 1962, pp. 133–140.

—— 2. 'Moses and Atomism', *J.H.I.*, 25, 1964, pp. 3–16.

W. Salmon (ed.) *Zeno's Paradoxes*, Bobbs-Merrill, Indianapolis and New York, 1970.

S. Sambursky 1. *The Physical World of the Greeks*, Tr. M. Dagut, RKP, London, 1956.

—— 2. 'Some References to Experience in Stoic Physics', *Isis*, 62, 1971, pp. 331–335.

—— 3. *Physics of the Stoics*, RKP, London, 1959.

—— 4. *The Physical World of Late Antiquity*, RKP, London, 1962.

J. A. Saveson 'Differing Reactions to Descartes among the Cambridge Platonists', *J.H.I.*, 21, 1960, pp. 560–567.

C. B. Schmitt 1. 'Experimental Evidence for and against a Void: the Sixteenth Century Arguments', *Isis*, 58, 1967, pp. 352–366.

—— 2. 'Towards a Reassessment of Renaissance Aristotelianism', *H.S.*, 11, 1973, pp. 159–193.

M. Schramm 'Aristotelianism: Basis and Obstacle to Scientific Progress in the Middle Ages', *H.S.*, 2, 1973, pp. 91–113.

M. Schwartz 'The Affirmation of Empty Space by an Eleventh-Century Mu'tazilite', *Isis*, 64, 1973, pp. 384–385.

J. F. Scott *The Scientific Work of René Descartes*, Taylor & Francis, London, 1952.

T. K. Scott 'Nicholas of Autrecourt, Buridan, and Ockhamism', *J.H.P.*, 9, 1971, pp. 15–41.

W. Scott (ed.) *Hermetica*, Vol. 1 (Texts and Translation), Clarendon, Oxford, 1924.

W. L. Scott 1. 'The Significance of "Hard" Bodies in the History of Scientific Thought', *Isis*, 50, 1959, pp. 199–210.

—— 2. *The Conflict Between Atomism and Conservation theory, 1644–1860*, MacDonald, London, 1970.

W. Sellars 'Substance and Form in Aristotle', *Journal of Philosophy*, 54, 1957, pp. 688–699.

Seneca *Quaestiones Naturales*, Tr. J. Clarke, Macmillan, London, 1910.

Sextus Empiricus *Against the Physicists*, Works, vol. 3, Tr. R. G. Bury, Heinemann, London, 1936.

D. Shapere *Galileo: A Philosophical Study*, University of Chicago Press, Chicago, 1974.

L. Sharp 'Walter Charleton's Early Life, 1620–1659', *A.S.*, 30, 1973, pp. 311–340.

W. R. Shea 'Galileo's Atomic Hypothesis', *Ambix*, 17, 1970, pp. 13–27.

H. J. Sheppard 'The Ouroboros and the Unity of Matter in Alchemy: A Study in Origins', *Ambix*, 10, 1962, pp. 83–96.

T. P. Sherlock 'The Chemical Work of Paracelsus', *Ambix*, 3, 1948, p. 33–63.

D. W. Singer *Giordano Bruno: His Life and Thought*, containing a translation of 'On the Infinite Universe and Worlds', Schuman, New York, 1950.

D. Skabelund & P. Thomas. 'Walter of Odington's Mathematical Treatment of the Primary Qualities', *Isis*, 60, 1969, pp. 331–350.

A. M. Smith 'Galileo's Theory of Indivisibles: Revolution or Compromise?', *J.H.I.*, 37, 1976, pp. 571–588.

A. J. Snow *Matter and Gravity in Newton's Physical Philosophy*, Arno, New York, 1975.

R. Sokolowski 'Matter, Elements and Substance in Aristotle', *J.H.P.*, 8, 1970, pp. 263–288.

F. Solmsen 1. 'Epicurus on Cosmological Heresies', *American Journal of Philology*, 72, 1951, pp. 1–23.

—— 2. 'Epicurus on the Growth and Decline of the Cosmos', *American Journal of Philology*, 74, 1953, pp. 34–51.

—— 3. 'Aristotle and Prime Matter: A Reply to Hugh R. King', *J.H.I.*, 19, 1958, pp. 242–252.

—— 4. *Aristotle's System of the Physical World*, Cornell University Press, Ithaca, New York, 1960.

—— 5. 'The Tradition about Zeno to Elea Re-examined', in Mourelatos (ed.), pp. 368–393.

J. S. Spink *French Free-Thought from Gassendi to Voltaire*, Greenwood, New York, 1969.

T. Sprat *History of the Royal Society*, RKP, London, 1959.

H. E. Stapleton 'The Antiquity of Alchemy', *Ambix*, 5, 1953, pp. 1–43.

C. A. Staudenbaur 1. 'Galileo, Ficino, and Henry More's Psychathanasia', *J.H.I.*, 29, 1968, pp. 565–578.

—— 2. 'Platonism, Theosophy and Immaterialism: Recent Views of the Cambridge Platonists', *J.H.I.*, 35, 1974, pp. 157–169.

M. C. Stokes *One and Many in Presocratic Philosophy*, Harvard University Press, Cambridge, Mass., 1971.

G. B. Stones 'The Atomic View of Matter of the XVth, XVIth and XVIIth Centuries', *Isis*, 10, 1928, pp. 444–465.

C. L. Stough. 'Parmenides' Way of Truth, B8 12–13', *Phronesis*, 13, 1968, pp. 91–106.

E. W. Strong 1. 'Newton's "Mathematical Way" ', *J.H.I.*, 12, 1951, pp. 90–110.

—— 2. 'Newton and God', *J.H.I.*, 13, 1952, pp. 147–167.

—— 3. 'Newtonian Explications of Natural Philosophy', *J.H.I.*, 18, 1957, pp. 49–83.

—— 4. 'Barrow and Newton', *J.H.P.*, 8, 1970, pp. 155–172.

P. Suppes 1. 'Descartes and the Problem of Action at a Distance', *J.H.I.*, 15, 1954, pp. 146–152.

—— 2. 'Aristotle's Concept of Matter and its Relation to Modern Concepts of Matter', *Synthese*, 28, 1974, pp. 27–50.

L. Sweeney *Infinity in the Presocratics*, Martinus Nijhoff, The Hague, 1972.

P. Tasch 1. 'Quantitative Measurements and the Greek Atomists', *Isis*, 38, 1947, pp. 185–189.

—— 2. 'Diogenes of Apollonia and Democritus', *Isis*, 40, 1949, pp. 10–12.

A. E. Taylor *Epicurus*, Books for Libraries Press, Freeport, New York, 1969.

C. C. W. Taylor 'Pleasure, Knowledge and Sensation in Democritus', *Phronesis*, 12, 1967, pp. 6–27.

F. S. Taylor 1. 'The Origins of Greek Alchemy', *Ambix*, 1, 1938, pp. 30–47,

—— 2. *The Alchemists, Founders of Modern Chemistry*, Heinemann, London, 1951

—— 3. 'An Alchemical Work of Sir Isaac Newton', *Ambix*, 5, 1956, pp. 59–84.

O. Temkin 'Physics in Aristotle', *H.S.*, 2, 1963, pp. 135–139.

Texas Quarterly *The Annus Mirabilis of Sir Isaac Newton*, Tricentennial Celebration, *Texas Quarterly*, vol. 10, No. 3, Autumn 1967.

A. Thackray 1. ' "Matter in a Nut-Shell": Newton's Opticks and Eighteenth-Century Chemistry', *Ambix*, 15, 1968, pp. 29–53.

—— 2. *Atoms and Powers*, Harvard University Press, Cambridge, Mass., 1970

K. Thomas *Religion and the Decline of Magic*, Weidenfeld & Nicholson, London, 1971.

L. Thorndike *A History of Magic and Experimental Science*, 8 Vols., Macmillan, London, 1923–1958.

S. S. Tigner 'Empedocles' Twirled Ladle and the Vortex-Supported Earth', *Isis*, 65, 1974, pp. 433–447.

E. M. W. Tillyard *The Elizabethan World-Picture*, Chatto & Windus, London, 1943.

S. Toulmin 'Criticism in the History of Science: Newton on Absolute Space, Time, and Motion', 2 parts, *P.R.*, 68, 1959, pp. 1-29, 203-227.

S. Toulmin & J. Goodfield 1 *The Architecture of Matter*, Hutchinson, London, 1962.

—— 2. *The Fabric of the Heavens*, Penguin, Harmondsworth, 1963.

A. T. Tymieniecka *Leibniz's Cosmological Synthesis*, Van Gorcum, Assen, 1964.

A. G. Van Melsen *From Atomos to Atom*, Harper, New York, 1960.

Vitruvius *The Ten Books of Architecture*, Tr. M. H. Morgan, Dover, New York, 1960.

G. Vlastos 1. Sum of an Infinite Series', *Gnomon*, 31, 1959, pp. 193-204.

—— 2. 'The Physical Theory of Anaxagoras' in Mourelatos (ed.), pp. 459-488.

—— 3. 'Minimal Parts in Epicurean Atomism', *Isis*, 56, 1965, pp. 121-147.

—— 4. 'Zeno's Race Course', *J.H.P.*, 4, 1966, pp. 95-108.

—— 5. 'A Note on Zeno's Arrow', *Phronesis*, 11, 1966, pp. 3-18.

—— 6 'A Zenonian Argument against Plurality' in Anton & Kustas (eds.), pp. 119-144.

—— 7. *Plato Universe*, University of Washington Press, Seattle, 1975.

D.P. Walker *Spiritual and Demonic Magic from Ficino to Campanella*, Warburg Institute, London, 1958.

W. A. Wallace 1. 'The Enigma of Domingo de Soto', *Isis*, 59, 1968, pp. 384-401.

—— 2. 'Mechanics from Bradwardine to Galileo', *J.H.I.*, 32, 1971, pp. 15-28.

—— (ed.) 3. *Galileo's Early Notebooks: The Physical Questions*, University of Notre Dame Press, Notre Dame, Indiana, 1977.

—— 4. 'Causes and Forces in Sixteenth Century Physics', *Isis*, 69, 1978, pp. 400-412.

—— 5 *Prelude to Galileo: Essays on Medieval and Sixteenth-Century Sources of Galileo's Thought*, Reidel, Dordrecht, 1981.

M. T. Walton 'Boyle and Newton on the Transmutation of Water and Air', *Ambix*, 27, 1980, pp. 11-18.

R. A. Walton *The Downfall of Cartesianism*, Martinus Nijhoff, The Hague, 1966.

C. Webster 1. 'Water as the Ultimate Principle of Nature: The Background to Boyle's Sceptical Chemist', *Ambix*, 13, 1966, pp. 96–107.

—— 2. 'Henry Power's Experimental Philosophy', *Ambix*, 14, 1967, pp. 150–178.

—— 3. 'Henry More and Descartes: Some New Sources', *B.J.H.S.*, 4, 1969, pp. 359–377.

J. R. Weinberg 1. *Nicholas of Autrecourt*, Princeton University Press, Princeton, New Jersey, 1948.

—— 2. *Ockham, Descartes, and Hume*, University of Wisconsin Press, Wisconsin, 1977.

J. A. Weisheipl 'The Principle "Omne quod movetur ab aliquo movetur" in Medieval Physics', *Isis*, 56, 1965, pp. 26–45.

M. West 'Notes on the importance of Alchemy to Modern Science in the writings of Francis Bacon and Robert Boyle', *Ambix*, 9, 1961, pp. 102–114.

R. S. Westfall, 1. 'Unpublished Boyle Papers relating to Scientific Method', 2 papers, *A.S.*, 12, 1956, pp. 63–73, 103–117.

—— 2. 'The Foundations of Newton's Philosophy of Nature', *B.J.H.S.*, 1, 1962, pp. 171–182.

—— 3. 'Hooke and the Law of Universal Gravitation', *B.J.H.S.*, 3, 1967, pp. 245–261.

—— 4. *Force in Newton's Physics*, MacDonald, London, 1971.

—— 5. *The Construction of Modern Science: Mechanisms and Mechanics*, CUP, Cambridge, 1977.

P. J. White 'Materialism and the Concept of Motion in Locke's Theory of Sense-Idea Causation', *S.H.P.S.*, 2, 1971, pp. 97–134.

P. Wiener 'The Experimental Philosophy of Robert Boyle, 1626–1691', *P.R.*, 41, 1932, pp. 594–609.

J. Wild 'The Cartesian Deformation of the Structure of Change and Its Influence of Modern Thought', *P.R.*, 50, 1941, pp. 36–59.

R. S. Wilkinson 1. 'George Starkey, Physician and Alchemist', *Ambix*, 11, 1963, pp. 121–152.

—— 2. 'The Problem of the Identity of Eirenaeus Philalethes', *Ambix*, 12, 1964, pp. 24–43.

—— 3. 'The Hartlib Papers and Seventeenth-Century Chemistry, 1', *Ambix*, 15, 1968, pp. 54–69.

—— 4. 'The Hartlib Papers and Seventeenth-Century Chemistry, 2', *Ambix*, 17, 1980, pp. 85–110.

B. Willey *The Seventeenth Century Background*, Chatto & Windus, London, 1934.

C. Wilson 1. 'Kepler's Derivation of the Elliptical Path', *Isis*, 59, 1968, pp. 5–25.

—— 2. 'Newton and Some Philosophers on Kepler's "Laws" ', *J.H.I.*, 35, 1974, pp. 231–258.

H. A. Wolfson 1. *Crescas' Critique of Aristotle*, Harvard University Press, Cambridge, Mass., 1929.

—— 2. *Repercussions of the Kālām in Jewish Philosophy*, Harvard University Press, Cambridge, Mass., 1979.

N. Wood 'The Baconian Character of Locke's Essay', *S.H.P.S.*, 6, 1975, pp. 43–84.

M. J. Woods 'Problems in Metaphysics Z Chapter 13' in J. M. E. Moravcsik (ed.), *Aristotle*, Anchor, New York, 1967.

R. Woolhouse, 'Locke's Idea of Spatial Extension', *J.H.P.*, 8, 1970, pp. 313–318.

F. A. Yates 1. *Giordano Bruno and the Hermetic Tradition*, RKP, London, 1964.

—— 2. *The Rosicrucian Enlightenment*, RKP, London, 1972.

R. M. Yost 'Locke's Rejection of Hypothesis about Submicroscopic Events', *J.H.I.*, 12, 1951, pp. 111–130.

P. Zetterberg 'Hermetic Geocentricity: John Dee's Celestial Egg', *Isis*, 70, 1979, pp. 385–393.

E. Zilsel 1. 'Copernicus and Mechanics, *J.H.I.*, 1, 1940, pp. 113–118.

—— 2. 'The Genesis of the Concept of a Physical Law', *P.R.*, 51, 1942, pp. 245–279.

ALSO AVAILABLE FROM THOEMMES PRESS

KEY ISSUES

Series Editor: **Andrew Pyle**, *University of Bristol*

The *Key Issues* series makes available the contemporary reactions that met important books and debates on their first appearance.
Examining the range of contemporary literature – journal articles and reviews, book extracts, public letters, sermons and pamphlets – *Key Issues* gives the reader an essential insight into the historical, social and political context in which a key publication or particular topic emerged.
Each text has a new editorial introduction to supply the necessary historical background.

Liberty
Contemporary Responses to John Stuart Mill
Edited and introduced by **Andrew Pyle**, *University of Bristol*

PHILOSOPHY, POLITICS
ISBN 1 85506 244 5 : 466pp : Hb : 1994 : £45.00 **$24.95**
ISBN 1 85506 245 3 : 466pp : Pb : 1994 : £14.95 **$72.00**

Population
Contemporary Responses to Thomas Malthus
Edited and introduced by **Andrew Pyle**, *University of Bristol*

ECONOMICS, POLITICS, SOCIAL HISTORY
ISBN 1 85506 344 1 : 320pp : Hb : 1994 : £40.00 **$24.95**
ISBN 1 85506 345 X : 320pp : Pb : 1994 : £13.95 **$72.00**

Group Rights
Perspectives Since 1900
Edited and introduced by **Julia Stapleton,** *University of Durham*

SOCIOLOGY, POLITICS
ISBN 1 85506 403 0 : 360pp : Hb : 1995 : £45.00 **$24.95**
ISBN 1 85506 402 2 : 360pp : Pb : 1995 : £14.95 **$72.00**

Agnosticism
Contemporary Responses to Spencer and Huxley
Edited and introduced by **Andrew Pyle,** *University of Bristol*

PHILOSOPHY, THEOLOGY
ISBN 1 85506 405 7 : 328pp : Hb : 1995 : £45.00 $24.95
ISBN 1 85506 404 9 : 328pp : Pb 1995 : £14.95 $72.00

Leviathan
Contemporary Responses to the Political Theory of Thomas Hobbes
Edited and introduced by **G. A. J. Rogers,** *Keele University*

PHILOSOPHY, POLITICS
ISBN 1 85506 407 3 : 317pp : Hb : 1995 : £45.00 $24.95
ISBN 1 85506 406 5 : 317pp : Pb : 1995 : £14.95 $72.00

The Subjection of Women
Contemporary Responses to John Stuart Mill
Edited and introduced by **Andrew Pyle,** *University of Bristol*

PHILOSOPHY, POLITICS, HISTORY OF FEMINISM
ISBN 1 85506 409 X : 340pp : Hb : 1995 : £45.00 $24.95
ISBN 1 85506 408 1 : 340pp : Pb : 1995 : £14.95 $72.00

The Origin of Language
Edited and introduced by **Roy Harris,** *University of Oxford*

LANGUAGE, ANTHROPOLOGY
ISBN 1 85506 438 3 : 344pp : Hb : £45.00 $24.95
ISBN 1 85506 437 5 : 344pp : Pb : £14.95 $72.00

Pure Experience
The Response to William James
Edited and introduced by **Eugene Taylor** and **Robert H. Wozniak,**
Harvard University & Bryn Mawr College, Pennsylvania

PSYCHOLOGY, PHILOSOPHY
ISBN 1 85506 413 8 : 294pp : Hb : £45.00 £24.95
ISBN 1 85506 412 X : 294pp : Pb : £14.95 $72.00

Gender and Science
Late Nineteenth-Century Debates on the Female Mind and Body
Edited and introduced by **Katharina Rowold**,
Wellcome Institute, London

GENDER, HISTORY OF SCIENCE
ISBN 1 85506 411 1 : 310pp : Hb : £45.00 **$24.95**
ISBN 1 85506 410 3 : 310pp : Pb : £14.95 **$72.00**

Free Trade
The Repeal of the Corn Laws
Edited and introduced by **Cheryl Schonhardt-Bailey**, *LSE*

POLITICS, ECONOMICS, BRITISH HISTORY
ISBN 1 85506 446 4 : 340pp : Hb : £45.00 **$24.95**
ISBN 1 85506 445 6 : 340pp : Pb : £14.95 **$72.00**

Hume on Miracles
Edited and introduced by **Stanley Tweyman**, *York University, Toronto*

PHILOSOPHY, THEOLOGY
ISBN 1 85506 444 8 : 190pp : Hb : £45.00 **$24.95**
ISBN 1 85506 443 X : 190pp : Pb : £14.95 **$72.00**

Hume on Natural Religion
Edited and introduced by **Stanley Tweyman**, *York University, Toronto*

PHILOSOPHY, THEOLOGY
ISBN 1 85506 451 0 : 352pp : Hb : £45.00 **$24.95**
ISBN 1 85506 450 2 : 352pp : Pb : £14.95 **$72.00**

Herbert Spencer and the Limits of the State
Contemporary Responses to Spencer's The Man Versus the State
Edited and introduced by **Michael Taylor**, *London Guildhall University*

POLITICS, PHILOSOPHY, SOCIOLOGY
ISBN 1 85506 453 7 : 296pp : Hb : £45.00 **$24.95**
ISBN 1 85506 452 9 : 296pp : Pb : £14.95 **$72.00**

Race: The Origins of an Idea, 1760–1850
Edited and introduced by **Hannah Augstein**,
Wellcome Institute, London

ANTHROPOLOGY, POLITICS
ISBN 1 85506 455 3 : 294pp : Hb : £45.00 $24.95
ISBN 1 85506 454 5 : 294pp : Pb : £14.95 $72.00

Religious Scepticism
Contemporary Responses to Gibbon
Edited and introduced by **David Womersely**, *Jesus College, Oxford*

THEOLOGY, PHILOSOPHY, HISTORY, LITERATURE
ISBN 1 85506 509 6 : 250pp : Hb : £45.00 $29.95
ISBN 1 85506 510 X : 250pp : Pb : £14.95 $75.00

John Locke and Christianity
Contemporary Responses to the Reasonableness of Christianity
Edited and introduced by **Victor Nuovo**, *Middlebury College, Vermont*

ANTHROPOLOGY, POLITICS
ISBN 1 85506 539 8 : 250pp : Hb : £45.00 $24.95
ISBN 1 85506 540 1 : 250pp : Pb : £14.95 $72.00

Mill and Religion
Contemporary Responses to Three Essays on Religion
Edited and introduced by **Alan P.F. Sell**,
United Theological College, Aberystwyth

ANTHROPOLOGY, POLITICS
ISBN 1 85506 541 X : 250pp : Hb : £45.00 $24.95
ISBN 1 85506 542 8 : 250pp : Pb : £14.95 $72.00